THE UNIVERSE

THE UNIVERSE

COLIN A. RONAN

MACMILLAN REFERENCE USA
An imprint of the Gale Group
Detroit • New York • San Francisco • London • Boston • Woodbridge, CT

CONTENTS

A MARSHALL EDITION

This book was conceived, edited and designed by
Marshall Editions, The Orangery, 161 New Bond Street,
London W1S 2UF

First published in the U.S.A. in 2000 by
Macmillan Reference USA, an imprint of the Gale Group

Copyright © 1993/2000 by Marshall Editions Ltd

Consultant Editor, Carole Stott

The right of Colin A. Ronan to be identified as the author of this work has been asserted in accordance with sections 77 and 78 of the Copyright Designs and Patents Act 1988

All rights reserved. No part of this book may be reproduced or transmitted in any form or by any means, electronic or mechanical, including photocopying, recording, or by any information storage and retrieval system, without permission in writing from the Publisher.

Macmillan Reference USA
1633 Broadway
New York, NY 10019

Library of Congress Card Number: 00-105270

ISBN 0-02-865591-5

10 9 8 7 6 5 4 3 2 1

Printed and bound by Imago, Singapore

Introduction	7
Creation of the Universe	**8**
The Scale of the Universe	10
Mathematics and Reality	14
Space and Time	16
Curved Space-Time	18
Relativity Observed	20
Physics of the Atom	22
The Subatomic World	26
The First Moment	30
Building Matter	34
The Cooling Universe	36
Protogalaxies	38

The Grand Design	40	**The Living Universe**	148
Observing the Universe	42	Earth	150
Telescopes of Today	44	The Nature of Life	154
Yardsticks of the Cosmos	46	Chemistries of Life	156
Superclusters	48	The Beginning of Life	158
Clusters of Galaxies	50	Molecules of Life	160
Spiral Galaxies	52	Evolution on Earth	162
The Milky Way	56	Universal Communication	166
Colliding Galaxies	58	The Future of the Universe	168
Elliptical Galaxies	60	Obituary for the Universe	170
Black Holes	62	Hidden Dimensions	172
Active Galaxies	66	Quantum Space	176
Quasars	70	Big Bang in Question	178
Nebulae	74	The Anthropic Principle	180
Starbirth	78	Creating the Universe	184
Stellar Companions	80	Wormholes	186
The Chemistry of the Stars	82	Star Charts	188
Lives of the Stars	84	Glossary	192
Variable Stars	86	Biographies	204
Death of a Star	88	Bibliography	205
Pulsars	92	Index	206
The Sun	94	Acknowledgments	212
Birth of a Solar System	102		
Other Solar Systems	104		
Our Solar System	106		
Mercury	108		
Venus	110		
Earth	114		
Mars	116		
Jupiter	120		
Saturn	124		
Uranus	128		
Neptune and Pluto	130		
Satellites	134		
Asteroids	142		
Comets	144		
Meteors	146		

Frontispiece: *A detail of the largest pillar of gas and dust in the Eagle Nebula. The silhouetted "fingers" of material are young stars in the making.*

Title page: *Part of the huge nebula in the constellation Carina, in the southern skies.*

Left: *Io, the volcanic, sulphur-covered satellite of Jupiter.*

Above: *The fine structure of Saturn's rings, revealed in a false-color image.*

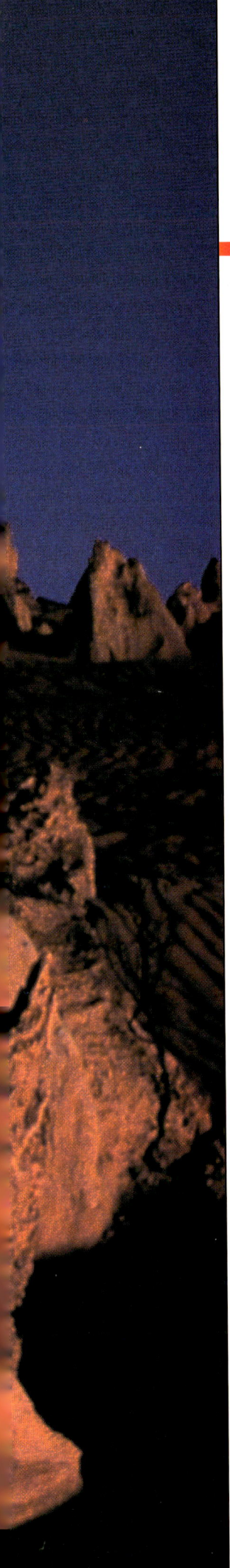

INTRODUCTION

From time immemorial, the skies have intrigued the human inhabitants of Earth. The pageant of sunrise and sunset, the changing phases of the Moon, and the silent procession of the stars across the black dome of heaven have long provided both a spectacle and a puzzle. The spectacle has inspired the artist, the musician, and the poet; the puzzle has intrigued philosophers and scientists. What is the meaning of it all? Why is it as it is, and in what manner was it formed?

To the earliest civilizations, the sky was a kind of dome which no human could reach. On it were stars, fixed in patterns which they recognized as familiar objects and characters of myth and legend. Yet as time passed, what intrigued many astronomers in the West was why a few of these stars wove paths among the rest. The motions of these "planets," or wanderers, were first investigated in detail in Mesopotamia.

Later, in ancient Greece, the challenge was resumed, and a new view taken of the universe. For aesthetic and mathematical reasons, it was expanded from a dome to a sphere. In doing so, the Greeks took the first step toward realizing that the universe is larger than it appears. They also developed an elaborate mathematical way of describing the cyclic motion of the Moon and planets around the Earth, which, on what seemed to be good evidence, appeared fixed at the center of the universe. In fact, their plan was so satisfactory that it was used by philosophers and astronomers for more than two thousand years. Not until the 1540s did mathematical reasoning and aesthetic considerations again bring about a change. This time, it turned into an intellectual watershed.

The Sun, not the Earth, was put at the center of the universe, the planets were set in orbit around it, and human beings were dethroned from their privileged position at the center of all creation. With subsequent mathematical work on planetary motion by Isaac Newton, the foundations of a new scientific universe were laid: a universe whose boundaries seemed infinite.

Since Newton's time, the march of science has accelerated. Now, more than three centuries later, its unceasing progress has finally brought us ways of observing the universe which would have astounded earlier astronomers. Yet this is not all. Our understanding of the nature of things has grown immensely, and new theoretical tools have been forged. Mathematics has been developed to an astonishing degree, and our knowledge of the world of physics has leaped ahead in a way that would leave previous generations gasping. As a result, our up-to-date picture is truly amazing compared with anything conceived in previous ages.

A monument to the profound imagination of 20th-century science, it calls on the most advanced and aesthetically elegant concepts of quantum theory to describe the very particles of the material world of which everything in the universe is made. Together with the almost incredible doctrines of relativity, and the stupendous observations brought back from space itself, we have evidence never previously available.

The astonishing picture that emerges is traced in the account that follows. It looks at the universe from its very beginnings to its ultimate end, and assesses the place of humankind in today's totally new scheme of space and time.

A full Moon shines down on bleak pinnacles of rock in the Australian desert. Earth's companion world is only a step away compared with the immensity of the cosmos known to modern astronomy.

CREATION OF THE UNIVERSE

The universe began in a colossal explosion, in which energy, space, time and matter were created. There is little doubt of this among scientists. The evidence that makes them so certain that the story is correct, in broad outline at least, comes from the discoveries of astronomy and subatomic physics, and from relativity and quantum theory, the two revolutionary theories that lie at the heart of modern physics. With the aid of these, theorists can trace the history of the universe to within a minute fraction of a second after the beginning.

There is doubt about precisely when the Big Bang occurred. It was at least 11 billion years ago, and perhaps as much as 18 billion. (Here, as throughout this book, "billion" means "thousand million.") The extreme conditions of the Big Bang do not exist today, but on both the astronomical and the subatomic scales the universe displays some extremely strange aspects. For instance, it almost certainly contains black holes, whose interiors are cut off from the surrounding cosmos. There are stars that spin hundreds of times a second and are made of material 35 billion times as dense as lead. There are stars, and even whole galaxies, that explode.

As for the particles at the heart of the atom, they are composed of still smaller particles called quarks, with such peculiar properties that language has to be stretched to describe them. Words like "color" and "charm" are pressed into service with new senses – though color is a meaningless concept at this level and the particles charm the physicist only because they are so puzzling.

Among the astronomical observations supporting the Big Bang theory, three pieces of evidence are specially important. The first is that the galaxies – the vast systems into which stars, gas, and dust are grouped – are all moving away from each other. We find that we are living in an expanding universe. A primordial Big Bang can explain this.

The second piece of observational evidence is the discovery of radiation reaching us from every direction of the universe. Moreover, this radiation is of equal intensity from every part of the sky. This fits in well with the idea of a hot Big Bang; what we are seeing is the glow of the primordial universe as it was at a very early date, a few hundred millennia after the beginning. Now, about 13 billion years later, this radiation has cooled to a few degrees above the absolute zero of temperature. This is just the temperature to be expected today if the radiation had originated in a hot Big Bang.

The third item of evidence for the Big Bang comes from nuclear physics. Studies of how the chemical elements would evolve after a Big Bang suggest that in the present-day universe we should find a particular ratio between the amounts of deuterium (a form of hydrogen) and of helium. Astrophysicists have verified that the existing ratio is what the theories predict.

So the hypothesis of a primordial explosion is extremely well founded. Yet there are variations of the theory, and outright alternatives to it, that will merit consideration at the end of this book. Nevertheless, however these unconventional theories fare in the future, the Big Bang theory will remain one of the grandest constructs of 20th-century scientific thinking. It tells a story that spans the microcosmos and the macrocosmos and leads from the beginning of the universe to an ending that can still only be guessed at.

The tracks of subatomic particles form an intricate pattern in a bubble chamber filled with liquid hydrogen. The reactions brought about by modern particle accelerators mimic processes that occurred in the first split second of the universe.

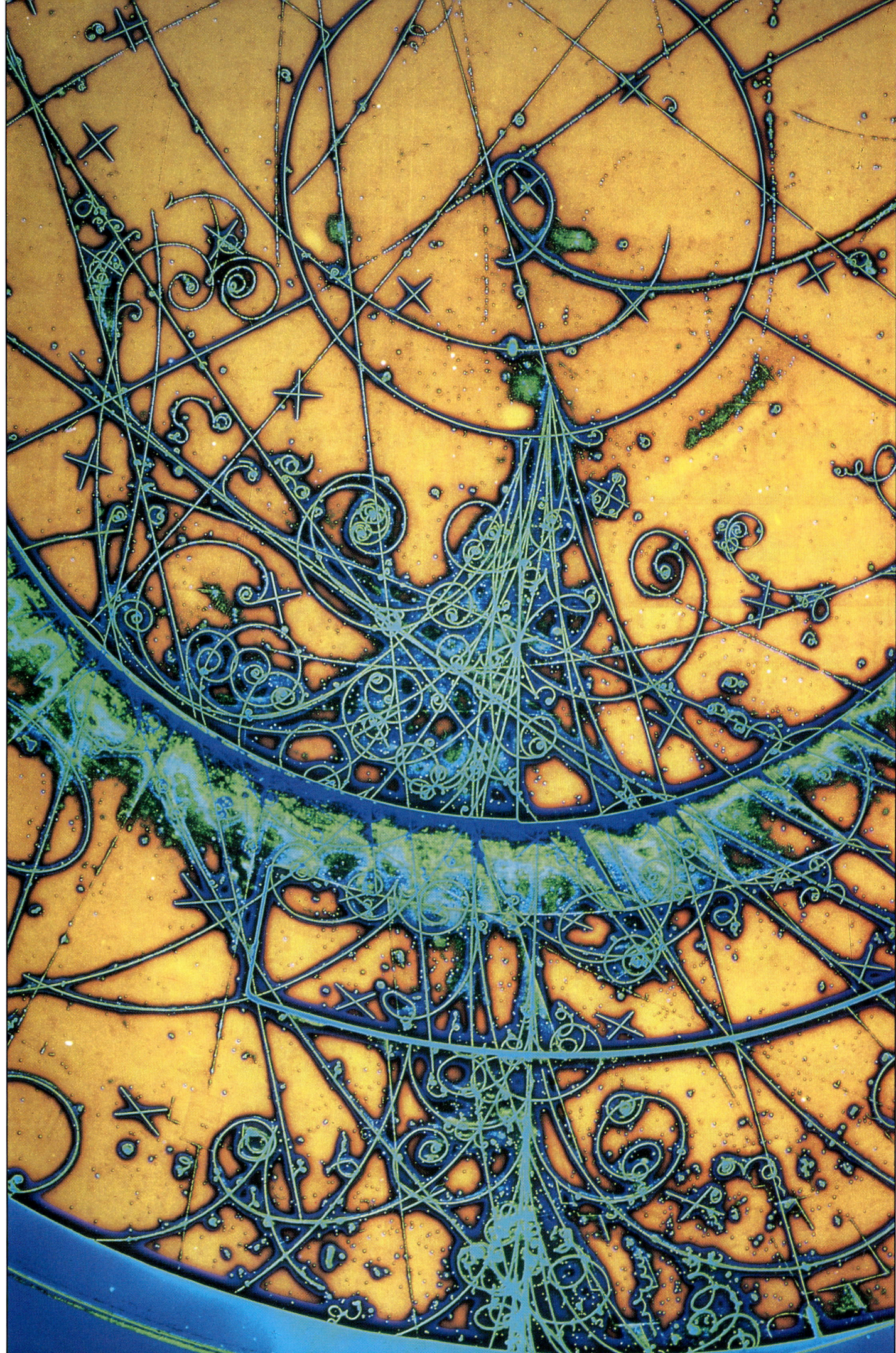

THE SCALE OF THE UNIVERSE
• *From quarks to superclusters*

The universe that has been opened up to inquiry by development of modern scientific instruments ranges from the wastes of intergalactic space to the interior of the atom. The largest entities of which we now have knowledge are the so-called superclusters – clusters of clusters – of galaxies. The smallest entities are the fundamental subatomic particles, such as quarks. Our knowledge of the latter is indirect, since they have never been observed singly, but is nonetheless firmly based. To space this vast range, a progressively increasing scale of length is shown here, each interval representing a distance 10 times larger than the one before. Scientists speak of a tenfold increase or decrease as a change of one "order of magnitude." The scale shown on these pages ranges over more than 40 orders of magnitude. Shorter than the wavelength of light, and therefore invisible, are atoms, molecules, and the smallest living entities, the viruses. Human beings are more than 10 million times, or seven orders of magnitude, larger.

In the same relation to the human body as the human body is to a virus are the gas giants, the largest planets of the solar system. In roughly the same proportion again are the planetary nebulae, the clouds of gas thrown off by dying stars. And in roughly the same ratio to these are the clusters of galaxies.

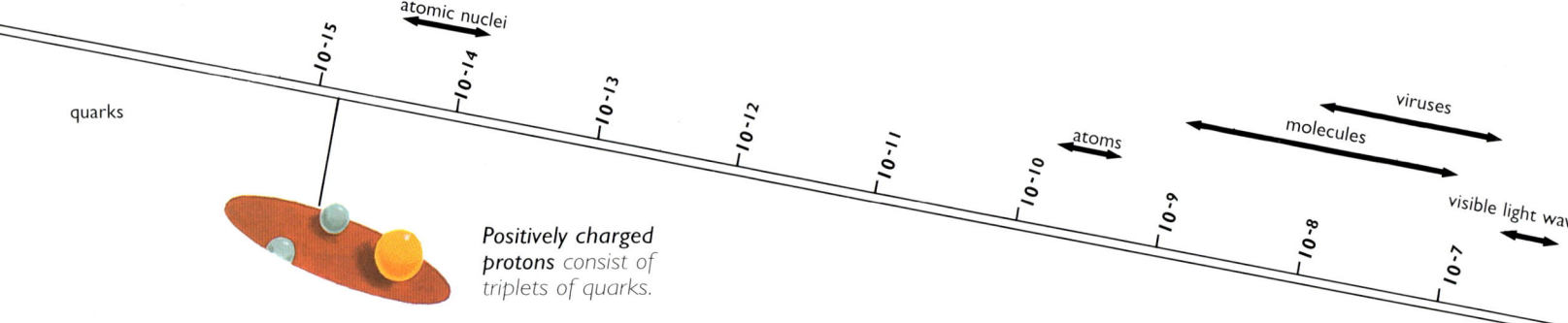

Positively charged protons consist of triplets of quarks.

Number shorthand

The scale running across these and the following two pages is logarithmic – that is, each division represents a tenfold increase in size over the one before. This convention enables sizes from the subatomic to the cosmic to be encompassed in a small space.

In a similar way, a simple notation enables very large and very small numbers to be written in a compact manner. To begin with everyday numbers: since $1,000 = 10 \times 10 \times 10$, it is written as 10^3. A number such as a billion billion, which is the result of multiplying 18 tens together and is normally written as 1 followed by 18 zeros, is written in this notation as 10^{18}.

Very small numbers are written in an analogous way. One millionth is 10^{-6} (that is, the result of dividing 1 by 10^6). The very brief period of time called the Planck time is written as 10^{-43} seconds – that is, as one second divided by 10^{43}.

The price of such compactness of expression is a distortion of scale. It is easy to forget the enormous difference represented by the easy step from, say, 10^2 to 10^4 light-years. But scientists and mathematicians find this so-called exponential or powers-of-10 notation indispensable.

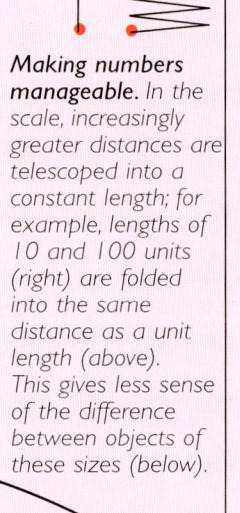

Making numbers manageable. In the scale, increasingly greater distances are telescoped into a constant length; for example, lengths of 10 and 100 units (right) are folded into the same distance as a unit length (above). This gives less sense of the difference between objects of these sizes (below).

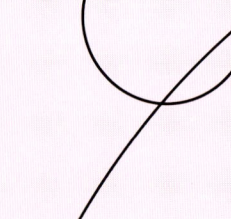

CREATION OF THE UNIVERSE

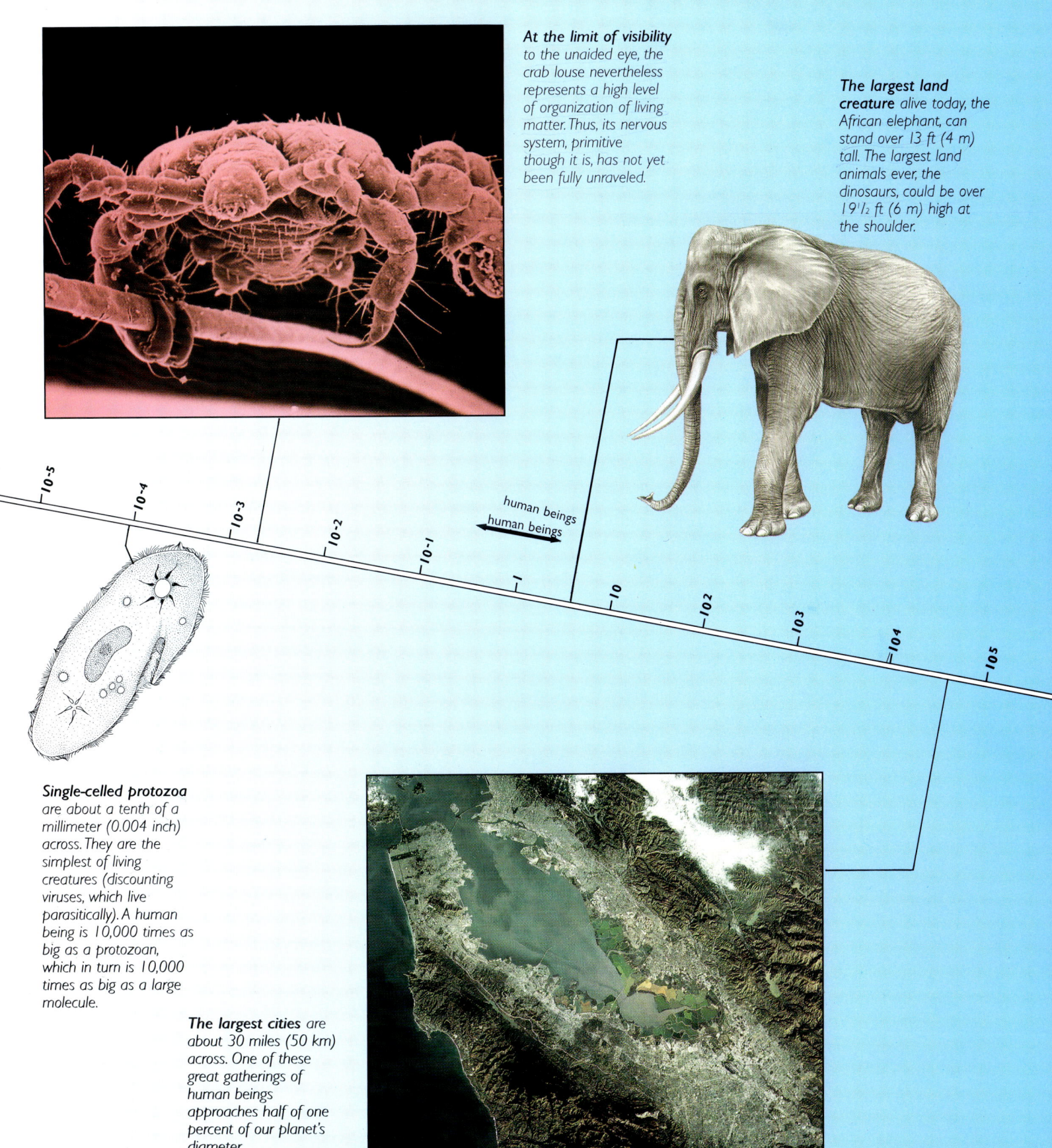

At the limit of visibility to the unaided eye, the crab louse nevertheless represents a high level of organization of living matter. Thus, its nervous system, primitive though it is, has not yet been fully unraveled.

The largest land creature alive today, the African elephant, can stand over 13 ft (4 m) tall. The largest land animals ever, the dinosaurs, could be over 19 1/2 ft (6 m) high at the shoulder.

Single-celled protozoa are about a tenth of a millimeter (0.004 inch) across. They are the simplest of living creatures (discounting viruses, which live parasitically). A human being is 10,000 times as big as a protozoan, which in turn is 10,000 times as big as a large molecule.

The largest cities are about 30 miles (50 km) across. One of these great gatherings of human beings approaches half of one percent of our planet's diameter.

11

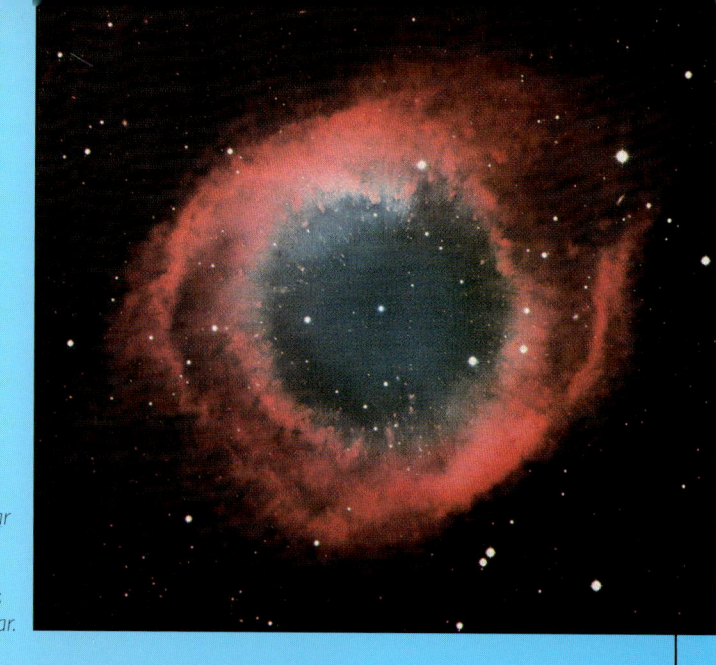

A planetary nebula, because of its vast size, appears in a telescope as a disk, whereas a star remains merely a point of light. This type of nebula is a globe of gas puffed off by a dying star.

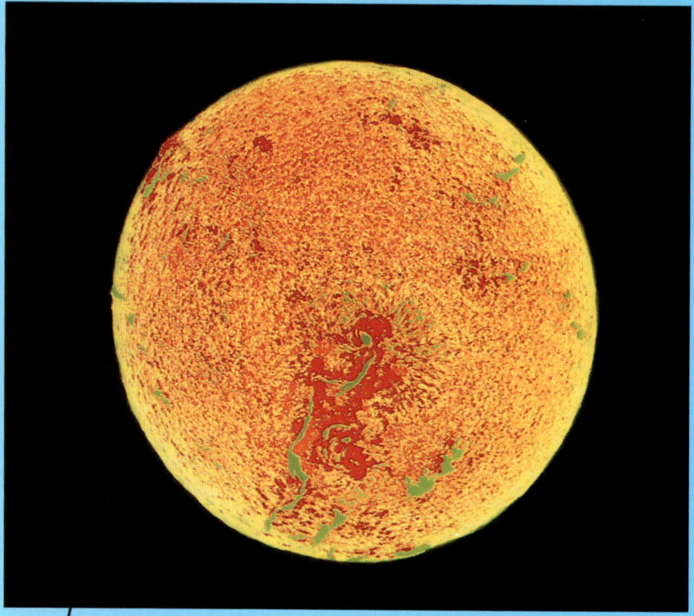

Dominating the solar system, the Sun has a diameter over 100 times that of the Earth. It is a very ordinary star – many others have diameters hundreds of times as great again.

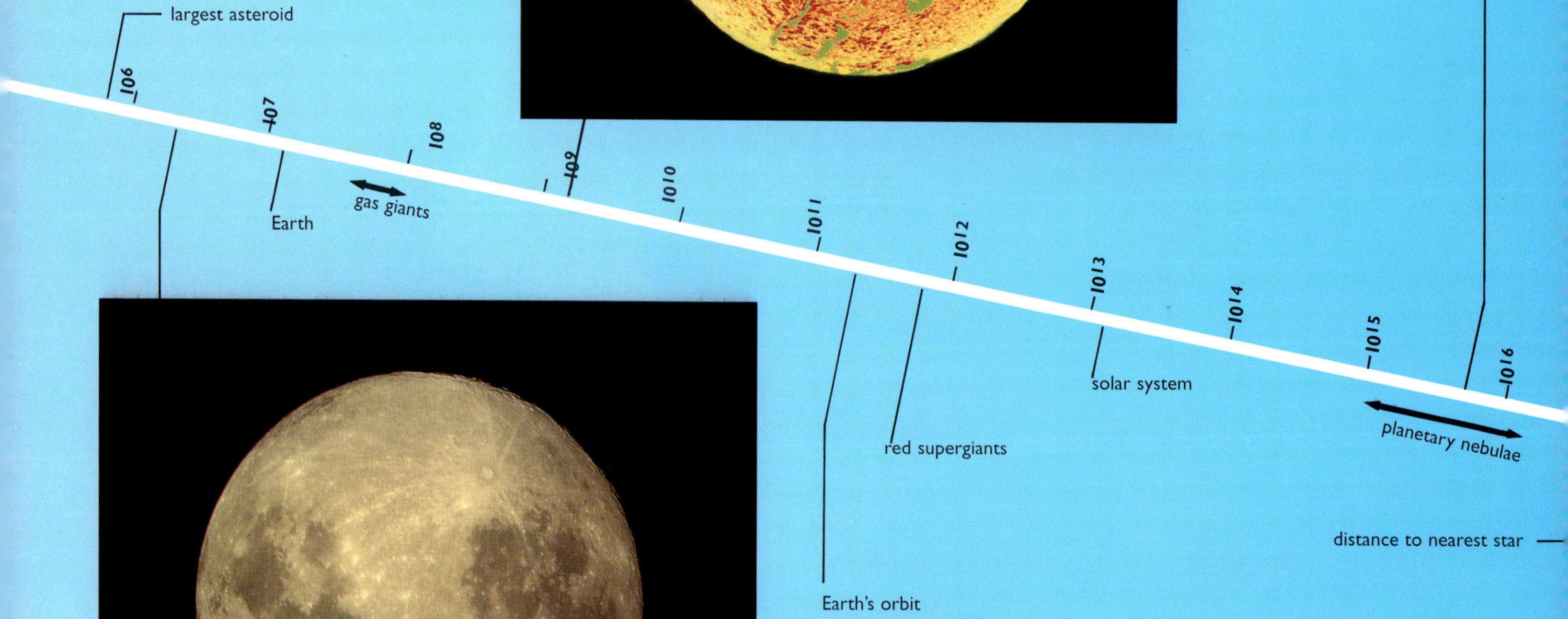

largest asteroid

10^6 10^7 10^8 10^9 10^{10} 10^{11} 10^{12} 10^{13} 10^{14} 10^{15} 10^{16}

Earth

← gas giants →

red supergiants

solar system

← planetary nebulae →

distance to nearest star

Earth's orbit

The Earth's satellite, the Moon, is 2,160 miles (3,476 km) across. It is one of the largest satellites in the solar system, larger than the planet Pluto.

CREATION OF THE UNIVERSE

Galaxies are huge assemblages of stars, gas, and dust, bound together by gravitation. Many are 100,000 light-years across. A galaxy bears roughly the same proportion to the Earth's orbit that the human body does to an atom.

The birthplaces of the stars are the diffuse nebulae, scores of light-years across. These masses of hydrogen gas and grains of dust are more rarefied than many laboratory vacuums.

Clusters of galaxies may have millions of members. Even these are not the largest unit into which matter organizes itself: clusters form superclusters, which can be hundreds of millions of light-years across.

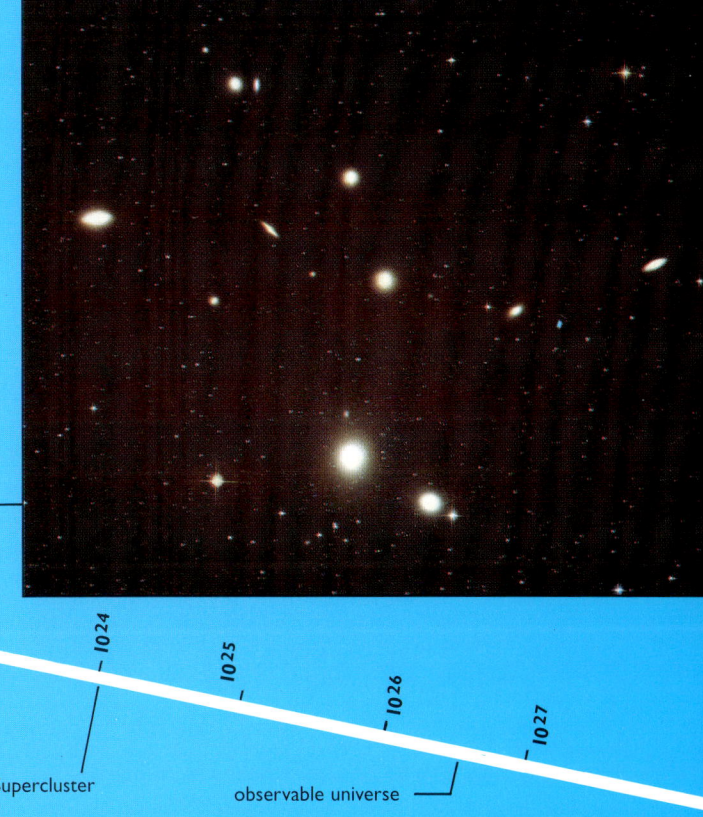

10^{17} 10^{18} 10^{19} 10^{20} 10^{21} 10^{22} 10^{23} 10^{24} 10^{25} 10^{26} 10^{27}

galaxies

radio galaxies (across lobes)

Local Supercluster

observable universe

13

MATHEMATICS AND REALITY
- *Insight through symbols*

The theory of the Big Bang has much observational evidence to support it. Yet there is another side to the study of the universe. Mathematics often enables cosmologists, like other scientists, to work out in theory what nature should be like *before* experimentalists and observers have confirmed that such is the case.

Mathematical reasoning is the only way to grasp the fundamentals that lie behind what we observe. This is so because mathematics is a language in which ideas can be formulated in a precise way and which allows the mind to work out logical consequences at profound depths, where mere words would present quite insurmountable obstacles. Time and again mathematical reasoning has provided insights available in no other way.

Unfortunately, many people suffer a mental block immediately when they see a simple equation; they find even one as simple, and as fundamental, as $E = mc^2$ to be elusive. So what does this particular equation mean?

The relationship was first derived by Einstein in the early years of this century in connection with his theory of relativity (pp. 16–17). It concerns the energy obtainable by the complete annihilation of a quantity of matter, which he derived by mathematical analysis. In the equation, m refers to the mass of the material, E refers to the energy, and c is the velocity of light. The latter is an enormous quantity – 186,420 miles (300,000 kilometers) per second. The equation asserts that, in suitable units, the energy is equal to that mass multiplied by c^2 – that is, c multiplied by itself.

The equation therefore specifies precisely the quantity of energy equivalent to a given mass. Since c is such a huge number, it implies that the amount of energy released is immense. This explains why an atomic bomb creates such a tremendous explosion and why the Sun shines so intensely – because the Sun, and all the stars, shine by converting mass into energy.

Expressing the meaning of $E = mc^2$ in words is much more long-winded than writing it in the form of an equation. Furthermore, once you understand the symbols, the equation's meaning is far more readily understood than the words.

Yet there is still more to the use of symbols. From everyday algebra we can write a new equation thus: $m = E/c^2$. This tells us how much mass is equivalent to a given amount of energy, and further suggests that mass can be created from energy. This is highly significant for cosmologists, because it tells them that all the matter in the universe could have come from the energy of the hot Big Bang.

With good reason, it has been said that mathematics is an art, and mathematicians certainly like to see symmetry, elegance, and simplicity in their equations. Furthermore, these aesthetic features seem to guide scientists in discovering facts about the real world.

A classic example of the power of symbols is afforded by the equations derived by the 19th-century Scottish physicist James Clerk Maxwell. Maxwell was intrigued by the electrical research of the brilliant self-taught experimentalist Michael Faraday. Faraday was no mathematician, and Maxwell set himself the important task of expressing in mathematical form the relationships Faraday had discovered experimentally.

Faraday had introduced the idea of fields to explain how electric and magnetic effects could be detected over a distance. A field was the pattern of electrical or magnetic influence filling space around a magnetized or electrically charged object. The field was depicted by drawing lines representing the direction of the electrical or magnetic force at each point of space. Maxwell was able to derive equations that expressed the observed effects, and he related them to the characteristics of the magnets and electric currents.

On examining these equations, Maxwell recognized that they gave results of exactly the same form as those expressing the motion of waves in fluid. He therefore concluded that there are waves in electric and magnetic fields. From the known values of certain quantities that had been measured in laboratory experiments, Maxwell was able to calculate the speed of the waves. It turned out to be 186,420 miles (300,000 kilometers) per second, the speed of light.

Maxwell concluded that light consisted of electrical and magnetic waves. Furthermore, there should be a whole range of such waves, some longer and some shorter than the familiar waves of visible, infrared, and ultraviolet light.

About 30 years later, Heinrich Hertz in Germany succeeded in producing electromagnetic waves about 1 ft (30 cm) long. Hertz showed that these radio waves, as we now call them, behaved as Maxwell's equations had predicted; they traveled at the velocity of light and were reflected or refracted (bent) in the same way as ordinary light waves.

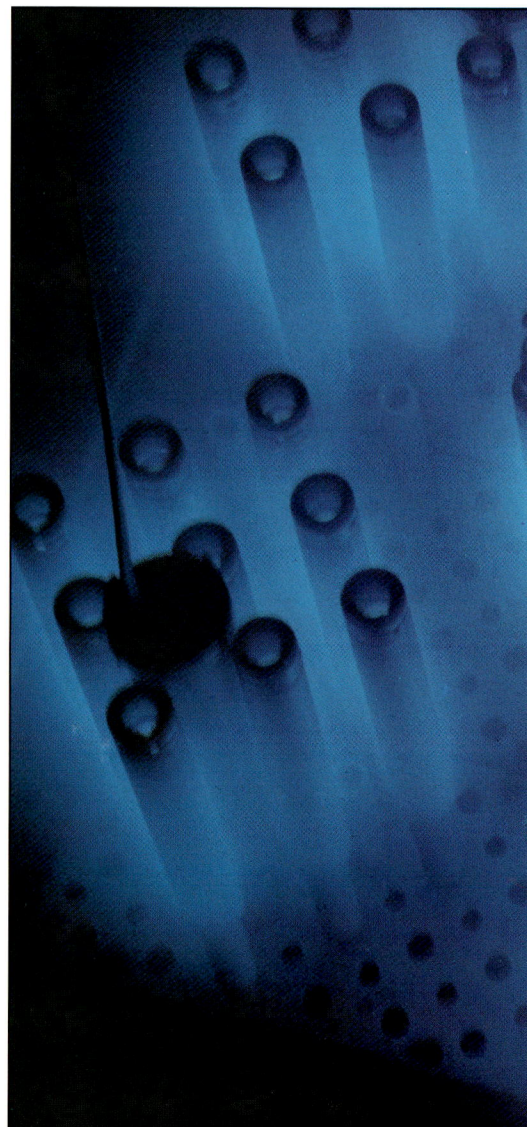

Thus mathematics, devised to express already existing results of experiments, had pointed to totally new experiments and discoveries – a feat that words alone would almost certainly have been quite unable to achieve.

Furthermore, once he had formulated his equations, Maxwell went on to make them more self-consistent, elegant, and symmetrical. He introduced an additional term, now known as the displacement current, which proved to play a part in the growth and contraction of a magnetic field associated with changing electric field. The urge to achieve mathematical symmetry had important consequences. Today it plays an equally crucial part in helping cosmologists fathom what happened during the earliest moments of the universe.

A blue glow reveals that energy is being released from nuclear fuel rods stored underwater (below). The energy is emitted when heavy atoms of uranium or plutonium fission (split) into smaller atoms. The combined masses of the fragments are less than the mass of the original atom. The missing mass is converted into energy in accordance with Einstein's equation $E = mc^2$.

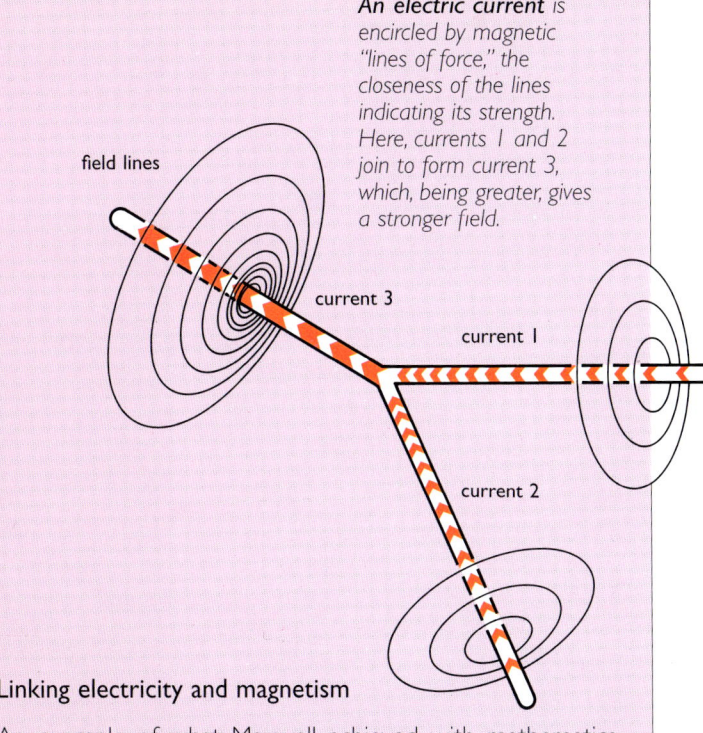

An electric current is encircled by magnetic "lines of force," the closeness of the lines indicating its strength. Here, currents 1 and 2 join to form current 3, which, being greater, gives a stronger field.

Linking electricity and magnetism

An example of what Maxwell achieved with mathematics may be illustrated by the first two of his four equations. The first equation reads:

$$\text{curl } \mathbf{E} = -\frac{\delta \mathbf{B}}{\delta t}$$

This is written in modern vector notation. A vector is a quantity, such as velocity or force, which has both magnitude and direction. The vector **B** represents the magnetic field, **E** the electric field. The symbol t refers to time, and $\frac{\delta \mathbf{B}}{\delta t}$ means "rate of change of **B**." The equation says that **curl E** – a certain property of **E**, which may be called its "rotation" – is proportional to the rate at which **B** changes.

Maxwell's second equation reads:

$$\text{curl } \mathbf{H} = \mathbf{J} + \frac{\delta \mathbf{D}}{\delta t}$$

This says that the "rotation" of a magnetic field **H** produces an electric current **J** and vice versa. The new term $\frac{\delta \mathbf{D}}{\delta t}$ added a certain symmetry previously lacking in the first equation. The equation shows that the magnetic field depends not only on the current **J** – which was well known before Maxwell – but also on the rate of change of a hitherto unrecognized "displacement" current, **D**, spreading throughout space. This asserted the distribution of electromagnetism throughout space. Later experiments showed Maxwell to be correct in this matter.

SPACE AND TIME
• *The meaning of relativity*

The laws expressed by James Clerk Maxwell were to play a key role in the scientific revolution that culminated in the work of Albert Einstein in the early years of the 20th century. It was a revolution that shook our very ideas of the nature of space and time. The classical physics that had developed from Newton's work had held that events everywhere took place within a shared universal time and an absolute and universal space. But by the beginning of the century it had become evident that mass, distance, energy, and the passage of time itself should vary according to the motion of the observer.

When Einstein first examined these ideas, he considered only observers moving relative to each other at unchanging velocities – at constant speeds in straight lines. This first version of his theory, called the special theory of relativity, was put forward in 1905. It was founded on the postulate that the laws of physics must be identical for every observer moving in this way – that is, for every frame of reference or viewpoint from which the measurements of physics are made. No frame of reference could be picked out as being at absolute rest.

In particular, Maxwell's laws of electromagnetism (pp. 14–15) must be the same in all frames of reference. Since the speed of light (as measured in a vacuum) is determined by these equations, the speed of light must also be the same for all observers.

Paradoxical as this assertion seems, it already had experimental backing, for no variation in the speed of light had ever been detected. Light from double stars, for example, which orbit each other at significant speeds, is neither delayed nor advanced by the approach or recession of the source stars. If it were, we should sometimes see such stars in more than one place at once. And a classic experiment by A. A. Michelson, later with the help of E. W. Morley, was designed to measure the speed of the Earth through space by its effect on light waves. But it completely failed to detect any of the expected effects.

The consequences that Einstein was able to deduce from the principles of relative motion and the constancy of the speed of light were astounding. First of all, he showed that the length of a body was relative to the frame of reference in which it was measured. If the body were moving at high speed in relation to the observer, it would appear to contract in the direction of its motion, while remaining unaffected in other directions.

The mass of a body also proved to be relative to the frame of reference adopted. (Mass is the "quantity of matter" in a body: the greater its mass, the greater its resistance to changes in its motion.) As a body moves faster, its mass increases relative to an observer at rest. As it approaches the speed of light, more and more energy is needed to give it an extra unit of speed, with the result that no material body can ever reach the speed of light.

Even more astonishing was the discovery that time also is relative. On an object that is moving fast relative to the observer, it seems to pass more slowly. So there is no universal standard of time. Our time literally does not pass at the same rate as that of another observer in a different frame of reference. There is no absolute simultaneity in the universe.

By 1915 Einstein had made another enormous advance: he had achieved his general theory of relativity, which was able to deal with changing velocities – that is, with frames of reference in accelerated motion. And because gravity makes bodies accelerate, this new theory became a theory of gravity as well.

The new theory showed that the motions of bodies have to be considered as motions through space-time: a four-dimensional amalgam of the three dimensions of space and the single dimension of time, all intimately connected. From the mathematics of the theory came the result that space-time is distorted – curved – around any mass, a distortion that compels other bodies to follow curved paths in response. Gravitational force is replaced by curvature of space-time.

Einstein followed a clue offered by what had seemed to be an isolated fact in Newton's theory of gravity. There are

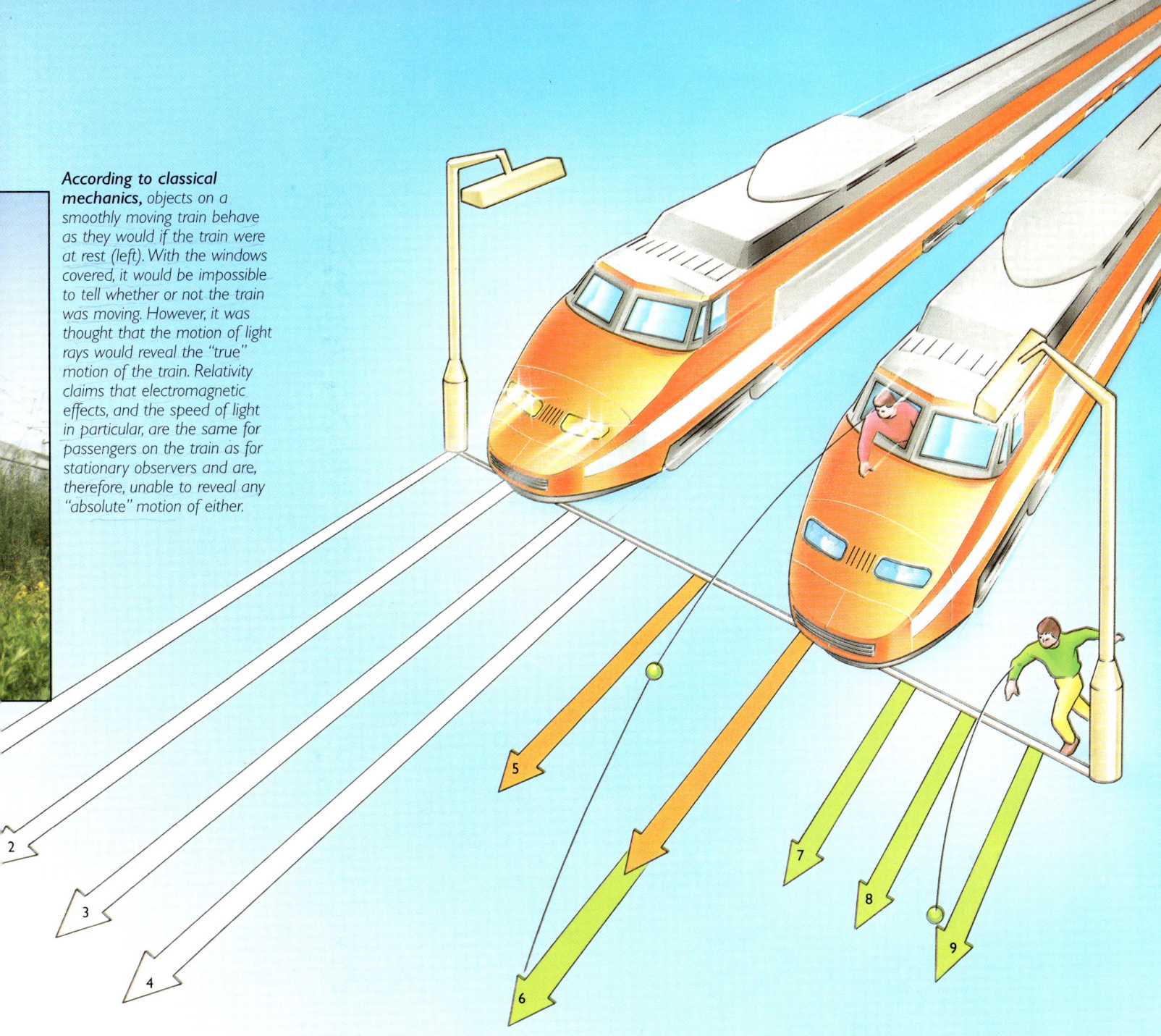

According to classical mechanics, objects on a smoothly moving train behave as they would if the train were at rest (left). With the windows covered, it would be impossible to tell whether or not the train was moving. However, it was thought that the motion of light rays would reveal the "true" motion of the train. Relativity claims that electromagnetic effects, and the speed of light in particular, are the same for passengers on the train as for stationary observers and are, therefore, unable to reveal any "absolute" motion of either.

The speed of light has an enormous value and, for a stationary observer, is always the same from any fixed source (1), whereas the speed of a material object, such as a moving ball, is variable (9). Remarkably, the speed of light from a moving source, such as a train, is the same for a stationary observer (2), for an observer on the train (3), and for an observer moving in relation to the train (4). This contrasts markedly with the observed speed of the ball. If thrown from the train, its speed relative to a stationary observer (6) is the sum of the speed of the train (5) and the speed of the object relative to the train (7). The latter is, in turn, equal to the speed of the object measured in relation to the person throwing it (8).

two ways of measuring mass: by its inertia – its resistance to being accelerated – and by its weight. Inertial mass is always found to be proportional to gravitational mass, and this is why heavy bodies fall at the same rate as light ones: if one body weighs twice as much as another, it will also have twice the inertia, and the one exactly compensates the other. The theory at which Einstein arrived made it seem natural that a heavy and a light object should move together in a gravitational field; whatever the mass of a body, its movement depends only on the local curvature of space.

Einstein's most famous discovery emerged from the special theory. This was the equivalence, and interconvertibility, of mass and energy, expressed in the equation $E=mc^2$ (pp. 14–15). The equation was established in the popular mind, firmly linked with Einstein, by the advent of nuclear bombs and the development of atomic energy. It remains a cornerstone of physics and cosmology today.

CURVED SPACE-TIME
● *The geometry of relativity*

The general theory of relativity is a theory of gravitation, which is viewed as a distortion of the space and time surrounding a body. In this sense, relativity is a geometrical theory because the mathematical study of space, whether curved or flat, is geometry.

The type of geometry with which most people are familiar is "Euclidean" geometry, named after the Greek philosopher Euclid, who lived about 290 BC. Euclid systematized all the geometrical knowledge of his time in a book now known as the *Elements*, which remained the undisputed touchstone of all geometrical knowledge until the late 19th century. Euclidean geometry was believed to be the only way in which to describe the space that everyone experiences.

Not until 1823 did Euclidean geometry begin to lose its unique position. In that year, the Hungarian mathematician Janos Bolyai discovered that there could be totally consistent geometries different from Euclid's. One of these later proved to be closely linked with general relativity.

In non-Euclidean geometry, the concept of a straight line is replaced by the more general idea of a geodesic, the line that follows the shortest distance between two points. On the curved surface of the Earth, any section of the equator or a line of longitude (a circle passing through both poles) is a geodesic. In Euclidean geometry – the geometry of flat surfaces – the geodesics are straight lines.

Consider a three-sided figure whose sides are geodesics. In Euclidean geometry this is a triangle whose internal angles always add up to 180°. But in the geometry discovered by Bolyai (and independently by the Russian Nikolai Lobachevsky) called hyperbolic geometry, the internal angles add up to less than 180°. The smaller the triangle is, the closer the sum approaches 180°, but a very large triangle can have very small internal angles.

The second alternative to Euclidean geometry was worked out in the 1850s by a Swiss mathematician, Ludwig Schläfi, and a German, Bernhard Riemann. In such a "Riemannian" geometry, as it has become known, the internal angles of a triangle always add up to more than 180°. Again, the difference from 180° increases with the size of the triangle.

To visualize this new geometry, Schläfi described it as being the geometry of the surface of a hypersphere – the analog of a sphere drawn in four dimensions. The geodesics then actually do appear as straight lines. (The subject of the fourth dimension – and higher ones – is pursued on pp. 172–75).

A simple analogy with hyperbolic and Riemannian geometries can be obtained by considering them in two dimensions only – that is, by concentrating on just the surfaces of solid bodies. Hyperbolic geometry is, then, the geometry of a saddle-shaped surface, while Riemannian geometry is that of a spherical surface. In the latter case, for example, the angles of a triangle clearly grow larger as the triangle increases in size.

Riemannian geometry extends our ideas of space. In fact, there can be some striking differences from the familiar space of Euclid; a geodesic can, so to speak, curve over on itself. Riemannian space can be finite in volume, though lacking any boundary. (This can be grasped by analogy with the surface of a sphere, which is finite but boundless.) Space extends outward without limit – we can travel up and down, or side to side, as far as we like, and we shall never reach a boundary. Yet this space is finite in volume.

Furthermore, travelers continuing in a straight line, without deviation, will eventually return to their starting point from the direction opposite to that in which they set out – just as an imaginary creature confined to the surface of a sphere would do.

Einstein maintained that not only matter but light would follow a curved path through space-time distorted by the presence of matter. This prediction of the bending of light by gravity was triumphantly vindicated by observations of an eclipse of the Sun in 1919. The new theory also accounted for anomalies in the motion of Mercury.

People have great difficulty in grasping the idea of a beginning of the universe. They want to know what was there before it started. And when they consider the expanding universe, they find themselves asking the question, "What lies outside it?" In the Riemannian geometry of relativity, it is possible to answer that space and time did not exist before the Big Bang, and that nothing exists outside the present expanding universe because there is no

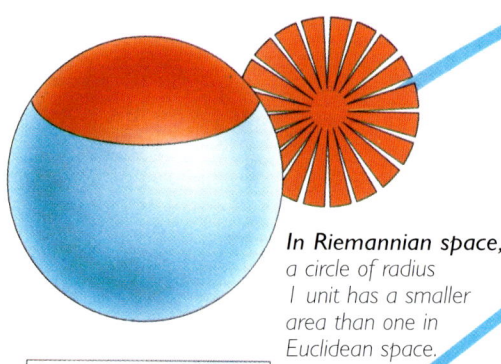

In Riemannian space, a circle of radius 1 unit has a smaller area than one in Euclidean space.

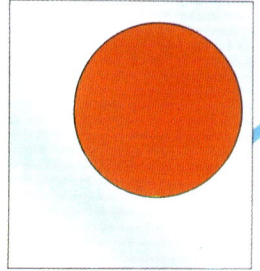

In Euclidean space, a circle's area is proportional to the square of its radius. Doubling the size of a circle will, therefore, increase its area four times.

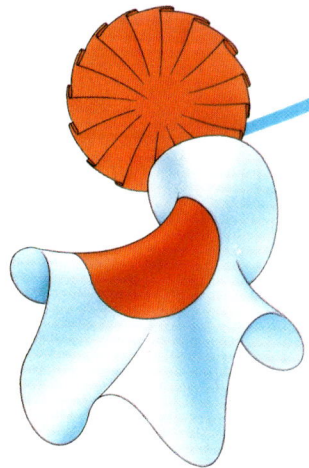

In hyperbolic space, the area of a circle is greater than that of a Euclidean circle of the same radius. The larger the circle that is considered, the greater the disparity.

Observations on an astronomical scale can, in principle, reveal whether the overall geometry of our universe is hyperbolic, Riemannian, or Euclidean.

able to use probability waves to explain the energy levels in the Bohr atom. The waves associated with the electrons were regarded as being standing waves, or resonances, formed much as vibrations of certain frequencies are set up in violin strings. The well-defined paths of electrons in the Bohr atom were replaced by "clouds of probability."

An electron is more probably to be found where the cloud is dense than where it is thin, but we cannot say precisely where and with precisely what speed or energy. It is like being told that a wave of burglaries is affecting a certain town: it is impossible to say whether a break-in will happen at a particular house on a specified day; we can only assess the probability that one will occur in a given period over a given area. Such statistical probabilities are real enough, as insurance companies know. Although the electrons cannot now be viewed as having definite orbits, they do have definite energies.

The fact that particles are wavelike has other enormously important consequences. It means, for example, that in general there cannot be complete certainty about both the position *and* the speed of a particle at a particular instant. We can specify only one factor at a time – the other one remains uncertain. If a physicist performs an experiment to determine the particle's precise position, he will disturb it so that its motion will become completely uncertain. And if he tries to determine the speed precisely, he will disturb the particle so that its position becomes unknown.

This crucial feature of the new physics, first stated by the German physicist Werner Heisenberg, has become known as the uncertainty principle. Another aspect of it is that we cannot specify both the energy involved in a certain event *and* the time of occurrence of that event to arbitrary accuracy. Such uncertainties are a consequence of the nature of the atomic world itself, not of any inability on our part to measure precisely enough.

In 1925, the Austrian-American physicist Wolfgang Pauli proposed his exclusion principle, which states that in any system, only one electron can be in a given quantum "state." The state is defined by such quantities as energy and position, and also by spin, which is in some ways analogous to the spin of a

According to quantum mechanics, position and velocity cannot be precisely specified at the same time. An observation that reveals the speed and direction of movement of a particle "smears" its position – rather as the time exposure (above) reveals the dancer's motion but makes his position less definite. An observation that fixes a particle's position leaves its motion completely uncertain – just as a high-speed exposure (right) makes the dancer's position clear and precise, while giving no information at all about movement.

macroscopic object, but with some differences which show that subatomic particles "see" a different world from our own. Spin comes in discrete values, intrinsic to a given particle. Some particles have whole-number spin values, labeled 0, ± 1, ± 2 and so on. Others, including the proton, neutron, and electron, have fractional, or non-integral, values: $\pm 1/2$, $\pm 3/2$, and so on.

An electron in the innermost orbit of an atom can be in either of two states, according to the direction of its spin. Therefore, two electrons can occupy this orbit. Limited numbers of states are permitted in other orbits. This is why electrons do not all congregate at the lowest energy state.

The creation of quantum mechanics led to an understanding of the architecture of the atom's outermost layers. But as physicists probed deeper, into the nucleus, they were to encounter new forces and new particles.

THE SUBATOMIC WORLD
• A profusion of particles

A bewildering variety of subatomic particles has been discovered by modern physicists. Every advance in the power of "atom-smashers," or particle accelerators (see box), leads to the production of yet more, so that today thousands have been cataloged. But in recent years, theorists have made great strides in bringing order into this seeming chaos.

Every type of particle is now known to have its own antiparticle (except for a few cases where particle and antiparticle are identical). The properties of an antiparticle are opposite to those of its corresponding particle. Thus, the antiparticle of the ordinary negatively-charged electron is the positron, which has the same mass but a positive charge equal to the electron's negative charge.

This practically halves the number of distinct types of particle. One way of simplifying things further is to consider how particles interact with each other via "messenger particles."

Consider the electromagnetic force. When an electron in an atom jumps from a higher-energy outer orbit to a lower-energy one closer to the nucleus, the energy difference escapes in the form of a photon – a quantum of electromagnetic radiation. The photon can be considered to be a messenger which carries the energy away.

Now imagine what happens when moving electrons pass near each other. Both have negative charges and will therefore repel each other. In quantum physics this force is regarded as being due to the emission of a virtual photon by one and its absorption by the other. (A virtual particle is one so short-lived that its existence can be inferred only indirectly.)

An analogy in the macroscopic world occurs when one person throws a ball to another. As the ball is released, the thrower experiences a reaction and is impelled backwards. The catcher experiences a force, too, also pushing backwards. In the quantum world the virtual photons act like balls – messengers that carry electromagnetic forces.

This kind of explanation can account for other subatomic forces. One of these is the weak nuclear force. An example of a process involving the weak force is the decay of a neutron into a proton, an electron, and an antineutrino.

The messenger particles of the weak nuclear force are know as W and Z particles. These particles can be detected if they are given enough energy in a particle accelerator to enable them to shoot away and cease being virtual.

However, physicists need another force to explain what holds the particles of an atomic nucleus together, overcoming the repulsion between the positive electric charges of the protons. The weak force is billions of times too weak, and gravity still more so. So the additional force is called the strong nuclear force but, although it is powerful, it extends only a very short distance – no more than 10^{-12} millimeters – and so it is limited to the atomic nucleus.

To understand more about the messenger particles of the strong force, it is necessary to know more about the particles on which they act. In 1963, the American physicists Murray Gell-Mann and George Zweig proposed that protons and neutrons are not basic particles, but are made of smaller entities. They called these quarks, a word coined by the Irish writer James Joyce in his novel *Finnegan's Wake*. According to Gell-Mann and Zweig, particles such as the proton and neutron are composed of three quarks. These are called baryons ("heavy particles"). Mesons are each made up of a quark and an antiquark.

When particles take part in reactions via the strong force, it is actually the quarks of which they are composed that are interacting. The messenger particles of the strong force are called gluons. They are responsible for the remarkable fact that quarks are never observed singly. The strong force grows more and more powerful as two quarks are pulled apart, until it overwhelms the separating force.

Despite the great differences between the basic forces, physicists are seeking to unify them – to show how they could be derived from a single fundamental force. Nor is this merely a theoretical exercise. They believe that in the first moments of the universe, the basic forces were literally merged into one.

In the heart of the atom

Part of the Large Electron-Positron collider at CERN, Switzerland. The 17-mile-long tube of the machine is 330 ft underground in a tunnel straddling the Swiss-French border. Rapidly oscillating electric and magnetic fields, created by electromagnets along the tube, give the circulating particles energy equivalent to a single boost of 50 billion volts.

Fundamental particles are probed by smashing them into each other at high speed and studying the fragments that result from the collisions. Nearly all the most powerful accelerators are synchrotrons – ring-shaped tubes, up to 17 miles long, with a high vacuum inside. An initial burst of electrons or protons (or their antiparticles) is injected into the synchrotron. Powerful electromagnets create rapidly varying fields that whirl the particles around the ring thousands of times at close to the speed of light.

The high-speed particles may then be flung at a stationary target, such as a block of aluminum. Higher energies of impact are

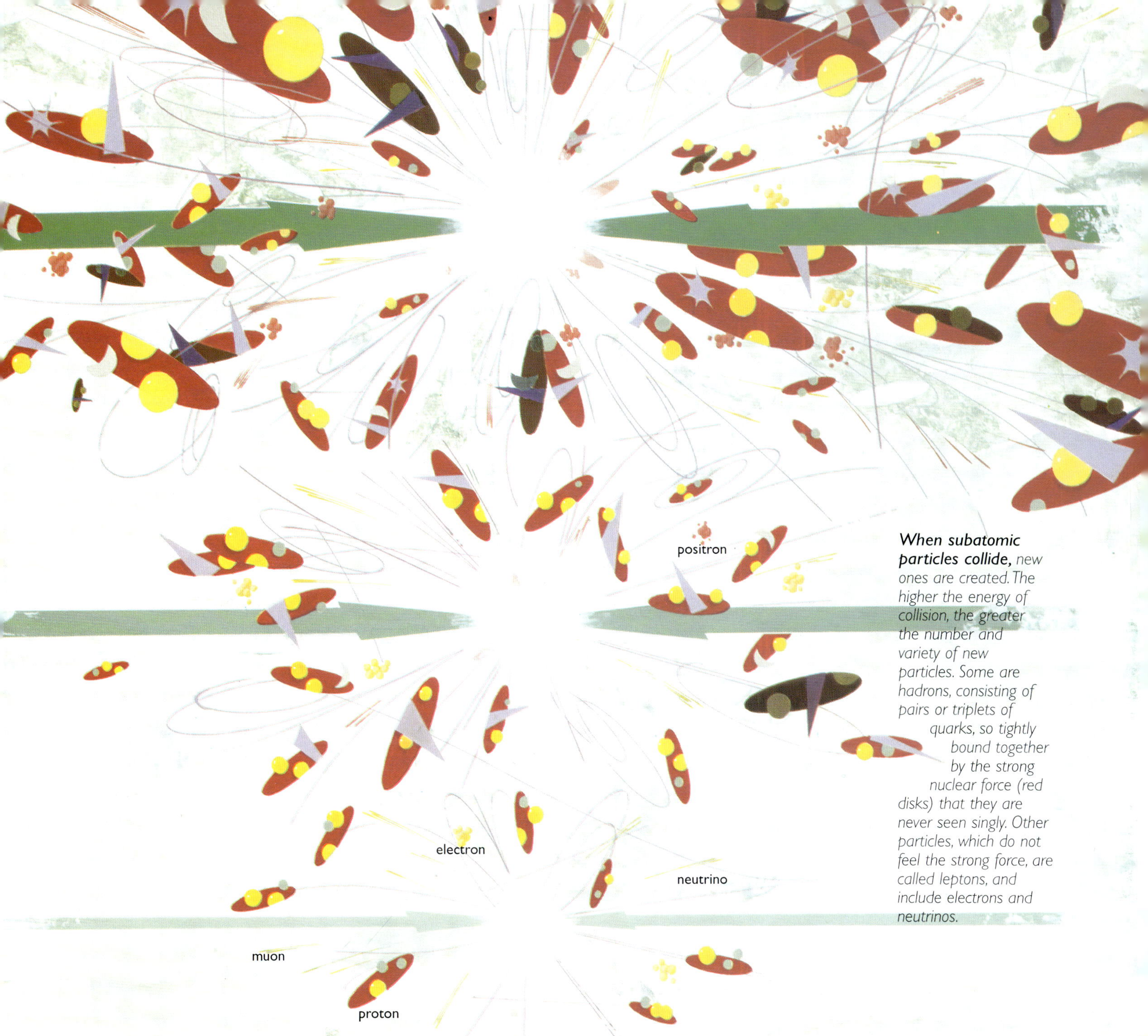

When subatomic particles collide, new ones are created. The higher the energy of collision, the greater the number and variety of new particles. Some are hadrons, consisting of pairs or triplets of quarks, so tightly bound together by the strong nuclear force (red disks) that they are never seen singly. Other particles, which do not feel the strong force, are called leptons, and include electrons and neutrinos.

achieved by accumulating two streams of particles that circulate opposite ways in storage rings, and then letting the two streams collide head-on.

Much of the energy of the collision is converted into a cascade of particles, many of which come into existence at the moment of the interaction. Nearly all the particles are ephemeral. The charged particles among them leave tracks in special detectors (the presence of uncharged particles has to be inferred indirectly).

Nearly all known particles fall into two main groups: hadrons and leptons. Leptons seem to have no structure and do not experience the strong force. They include electrons, neutrinos, and muons.

Hadrons, on the other hand, consist of pairs or triplets of quarks and are subject to the strong nuclear force. Quarks come in six "flavors," and each of these in three "colors." The two building blocks of atomic nuclei, the proton and neutron, are triplets of quarks. The neutron consists of one up and two down quarks. In isolation it rapidly breaks down into a proton, electron, and antineutrino, but in the nucleus it is stable. The proton consists of two up quarks and one down quark. Its lifetime is enormously long, possibly infinite.

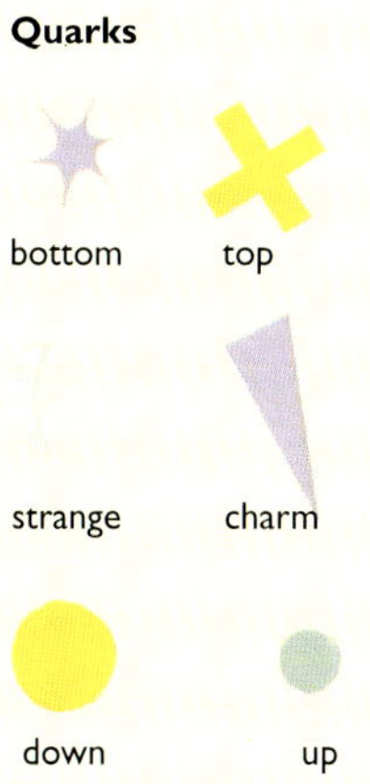

Quarks

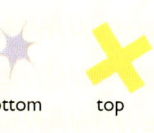
bottom   top

strange   charm

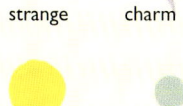
down   up

The six types of quark believed to exist. The top quark is also called the "truth" quark. Strikingly, they have electric charges that are either $1/3$ or $2/3$ of the electron charge – once thought to be fundamental. Quarks always combine to give charges that are either zero or whole-number multiples of the electron's electric charge.

UNIFIED FORCES
Toward a "theory of everything"

The great goal of modern physics is to create a unified theory of the four fundamental interactions that govern the material world: gravitation, electromagnetism, and the strong and weak nuclear forces. The theorists' first successes have linked the weak and electromagnetic forces.

The range of the weak force is less than 10^{-14} millimeters – a thousandth that of the strong force. This short range is due to the fact that its messenger particles are very heavy. Sheldon Glashow of Harvard took the first steps toward unifying the weak and electromagnetic forces in 1974. He discovered that an electrically neutral particle, which he called the Z^0, was required, in addition to two electrically charged particles, which he called the W^+ and W^-.

Although this was an advance, the fact remained that all three particles must be very massive, whereas photons, the messengers of the electromagnetic field, have no rest mass. This seemed to spoil the so-called gauge symmetry between the two forces.

Mathematicians maintain that something possesses symmetry when it is unchanged by some operation. Thus, a square is said to have a fourfold rotational symmetry, because in one rotation there are four positions in which it is unchanged. Symmetries can involve physical quantities, not just geometrical shapes. For example, suppose temperature is raised by 10°C (50°F): for many processes it does not matter if we start from a temperature of 5°C (41°F) or 50°C (122°F) – the effects of the rise are the same. The process is symmetrical with respect to these changes of starting temperature. To put it another way, the gauge (roughly – the position on the scale) can be altered without affecting the change. None of the original symmetry will be lost.

The mathematics of the electromagnetic field reveals that there is a gauge symmetry connected with the zero mass of the photon. The weak force, having massive messenger particles, lacks this symmetry. But Abdus Salam working at Imperial College, London, and Steven Weinberg at the Massachusetts Institute of Technology independently showed how the two fields could, nonetheless, possess a mutual symmetry, which is spontaneously broken in the present-day universe.

Spontaneous symmetry-breaking happens in the everyday world as well as on the atomic scale. Stand a pencil on its end. It possesses rotational symmetry: if rotated on its vertical axis, it looks the same. Let it fall over and lie flat, and the rotational symmetry is broken: if it is rotated on a vertical axis, it looks different because its orientation changes. The symmetry is present but hidden: stand the pencil up again and the symmetry is once more apparent.

So it is in the atomic world. Salam and Weinberg suggested that the huge mass of the W and Z particles appears when the underlying symmetry of the electromagnetic and weak forces is spontaneously broken. They found that this could happen if they introduced a new field of force – the Higgs field, devised by Peter Higgs of Edinburgh University. In our present-day, comparatively low-energy universe, the Higgs field is present everywhere and couples with the weak field. It acts like molasses, making the W and Z particles sluggish, and giving them a large effective mass. Photons are unaffected.

At high energies, however, the Higgs field vanishes. The W and Z particles behave like photons and can no longer be distinguished from them. The symmetry between the weak and electromagnetic fields becomes manifest. In 1983, the W and Z particles were actually observed in the proton-antiproton collider at CERN, confirming the existence of the unified electroweak force.

The next step is the unification of the electroweak force with the strong force. At the temperatures prevailing at the beginning of the Big Bang, quarks would have been packed closely together. The strong nuclear force becomes weaker at shorter distances (p. 26), so there was a temperature at which it was equal in strength to the electroweak force. The two forces could have been unified at that time.

This so-called grand unification requires that leptons and quarks be able to turn into each other. They could do so by exchanging a new messenger, the X particle, which is very massive – 10^{15} times heavier than the proton.

Even at low energies a few changes caused by X particles are expected to occur. For example, in a large quantity of matter protons are expected to break up very occasionally. A watch for such events has been mounted, using equipment deep in mines to avoid contamination of the results by cosmic rays. But firm evidence has yet to be found.

Another prediction of grand unified theories is the magnetic monopole. In the everyday world magnets always have two poles, a north and a south. If a magnet is cut in half, each piece has a north and a south pole. But GUTs (grand unified theories) predict that in the early moments of the universe, single magnetic poles should have been created. They should survive today, and should be as massive as the X particle. However, experiments to find magnetic monopoles have so far been inconclusive. According to the theory of inflation (pp. 30–31), monopoles should have been so widely dispersed in the Big Bang that this is not surprising.

What physicists call a "theory of everything" will go farther than the grand unified theories: it will unify gravity with the other forces. This is the next great prize for theorists.

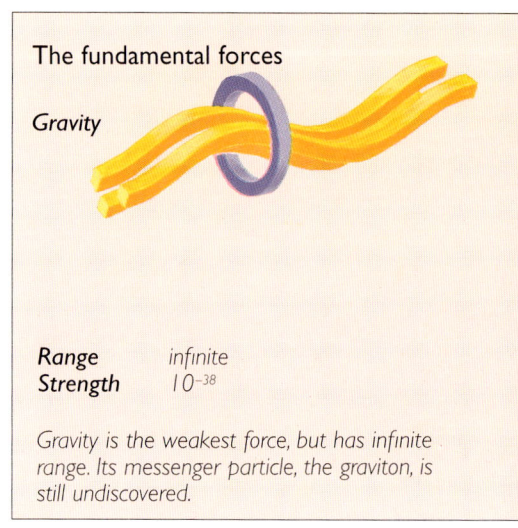

The fundamental forces

Gravity

Range	infinite
Strength	10^{-38}

Gravity is the weakest force, but has infinite range. Its messenger particle, the graviton, is still undiscovered.

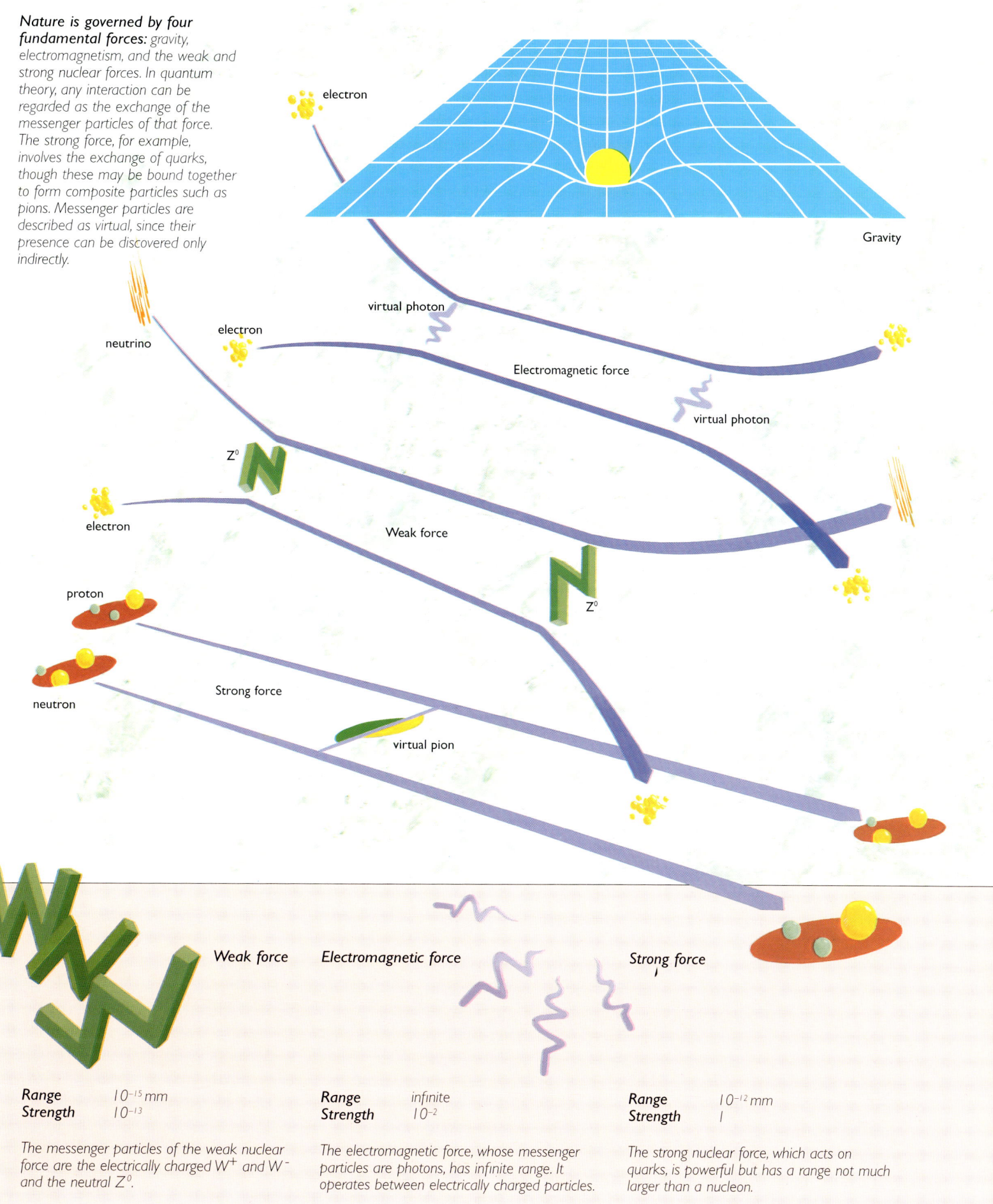

Nature is governed by four fundamental forces: gravity, electromagnetism, and the weak and strong nuclear forces. In quantum theory, any interaction can be regarded as the exchange of the messenger particles of that force. The strong force, for example, involves the exchange of quarks, though these may be bound together to form composite particles such as pions. Messenger particles are described as virtual, since their presence can be discovered only indirectly.

Weak force
Range 10^{-15} mm
Strength 10^{-13}

The messenger particles of the weak nuclear force are the electrically charged W^+ and W^- and the neutral Z^0.

Electromagnetic force
Range infinite
Strength 10^{-2}

The electromagnetic force, whose messenger particles are photons, has infinite range. It operates between electrically charged particles.

Strong force
Range 10^{-12} mm
Strength 1

The strong nuclear force, which acts on quarks, is powerful but has a range not much larger than a nucleon.

THE FIRST MOMENT
- *The birth of space, time, and matter*

At the birth of the universe, all matter and energy were compressed into a fiery mass of unimaginable density. The earliest moment that can be spoken of with certainty came after a period called the Planck time — the incredibly brief interval of 10^{-43} seconds.

The entire universe that is observable today — which may be only part of some unknown whole — occupied a space 10^{-20} times smaller than an atomic nucleus. And it is believed that at an even earlier moment, the four basic forces — gravitation, the strong and weak nuclear forces, and electromagnetism — were unified into a single force (p. 28). However, present theories break down at this point because it is not yet known how gravitation and the other forces were united.

Present theories can be tentatively applied after the Planck time. But detailed information about the state of the universe at this time was irrevocably altered by the next major event in the history of the universe. When it was 10^{-35} seconds old, the era of inflation began: a period of fantastically rapid increase in size, during which the universe swelled up to at least 10^{50} times its previous size. Although the characteristics of inflation are incredible, theorists believe this idea — devised originally by Alan Guth in the US — must be correct, since it explains a number of puzzling features of today's universe. First, it accounts for the fact that the closely-packed infant universe expanded neither so slowly that gravitation could crush it back into nothingness, nor so fast that it could thin out before galaxies and stars could form. Relativity tells us that space is in general curved, and that the amount of curvature depends on the amount of mass present in unit volume — the density. Too great a density and the universe would close around on itself and collapse; too low a density and space would open out uncontrollably.

Mathematical analysis shows that to account for this fine balance, the density of the universe must have had a particular, or critical, value, and no other, at a very early time. In fact, at 10^{-33} seconds

Matter, energy, space, and time were created in an inferno of particles and forces unknown in the modern universe. Within a billionth of a second, the initial force affecting all matter had frozen out into the forces known today. It took three minutes for the kaleidoscope of ephemeral particles to give way to stable atomic nuclei. Millennia were to pass before complete atoms were formed.

after the Big Bang, it could not have differed from the critical value by more than one part in 10^{-49}. Such a precise agreement is predicted by the theory of inflation, and no other way of explaining it is known.

Inflation also solves the "horizon" problem. In our present universe, we see galaxies moving away from each other into the depths of space. There is a horizon beyond which we cannot see, since galaxies at that distance are receding at the speed of light. The horizon is different for each galaxy. Consider two galaxies, close to our horizon but lying in opposite directions from us: they will both be able to see us, lying at *their* respective horizons, but not each other.

THE FIRST MOMENT

The problem that arises is this: such galaxies, which cannot receive signals from each other, are very similar in what they contain, and in the density and distribution of their matter. The cosmic background radiation, too, is at exactly the same temperature, whichever part of the sky it comes from (p. 36). Why should the universe be so homogeneous when each part of it sees only a small part of the whole?

Certainly, as we go back ever closer to the Big Bang itself, these widely separated parts of the universe were closer together than they are now. But the time that had elapsed from the beginning was shorter, too. There still had not been time for radiation to travel between them, evening out any possible differences. On the basis of the galaxies' present motions, there would never have been any contact between these regions.

But the theory of inflation solves the problem. Before inflation occurred, the presently observable universe filled a tiny volume, much smaller than the horizon distance of that time. All parts of it reached temperature equilibrium, with any differences evened out. We see this equilibrium today, long after the most widely separated parts of the universe have ceased to influence each other.

The universe we see now is, in fact, very "smooth" on the large scale. Inflation shows how any initial irregularity in the density of matter and energy would be ironed out to a vanishingly low value. The problem now is to explain how sufficient irregularity survived inflation to make it possible for matter to clump into galaxies and stars.

Lastly, physicists have discovered that presently the universe contains between 100 million and a billion photons for each atom in the universe. Why it should be this number rather than any other is explained by the idea of inflation.

Inflation started when the strong force was beginning to separate from the electroweak (combined electromagnetic and weak) force, but before their separation was complete. By the time this had happened, the temperature of the universe had dropped to a ten-thousandth of what it had been at the end of the Planck time, although it was still at the enormous value of 10^{28} K. Inflation was triggered by extreme supercooling of the infant universe.

Supercooling occurs when a system falls below a temperature at which it normally changes state, yet fails to undergo that change. For example, steam normally changes to water when it is cooled below 100°C (212°F). This is because its particles move increasingly slowly as the steam cools, until they begin to coalesce. In practice, this happens only when grains of dust or ions are around to act as centers on which liquid drops can grow. If there are no such centers, the steam can be supercooled below 100°C (212°F) while remaining in its gaseous state.

At 10^{-35} seconds, the universe was on the brink of a change of state marked by the separation of the strong and electroweak forces. But it continued to cool until it was about 10^{-32} seconds old before this transition occurred. While it was in its supercooled condition, a false vacuum was formed, the properties of which were very different from those of the true vacuum.

Usually the density of energy in any system – whether in the form of radiation or of matter – decreases when the volume of the system increases and its particles become less densely packed. But the false vacuum state is one in which the energy density stays constant during expansion. Relativity theory shows that the existence of the false vacuum, with its constant energy density, would have caused a vast force of repulsion. Inflation occurred.

While the inflation period lasted, the universe doubled in size every 10^{-35} seconds. It doubled hundreds of times, with the result that its volume grew at least 10^{50} times. The temperature plummeted from 10^{28} to 10^{23} K (see below).

Inflation ceased when the change of state had occurred and the strong and electroweak forces had separated. The energy density of the false vacuum was released, much as the latent heat stored in steam is released when it changes state to become water. This burst of energy created a vast number of atomic particles, which reheated the universe to the temperature it had been at when inflation began. In this chaos of radiation and exotic particles began the building of matter as we know it today.

The meaning of temperature

The temperature of matter is judged by the movement of the particles of which it is composed. At high temperatures they move swiftly, with great energy; at low temperatures they slow down and their energy is less.

There is a relationship between the temperature, volume, and pressure of a gas. At a given temperature, the pressure is greater the smaller the volume; and at a given pressure, cool gas occupies a smaller volume than it does when at a higher temperature.

In fact, for every drop in temperature of 1°C (34°F), a gas shrinks by $1/273$ of the volume it occupies at 0°C (32°F), the freezing point of water. In the 19th century, the British physicist Lord Kelvin argued that at −273°C (−459.67°F), the energy of movement of molecules would be zero. Actually, the molecules would still possess *some* energy of motion – they would vibrate a little – but they would have no spare energy to impart to any other substance.

The absolute scale of temperature is based on this fact. It begins from "absolute zero," which is actually −273.18°C (−459.67°F), and each unit, called a kelvin, is by definition equal to one degree of the Celsius scale. Thus 0°C is equivalent to 273.18 K.

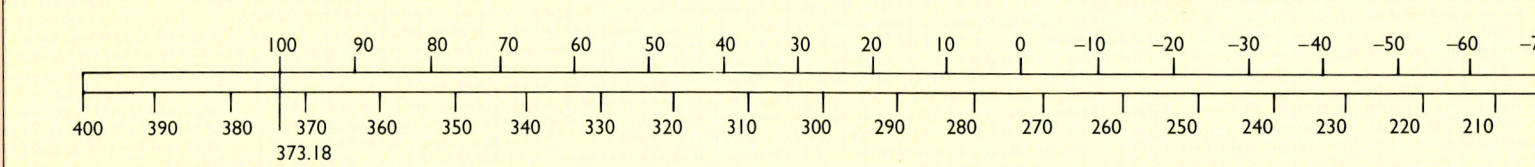

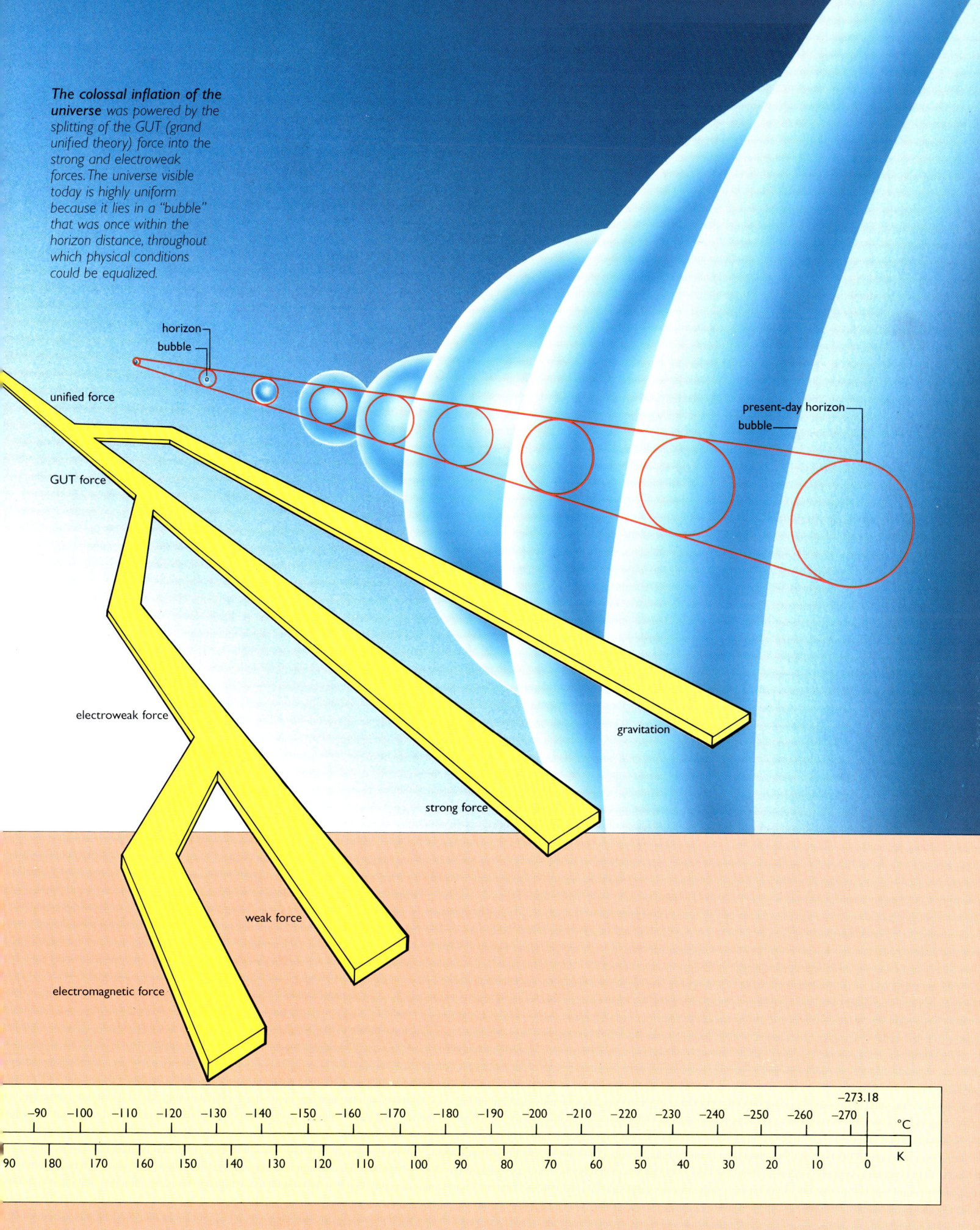

BUILDING MATTER
• *Stable atoms make their appearance*

The universe emerged from its burst of inflation at the colossal temperature of 10^{27} K. It was a seething mass of particles appearing fleetingly and disappearing again. In this "quark soup," quarks and antiquarks were constantly colliding with each other and being annihilated, giving out a flash of radiation. At first there was sufficient energy to produce quarks and antiquarks to balance this destruction, but as the universe expanded and cooled, production of these particles ceased. But because there had been a slight preponderance of quarks over antiquarks, these survived to form the basis of matter today.

Later, the temperature dropped far enough to ensure that quarks and leptons could not turn into each other. There was an abundance of W and Z particles, so the weak force was still united with the electromagnetic force (p. 28). But when the universe was a hundred-millionth of a second old, and its temperature had fallen to about 10^{14} K, the electroweak force separated into the weak and electromagnetic forces, as they are known today. This change did not, however, have such dramatic effects as the separation of the strong force, which powered inflation.

Then, when the universe was still only a millionth of a second old, the quarks began to bind together to produce hadrons, particles sensitive to the strong nuclear force. These included protons and neutrons, consisting of triplets of quarks, and mesons, consisting of quark pairs. Theory shows that some of the quarks may still exist singly, and could occasionally reach the Earth in cosmic rays. However, none have definitely been observed so far.

The hadrons present at this time included equal numbers of protons and neutrons. They were constantly being converted into each other by high-energy reactions. Protons and electrons combined to form neutrons, also giving out neutrinos. At the same time, neutrons were colliding with positrons to form protons and antineutrinos.

However, these reactions required the presence of a great number of electrons and positrons. These were produced in pairs by the annihilation of high-energy photons, but when the universe was about one second old, such pair production ceased, and with it the equilibrium between the numbers of neutrons and protons. There were fewer neutrons than protons, because slightly more energy is required to produce a neutron from a proton than vice versa.

Nevertheless, there were still a good number of neutrons in existence – one to every six protons. By themselves neutrons are unstable: they have a 50 percent chance of decaying after 15 minutes. So now the free neutrons present began to decay into protons and two kinds of lepton: electrons and neutrinos. The leptonic era had commenced.

Yet, there was time for neutrons to combine with other particles. In other words, they took part in the creation of the elements by nuclear fusion, the process that releases energy in a hydrogen bomb and in the stars.

Nuclear fusion became significant when the universe was one minute old. To begin with, one neutron and one proton combined to produce nuclei of heavy hydrogen, or deuterium. Each deuterium nucleus had one proton, which determined its identity as a type, or isotope, of hydrogen. But unlike ordinary hydrogen nuclei, it also contained a neutron. Next, some deuterium nuclei collected a second neutron to produce nuclei of tritium, an even heavier isotope of hydrogen.

Reacting with another proton, the tritium then built a nucleus of a new element, helium, containing two protons and two neutrons. Virtually all the neutrons were incorporated into helium nuclei, so it is possible to work out that at this time there were 10 hydrogen nuclei to every helium nucleus. Other nuclei were also built up in small proportions by a variety of reactions. They included helium 3 (two protons, one neutron), beryllium 7 (four protons and three neutrons), and lithium 7 (three protons and four neutrons).

Radiation still played the dominant role in the universe when it was a few

In the first millionth of a second of its life, the universe was populated by photons and free quarks. Then quarks combined to form hadrons – protons and neutrons, consisting of triplets of quarks – and mesons, consisting of pairs of quarks.

minutes old. This radiation era lasted from about one minute to about 10,000 years. At first, much of the radiation consisted of high-energy gamma rays produced during the breakdown of deuterium and build-up of light nuclei. The radiation lost energy as the universe continued to expand, although photons could still knock particles around and be scattered in turn. It would have been impossible to see far in any direction.

As the universe continued to expand and cool, the photons lost energy. The masses of the nuclear particles did not change, and 10,000 years after the Big Bang, their mass came to dominate the energy of the radiation, even though there were still about 10 billion photons to every proton. The universe had entered the matter era.

Even so, the photons were still moving at high enough speeds to break up any atoms that might briefly form when a nucleus and an electron came together. But after 300,000 years, the energy of even the most energetic photons had dropped sufficiently to allow atoms to form and survive. Atoms and photons could now coexist, pursuing their respective paths, since the electrically neutral atoms did not significantly

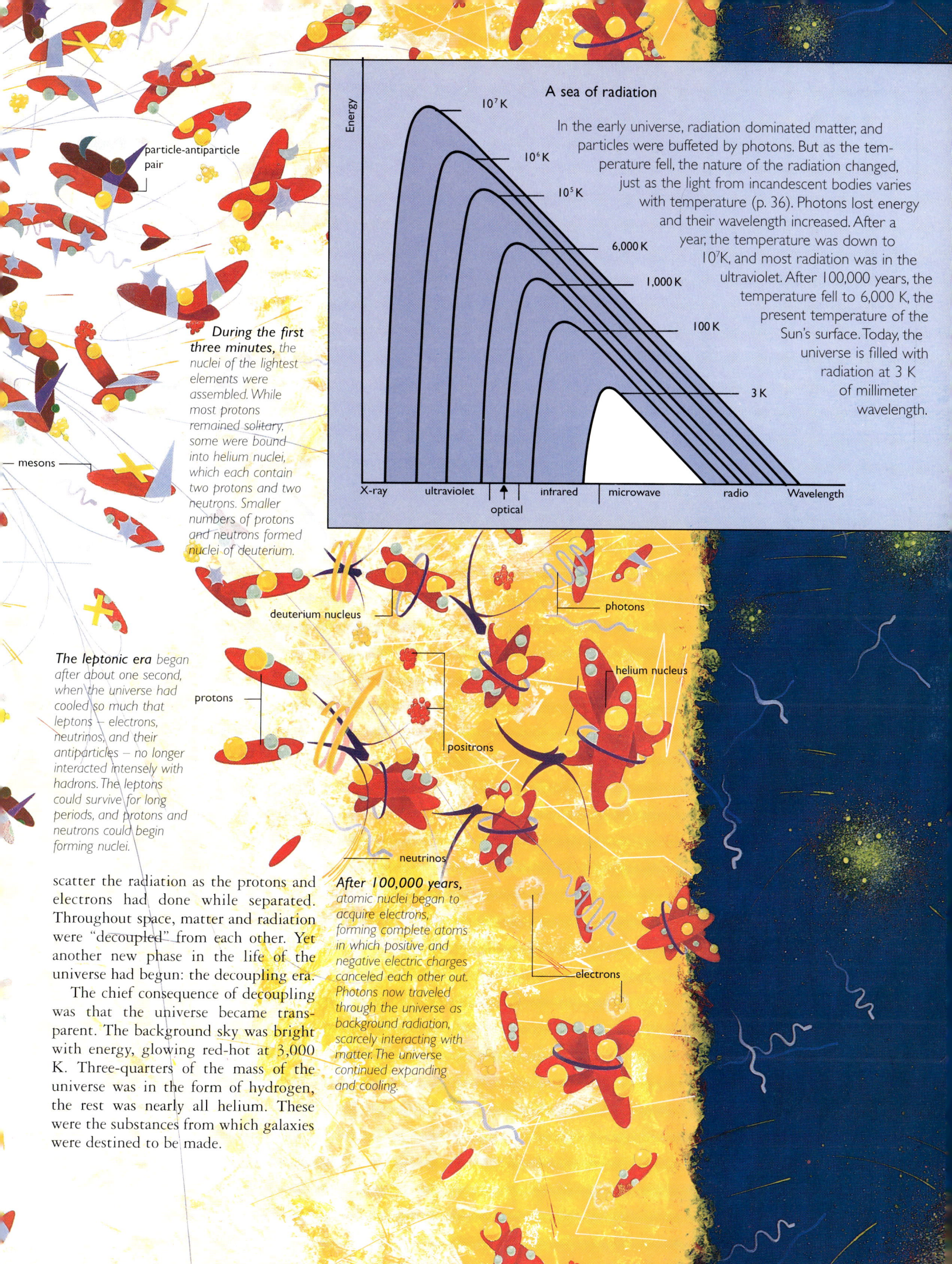

A sea of radiation

In the early universe, radiation dominated matter, and particles were buffeted by photons. But as the temperature fell, the nature of the radiation changed, just as the light from incandescent bodies varies with temperature (p. 36). Photons lost energy and their wavelength increased. After a year, the temperature was down to 10^7 K, and most radiation was in the ultraviolet. After 100,000 years, the temperature fell to 6,000 K, the present temperature of the Sun's surface. Today, the universe is filled with radiation at 3 K of millimeter wavelength.

During the first three minutes, the nuclei of the lightest elements were assembled. While most protons remained solitary, some were bound into helium nuclei, which each contain two protons and two neutrons. Smaller numbers of protons and neutrons formed nuclei of deuterium.

The leptonic era began after about one second, when the universe had cooled so much that leptons – electrons, neutrinos, and their antiparticles – no longer interacted intensely with hadrons. The leptons could survive for long periods, and protons and neutrons could begin forming nuclei.

scatter the radiation as the protons and electrons had done while separated. Throughout space, matter and radiation were "decoupled" from each other. Yet another new phase in the life of the universe had begun: the decoupling era.

The chief consequence of decoupling was that the universe became transparent. The background sky was bright with energy, glowing red-hot at 3,000 K. Three-quarters of the mass of the universe was in the form of hydrogen, the rest was nearly all helium. These were the substances from which galaxies were destined to be made.

After 100,000 years, atomic nuclei began to acquire electrons, forming complete atoms in which positive and negative electric charges canceled each other out. Photons now traveled through the universe as background radiation, scarcely interacting with matter. The universe continued expanding and cooling.

THE COOLING UNIVERSE
• Big Bang's lingering glow

The temperature of the universe has played a significant role throughout its development. The electromagnetic radiation that bathed matter from the moment of the Big Bang has cooled along with matter itself. It survives today, providing perhaps the most powerful evidence for the occurrence of the Big Bang.

A basic concept used in studies of radiation is that of a black body. This is an ideal object that, by definition, absorbs all electromagnetic radiation falling on it and reflects none whatsoever, which is why it is called a black body. Nevertheless, such a body does radiate energy, in a manner that varies depending on its temperature. When it is cool, it emits most energy at long (radio) wavelengths, less at others. As energy is fed into it and its temperature rises, its total emission increases, while its peak emission shifts progressively to shorter and shorter wavelengths.

The peak moves first through the infrared region, and then into visible wavelengths. The object gradually becomes bright red and then yellowish. At about 6,000°C (10,832°F), it becomes white-hot and its light resembles sunlight. As its temperature rises further, its radiation peaks successively at blue, violet, ultraviolet, and eventually X-ray and gamma-ray wavelengths.

The universe can be regarded as a black body, cooling as it expands from the Big Bang. At each moment, most of its radiation is concentrated at the wavelength corresponding to its temperature, but a drop in temperature is to be expected as its energy spreads out more and more thinly. Space should be filled today with radiation at the average temperature of the present universe.

This fact was first recognized by the theoretical astronomers Ralph Alpher, Robert Herman, and George Gamow, working in the United States as early as 1948. But their conclusion was largely forgotten since there was then no equipment to detect such radiation.

Only in 1965 did the situation change, when Arno Penzias and Robert Wilson at the Bell Telephone Laboratories in Holmdel, New Jersey, constructed a radio telescope to observe radiation from the Galaxy, and picked up a constant background "noise." This did not change in strength as they pointed their horn-shaped antenna to various parts of the sky, nor as the Earth rotated, as it would have been expected to do if its source was celestial.

It was at this juncture that they were contacted by Robert Dicke of Princeton University, who was trying to detect radiation left over from the Big Bang. It immediately became evident that this was just what Penzias and Wilson had observed. The wavelength of the noise detected by their equipment was 7.3 centimeters. The radiation has since been detected over a range of wavelengths, and it has been found that it peaks at about 1 millimeter, corresponding to a temperature of 2.7 K.

In 1965, there were alternative theories of the origins of the universe besides that of a hot Big Bang, but none of them had predicted the existence of such radiation and none could account for it. Now almost all astronomers have been persuaded that the background radiation is the strongest evidence for a high-temperature Big Bang about 13 billion years ago.

More remote "slices" of space contain more stars, although they are fainter. If the slices continue to infinity, they send us an infinite amount of light.

Olbers' paradox

The expansion of the universe and its finite age are relevant to a simply stated, yet profound puzzle: Why is the sky dark at night? In 1826, Heinrich Olbers, a German physician and astronomer, discussed this question, and his argument is now called Olbers' paradox, although the problem had been mentioned long before.

Olbers started by assuming that the true brightness of all stars is the same. This is an approximation, but it is satisfactory as a start. He also assumed that the stars are evenly distributed through space and that the universe is static.

Consider, then, a sphere with a radius of, say, 10 arbitrary units and a thickness of 1 unit, centered on the Earth. It is easy to calculate the total brightness of the stars lying within this shell. In another sphere of twice the radius, but the same thickness, the individual stars will be of only a quarter the brightness (since the brightness of an object falls to a quarter if its distance doubles). On the other hand, there will be four times as many stars to observe, since the volume of such a spherical shell will be four times as great. So the total amount of light coming from this shell will be the same as that from the first one. The argument can be continued by assuming further shells, each contributing the same amount of light. The brightness of the sky will be given by the total of the light from all these shells, which, by assumption, can be continued to infinity.

However, modern research has shown that the universe is not static, but expanding. As a result of the expansion, the radiation emitted from a distant object loses energy – the farther away it is, the greater the extra dimming. So in the argument above, the more distant spherical shells contribute less light.

More importantly, the universe is not infinitely old: it started at the time of the Big Bang. Even in an infinite universe, most light could not yet have reached the Earth. The observable universe is only a tiny fraction of what would be needed to make the sky bright.

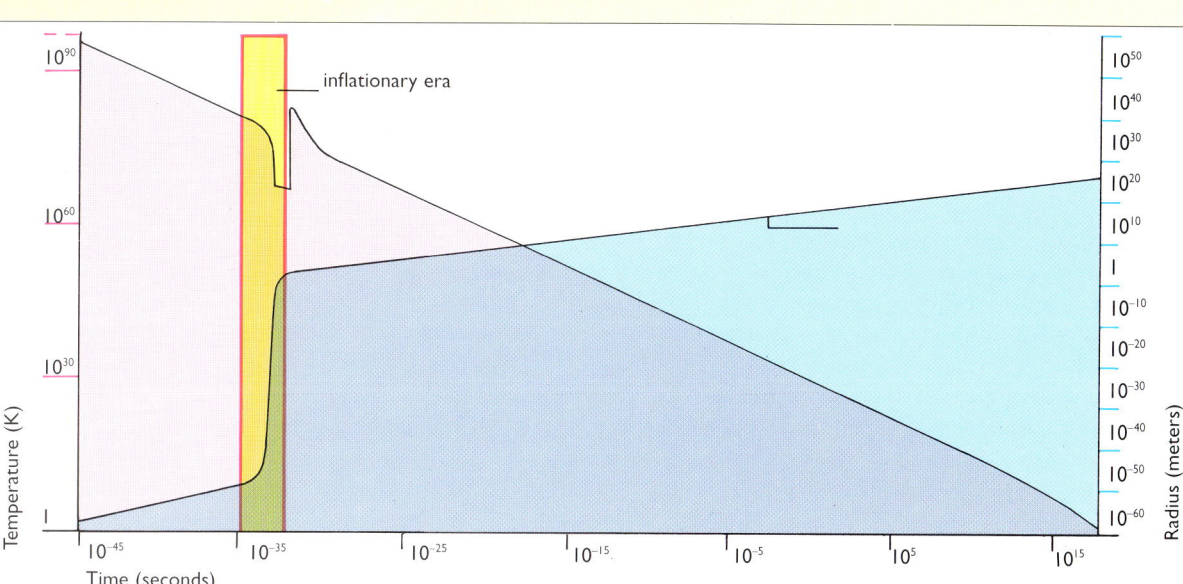

The temperature of the universe has fallen continuously since the Planck time, except for a brief period of reheating marking the end of inflation. It fell to levels at which quarks could no longer exist singly and formed protons and neutrons. These combined to make the lightest atomic nuclei. Later electrons attached themselves to the nuclei to form atoms. Today the temperature is down to about 3 K.

PROTOGALAXIES
● *The first celestial bodies*

The primordial gas seems to have expanded for two billion years before the next great event in the evolution of the universe occurred: the appearance of embryonic galaxies. While the gas was thinning and cooling everywhere, in some places it was slightly denser than elsewhere. Here the gas expanded slightly more slowly because of its own gravitational attraction, until the expansion stopped and was turned into contraction.

In such areas, the material of future galaxies slowly concentrated, probably condensing to form disks of material even much larger than the galaxy, or cluster of galaxies, that was to be born from it. The amount of material in these protogalaxies varied from one place to another, and observation of the universe today gives every reason to suppose that often a single cloud gave birth to one of the clusters of galaxies now visible.

There is a slight problem in understanding how the process of "seeding" galaxies could commence. For gas to begin to gather at one place rather than another, there initially had to be some minute differences in density from place to place. But after the "smoothing" of the universe by the process of inflation, it would seem that the distribution of matter throughout all space would have been too uniform to allow this process to begin. However, recent research indicates that "fossil" traces of inflation left their mark in the young universe. One of the most widely discussed concepts is that of cosmic strings.

Cosmic strings, if they exist, consist of tremendously dense matter and energy – in fact, surviving samples of the false vacuum that initiated inflation (p. 30) – trapped in long, closed loops. Their masses would be measured in thousands of billions of tonnes per centimeter, yet any that now survive will have dimensions of millions of light-years. But formerly, loops of all sizes would have existed, and would have vibrated very quickly – almost at the speed of light –, emitting energy in the form of gravitational waves as they did so. As it lost energy, each loop would shrink until it became a concentrated lump of very dense material, ready to act as a seed around which a galaxy or, perhaps, a cluster of galaxies could form.

Attractive though this suggestion is, it presents certain theoretical difficulties, and its originator, Neil Turok in the United States, has modified and developed the idea into another stimulating theory. He now suggests that after inflation, space was left with a "knot" texture. Each knot was a tiny region of space where energy was concentrated, distorting the geometry of space-time. When the supercooling of the universe that occurred during inflation had ended, the knots in space gradually became unraveled. As each one disappeared, a spherical blast wave of energy blew out from it. Where this met particles, it pressed them together, and such concentrated matter later produced embryonic galaxies.

In this theory, the ending of inflation may also have created strings of energy which would leave behind a wake as they moved, just as a speeding ship leaves a wake in the water. Such wakes would be an additional cause of clumping in the universe.

Whatever the processes by which density differences were created, only after about two billion years did they become marked. Where density was greater, protogalaxies formed, so that by the time the universe was seven billion years old, the process of expansion had made it possible for galaxies to form in considerable numbers.

It seems that relatively concentrated regions of gas may first have condensed into large clouds of cold, dark matter, which later broke up to produce galaxies. The resulting galaxies would have an immense range of sizes – from one hundred times the mass of our galaxy down to a hundred-thousandth of it. Clusters of galaxies would also grow by attracting new galaxies to them, and the largest clusters would even be attracted to each other to produce superclusters.

The possibility of galaxy formation from hot rather than cold matter has been considered, using computers to investigate the results. The outcome was not encouraging at first because a vast number of very large clusters of galaxies would be created, though observations now seem to favor this scenario.

What is more, a computer simulation of formation from cold, dark matter shows that on larger and larger scales, the clustering of galaxies decreases: there would be fewer extremely large clusters than might be expected if they formed from matter that was hot, but observation indicates otherwise. In short, material is not evenly spread through the universe on a large scale, and is also unevenly distributed on a smaller and smaller scale. Astronomers refer to this as hierarchial clustering.

Computer studies also demonstrate not only that the first condensations occurred about two billion years after the Big Bang, but also that smaller-scale condensations took place as the larger condensations merged. They make it

A million years after the Big Bang, the universe was a featureless sea of hydrogen and helium gas. It became differentiated into regions of higher and lower density that ultimately gave birth to the present clusters and superclusters of galaxies. Computer simulations attempt to reproduce the details of this process. The images shown here were produced by a simulation based on the idea that neutrinos, generally thought to be massless, may have a non-zero mass. The distribution of matter in the universe is shown at one-eighth and one-third of its present age, and finally at the present day. The organization of matter into strings separated by voids is a fair representation of the observed grouping of clusters and galaxies.

evident that nearly all traces of any earlier structure – knots, for instance – were lost as the previously separate masses merged. It seems, therefore, that while small clumps of hot matter formed first, their merging produced the hierarchy of galaxy sizes that astronomers see today.

There are still problems with this picture of the formation of galaxies. When astronomers study the way the brightness of the galaxies is distributed among the different types, they find that there are now more faint galaxies than bright ones. The reason may be that as embryonic galaxies gently collided and merged, energy was released and gas was driven off, limiting the sizes and hence brightnesses of the galaxies that formed. Only in a small proportion of cases would a sufficient number of gas clouds collide in a short enough time to build up those very large galaxies that we observe.

The universe had been dark for two billion years. Then, while the gas clouds were still contracting, they began to give birth to the first generation of stars. The scene was set for the evolution of the universe as it is known today.

THE GRAND DESIGN

The universe is an enormous hierarchy, ranging from vast clusters of galaxies to minute fragments of interplanetary debris. What is more, everything in it obeys the same laws. Even at the farthest distances that can be observed, there is the same kind of material, behaving just as it does in the nearer reaches of space.

The description of the universe that follows deals with those celestial objects that formed first and goes on to describe the others in the order in which they most probably came into being. It starts with clusters of galaxies, and even clusters of clusters, and then moves on to the galaxies themselves, then the stars and clouds of gas and dust of which the galaxies are composed, and so on to the planets and their moons.

Galaxies were once known as island universes because they are conglomerations of stars and gas, and sometimes a lot of dust as well, separated by great distances. They can be observed because they emit light and other radiation, which travels over vast distances of space before it reaches Earth. This radiation takes time to travel. Even though it covers almost 186,420 miles (300,000 km) each second, it takes billions of years to reach Earth from the most distant galaxies. So here, astronomical distances on the largest scale are being considered.

The next class of objects in the hierarchy of the universe is comprised of the gas clouds, or nebulae, within the galaxies. These are the birthplaces of the stars, which form the main visible content of galaxies. Billions of stars go to make up even the smallest galaxies. They shine by their own light, which is generated within them by nuclear fusion. Our Sun is just such a body, its central regions resembling a vast collection of exploding nuclear bombs.

Moving still farther down the hierarchical order is cool matter associated with the stars and derived from the material from which the stars themselves formed. Little is known of such material except in the case of our own solar system, where much of it has condensed to form lumps that orbit the Sun. These lumps are planets, some of which have smaller condensations – their moons – orbiting them.

The Sun's family is comprised of nine major planets and scores of moons, as well as thousands of fragments known variously as minor planets, planetoids or, more usually, asteroids – "little stars." Other material has condensed into small pieces of matter, the comets and meteors. The Sun's planetary system and interplanetary debris are all cold and are visible only because they reflect the Sun's light. There is good reason to believe that this system is not unique in our galaxy.

The matter and energy that poured forth from the Big Bang produced the hierarchy of the present-day universe. Minute irregularities in the infant universe led to the growth of structures on all scales. Largest of all were the superclusters – groupings of galaxy clusters. A map of the superclusters is the basis of this computer-generated image, expressing the dynamism of the universe revealed by modern cosmology.

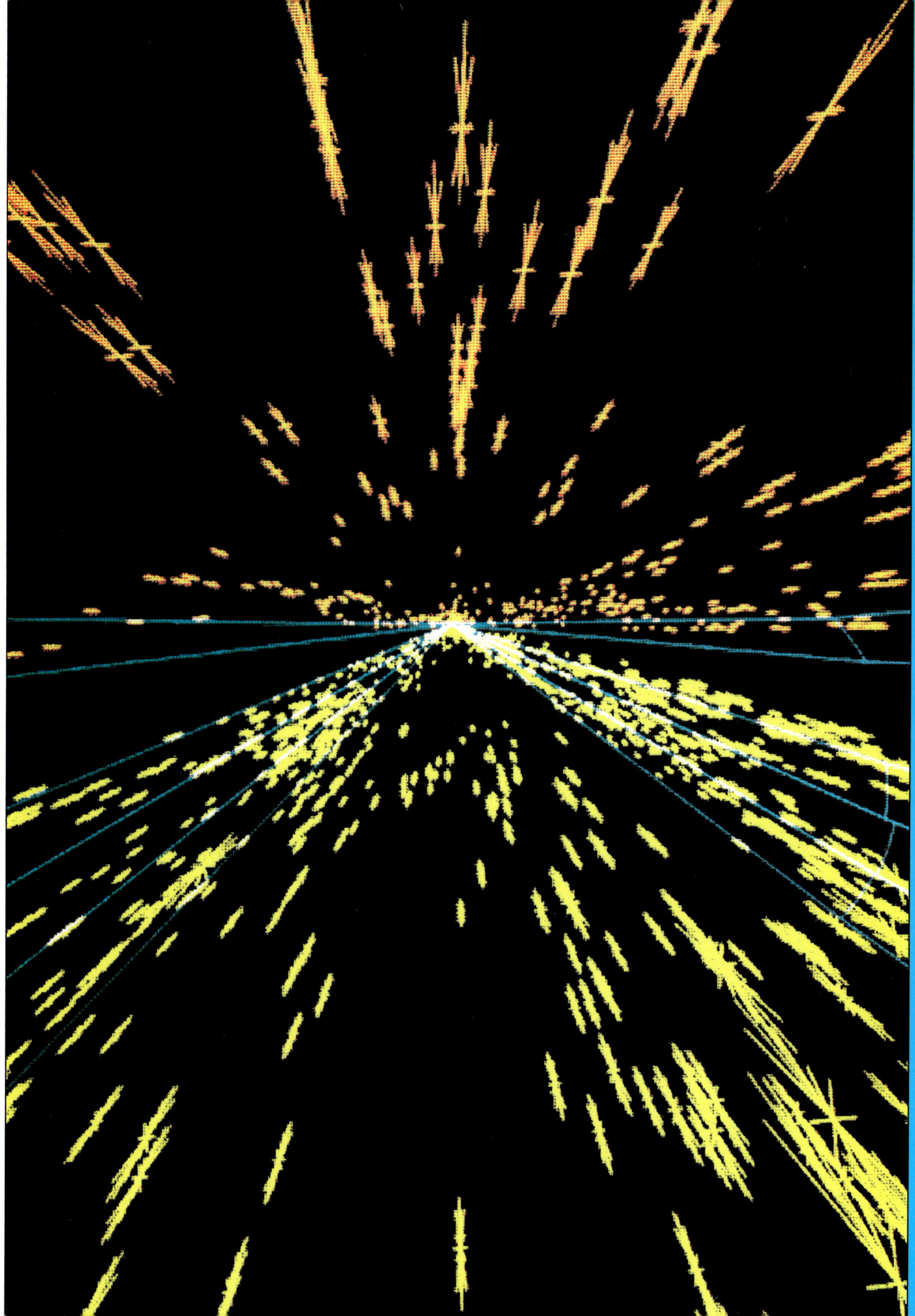

OBSERVING THE UNIVERSE
● *New windows on the cosmos*

From the earliest times until a little less than 400 years ago, the only way of observing the universe was simply to look at it. There were no optical instruments that could be used – only the eyes, and this brought with it certain restrictions. Above all, it meant that there was a limit to the faintness of the objects that could be detected, because the pupil of the eye is small and therefore allows only a small amount of light to fall on the light-sensitive retina.

The situation had changed by the early 17th century, as a result of the application of the telescope to astronomy by Galileo and others. Because the telescope gathers a greater amount of light than the human eye, thousands of faint stars suddenly became visible for the first time. Astronomers could now probe farther into space. Moreover, the telescope also magnified the objects that it viewed. While this made no difference to the images of stars, which remained point-like, the telescope revealed to the observer detail that the naked eye could not detect in extended objects, notably the Moon and planets.

With the new instrument, Galileo made discoveries that helped overthrow the theory of an Earth-centered universe. For example, he detected the phases of Venus, showing that the planet revolves around the Sun, and he observed moons revolving around Jupiter. Moreover, he found that the Moon was a mountainous world, not unlike Earth.

As the centuries passed, larger and more efficient telescopes were designed and built, and the bounds of the known universe were pushed still farther into space. In the 19th century, the power of the telescope to penetrate space was greatly enhanced by the introduction of photography. Whereas the eye's sensitivity is not increased by staring for a long time at something too dim for it to detect in the first place, a photographic plate will allow images to build up while it is exposed, and so enable dimmer objects to be detected.

Another vitally important tool of the astronomer is the spectroscope. It was in the mid-17th century that Isaac Newton showed that "white" light is a combination of light of all colors. He did this by passing sunlight through a prism to split it into its component parts, forming a colored band – a spectrum. The later spectroscope was a more sophisticated device designed to disperse the components more thoroughly.

Light was first passed through a narrow slit to produce a thin beam, before going to a prism or other light-scattering device. Each color in the light formed a separate line – an image of the slit. Although with many types of source the lines merged to form a continuous band of colors, gases at low pressure emitted certain very specific colors when incandescent, and absorbed the same colors when cool. These colors appeared as characteristic bright or dark lines crossing the spectrum.

Thousands of dark lines appear in the spectrum of the Sun. Investigations, notably by Gustav Kirchhoff and Robert Bunsen in Germany during the 1860s, proved that these were the "fingerprints" of chemical elements in the Sun's outer layers. Later still, it was found that all stars display such spectra, and that detailed examination of the lines could not only indicate the elements present, but also show much about temperatures and pressures in the stars.

It was found that visible light does not constitute the whole of the spectrum of sunlight. There is also a component lying beyond the violet end of the visible spectrum, and hence called ultraviolet radiation. It affects human skins by producing a tanning reaction. There is another component, which lies beyond the red end of the spectrum, called infrared radiation. It is also described as heat radiation, since it warms objects that absorb it and is the main type of radiation that is given out by objects at the normal temperatures of surroundings on Earth.

Although a few astronomical observations were made in the infrared and ultraviolet bands, they were not pursued, primarily because their significance was not appreciated. The work of the fourth Earl of Rosse, who as early as 1877 measured the Moon's temperature by observing its infrared radiation, was followed only sporadically for decades. Yet James Clerk Maxwell had shown (pp. 14–15) that even ultraviolet, infrared, and visible light make up only part of a wider-ranging spectrum of electromagnetic radiation. New fields of observation were awaiting the entry of astronomical pioneers using new techniques.

X-ray sources in the skies (right, above) were detected by the HEAO-1 X-ray observatory. They show a concentration of bright sources toward the center of the Galaxy, but away from the obscuring material in the galactic plane, a more uniform distribution of such remote objects is visible.

The form of the Galaxy at visible wavelengths is revealed (above) by the patterns of light and dark nebulae and the distribution of stars down to the 10th magnitude – 40 times fainter than the dimmest stars the naked eye can see. The best optical telescopes can see much fainter stars, but to penetrate the heart of the Galaxy, other wavelengths are required.

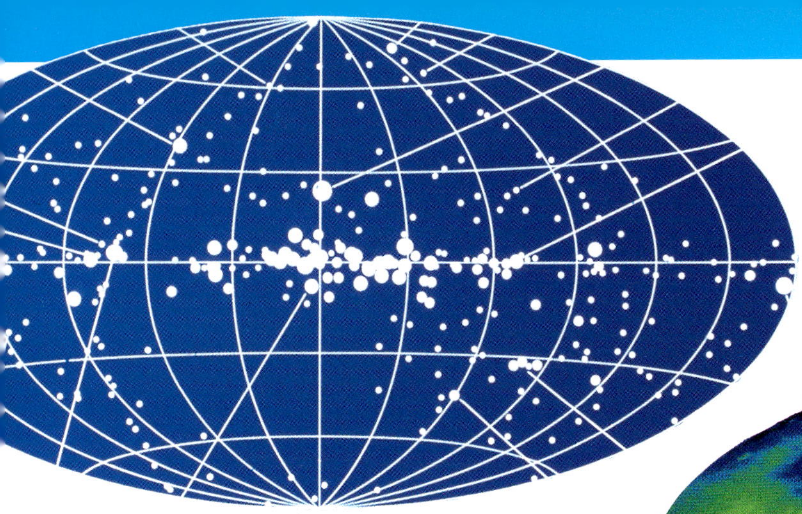

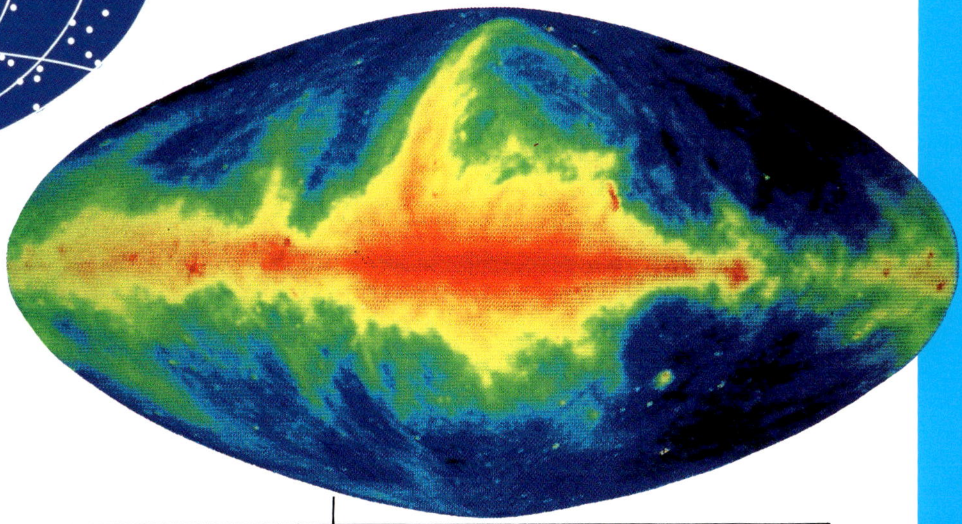

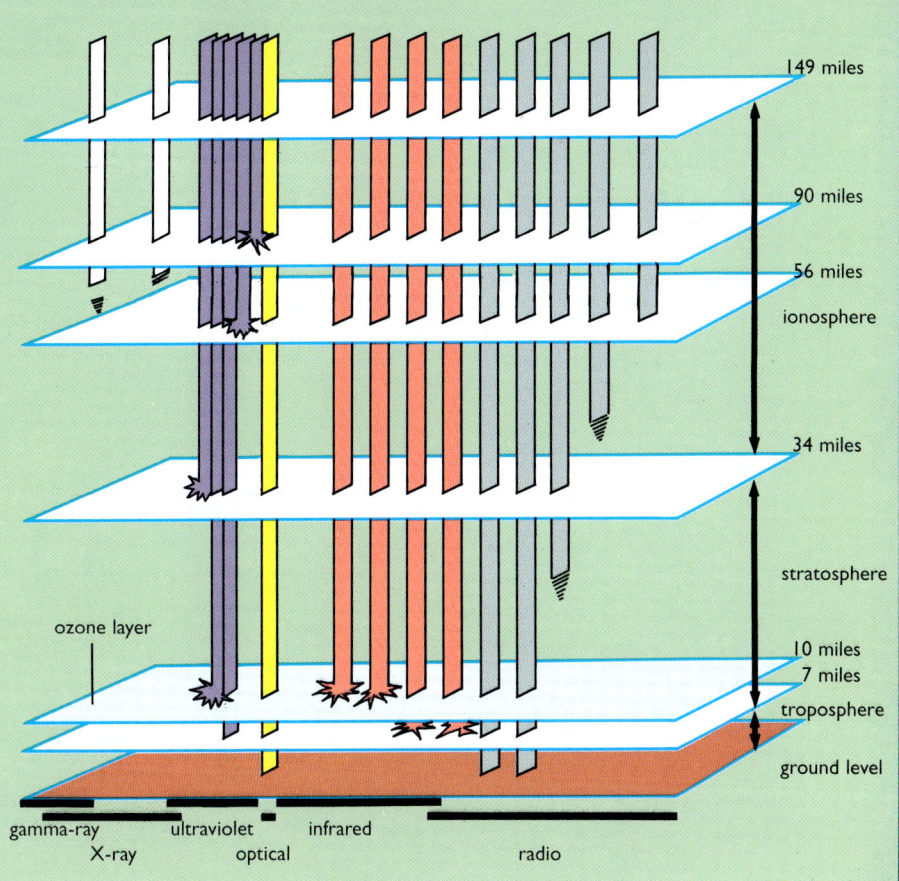

Longer-wave radio observations (right), while dominated by the disk of the Galaxy, reveal new structures – loops and streamers of active material extending high out of the galactic disk.

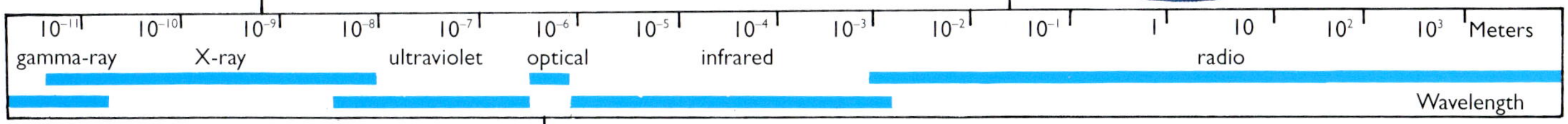

10^{-11} 10^{-10} 10^{-9} 10^{-8} 10^{-7} 10^{-6} 10^{-5} 10^{-4} 10^{-3} 10^{-2} 10^{-1} 1 10 10^{2} 10^{3} Meters
gamma-ray
X-ray
ultraviolet
optical
infrared
radio
Wavelength

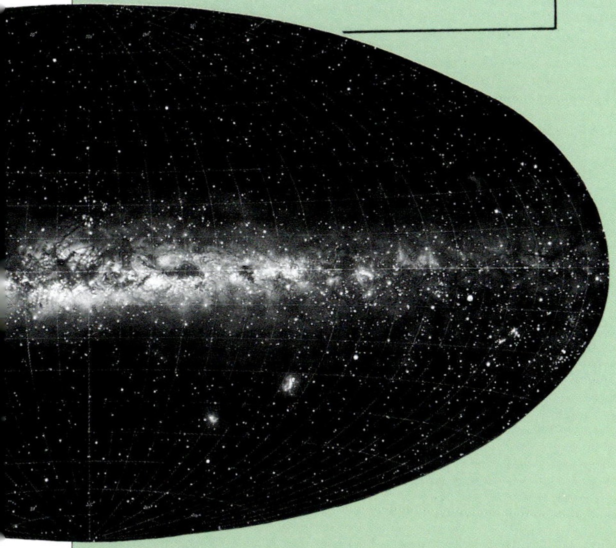

The barrier of the atmosphere

149 miles
90 miles
56 miles
ionosphere
34 miles
stratosphere
ozone layer
10 miles
7 miles
troposphere
ground level
gamma-ray
X-ray
ultraviolet
optical
infrared
radio

The thin skin of air that sustains life on Earth shields us from harmful radiations, while blocking the astronomer's view of the universe at most wavelengths. There are two major "windows": one allows visible light to reach the ground. The other lets through most radio wavelengths, but the ionosphere contains relatively high concentrations of ions which act like mirrors to some radio waves from above and below. Other wavelengths are absorbed at various heights. The ozone layer absorbs damaging ultraviolet rays.

TELESCOPES OF TODAY
● *New technology for the astronomer*

Optical astronomy *continues to advance despite the explosive growth of "invisible" astronomy. The Very Large Telescope (VLT), the world's biggest telescope, is in fact four 27-foot telescopes. Each component is housed in its own dome at the European Southern Observatory in Chile. A computer links the four to work as if they are one telescope with a mirror 54 ft in diameter.*

The dishes of the Very Large Array *(far right) make up the world's largest single radio telescope. There are 27 receivers, moving on railroad tracks that form a Y shape. Each arm of the Y is about 12½ miles long. The VLA is located at Socorro, New Mexico.*

The first major breakthrough in extending the wavelength range of astronomical observations came in 1932, when an American radio engineer, Karl Jansky, made the serendipitous discovery that radio waves were emanating from the Milky Way. His work was extended by another American, an amateur radio ham, Grote Reber. Yet only since the mid-20th century did radio astronomy really get under way.

The results have proved of vital importance in extending our understanding of the universe. For instance, radio telescopes have made it possible to detect molecules in space that are invisible at optical wavelengths; they have made it clear that galaxies are not the peaceful islands of gas, dust, and stars that they were once thought to be; and, above all, they have shown that the universe is bathed in background radiation that survives from soon after the Big Bang.

The desire to extend the observable spectrum farther led, in the 1960s, to the construction of special infrared telescopes, such as the United Kingdom Infrared Telescopes (UKIRT) set up on a mountain peak in Hawaii in order to observe in very clear, dry air. But even instruments placed as high as this are severely limited by water vapor in the Earth's atmosphere, which acts like a blanket, absorbing much of the infrared radiation from space. This limitation can only be overcome by Earth-orbiting space probes.

In 1983, the joint European and United States Infrared Astronomy Satellite, IRAS, was launched. It had many triumphs, discovering places where starbirth is proceeding in our galaxy, detecting cool stars that are invisible optically, and producing evidence of the formation of planetary systems around other stars.

The equipment on an infrared telescope, including that on IRAS, has to be refrigerated, because at normal temperatures the instrument and its surroundings emit vast quantities of infrared radiation that would swamp the images of faint celestial sources. To view the shorter infrared wavelengths, the component that detects the focused infrared radiation is cooled with liquid nitrogen to $-196°C$ ($-321°F$); to view longer wavelengths, it must be cooled to $2°C$ ($3.6°F$) above absolute zero with liquid helium.

Space technology has, of course, proved of immense importance at all wavelengths. Its most obvious applications have been the close-up examination of planets and moons, but of even wider significance has been the extension of observation into the extreme ranges of the electromagnetic spectrum. Although some ultraviolet radiation reaches the surface of the Earth, most is blocked by the atmosphere. Space satellites can detect this radiation, and also X-rays and the even shorter-wavelength gamma rays originating from space.

Great ingenuity is often used in designing modern astronomical equipment. For instance, since X-rays are not deflected when they penetrate a material, it might seem impossible to construct an X-ray telescope. However, if the X-rays graze the surface of a suitable metal, they can be reflected, and so can be made to form an image. Such grazing-incidence telescopes are now commonplace in X-ray satellites.

The optical telescope has gained a new lease on life with the advent of computer-controlled optics. This has permitted the building of large instruments that exploit the interference of light waves falling on separated mirrors, a

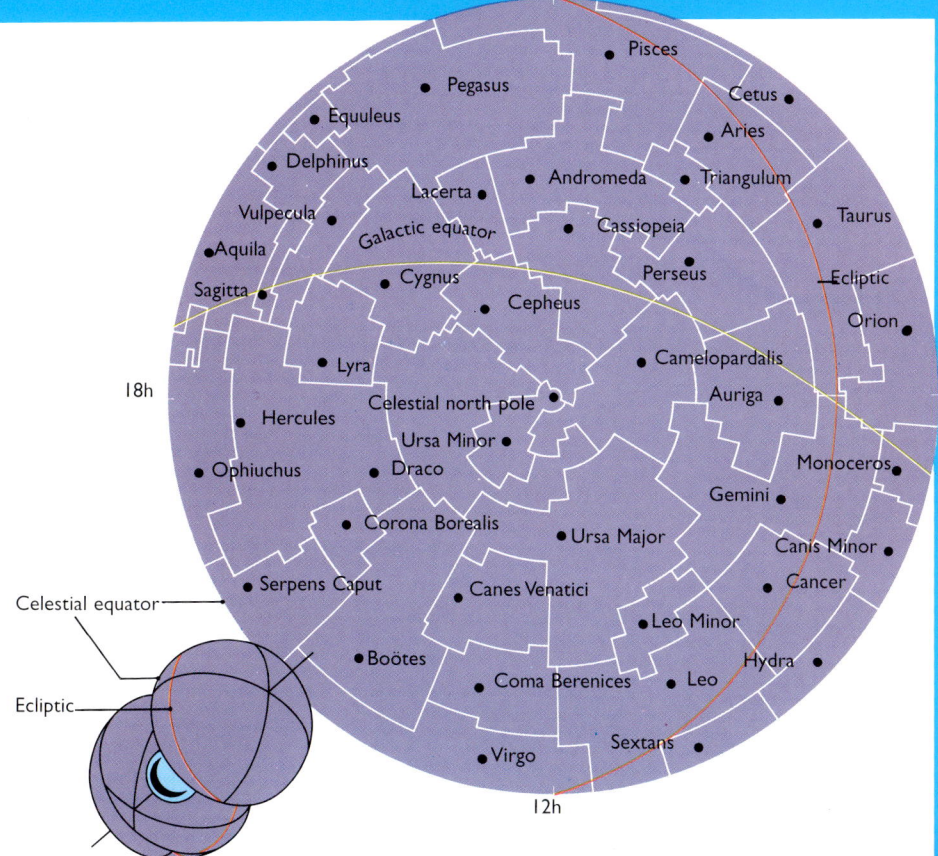

technique derived from radio astronomy. Such instruments include special telescopes that can measure the diameters of large nearby stars.

Another revolution has occurred in the processing of the information from telescopes, satellites, and space probes. Until the advent of the microchip and the computers based on it, the measurement of the positions of images on photographic plates and their analysis was a time-consuming process. For every night spent at the telescope, an astronomer might spend five days or more processing his results. Now automatic computer-controlled plate-measuring machines do the task in a few hours. Another computerized electronic technique that has proved extremely useful is the replacement of the photographic plate by more sensitive electronic detectors, known as charge-coupled devices (CCDs).

Extremely valuable also is the electronic process of image enhancement. Computer processing of the data from spacecraft — for example, the Galileo probe to Jupiter — is vital to obtain really good and detailed pictures. The astronomers can program enhancements of certain features to help the analysis of observations in each wavelength range. The resulting images make interpretation easier and more exact.

The sky is mapped by imagining it to be a huge sphere centered on the Earth (above). It is divided into areas, each representing a single constellation (top and below).

The celestial poles and equator lie above their equivalents on Earth (below). The ecliptic is the Sun's apparent path around the sky — where the plane of the Earth's orbit cuts the celestial globe.

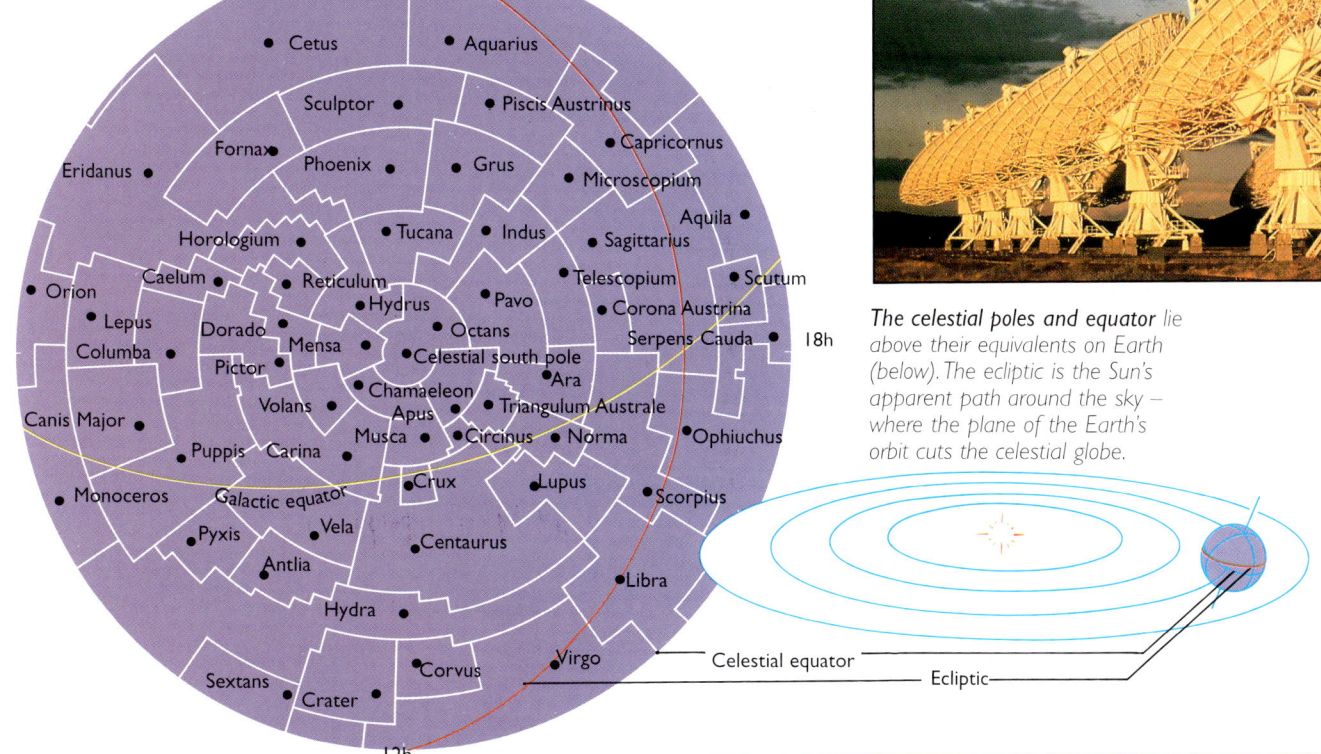

YARDSTICKS OF THE COSMOS
● *Techniques for measuring the universe*

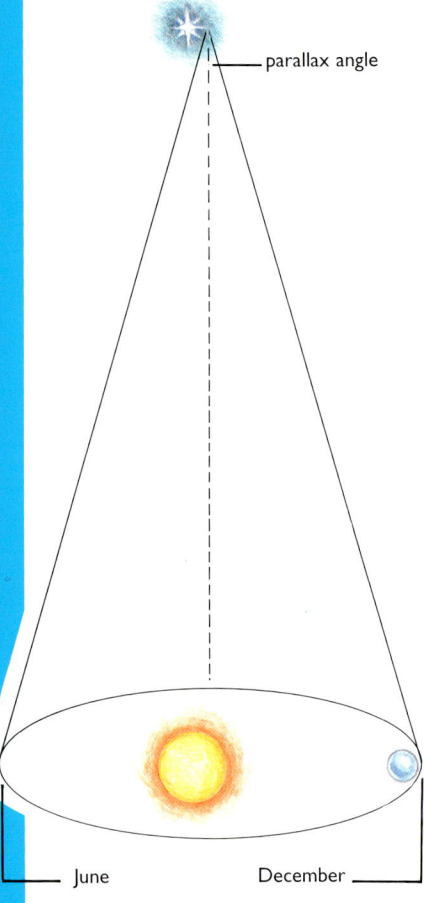

The distance of a nearby star can be found by observing its apparent yearly movement, caused by the Earth's motion around the Sun. Its position in relation to more distant stars is measured at intervals of six months, from opposite sides of the Earth's orbit. The star's parallax is defined as half the apparent angle of shift. Since the size of the Earth's orbit is known precisely, the star's distance can be calculated.

Estimates of the distances of the celestial bodies by the earliest astronomers were woefully small. But it is not surprising that they should have erred: the difficulty of measuring the distances of the stars, in particular, is such that the feat was not accomplished until 1839. Today, astronomers have a large repertoire of methods of measuring distance; they are of varying precision and each is limited to a certain range of distances.

Since the mid-20th century, the distances of the Moon, Sun, and planets, long established by older methods, have been obtained with unprecedented accuracy by means of three new techniques. Laser beams are bounced from the surface of the Moon and their return is carefully timed to yield the distance with such precision that it can be calculated to the nearest centimeter; radar pulses are used for determining the distance of the Sun; and distances for all the planets except Pluto have been determined by means of radio signals from visiting space probes. Finding the distances of the stars, however, is not quite as simple because of their sheer immensity.

The method used for the nearer stars is triangulation, which is also used by terrestrial surveyors. The astronomer selects a star that, because of its apparent brightness or comparatively large motion, is believed to be near the Earth, and measures its position in relation to other stars that are more distant. Six months later, when the Earth is at the other side of its orbit around the Sun, the star's apparent position is measured again. The star will seem to have moved in relation to the background. The star's annual parallax is defined as equal to half the angle through which it has moved (see illustration). Using trigonometry, the actual distance of the star can be calculated, since the diameter of the Earth's orbit is known accurately.

When such an angle has thus been translated into a distance, the result is huge. To make the figures understandable, the speed of light provides a convenient yardstick. On this scale, the Moon is 1.3 light-seconds away from us and the Sun is 8.3 light-minutes away. But the nearest star, Proxima Centauri in the southern sky, is 4.3 light-years from us. The light-year is equal to the colossal distance of approximately 5.9 trillion miles.

Astronomers have found other, less direct, ways of determining distances beyond the range of parallax measurements. The most important means of taking larger strides into space is to use a form of celestial "lighthouse" – the Cepheid variable stars (pp. 86–87). These stars vary in brightness in a regular way, each repeating a pattern of variations over a fixed period measured in days or months. Their importance lies in the fact that their true brightnesses and their variation periods are intimately linked.

If the period of a Cepheid variable star is measured, its true brightness is known. From its apparent brightness the distance of the star can be determined. Cepheids have been detected in some of the nearer galaxies, thereby showing them to be millions of light-years distant. Yet what of those galaxies that are too far off for the Cepheids they contain to be detected?

One of the consequences of the Big Bang is that all the galaxies are moving away from each other. The farther a galaxy is from us, the faster its velocity of recession. Astronomers have detected this by the red shift of the tell-tale lines in a galaxy's spectrum: all the lines in the visible part of the spectrum are shifted toward the red, and lines in the red part of the spectrum are shifted into the infrared, and so on.

This happens because the crests and troughs of the electromagnetic radiation reach Earth less frequently the larger a galaxy's speed of recession is. Consider, for example, the conspicuous twin yellow lines produced by sodium in a stationary source of light: the wave crests reach Earth 500 trillion times a second. But in the same light coming from a galaxy receding at one-tenth the speed of light, each crest has farther to travel than the one preceding it. They therefore reach Earth only about 450 trillion times per second, and the light seems to us to have a lower frequency and longer wavelength – to be reddened, in fact. Conversely, if the light source were moving toward Earth at this velocity, the crests would arrive 550 trillion times per second, and a shift of the line toward the blue end of the spectrum would be observed.

By measuring red shifts, astronomers can

determine the speeds of recession of galaxies, and since they know that the farther off they are, the faster they move, they can calculate their distances. Of course, this is possible only if they know exactly how velocity increases with distance, and at the moment, there is uncertainty about this. The ratio of speed to distance – called the Hubble constant – could be as low as 9.3 miles per second per million light-years or as high as 18.6.

The value of the Hubble constant affects astronomers' ideas about the size of the universe: the smaller the constant is, the farther away a galaxy of a given speed of recession must be, and the larger the universe must be. And the larger the universe is, the longer it has taken to reach that size. At present, the upper value of the age of the universe is taken to be 18 billion years, but this value is highly dependent on the accuracy of the distance scale that astronomers have painstakingly constructed.

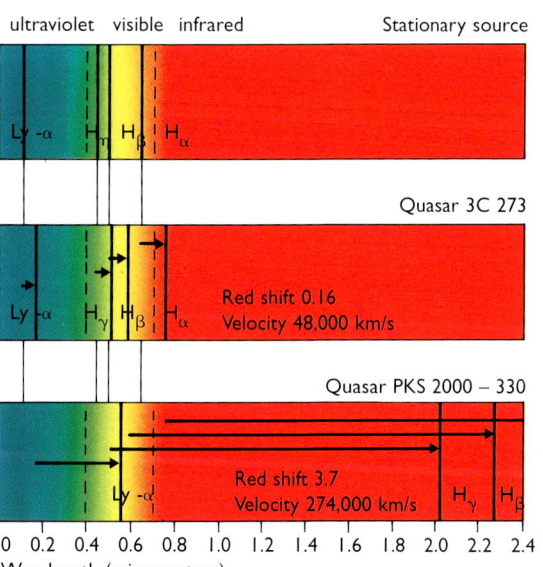

Some of the furthest galaxies yet observed appear in this deep field image. It was created from observations made over a 10-day period by the Hubble Space Telescope in December 1995. All but one of the bright lights in the image is a galaxy. (The exception is the pointed star at left of center.) There are over 3,000 of them, most so dim they are 4 billion times fainter than the faintest star the naked eye can see.

The remoteness of quasars is revealed by their hugely redshifted hydrogen lines. In light from 3C 273, H_α is shifted from the visible into the infrared; in that from PKS 2000 – 330, the normally ultraviolet Lyman-α line is shifted into the visible.

SUPERCLUSTERS
● *The largest structures in the universe*

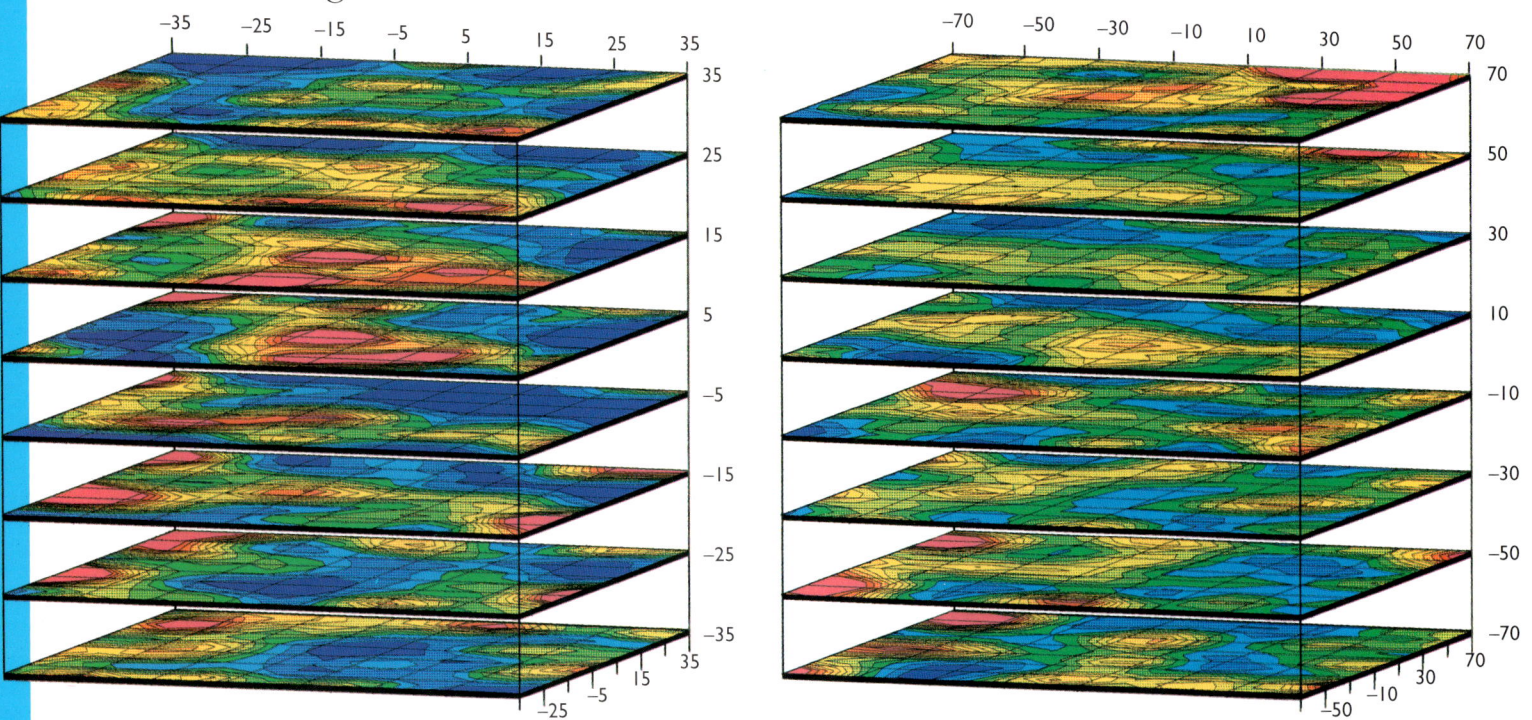

Theories of cluster formation *from cold matter do not appear to match newly-made observations. The blocks show the distribution of clusters of galaxies, with our galaxy at the center of each block. The scales are in megaparsecs (a megaparsec is equal to just over three million light-years). The first three blocks show data from the Infrared Astronomical Satellite (IRAS), while the fourth is a computer simulation based on the theory that the clusters formed from cold dark matter (CDM). In the observations at the largest scale (third block), very high and very low densities of galaxies are more apparent than in the CDM simulation and galaxy formation from hot matter must therefore be considered.*

In the evolution of the hierarchy of the present universe, the largest structures appeared first: protogalaxies were formed and then shrank to become the galaxies that can now be observed. These, however, were probably grouped from the first into clusters and superclusters, the largest units that can be discerned and the highest stages of the cosmic hierarchy.

In the early years of last century, it was already known from the work of the English astronomer William Huggins that nebulae (so called from the Latin word for "clouds") were of two kinds – hazy patches of gas, and spiral or elliptical shapes. Vesto Slipher at the Lowell Observatory in Arizona and Francis Pease at the Mount Wilson Observatory in California began an intensive study of the spectra of spirals and found that almost all displayed red shifts (pp. 46-47): in other words, virtually all were moving away from the Earth at great velocities. No such general recession was displayed by other kinds of nebulae. The question arose of whether the spirals were even part of our "star-island," the Galaxy.

In the 1920s, Edwin Hubble at Mount Wilson was able to detect variable stars, of the type known as Cepheids, in a few nearby spiral nebulae. Since the true brightness of Cepheids can be determined (pp. 86–87), he could thus obtain definite distances for the spirals, showing that they lay far beyond the confines of our own galaxy. They came to be called spiral galaxies, for they are the same kind of system as the Galaxy. It became clear, too, that some other types of nebulae are also galaxies beyond our own.

By 1930, it had been established not only that most galaxies are receding from Earth but that their velocities of recession are greater the fainter, and therefore farther away, they are. This was the first evidence to indicate an expanding universe.

In the 1930s, the Swiss astronomer Fritz Zwicky, who had settled in the United States, discovered that galaxies are grouped into clusters, some of them thousands strong and spreading tens of millions of light-years across space. Later it became clear that our own galaxy and the famous spiral in Andromeda are members, together with a number of other galaxies, of a small cluster now known as the Local Group. Members of a cluster move under the influence of their mutual gravitational attractions, and this provided a simple explanation for those few galaxies that display blue shifts in their spectra. They are all

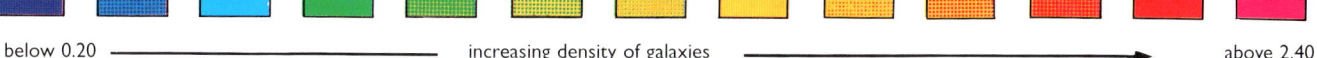

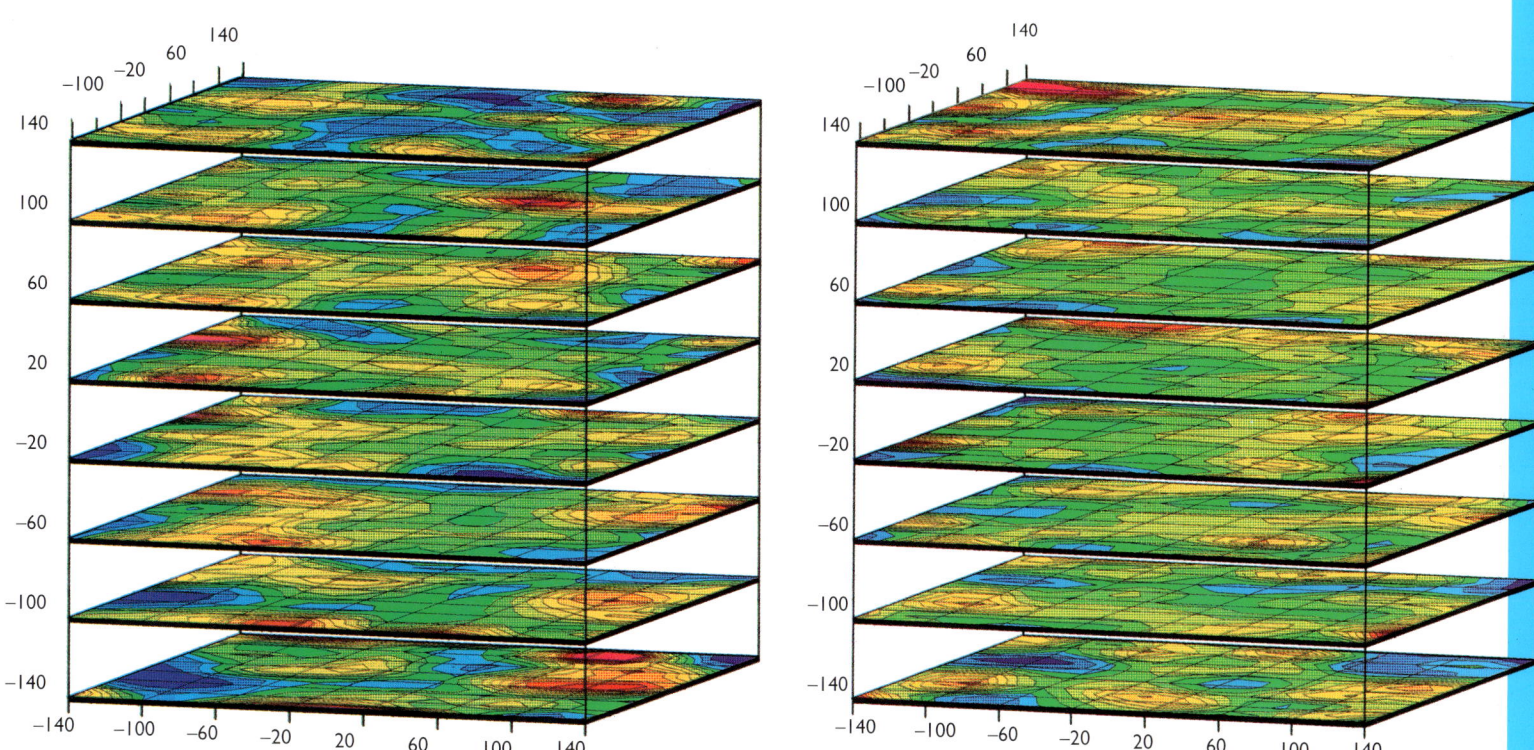

members of the Local Group that happen to be moving toward Earth at present.

Further work revealed a concentration of clusters of galaxies in a belt – a veritable Milky Way – around the northern and southern skies. Viewed from Earth, the center of this string appears to lie in a vast cluster of thousands of galaxies, the Virgo cluster – so called because its center appears to lie in the constellation Virgo. The string itself is therefore called the Virgo supercluster, or Local Supercluster. The Virgo supercluster is not a unique phenomenon; other superclusters also take the form of lines or strings. There seem to be no observations to lead astronomers to assume any structure larger than superclusters.

Astronomers are vigorously mapping out the superclusters. The Local Supercluster is comprised of not only the vast Virgo cluster, as already mentioned, but also our Local Group and dozens of smaller groups that can be seen in the constellations of Canes Venatici, Leo, and Crater. The supercluster's shape is a flattened ellipsoid, with one or two protruding sections, so that the whole requires a cube of 100 million light-years on a side to contain it. All the member clusters are attracted toward the central Virgo cluster: the Local Group is moving at about 155 miles per second toward a point about 20 to 30 degrees from its center.

There are other huge conglomerations – the Coma, Hercules, and Perseus superclusters, for example. Each has a large cluster at its heart, the cluster at the center of the Coma supercluster containing more than a thousand galaxies. The Hercules supercluster is centered on a much less dense cluster, which lies at a mean distance of 650 million light-years, and the Perseus supercluster is centered on a large cluster at a mean distance of about 235 million light-years. Another supercluster that deserves mention is the Southern Supergalaxy, centered on clusters in the constellations Fornax and Dorado.

Some astronomers believe that motions noted in the Virgo supercluster indicate the presence of an even larger supercluster farther out in space. It has been named the Great Attractor and lies around 150 million light-years away, in the direction of the constellations Hydra and Centaurus. The region contains a vast number of galaxies and dark matter, and extends for about 500 million light-years. This large concentration of mass exerts a powerful gravitational pull on our supercluster.

CLUSTERS OF GALAXIES
● *Galaxies that travel in concert*

Tenuous gas (in blue, below left) surrounds and links the galaxies in the cluster called Pavo 5. Where there is a significant amount of gas and dust between galaxies in a cluster, scientists reason that the cluster is relatively old, because the gas will have needed time to escape from the gravitation of the individual galaxies. Typical of the type of galaxy in the outer regions of a cluster is a spiral galaxy, such as NGC 6744 (below right) in Pavo (taken by the Anglo-Australian telescope). Toward the center of a cluster, large elliptical galaxies tend to predominate.

The size of clusters of galaxies ranges from mere pairs or triplets of galaxies to giants with thousands of members. Although there appear to be a few isolated galaxies drifting in space, they are probably former members of nearby clusters that have escaped.

All clusters consist of a variety of galaxies, of which there are three main types. Spirals have spiral arms with central bulges; ellipticals are approximately the same shape as a football, although they approach a spherical shape in varying degrees; irregulars, although fairly formless as their name implies, seem to be more closely allied to spirals than to ellipticals. Ellipticals and spirals usually predominate in a cluster. The galaxies vary greatly in size, from thousands to hundreds of thousands of light-years in diameter. The smallest are termed dwarf galaxies.

The cluster to which our own galaxy belongs, the Local Group, is a small one. How many galaxies it contains depends on whether certain outlying members are included. Certainly there are at least 26 chief members, and the figure is extended by general agreement to include a few other small, faint galaxies in the vicinity. These bring the total to at least 28, and no doubt there are further faint galaxies yet to be discovered.

All these galaxies are in motion: our own travels at a velocity of about 125 miles per second in relation to the group as a whole. The entire group is now known to have a diameter of almost four million light-years. No other galaxies are known within a distance of about 10 million light-years from our galaxy.

The nearest rich cluster to our Local Group lies in the constellation Virgo. This is a vast conglomeration of thousands of galaxies of every kind, of which 2,500 are fairly bright. Of these, 30 percent are elliptical galaxies. Prominent at the center of the Virgo cluster is a huge elliptical galaxy, M87. A similar feature is evident in many clusters.

The average distance of the Virgo cluster is 52 million light-years. It has a somewhat irregular shape, the longest dimension of which is about 8.8 million light-years. The members of the cluster are embedded in gas. Observations show that M87 is also a source of strong X-ray emissions.

The large cluster in Coma Berenices, the chief component of the Coma supercluster, can be examined in some detail because it lies above the gas and dust associated with the Milky Way. Containing way over a thousand galaxies, it lies at an average distance of 326 million light-years. It is approximately spherical and its diameter is estimated as at least 10 million light-years. It has a central core that is three times as dense as the rest of the cluster, containing a number of massive spiral and elliptical galaxies. Possibly, this is because the cluster is old, with

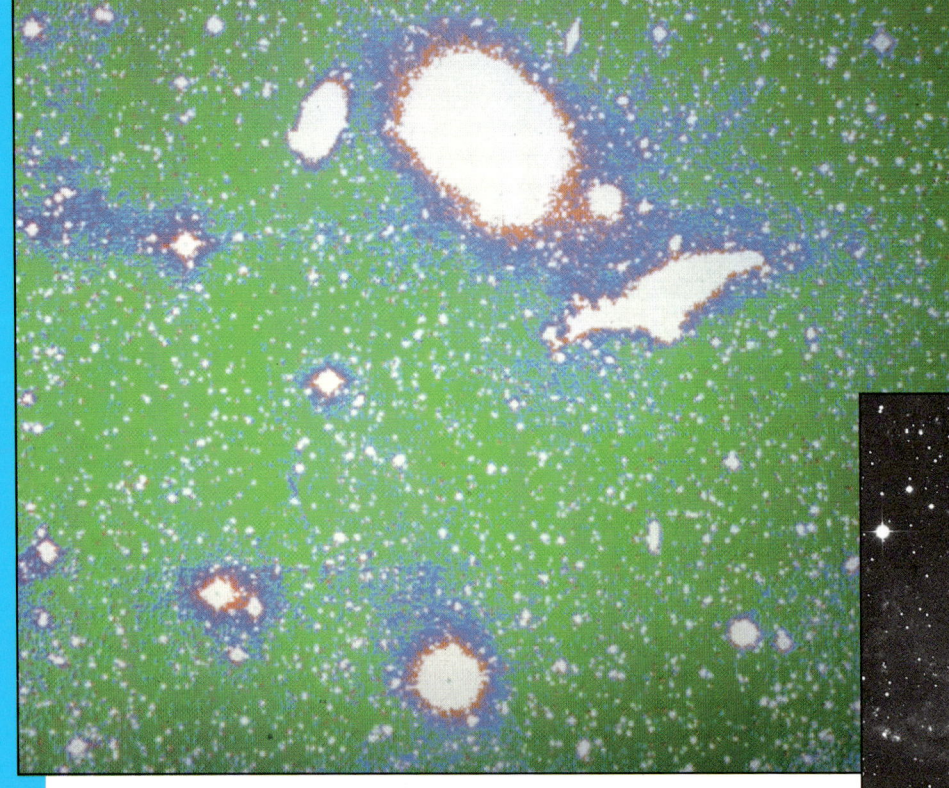

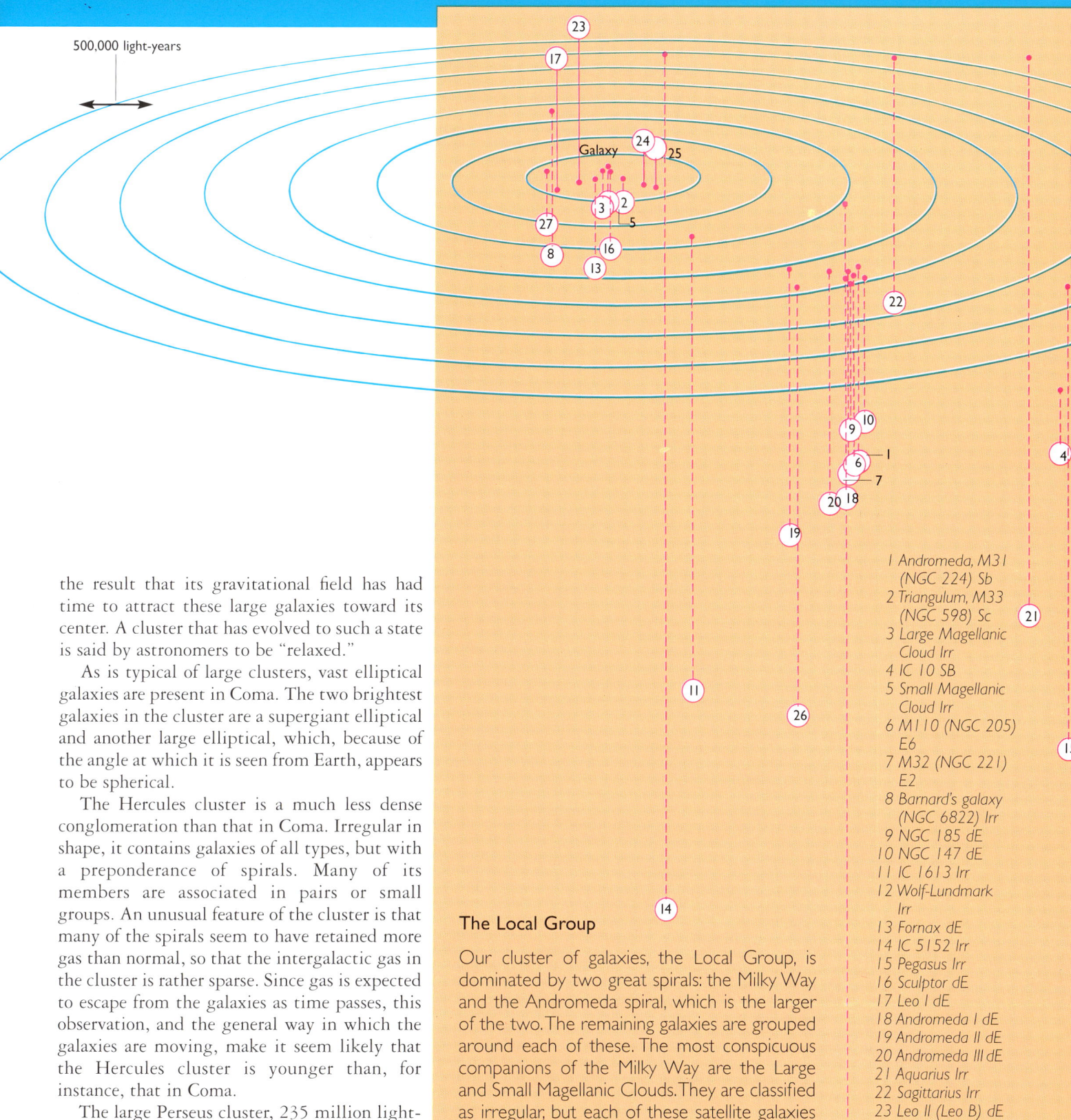

the result that its gravitational field has had time to attract these large galaxies toward its center. A cluster that has evolved to such a state is said by astronomers to be "relaxed."

As is typical of large clusters, vast elliptical galaxies are present in Coma. The two brightest galaxies in the cluster are a supergiant elliptical and another large elliptical, which, because of the angle at which it is seen from Earth, appears to be spherical.

The Hercules cluster is a much less dense conglomeration than that in Coma. Irregular in shape, it contains galaxies of all types, but with a preponderance of spirals. Many of its members are associated in pairs or small groups. An unusual feature of the cluster is that many of the spirals seem to have retained more gas than normal, so that the intergalactic gas in the cluster is rather sparse. Since gas is expected to escape from the galaxies as time passes, this observation, and the general way in which the galaxies are moving, make it seem likely that the Hercules cluster is younger than, for instance, that in Coma.

The large Perseus cluster, 235 million light-years from Earth, has a supergiant elliptical galaxy at its center. It also contains a number of radio sources — evidence of eruptive galaxies — and is rich in high-energy intracluster gas. Member galaxies have high velocities in relation to each other, and this, coupled with the evidence of the gas, may indicate that, like the Coma supercluster, the Perseus cluster is not particularly young.

The Local Group

Our cluster of galaxies, the Local Group, is dominated by two great spirals: the Milky Way and the Andromeda spiral, which is the larger of the two. The remaining galaxies are grouped around each of these. The most conspicuous companions of the Milky Way are the Large and Small Magellanic Clouds. They are classified as irregular, but each of these satellite galaxies seems to have a central bar of densely-packed faint stars, as do some spirals.

There is a third large spiral in the Local Group: M33. It possesses no companion galaxies. The remaining galaxies are dwarf ellipticals and small irregulars. It is not certain exactly how many faint, relatively nearby galaxies belong to the Local Group: the main members are shown here.

1 Andromeda, M31 (NGC 224) Sb
2 Triangulum, M33 (NGC 598) Sc
3 Large Magellanic Cloud Irr
4 IC 10 SB
5 Small Magellanic Cloud Irr
6 M110 (NGC 205) E6
7 M32 (NGC 221) E2
8 Barnard's galaxy (NGC 6822) Irr
9 NGC 185 dE
10 NGC 147 dE
11 IC 1613 Irr
12 Wolf-Lundmark Irr
13 Fornax dE
14 IC 5152 Irr
15 Pegasus Irr
16 Sculptor dE
17 Leo I dE
18 Andromeda I dE
19 Andromeda II dE
20 Andromeda III dE
21 Aquarius Irr
22 Sagittarius Irr
23 Leo II (Leo B) dE
24 Ursa Minor dE
25 Draco dE
26 LGS 3 Irr
27 Carina dE

dE = dwarf elliptical. Irr = irregular. For other symbols, see pp. 52 and 60. Leo III (Leo A) is too remote to be shown.

SPIRAL GALAXIES
● *Pinwheels of the cosmos*

The spiral galaxy, NGC 1232, in the constellation Eridanus, lies face-on to us. The arms are not continuous but have broken into segments. They are made of glowing gas and dark dust, and blue "knots" where new stars are being born. Older stars make up the yellow core.

The pinwheel shapes of spiral galaxies are among the finest of cosmic spectacles. It is not surprising that most people think of a spiral whenever the word "galaxy" is mentioned.

The essential features of a spiral are a central bulge and a disk that spreads out from the bulge, far into space. The arms almost always lie in the plane of the disk. Spirals are classified according to how tightly their arms are wound: from Sa, for the most tightly wound, through Sb to Sc, which have arms that are extremely widely spread. Generally, the central bulge is roughly spherical. However, in one class of spiral, the barred spirals, the central region is elongated to form a bar, from the ends of which the spiral arms protrude.

In addition to visible light, many spirals emit radio waves. These come from cold, dark hydrogen gas and from patches of glowing hydrogen. Furthermore, radio waves give evidence of halos of star clusters and gas around spirals. In fact, it is evident that, on account of their halos and gaseous envelopes, spirals are much larger and more massive than optical observations indicate. Some spiral galaxies are strong emitters of X-rays, which appear to emanate from the galaxies' central regions.

Spiral galaxies vary widely in size. The Andromeda spiral, M31, which we see almost edge-on, is a vast system, with a diameter of at least 124,000 light-years. Optical observations using techniques for detecting faint radiation indicate gas extending much farther out than this. Our own galaxy is also very large, with a diameter of not less than 100,000 light-years.

However, the other substantial spiral in our Local Group, M33 in Triangulum, has a diameter considerably less than either of these. Its main disk is only 52,000 light-years across, yet it is still large compared with most spirals. Even though it is smaller than the Andromeda, it can be easier to observe. It lies face-on to us and therefore appears comparatively large, and the nucleus and arms are easy to see.

Spiral galaxies contain myriads of stars. There are at least 200 billion in our own galaxy, and the Andromeda galaxy may have as many as a trillion. But these stars are not evenly distributed. The main concentration is in the central bulge. These are comparatively old stars, which collectively appear yellowish. Those in the disk and spiral arms are far younger: many are in fact being born out of the gas and dust of these regions at the present time, and they give both the disk and the arms a bluish color.

How did spiral galaxies originate? They could not form until there were collections of stars to act as nuclei that could attract other material gravitationally. The same processes that produced the protogalaxies (pp. 38–39) were also favorable to star formation.

Collisions between atoms and molecules in the early universe meant a loss of energy, with

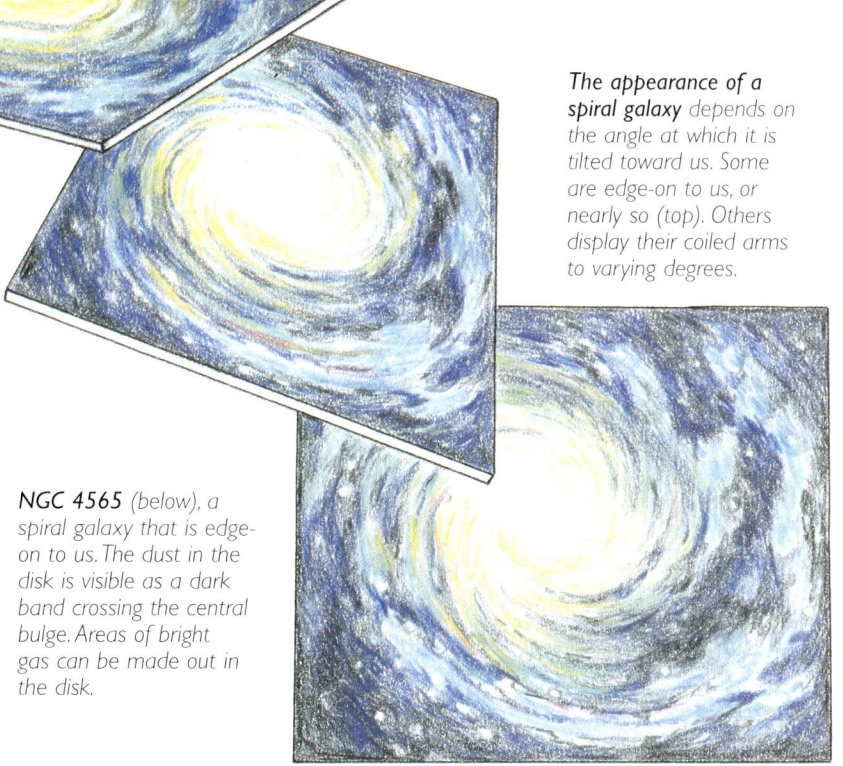

The appearance of a spiral galaxy depends on the angle at which it is tilted toward us. Some are edge-on to us, or nearly so (top). Others display their coiled arms to varying degrees.

the result that they slowed down and gravitation took over. The material gathered into small massive concentrations, the protogalaxies, which then began to contract rapidly under their own gravity to form central cores. At the same time, these broke into separate lumps of material, which contracted to become stars. Stars are still forming out of such concentrations of gas and dust today.

The material from which this central conglomeration of stars formed was in motion around the central core. Thus, as the stars formed they were concentrated into a rotating mass. The galaxy's rotation had the effect of throwing gaseous material outward to form a flattened disk surrounding the central regions. The disk would display differential rotation, with the slowest-moving material being closest to the rim. Initially, such a disk did not have spiral arms.

However, present theories suggest that density waves appeared. Due to odd disturbances in the gravitational field in the disk — caused, presumably, by the precise way in which material was distributed — stars would travel at different speeds. When stars passed each other, they caused mutual alterations in their orbits. These set up gravitational disturbances that traveled like waves through the material and produced spiral arms.

The density waves were influenced by a further factor: the rotation of the gaseous material itself. This rotation and the formation of arms together generated further shock waves. These caused gas molecules to crowd together into denser clouds. In due course, gravitational attraction between the molecules in each of these clouds caused stars to be born; it is these later-forming, bright young stars that give spiral arms their blue color.

The shock waves generated by these processes die away after a billion years at most. Nevertheless, if stars are to be born continually out of the disk material, as they seem to be, new shock waves must be set up all the time within spiral galaxies. Computer simulations of the conditions in spiral arms have shown astronomers how this could happen, and they also lead to other interesting results.

NGC 4565 (below), a spiral galaxy that is edge-on to us. The dust in the disk is visible as a dark band crossing the central bulge. Areas of bright gas can be made out in the disk.

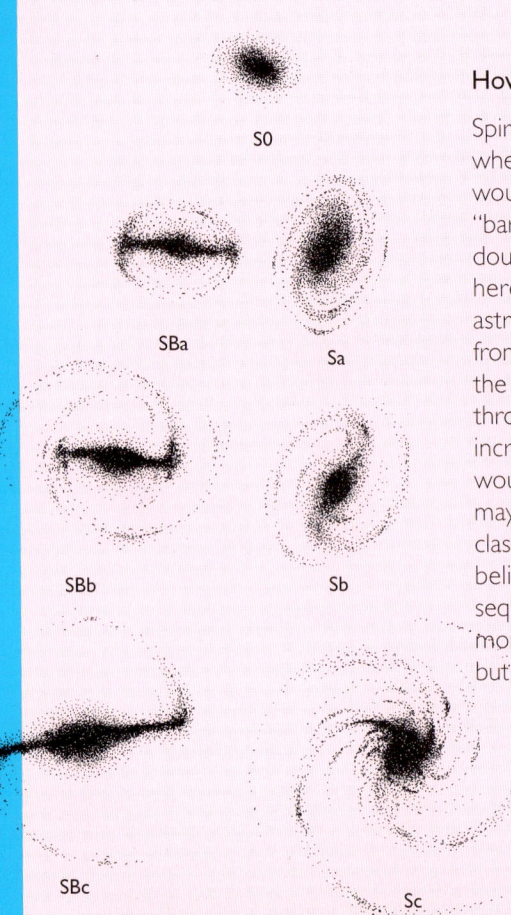

How spirals are classified

Spiral galaxies are classified according to whether their arms are tightly or loosely wound, and whether they have normal or "barred" cores. They can be arranged in a double sequence, as in the diagram shown here, first drawn by the American astronomer Edwin Hubble. They range from the lenticular galaxies S0, which bridge the gap between ellipticals and spirals, through Sa, Sb, and so on, which have increasingly well-developed and loosely wound arms. Intermediate classifications may be added. Barred spirals are similarly classified: SBa, SBb, and so on. It was once believed that these were evolutionary sequences – that spiral galaxies evolve from more open to more tightly wound forms – but this view is no longer held.

NGC 1365, an SBb barred spiral, is shown below. The bar is probably a feature that develops late in the life of a spiral galaxy.

The disk containing the spiral arms is always thin, its thickness being about one-fifteenth of its diameter. Calculations show that such disks are inherently unstable and, furthermore, that the individual random motions of stars will become larger as time passes.

The result of such an increase is to destabilize the disk even more, setting up new shock waves. The principal waves will not be spiral in form, but long and narrow. This means that a uniform disk will first form spiral arms and then, after the galaxy has rotated a few times, the arms will largely disappear and be replaced by a bar. Such bars may contain as much as 40 percent of the optically visible material of the galaxy.

This is how barred spirals form, and clearly they are a late stage in the evolution of spiral galaxies. However, the age at which a spiral galaxy changes into a barred spiral depends on its enveloping halo of material. The more

material there is, the more stable the disk, and the more slowly the galaxy will evolve.

This general scenario of the formation of spiral galaxies has been supported by the observation of a supergiant spiral in the Virgo constellation. There seems to be no doubt that this galaxy, called Malin 1, is a lens-shaped spiral at an immense distance. The galaxy is actually fainter than the night-sky background. The fact that we can see it at all is thanks to a powerful technique of photography which was developed by David Malin – for whom the galaxy is named – of the Anglo-Australian Observatory in New South Wales.

What is so surprising about the Malin 1 galaxy is its vastness. Its central region is dominated by a huge bulge resembling an elliptical galaxy. It shows bright lines in its spectrum, indicating the presence of a considerable amount of glowing gas. This region is surrounded by a huge disk of gas, with a diameter of not less than 490,000 light-years, or approximately five times that of our own, not inconsiderable, galaxy.

The whole vast disk of neutral hydrogen is rotating more slowly than our own galaxy. If Malin 1 later contracts, its speed of rotation will increase. In fact, its present velocity is to be expected if, as its discoveries suggest, this supergiant galaxy is a spiral in the early stages of formation.

If Malin 1 is a very young spiral, it fits in well with the development of spirals sketched above. It is just what we should expect to find if galaxies contract from more diffuse protogalaxies: a core of older stars and a disk of gaseous material that will in time split into spiral arms, due to shock waves within it. We see it as it was 800 million years ago, long after the first spiral galaxies were born, and it seems to be evidence for the continuing production of galaxies in the universe.

The spiral galaxy M82, 10 million light-years from the Earth, is seen in computer-enhanced colors in this photograph. It was long believed to be an exploding irregular. Now, however, it is thought to be a normal spiral. The reddish filaments that appear to emerge from the core probably belong to a cloud of gas and dust through which the galaxy is moving.

THE MILKY WAY
• *Our galactic home*

Earth is a planet orbiting one of the 200 billion stars that make up a large spiral galaxy. The concept of this island of stars was first proposed in the 1780s by William Herschel, a pioneer of the study of the skies with large telescopes. On the basis of counting the stars visible in different directions, he concluded that the Sun lay at the center of a star system that was flattish and elongated in shape.

When viewed across its breadth, this system appears as the Milky Way. In fact, even a modest telescope shows that the Milky Way is composed of thousands upon thousands of stars. Fewer stars are visible when looking away from the plane of the system. The system has become known to astronomers as the Galaxy – from *galaxias*, which is what the Milky Way was called in ancient Greece.

In various places there appear to be vast "holes" in the Milky Way. These are not truly holes, as astronomers believed for a time, but are caused by the absorption of light by dust composed of carbon, iron, and other material (pp.74-77). Enormous clouds of gas are also present. Since Herschel's time, it has become still clearer that the stars making up the Milky Way and its associated gas and dust are all part of a vast disk of material spreading from the central regions of our star-island.

The observations of radio telescopes, which can penetrate dust and gas, have confirmed that the system is fully disk-like, and have also led to the detection of spiral arms. In fact, it is now known that there are four such arms. The one that reaches closest toward the center of the Galaxy, the Centaurus arm, can be seen in the area of the sky covered by the constellations Centaurus, Crux, and Carina. Two others curve through Sagittarius and Orion. The arm that extends farthest from the center lies across the area covered by the constellation Perseus. The diameter of the main part of the disk is 100,000 light-years, although it extends much farther than this in an increasingly tenuous state. The Sun lies about 30,000 light-years from the center. The thickness of the luminous disk proves to be only 2,000 light-years.

At the center of the Galaxy, as of all spirals, there is a concentrated core of stars. These form a bulge at least 20,000 light-years in diameter and about 6,000 light-years thick; they are accompanied by a comparatively small amount of gas – no more than 10 percent by mass. Viewed from Earth, this central bulge lies in the direction of Sagittarius. The spiral arms contain stars of all ages, including newly forming ones – known as stars of Population I –, but the central regions contain only old stars – those of Population II.

The Galaxy is rotating, with the outer parts of the disk moving more slowly than those nearer to the central bulge. The Sun, roughly halfway out toward the edge, moves at a velocity of 155 miles per second and takes no less than 200 million years to complete each rotation, a period sometimes called the cosmic year. For a star only 15,000 light-years from the center, rather than the Sun's 30,000 light-years, the rotation period would amount to about half this figure.

Since the late 1970s, satellites such as the two US High-Energy Astronomical Observatories have found that the central regions of the Galaxy, as of other spirals, are an intense source of X-rays. This supports the hypothesis that at the nucleus of the Galaxy there is a black hole (pp. 62–65) voraciously swallowing gas, dust, and old stars, and causing such material to emit high-energy radiation as it falls in. Such a black hole could perhaps have been formed by the convergence, due to gravity, of a number of the old stars in the central regions.

The entire Galaxy is embedded in a vast halo whose visible diameter is similar to that of the disk – about 100,000 light-years. In this halo there are hundreds of globular clusters, dense concentrations of tens of thousands of stars (pp. 80–81), and also countless individual stars. And it is probable that the Galaxy – and presumably all others like it – is cocooned in an extensive body of unobserved matter. This seems to be confirmed by the observed movement of stars in the disk of the Galaxy.

The presence of such matter around other galaxies would be of cardinal importance to astronomers' ideas of the past and future of the universe: on the one hand, it is virtually demanded by the inflationary theories of the Big Bang; on the other, it would slow the expansion of the universe – and perhaps even, some time in the remote future, turn it into a collapse.

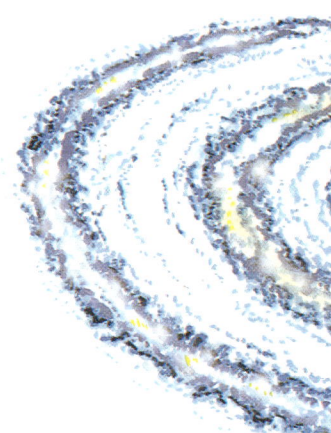

The disk of the Galaxy (below) appears to us as the Milky Way, a band of light that circles the sky and consists of thousands of stars too closely packed to distinguish. The Sun lies in the Orion arm, one of the Galaxy's four spiral arms.

The Galaxy cuts the celestial sphere (an imaginary globe centered on the Earth – right) in a circle, the galactic equator. Opening out the sphere makes a sky map (top). The celestial north and south poles, to which the Earth's axis points, are tilted away from the galactic poles.

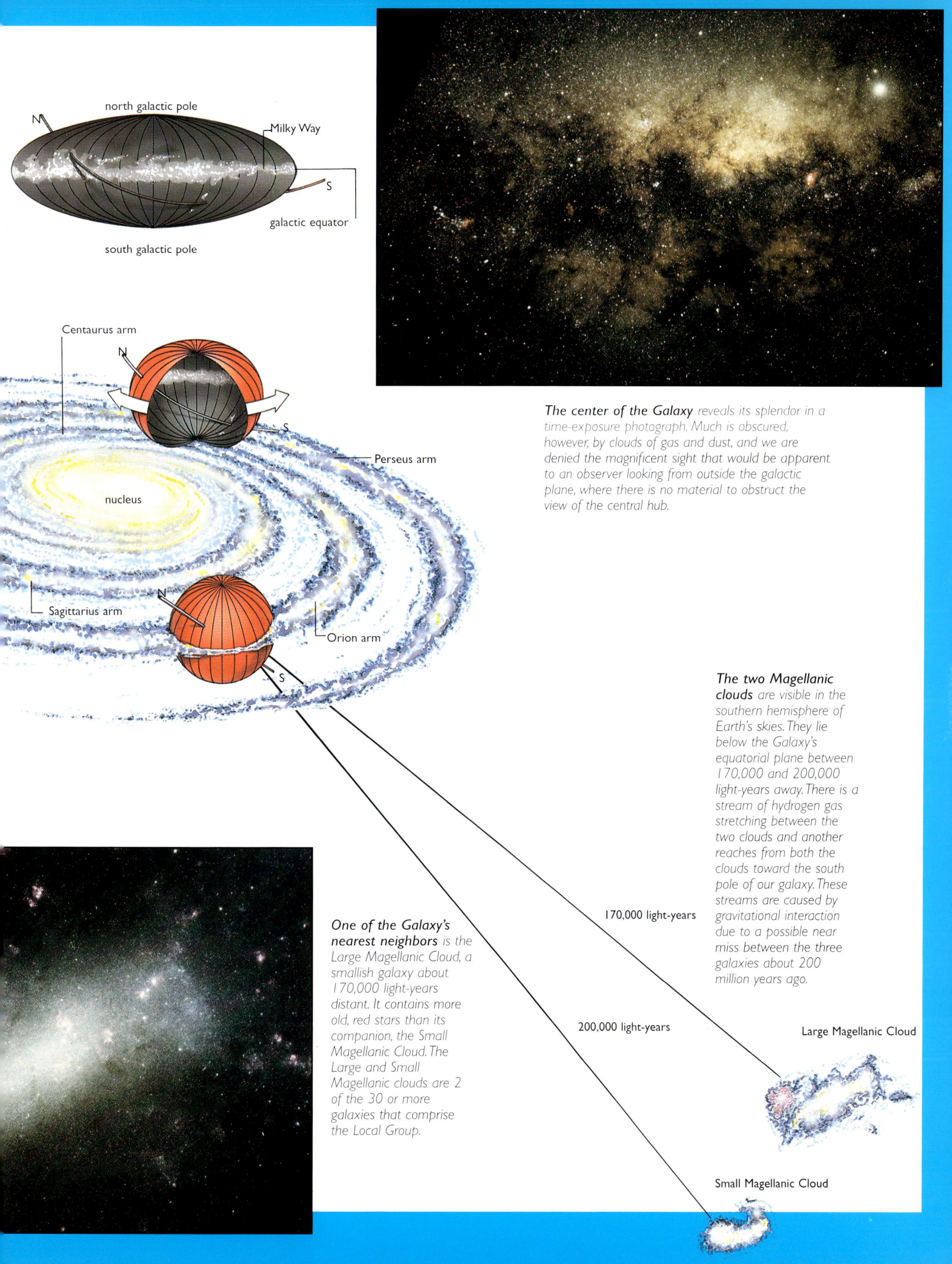

The center of the Galaxy reveals its splendor in a time-exposure photograph. Much is obscured, however, by clouds of gas and dust, and we are denied the magnificent sight that would be apparent to an observer looking from outside the galactic plane, where there is no material to obstruct the view of the central hub.

The two Magellanic clouds are visible in the southern hemisphere of Earth's skies. They lie below the Galaxy's equatorial plane between 170,000 and 200,000 light-years away. There is a stream of hydrogen gas stretching between the two clouds and another reaches from both the clouds toward the south pole of our galaxy. These streams are caused by gravitational interaction due to a possible near miss between the three galaxies about 200 million years ago.

One of the Galaxy's nearest neighbors is the Large Magellanic Cloud, a smallish galaxy about 170,000 light-years distant. It contains more old, red stars than its companion, the Small Magellanic Cloud. The Large and Small Magellanic clouds are 2 of the 30 or more galaxies that comprise the Local Group.

COLLIDING GALAXIES
● *Catastrophes in deep space*

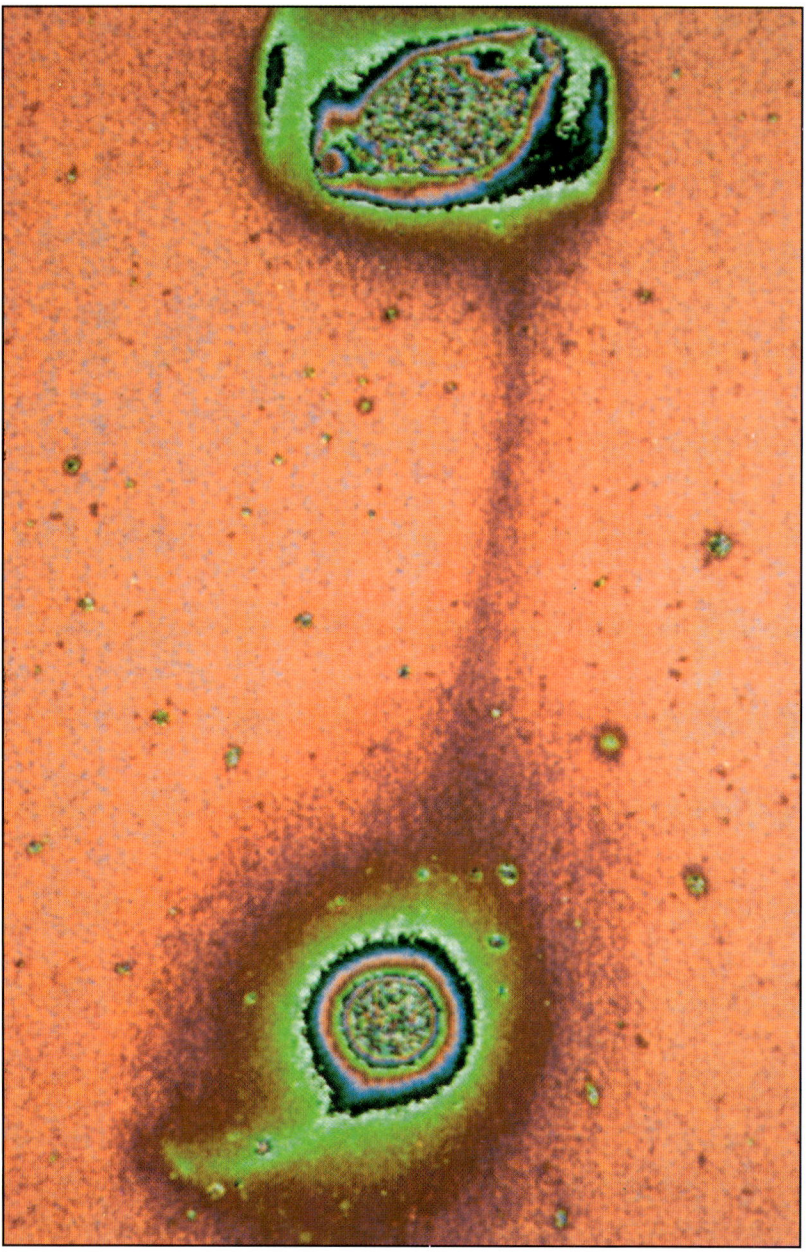

A "rope" of stars links the two galaxies NGC 5216 and NGC 5218. The rope was torn out of the two galaxies by their mutual gravitational pull. This image was obtained by the Isaac Newton telescope at La Palma Observatory, Tenerife, using a sensitive electronic detector in place of film.

Most galaxies are a long way apart: even within clusters, the average separation of the larger ones is about 1,500,000 light-years. Nevertheless, approaches sufficiently close to cause gravitational interactions and even collisions should occasionally occur, and observations confirm this.

The two Magellanic Clouds are in orbit around our own galaxy, completing a circuit in something like 500 million years. They have noticeable gravitational effects on the Galaxy: they distort its disk of stars, gas, and dust, pulling one side upward and the other downward. In one part of the disk, in a region that is a little farther from the center than our Sun, the disk is bent northward, while at an equal distance from the center on the opposite side, it is bent southward. This warp is not fixed in position: it revolves slowly around the disk, taking perhaps a billion years to complete a circuit. As for the Galaxy, it has tidal effects on the Magellanic Clouds, and an extended gas trail – the Magellanic Stream – has been found to circulate among the three galaxies.

Such mutual tidal effects are to be found in 8 out of 10 spiral galaxies, whose bent or distorted disks can be observed with modern techniques. The Andromeda galaxy suffers a similar distortion, although the cause in this case is not clear since its satellite galaxies seem to be too far away to do the warping. However, the halos of spiral galaxies very probably contain dark matter (pp. 168–69), and such material could react on the disks.

This material may take various forms: some may consist of WIMPs (weakly interacting massive particles), a hypothetical type of matter that is still being sought in the laboratory, and some may consist of MACHOs (massive compact halo objects). The latter would be the remnants of stars that have collapsed long ago, or else are brown dwarfs or jupiters – objects of too low a mass to have formed into stars.

Many interesting interactions between galaxies have now been observed. One example is the interaction between the two spirals NGC 5426 and NGC 5427 in Virgo, whose spiral arms are actually linked. Moreover, their disks are distorted by mutual tidal action, and seem forced to oscillate with a period in the order of 100 million years.

Sometimes more than two galaxies are concerned in an interaction, as in the case of the Magellanic Clouds and our galaxy. In the constellation of Serpens Caput there is a group originally named Stephan's Sextet for its discoverer. It had to be renamed when detailed studies showed that what appears to be a sextet of interacting galaxies is not quite what it seems. Only five of the objects are interacting; the sixth object, a spiral galaxy, merely happens to be in the same field of vision: it is in fact

THE GRAND DESIGN

five times as far away as the others.

The remaining five objects are certainly connected: they consist of four spirals and a large cloud of gas that has been ejected by one of them. Their masses and mutual separations show that they must be interacting gravitationally. Lanes of gas can actually be seen running between all five objects.

It is only a step from this type of situation to an actual collision between galaxies. A pair of spirals in Corvus seem to have collided and, as a result, to have ejected two vast streams of glowing gas, looking like the antennae of a giant insect. Computer simulations have shown just how this would occur if the disk of one spiral were to meet the disk of the other almost at right angles. The centers of these galaxies now lie about 65,000 light-years apart: the magnitude of the eruption that was generated may be judged from the fact that the ends of the antennae are now almost 500,000 light-years away from each other.

There are other examples of colliding galaxies that are even more spectacular in their effects. Two spirals called IG 29 and IG 30 have evidently collided. Again the disks seem to have met at right angles; what may seem surprising is that the two galaxies have passed right through each other and are still recognizably spirals. Nevertheless, they are still gravitationally bound, and there is a luminous bridge of gas joining their centers. One of the galaxies is surrounded by a ring of gas and bright stars; galaxies such as this are given the name ring galaxies.

Perhaps even more spectacular is the Cartwheel galaxy, lying at a distance of about 650 million light-years from us. This seems originally to have been a spiral galaxy, which developed into a ring galaxy in the aftermath of a head-on collision with a smaller galaxy that passed through it completely. The two are now over 250,000 light-years apart. The rim of the Cartwheel galaxy is an expanding shock wave that is triggering the generation of a vast number of bright, massive stars, which lead short lives before they explode as supernovae (pp. 90–91). The rate of production of such exploding stars is 100 times greater than in a normal spiral, which underlines the catastrophic scale of the collision.

A small elliptical galaxy passing through a spiral galaxy (similar to our own) attracts stars, gas, and dust from the larger galaxy toward the point of impact. After the elliptical galaxy has passed through, the material rebounds, forming rings rather like the ripples on a pond when a stone is thrown in it. The rings spread out and trigger star formation as they compress gas and dust.

The Cartwheel galaxy (below), in the constellation of Sculptor, has an outer ring of billions of new stars moving out from its center. The galaxy is the result of a head-on collision with a smaller galaxy – possibly, one of the two on the right of the image.

ELLIPTICAL GALAXIES
● *Remnants of galactic collisions?*

Elliptical galaxies have complex shapes. In general, the three main axes, which pass through the center at right angles to each other, are all of unequal lengths.

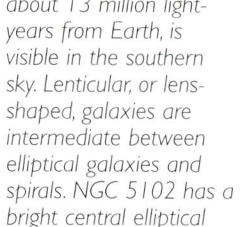

Lenticular galaxy NGC 5102, about 13 million light-years from Earth, is visible in the southern sky. Lenticular, or lens-shaped, galaxies are intermediate between elliptical galaxies and spirals. NGC 5102 has a bright central elliptical bulge, about 3,400 light-years wide by 8,000 light-years long. The elongated disk is 33,000 light-years across.

Elliptical galaxies seem, at optical wavelengths, to be composed entirely of old, reddish stars. They do not possess the disks, consisting of gas, dust, and young stars, that characterize spiral galaxies. Their shapes are classified from E0, spherical, to E7, the most elongated. The last shape merges into the S0 galaxies, sometimes referred to as lenticular galaxies, which seem to form an intermediate class of galaxy between ellipticals and spirals.

It was once thought that the ellipticals had the shape of a football or rugby ball (apart from some that happened to be spherical). Such a galaxy would look like an E7 if it presented a side view to Earth, but like an E0 galaxy if it were end-on to Earth. However, since the 1970s, observational techniques have developed to a stage where astronomers can closely examine the orbital speeds of stars in galaxies. The results have been something of a surprise, because they have shown that in general, all three of the axes of an elliptical are unequal.

It was also once thought that elliptical galaxies evolve into spirals, but the standard view now is that all galaxies were formed at about the same time. Whether a protogalaxy (pp. 38–39) produced a spiral or an elliptical was a consequence of the speed of its rotation. A high rotation velocity resulted in condensation into a spiral galaxy; a slower one led to the formation of an elliptical galaxy.

In elliptical galaxies, the formation of stars was largely completed as the protogalaxy contracted, and there is now an absence of young stars. This is very striking, since radio observations have shown that ellipticals contain a considerable amount of neutral hydrogen, which gives no signs of its presence at optical wavelengths, but is very evident at the radio wavelength of 21 centimeters. Furthermore, modern optical detectors have made it clear that ellipticals contain considerable quantities of dust. Despite these facts, it remains true that elliptical galaxies consist predominantly of older stars. Spirals also contain aging stars in their cores and halos; it is in their disks that the process of starbirth from gas clouds continues.

However, very massive ellipticals are found in the central regions of clusters of galaxies (pp. 50–51), and it now appears that these may have been formed more recently in a different manner. They may indeed have been formed from spirals, but not by a process of evolution, as was once supposed.

One clue is provided by the study of the way the brightness falls off toward the edge of an elliptical galaxy. The form of this brightness distribution is called de Vaucouleurs' law. Since 1987, it has gradually become clear to astronomers that this law is also applicable to another class of objects, known as starburst galaxies. These are irregular galaxies that contain a great deal of dust spread throughout them in a seemingly chaotic way. In these galaxies, stars are being formed in abundance.

Few starburst galaxies are found in isolation. Most are interacting with other galaxies, and there is also an extensive class formed by the actual collision of two or even more galaxies.

THE GRAND DESIGN

These are now observed as single objects, surrounded by trails of gas and dust and other remnants. Optical observations, supplemented by infrared studies, often show that the movements of the older stars as they were before the collision is of the kind seen in elliptical galaxies.

In addition, computer studies have shown that the violent disturbances that would occur during collisions of two similar spirals with large dark halos would produce two ordinary ellipticals to which de Vaucouleurs' law would apply. Observations of the galaxy NGC 7252 in Aquarius seem to confirm this.

The study of colliding galaxies has been taken further by Gillian Wright and her colleagues, using the United Kingdom Infrared Telescope in Hawaii. They have examined galaxies that emit vast amounts of infrared radiation, with very long wavelengths up to one-tenth of a millimeter. These include the starburst galaxy Arp 220, one of the most luminous known sources at these wavelengths.

Such galaxies display the wreckage to be expected if two spiral galaxies had collided. Optical pictures of Arp 220 show it to be accompanied by a great deal of dust, some of it in the form of a lane running across the image. This is the kind of dust associated with the formation of new stars. The infrared observations show not only large amounts of hot dust, but also the presence of young stars and complex gas molecules.

These observations prove that elliptical galaxies with active centers are still forming. At earlier times in the history of the universe, when galaxies were closer together, collisions must have been more frequent. It is therefore possible that, in contrast to the predominant view, most or even all ellipticals are in fact the result of collisions between spirals.

Centaurus A is an intense radio source, the third strongest in the sky. The dark band seen here is a disk of gas and dust, lying at the center of a much more extensive elliptical galaxy, which needs special photographic techniques to reveal it. The disk is a site of star formation.

BLACK HOLES
• *Cut off from the universe*

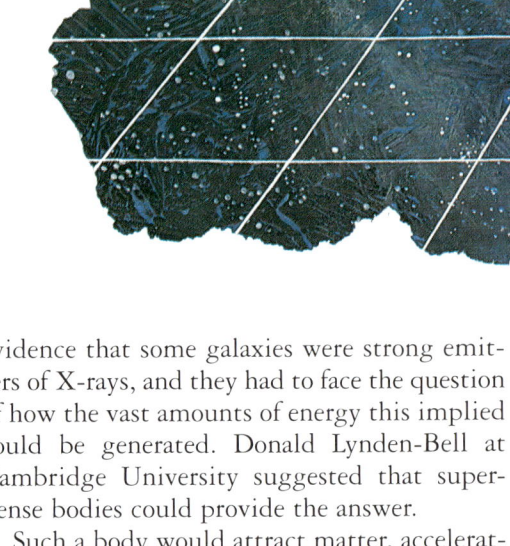

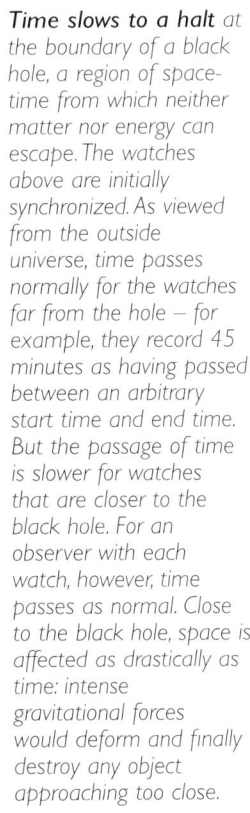

Time slows to a halt at the boundary of a black hole, a region of space-time from which neither matter nor energy can escape. The watches above are initially synchronized. As viewed from the outside universe, time passes normally for the watches far from the hole – for example, they record 45 minutes as having passed between an arbitrary start time and end time. But the passage of time is slower for watches that are closer to the black hole. For an observer with each watch, however, time passes as normal. Close to the black hole, space is affected as drastically as time: intense gravitational forces would deform and finally destroy any object approaching too close.

As long ago as 1783, John Michell, an astronomer and geologist, suggested that gravity could act on light as well as on matter. At that time, Newton's theory that light was composed of small particles or "corpuscles" was still current, so Michell's idea did not seem too far-fetched. He drew the interesting conclusion that some stars might be so large as to have escape velocities greater than the speed of light, thus preventing any light corpuscles that they radiated from escaping into space. To an observer such massive stars would therefore be black, and invisible against the night sky.

Pierre Laplace, the great French mathematical physicist, put forward similar ideas independently. Yet to most scientists, they appeared to be no more than a curiosity. Moreover, they were suspect because there was no reason to suppose that light could have one and only one velocity in all circumstances. In the early 19th century, when Newton's corpuscular theory gave way to the idea that light is a wave disturbance, it seemed even more unlikely that gravity would be able to trap light.

Only in 1915, with the advent of Einstein's general theory of relativity, was the situation radically altered. The bending of light was one consequence of the theory, and once the observations made at the total eclipse of 1919 had confirmed that light is bent by the Sun, the whole subject of the effects of gravity on light was revived.

The English physicist Oliver Lodge calculated that a body of "reasonable" mass could trap light completely only if it were of a density that could not be attained by matter as it was conceived of in his day. But later, the concept of "degenerate" matter was proposed (p. 88). This is extraordinarily dense, and put a different complexion on the idea of a "black star."

But interest in the subject did not really awaken until 1969. By then, astronomers had evidence that some galaxies were strong emitters of X-rays, and they had to face the question of how the vast amounts of energy this implied could be generated. Donald Lynden-Bell at Cambridge University suggested that super-dense bodies could provide the answer.

Such a body would attract matter, accelerating it to huge velocities as it fell in. An accretion disk would form, consisting of matter revolving around the black hole before its final disappearance. The infalling material would move at immense velocities, emitting X-rays in huge quantities.

Relativity now enters the scene. The extreme density of such a body would create an intense gravitational field. This would mean that space-time around the body would be so strongly curved as to cause the interior to be completely closed off from the outside universe. Nothing could escape from it. This is why such objects are now called black holes.

These holes in space-time not only absorb whatever falls into them: they have other astounding effects. As a body approaches a black hole, the immense gravitational field causes the passage of time for that body to slow down, as measured by an outside observer. The frequency of the light that it emits falls lower and lower – that is, it is redshifted – and it becomes weaker. The object approaches, but never quite reaches, the so-called event horizon surrounding the black hole, and finally seems to hover there. In consequence, the outside observer never sees it actually fall into the hole.

In the frame of reference of the body itself, however, there will be no apparent change in the rate at which time passes. But the body will be subjected to immense forces that will tear apart any macroscopic body, and certainly make survival impossible for any living thing. In fact, all structure is destroyed, so that matter inside the black hole loses all individuality – all "memory" of what it was before. It falls toward a center known mathematically as a singularity, where the density of matter becomes infinite and space-time is reduced to a mere point.

To understand these and other facts about black holes, it is helpful to draw space-time diagrams. The vertical dimension is taken to represent not a space dimension, but time. A vertical line directed upward represents the position of a stationary object as it moves into the future. A line that slants, or is curved, shows that the object is moving. Such a line is called its world line.

Waves of light emitted from some event spread out across space. In such a diagram, it is represented by rings increasing in diameter up the page. We end up with a light-cone, its apex representing the original event.

Now imagine that a star falls into the black hole. Successive wavefronts of the star's light

start time

end time

63

red giant

white dwarf

light rays

neutron star

light rays

black hole

trapped light

Black holes are formed *when matter reaches sufficiently high density. Light from a star of low density, such as a red giant, is scarcely deflected by the gravitational field. A white dwarf, which has a density of 18 tons per square inch, also does not trap light, although the light rays are appreciably curved. They are even more strongly curved by the gravity of a neutron star, which is about a hundred million times more dense than a white dwarf. An even denser stellar remnant, left after the death of a very massive star, bends light rays so strongly that they cannot escape at all, and the body is a black hole.*

remain circular in the diagram but are emitted from a point that is falling inward with increasing speed. The result is that the light-cone appears to be tipped over toward the black hole. At the event horizon, the light-cone is tilted so far that its outer edge is vertical, representing the fact that the light cannot escape from the event horizon, and we see the star apparently hovering there forever.

This is probably an oversimplified picture. It seems likely that in the real universe, black holes, being formed from material that is rotating, are themselves rotating. The British cosmologist Stephen Hawking and his colleagues have established that such black holes can be described by equations first discovered by the New Zealander Roy Kerr.

The area around the event horizon of a rotating black hole is known as the ergosphere. It represents that part of space-time in which the black hole interacts with the rest of the universe. For instance, the Kerr equations tell us that as matter falls through the ergosphere and into a rotating black hole, the area of its surface increases. But strangely enough, this infalling matter can cause a *decrease* in the mass, provided that the spin of the black hole is reduced as well.

Despite all that has been said above, Stephen Hawking has found that black holes can lose energy (see box). He has applied the laws of quantum theory and thermodynamics to black holes, with interesting results. The first is that black holes have a temperature, which is larger, the smaller the mass of the hole. Even so, the temperatures are minute by everyday standards: even a comparatively small black hole with the mass of a star would be no hotter than a ten-millionth of a kelvin.

However, Hawking believes that minute black holes could have been formed from very dense matter crushed together at an early stage of the Big Bang. Being so small, their temperature would be very high, and some of them would be evaporating now, in explosions that would put an artificial nuclear explosion to shame. But no radiation from such miniholes has been observed to date, and it seems probable that the Big Bang was too uniform to allow them to form.

But black holes of larger masses probably exist. Where are they? It seems likely that some reside in the central regions of spiral and elliptical galaxies. They also form after the catastrophic collapse of very large stars, and also, possibly, after the deterioration of weird supermassive objects called spinars. Astronomers are now convinced such black holes exist, even though they are difficult to track down. Some are thought to have a mass millions or billions that of the Sun, crammed into a small space. Such an object is believed to lurk at the centers of active galaxies. If not for their rapid rotation, they would collapse immediately. When the energy of their rotation is spent, they form black holes.

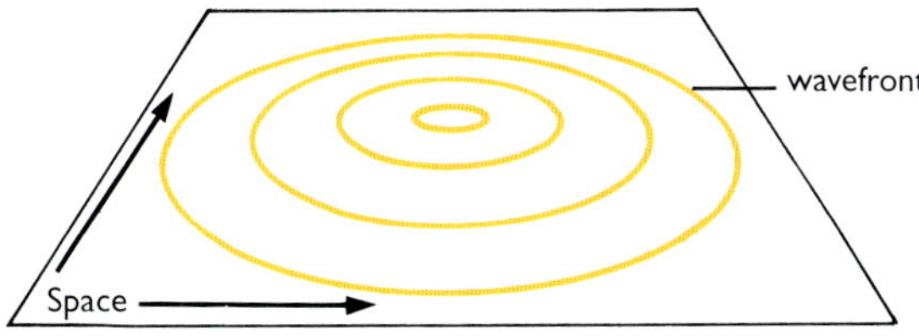

In a two-dimensional cross-section, spreading light waves are represented as a series of expanding circles, called wavefronts (right).

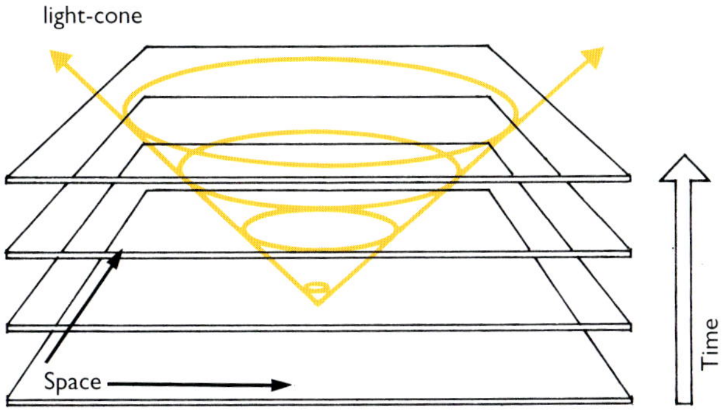

In a space-time diagram (above), only two dimensions of space are represented, while the third dimension represents time. The spreading wavefronts of light in the top diagram become the surface of a conical figure called a light-cone.

Stars falling into a black hole are represented by this space-time diagram (right). The stars' world lines, or paths through space-time, curve in toward the center of the black hole, which is a singularity – a point where known laws break down. Light-cones from the falling stars tilt until within the event horizon, no light is able to escape.

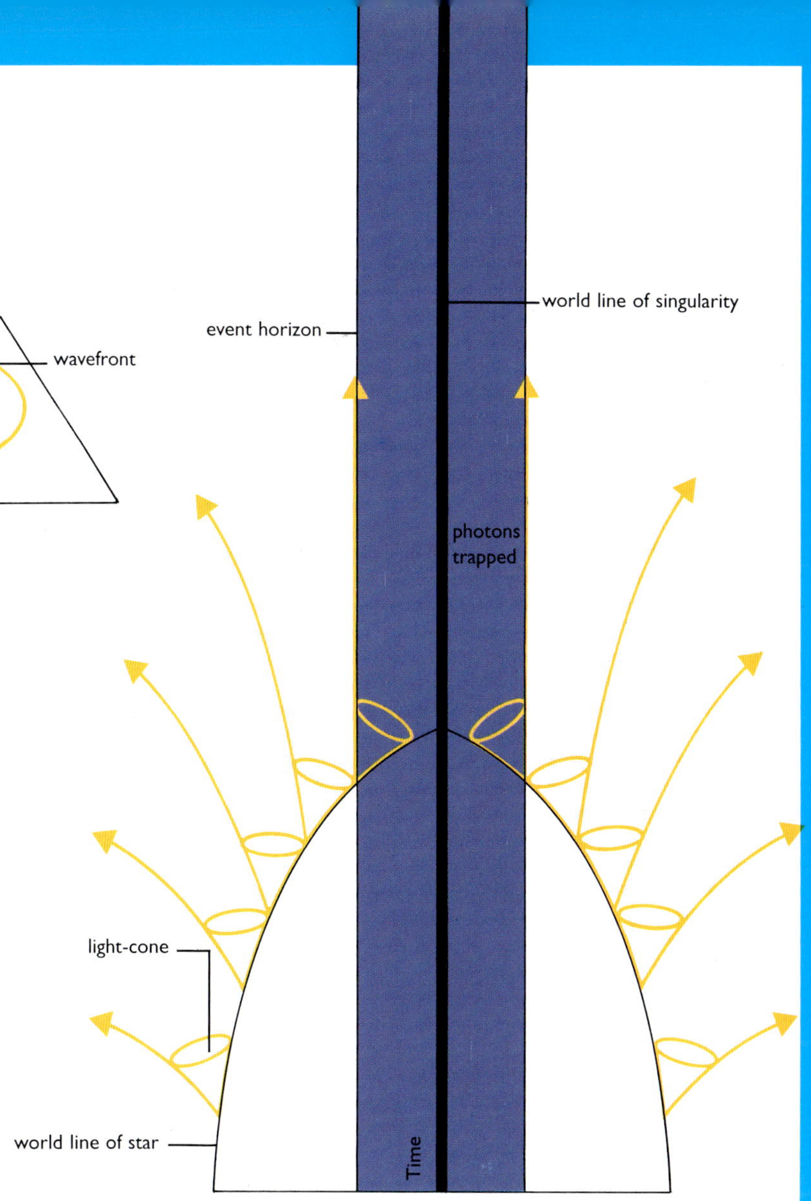

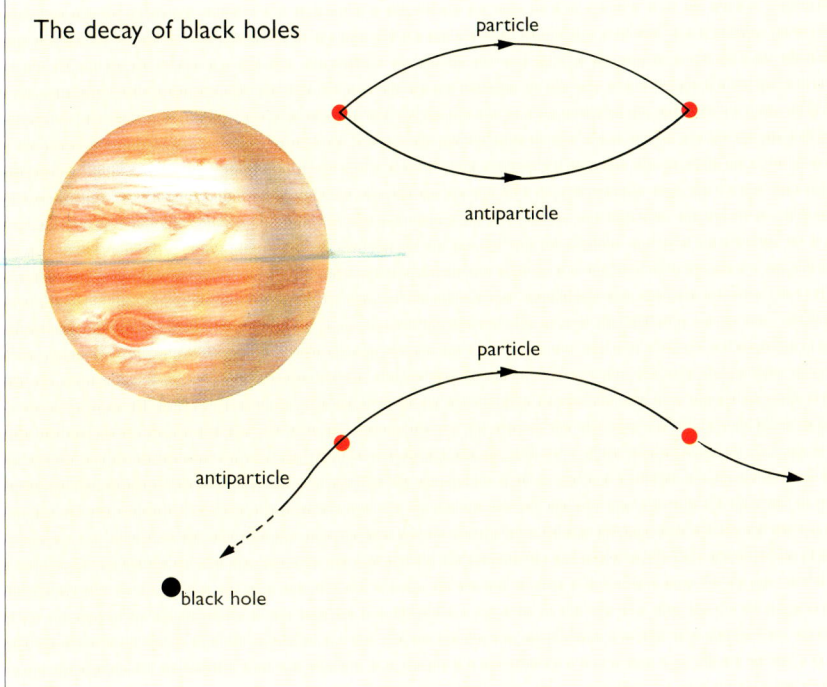

The decay of black holes

Throughout space, virtual particles (p. 26) are constantly appearing fleetingly, and then disappearing. They appear in particle-antiparticle pairs, so that there is no net creation of electric charge or spin. Pair production is stimulated by the gravitational field of any massive body (upper diagram), but when it happens near a black hole, one member of the pair can fall into the black hole, while the other escapes. The one that escapes carries energy away, while the one that is captured carries negative energy into the black hole. The black hole appears to radiate energy with a spectrum resembling that of a body with a definite temperature. This process of decay is extraordinarily slow. A black hole with the mass of the Sun would take 10^{56} times the present age of the universe to evaporate.

65

ACTIVE GALAXIES
• *Star systems in turmoil*

Once galaxies were recognized as star systems lying far outside our own, astronomers generally gained the impression from photographs that they were staid, well-behaved objects. Only in the early 1950s did it suddenly become clear that this might not be true. Optical and radio astronomers working together established that two powerful radio sources, called Virgo A and Centaurus A, were actually galaxies. Virgo A turned out to be the E0-type elliptical galaxy M87 and Centaurus A the S0-type galaxy NGC 5128. Optically, neither appeared remarkable except that Centaurus A was crossed by a band of dark material. The name "radio galaxies" was coined.

As the detail that could be "seen" with radio telescopes was improved, it was found that there were many radio galaxies. On optical photographs, they appeared as galaxies might be expected to, but the radio observations gave a very different picture. Centaurus A was shown in detailed radio pictures to be a triple source. In the center is a small intense radio source, and on each side of it are two large and strongly emitting radio "clouds."

At some time in the past, the central regions of the galaxy have suffered a vast explosion, causing two jets of matter to be thrown out in opposite directions. The jets are composed of very fast-moving particles, reaching velocities of about 37,300 miles per second (one-fifth of the speed of light). Their energy is so great that they burn their way through the surrounding intergalactic gas, setting up pressure waves rather like the bow waves of a ship.

The particles are ions (electrified atoms) and electrons. Their motion creates magnetic fields that restrict the volume of space through which the particles can spread. As a result, the particles pile up, forming lobes at the end of each jet. When the galaxy stops emitting jets, the two lobes of plasma continue to expand, forming the clouds we now observe. As time goes on, their radio "brightness" diminishes.

In the case of Centaurus A, there is also a radio halo, somewhat elongated in shape and about one million light-years in extent. The clouds lie within this, two at a distance of about 30,000 light-years from the galaxy's center, one at a distance of 100,000 light-years. The two inner lobes form a ring around the galaxy, and were formed from two jets. The single, more distant cloud is presumably the result of a single jet of atomic particles. The amounts of energy emitted by these clouds are prodigious. The radio emission is anything up to a hundred times the optical output of a typical galaxy, and lasts for between one million and 100 million years.

Not all radio galaxies display lobes. Some, such as M87, emit only a single jet of material, as Centaurus A seems to have done once. The jet can be seen optically, as well as detected at radio wavelengths. Radio galaxies also emit at X-ray and gamma-ray wavelengths. Most of this very energetic radiation comes from a small region in the nucleus of the galaxy.

Clearly, radio galaxies are undergoing enormous eruptions. What is more, radio studies show that our own galaxy and M31 in Andromeda, which at first sight appear to be so quiet, also have active centers, although on nothing like the same scale.

Certain active galaxies have become known as blazars, or BL Lac objects. The name is derived from the first object of this kind, which was originally thought to be a star and was designated as BL Lacertae (that is, variable star BL in the constellation Lacerta, the Lizard). It is now known to be a star-like object within the nucleus of a galaxy surrounded by very strong magnetic fields. These fields vary, like BL Lac's radiation output, occasionally becoming 100 times greater than usual. Moreover, the object itself is no larger than a few light-months in diameter. About 100 BL Lac objects are now known, and it is evident that each is part of the radiating area of an active galactic nucleus.

What, then, is happening in the central regions of such explosive and active galaxies? Certainly, atomic particles are being emitted at immensely high speeds. In fact, sometimes the clouds appear to be traveling faster than light. Such superluminal velocities would contradict the theory of relativity if they were real (pp. 16-17), but they are in fact illusory. Such cases occur when the radiation-emitting jet is coming from the galactic nucleus in a direction very close to the line of sight.

The jet from an active galaxy comes from a source about 3 to 30 light-years away from its center. Although we cannot directly observe this central powerhouse, we can come to some

The structure of an active galaxy, according to modern theory. The source of its energy is a black hole at its heart. Gas, dust, and even stars are continually drawn into the black hole. As they spiral in, they form an accretion disk of orbiting matter. As the material is swallowed, energy equivalent to 10 percent or more of its mass may be emitted. This outpouring of energy creates jets of charged particles that escape along the black hole's axis of rotation.

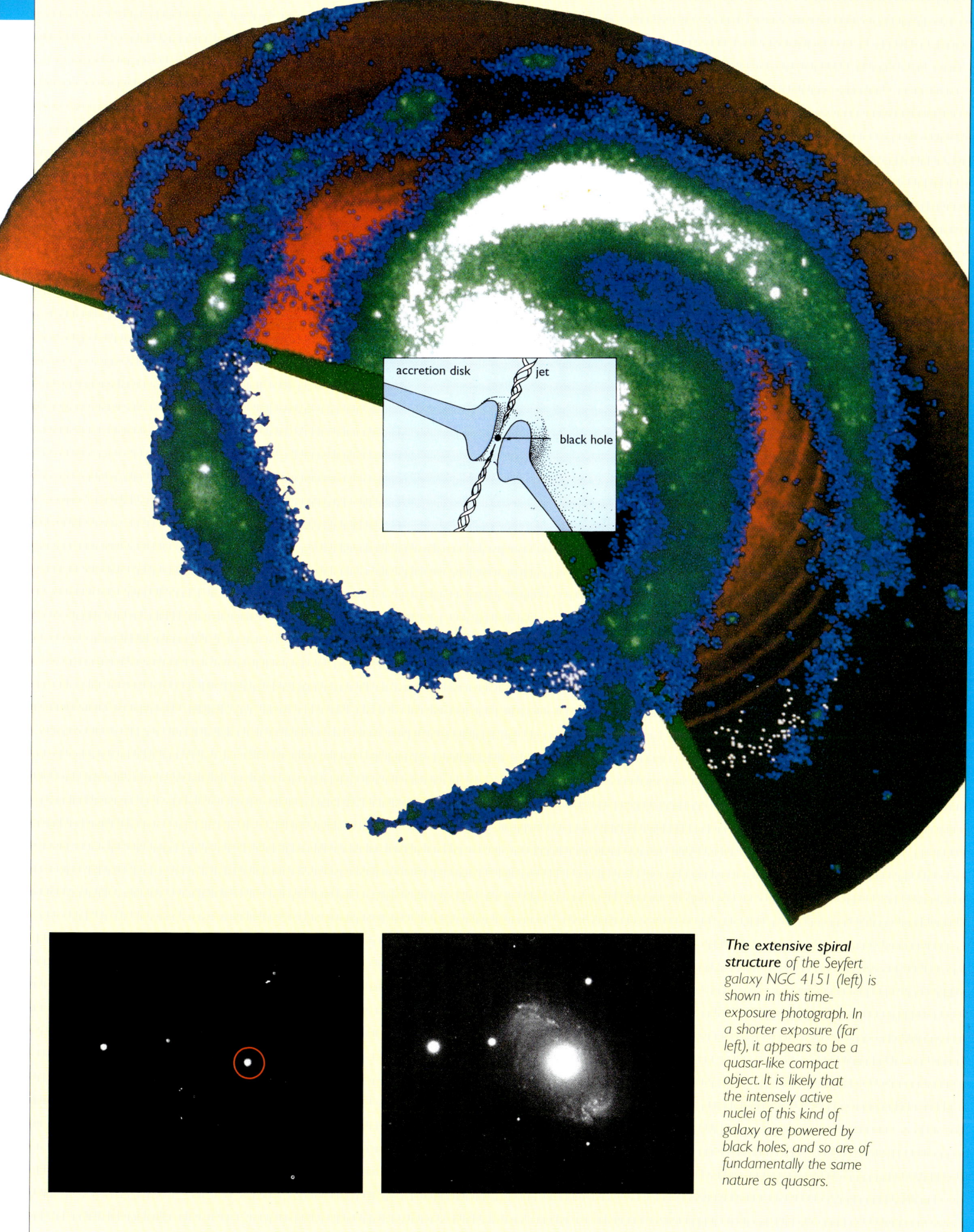

The extensive spiral structure of the Seyfert galaxy NGC 4151 (left) is shown in this time-exposure photograph. In a shorter exposure (far left), it appears to be a quasar-like compact object. It is likely that the intensely active nuclei of this kind of galaxy are powered by black holes, and so are of fundamentally the same nature as quasars.

An intense jet of charged particles streams from an active elliptical galaxy. The low density of interstellar gas and dust in the galaxy permits the jet to escape and create radio-bright lobes where it collides with intergalactic matter.

The active galaxy Cygnus A (below) is the brightest radio source in the sky. Most of its radio emission comes from two lobes of gas 120,000 light-years from the galaxy. These are sending out a million times as much radio energy as our Milky Way galaxy.

definite conclusion about it. Astronomers find that it must be associated with a disk of gas at the galaxy's center. The disk is probably thinner at its center than at the edge. In that event, material driven outward by the violent energies at the center would find it difficult to escape through the thickness of the disk; instead, the material would stream away from the galaxy at right angles.

The disk is formed by accretion of material orbiting a tiny object at its center; around this core, the disk bulges. In addition to the beams of particles ejected, the disk itself would emit high-energy ultraviolet and X-radiation. If the disk is surrounded by very dense gas clouds, these would be heated by this bombardment and would give out radiation with the kind of spectrum observed in many active galaxies.

In such a scenario, there has to be something very active at the center of the galactic nucleus, something powerful enough to cause the accretion disk to blast away atomic particles at a sizable fraction of the speed of light and produce super-energetic synchrotron radiation, X-rays, and short-wavelength ultraviolet.

The only viable candidate is a black hole. Black holes can be caused by the collapse of very large stars; at the centers of galaxies, there are concentrations of stars, among which at least some massive examples are to be expected. Moreover, once a black hole — even a small one — has formed, it will gather other material into its embrace. This will revolve around the black hole, and will produce an accretion disk.

Matter near the center will be heated intensely, emitting X-rays and ultraviolet radiation before plunging into the black hole. What is more, although formed from massive stars and a lot of other material, the black hole would be very small by cosmic standards.

The presence of black holes at the centers of active galaxies makes it seem likely that black holes lurk at the center of our own and perhaps in the centers of all galaxies, because they all have dense central cores of old stars. What is more, active galaxies do

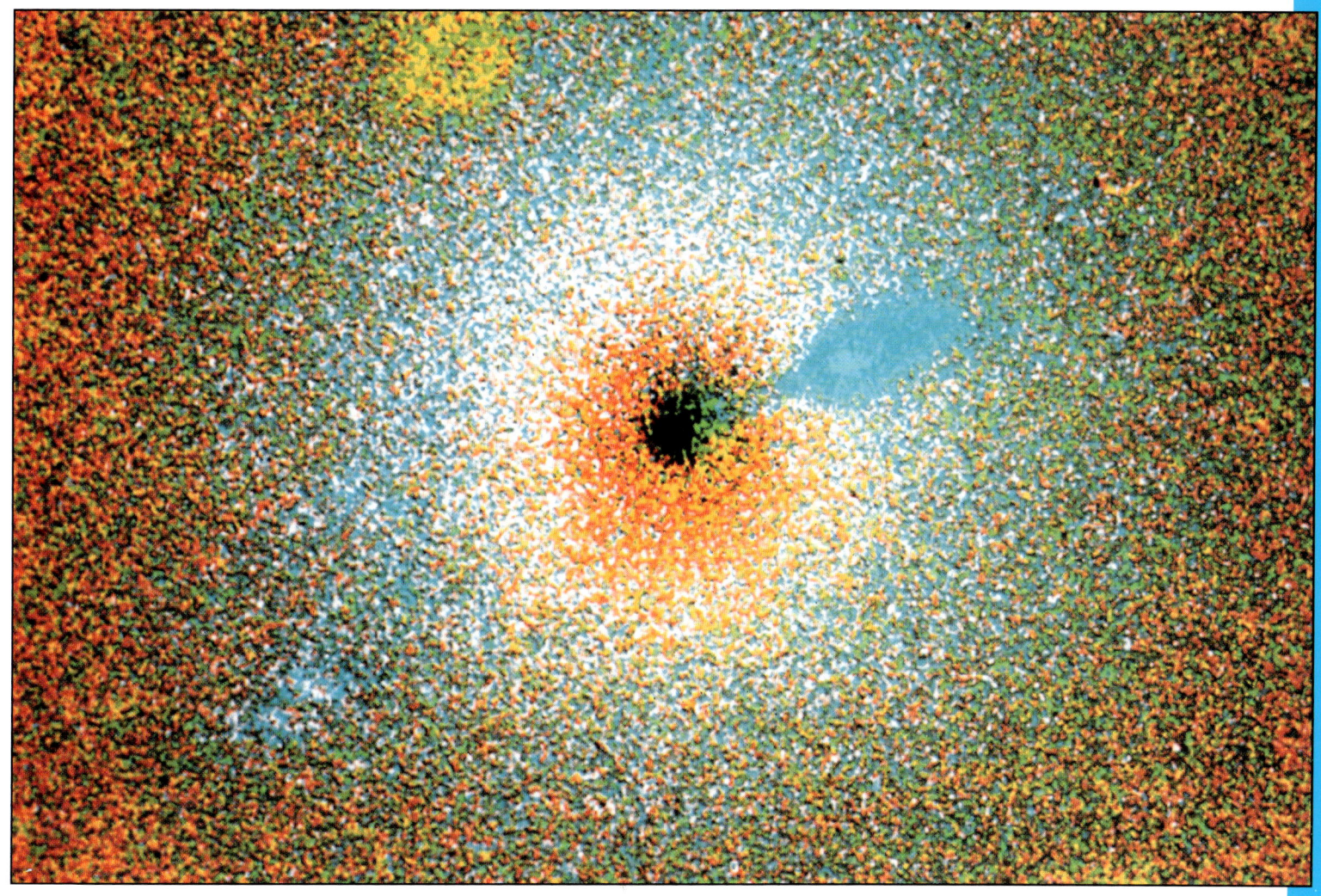

not stay wildly active permanently, and may eventually settle down. The more sedate galaxies may once have been active themselves.

Even galaxies that do not possess violently energetic cores and do not puff out lobes of radio-emitting ions vary among themselves in their level of activity. They differ especially in the vigor with which they form stars. Some dwarf galaxies appear to be active in this way.

Dwarf galaxies are generally classified as dwarf ellipticals (dE), dwarf irregulars (dI), or blue compact dwarfs (BCD). Dwarf ellipticals possess very little unused hydrogen and show little evidence of current star formation. The light from their stars shows that they contain a comparatively high proportion of metals, which means that they have existed for many stellar life cycles. Dwarf ellipticals are old and some, perhaps most, were formed directly from the protogalaxy material. However, it also seems possible that some dE galaxies evolved later from dwarf irregulars.

Dwarf irregular and blue compact dwarf galaxies are very different. Both are roughly disk-shaped. The irregulars contain a great deal of hydrogen and give evidence that star formation is going on at present. When their hydrogen is depleted, they have a good chance of becoming dwarf ellipticals. As far as the BCD galaxies are concerned, their color alone shows that they are rich in young blue stars.

The chief difference between dI and BCD galaxies is that the dIs show only isolated patches of star formation, while the BCDs seem to have such activity everywhere. In consequence, BCDs are brighter than dIs. However, it may be that BCDs and dIs are really the same kind of dwarf galaxy. The apparent differences would be due to star formation proceeding in cycles. The BCDs would be those galaxies that happen to be at the more active parts of their cycles.

A jet of gas emanates from the giant elliptical galaxy M87, 50 million light-years from us in the Virgo cluster of galaxies. Radio waves and X-rays from fast-moving electrons come from the jet and the galaxy. The jet is probably emitted by a massive black hole at the center of the galaxy. The black area at the center of this false-color computer-processed picture represents a region of closely packed, fast-moving stars, which are probably moving under the influence of the black hole's intense gravity.

QUASARS
●Celestial powerhouses

In 1963, radio astronomers were watching intently as the Moon passed in front of the radio-bright source of 3C 273. They were interested in this object because it had been tentatively identified with an optically dim blue star-like object with a peculiar spectrum. The identification remained uncertain, however, because radio telescopes could then fix the positions of sources only to an accuracy of five seconds of arc. When the Moon's edge blotted out the radio waves from 3C 273, it revealed the precise location of the radio source, and showed that it was indeed the suspected "star."

The occultation, as such events are called, also showed the 3C 273 was not a simple source, but consisted of two parts. Detailed optical studies later showed that although one part was certainly a compact source, the other was associated with a faint jet.

The optical spectrum of the star-like object was puzzling because it seemed to display no trace of hydrogen, which is the main constituent of stars. However, these optical observations held an even greater surprise in store. It was realized late in 1963 that the lines characteristic of hydrogen were present: the lines normally seen in the optical region had been shifted into the infrared, while the lines that were now in the visible range were ones normally visible in the ultraviolet. In other words, the object had an unprecedented red shift, far larger than could be expected in a star.

What was true of 3C 273 proved to be true also of 3C 48, which had first been tentatively identified with a blue "star" in 1960. Because of their star-like appearance, the two objects became known as quasi-stellar radio sources, or quasars. Later, it turned out that there were many more such objects, all showing strong red shifts, although most did not have their peak emission at radio wavelengths. They were given the general name quasi-stellar objects, or QSOs, but the name quasar has stuck.

But what were the quasars? If the immense red shifts indicated a recession, as those of galaxies were assumed to do, then 3C 273 must lie at a distance of no less than 1,500 million light-years, and 3C 48 must be twice as far away. Yet if this were so, the optical radiation they were emitting must be more than 1,000 times greater than that of a galaxy such as our own.

There was another problem. The radiation from quasars is not only extremely intense, but also varies in strength, often over a few weeks. This implies that the source can be no more than a few light-weeks across; otherwise, radiation from the farthest parts of it would reach us with a delay sufficient to "smear" the variation over a longer period.

In 1963, no mechanism was known that could produce such immensely strong radiation in so tiny a region of space. Other explanations of the huge red shifts were therefore sought. Could the quasars be nuclei ejected from their parent galaxies, so that their high speeds had nothing to do with the expansion of the universe, and the objects were in fact relatively nearby? Or could their red shifts be caused by intense gravitational fields, which is possible according to relativity theory?

But neither of these explanations fitted all the observations, and consequently these suggestions had to be rejected. In fact, there seemed to be only one satisfactory answer, and that was that the red shifts were the consequence of distance: they were truly "cosmological." This explanation is now generally accepted, with only a few dissenters (pp. 178–79).

A clue to their nature comes from the fact that quasars are so far away. We are observing objects formed in the comparatively early days of the universe, at the stage when the galaxies were coming into being. Their spectra display a continuous background on which are superimposed broad lines. The continuous part probably comes from a compact source, and the lines from clouds of hot gas surrounding this core.

Quasars' spectra are in fact similar to those of galaxies of the kind known as Seyfert Type I, which have extremely active cores. In fact, Seyfert galaxies and quasars seem to be the same kind of object, with Seyfert Type I being at a slightly later stage of evolution than quasars. Type II Seyferts, which have less active cores, are at a still later stage of development. Neither Seyfert galaxies nor quasars are permanently present in the universe, but instead mark different phases in its evolution.

Yet, an explanation is still required for the intensity of these objects, since their emission of radiation is truly astounding. In 1969, it was

Hot gas surrounds the core of a quasar. At the center, enormous quantities of X-rays and ultraviolet radiation are released. This energy is absorbed and re-emitted by the surrounding matter. Within about one light-year of the core small, hot clouds move at almost the speed of light, both toward us (blue) and away from us (red). Cooler gas extends thousands of light-years beyond this (outer zone).

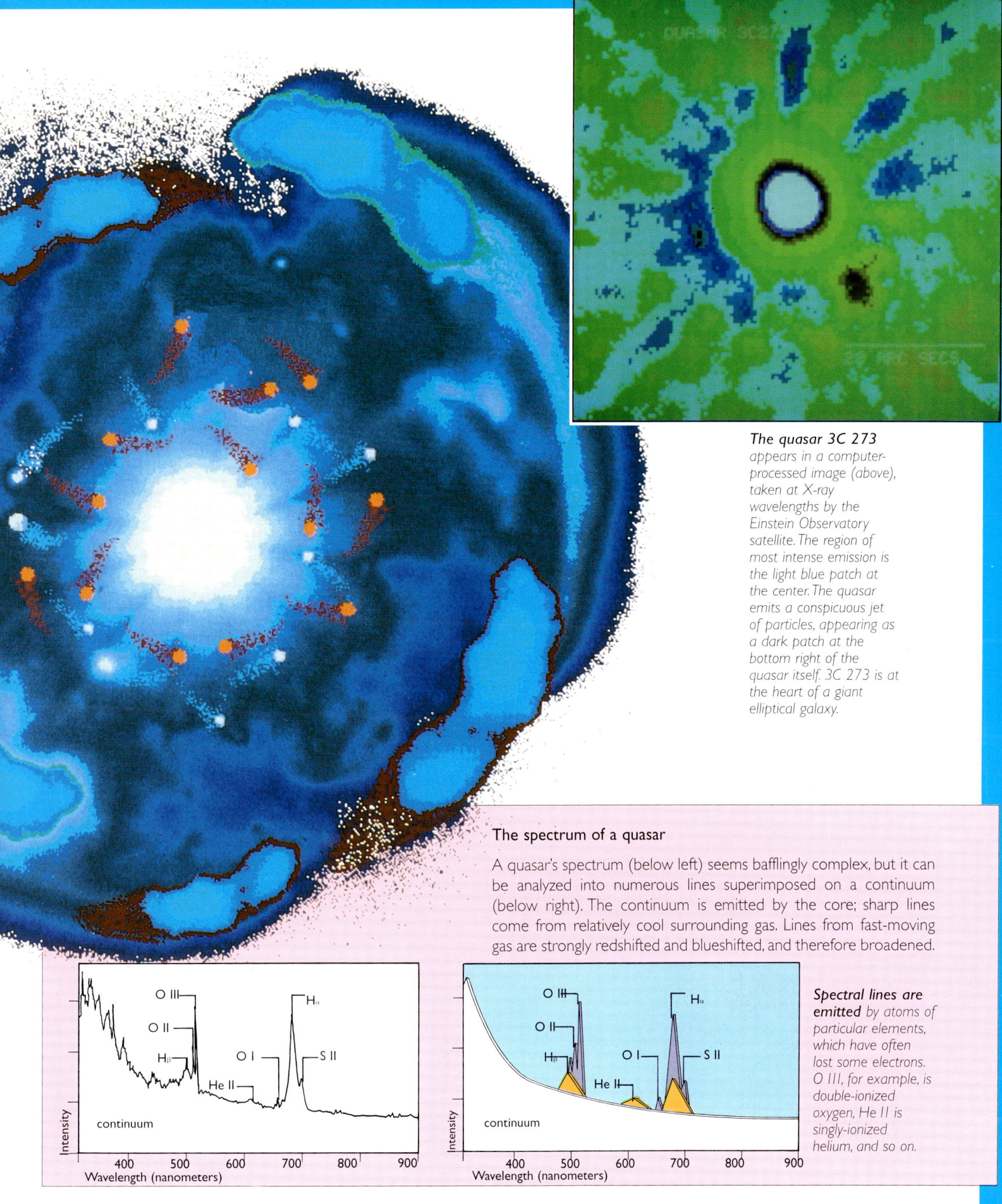

The quasar 3C 273 appears in a computer-processed image (above), taken at X-ray wavelengths by the Einstein Observatory satellite. The region of most intense emission is the light blue patch at the center. The quasar emits a conspicuous jet of particles, appearing as a dark patch at the bottom right of the quasar itself. 3C 273 is at the heart of a giant elliptical galaxy.

The spectrum of a quasar

A quasar's spectrum (below left) seems bafflingly complex, but it can be analyzed into numerous lines superimposed on a continuum (below right). The continuum is emitted by the core; sharp lines come from relatively cool surrounding gas. Lines from fast-moving gas are strongly redshifted and blueshifted, and therefore broadened.

Spectral lines are emitted by atoms of particular elements, which have often lost some electrons. O III, for example, is double-ionized oxygen, He II is singly-ionized helium, and so on.

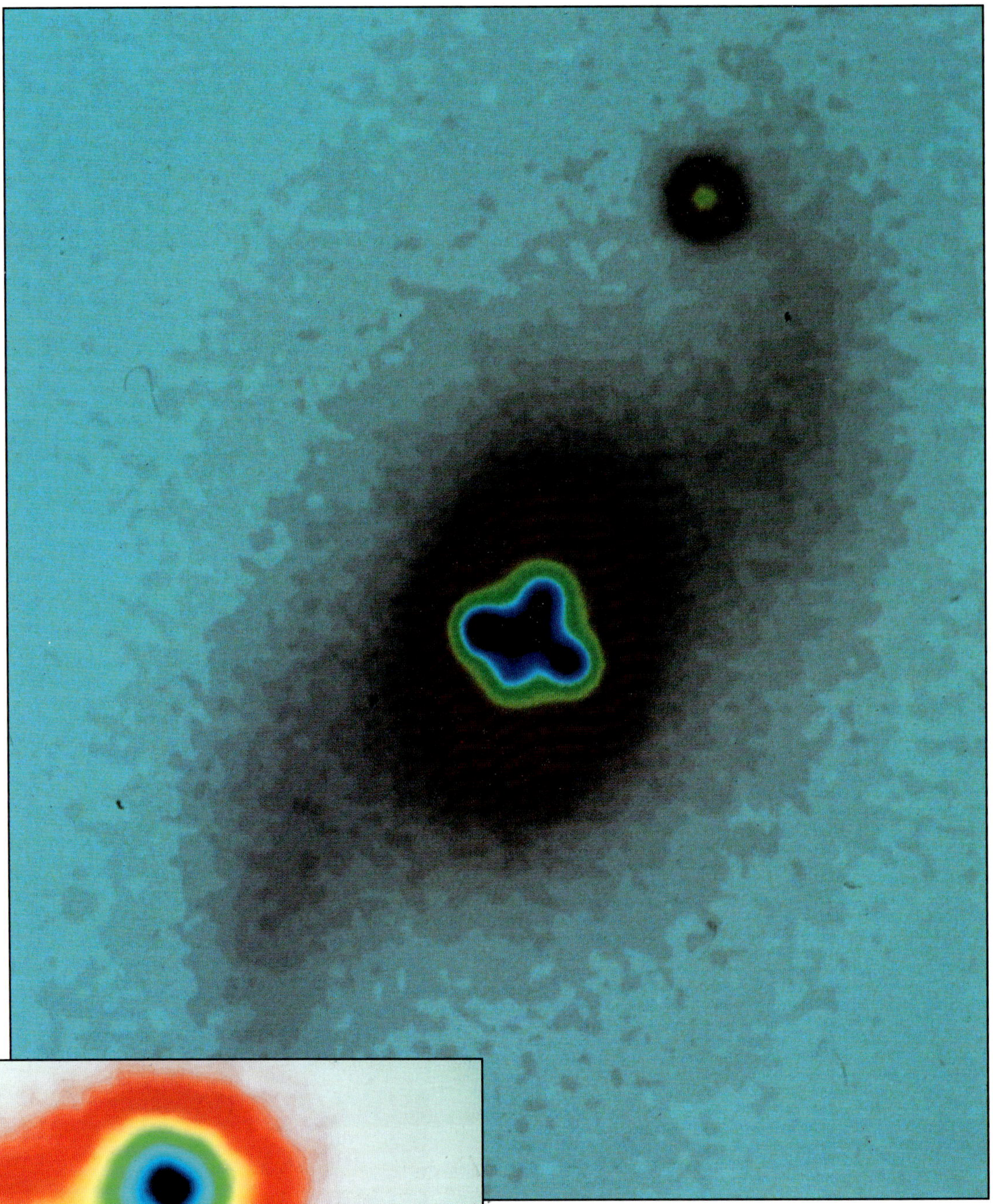

QUASARS

A cloverleaf image of a quasar is seen, apparently superimposed on a spiral galaxy (right). In fact, the galaxy is 500 million light-years away, whereas the quasar is billions of light-years distant. The cloverleaf effect is produced by "microlensing," which is gravitational focusing by individual stars, rather than by entire galaxies. When the "cloverleaf" is magnified (below), it proves to consist of four images of the quasar around the galaxy's bright core.

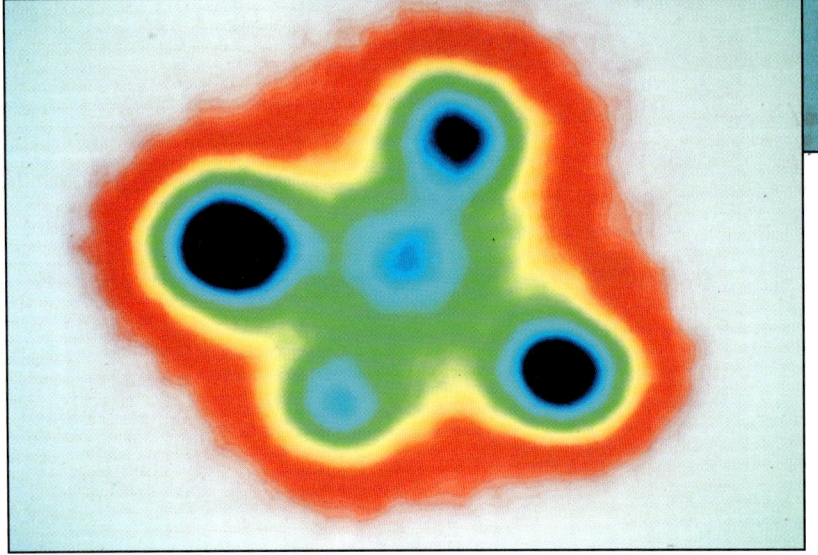

suggested that black holes (pp. 62–65) are the power sources for quasars. The amount of energy emitted when gas or other material falls into a black hole is a high proportion of that which would be produced by the complete annihilation of the matter, as given by Einstein's equation $E = mc^2$. This is about 1,000 times what the same amount of matter could yield if it fueled nuclear reactions inside a star (pp. 82–83), and is of the right order of magnitude to account for a quasar's energy.

Such quantities of radiation would be

generated if the black hole at the center of a quasar attracted material at relativistic speeds, producing intense synchrotron radiation (electro-magnetic waves emitted whenever charged particles are whirled around by magnetic fields). A large amount of the energy given out by quasars is in the form of synchrotron radiation. What is more, strong "winds" of ionized particles would also blow outward. This is confirmed by observations showing that the central regions of quasars have been swept clear of dust. From every point of view, black holes seem to be the cause of quasars.

Calculations based on the observational evidence indicate that the masses of black holes could span an enormous range. At the lower end of the scale, the holes might be the equivalent of no more than 100 million Suns, while at the top end they might be as massive as a trillion Suns.

Because they are so bright, yet so remote, quasars are uniquely valuable as probes of the distant, early universe. That they are truly distant seems confirmed by additional observations, which show the gravitational deflection of their light by intervening galaxies.

From the general theory of relativity, it can be calculated that the light from, say, a quasar, or other distant source lying almost behind a nearer massive body, should be deflected by the distortion of space-time. The image of the quasar should not only be shifted, but should in general also be broken up into several images, or spread out into arcs or an "Einstein ring" apparently centered on the nearer body.

Such gravitational lensing was first detected by the Nuffield Radio Astronomy Observatory at Jodrell Bank, England, in what appeared to be a double source. In 1979, several years after the first radio observations, the pair was studied optically. It was found that the objects were quasars, and that they had identical spectra. This ruled out the possibility that they consisted of two clouds thrown out by an active galaxy lying between them. The two images were clearly of the same object, about 10 billion light-years from Earth. The cause of this double image is a giant elliptical galaxy that has been detected by radio and photographed. It is less than half as far away as the quasar whose light it affects.

A number of Einstein rings and arcs have been detected. Two arcs associated with the quasar MG 1654 + 1346 in Hercules are a fine example of gravitational lensing at radio wavelengths. The quasar possesses two lobes, in addition to the emission from its center. The southern lobe is directly in line with a large intervening elliptical galaxy, which splits the image of the lobe into two arcs. The galaxy's mass can be calculated from the lensing it causes: it is equal to 89.8 billion solar masses, which makes it large as elliptical galaxies go.

Light can be focused by the gravitational fields of single stars, an effect known as "microlensing." There is little chance of observing this phenomenon in the case of stars in our own galaxy, but it has been detected in some external galaxies. Some astronomers claim that increased numbers of quasars can be seen close to the lines of sight to nearby galaxies; they believe this is due to the focusing, and therefore brightening, of their light by stars in the galaxies.

Not all double quasars are the product of gravitational lensing. Some are true binary quasars. In 1989, it was reported that quasar PHL 1222 in the southern sky is double. Its two components are very close — no more than 100,000 light-years apart — but are not identical. The system is calculated to be 12 billion light-years away, so what we are probably observing here is a pair of quasars being born.

A partial Einstein ring is seen in this image taken by the Hubble Space Telescope. A cluster of galaxies is in the center of the image. A more distant galaxy lies along almost the same line of sight as the cluster. The cluster acts as a gravitational lens and bends the light of the distant galaxy. A mirage of the galaxy, the blue arcs around the cluster, result. If the cluster and galaxy were lying exactly on the same line of sight, a complete ring would be seen.

NEBULAE
• Cradles of the stars

Our galaxy contains not only myriads of stars, but also an abundance of gas and dust. The gas and dust are referred to as interstellar material, and take the form of bright and dark clouds (nebulae) which can be observed at many wavelengths. They can be wonderful sights – especially in optical photographs, in which their colors can be detected (the light of a nebula is too dim to activate the color receptors of the eye). The Great Nebula in Orion, M42, which lies just a little south of the central star of Orion's belt, shows no more to the eye than a hazy patch of apparently colorless light in any telescope. Yet in time-exposure photographs, spectacular colors are revealed.

The Orion Nebula is a grand sight because it is a mixture of bright and dark clouds. There are bright pinkish regions, which seem to be divided into two parts: a small nebula (M43) to the north and a vast glowing patch (M42) to the south. Between them there appears to be a dark, empty space, but in fact, this is an area of dust, blocking the light of the stars beyond. Such dark nebulae can be seen in various other places against the background of both these bright nebulae. Both M42 and M43 lie at a distance of about 1,500 light-years.

The dark nebulae are dark because they contain dust, which does not emit visible radiation – dust that is dense enough to prevent light reaching us from the bright nebulae and stars beyond. The most spectacular dark nebula in Orion is the famous Horsehead, named for its shape. Another dramatic example is the Coalsack in the constellation of the Southern Cross. Interstellar dust particles are composed of many substances, including carbon, silicon, and silicates of iron, magnesium and aluminum, as well as some other molecules.

Bright gas clouds are of two types: reflection nebulae and the more numerous kind that consist of glowing gas. Reflection nebulae contain a great deal of dust as well as gas, and happen to lie close to hot stars; it is the dust that reflects the starlight. Such nebulae appear blue because the dust scatters only the shorter wavelengths. A similar process operates on Earth, where the blue component of sunlight is scattered by the molecules of the gases in our atmosphere, making the sky blue.

Nebulae that consist of glowing gas also receive their energy from starlight, but by a different process. There are hot bright stars within or behind these gas clouds. Such stars are strong emitters of high-energy, short-wavelength ultraviolet radiation. When this falls on the cloud, it ionizes the gas atoms, stripping electrons from them. When the electrons recombine with the atoms, they give out lower-energy, longer-wavelength visible light. This process is called fluorescence.

In M42, the stars responsible for the nebula's glow lie in a very bright patch of gas toward the northwest; there are four in a cluster known to astronomers as the Trapezium. The clouds here are composed primarily of hydrogen, which gives them their characteristic pinkish color. However, above the Trapezium lies some darker gas; this does not glow because it is too far away from the Trapezium, and the ionizing radiation is too weak by the time it reaches the cloud. The gas here, therefore, remains neutral (not electrically charged).

Yet, dark hydrogen does emit radiation – at the radio wavelength of 21 centimeters. Because of this, it is possible to trace the distribution of hydrogen not only in the Orion Nebula but in other nebulae and elsewhere throughout interstellar space.

Hydrogen is not the only gas in interstellar space, although it was not until the advent of radio astronomy that the presence of other material became known. In particular, carbon and oxygen are prevalent, but atoms of nitrogen, silicon, and sulphur have also been detected. Such atoms do not appear singly, but are grouped into molecules.

Some of these are organic molecules (complex molecules based on carbon). They include formaldehyde and formic acid. Even ethyl alcohol has been discovered in prodigious quantities; in fact, there is enough comparatively near to us to satisfy humankind's alcohol consumption for countless generations. It may be that even more complex molecules are yet to be discovered.

The presence of carbon-containing molecules has important implications for the possibility that life can be generated in space (pp. 158-59). Here it should be noted that they can exist only in particularly cool areas of

The spectacular M42, part of the Orion Nebula, is stimulated to glow by the ultraviolet radiation of hot young stars.

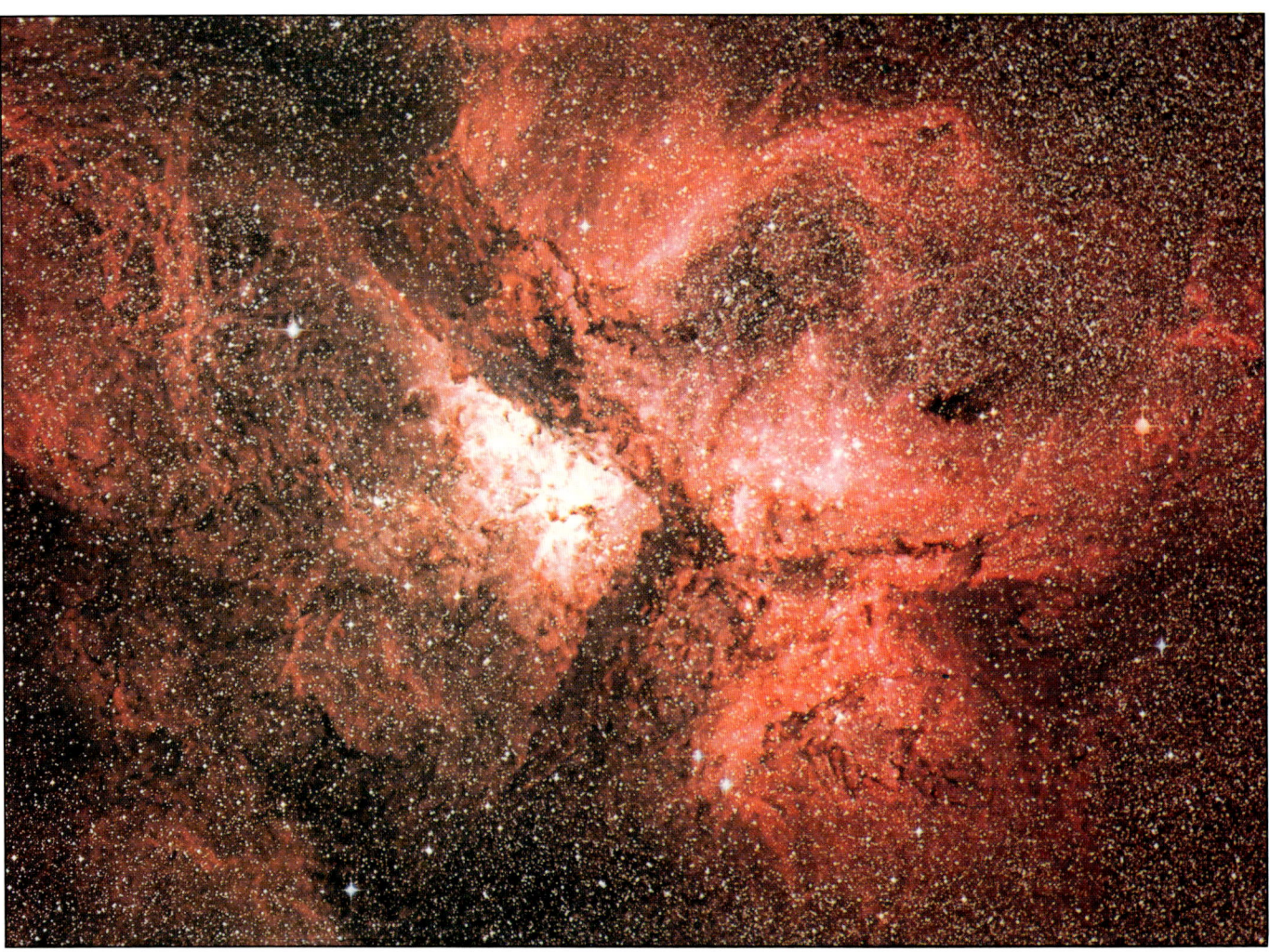

The magnificent nebula in the constellation Carina (the Keel of the Ship), in the southern skies. Dark lanes of dust cross the bright regions of gaseous matter. At the center of the nebula lies a denser knot of gas and dust, swathing the remarkable variable star Eta Carinae. In the 1840s, it suddenly brightened: having been a second-magnitude star, it became the second-brightest star in the sky. In a few decades it faded, to the extent that it is now invisible to the naked eye. It may have been 200 times as massive as the Sun when it was born. It has shed a lot of of its matter into space, but it still ranks as a supermassive star.

nebulae. This is because they cannot survive unless protected from the damaging effects of ultraviolet and other high-energy radiation, which would break up such complex molecules into their constituent atoms.

Molecules of water and hydroxyl are also present. In the Orion Nebula, M42, they are part of a vast molecular cloud, contracting under the influence of gravity. The nebula possesses two strong infrared sources. Almost certainly one of these is a star in the process of being born. The evidence for this comes from the water and hydroxyl molecules, which act as "masers." (This name is derived from the initials of the phrase "*m*icrowave *a*mplification by *s*timulated *e*mission of *r*adiation.") Radio waves emitted by the nascent star stimulate the molecules to radiate, providing a clue to its existence.

The material in a nebula is extraordinarily thinly spread. Generally, there are no more than 164,000 atoms per cubic inch, even at the center – billions of times less than the density of a puff of smoke. Some gas clouds are very much less dense than even this. As far as the dust is concerned, there is on average one grain per 3.5 million cubic feet – the volume of an average concert hall. This, too, represents a very small quantity of matter, for typical dust grains are no larger than one 25-thousandth of an inch, and may be only one-tenth of this size. They are composed of carbon or, less frequently, silicon. However, they are sometimes coated with ice or with large, complex molecules.

It may seem astounding that such tenuous clouds can be observed at all, until one realizes their vast extent. As an example, the optically visible regions of the Orion Nebula stretches 20 light-years across space. It would take over 10,000 solar systems in a line to span it. Such vastness makes it easier to understand how the incredibly thin dust of a nebula can dim and even block out the light of the stars beyond it.

Like many other objects in the universe, nebulae are in constant motion. They share in the general rotation of the Galaxy. And when new stars form within them, the newly emitted

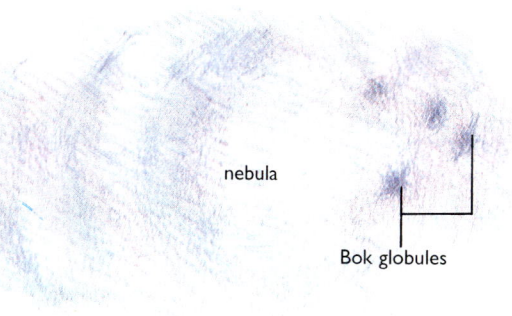

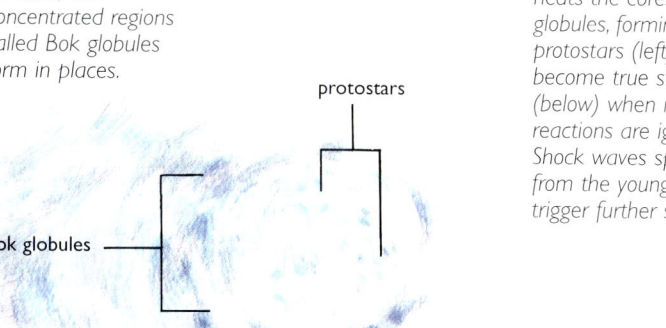

A nebula that is sufficiently dense is destined to form stars (left). It begins to contract, and concentrated regions called Bok globules form in places.

The collapse of the gas heats the cores of the globules, forming protostars (left). These become true stars (below) when nuclear reactions are ignited. Shock waves spreading from the young stars trigger further starbirth.

The Cone Nebula (below), in Monoceros the Unicorn, is the scene of starbirth today. There are many young, bright T Tauri stars nearby, which vary erratically in brightness because they are still surrounded by swirling remnants of the gas and dust from which they formed.

radiation disturbs the cloud, blowing away the gas and dust. It is estimated that the presently visible regions of M42 in Orion will be dispersed within the next 10,000 years.

Yet, the disappearance of this material will not leave a gap, because new clouds of gas and dust will form in the Orion spiral arm, in which the nebula lies. This matter is at present spread out along the spiral arm, but it will gather into nebulae by the processes mentioned in the discussion of the formation of spiral galaxies (pp. 54–55). Thus, new nebulae will appear, some of them lit up by stars that are still unborn.

The presence of nebulae and diffuse interstellar matter poses problems for the observational astronomer. They can dim the stars and even hide the most distant objects from view. The small size of the dust particles means that they scatter the shorter optical wavelengths, and therefore produce an apparent reddening of distant objects. This makes it difficult to assess the true brightness and color of distant stars and galaxies.

Astronomers can allow for the drop in intensity with distance, provided that they have some standard by which to assess the diminution. Some galaxies provide such a standard because they are of known brightness. The true brightness of certain stars can also be determined; for example, the brightness of a Cepheid variable can be determined with considerable precision from its period of variation (pp. 86–87). Such compensation has to be applied regularly when observing across interstellar space.

However, as was mentioned above, neutral hydrogen radiates at a radio wavelength of 21 centimeters. This allows radio astronomers to penetrate some of the dark clouds and to "see" farther into space in certain regions than is possible optically. This is one of the reasons why knowledge of the universe has expanded so greatly in recent decades.

STARBIRTH
• *Lighting the nuclear fire*

If you look very closely at photographs of bright nebulae, you will find that many of them are pitted with tiny dark blobs. These are called Bok globules, for Bart J. Bok, the American astronomer who first observed them. The infrared and radio waves coming from them show that they are the birthplaces of stars.

Stars could not form in the very early universe, when matter was still at temperatures of hundreds of thousands of degrees. However, with the continuing expansion of the universe, the hydrogen gas cooled. About two billion years after the Big Bang, the protogalaxies formed (pp. 38–39). Then local condensations of gas appeared within the protogalaxies, forming nebulae. In some places, the density of gas in these reached billions of molecules per cubic meter. This density, though far thinner than the best laboratory vacuum, enabled gravity to pull the material into ever more concentrated conglomerations.

As this happened, the central regions of each concentration heated up, somewhat like air heats up as it is pumped into a tire. The temperature in each knot of matter rose until the molecules were broken up into individual atoms, and then still further, until the atoms were ionized, losing their outer electrons. Pressure rose as fresh material was pulled in, pressing down on the central regions.

Although nuclear reactions had not yet begun, this protostar was already producing copious amounts of energy. But no visible light could penetrate the gas and dust that enveloped it. Only infrared radiation could escape.

In due course, the situation at the heart of the protostar became critical. The density had increased billions of times, and the temperature had risen to 10 million K, or even more. The positively charged hydrogen nuclei, now stripped of all their electrons, were pushed so close together that they began to collide, overcoming the strong repulsion of their electrical charges. Hydrogen nuclei began to fuse to form helium nuclei. The protostar became a true star.

Each helium nucleus had slightly less mass than the hydrogen nuclei that formed it. The mass that vanished was converted into huge amounts of energy, according to the relationship $E = mc^2$ discovered by Albert Einstein (pp. 14–15). Each gram of hydrogen that reacted produced as much energy as 600 million single-bar electric heaters radiate every second. The temperature at the center of the protostar rose further as these vast amounts of energy tried to escape. The new star would have exploded were it not for the huge weight of material of which it was composed.

As radiation made its way outward, it set up convection in the outer layers of the star. The radiation heated the deep-lying gas, which rose, cooled at the surface, and then sank, to begin the cycle all over again. (The same sort of movement can be seen in a pan of milk heated on a stove.) The light and heat from the young star blew the clouds of gas and dust away and it became visible to the universe beyond.

The same process of starbirth is going on today. Infrared astronomers think they can glimpse it in parts of some nebulae, such as the nebula in Orion. But not every concentration of gas and dust forms into stars. If there is insufficient material in a gas cloud, its gravitation will be too weak to make it condense sufficiently, and its temperature will not rise to the critical level at which nuclear fusion begins. The "star" that forms in this way will merely be a warm body, detectable mainly because of its infrared radiation.

Many such objects seem to exist in our own region of the Galaxy, particularly within the spiral arms. They have become known as brown dwarfs. Some failed stars are so small, they are only about the size of the planet Jupiter, and accordingly are called jupiters.

Fully developed stars bear signs of the epoch in which they were born. The earliest stars – the first to be formed after the Big Bang – condensed from the primal material: hydrogen with a proportion of helium. Later generations of stars formed from material that had been mixed with the remnants of exploded stars. These included elements with nuclei heavier than that of helium, for these had been synthesized in the cores of the stars before they exploded (pp. 82–83). The Sun contains such elements, which shows that it is not a first-generation star. Furthermore, our theories of stellar life cycles tell us that the Sun is only five billion years old – making its origin more recent than the first period of star formation.

A cloud of interstellar matter begins to shrink, its central portions collapsing faster than its outer parts. The core warms up, while the cloud begins to rotate.

After hundreds of thousands of years of contraction, the cloud is rotating faster and its globular shape has begun to flatten. Its central regions have formed a hot protostar, emitting several times as much radiation as today's Sun. But only infrared radiation can escape from the surrounding cloud.

The bulk of the circling gas and dust has become concentrated into an accretion disk. Matter continues to fall onto the protostar, which is so hot that it violently ejects matter. This escapes at the poles, where there is least obstructing matter. This outflow sweeps away much of the gas and dust surrounding the protostar.

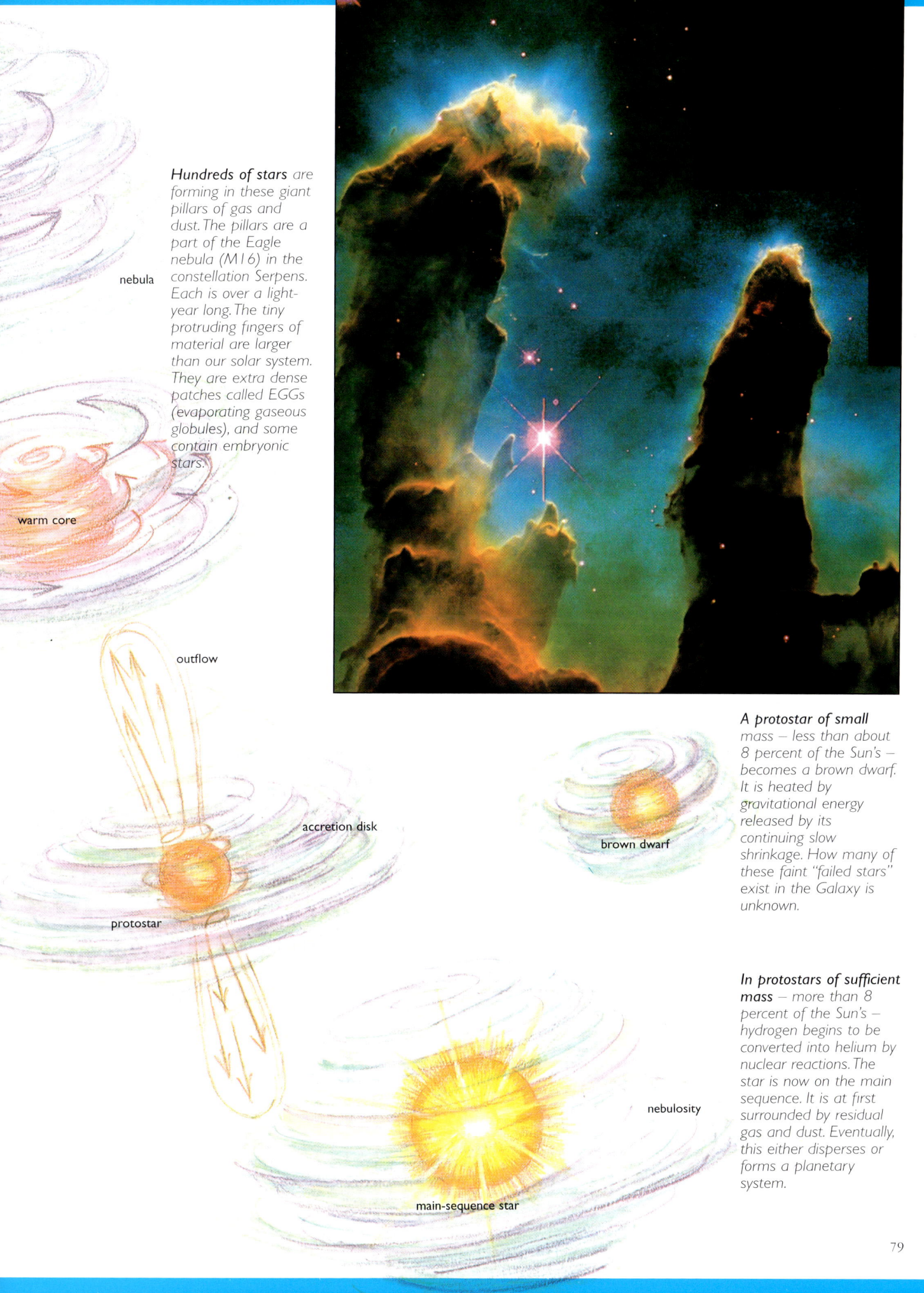

Hundreds of stars are forming in these giant pillars of gas and dust. The pillars are a part of the Eagle nebula (M16) in the constellation Serpens. Each is over a light-year long. The tiny protruding fingers of material are larger than our solar system. They are extra dense patches called EGGs (evaporating gaseous globules), and some contain embryonic stars.

A protostar of small mass — less than about 8 percent of the Sun's — becomes a brown dwarf. It is heated by gravitational energy released by its continuing slow shrinkage. How many of these faint "failed stars" exist in the Galaxy is unknown.

In protostars of sufficient mass — more than 8 percent of the Sun's — hydrogen begins to be converted into helium by nuclear reactions. The star is now on the main sequence. It is at first surrounded by residual gas and dust. Eventually, this either disperses or forms a planetary system.

STELLAR COMPANIONS
• *Star clusters and multiple stars*

Stars form not only singly but also in groups; this occurs when a number of protostars are born close together. They form a gravitationally linked cluster, with a collective motion on which their individual motions are superimposed.

Clusters with a few hundred members, loosely grouped, are termed open clusters. The conspicuous Pleiades, in the northern constellation Taurus, are of this kind. The nearest open cluster, however, is the group called the Hyades, also in Taurus, only about 130 light-years away. Its age is about two billion years.

In the southern skies, the most spectacular open cluster is the Jewel Box, NGC 4755, around 7,800 light-years away. It derives its name from the fact that it consists of a group of about 50 bright young blue stars together with a bright red star, Kappa Crucis.

Even looser groupings of stars called stellar associations occur, but they are difficult to identify. They are young stars often with an open cluster at their center.

On the other hand, it may happen that thousands or hundreds of thousands of stars form together — not enough to form an elliptical galaxy, but enough to make a roughly spherical grouping known as a globular cluster. Such clusters are always old, and it seems that they formed around the same time as the galaxies with which they are associated.

Our own galaxy has a halo of about 150 globular clusters evenly distributed around the galactic center, and up to 60,000 light-years from the galactic plane. The most spectacular is Omega Centauri, in the southern skies, only 16,500 light-years away. Through a large telescope, it is seen to cover more than a degree of arc — twice the apparent size of the full Moon. It is probably the most massive globular cluster in the Galaxy, spreading over approximately 65 light-years. It is not spherical, but ellipsoidal, apparently because of its rapid rotation.

Single stars like our Sun account for only about half the stars we see: the other half are binary or multiple stars. Some apparent doubles do not consist of genuinely associated stars, but merely of ones that happen to lie in the same line of sight from the Earth. For example, Mizar, the middle star of the handle of the Big Dipper, or Plough, has a fainter "companion," Alcor. Yet Mizar is about 58 light-years from the Earth, while Alcor is 81 light-years away, and the two are not physically connected, although each is a double itself.

However, most multiple star systems are genuinely associated stars. About 30 percent have more than two members; about one in nine have four or even more components. In the overwhelming majority of such systems, the components were formed at the same time, but in a few cases, a massive star has captured a companion.

All the members of a multiple star system orbit around their common center of gravity, and have a complex motion. The members of some systems like this cannot be separated in a telescope, but their combined spectrum shows distinctive behavior. While one star is moving toward the Earth, its light will be blueshifted (pp. 46-47). At the same time, its companion will be moving away from the Earth, and its light will be redshifted. The result will be that lines in the combined spectrum will periodically split and recombine, or at least become broader and narrower. Thus, the multiple nature of the light source can be recognized.

A globular cluster beyond our galaxy (left). This object orbits the center of the Large Magellanic Cloud, a companion galaxy to the Milky Way. Globular clusters are present in all galaxies. Although they contain hundreds of thousands of stars, these are widely spaced even near the center of the cluster — but by light-months, rather than the light-years that separate stars in our own stellar neighborhood.

The brightest members of the Pleiades in Taurus make a fine naked-eye cluster, and form a spectacular grouping in a telescope. There may be 500 stars in all.

Cocooned in clouds of cold gas and dust, the cluster spreads across space for about 30 light-years, and is at a distance of 400 light-years. The cluster is young, having an age of no more than 50 million years, and its stars are bluish. The nebulosity shines by reflected starlight.

THE CHEMISTRY OF THE STARS

● *Sources of stellar energy*

In the nuclear powerhouse at the heart of a star like the Sun, hydrogen nuclei are smashed together and welded into heavier nuclei. At the same time, energy in the form of lighter particles and high-energy gamma rays is released. The hydrogen nucleus is a single particle, the proton, and the dominant process, shown here, is called the proton-proton cycle.

Protons are constantly colliding to form nuclei of deuterium, called deuterons. These eventually collide with protons to form nuclei of helium 3. The collision of two helium 3 nuclei forms one nucleus of helium 4, while releasing two protons. This ends the chain.

The fundamental power source of all stars is the fusion of hydrogen into helium. But this conversion can be achieved by any of a variety of processes. However, in stars of the Sun's size, it proceeds almost wholly by one process, the hydrogen, or proton-proton cycle.

The sequence of events begins when two hydrogen nuclei combine. The nucleus of a hydrogen atom is the simplest of all nuclei. It consists of a single positively-charged particle, the proton. Since protons have like electrical charges, they repel each other. Two of them can be forced together by the high temperatures and pressure of a stellar interior, but they can stay together only if one of them loses its charge. It can do this through reactions that turn it into a neutron, which has slightly more mass than a proton but is electrically neutral. The combination of proton and neutron thus formed is called a deuteron.

In the process, a positron (positively-charged electron) and a neutrino are given off. The neutrino has no electrical charge and no rest mass – or such a tiny mass that it has not been measured yet – and it has a tremendous penetrating power. It carries away a substantial amount of energy.

At Earth-like temperatures and pressures, a deuteron would soon pair up with a single electron, which would balance the deuteron's single proton. The combination would be an atom of deuterium. Deuterium is an isotope of hydrogen: since its atom has only one electron, it has the chemical properties of a hydrogen atom, but roughly twice the weight.

In due course, the deuteron will collide with another proton, again giving out energy – this time in the form of a gamma-ray photon. The result will be a nucleus of two protons and one neutron – an isotope of helium. Two of these nuclei later collide and produce a nucleus of ordinary helium (two protons and two neutrons), together with two protons.

So the net result of the hydrogen cycle is the building up of a nucleus of helium from four protons. Part of the energy given out in the successive steps of the reaction goes into heating the star; part escapes from the star. Most of the neutrino's energy is lost, because these ghost-like particles can slip through thousands of miles of matter as if it did not exist. The energy carried by positrons, on the other hand, is soon converted into radiation when the positrons are annihilated in collisions with electrons. The radiation is absorbed by the star, thus maintaining its temperature.

In stars much more massive than the Sun, the temperature of the central regions is above 27 million degrees Fahrenheit. At these temperatures, heavier nuclei move at the speeds

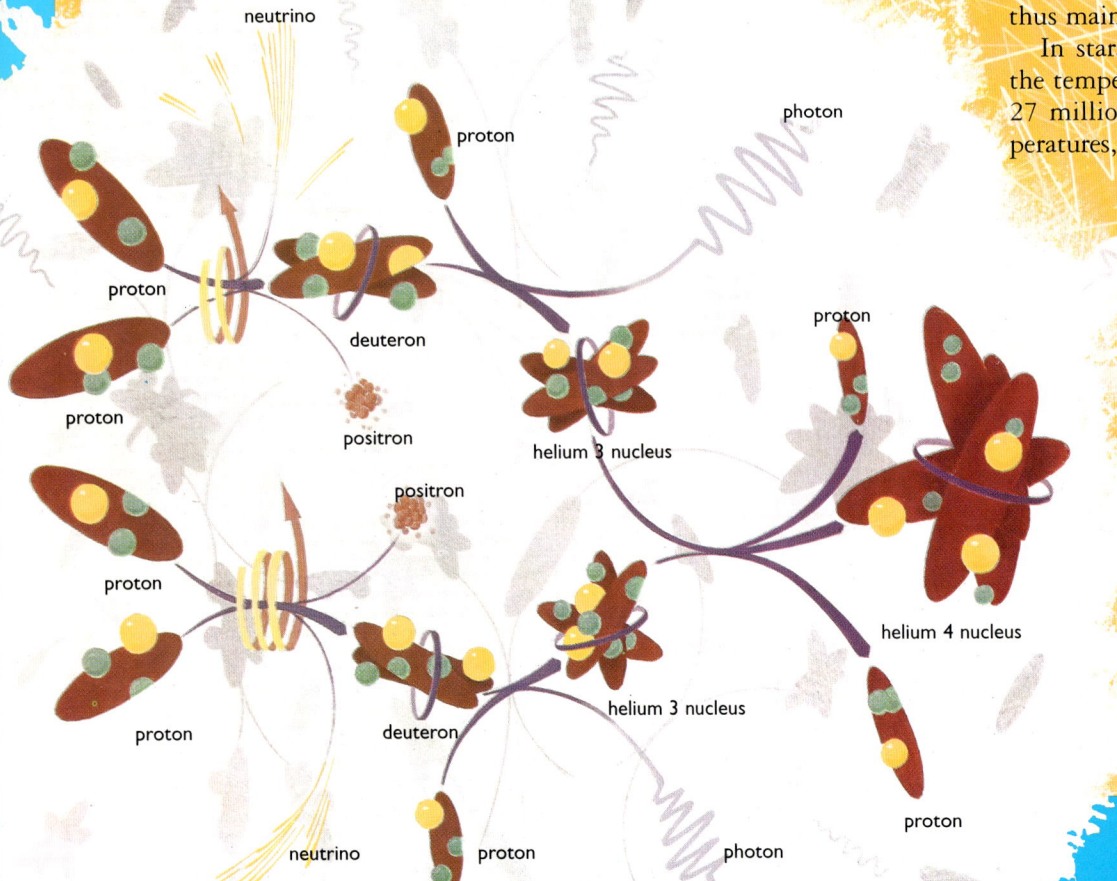

necessary to enable them to react. Carbon nuclei, in particular, take part in a sequence of reactions called the carbon cycle. The carbon nuclei emerge at the end of the carbon cycle unchanged – in the chemist's terminology, they are catalysts:

1: A carbon nucleus (six protons and six neutrons) combines with a proton, forming a nitrogen nucleus (seven protons and six neutrons) and giving out a gamma-ray photon. This form of nitrogen is unstable, since there are not enough neutrons present; in all long-lived atomic nuclei, the number of neutrons equals or exceeds the number of protons. Thus, the nitrogen emits a positron (plus a neutrino) so that one of its protons turns into a neutron. The nucleus is now a carbon nucleus, although it has seven neutrons instead of six.

2: When this carbon nucleus encounters another proton, it forms a nitrogen nucleus again – more stable this time, since its seven protons are combined with seven neutrons. Another gamma-ray photon is given out as the nitrogen nucleus is formed.

3: When this nitrogen meets another proton, they combine to form an unstable isotope of oxygen, with eight protons and seven neutrons (eight neutrons are present in ordinary oxygen). At the same time, another gamma is emitted. The oxygen soon "decays" by sending out a positron (with, for good measure, a neutrino as well), changing to yet another form of nitrogen (seven protons, eight neutrons).

4: The nitrogen nucleus now meets another proton, and they form a nucleus of helium (two protons, two neutrons) and one of carbon (six protons, six neutrons), identical to that which started the cycle.

The net result of the carbon cycle is the production of helium nuclei – one for each of the four protons that are used up – and the release of energy in the form of neutrinos, gamma-ray photons and positrons.

In massive stars, the carbon cycle is faster than the proton-proton cycle at converting hydrogen into helium. It is also an effective producer of energy, since during each cycle three gamma-ray photons are brought into being. The carbon cycle is one of the main reasons for the much greater energy production of massive stars.

In giant red stars, yet another process, the triple-alpha reaction, occurs. The process is so called because three helium nuclei are involved, and alpha-particle is another name for the helium nucleus. To begin with, two helium nuclei combine to make an isotope of the element beryllium (four protons, four neutrons), with the emission of a gamma-ray photon. Then another helium nucleus fuses with the beryllium; this produces a stable isotope of carbon, with six protons and six neutrons. Again, a gamma-ray photon is emitted. So the net result of this reaction is the conversion of helium into carbon.

There are still further reactions that lead to the synthesis of heavier chemical elements. For instance, two carbon nuclei can fuse to produce a magnesium nucleus. Under immensely hot conditions, such heavier nuclei can combine to make elements as heavy as iron. Such reactions occur only in the most massive stars, as they approach their deaths.

Energy travels outward from the Sun's energy-producing core in the form of high-energy radiation. Photons travel a zigzag path as they are absorbed and re-emitted by nuclei, taking thousands of years to battle their way to the surface. In the outermost layers, however, convection becomes more important in transferring heat. Here, hot gas, being less dense, rises to the surface of the Sun. There it radiates light and heat, cools, becomes denser, and sinks. The gases circulate in convection cells (darker arrows above).

83

LIVES OF THE STARS
● *Paths of stellar evolution*

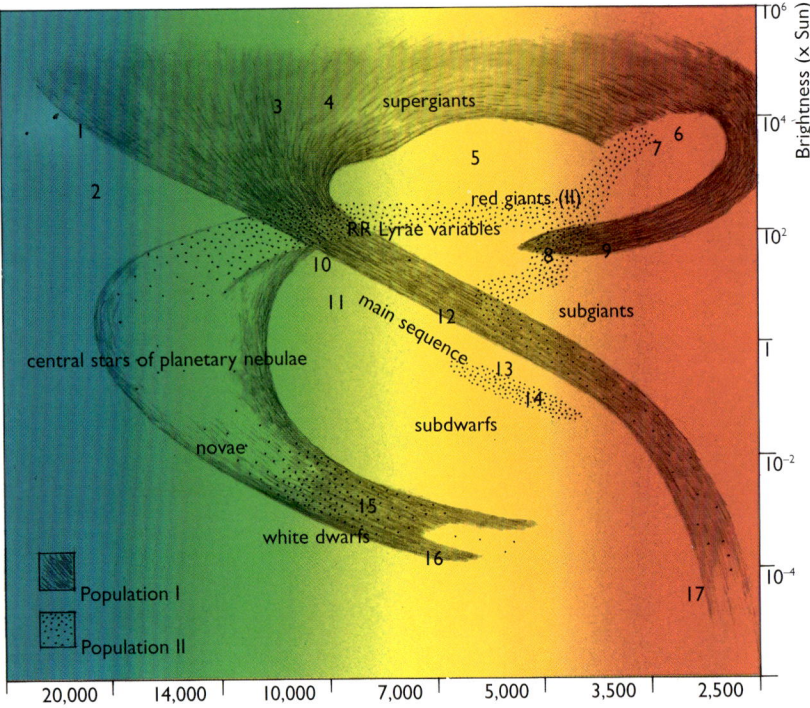

In the Hertzsprung-Russell diagram, luminosity is plotted against temperature. Most stars fall in the central band called the main sequence. Population II stars are relatively old, and many have evolved away from the main sequence. Only the most massive and fastest-evolving of the younger Population I stars have done so yet.

1 δ Orionis
2 Spica
3 Rigel
4 Deneb
5 Polaris
6 Betelgeuse
7 Antares
8 Arcturus
9 Aldebaran
10 Vega
11 Sirius A
12 Procyon A
13 The Sun
14 τ Ceti
15 Sirius B
16 Procyon B
17 Proxima Centauri

Nuclear reactions determine the complex life cycles of the stars. An essential tool in describing stellar evolution is called the Hertzsprung-Russell, or H-R diagram. It was first devised during the second decade of this century by Ejnar Hertzsprung and Henry Norris Russell.

An H-R diagram carries a scale of brightness – the true brightness, or absolute magnitude, of a star – running upward, from faintest at the bottom to brightest at the top. Horizontally it carries a scale of temperature, or of types of spectra that correspond to temperatures. These classes bear the traditional labels O, B, A, F, G, K, M. (All English-speaking astronomers remember this sequence with the mnemonic "Oh Be A Fine Girl, Kiss Me!")

At the high temperature end, O-type stars are blue and have temperatures in excess of 25,000 K, B-type stars are a little cooler, having temperatures of 11,000 – 25,000 K, but are still bluish, and so on. The Sun is a G-type star, with a surface temperature of about 6,000 K, described by astronomers as yellowish. M-type stars are red and have temperatures below about 3,500 K.

When the brightnesses and temperatures of stars are plotted on the H-R diagram, it is found that some regions are occupied and others are not. The majority of stars lie on a band called the main sequence, running from bottom right to top left. It has become clear that the pattern of the H-R diagram is related to the evolution of stars: it is a plan of the way they spend their lives.

When a star has just been born in a Bok globule, and once nuclear reactions begin in its core, it appears close to the main sequence. Precisely where it enters the sequence depends on its mass.

Very small stars, of about a quarter the mass of the Sun, enter as M-type red dwarfs. The more massive Sun entered farther up, while even more massive stars join still farther along the sequence. All stars spend the greater parts of their lives on the main sequence, their positions changing very little, while the hydrogen in their central regions lasts.

A large central core of nonreactive helium "ash" gradually builds up. This core contracts and its temperature rises, but a surrounding shell of hydrogen still "burns." Now the star moves off the main sequence.

The length of life of a star depends on its mass, and thus on its position on the main sequence. A dim red dwarf evolves so slowly that it will take 200 billion years before it leaves the main sequence; the Sun will do this after about 10 billion years; a star of five solar masses takes only 70 million years.

As a sun-like star evolves away from the main sequence, it expands to something like 50 times its previous size. It becomes cooler and redder, and therefore moves to the right of the H-R diagram. But because of its huge increase in size, it also grows brighter, and therefore moves upward in the diagram, becoming a red giant.

But the core shrinks and its temperature rises still farther. When it reaches 100 million K, the helium begins to take part in nuclear reactions. The outer regions contract and the star ceases to be a red giant.

By this time, the core of the star is mainly composed of carbon and oxygen which have been formed by the helium-burning (pp. 82–83). It now reaches the final stages of its life. At first, the energy output becomes less and the star contracts. However, the interior of the star expands again, and for a short time the star becomes a red giant once more, with helium as well as hydrogen burning around the core.

Suddenly, there is a change: the energy generated around the core blows off the outer envelope, and for a time the star is surrounded by a shell of gas. It has become a planetary nebula – so called because in a telescope it appears as a hazy disk, similar to that of a planet. After this, the star contracts and all that remains is a superdense core, still burning nuclear fuel in its outer regions. The star ends as a white dwarf, cooling and fading (pp. 88–89).

After leaving the main sequence, a Sun-like star will spend about 10 billion years evolving toward the red giant stage. More massive stars live their lives faster because their nuclear reactions proceed at a higher rate. Stars of five times the Sun's mass will take only 70 million years to evolve to the red giant stage; those 15 times as massive will last only 10 million years.

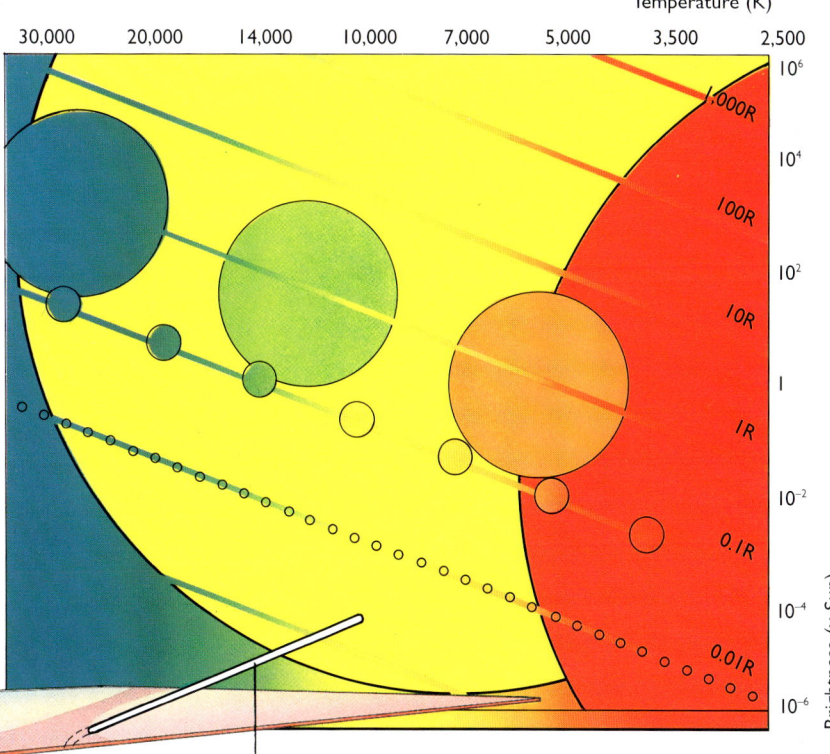

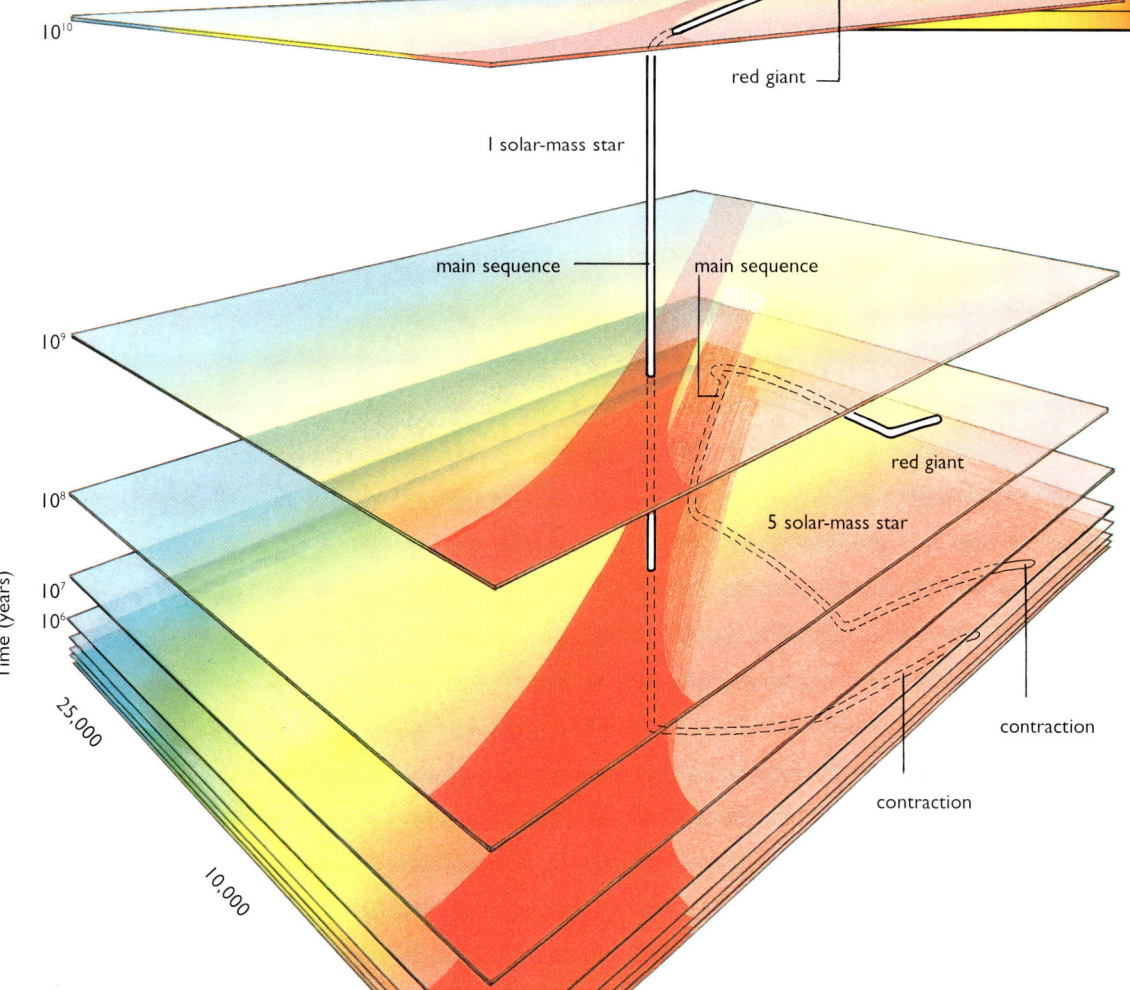

The sizes of the stars can be calculated from their temperature and brightness, as shown by their positions on the H-R diagram (above). Stars of a given size lie on the straight lines shown here; thus, on the line labeled 10R, they have a radius 10 times that of the Sun.

The evolution of a star is represented by a track through the H-R diagram. Newly born stars join the main sequence as nuclear reactions begin, and spend most of their lives there. As they age, they swell, become hot and bright, and move into the red giant region of the diagram. The star of five solar masses shown here completes its life cycle about a hundred times as fast as a star like the Sun.

VARIABLE STARS
● *Shining with inconstant light*

Stars vary enormously in brightness in the course of their lives. But these changes occur over millions and even billions of years. Some stars, however, lead a less sedate existence in which their radiation output varies far more rapidly. Study of such variable stars brings new evidence about their internal processes.

One type of variable star does not truly vary at all. Such are binary stars with orbits that happen to be edge-on to us, so that we see each component eclipsing the other in turn. But in most variables, there are intrinsic changes in the radiation emitted. This is the case with the irregular and eruptive variables. These are generally stars that are very young and still associated with the nebulosity from which they were born. They vary in brightness because from time to time, huge "flares" explode on them, blowing masses of hot gas into space.

Then there are the periodic variables, which brighten and dim in a regular fashion. The surfaces of some are not uniformly hot, but have brighter and dimmer spots. As a result, their apparent brightness varies as they rotate, with periods from about 12 hours to a few hundred days. These are mainly hot stars, on or close to the main sequence. Their variations in surface temperature are probably caused partly by uneven mixing of the gases in their outer regions, and partly by an uneven magnetic field near the surface.

A quite different type of periodic variable is named after the star R Coronae Borealis; these are stars that suddenly dim by a factor of 10,000 times and then recover. They are old red supergiants, rich in carbon that periodically forms clouds of "soot" around the star.

A very important class of periodic variables comprises the Cepheids, named after the prototype Delta Cephei. The outer gaseous envelopes of these stars pulsate outward and inward. This is caused by a complex process in which atoms in a layer below the star's atmosphere first lose some of their electrons and then, as the star contracts, lose even more, so that the layer becomes opaque to radiation from the interior. Pressure builds up, the star expands and cools, ionization is reduced, and radiation escapes. The cycle is then repeated.

A Cepheid's brightness varies by 10-20 percent of its maximum brightness, and this maximum is related to the period of the variations: the brighter the star, the longer its period. Most Cepheids are of Type I, massive young yellow supergiants. The pulsation periods of most of these range from about one day to about 50 days.

Type II Cepheids – sometimes called W Virginis variables – are old stars. They are found in globular clusters and toward the central regions of the Galaxy. The range of their variation periods is a little more restricted than those of Type I Cepheids, and they are about six times fainter.

Once the period (and the type) of a Cepheid is known, its true brightness can be worked out. By comparing this with the apparent brightness, it is possible to calculate the star's distance (pp. 46–47).

There are other kinds of pulsating variables, among them the RR Lyrae stars, whose periods range from hours to about 1½ days. They are all old stars, originally deficient in carbon and oxygen, which later turn into Type II Cepheids. They are of about the same absolute

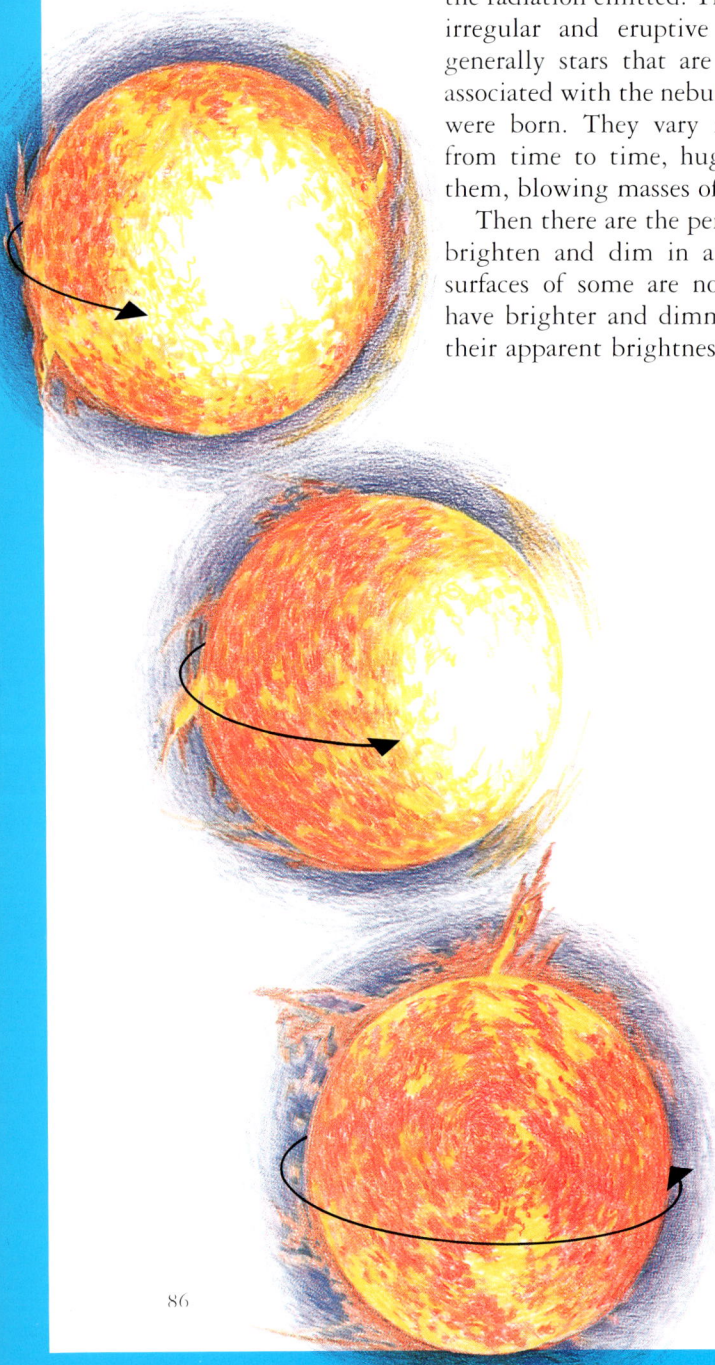

One type of variable star has a surface of uneven temperature, and hence brightness. The changes in apparent brightness are solely due to the fact that the star presents different parts of its surface to us during each rotation.

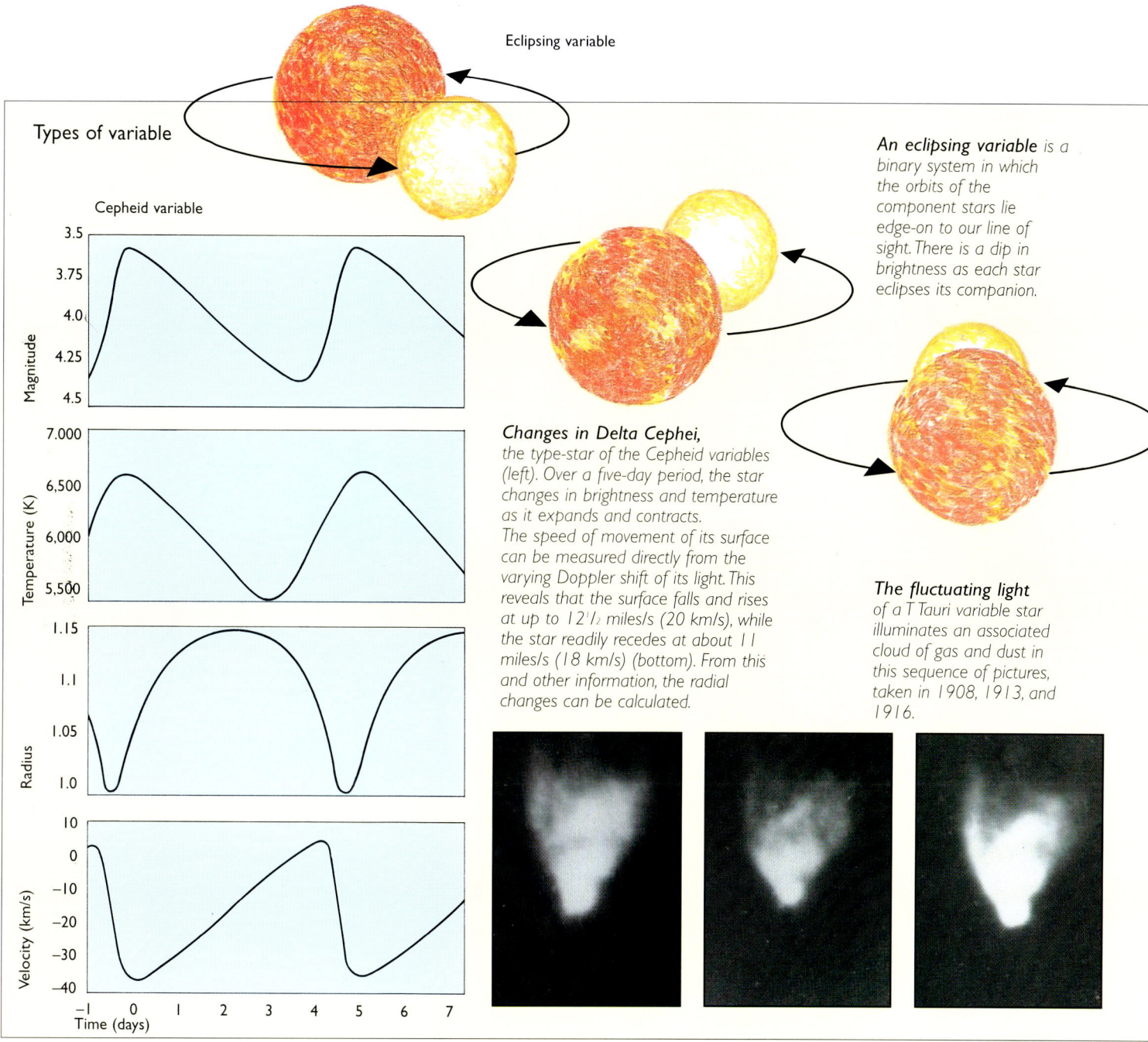

An eclipsing variable is a binary system in which the orbits of the component stars lie edge-on to our line of sight. There is a dip in brightness as each star eclipses its companion.

Changes in Delta Cephei, the type-star of the Cepheid variables (left). Over a five-day period, the star changes in brightness and temperature as it expands and contracts. The speed of movement of its surface can be measured directly from the varying Doppler shift of its light. This reveals that the surface falls and rises at up to 12½ miles/s (20 km/s), while the star readily recedes at about 11 miles/s (18 km/s) (bottom). From this and other information, the radial changes can be calculated.

The fluctuating light of a T Tauri variable star illuminates an associated cloud of gas and dust in this sequence of pictures, taken in 1908, 1913, and 1916.

brightness (about 100 times that of the Sun), which makes it easy to determine their distances from their apparent brightnesses.

Novae are yet another kind of variable star. "Nova" is the Latin word for "new"; in a few hours, a nova brightens by about 10,000 to a million times. It then sinks back to its original luminosity in a couple of months. A star becomes a nova when it suddenly throws off a shell of matter, which may comprise about one-hundred-thousandths of its total mass. (Supernovae are even more extreme outbursts: they are much rarer and of a very different nature (pp. 90–91).)

Many novae are observed to flare up repeatedly, at periods from 10 days to tens of years, and many more probably have longer periods. Such recurrent novae are close binary systems containing a white dwarf, which draws matter from its companion. This falls onto the dwarf and is violently burned in nuclear reactions, which create the outburst. A star that was invisible may become visible to the naked eye, truly appearing to be a new star.

DEATH OF A STAR
● *When the fuel runs low*

main-sequence star

shrinking core

red giant

When an aging star runs low on hydrogen fuel, its core shrinks and it begins to "burn" helium. But hydrogen-burning continues in the outer layers, which swell up to giant size. The surface cools and reddens, but because of the star's huge size its overall brightness increases.

hot central star

planetary nebula

The giant star becomes unstable and throws off its outer layers, forming an expanding planetary nebula. The core of the star, intensely hot and therefore bluish-white, is exposed. Its invisible ultraviolet radiation makes the nebula glow like a fluorescent lamp.

white dwarf

Death is a slow fading away for a star of about the Sun's mass. When all the helium in the core has been used up, the star shrinks to become a white dwarf, about the size of the Earth. It grows fainter as it radiates its energy away.

Some stars end their lives spectacularly, torn apart by colossal explosions. Others, after suffering less violent disturbances, quietly fade from view over millions of years. What are the factors that determine which path a star takes at the end of its life?

A star leaves the main sequence when it has aged to the point at which helium-burning processes start up within it (p. 83). It begins to suffer from "middle-age spread" as it swells into a red giant or, if it is very massive, into a supergiant. Eventually, it becomes a variable, and throws off its outer layers, forming a planetary nebula around itself. This heralds the onset of death.

The simplest case is that of a star like the Sun, which is neither very large nor very small. Such a star is "well behaved" and lives its life at a modest rate. By the time it reaches the planetary nebula stage it has shrunk and become fairly hot, since it has used up its helium. After the nebula has formed, the central star cools and shrinks further.

What happens then was analyzed in the 1920s by the Indian astrophysicist Subrahmanyan Chandrasekhar, when he developed his theory of white dwarf stars. He proposed that when a star shrinks, the gravitational pressure at its center grows so intense that the matter becomes far more closely packed than normal. It is forced into a state called degeneracy.

Degenerate matter was undreamed of before quantum theory. Normal matter is made of atoms, each consisting of a nucleus circled by one or more electrons. How many electrons there are depends on the chemical identity of the atom. The Pauli exclusion principle (p. 25) states that in a given region no two electrons can be in the same state – having the same energy, spin, and so on. This holds the electrons apart by forcing them to occupy different energy levels. The principle is responsible for preventing atoms from collapsing. And it keeps the density of everyday matter below a certain maximum, about 90 times that of water.

Inside a star, the extremely high temperature causes the atoms to be completely ionized – separated into atomic nuclei and electrons. In this state the matter can be crushed together, becoming much denser. It becomes denser still in the heart of a star that is shrinking as it ages. Nevertheless, the Pauli exclusion principle again sees to it that no two electrons can occupy the same state. As the electrons are squeezed closer together, they are forced to move faster and faster, and thus create a pressure that opposes the crushing force of gravitation.

In a star of moderate mass (up to 40 percent greater than the Sun's) the electron pressure becomes great enough to prevent further contraction when the density at the center of the star is about 18 tons per square inch – 10,000 times heavier than the densest material met with on Earth. This is the first stage of degeneracy, and the pressure supporting the star is termed degeneracy pressure.

By this stage, the envelope of the star has been thrown off into space. The core of the star is exposed and is so hot that it shines with a white light. The star is now a white dwarf. However, its internal temperature is not high enough for more complex nuclear reactions to occur, and so it shines merely by radiating its internal energy. It cools as it does so and gradually fades to become a black dwarf.

Observation has shown that white dwarfs do exist. In 1844, it was realized that wobbles in the path of Sirius across the sky reveal the presence of an unseen companion. The companion was observed in 1862. From the strength of its pull on Sirius, its mass was found to be about equal to the Sun's. Yet analysis of its light showed that it could be no more than five times as big as the Earth. Sirius B, as it is called, is a white dwarf. Several hundred white dwarfs have since been discovered, and thousands more "suspects" await confirmation of their nature.

The above account applies, as we have seen, to any star with a mass up to 40 percent greater than the Sun's. (This figure is called the Chandrasekhar limit.) If the star is above this limit, the temperatures reached in the core will be higher, and new, more complex, nuclear reactions will occur. In these reactions, nuclei as heavy as those of iron are synthesized.

But once the core has been converted to iron, no further nuclear reactions are available to release energy. The pressure needed to prevent further collapse cannot be maintained. The force of gravity in such a star overwhelms even

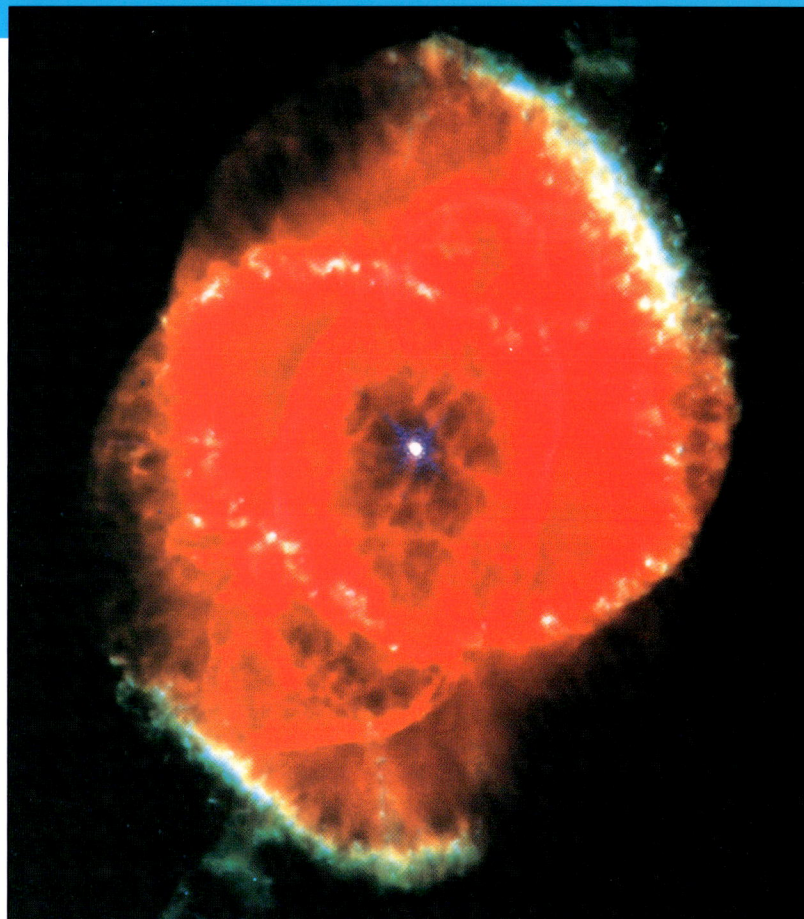

the electron degeneracy pressure and the core collapses catastrophically, to a density much greater than that at the center of a white dwarf. Electrons crash into protons and combine with them to form neutrons. As a result, a neutron gas is generated.

Collapse continues until the neutrons are moving so fast that they exert enough degeneracy pressure to prevent any further collapse. Since neutrons are 2,000 times heavier than electrons, the neutron gas can exert much greater pressures than an electron gas. The core is now in a superdense state.

The collapse of the core is the trigger for a supernova explosion (pp. 90–91). But there is a further stage of collapse that can be reached by the most massive stars – those that are more than five times as massive as the Sun. When these stars collapse, the gravitational force of their overlying material crushes even the dense core of neutrons. Although the outer parts of the star are torn apart by the supernova explosion, a dark remnant of the core is left.

Since matter and energy cannot escape from the star, it becomes a black hole (pp. 62–65). Such objects are what Chandrasekhar called "the most perfect macroscopic bodies in the universe" because, as he put it, "the only elements in their construction are our concepts of space and time."

The Cat's Eye Nebula (NGC 6543) lies in the constellation of Draco. It is a planetary nebula, a late stage in the evolution of a dying star. It is believed to be about a thousand years old. The nebula's unusual shape is explained by the presence of two stars. Shells of ejected gas are moving away from a dying red star in the center. The gravity of a companion star, in orbit around the dying star, is possibly pulling the gas into this extraordinary pattern.

DEATH OF A STAR

Nearly a thousand years ago, in AD 1054, Chinese astronomers noticed that a new star had suddenly appeared in the constellation we now know as Taurus. It was so bright that it could be seen even in daylight. The wreckage of its explosion, an expanding mass of gas and dust, is visible today as the Crab Nebula. Such extremely bright objects are rare in our galaxy – none has been seen since 1604. Hundreds have been observed in other galaxies, however.

Some of these immense catastrophes occur in binary systems in which one member is a white dwarf, the dying remnant of a star (pp. 88–89). Like all white dwarfs, it has consumed almost all its hydrogen, but does contain elements built up from hydrogen in its vigorous youth: carbon, calcium, magnesium, oxygen, silicon, and sulphur. The white dwarf attracts material from the outer layers of its companion star, which "burns" when it falls on to the dwarf's outer regions.

At the same time, the dwarf's core becomes extremely hot, reaching a temperature of at least 100 million K. In consequence, carbon-burning takes over in the core and a runaway reaction occurs; the white dwarf is consumed and its remains are thrown out in a vast cloud of gas containing elements with heavy nuclei, which have been synthesized in the great heat.

This is Type I supernova. On average, one occurs in a particular galaxy every 140 years. Because such supernovae are always of about the same brightness, observing them allows astronomers to determine the distances of very remote galaxies.

A Type 2 supernova is very different. It is the result of the sudden collapse of a lone star. Such an event occurs on average once every 91 years in a particular galaxy and, by a stroke of good fortune for science, one happened comparatively close to Earth in 1987, in the Large Magellanic Cloud.

Before its explosion, this supernova – known as SN 1987A – was a star about 15 times as massive as the Sun. It had passed through its short life of about 10 million years on the main sequence, followed by less than a million as a red giant, before entering the blue super-giant stage. Calculations show that at the end of its life, iron was synthesized in its core. The core was tiny – about half the diameter of the Earth – yet so dense that it accounted for one-tenth of the mass of the entire star.

The supernova explosion occurred when this core shrank catastrophically. Because the core was massive, the collapse did not stop at the density levels characteristic of a white dwarf, but went farther. Two things therefore happened at the same time: the newly-made iron nuclei were broken down again into lighter elements; and electrons and protons combined to form neutrons. The core turned into a ball of neutrons – a neutron star.

The rest of the star's material then fell in on to the core under its own weight. In doing so, it reached about one-tenth of the speed of light, and hit the core with such force that it bounced back outward again. It spread out into space, emitting vast amounts of energy as it did so. This gaseous material was moving at prodigious velocity: within 10 hours it had swollen to the diameter of the Earth's orbit around the Sun, about 186.4 million miles.

The core of SN 1987A remains as a tiny neutron star, its diameter no more than $12\frac{1}{2}$ miles. Yet its mass is comparable with that of the Sun. That means its density is huge – about 5,400 million tons per cubic inch (US). Its gases continue to spread outward into space, just as those that now form the Crab Nebula have been doing for over 900 years.

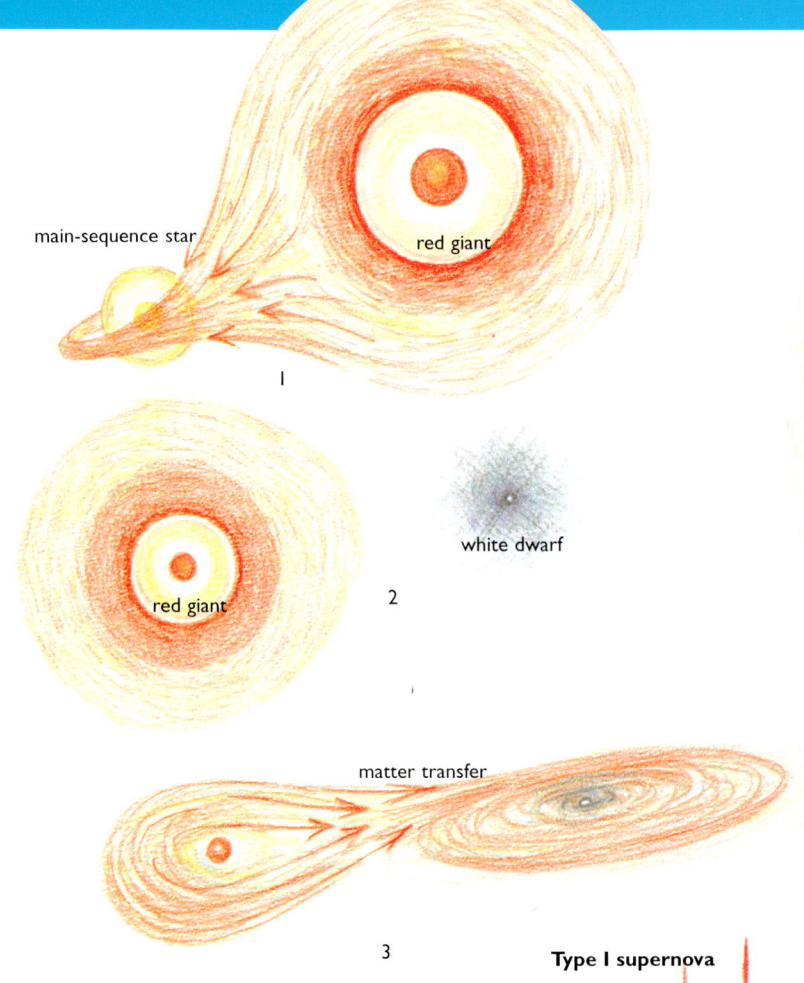

A Type I supernova results from a star belonging to a binary system. When it becomes a red giant, it loses matter to its less massive, slower-evolving companion (1) and becomes a white dwarf. The other star evolves to the giant stage (2), sheds matter on to the dwarf (3), and triggers an enormous explosion (4).

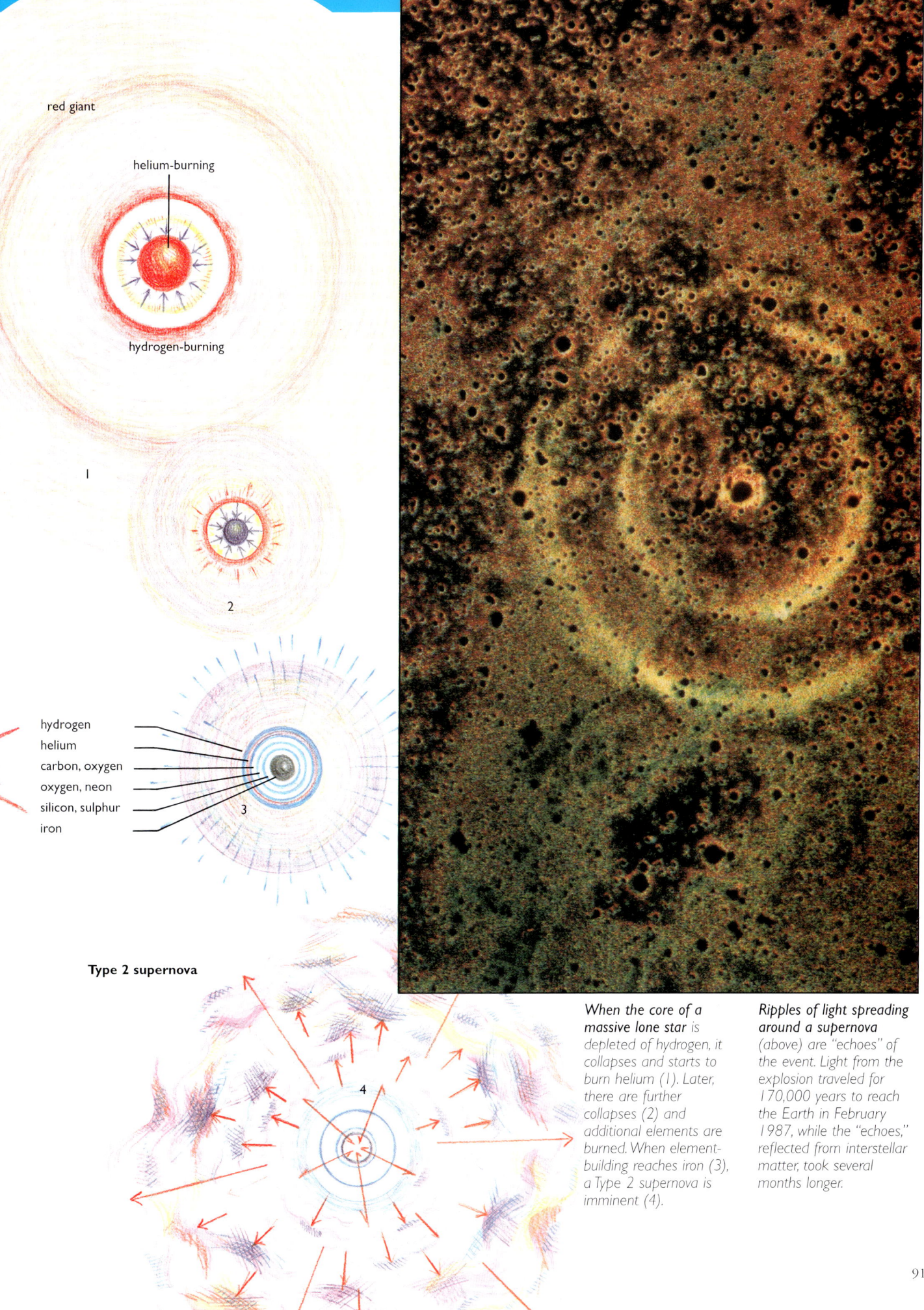

When the core of a massive lone star is depleted of hydrogen, it collapses and starts to burn helium (1). Later, there are further collapses (2) and additional elements are burned. When element-building reaches iron (3), a Type 2 supernova is imminent (4).

Ripples of light spreading around a supernova (above) are "echoes" of the event. Light from the explosion traveled for 170,000 years to reach the Earth in February 1987, while the "echoes," reflected from interstellar matter, took several months longer.

PULSARS
● *Cosmic radio beacons*

In 1967, a new research program began at the Mullard Radio Astronomy Laboratory of Cambridge University. It was humorously referred to as the only research project carried out with a sledgehammer. In fact, the sledgehammer was needed to put up posts carrying 2,048 radio antennae resembling clotheslines. These made up a radio telescope covering no less than 172,220 square feet.

When observations began Jocelyn Bell, a postgraduate student, noticed that the telescope recordings displayed some surprising signals. They consisted of pulses about a twentieth of a second in length that recurred at precise intervals of about $1\frac{1}{3}$ seconds. She discussed this with her supervisor, Antony Hewish, and they decided to investigate in more detail, using a special high-speed recording instrument.

They soon ruled out the idea that the signals could be due to interference from electrical equipment near the observatory: they were definitely celestial because they moved across the sky at the same rate as the stars.

At the time this work was being carried out, there was considerable discussion of the possibility of civilizations elsewhere in the universe. Could the Mullard Observatory's pulses be signals from alien beings? The question was rightly treated seriously, even if the signals were for a while flippantly referred to as LGMs (for "little green men"). In fact, this name was used as a precaution against the media getting wind of the discovery at this early stage of the investigation.

However, the signals varied in intensity in what appeared to be a completely random way, and no intelligible code could be derived from them. What is more, they did not show Doppler shifts in frequency (p. 46) of a kind that would be expected if they originated from a planet in orbit around a distant star.

Nevertheless, they proved to be at the right sort of distance to be from a star of some kind. Longer wavelengths were slightly delayed compared with shorter ones, due to the scattering of the radio waves by electrons in interstellar space, and calculations showed that the source therefore lay at a distance of 400 light-years.

By 1970, no fewer than 50 of these pulsars had been discovered, and today more than a thousand are known. At first, there was no optical identification. But it soon became clear that the pulses must come from bodies that were oscillating, pulsating, or rotating. They must also be very compact to give such sharply-defined pulses.

In 1968, the British cosmologist Thomas Gold suggested that pulsars could be rapidly spinning neutron stars, the superdense remnants of massive stars that had exploded (pp. 88–91). The central star of the Crab Nebula, the remnant of the supernova of 1054, was identified as a pulsar, and it was found to flash at optical as well as radio wavelengths. The remnant of the 1987 supernova in the Large Magellanic Cloud is also a pulsar. Thus, the mystery of Hewish and Bell's LGMs has received a purely astronomical explanation.

The nature of pulsars

Current ideas about the structure of a pulsar are deduced purely theoretically from fundamental physics. They show that neutron stars probably have a solid core of neutrons. Around this is a "superfluid" layer, consisting of neutrons, protons, and electrons. (A superfluid is matter in a state in which it flows with no resistance. Since electrically-charged protons and electrons are present as well, the superfluid in the neutron star is also a superconductor, which means that it possesses no electrical resistance.)

Outside this layer lies a very thin crust, probably only about 2,000 feet thick, composed of neutrons also in a superfluid state. Finally there is a solid outer crust, only half as thick as the inner crust, containing atomic nuclei and electrons as well as neutrons.

The star from which the neutron star formed would certainly have possessed a magnetic field. During its final collapse, this field would have been "frozen into" the neutron core, and would have become billions of times more intense. Charged particles – protons and elecrons – are constantly given off from the surface of the neutron star. The intense magnetic field whirls these around, generating radio waves that are channeled into beams emerging from the magnetic poles.

The neutron star is spinning very fast – at a rate ranging from once every few seconds to about 1,000 times per second – and the two beams from the magnetic poles swing around. If one happens to point in the right direction, it is detected on Earth as a series of pulses.

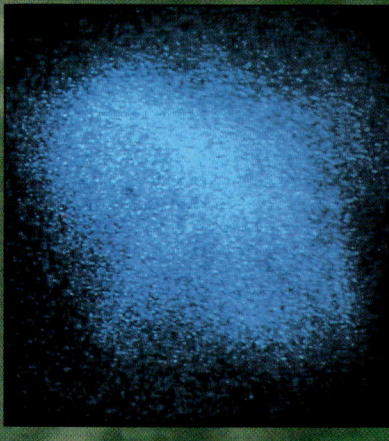

THE GRAND DESIGN

The Crab Nebula is the wreckage of a supernova, which has been spreading outward since the explosion was observed over 900 years ago. At its heart is the collapsed core of the star that exploded, now a pulsar giving out flashes of radio waves, visible light, and X-rays. As it ages, it will slow down and radiate only at the lower-energy radio wavelengths.

The Crab Nebula pulsar flashes on (above) and off (left) 30 times per second.

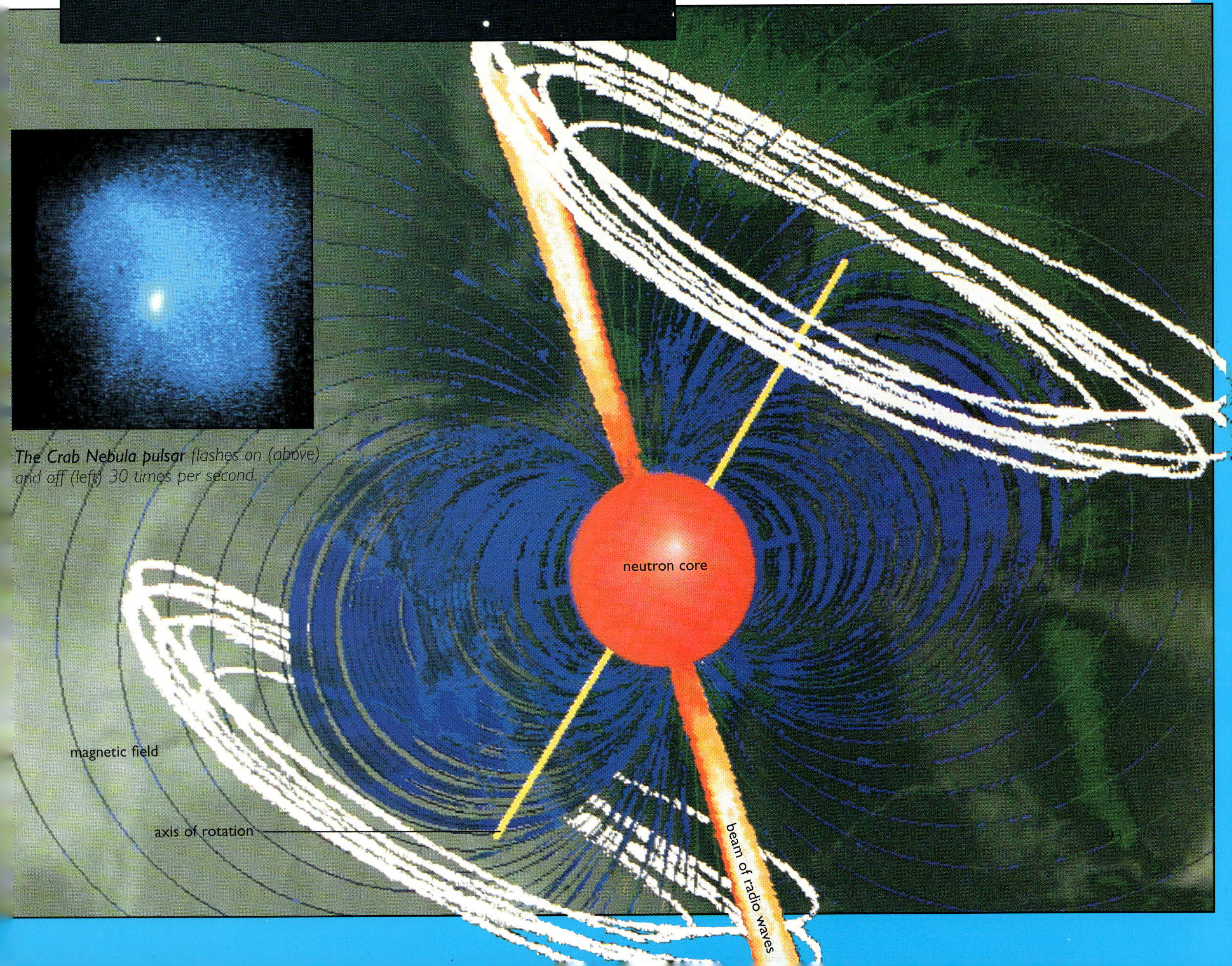

neutron core

magnetic field

axis of rotation

beam of radio waves

THE SUN
● *Nearest of the stars*

The Sun lies in the Orion arm of the Galaxy, which is marked by dark and bright nebulae from which new stars are constantly born.

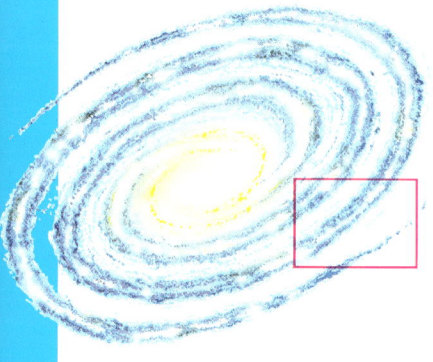

A representative sample of stars is provided by the Sun and its neighbors in our small area of the Galaxy. Although these stars are near us on the galactic scale, they are a very long way off on the scale of the solar system. Whereas the Sun's average distance from Earth is 8.3 light-minutes, the distance of the nearest star, Proxima Centauri, is 4.28 light-years – over 250,000 times greater.

Proxima Centauri is part of a triple star system; its companions are the two components, A and B, of Alpha Centauri, the brightest star of Centaurus, in the southern sky. Alpha Centauri A is a yellowish star like the Sun; the other component, B, is whiter, while Proxima is a dim red dwarf.

Notable among the stars in our vicinity is Sirius, lying 8.8 light-years away. It is the brightest star within 20 light-years and also has the greatest apparent brightness as viewed from the Earth. Sirius is a binary, its companion being a white dwarf.

Another nearby star, 61 Cygni, is of historical significance because it was the first star – other than the Sun – whose distance was measured. This was achieved in 1839, by the German astronomer Friedrich Bessel, using the parallax method (p. 46).

Another multiple system is Procyon, brightest star of Canis Minor. Procyon A is a bright yellowish-white star, orbiting a white dwarf, B. Even though Procyon is at a distance of 11.4 light-years, it would be a disaster for us if B, as a result of interactions with A, were to become a supernova; however, the distance between the two stars may preclude this. It would be an even more serious state of affairs if Sirius B were to become a supernova, but fortunately

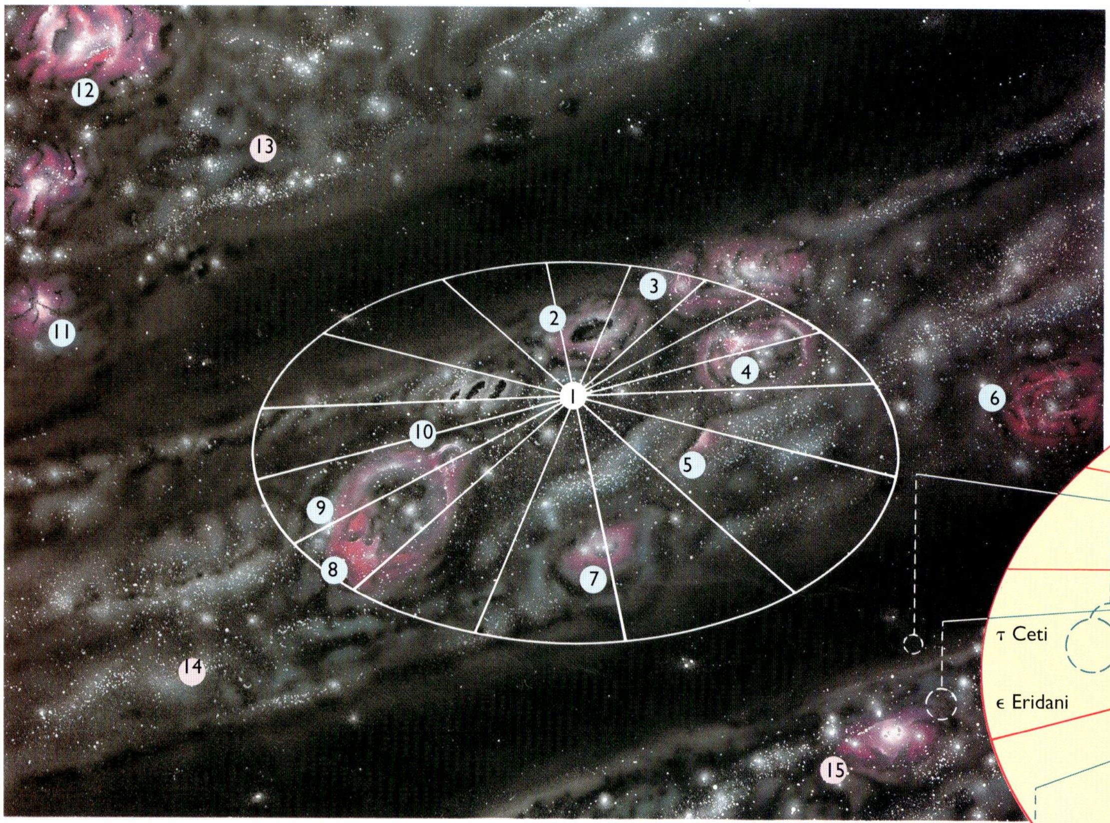

1 Sun
2 Coalsack
3 Vela Nebula
4 Orion Nebula
5 California Nebula
6 Rosette Nebula
7 IC 1805
8 North America Nebula
9 Pelican Nebula
10 Cygnus Loop
11 Trifid Nebula
12 NGC 6164-5
13 Sagittarius arm
14 Orion arm
15 Perseus arm

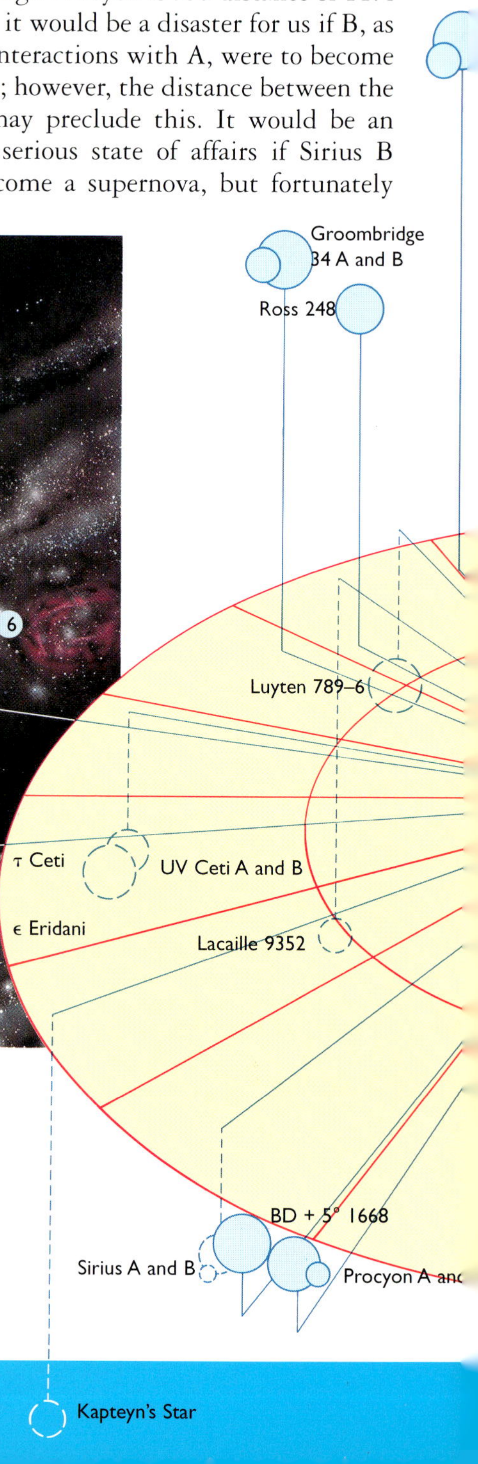

THE GRAND DESIGN

More than 20 star systems lie within 13 light-years of the Sun. Only four of these are visible to the naked eye. The nearest system, Alpha Centauri, is triple. Seven others are binary stars. The circles here are at 5 and 10 light-years from the Sun.

Kruger 60 A and B
Σ 2398 A and B
61 Cygni A and B
Barnard's Star
Lalande 21185
Ross 154
Lacaille 8760
Proxima Centauri
Wolf 359
α Centauri A and B
ε Indi
Ross 128

for us the same argument applies.

The nearest star after Alpha Centauri is Barnard's star, a red dwarf 5.8 light-years from us in the constellation Ophiuchus. Edward Barnard drew attention to it in 1919 because it appears to have a faster individual motion across the sky than any other star. Nevertheless, this amounts to traveling only half a degree – the Moon's angular diameter – every 180 years. Part, but by no means all, of this speed is due to the star's proximity to us. Barnard's star is also interesting because it may have planets in orbit around it (pp. 104–5).

All in all, these are the kinds of star that might be expected in a random sample taken from a spiral arm of a galaxy such as ours. Not only is the Sun an ordinary star, but so are its nearest companions.

Yet if the Sun is undistinguished among its celestial neighbors, it is immense by terrestrial standards. Its diameter is 865,000 miles, more than 109 times the equatorial diameter of the Earth; accordingly, its volume is 1,303,600 times that of our planet.

Although the Sun displays a disk with a firm edge when seen through mist, fog, or thin cloud, it is a gaseous body and has no definite boundary, as highly-magnified photographs show. It looks so bright merely because it is so near to us, appearing 400,000 times brighter than the full Moon; yet it has only one twenty-fifth the brightness of Sirius. Nevertheless, it is vital never to look at the Sun directly, either with the naked eye, through dark glasses, or through binoculars or a telescope. Doing so may cause blindness.

Generally, the Sun is a well-behaved star and radiates energy steadily. It probably took more than 3.7 billion years from the onset of nuclear reactions in its core before it settled down to its present form, and a further 800 million years before it reached its present brightness and temperature. Today, it consists of about 60 percent hydrogen by weight, and no notable change in its brightness is to be expected for at least another 1.5 billion years. It will be several billions of years after that before it becomes a red giant. By the time it fades to become a white dwarf it will have been in existence for about 10 billion years.

The temperature of the Sun's core, where nuclear reactions take place, is about 15 million K. Matter here exists in the form of bare atomic nuclei and the electrons that have been stripped from them. Energy escapes from the core by a complicated route. When first produced, the radiation is mainly in the form of high-energy gamma-ray and X-ray photons. These are soon absorbed by the dense gas of the interior, only to be re-emitted. This happens again and again, the distance between one absorption and the next near the Sun's center being no more than about 0.39 inch (1 cm). It probably takes anything from 8,000 to 80,000 years for energy to make its way through the "radiative" zone by this tortuous path.

At about 373,000 miles from the center (85 percent of the Sun's radius) the gas is sufficiently cool for the atoms to retain their electrons, and they are opaque to radiation. The radiative zone ends and heat now escapes in a different manner: masses of gas rise to the surface, radiate away both light and heat, then cool and sink back again. There the gas is warmed again, and the process – termed convection – is repeated.

Large cells of gas circulate at lower levels,

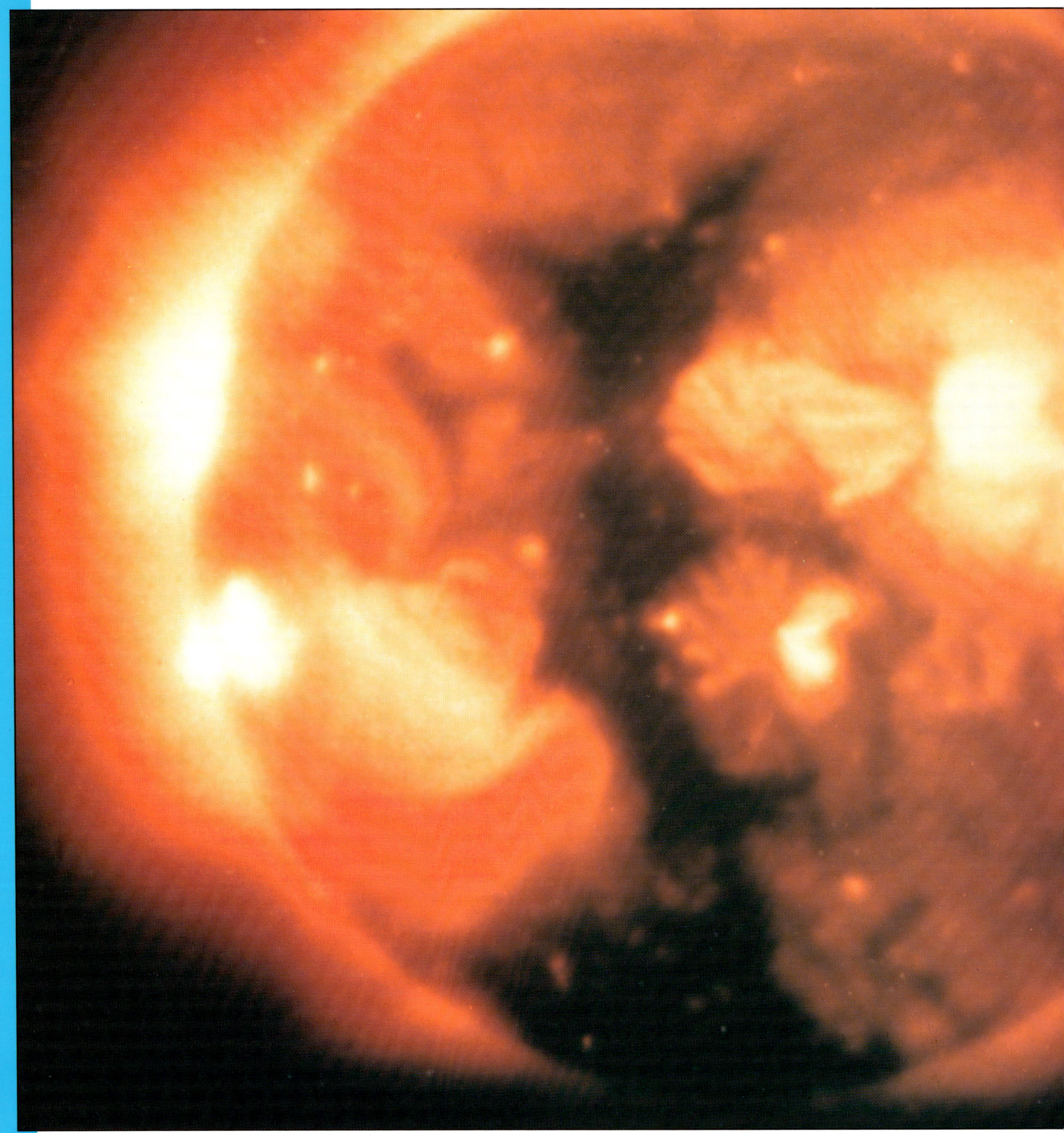

An X-ray image of flares on the Sun, taken from the Skylab space station, conveys the turmoil of its surface.

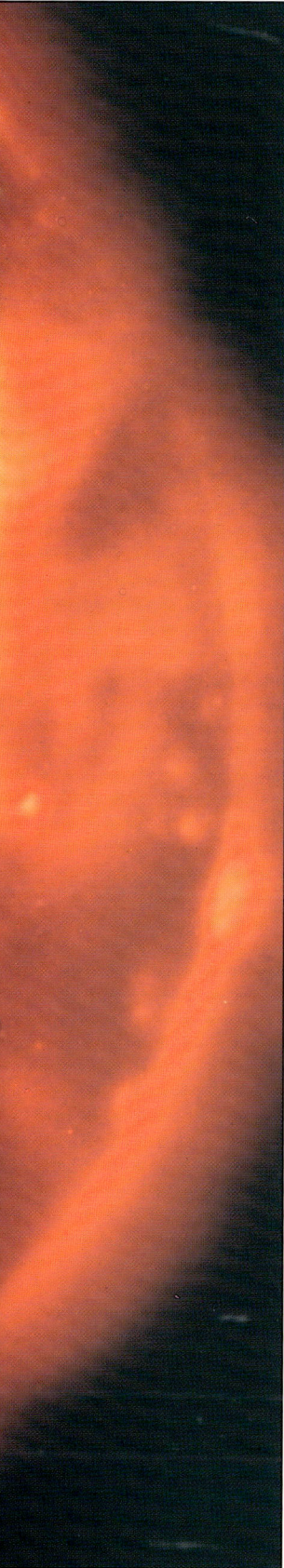

while smaller cells do so at higher levels. Viewed under very clear conditions through a telescope, the topmost cells can be seen at the surface as incessantly moving bright granules. This visible surface is called the photosphere ("sphere of light") and has a temperature of about 6,000 K.

The composition of the outer layers of the Sun can be directly determined by analyzing sunlight spectroscopically. It consists of 73.5 percent hydrogen and 25 percent helium, some of which was formed in the first few minutes after the Big Bang, and the rest of which is the product of the nuclear reactions in the Sun. However, there are also traces of elements with heavier atoms, built up in earlier generations of stars that exploded, scattering the elements through the interstellar matter from which the Sun was formed. These include oxygen, carbon, iron, magnesium, and so on. However, none of these elements accounts for even 1 percent of the Sun's weight.

It has been discovered that the outer skin of the Sun, 6,200 miles thick, oscillates, its depth increasing and decreasing by about 15½ miles. In fact, the Sun is vibrating like a ringing bell, with an oscillation period of only five minutes.

The cause of this oscillation may be a slight periodic fluctuation in solar transparency, causing a corresponding change in the pressure exerted by the radiation from the interior. Alternatively, the oscillation may be due to pressure waves triggered by turbulence in the convective region beneath. This process is similar to the one that causes the far greater pulsations of the outer layers of Cepheid variables (pp. 86–87).

There is also another oscillation, with a period of 2 hours 40 minutes. This could be due to variations in the rate of the proton-proton reactions in the core (pp. 82–83). But to account for it in this way would mean that the central temperature of the Sun must be 10 percent less than has been calculated. Yet such a lower temperature would result in a Sun that is not as bright as is actually observed. It appears that our present understanding of what is going on in the Sun's core needs reconsidering.

Another clue to this involves the output of neutrinos from the nuclear reactions. Only one-third of the theoretically expected neutrinos are observed. This would be accounted for if the temperature were 10 percent below the figure that is generally accepted.

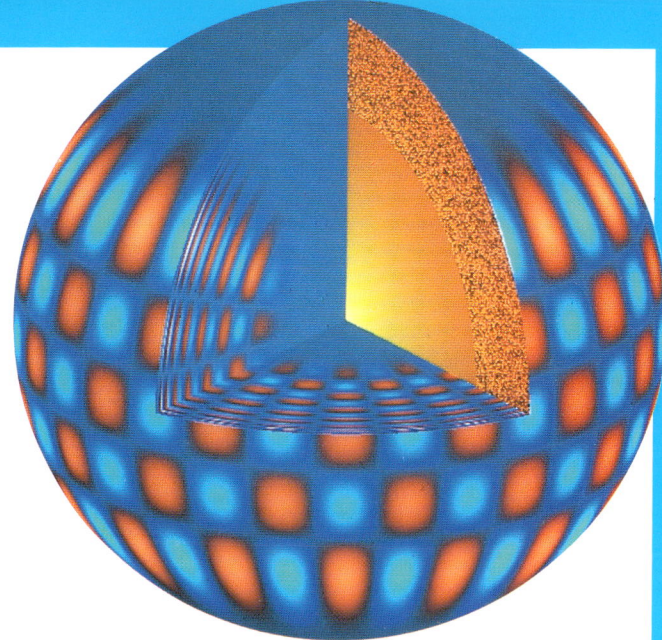

Power is generated within the innermost 30 percent of the Sun's radius. Energy is carried most of the way to the surface by radiation, but in the outer layers by convection. One oscillation pattern is shown: red areas are receding, blue are approaching.

Lying above the photosphere, and visible only when the brilliant light of the latter is cut off in an eclipse or by special instruments, is a layer of reddish-pink gas, the chromosphere ("color sphere"), most of which is only about 3,100 miles thick. Its temperature rises from about 4,000 K in its lower regions to some 50,000 K at its top. This upper surface is composed of spikes, or spicules, vertical columns of gas 620 miles broad and 6,200 miles high. The gas in the spicules shoots upward at velocities of about 9½–18½ miles per second and then falls back into the chromosphere.

The chromosphere also contains fibrils, horizontal strands of gas, appearing dark by contrast with their surroundings, which last for only 10–20 minutes. They are about 6,200 miles long and 620–1,250 miles thick. They are associated with active regions in the photosphere beneath.

The most obvious phenomena of the photosphere are sunspots, first observed and recorded by Chinese astronomers over 2,000 years ago. They were only studied seriously in the West after 1610, when they were first observed with the telescope.

All sunspots display a relatively dark central region, the umbra, and a lighter surrounding area, the penumbra, which is nevertheless darker than the surrounding photosphere. Sunspots are not really dark; they appear so only by contrast with their surroundings. The temperature of a spot's penumbra is about

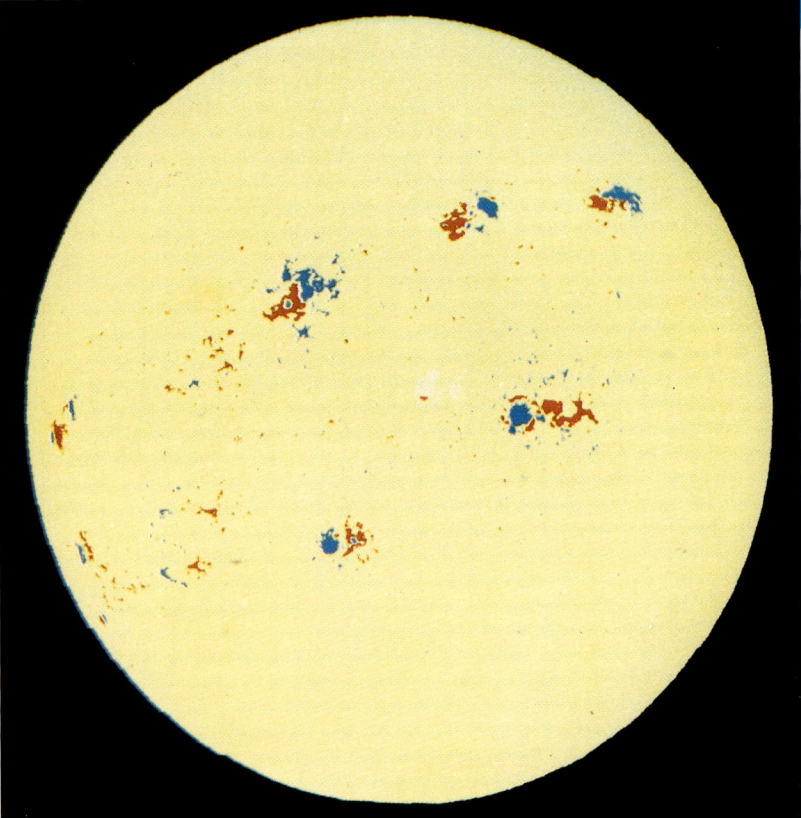

The magnetic fields of sunspot groups are displayed in this computer-colored image. The north and south magnetic poles are shown in red and blue, respectively. In one hemisphere, north magnetic polarity leads in each sunspot pair, while in the other, the south pole leads. In the next sunspot cycle, the leading spots will have the opposite polarity.

5,600 K, and of the umbra about 4,000 K, compared with the 6,000 K of the photosphere. It has been calculated that if the umbra of a sunspot could be seen in isolation, but still at the Sun's distance, it would be about 50 times brighter than the full Moon.

The movement and appearance of sunspots show that the Sun rotates at a speed that varies according to latitude. Studies of Dopler shifts show that near the solar equator it takes about 26 days to complete one revolution; at latitudes of 30 degrees the time taken is over 28 days. Near the poles, the rotation period is about 37 days.

Sunspots vary enormously in size: some are no more than 620 miles across, while others greatly exceed the Earth in diameter. They are usually to be found in pairs, at latitudes rarely exceeding 45 degrees north or south.

Sunspots display a cycle lasting 11 years. At the beginning of a cycle, the Sun is almost free of spots. A few appear at high northern and southern latitudes and then, as the cycle progresses, they appear in greater numbers at lower latitudes. They reach their maximum at latitudes of 15 degrees at the peak of a cycle. Their numbers then dwindle, although new ones continue to form closer to the equator.

This is the general rule, but the maximum number of spots varies from one cycle to another. What is more, there is sometimes a dearth of spots. During the period from 1645 to 1715 – the "Maunder minimum" – virtually none were visible.

When sunspots are observed as they approach the Sun's edge, or limb, the umbra appears to be lower than the photosphere, but this is an optical illusion. Gas can often be observed being emitted from the umbra and falling back again into the companion spot. The route of the gas always follows the lines of force of the magnetic field associated with every spot.

Although details of the sunspot phenomenon are still a matter for debate, an explanation proposed by Horace Babcock and Robert Leighton is now generally accepted. Their model involves the Sun's magnetic field, which at the beginning of a solar cycle is believed to lie in the lower part of the convective layer. Magnetic field lines stretch between the Sun's magnetic north and south poles and, being electrically highly conductive, the gas traps the field. Because different parts of the Sun rotate at different rates depending on their latitudes – faster near the solar equator and slower near the poles – the field lines become wrapped around

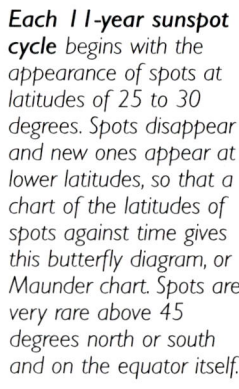

Each 11-year sunspot cycle begins with the appearance of spots at latitudes of 25 to 30 degrees. Spots disappear and new ones appear at lower latitudes, so that a chart of the latitudes of spots against time gives this butterfly diagram, or Maunder chart. Spots are very rare above 45 degrees north or south and on the equator itself.

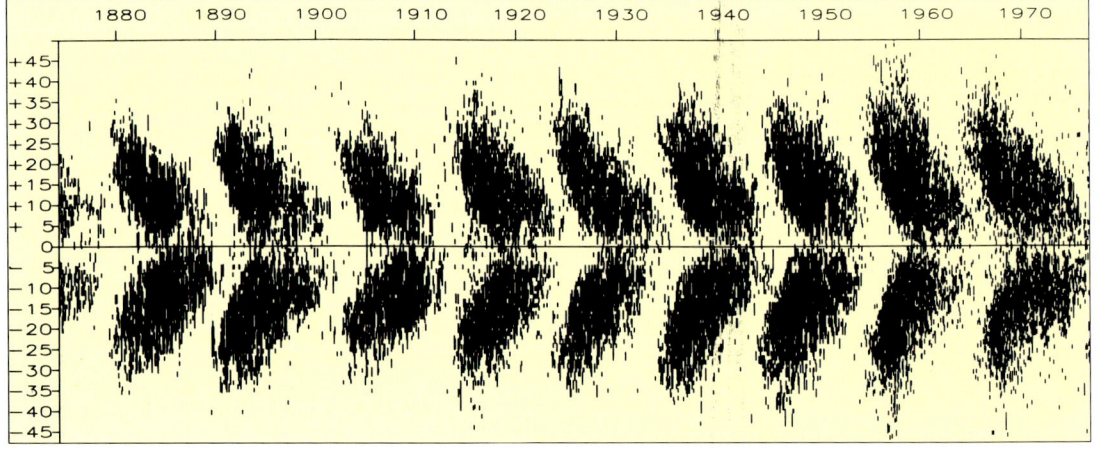

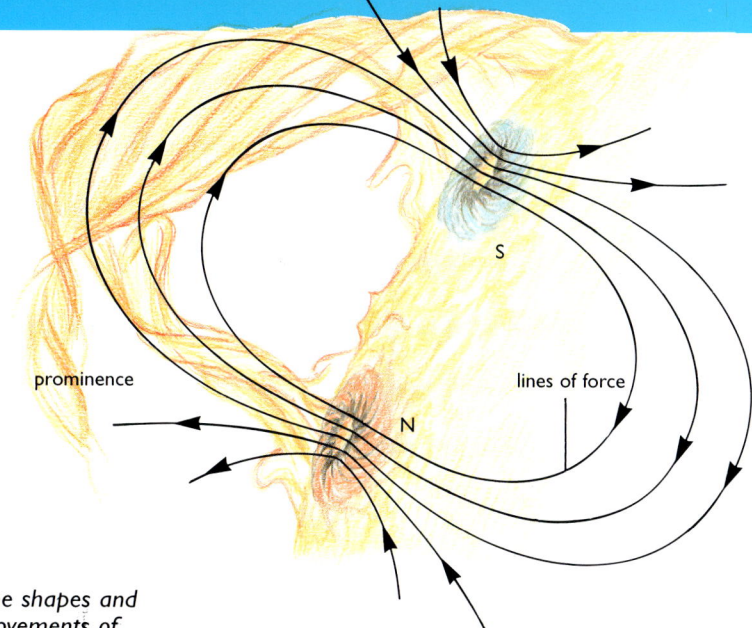

the Sun. Wherever the lines come close together, the strength of the field is increased, and here "tubes" of gas form, threaded by intense magnetic fields.

Because of turbulence, the magnetic tubes also become wound up like the strands in a rope. They also tend to rise, and a time comes when the upward force on these magnetic ropes becomes so great that they float up to the photosphere, where kinks in them break through the surface. The magnetic field lowers the temperature of the electrified gas; accordingly, a pair of spots will form, one where the field lines emerge, the other where they re-enter the photosphere.

The spots are effectively magnetic poles – one a north pole where the lines emerge, the other a south pole. This has long been known from study of the effect of the magnetic field on the sunlight from the spots. In one hemisphere, the leading spot in each pair is a north pole in one sunspot cycle, and a south pole in the next; the leading spots in the other hemisphere are always of the opposite polarity.

The disruption in the photosphere caused by sunspots is accompanied by other effects. Brighter patches in the upper part of the photosphere, called faculae, can be seen at places where sunspots are about to appear, and for a short time after they have arrived. The spots also produce areas of increased brightness called plages (the French word for beaches) and filaments of dark light-absorbing gas between the sunspot pairs. These phenomena generally disappear two to four weeks after the sunspots have gone, although sometimes they will last for several months.

The Sun possesses an outer atmosphere known as the corona (Latin for crown), which may have been discovered as early as the first century AD. Because it is faint, it is ordinarily invisible to the unaided eye, but becomes visible during total solar eclipses. In fact, the first knowledge of the less obvious properties of the Sun was afforded by observations made during such eclipses.

A solar eclipse occurs whenever the Moon comes between the Sun and the Earth. The apparent size of the Moon is, by coincidence, almost exactly the same as that of the Sun. If the Moon passes centrally in front of the Sun, it normally just covers the latter's disk, causing a total eclipse. If the Moon happens to be at its greatest distance, however, the rim of the Sun

The shapes and movements of prominences are dominated by the magnetic fields of sunspots. Magnetic field lines loop between the two members of a spot pair, emerging from a spot that has north polarity and re-entering a spot with south polarity. Despite appearances, the gas in a prominence is usually moving downward. It often forms an arch following the field lines, and may show braiding and twisting effects.

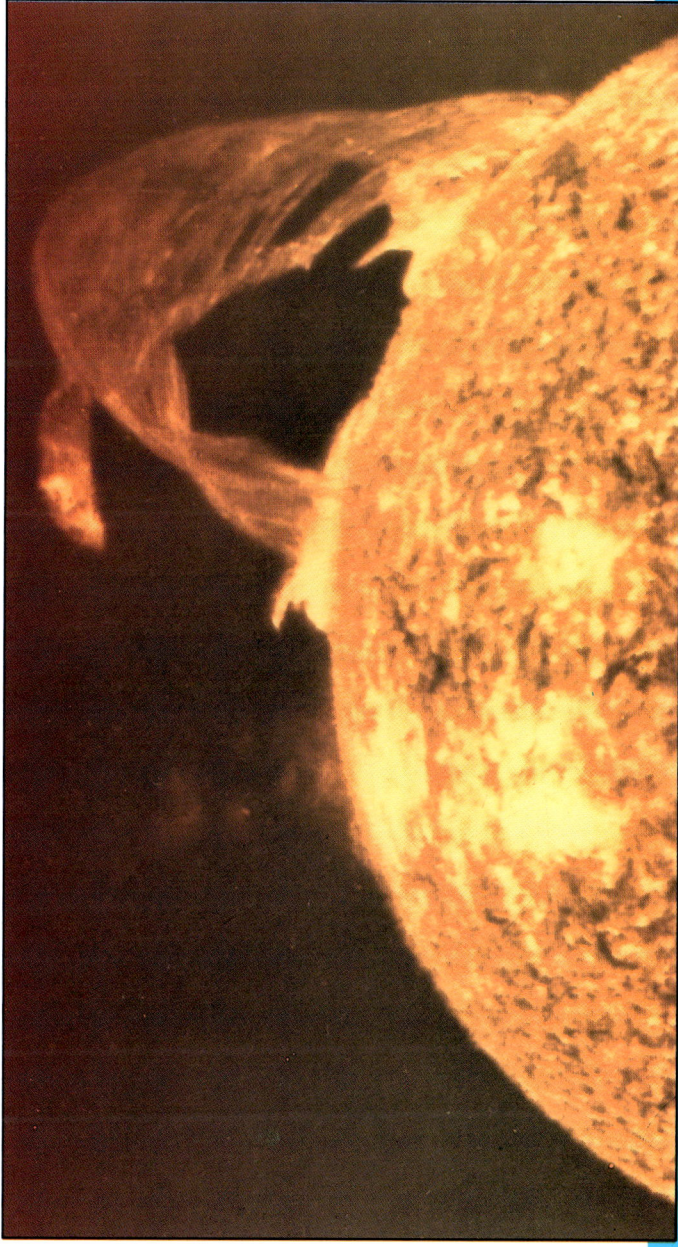

A solar prominence is seen in ultraviolet light by instruments on board Skylab on December 19, 1973. The arc of ionized helium extends 365,400 miles – 45 times the diameter of the Earth – across the surface of the Sun. Before expanding to this size, it had the appearance of a quiescent prominence, apparently destined to last for months while showing little activity.

During a total eclipse the pearly white corona, or atmosphere, of the Sun becomes visible. The corona is extremely tenuous but intensely hot, its outer regions reaching temperatures of millions of degrees.

Blotting out the Sun

The Moon as viewed from the Earth is just big enough to cover the Sun's disk when the three bodies line up. The eclipse is total only in the central part of the shadow, which is at most 167 miles across on the Earth's surface. Outside this region, the eclipse is partial, with part of the Sun visible around the edge of the Moon.

Solar eclipses occur at least twice a year, and at most five times. They would be far more common if the Moon's orbit were not tilted in relation to the Earth's orbit around the Sun; the Moon would then pass between the Earth and the Sun every month.

remains visible around the Moon, giving an annular, or ring, eclipse.

Total eclipses occur about six times in a decade. Each eclipse is total only in a narrow band, never more than 167 miles wide and a few thousand miles long. To each side of this band, the eclipse is partial.

A total solar eclipse is very impressive: the sky darkens, stars appear, animals lie down as if it were night, and buildings look strangely flat, like a stage set. The temperature drops noticeably. Because the Moon moves quickly, the shadow it casts never stays long in a particular place: the maximum duration of a total eclipse is 7 minutes and 31 seconds. In these brief moments, some of the most spectacular phenomena on the Sun can be seen.

During an eclipse, huge, flamelike clouds of hot gas called prominences may be seen. Prominences look as if they are emerging from the chromosphere. They have temperatures of around 10,000 K and radiate in the ultraviolet and X-ray regions of the spectrum as well as at visible wavelengths. Some "quiescent" prominences are vertical masses of hot gas, over 99,400 miles long and 3,000 to 5,000 miles thick, hovering more than 186,000 miles above the photosphere. They last a matter of months, or in some cases as much as a year.

Occurring in the magnetically neutral areas between the two members of pairs of sunspots, quiescent prominences look as if they consist of gas moving upward from the chromosphere, especially since they display the same reddish-pink color. This is true of some, although most are actually material from the corona, which becomes more concentrated as it descends.

Active prominences are smaller than quiescent ones, being about 40 percent as long. Their activity is measured in minutes rather than months. Some take the form of loops or arches; these are composed of material descending from the corona and following the local magnetic field associated with sunspots.

Other active prominences clearly consist of material thrown up from the chromosphere and sometimes ejected from it, at speeds of 62 to 125 miles per second. Such prominences are related to flares, sudden releases of energy that occur in active regions, again usually in the neutral regions between associated sunspots. The frequency of flares is closely connected with the numbers of sunspots present on the Sun at the time. The flare energy is released in the form of radiation, masses of ejected gas, and fast-moving electrons and protons.

The corona is variable in form and extent. Above active sunspot regions coronal streamers may travel 87 million miles into space – almost reaching the Earth. The part of the corona seen at visible wavelengths is large and spread around the Sun fairly evenly at sunspot maximum, whereas at minimum it is generally smaller, but stretches in long streamers from the Sun's equatorial regions. At the poles it forms spreading, curved lines with a "brushed" appearance, revealing the form of the Sun's magnetic field.

The corona is extremely tenuous – at its densest it is 10,000 times less dense than the photosphere. Nevertheless it is very hot, its temperature ranging in different regions between 1 million and 5 million K. Yet because of its thinness, it contains very little energy – it would not, for example, melt a spacecraft within it.

In pictures taken at ultraviolet and X-ray wavelengths, the corona displays conspicuous dark regions, which have become known as coronal holes. These occur where the magnetic field of the Sun is weak and the field lines rising from one pole of the Sun do not return to the other but break away and trail into space. Here hot matter escapes into the solar wind. Such regions are cooler than the surrounding corona, which is why they seem dark at these energetic wavelengths.

As early as 1900, Sir Oliver Lodge suggested that electrified gas might be ejected from the Sun, but it was not until 1958 that it was realized that the matter in the corona is continually expanding, becoming the solar wind as it moves outward. The electrified material consists of protons, electrons, and ionized heavier atoms, ejected at average speeds of 280 miles per second.

The solar wind carries with it a magnetic field which affects the magnetic fields of the planets, including the Earth; when charged particles enter the atmosphere, they produce the beautiful aurorae, or polar lights. What is true of the Sun is true also of other stars: spots have been detected on some of the nearer ones, and others show evidence of stellar winds, consisting of electrified particles.

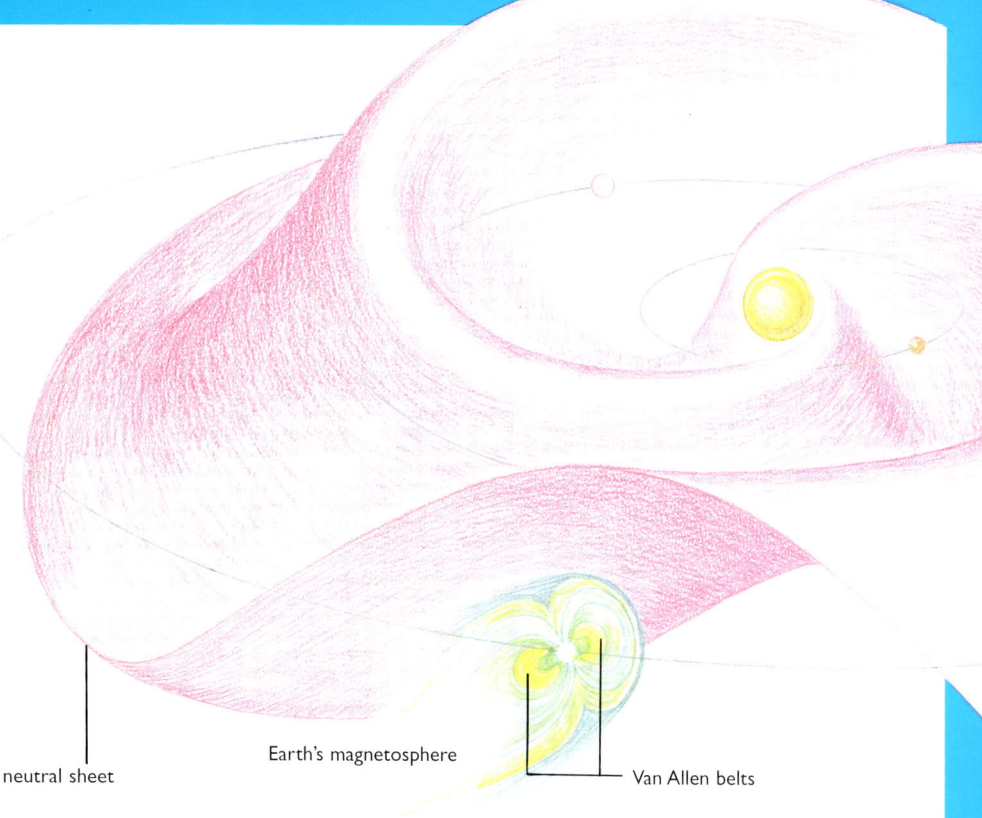

The solar wind distorts the magnetic fields of the Sun and the Earth. Fluctuations in the wind – a stream of electrically-charged particles from the Sun – make the Sun's field wave up and down, as revealed by the buckling of the central, neutral sheet. (Above this sheet, the field points away from the Sun, below it in the opposite direction.) The solar wind blows the Earth's magnetosphere into a long tail, and some of its particles are trapped in the Van Allen belts.

Lights in the sky

Aurorae are luminous displays that appear in the night sky at high latitudes, taking the form of colored arcs, streamers or draperies. They are caused by charged particles from the Sun, such as electrons and protons, interacting with atoms and molecules in the Earth's atmosphere, usually more than 62 miles up. The particles arrive in the solar wind and succeed in penetrating the Earth's magnetosphere, the region occupied by our planet's magnetic field. They are trapped for a time in the Van Allen belts, but they gradually escape and follow the Earth's field down toward the magnetic poles.

The influence of the solar wind on the Earth shows up clearly in this colorful aurora. The best chance of observing aurorae comes when the Sun is active, because then the solar wind is the most intense.

BIRTH OF A SOLAR SYSTEM
● *From planetoids to planets*

When the Sun formed from gaseous material about 4.56 billion years ago, something like 99 percent of the gas cloud formed the protosun before condensing down into the Sun itself. The remaining material formed a disk – often referred to as the solar nebula – around the condensing star. This solar nebula was not composed only of the lighter gases such as hydrogen and helium; there were other, heavier elements present as well.

It seems that the Sun formed in company with a number of other stars, some of which were probably very large. The comparatively slow contraction of the Sun continued while these large stars quickly passed through their life cycles to reach the supernova stage.

Moreover, some of an earlier generation of stars within our galaxy had reached that stage too. When stars become supernovae, they manufacture and then spew out heavier elements. Thus, there were two sources to provide heavy elements in the solar environment. Some of these elements were swept up into the Sun, but 1 percent, at least, remained in the solar nebula.

To begin with, the solar nebula was hot and opaque. Gravitation made it contract, causing its temperature to rise, and its opaqueness helped retain infrared radiation in its central regions. Yet although the central regions, at least, became very hot, the heat did radiate away slowly, and the nebula began to cool.

As the solar nebula cooled, various chemical substances condensed out. Once the temperature was less than 2,000 K, compounds of aluminum, calcium, magnesium, and titanium formed; below 1,000 K, compounds of silicon and oxides of metals appeared. At 180 K ice formed from water vapor, and in the outer parts of the nebula, where the temperature would have dropped to 20 K, methane solidified. Such chemical condensation resulted in the formation of tiny grains.

Mathematical studies of what happens in thin disks of condensing material show that the material soon ceases to be evenly distributed throughout and pockets of denser material form. In the solar nebula, the grains of the various compounds began to collect together to form lumps of material that measured a few miles across. By cosmic standards, the whole cooling, condensing, and coalescing process described so far did not take long – only about 1,000 years.

Thus, at the end of that period, the solar nebula had condensed and accumulated into a rotating disk composed mainly of small lumps or planetoids. Two processes then took place.

In the first, collisions occurred between planetoids. In some cases, when the collisions were fast enough to be disastrous, the planetoids broke up. In others, where the collisions amounted to more gentle encounters, fragments coalesced and fused, due to mutual gravitational attraction, to form larger units, perhaps as much as 620 miles across. This left the solar nebula containing a considerable number of protoplanets. The second process was the coalescence and fusion, again due to gravitational attraction between them, of protoplanets into true planets.

Both these processes swept up a considerable amount of the planetoid material, and the resulting orbits of the planets so formed were almost circular. Thus, the newly-formed planets became isolated from each other, each pursuing its own orbit. These two processes took in the order of 100 million years. Planetary satellites formed at the same time, as part of the planetary formation process. Material not used in the planets fused to form the satellites in the equatorial planes of the planets.

Our Moon, however, has a different origin. It is believed that a Mars-sized body collided with young planet Earth. Material from the two bodies formed a ring around Earth and then coalesced into one body – the Moon.

Planetoid material that did not coalesce into larger units formed the asteroids or minor planets (pp. 142–43). Moreover, calculations show that some of the smaller primary planetoid material did not coalesce at all. Instead it formed icy conglomerates, sometimes called planetesimals, which became comets (pp. 144–45) and meteors (pp. 147–48). These objects are particularly interesting from the point of view of the history of the solar system because they provide astronomers with evidence as to what chemical elements made up the material in the very early solar nebula.

A planetary system is born from the gas and dust remaining after the formation of a protostar (top). The first stage is the building up of countless rocky lumps called planetoids by the accretion of grains of dust (center). In further collisions, the planetoids break down and re-form, finally forming a small number of protoplanets (bottom). Nuclear reactions begin in the star, while its radiation blows away remnants of the cloud. Light gases, such as hydrogen, helium, methane, and ammonia, cannot be held by the small protoplanets in the hotter inner regions of the system, but are retained as deep atmospheres around the larger, colder bodies farther out.

THE GRAND DESIGN

103

OTHER SOLAR SYSTEMS
● *Do they exist?*

A system of planets in the making may be revealed in this picture taken at infrared wavelengths by the IRAS satellite. A disk of gas and dust is seen almost edge-on, reaching thousands of billions of miles from the star Beta Pictoris, the light of which is blocked out by a dark mask at the center.

The prospect of other stars possessing planetary systems in orbit around them has long intrigued astronomers. The debate about whether they have or not has, at any given time, been greatly influenced by the prevailing opinion about how our own solar system was formed.

When it was thought that the planets were the result of a rare set of circumstances, it was considered extremely doubtful that many stars possessed a set of orbiting planets. In the 1940s and even earlier, the generally accepted scenario was that a chance close encounter by the Sun with another star was the cause.

In such a case, according to theory, mutual gravitational attraction would have drawn out material from both stars into an elongated filament, which would have been fatter in the center than at the ends. This would have broken up in time, leaving one part in orbit around the Sun. It was claimed that gradually this would condense into planets, the terrestrial ones being those which formed at the thin end near the Sun, with the gas giants forming from the more distant bulge.

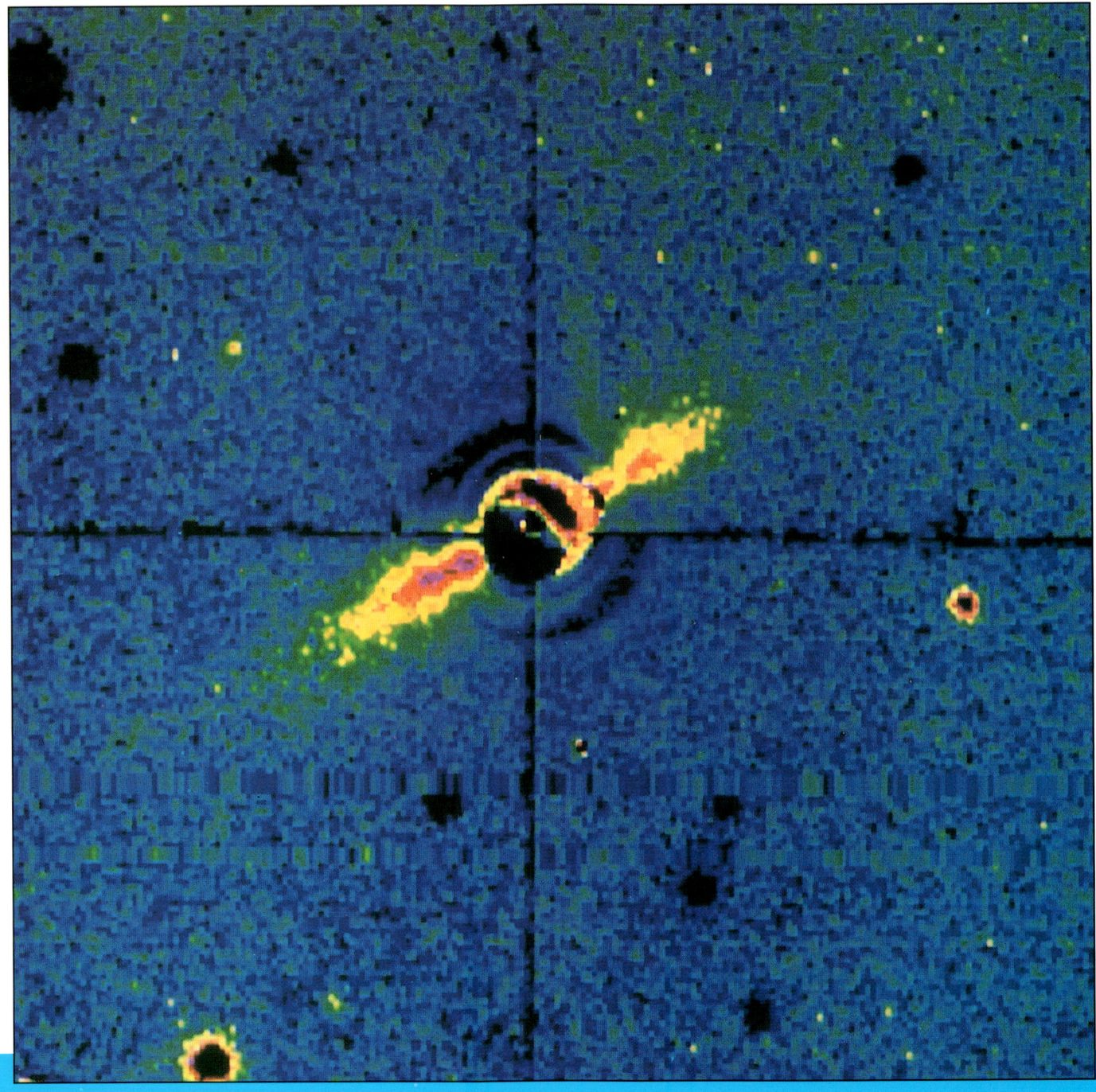

However, later mathematical analysis has shown that such a scenario would not produce a solar system such as ours. Furthermore, increased knowledge about how stars are born makes the accretion theory (pp. 102–3) more likely. Mathematical studies confirm its validity but, above all, observational evidence now makes it almost certain.

Evidence indicating, and then proving the existence of planets around other stars has been gathered in the past two decades. Observations of Barnard's star revealed that the star "wobbled" as it moved across the sky against the background of more distant stars. This wobble is just what one might expect if such a star is orbited by two planets, each about the size of Jupiter.

Accepting accretion in a solar nebula as the explanation of how the solar system was formed opened the way to seeking proof by direct observation. Put simply, astronomers wanted to see a solar nebula around some not-too-distant stars.

During 1983, the Infrared Astronomical Satellite (IRAS) did, in fact, detect a bright radiating dust structure around the star Vega (Alpha Lyrae) – a white, class A0 star about 60 times as bright as the Sun. This structure extends out almost half a light-day, or a distance 2.8 times that of Neptune from the Sun, and its mass is about the same as that of our own solar system.

But observationally, the more exciting was Beta Pictoris, only 60 light-years away. Examination by IRAS showed it had a solar nebula and astronomers managed to obtain a picture. Tilted about 7.5 degrees to our line of sight, the disk-shaped nebula has been traced out to a distance of about 28 thousand billion miles from the parent star.

There is yet another group of celestial bodies which appears to possess solar systems in the making. These are the T Tauri stars, named after a typical example in the constellation Taurus. They are extremely young stars, no more than a few million years old, settling themselves into the main-sequence stage of their life cycles and varying irregularly in brightness as they do so. During 1989 and 1990, astronomers analyzed radiation from T Tauri stars at various wavelengths and con-

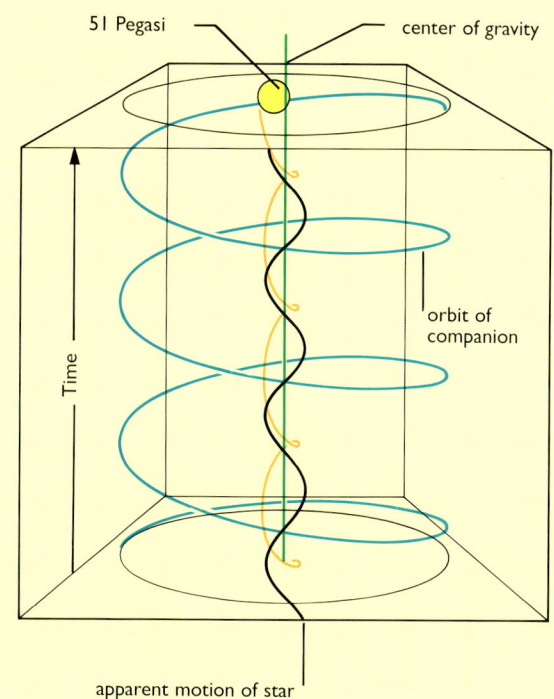

Unseen companions

A "wobble" in the movement of the star 51 Pegasi suggests the presence of an invisible companion. As the companion orbits 51 Pegasi, its gravity pulls on the star and makes it wobble. This wobble is detected in the star's light. The shift of lines in the star's spectrum indicate the forward and backward movement of the star. The wobble-effect, however, is only indirect proof of an orbiting planet. The same effect could be achieved by a pulsating star. An observation of a planet crossing in front of a star would confirm the existence of the orbiting body. The first observation of this kind was made in November 1999 when a planet about 30% larger than Jupiter crossed the face of the star HD 209458.

cluded that these stars are surrounded by flattish disks of material, with masses similar to those of our solar system.

The real breakthrough came in the late 1990s. The first extra-solar planet was discovered in 1995. Swiss astronomers at Geneva Observatory detected a "wobble" in the star 51 Pegasi which they concluded was caused by a planet in orbit around the star. By the close of of the 20th century, 28 extra-solar planets had been detected. Others have already been found and the search for more continues. These indirect discoveries (made by detecting the "wobble" of a star) are being followed up by a direct proof, by observing a planet's transit around its star, and by observation of light from a planet.

OUR SOLAR SYSTEM
● *The Sun's family*

The orbits of all the major planets display characteristics which are to be expected if the solar system was formed by accretion from a disk-type solar nebula (pp. 102–3). For example, the orbits all lie in the same plane, or very nearly so. There are exceptions, however. Venus, for instance, has an inclination of 3.4 degrees and that of Mercury, the closest planet to the Sun, is 7 degrees. All the others have inclinations less than this, except for one maverick, Pluto, the farthest planet from the Sun, with an inclination of just over 17 degrees.

This fact, along with other characteristics, leads astronomers to believe that Pluto is not actually a major planet (p. 133). Pluto aside, the orbits of the major planets of the solar system really do lie virtually in the same plane.

This is not so true, however, of asteroids, comets with their associated streams of meteors, and other small items of debris. Cometary orbits, in particular, vary widely in inclination. Some are inclined so much that they move around the Sun in a retrograde fashion; in other words, they move backwards compared with the major planets and the asteroids. Halley's Comet is a good example because its inclination is 162 degrees.

The second significant characteristic of the orbits of the major planets is that they are stable and approximately circular. Again there are exceptions. Mercury, for example, orbits in a pronounced ellipse with an eccentricity of 0.206; this is five times greater than the average elliptical eccentricity of the other major planets. But even this is not large.

A third characteristic is the observed fact that all the major planets orbit the Sun in the same direction: counterclockwise when viewed from above. This is another indication that they were formed in a rotating solar nebula.

The major planets can conveniently be divided into two groups. Closest to the Sun lie the terrestrial planets, so-called because they have some similarities with Earth. These are Mercury, Venus, Earth, and Mars. The second group, comprising the so-called gas giants, is farther away, beyond Mars. This group is made up of Jupiter, Saturn, Uranus, and Neptune.

The gas giants all have solid cores surrounded by vast, cold atmospheres where methane, ammonia, helium, and hydrogen are found in quantity. Such light gases were present originally on the terrestrial planets, but were lost.

Because the terrestrial planets are much closer to the Sun than the gas giants, they received much more heat and the light gas molecules began to move very fast. Since the terrestrial planets are less massive than the gas giants, each had an insufficient gravitational pull to retain such quickly moving molecules.

The most distant giant planet, Neptune, lies at an average distance of 2,794 million miles or just over 4 light-hours from the Sun. As the nearest independent star is Proxima Centauri, at 4.3 light-years, the scale of the solar system seems minute by interstellar standards. However, the solar system does not consist only of the major planets; there are also many small bodies such as asteroids and comets.

The asteroids are in the planetary section of the solar system. More than 90% of them orbit in a belt between Mars and Jupiter. The comets form the spherical Oort cloud which surrounds the system and marks its outer edge about 2.4 million million miles away. In between is the Kuiper Belt, made of tens of thousands of comet-like objects.

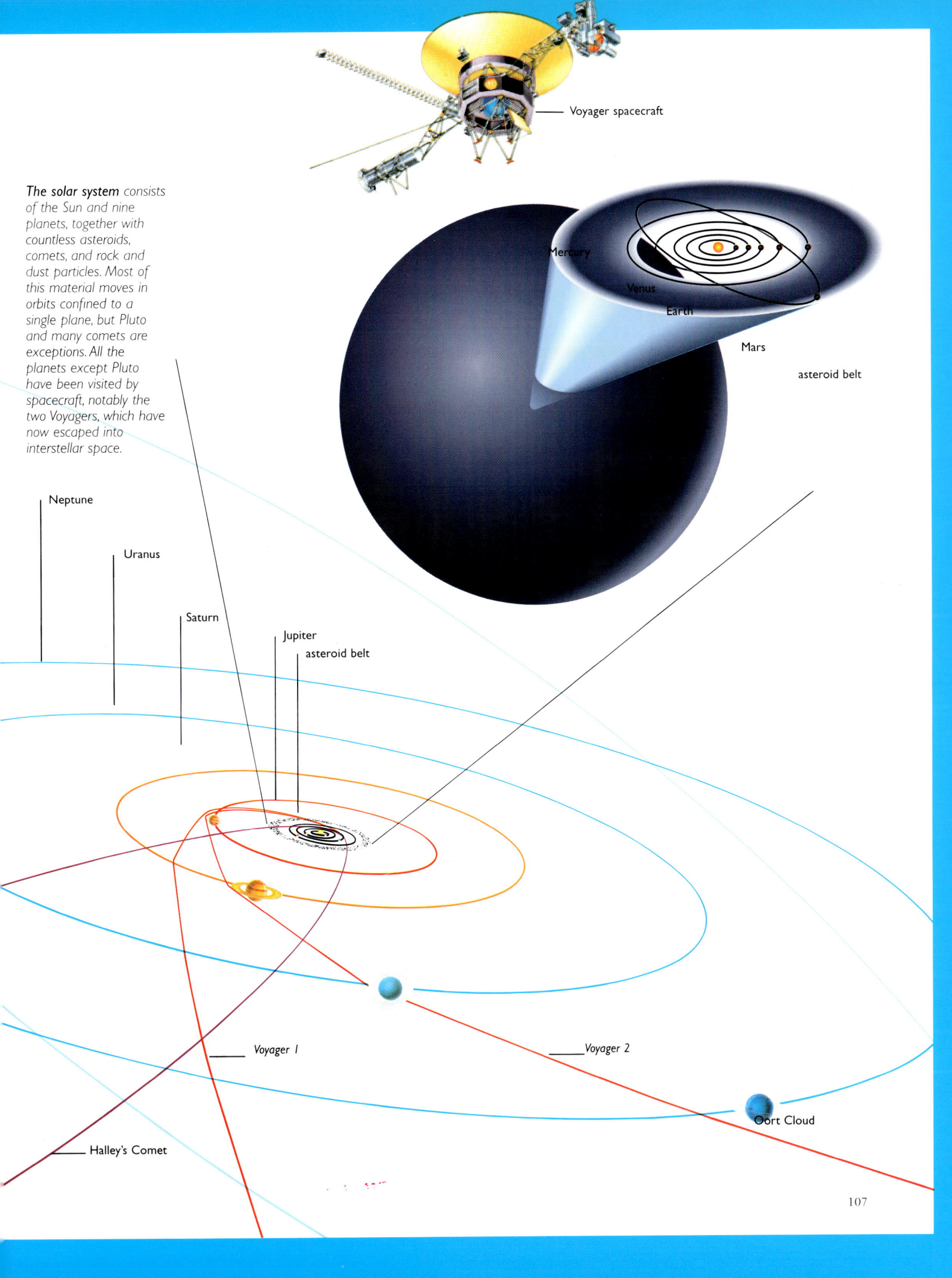

The solar system consists of the Sun and nine planets, together with countless asteroids, comets, and rock and dust particles. Most of this material moves in orbits confined to a single plane, but Pluto and many comets are exceptions. All the planets except Pluto have been visited by spacecraft, notably the two Voyagers, which have now escaped into interstellar space.

MERCURY
• Attendant of the Sun

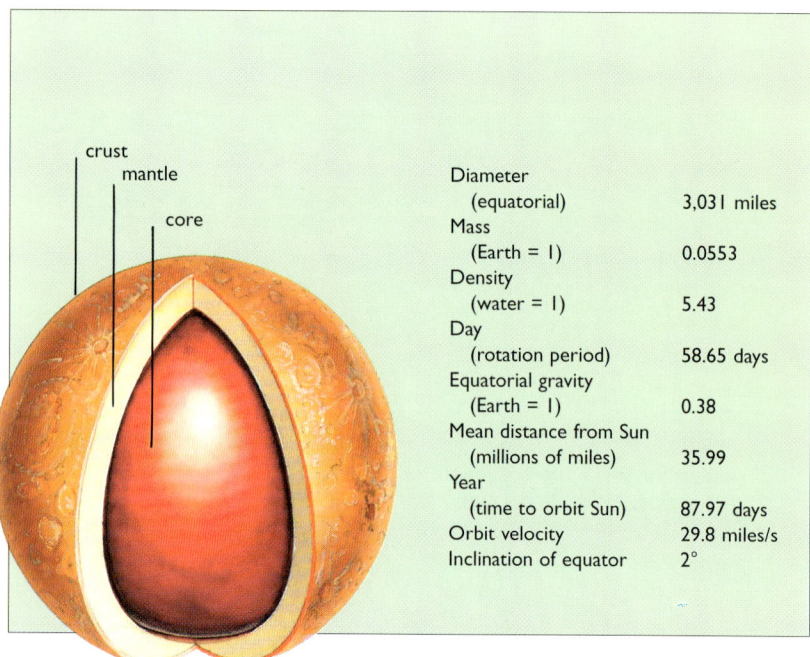

Diameter (equatorial)	3,031 miles
Mass (Earth = 1)	0.0553
Density (water = 1)	5.43
Day (rotation period)	58.65 days
Equatorial gravity (Earth = 1)	0.38
Mean distance from Sun (millions of miles)	35.99
Year (time to orbit Sun)	87.97 days
Orbit velocity	29.8 miles/s
Inclination of equator	2°

A double dawn is seen at some places on Mercury. Such a location (red dot) experiences its first dawn (1) when Mercury is accelerating as it nears the Sun. The planet turns through only a small angle (2–3) before it swings past the Sun, so night overtakes the given spot (4). The second dawn comes a little later (7). Daylight (8–13) lasts almost a Mercurian year.

Mercury, the closet planet to the Sun, can be observed either just before dawn or after sunset as a bright, silvery, starlike object. Planetary scientists have discovered that it is a parched, airless, cratered wilderness, baking in the fierce heat of the Sun.

Since its orbit lies within the Earth's, Mercury presents phases similar to our Moon. So when at its nearest to Earth, Mercury is seen merely as a relatively featureless thin crescent, even through the largest telescope. The whole disk is seen only when it is farthest away on the opposite side of the Sun. This has made it very hard for ground-based observers to examine the planet.

The challenge of observing Mercury made it difficult for astronomers to make accurate measurements and observations of the surface. Only in 1965, when radar pulses were sent out and successfully received back on Earth after bouncing off the planet's surface, did it first become possible to get an accurate fix on Mercury's rotation period. But it was not until a decade later that the value could be confirmed at 58.6461 days by the *Mariner 10* spacecraft, which passed close to Mercury in March and September, 1974 and March, 1975.

The surface of Mercury is crater pitted and looks just like our Moon, but Mercury does not have the large lava plains, or "seas," which dominate parts of the lunar landscape. Of the multitude of Mercurian craters, well over 230 have been named. The largest – called Beethoven – is 388 miles across. Planetary astronomers believe that craters are still being formed on Mercury by meteorite bombardment, as they are on the Moon. But, because of the general appearance of Mercury's surface, the rate of impact there is estimated to be about 20 percent greater.

Mercury is never farther from the Sun than 43.3 million miles and, with its elliptical orbit, it gets as close as 28.5 million miles. Thus, the Sun dominates the planet. Because it is so close, it gets at least 4.7 times more heat, light, and other radiation per unit area than the Earth. Because of this, its surface temperature can reach as high as 467°C (873°F).

The heat, combined with Mercury's low gravitational field, caused the planet's original atmospheric gases to boil off into space aeons

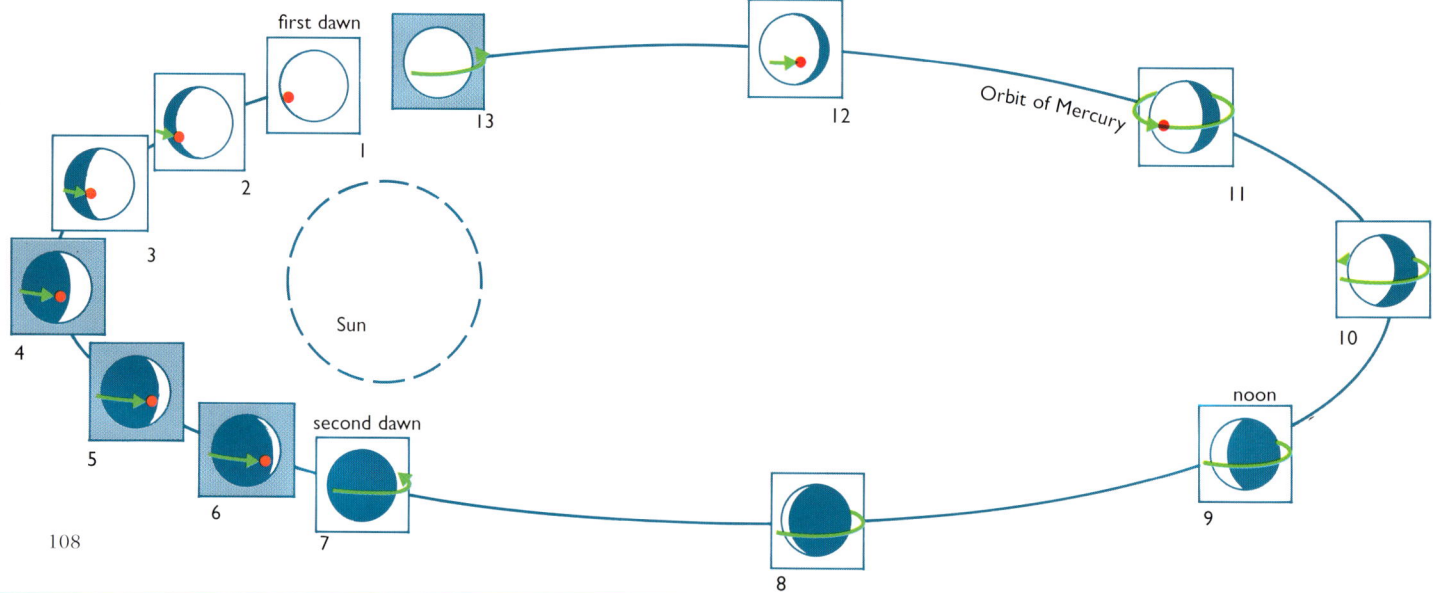

ago. Today, such atmosphere as there is consists of hydrogen and helium, gases from the solar wind briefly held back as it streams past the planet. But at only one million-billionth the density of Earth's atmosphere, it is much too thin to have any measurable effect on surface conditions.

Mercury's low mass contributes to its inability to retain an atmosphere. It is only 5.5 percent as heavy as Earth, so its escape velocity is 2.6 times less. Although only 1.4 times larger than our Moon in diameter, and apparently quite like it, with its cratered surface, Mercury is 1.7 times as dense. This is 5.4 times the density of water, almost the same as that of the Earth. Planetary scientists have concluded that Mercury must resemble the Earth in its internal structure, with a central iron-nickel core.

But to fit the theory, it is calculated that its core must contain about twice as much iron as Earth's and have a diameter of 2,240 miles. So it seems that Mercury's very dense solid core is actually larger than our entire Moon. Outside this huge core there is thought to be a relatively thin rocky mantle about 373 miles deep. Over the mantle is a light crust that at its deepest is no more than 41 miles thick.

Despite its bleak surface, Mercury is not totally inert. It has "hot" regions, caused by heat from volcanic activity below the surface, which are usually known as "hot poles" because they lie on opposite sides of the planet. *Mariner 10* also showed that Mercury has a weak magnetic field, about one-hundredth the strength of Earth's. However, it is stronger than those of the Moon, of Venus, and of Mars. As with the Earth, the magnetic poles of Mercury do not coincide with its poles of rotation; they are about 11 degrees away. But the existence of the field does not tie in with the idea that the core is solid iron, since magnetic fields are thought to be generated by molten iron cores.

A magnetic field does offer some protection against the radiation from the Sun but the weak field of Mercury, and the virtual absence of atmosphere, mean that the surface is continually bombarded by dangerous ultraviolet radiation and X-rays. This makes Mercury one of the most inhospitable members of the Sun's family of planets.

Mercury is covered with craters (above), caused by a heavy meteorite bombardment very similar to that which the Moon has experienced. The spacecraft Mariner 10 obtained this mosaic of views of Mercury during March 1974.

A relatively new crater (left), about 7½ miles across, lies within an older crater basin. This picture was taken from a distance of about 12,860 miles.

VENUS
● *The hothouse world*

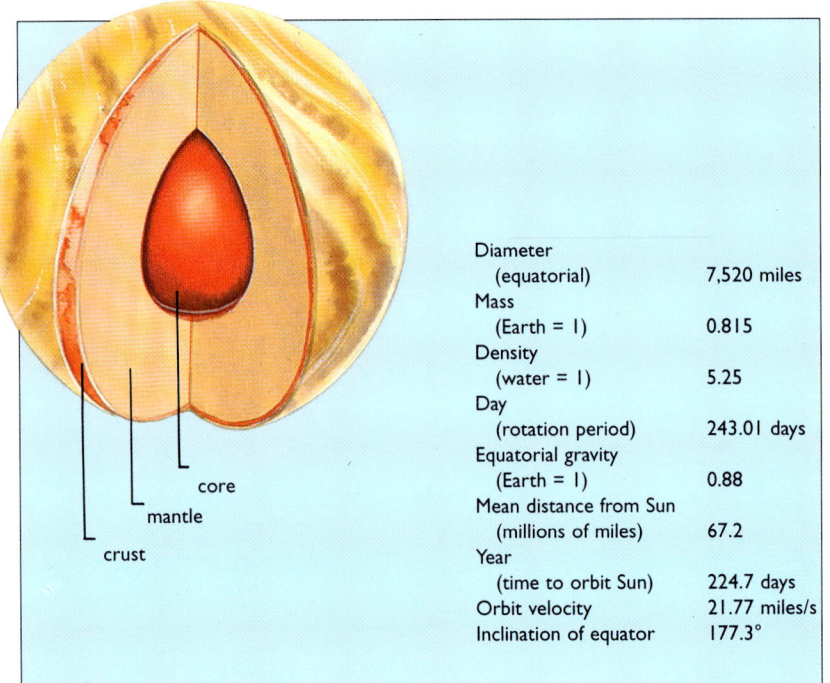

Diameter (equatorial)	7,520 miles
Mass (Earth = 1)	0.815
Density (water = 1)	5.25
Day (rotation period)	243.01 days
Equatorial gravity (Earth = 1)	0.88
Mean distance from Sun (millions of miles)	67.2
Year (time to orbit Sun)	224.7 days
Orbit velocity	21.77 miles/s
Inclination of equator	177.3°

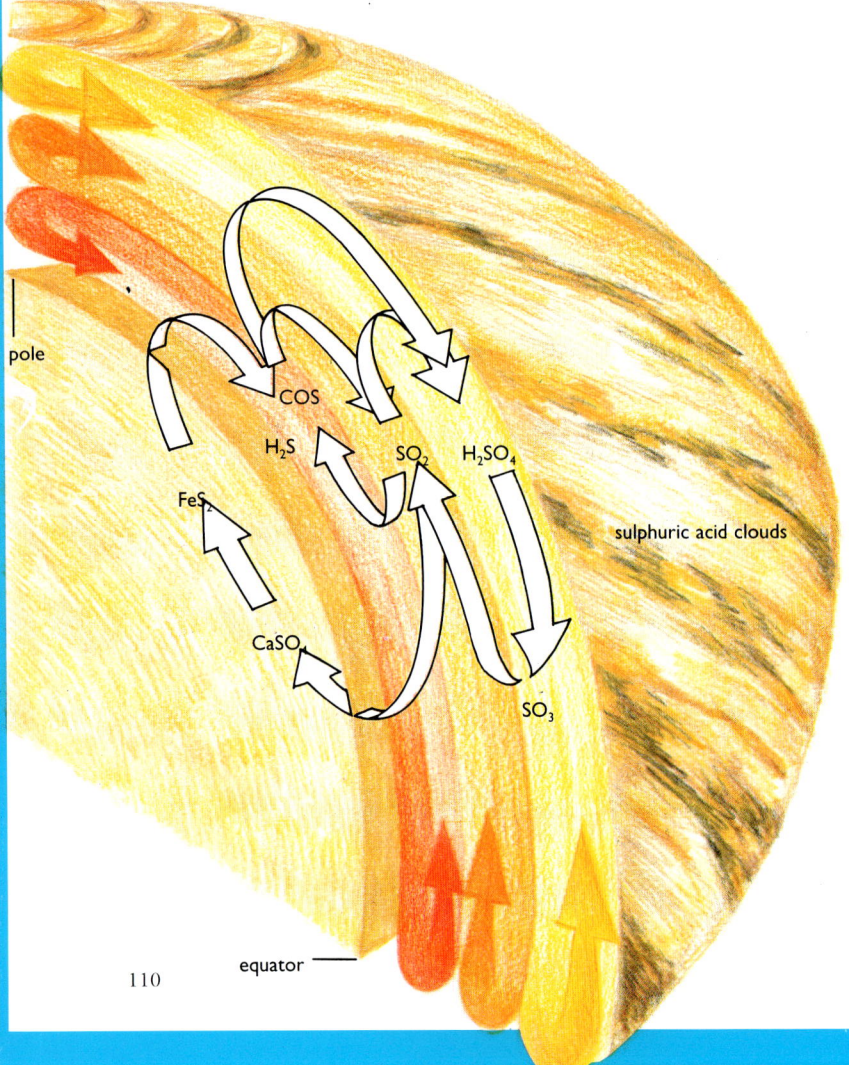

Apart from asteroids, the Moon, and the occasional comet, Venus can come closer to the Earth than any other object in the solar system. It is also close to the size, weight, and composition of the Earth, and its orbit, although inside Earth's, is not much different in shape. But while the Earth is temperate and supports life, Venus is unbearably hot with a choking, arid, crushing atmosphere of carbon dioxide, with dust and droplets of sulphuric acid (the pressure at the surface is greater than 1,280 lbs per sq inch).

Venus, like Mercury, lies closer to the Sun than the Earth. In consequence it, too, displays phases and can only be seen by Earth-based observers just before sunrise and for a short while after sunset. Thus, observations of its planetary disk are also fraught with difficulties, made more severe because its surface is covered by a dense, opaque atmosphere.

Since Venus is about the same size as the Earth — its diameter is only a little over 404 miles smaller — its internal construction is also like Earth's and it has a similarly-sized, dense nickel-iron core, part of which is probably in a liquid state. Over the Venusian core lies a mantle of rock, also about the same size as Earth's. The outer crust of Venus is 37 miles thick, about double that of Earth.

Like the Earth, Venus is covered with an appreciable atmosphere, but it is much denser, completely shrouding its surface. This means that Earth-based observers can see only the upper cloud layers. Because of such limitations, the rotation period of Venus was, for a long time, difficult to determine.

As with Mercury, it was from radar measurements made at the Arecibo Radio Astronomy Observatory in 1965 that a rotation period for the solid body of the planet, of about 243 days in a retrograde (east to west) direction, was deduced. This period is, incidentally, close to the Venusian orbital period of 224.7 days.

Since the 1960s, Venus has been observed by many spacecraft. The Soviet Union has sent a series of probes, from *Venera 1* in 1961 to *Venera 16* in 1983, and *Vega 1* and *Vega 2* in 1984. The United States has dispatched *Mariners 2, 5,* and *10*, as well as the two Pioneer craft in 1978 and *Magellan* in 1990. These have all revealed different facts about the planet.

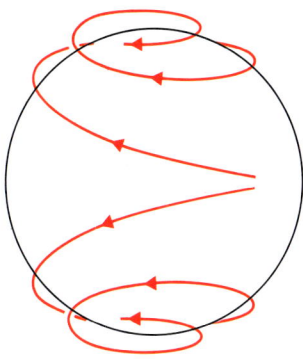

Swirling dense clouds shroud the baking surface of Venus (left). The mottled areas near the equator are thought to be convective cells due to heating by the Sun. The clouds move rapidly around the planet (below) at speeds up to 330 feet per second.

The sulphuric acid clouds of Venus are produced by complex reactions (far left). In the fast atmospheric cycle, sulphur dioxide (SO_2) is converted by sunlight into sulphuric acid (H_2SO_4). In the slow atmospheric cycle, hydrogen sulphide (H_2S) and carbon oxysulphide (COS) are converted into H_2SO_4. Both reactions utilize carbon dioxide and water vapor. In the lower atmosphere, H_2SO_4 breaks down to form, among other things, sulphur trioxide (SO_3).

In the crustal cycle, iron pyrites (FeS_2) react with water vapor and carbon dioxide to produce H_2S and COS, which go on to react with oxygen to form SO_2. Excess atmospheric sulphur dioxide forms calcium sulphate ($CaSO_4$), which with iron oxide and carbon dioxide produces iron pyrites once more.

The Venusian atmosphere is crossed by winds that blow from east to west, in the same direction as the planet's axial rotation. In addition, the whole atmosphere circulates from south to north and back again. With the close approach to the planet of *Mariner 10*, in 1974, and the Pioneer Venus Orbiter in 1979, direct measurements showed that the clouds high in the atmosphere rotate very fast. They reach speeds at the upper boundaries of the atmosphere of 330 feet per second (223 miles per hour) and take only four days to complete a circuit around the planet. But near ground level these winds travel comparatively slowly at only about 3.3 feet per second.

The Venusian atmosphere is very unlike that of Earth. It is dominated by carbon dioxide, which makes up 96 percent of the gases present. Next comes nitrogen at 3.5 percent – the remaining 0.5 percent is made up mainly of the following gases in order of decreasing quantity: sulphur dioxide, water vapor, argon, and carbon monoxide.

The sheer amount of carbon dioxide gives the atmosphere great depth and density, so the atmospheric pressure at ground level is a crushing 90 times greater than Earth's. Droplets or particles in the atmosphere are composed of sulphuric acid and occur in just two layers. The lower layer is between 30 and 50 miles above ground level. In the higher, cooler layer, water vapor combines with sulphur to form cloud droplets of sulphuric acid.

Because of the great density of the Venusian atmosphere and the prevalence of carbon dioxide, the planet suffers from a marked greenhouse effect; solar radiation is trapped by the atmosphere, strongly heating the planet. Thus while the Earth's atmosphere is at an average temperature of 27°C (87°F) at the surface, on Venus it is 457°C (855°F), well above the melting point of lead.

VENUS

Before surface details were revealed, the physical features of Venus were the subject of much speculation. Most theories were based on similarities in size between Earth and Venus, on the fact that, since it orbits much nearer to the Sun, Venus receives twice as much solar radiation as the Earth does, and on the fact that the albedo (reflectivity) of the planet is relatively high. Scientists deduced that vast oceans covered much, if not all, of the Venusian surface. This view remained popular until the spacecraft showed otherwise.

So far there have been two types of spacecraft investigations of Venus: landers and orbiters. Thirteen probes – some more successful than others – have landed on the planet. Some of these took measurements of the atmosphere on the way down before crashing into the surface, while others made soft landings and sent back video images of the surface and analyzed the soil. Most did not transmit for long – a couple of hours at the most.

Both *Venera 13* and *Venera 14* – which arrived at Venus in 1982 – made soft landings. Each craft sent back video images of a complete panorama (180 degrees) of a different part of the surface. The first, *Venera 13*, showed a relatively smooth terrain, with a surface which seemed to have a sandy covering of small grains, probably eroded – some scientists suspect originally by water – and then cemented together by atmospheric droplets and gases. The surface was also strewn with debris of small rocky pieces of various sizes.

Venera 14, which landed almost 590 miles from Venera 13, showed a slightly different scene. Again, some small rocks and pebbles appeared scattered on the surface, but otherwise the terrain was covered by flat pieces or plates of material, which showed signs of having suffered erosion followed by cementation. The plates had a sharp, angular appearance, but without any sandy or granular covering. Planetary geologists suggest that this is an area that was formed comparatively recently, probably about 10 million years ago. Analysis of the soil itself showed it to be a kind of basalt, similar to that found on the seabed on Earth, though with a stronger concentration of potassium.

The second method of investigating the surface by spacecraft has been by radar measurements made during close approaches to the planet. Short-wavelength radio pulses penetrated the cloudy atmosphere and reached the surface of Venus, from which they

Maat Mons, one of the largest volcanoes on Venus, rises 5¹/₂ miles above the surrounding land. The light-colored slopes of the volcano are lava flows believed to be not more than 10 million years old. The data for this image was collected by the Magellan space probe which orbited Venus between 1990 and 1994. It used radar beams to survey the planet, strip by strip. The result was that Magellan mapped 98% of the Venusian surface. The data was then combined and computer-enhanced to produce maps and landscape images of the planet. Areas like this one are not the norm, however, only about 15% of Venus is highland, most is lowland volcanic plain.

were reflected back and received by the same radar equipment. Computer analysis of the results has made it possible to chart the surface; the Magellan spacecraft in particular has provided astonishingly detailed pictures.

Most of the surface, about three quarters of the planet, is composed of flat plains. Additional land is depressed below the level of the plains. The balance consists of highlands, although these are mainly concentrated in two areas, one in the north and the other almost in line with the Venusian equator. The name given to the northern region is Ishtar Terra, with its Maxwell and Akna mountain ranges in the east and toward the west of the high region respectively. Ishtar Terra is not all highland and covers a large area, larger than the continent of North America. Most of it lies a few miles above the general flat surface, but the Maxwell mountains tower $7^1/_2$ miles above that already raised surface, while a large flat plateau to the west covers an area over 1,553 miles across.

The highland area on the equator is known as Aphrodite Terra and it covers an area half the size of Africa. It seems rougher and also more complex than Ishtar, with some deep straight canyons in the eastern central region. The latter are hundreds of miles wide and over 622 miles long; some are almost 2 miles deep.

Much of the Venusian landscape was molded by volcanic activity. The plains exhibit extensive lava flows, and the highlands feature giant volcanoes. Over 150 of the volcanoes are more than 62 miles across, and there are hundreds more smaller ones. These include the unusually-shaped pancake domes and the spider-like arachnoid volcanoes.

Here and there are some slight circular depressions. These range in size from 1 to 175 miles in diameter. They are impact craters; over 900 have been identified.

This global image of Venus was achieved by combining Magellan images. North is at the top. The light-colored wispy band that straddles the equator is Aphrodite Terra, the most extensive highland region on Venus. It is home to several large volcanoes, including Maat Mons (see image, p. 112).

EARTH
• The Living planet

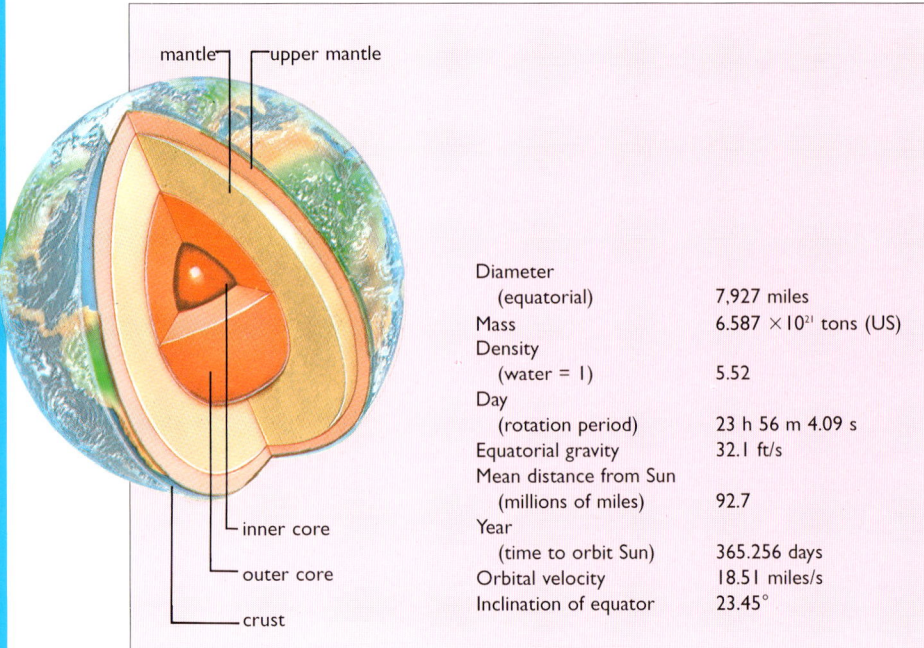

Diameter (equatorial)	7,927 miles
Mass	6.587×10^{21} tons (US)
Density (water = 1)	5.52
Day (rotation period)	23 h 56 m 4.09 s
Equatorial gravity	32.1 ft/s
Mean distance from Sun (millions of miles)	92.7
Year (time to orbit Sun)	365.256 days
Orbital velocity	18.51 miles/s
Inclination of equator	23.45°

Naturally enough, scientists know more about our beautiful, temperate world than any other in the universe. They are on hand to experiment directly, to interpret the evidence and to draw conclusions.

From the available evidence, planetary scientists are now reasonably certain that at the center of the Earth there is a nickel-iron core. This is divided into two parts: a metallic center 1,550 miles in diameter and, surrounding it, a layer where the iron and nickel are in a liquid state. This liquid section is about 1,370 miles thick.

Above the core, the Earth has a thick, rocky layer about 1,800 miles deep, called the mantle. The lower mantle is fairly rigid; the upper mantle consists of a somewhat more plastic layer known as the asthenosphere and a thin, rigid outermost layer, about 62 miles thick, the lithosphere. The topmost part of the lithosphere is the Earth's crust – the ground on

The Earth's crust

Geologists now think that the Earth's crust is divided up into six major, and a number of minor, rigid plates, floating on a "plastic" layer of the mantle. These plates are moved around by rising convection currents, which bring heat from the Earth's molten core to the surface.

Wherever this happens, usually in an ocean bed, new ocean floor is created, pushing the crust at each side outward. The ocean bed thus created is made of the comparatively dense rock, basalt. At present, the rate of drift of Europe and North America means they are moving apart by about 0.8 inch a year. Where new material is being created there is an underwater ridge at which red-hot new crust material comes into contact with icy cold sea water.

When magma wells up beneath a continent from the hot regions of the mantle (right, above), it can force the crust to separate. The thinner crust at this point may form new ocean floor. Where two plates collide (right, below), one may dive beneath the other, creating a trench.

The crustal plates have moved over the Earth in the course of time to create the pattern of oceans and continents we know today. About 300 million years ago, there was just one vast continent, known as Pangaea.

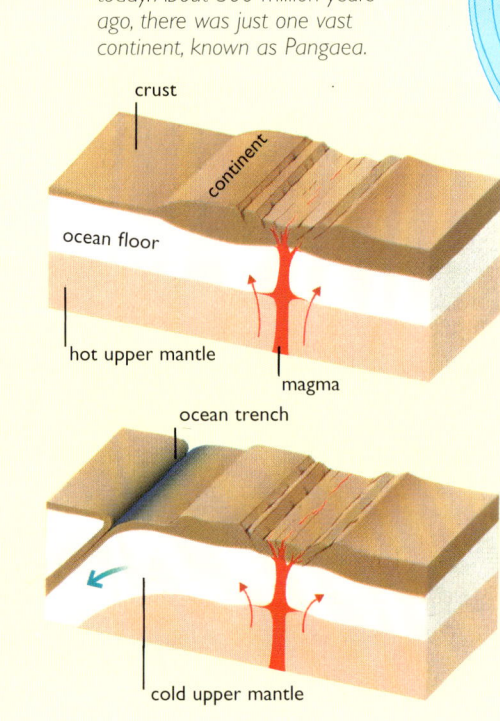

Pangaea split into *Laurasia*, made up of North America and Eurasia, and *Gondwana*, made up of South America, Australia, Africa, Antarctica, and India. It is thought that the original continental plate was so large that an upwelling convection current formed underneath and broke it apart, starting the movement of smaller plates.

114

which we live. This varies in thickness; the continental crusts are between 18½–25 miles thick, while the crust of the seabed is only 3 miles thick.

The lithosphere layer is constructed of many plates, which move across the mantle. Evidence for large-scale movement of continental land masses and of the sea floor comes from examination of rock formations, fossils, the magnetic polarity of rocks and the shapes of the continents. This evidence also shows that at an earlier stage of the Earth's history – probably about 160 million years ago – the world's continents formed two vast land masses. These gradually split apart to make the continents with which we are familiar today. The drifting of the continents still continues.

Above the Earth's surface lies the atmosphere, which is absolutely vital for the existence and maintenance of life on Earth. It is made up of a number of gases: nitrogen (77 percent); oxygen (21 percent); water vapor (1 percent); and the inert gas argon (0.93 percent). There are also some traces of carbon dioxide, neon, helium, and sulphur.

One highly important feature of the atmosphere is that it acts like a blanket around the Earth, preventing heat from the Sun's warming from being radiated away into space. This is the so-called greenhouse effect, whereby certain gases in the atmosphere reflect back infrared radiation instead of allowing it to escape. These gases include nitrous oxide, methane and, by far the most important, carbon dioxide.

Since the start of the Industrial Revolution, it has been estimated that the amount of carbon dioxide in the atmosphere has nearly doubled as a result of burning "fossil" fuels, such as coal, oil, and gas, and of large-scale deforestation. Scientists currently forecast that the average temperature of the Earth's atmosphere might rise significantly because of the increase in carbon dioxide. The effects of this are unpredictable and, in the worst scenario, could be disastrous.

The weather we experience is caused by the complex circulation of the atmosphere driven by heat from the Sun. Warm, moist air rises over the tropics and moves in the direction of the poles. At around 30 degrees latitude in both hemispheres, it cools and sinks, then returns to the equator. There are also eddies, generated by temperature differences between oceans and continental land masses. To complicate things further, the comparatively rapid rotation of the Earth whirls these currents and eddies around into an involved and ever-changing pattern.

In principle, however, the circulation of the Earth's atmosphere can be described fairly simply. East winds, coming from the poles, meet west winds from the temperate latitudes; while in the tropics there are trade winds, from the northeast in the northern hemisphere and the southeast in the southern. The trade winds are separated by low-pressure "depressions" in the equatorial regions.

To an observer from space, the Earth's constantly changing weather patterns contrast strongly with the fixed surface features of the continents and the oceans. A storm occurs when a heated mass of air rises. As it does so, it cools and water vapor condenses, causing clouds and rain. New air is sucked in to replace that which is rising. Because of the Earth's rotation it spirals in, creating the characteristic cloud patterns of a storm.

MARS
● *The red planet*

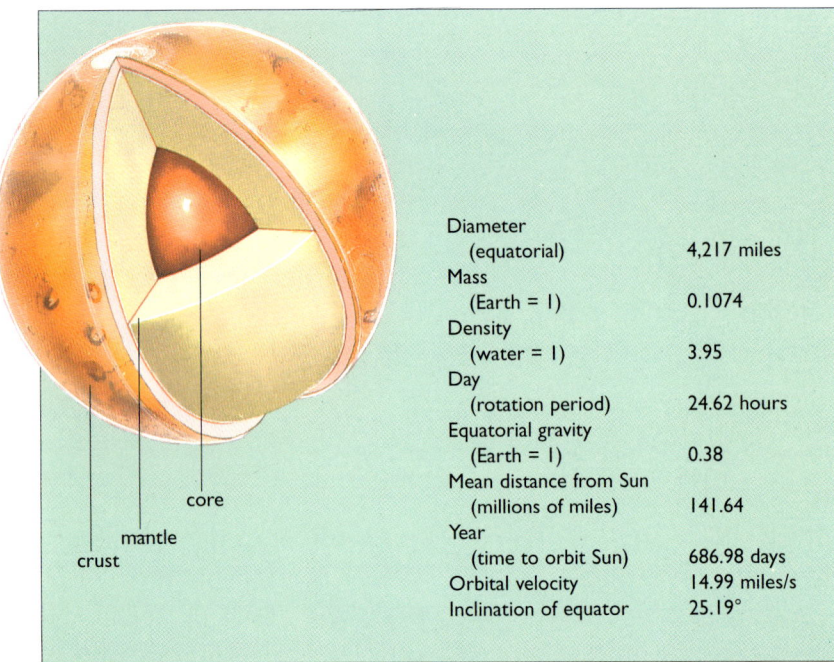

Diameter (equatorial)	4,217 miles
Mass (Earth = 1)	0.1074
Density (water = 1)	3.95
Day (rotation period)	24.62 hours
Equatorial gravity (Earth = 1)	0.38
Mean distance from Sun (millions of miles)	141.64
Year (time to orbit Sun)	686.98 days
Orbital velocity	14.99 miles/s
Inclination of equator	25.19°

Mars, the brilliant red planet with white poles, has long intrigued humankind. This, the last of the terrestrial planets, has exerted its fascination especially through comparisons with features of our own world – including the possibility of life. But Mars has become an object for serious scientific investigation only in the last two centuries. The most significant of modern studies are the revelations of space missions, which contradict the conclusions of earlier, Earth-based observations.

The diameter of Mars is only just over half that of the Earth, and its average density about 30 percent less. Since most of the mass of a terrestrial planet is concentrated in the core, that of Mars cannot, therefore, be large. Astronomers now favor a model of the interior similar to that of the other terrestrial planets, namely a core of iron and iron compounds with a diameter of about 1,865 miles. Outside the core is a mantle of silicate materials about 1,120 miles thick, overlaid by a crust a little more than 62 miles thick.

Taking 686.98 days – almost 1.9 Earth years – to complete an orbit around the Sun, Mars rotates on its axis once every 24.623 hours, giving a day length just a little more than the terrestrial one. But because its orbit is five and a half times more eccentric than the Earth's, Mars undergoes more marked changes in the length of its seasons.

Seasons on Mars are similar to those on Earth. This is due to the inclination of its axis

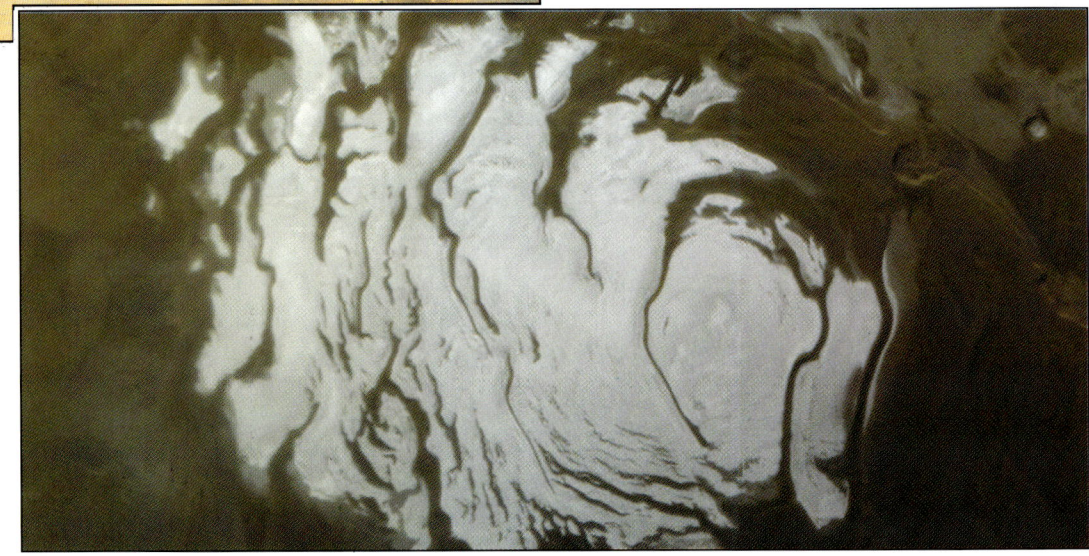

Valles Marineris, the giant Martian canyon that seems to show evidence of past erosion by flowing water, is captured above by the cameras on board Viking 1, which visited Mars in 1976. At right is a close-up view of the south polar ice cap. It shows the cap in summer when it has shrunk to about 400 km across. Frozen carbon dioxide has evaporated leaving behind a cap made mainly of water ice.

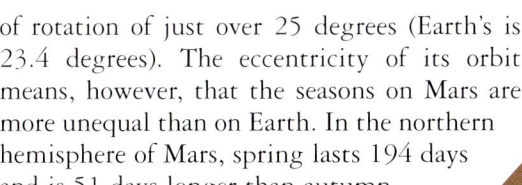

of rotation of just over 25 degrees (Earth's is 23.4 degrees). The eccentricity of its orbit means, however, that the seasons on Mars are more unequal than on Earth. In the northern hemisphere of Mars, spring lasts 194 days and is 51 days longer than autumn.

The seasonal changes on Mars have engaged astronomers' attention since the 19th century, for two very noticeable associated effects can readily be observed from Earth. The first is a discoloration of the reddish surface; areas that are less red – and sometimes even gray or green – begin to spread during the Martian spring. The second effect is that, during the spring in each hemisphere, the white polar cap starts to shrink. Early astronomers linked the two phenomena, and it became widely accepted that the discoloration was due to areas of vegetation spreading as they were irrigated by water released by the partial melting of a polar cap.

Like Venus, Mars has an atmosphere composed mainly of carbon dioxide (95 percent), with 2.7 percent nitrogen and 1.6 percent argon. Oxygen accounts for 1.3 percent and water vapor for no more than 0.3 percent. But the Martian atmosphere is much thinner than that of Earth. At ground level, it exerts a pressure only 0.7 percent of Earth's atmosphere. And because carbon dioxide is a very efficient radiator of heat (infrared) radiation, temperatures near ground level can drop at night to way below −53°C (−127°F), and to as low as −133°C (−271°F) at the winter pole.

Because Mars has no oceans, its surface responds quickly to temperature changes which lead to the initiation of strong winds that follow the hotter air warmed by the Sun. These Sun-following, or tidal winds can reach velocities of 150–300 feet per second at the Martian surface.

At such speeds, these winds can make the sand grains skip along the ground, and send dust up into the atmosphere. The dust grains are so small – about one-hundredth of a millimeter across – that they stay suspended in the atmosphere for months. Such dust storms are mainly local, but twice in each Martian year the disturbances are so widespread that great tracts, or even the whole, of the Martian surface become invisible from Earth.

Under good observation conditions it is possible to see from Earth long straight streaks on the Martian surface. The astronomer Giovanni Schiaparelli observed and, in fact, plotted a network of these. He called them *canali* – Italian for channels – but some astronomers linked the word to artificial canals, partly because such networks of intersecting straight lines gave every appearance of an artificial construction.

Mars viewed from above the northern hemisphere. The extensive canyon system, Valles Marineris, which lies along the planet's equator, here cuts across the lower part of the image. At its upper left are the giant volcanoes of the Tharsis region.

117

MARS

The notion of an "unnatural" canal system led to the implication that intelligent beings must exist on Mars. These creatures would be partially supported, presumably, by cultivation of plants, of which the gray-green areas seemed to be evidence. The greatest protagonist of this widely-held view was the astronomer Percival Lowell, who at his personal foundation, the Lowell Observatory in Flagstaff, Arizona, plotted the canals with diligence during the 1890s.

Until the space probes of the 1960s and '70s, most astronomers accepted the existence of long straight steaks on Mars and were happy to call them "canals," although almost no one thought that they were certain evidence for the existence of Martians. Attitudes remained unchanged until 1965, when the *Mariner 4* spacecraft took some closes-up pictures of the Martian surface. Although not detailed, they revealed conclusively that there were no canals; instead, Mars was a barren land, with craters scattered over it. It looked more like the Moon than the Earth.

Other Mariner probes provided similar visual evidence, but in 1971 when *Mariner 9* took pictures showing details as small as 330 feet across, it became clear that the Martian surface displayed not only impact craters but also vast canyons, volcanoes, and what appeared to be dried-up river beds. Five years later, two Viking spacecraft landed on the surface. There, they took close-up pictures of Martian rocks and made a chemical analysis of the ground itself.

Viking 1 landed on a fairly cratered area in the Chryse Planitia (Chryse Plain). Spacecraft pictures showed that rocks were strewn about nearby, while on the horizon a number of craters could be seen; these reached up to 1,980 feet in diameter. The surface was "sandy."

The second spacecraft, *Viking 2*, landed 5,497 miles to the northeast on a flat surface with fractured features. About 124 miles to the south was a large crater now named Mie, with a diameter of about 62 miles. It seems that *Viking 2* came to rest on the crater wall. The rocks near the Viking landings were pitted with holes, possibly evidence of the escape of gas once contained in them, or perhaps of ongoing erosion by windswept dust.

The surface of Mars, as revealed in close-up, is certainly red. Two-thirds of the sand particles consist of silicon and iron, and there is a strong concentration of sulphur — more than 100 times that found in terrestrial material. The red coloration is due, quite simply, to rust (iron oxide with other iron impurities, notably iron sulphide). Other evidence shows the surface to be an "iron-rich clay." The discoloration once thought to be vegetation is now known to be merely a chemical reaction on the rock-strewn surface. And both Vikings showed that the Martian sky is pink, not blue, a result of the suspension of fine dust particles of iron oxide.

The orbiter sections of the Viking spacecraft circled the planet and mapped the Martian surface down to details as small as 495 feet across, with some selected places at a resolution of $26^1/_2$ feet. The Vikings revealed some notable features, including the giant volcano Olympus Mons, and the Valles Marineris, a system of vast canyons in the equatorial regions which is well over 3,100 miles long and about $4^1/_2$ miles deep. Many other canyons have also

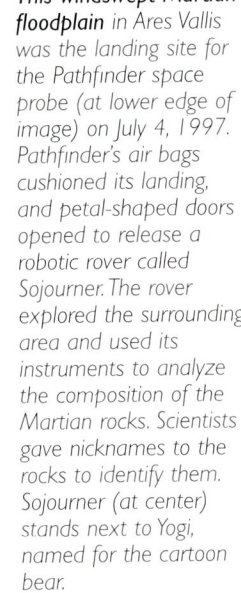

This windswept Martian floodplain in Ares Vallis was the landing site for the Pathfinder space probe (at lower edge of image) on July 4, 1997. Pathfinder's air bags cushioned its landing, and petal-shaped doors opened to release a robotic rover called Sojourner. The rover explored the surrounding area and used its instruments to analyze the composition of the Martian rocks. Scientists gave nicknames to the rocks to identify them. Sojourner (at center) stands next to Yogi, named for the cartoon bear.

been mapped, two of them over 622 miles long.

Arguably, the most intriguing fact brought to light by the Vikings is a confirmation that the surface of Mars does contain features which seem to be dried up water channels. (These are *not* the "canals" previously studied; those have been shown to be an optical illusion). There is no liquid water on the Martian surface now, but the channels seem to support the idea that, early in the history of the solar system, during the time when the Sun was developing, it underwent a brief period of intense radiation. If so, this could have removed the original planetary atmosphere, which might have been rich in water vapor and carbon dioxide, therefore allowing the presence of liquid water on the planet.

Another theory is that the channels may have been formed by the melting of ice during previous periods of intense volcanic activity. What seems certain, however, is that there is no evidence of life forms on Mars.

Olympus Mons (above), a vast shield volcano built up from successive eruptions of lava, is 375 miles across and more than 15 miles high. The largest equivalent structure on Earth is Mauna Kea in Hawaii, which stands only 6 miles above the seabed.

JUPITER
• Lord of the planets

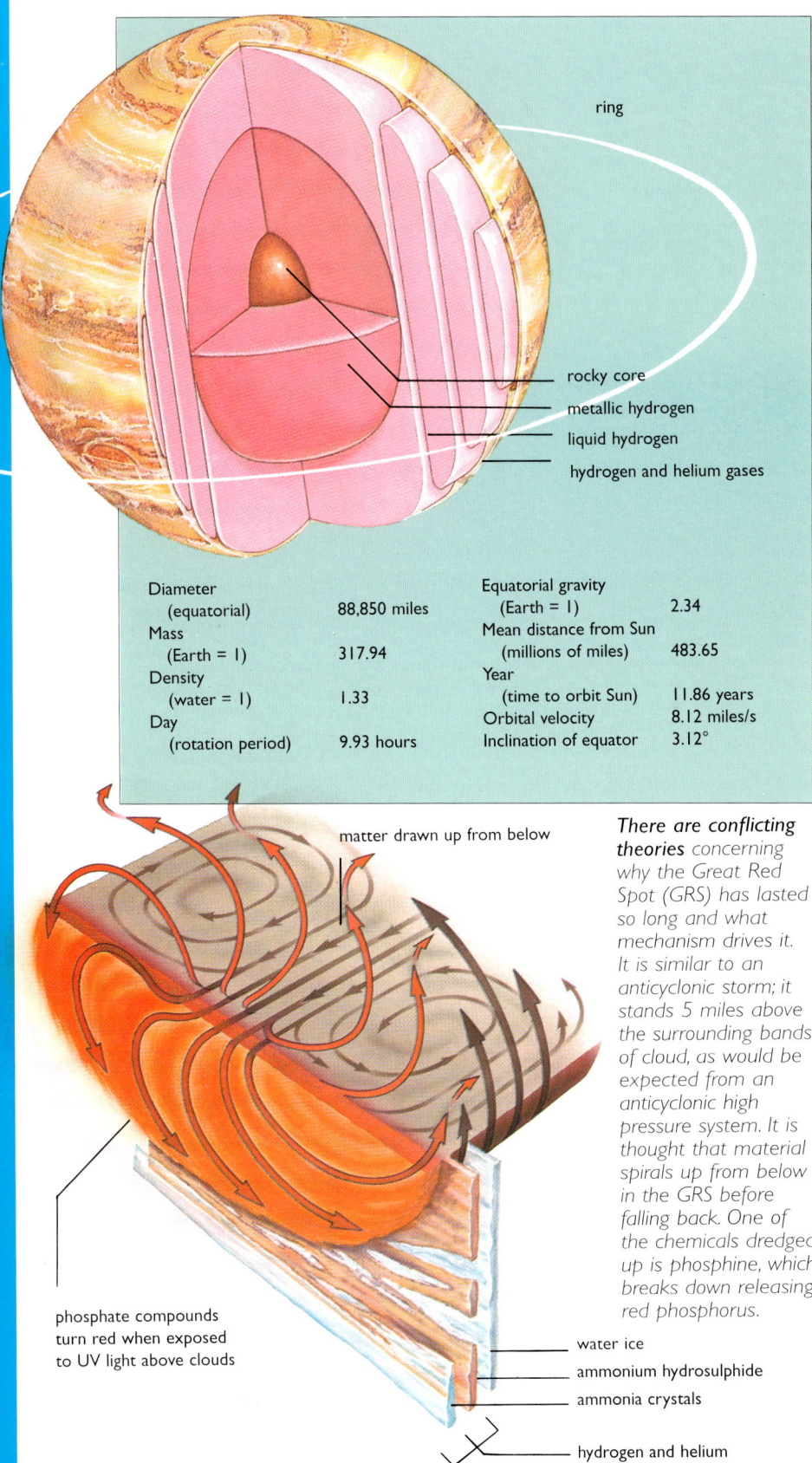

Diameter (equatorial)	88,850 miles	Equatorial gravity (Earth = 1)	2.34
Mass (Earth = 1)	317.94	Mean distance from Sun (millions of miles)	483.65
Density (water = 1)	1.33	Year (time to orbit Sun)	11.86 years
Day (rotation period)	9.93 hours	Orbital velocity	8.12 miles/s
		Inclination of equator	3.12°

There are conflicting theories concerning why the Great Red Spot (GRS) has lasted so long and what mechanism drives it. It is similar to an anticyclonic storm; it stands 5 miles above the surrounding bands of cloud, as would be expected from an anticyclonic high pressure system. It is thought that material spirals up from below in the GRS before falling back. One of the chemicals dredged up is phosphine, which breaks down releasing red phosphorus.

Jupiter is the closest to the Sun of the solar system's gas giants — huge planets possessing vast, dense atmospheres extending thousands of miles into the space around them. They contain the major proportion of the system's orbiting planetary mass and Jupiter alone accounts for more than 71 percent of this.

In essence, Jupiter is a vast ball of gas with a dense central core, with a diameter of about 18,640 miles and a temperature between 20,000 to 30,000 K at the center. According to calculations, the core appears to be mainly a mixture of iron and silicates, with some ices of water, ammonia, and methane converted into metallic form by the immense pressure of the overlying material. The pressure at the surface of the core is 45 million times the atmospheric pressure on Earth, or 6,400 million lbs. per sq inch.

Outside the central core is a zone of hydrogen, also under great pressure, in this instance, 2 million times our atmospheric pressure — over 28.4 million lbs. per sq inch. Under such pressure hydrogen becomes metallic, and its density is four times what it would be if it were in a gaseous state. This outer core extends for about 18,640 miles above the central core.

Outside the metallic hydrogen lies another zone of hydrogen which, while still under enormous pressure, is not metallic but takes the form of liquid molecular hydrogen. This extends out a further 15,535 miles. Above this is a gaseous, hydrogen-rich atmosphere 625 miles thick. It is this outer layer that can be seen from Earth and has been seen at close range by the Galileo spacecraft.

Even with a relatively small telescope, Jupiter is a fine sight with its banded atmosphere and four brilliant satellites (pp. 138-39). What can be seen of Jupiter is a squashed, or oblate, disk, wider at the equator than at the poles. This form of the planet is due mainly to its largely fluid nature and to its fast axial rotation; a Jovian day lasts not 24, but only 9.8 hours.

The observable Jovian atmosphere is not only banded but displays many features which show that there is a strong circulation pattern in the outer gaseous regions. The circulation is most rapid in the equatorial regions and slower toward the poles. Because of the different rotation speeds, the atmosphere presents an ever-changing series of features. Sometimes it is

enlivened by the appearance of the Great Red Spot (GRS), which has been observed through ground-based telescopes ever since the 1660s. In addition, the four larger satellites can be observed as they orbit the planet, becoming eclipsed as they move behind it and then crossing the disk, casting shadows on the cloudy surface.

Yet if Jupiter is spectacular using Earth-based telescopes, it proved to be even more striking when observed close up by the *Voyager 1* and *Voyager 2* spacecraft in March and July 1979. From the evidence of these observations, we now know that the surface clouds are 90 percent hydrogen. Helium accounts for almost all the remaining 10 percent, but there are traces, in descending order, of ammonia, methane, and water vapor.

The continual circulation of the bands of clouds effectively distributes the heat from the interior of the planet. The wind velocities, measured with reference to the eastward rotating mass of the atmosphere, range from easterlies traveling at up to 396 feet per second to westerlies (where the winds are slower than the general rotating mass) traveling at over 165 feet per second.

Using cameras and infrared detectors, the Voyager spacecraft have probed a little way below the outer cloud surface and also up into the atmosphere above this layer. Just above the clouds, where the pressure is five times that of Earth at sea level, are more brown clouds thought to be made up of water vapor and some as yet unknown compounds, probably sulphur. The temperature here is 7°C (47°F). About $18^{1}/_{2}$ miles higher, there are reddish-brown clouds of ammonium hydrogen sulphide along with other unidentified compounds. Here, the temperature has dropped to −73°C (−99.4°F).

At about $40^{1}/_{2}$ miles above the brown clouds, there are wispy cirrus clouds of ammonia and the temperature is −133°C (−207.4°F). However, at 56 miles – at Jupiter's tropopause – the temperature begins to rise again; but

The Great Red Spot *is the dominant feature of the southern hemisphere of Jupiter. This color-enhanced Voyager 1 picture clearly shows the circulating nature of the GRS and shows small puffy features within the spot itself. It rotates counterclockwise with a period of approximately 6 days and was about 16,280 miles long by 8,575 miles wide at the time of Voyager's encounter.*

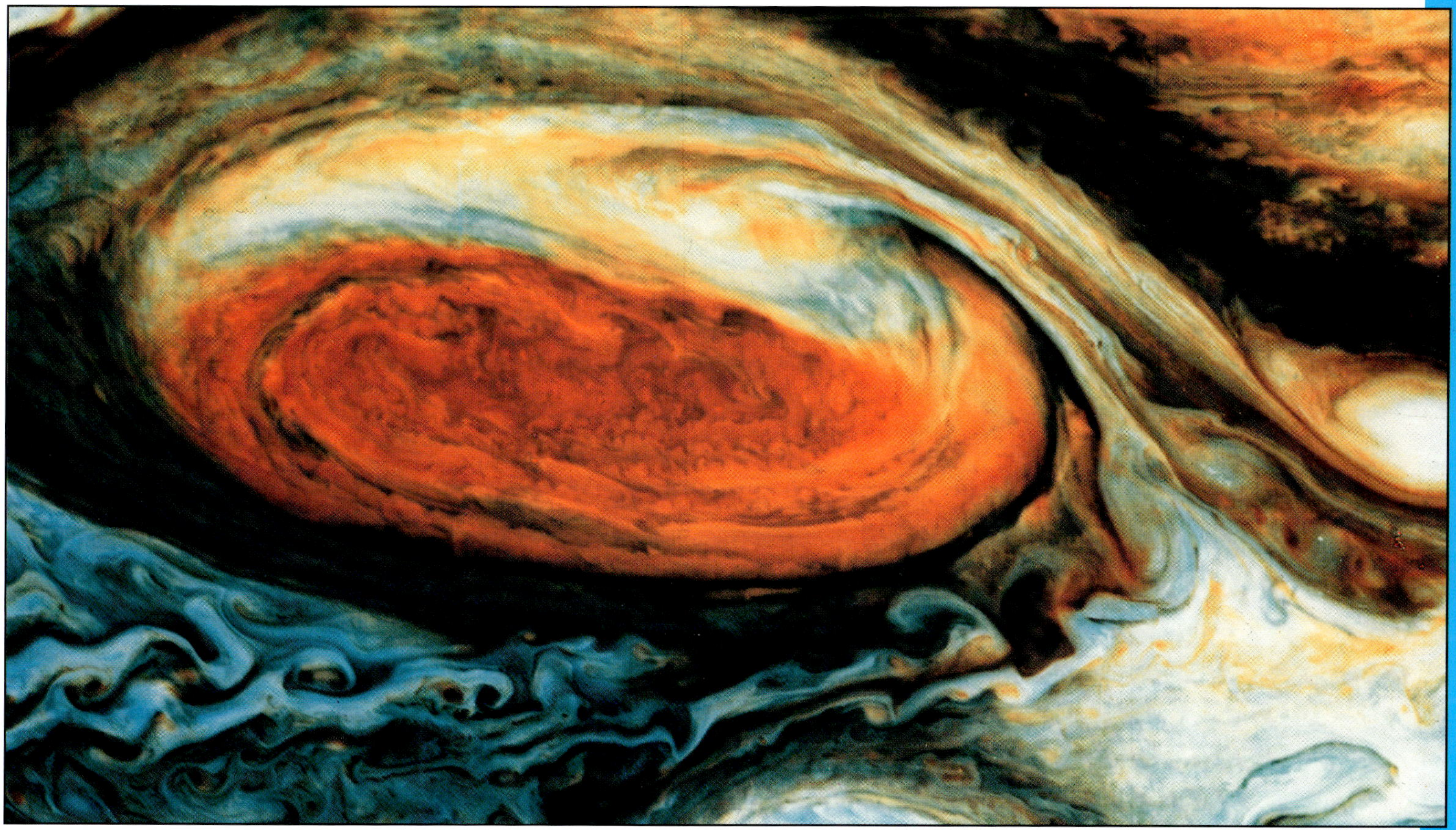

Surrounding Jupiter in a huge belt is a plasma ring, or torus (below). It consists of electrically-charged particles left behind in a trail by volcanic activity on the moon, Io. The particles are trapped by the lines of force of Jupiter's magnetic field. The torus is approximately the same shape and size as Io's orbit but is inclined so that it lies in the plane of Jupiter's magnetic equator.

even at 93 miles it never reaches more than −113°C (−171.4°F) because Jupiter is 485 million miles away from the warming radiation of the Sun.

Below the clouds, within the molecular hydrogen layer, planetary scientists think there may be a series of cylindrically-shaped envelopes lying one outside the other. Each has its own period of rotation which produces the wind currents observed in the upper cloud layer at the differing latitudes.

The Great Red Spot (GRS) is only one of a number of long-lasting oval patches in the cloudy Jovian atmosphere which can persist for months or even years. But since the GRS has been visible for about 340 years, its longevity seems exceptional. Moreover, it is unusual in that it rises about 5 miles above the surrounding cloud mass. Exceptional also is the size of the GRS. Its size varies, but at its largest, it can be 25,000 by 8,700 miles.

The ovals, including the GRS, keep to the same latitude but drift in longitude, seeming to roll between different layers of the high-velocity clouds. Voyager close-up pictures of the GRS, showing details down to 18½ miles across, reveal that it looks like a form of cyclone, with a complex pattern of swirling atmospheric motions. Currently, planetary meteorologists are uncertain as to the causes of such atmospheric features in general and of the GRS in particular.

Jupiter emits radio waves and has a large and strong magnetic field. Like that of the Earth, this field is a dipole, similar to a bar magnet. The planet's magnetic poles are inclined at 11 degrees to Jupiter's axis of rotation and, in addition, the magnetic axis is offset by one-

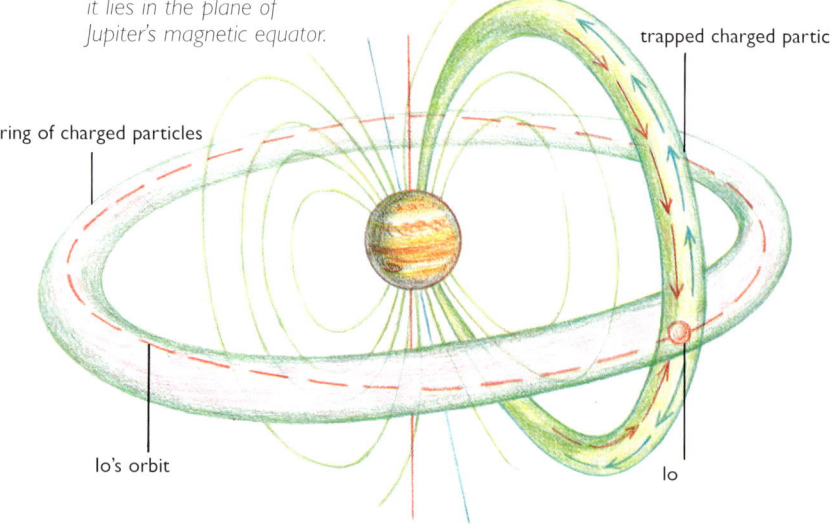

THE GRAND DESIGN

The complicated nature of the cloud belts on Jupiter is shown by a Voyager 1 photograph (left) taken in February 1979 when the craft was 17.6 million miles from the planet. Also visible are the inner moon, Io, and the moon Ganymede.

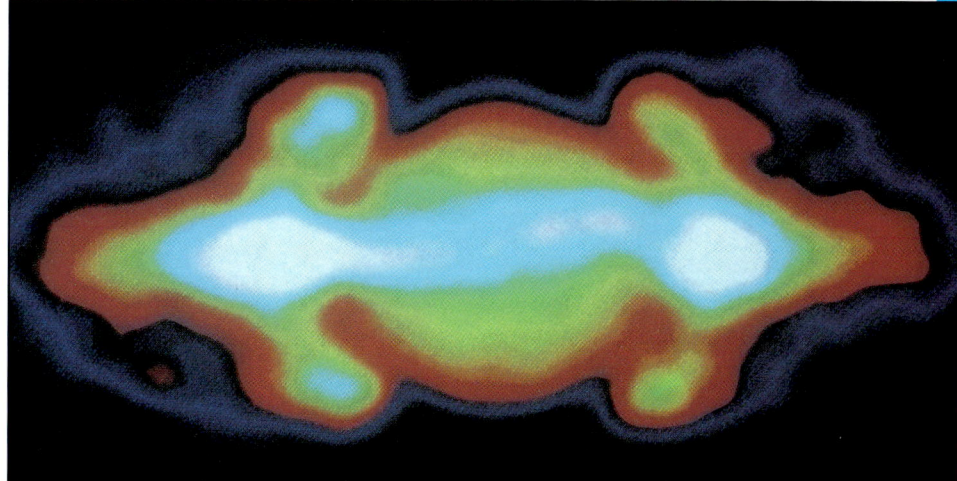

Jupiter's radiation belts shown in a false-color image taken by the Very Large Array radio telescope in New Mexico. Jupiter's strong magnetic field traps electrons in belts equivalent to the Earth's Van Allen belts. These electrons radiate synchrotron radiation at radio frequencies. This image was taken at a wavelength of 21 cm.

tenth of Jupiter's radius – that is, about 44,370 miles – from the planet's center. As a result, the magnetic field at Jupiter's cloud surface is not equal at every point; it differs most notably between the northern and southern hemispheres.

Because of its magnetic field, Jupiter possesses a magnetosphere which extends into space around the planet. Although it has a dipolar field, Jupiter's magnetosphere is very different from that of the Earth. Two main factors contribute to this. First, Jupiter's magnetic field is about 100 times larger than the Earth's and, second, the effect on it of the solar wind is about 25 times less because Jupiter is so much farther from the Sun.

On the sunward side of Jupiter, the magnetosphere extends about 1,243,000 miles above the planet, but on the side away from the Sun it is more extensive. Its intense central region has a diameter of 2,485,600 miles. The magnetosphere extends farther than this and is elongated away from the Sun, "blown" in that direction by the solar wind. This elongation is immense, and it is possible that the tail of Jupiter's magnetosphere extends as far as the orbit of Saturn, more than 373 million miles away. The magnetosphere on the sunward side is also drawn out a little toward the Sun, but not nearly as far as that on the opposite side.

The solar wind and electrified particles in the Jovian magnetosphere together produce auroral displays, as they do on Earth, and an auroral arc in the Jovian sky was, in fact, observed by *Voyager 1* on its close approach to the planet. Such displays occur over the whole surface of Jupiter, not just in polar regions as on Earth, possibly due to low-speed electrons which have leaked through to react with the Jovian ionosphere.

Jupiter emits more radiation than it receives from the Sun. This is due to the heating caused by shrinkage of the planet's core, a process which may still be continuing, and to radioactive heating. Because of thermal radiation from inside the core, Jupiter transmits at radio wavelengths. Some of these are centimeter wavelengths, due to the motions of electrified particles in the magnetosphere; others are longer – between 10 meters and 3 kilometers. These transmissions are disturbed by one of Jupiter's nearest satellites, the eruptive Io (pp. 138–39), which passes within 217,500 miles of the upper cloud layers of the planet and thus close enough to disturb the Jovian magnetosphere.

Jupiter also has a system of thin rings. The outer edge of the main ring is 31,070 miles above the cloud surface, with a very tenous ring extending farther out.

SATURN
• *The ringed world*

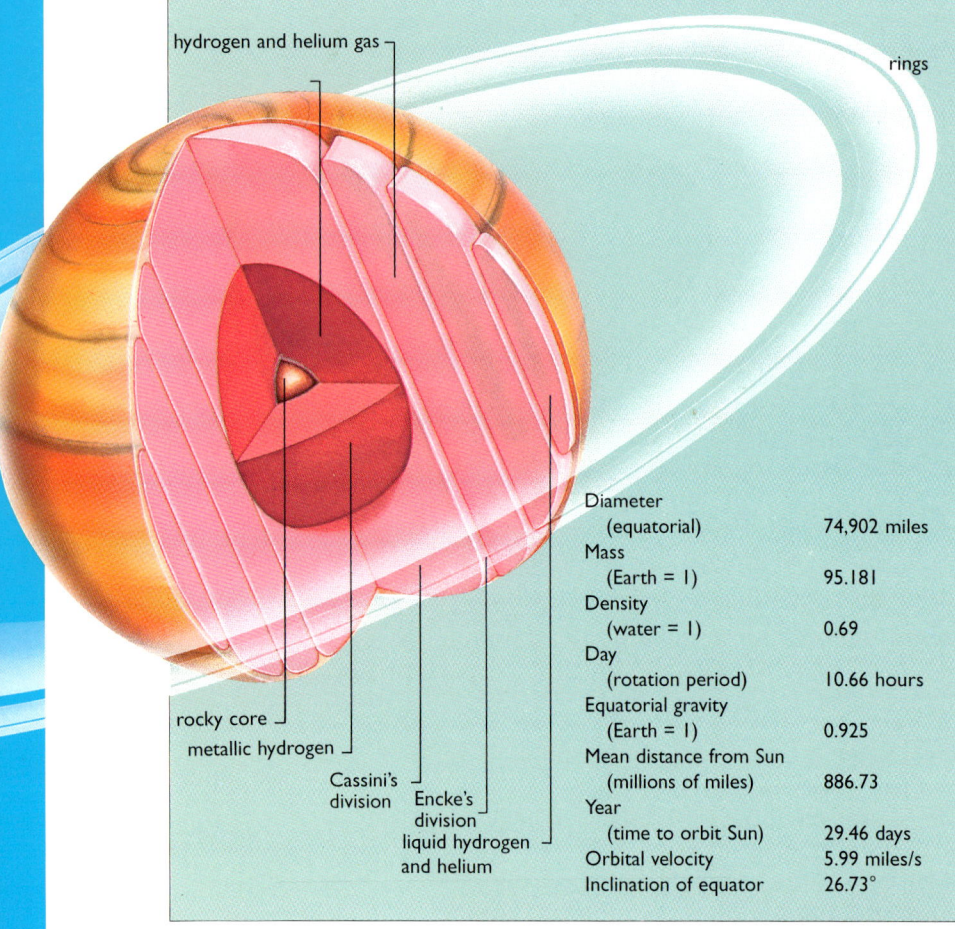

Diameter (equatorial)	74,902 miles
Mass (Earth = 1)	95.181
Density (water = 1)	0.69
Day (rotation period)	10.66 hours
Equatorial gravity (Earth = 1)	0.925
Mean distance from Sun (millions of miles)	886.73
Year (time to orbit Sun)	29.46 days
Orbital velocity	5.99 miles/s
Inclination of equator	26.73°

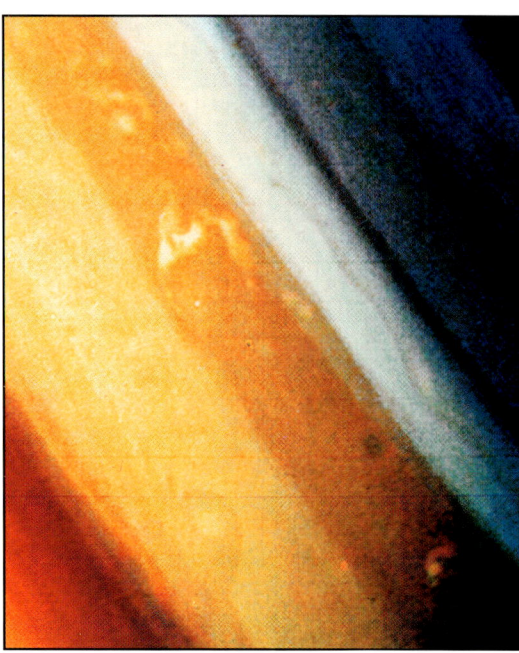

The visible features on Saturn consist of bands and other cloud patterns — ovals, eddies, and interacting spots. This false-color image of Saturn's northern hemisphere, taken by Voyager 1 in 1980 at a range of 5.6 million miles, shows an isolated convective cloud in the light brown region and a thin longitudinal wave in the blue belt. The smallest features discernible in the picture are 109 miles across.

Of all the planets in orbit around the Sun, Saturn, with its ring system, is the most spectacular. Yet Saturn is not alone in being surrounded by rings; Jupiter, Uranus, and Neptune also possess ring systems. But none of these rivals the Saturnian system either in extent or variety, and only Saturn's rings can be directly observed by telescope from Earth.

Though not as large as Jupiter, Saturn is another gas giant and it accounts for 21 percent of the total mass of all the planets. It is, however, less dense than Jupiter; while the average density of Jupiter is 1.33 times that of water, Saturn's average density is only 0.69. Thus, if Saturn could be put into a big enough bath of water, it would float.

Calculations indicate that Saturn has a central core about 15,535 miles across, composed of silicates, minerals, and various types of ice. The temperature of the core is about 14,000 K. The overlying pressure is 10 million times the atmospheric pressure on Earth.

Above the core, stretching up about 7,120 miles, lies a shell of metallic hydrogen, less than a tenth of the thickness of Jupiter's metallic hydrogen region. Beyond the metallic hydrogen is a layer, about 2,610 miles thick, of helium droplets. Beyond this again lies the remainder of the Saturnian material, extending outward for around 12,000 miles. Being composed of 93 percent molecular hydrogen and 7 percent helium, it is light in weight. This forms the outer layer of Saturn, which is visible from Earth and which was photographed by the Voyager spacecraft.

Like Jupiter, Saturn is oblate (flattened at the poles), though because of its lightness and a 10-hour rotation period, its degree of flattening is more than 50 percent greater than Jupiter's. Saturn is, in fact, the most oblate of all the planets. The surface we see is crossed by bands of clouds, although these appear more curved than those on Jupiter.

The cloud patterns are, nevertheless, similar, with jet streams of fast-moving clouds producing, as on Jupiter, eddies and even, at a latitude of 55 degrees south, a permanent, oval reddish spot. This is similar to the Great Red Spot (GRS) of Jupiter, but it is much smaller, with a length of only 3,730 miles in an east-west direction, compared with 12,500–25,000 miles

for the GRS. Saturn also has other ovals at different latitudes, mainly in the northern hemisphere. On Saturn, when spots and ovals meet, they move around each other and do not merge as such features do on Jupiter.

Saturn also possesses a permanent, distinctive dark wavy line in its cloud cover; this lies at 45 degrees latitude north and extends about 3,100 miles in an east-west direction.

Much has yet to be learned about the behavior of Saturn's cloud layer and the atmosphere above it, although it is now known that, as on Jupiter, the bands of cloud circulate at high velocities. Just above and below Saturn's equator, these reach speeds of about 1,585 feet per second; yet near its poles the speeds reduce to zero, which means that they merely keep pace with the planet's axial rotation.

Above the 186 ft-thick cloud surface, the atmosphere of Saturn extends outward and, at a height of 18½ miles or so, there are ammonia clouds. It is this hazy atmosphere that obscures the deeper cloud layers and gives the planet a more muted and less dramatic appearance than Jupiter.

Saturn, like Jupiter, emits more radiation than it receives from the Sun. In Saturn's case, this is 1.76 times more than it receives, and the internal heat, or energy, is thought to be caused by contraction of Saturn's core or by helium droplets condensing and falling deep into the planet's interior.

Saturn has a magnetic field generated by electric currents in the metallic hydrogen of the core. The field's magnetosphere – much like Jupiter's magnetosphere, although less extensive – stretches outward into space, trailing away in a direction opposite to that of the Sun but with a bulge on the sunward side. This bulge is elongated, extending 621,400–932,100 miles. But Saturn's magnetic poles are very close to its poles of axial rotation, making

Saturn and its magnificent ring system are here seen in full view by Voyager 2's cameras. The picture was taken four and a half days after the spacecraft made its closest approach to the planet. Saturn can never be seen in half-phase like this from Earth, because from our standpoint the Sun illuminates virtually the entire disk of the planet.

Saturn's rings are no more than 3,300 feet thick. How the rings were formed is something of a mystery. There are two competing theories. The first is that they are the debris left behind after an asteroid was torn apart by the gravitational forces of Saturn. The second is that they consist of material that failed to coalesce into an outer asteroid at the time the planet was forming. The second theory is now thought to be most likely.

its magnetosphere almost symmetrical.

The magnetic field and magnetosphere point to Saturn's having an ionosphere, which proves to be the case. It lies thousands of miles above the visible cloud surface, and, since it is composed mostly of ionized atoms of hydrogen, it reflects radio waves, although it is less concentrated than the Earth's ionosphere.

Saturn also emits electromagnetic radiation at radio wavelengths of between 15 and 300 meters. The emissions are intense and come from two sources, the stronger one only 10 degrees from the north pole on the sunlit side of the planet, the other from a similar spot in the southern hemisphere. The strength of both changes as Saturn rotates on its axis, and they reach a maximum every 10.65 hours. It seems likely that the sources are linked to, and indicate, the period of rotation of the magnetic field deep within the planet.

The cause of these transmissions is not fully explained, but it seems likely that they are associated with particles from the solar wind that are accelerated inside the magnetosphere, because they coincide with the appearance of aurorae at Saturn's poles. These auroral displays, which are widespread over the planet's surface, are caused by low-speed electrons interacting with the Saturnian ionosphere. Six of Saturn's satellites orbit within the magnetospheric regions, and so does the famous ring system.

When Saturn's ring system is observed from

THE GRAND DESIGN

Earth, its precise appearance depends on how the planet is situated in the sky with reference to the Earth's orbit. Every 14 or 15 years, the rings seem to disappear; this is because we are viewing them edge-on and, since they are extremely thin, they merge with the background of the sky. Between the occasions when they are visible in an edge-on aspect, they are seen from so much above or below their plane that they blot out large areas of the planet.

Saturn's ring system was first seen in July 1610 by the famous Italian physicist and astronomer, Galileo, with his newly-developed telescope. His instrument was not good enough to show the rings clearly, even though they were at their most open aspect at the time. All Galileo could report was that Saturn appeared to be a triple planet. When he was observing it about seven years later, it was near its edge-on aspect and the rings were invisible to him. Saturn, he said, seemed to have swallowed its own children. Because of the poor optical quality of the telescopes then, astronomers could not explain the phenomenon.

The mystery was finally cleared up in 1655, by the Dutch astronomer Christiaan Huygens, who managed to observe that there was, indeed, a ring around Saturn. But the true nature of Saturn's rings was not discovered until two centuries later, in 1856, when James Clerk Maxwell analyzed the evidence and showed that the gravitational field of Saturn would tear any sold ring to pieces. Maxwell concluded, therefore, that the rings could be composed of tiny particles in orbit around the planet. Subsequent studies, including results obtained from the Voyager probes, confirm Maxwell's conclusion.

Despite the fact that Earth-based observers had detected a number of separate rings around the planet, no one expected there to be as many as the astounding photographs from the Voyagers revealed. These photographs also confirmed astronomers' suspicions that the ring system was extremely thin. The Voyager measurements made it clear that the rings are no more than 0.6 miles thick, much thinner than previously believed. For their size, they are far and away the thinnest rings of any planet in the solar system.

The extreme thinness of the rings is thought to be caused by the gravitational effects of some small satellites, called "shepherd satellites," which orbit Saturn close to the plane of the rings. These satellites prevent the rings from spreading out above and below.

The rings are a magnificent spectacle and extend outward from 4,350 miles above the cloudy surface of Saturn to more than 46,000 miles. A broad faint outer ring stretches out further, to 298,300 miles.

Voyager pictures have also revealed dark patches, like the spokes of a wheel, in the ring system. These patches, which are more than 6,200 miles long and about 1,243 miles wide, rotate around Saturn.

The spaceprobe, Cassini, will soon show us Saturn, its rings and moons, in close-up once again, and offer us the opportunity to learn more about their nature and formation. The probe, launched in 1997, will reach Saturn in 2004, and spend the next four years orbiting the planet and its moons.

Dark radial markings resembling spokes constantly dissolve and re-appear, revolving with Saturn's rings. They seem to consist of clouds of electrically-charged particles, about a thousandth of a millimeter in size, suspended above the plane of the rings and moving under the influence of Saturn's magnetic field. They have been seen to form in a matter of minutes in places where the rings emerge from Saturn's shadow, and they survive for only one or two revolutions of the planet before vanishing.

URANUS
• *The tilted planet*

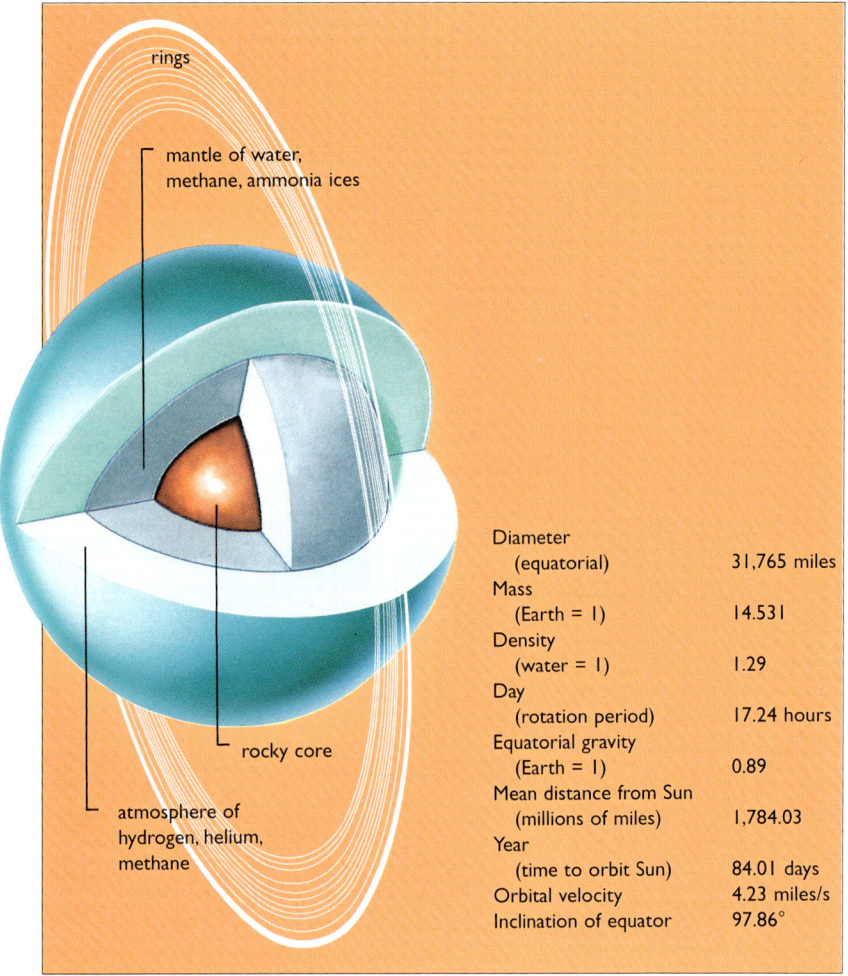

Diameter (equatorial)	31,765 miles
Mass (Earth = 1)	14.531
Density (water = 1)	1.29
Day (rotation period)	17.24 hours
Equatorial gravity (Earth = 1)	0.89
Mean distance from Sun (millions of miles)	1,784.03
Year (time to orbit Sun)	84.01 days
Orbital velocity	4.23 miles/s
Inclination of equator	97.86°

Uranus, another of the gas giants, was the first major planet to be discovered in relatively modern historical times; the others – Mercury to Saturn – have been known since earliest antiquity. Uranus was discovered in 1781 by William Herschel, probably the best visual observer in the history of astronomy.

Since then, knowledge about the planet has increased by leaps and bounds, especially with the evidence from *Voyager 2*. Calculations show that Uranus probably has an iron silicate core about 9,010 miles in diameter; its core is, therefore, a little larger than the entire Earth. Outside the core is a mantle, just over 6,215 miles thick, which is thought to be composed of water ice along with ammonia and methane, also in an icy, or possibly liquid, state.

Beyond this is a 5,600-mile-deep layer of molecules of hydrogen, helium and, perhaps, some methane. Methane certainly forms the planet's thick atmosphere, which is what we observe from Earth, and which was photographed by *Voyager 2* when it approached the south pole of the planet in 1986. Clearly, the structure of Uranus differs from that of Jupiter and Saturn, as does its internal temperature of about 7,000 K. Despite the fact that it emits slightly more heat – 0.1 percent – than it receives from the Sun, the core of Uranus appears to have few elements to give it an internal source of heat.

The pressure at the center of the planet's core, although equivalent to 20 million times the Earth's atmospheric pressure, is not high enough to liquefy hydrogen (which would make it electrically conductive). Although its mass is just under 5 percent of Jupiter's, Uranus has about the same average density as Jupiter, about 1.3 times that of water, and is, therefore, denser than Saturn.

The visible surface of Uranus rotates in slightly less than eight hours, which is fairly slow compared with Jupiter and Saturn. But Uranus' axis of rotation is unique, because it is tilted over so that it is almost in the plane of the planet's orbit. The north pole is, in fact, just below the plane of the orbit, which means that Uranus rotates on its axis in a retrograde, or backward, direction compared with the other planets. This is thought to have been caused by a collision which Uranus suffered with another substantial body early in its history.

In spite of having no electrically-conducting liquid hydrogen in its interior, Uranus does possess a magnetic field, most probably generated in the mantle. The field is dipolar and inclined at 60 degrees to the planet's axis of rotation, the most inclined magnetic field of any major planet in the solar system.

The upper cloud layer of Uranus is banded and rotates in the same retrograde (east to west) direction as the solid body of the planet, but faster. In fact, the nearer the clouds are to the poles the faster they rotate; this is opposite to what happens on Jupiter and Saturn.

The clouds in the upper atmosphere of Uranus are similar to those on the two larger gas giants. Unfortunately, details of the planet's weather system have so far been hidden, even from *Voyager 2*, because a great deal of the upper

Uranian atmosphere is shrouded by haze caused by sunlight acting on acetylene and ethane present there.

Uranus has a ring system, discovered in March 1977, even before the arrival of *Voyager 2*. When Uranus passed in front of a ninth-magnitude star, Earth-based observers noticed that the star "winked," or suffered very brief eclipses, five times before being blotted out by the disk of Uranus. From this evidence, it was deduced that Uranus is surrounded by a ring system, and this was confirmed by *Voyager 2*.

Uranus' rings show some similarities to Saturn's, although they are much less spectacular. We know now that there are eleven rings – not five as was assumed in 1977. They lie between 26,100 and 32,315 miles from the center of Uranus, which means they are 9,950 to 16,150 miles above its cloudy surface. The rings are extremely narrow, with most having an average width of less than 6 miles.

Some parts of them vary quite noticeably from this figure: the outermost, or Epsilon, ring is between 12.5 to 62 miles thick. This ring also has an elliptical shape, bringing it 500 miles closer to the planet at one point. The majority of the remaining rings are also elliptical; the rest are circular. Only the elliptical rings vary in width, but all of them, elliptical and circular, are very thin.

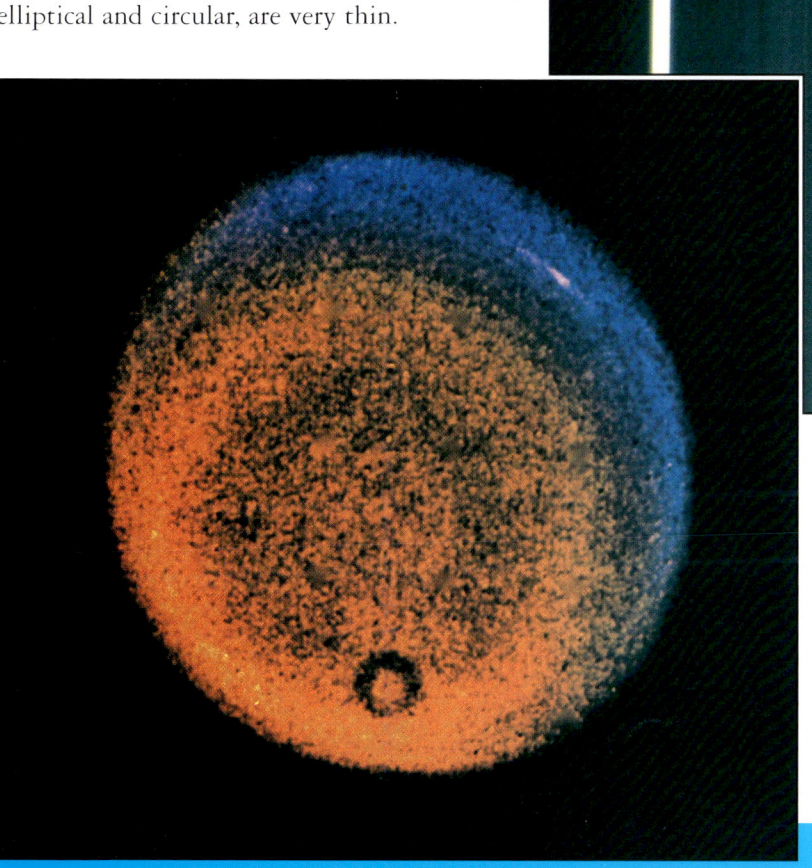

A methane haze shrouds the planet (left), obscuring the banded cloud systems. But the false colors of this computerized image enhance details such as the high-altitude cloud at top right.

The rings of Uranus (above) are made up of particles in two distinct size ranges. Many are between inches and yards across, but most are no more than a few thousandths of an inch across.

NEPTUNE AND PLUTO
● *Frontiers of the solar system*

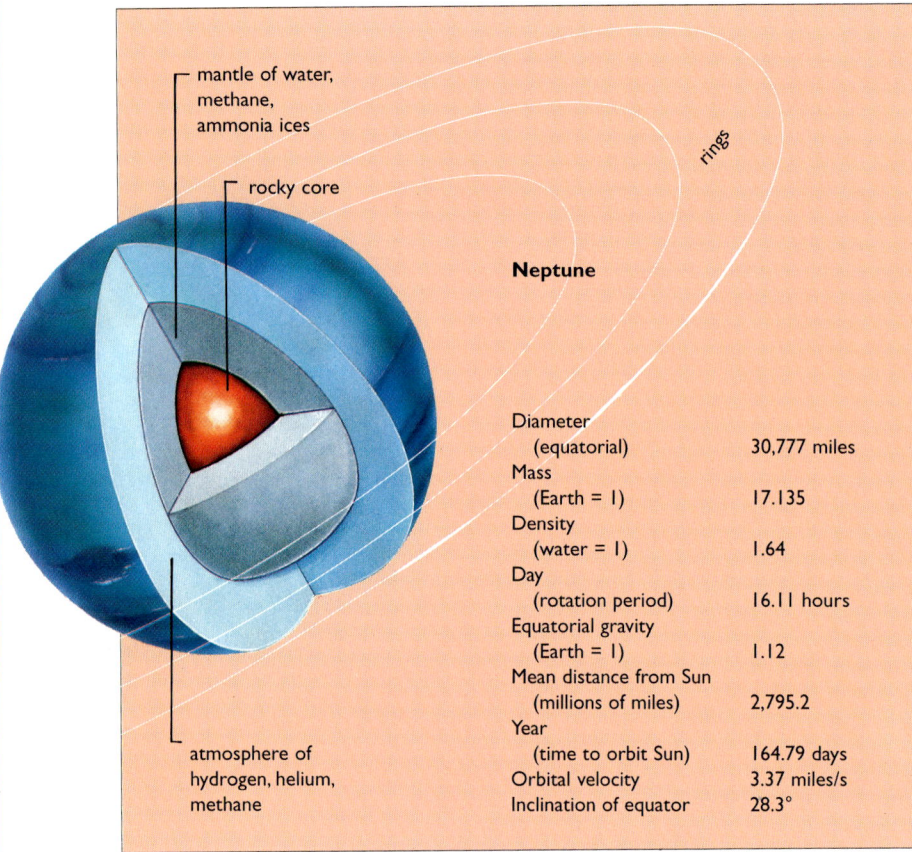

Neptune

Diameter (equatorial)	30,777 miles
Mass (Earth = 1)	17.135
Density (water = 1)	1.64
Day (rotation period)	16.11 hours
Equatorial gravity (Earth = 1)	1.12
Mean distance from Sun (millions of miles)	2,795.2
Year (time to orbit Sun)	164.79 days
Orbital velocity	3.37 miles/s
Inclination of equator	28.3°

The existence of Neptune, the last known gas giant out from the Sun, was predicted in the 19th century by two astronomers: John Couch Adams in England and Urbain Jean Joseph Leverrier in France. Both had been stimulated to search for an eighth planet by their attempts to account for apparent irregularities in the orbit of Uranus.

Like all astronomers of the time, these scientists assumed that Newton's law of universal gravitation was precisely correct and, using Newton's formulae, both independently concluded that the cause of Uranus' orbital irregularities must be the gravitational attraction of another planet beyond its orbit. Proving their case was no easy task: to discover where such a planet might lie in the sky they had to assume that it had a certain mass and orbit and then work out the consequences. A thorough knowledge of solar system dynamics helped reduce the possible choices, but to most astronomers the challenge was daunting.

In 1843, however, Adams cracked the problem, in his first research project after obtaining his first degree. At the age of 24, he was considered too young, as well as too inexperienced, and therefore failed to have his result taken seriously. And at that time, no observatory in Britain possessed an up-to-date chart of the region of the sky in which Adams' calculations showed that the planet might be. Without such a chart, checking would have been a long and tedious task.

Leverrier, who was eight years older than Adams, met with less resistance from colleagues when he announced similar findings in 1846. Fortunately, the Berlin Observatory had recently completed a new chart of the correct region of the heavens, and the director, Johann Galle, set a search in motion. Neptune was first observed on September 23, 1846, less than one degree from its predicted position.

Orbiting at an average distance of 2,794 million miles from the Sun, Neptune is not only the most distant known gas giant of the solar system but also very difficult to observe from Earth. Our knowledge of the planet has been revolutionized by information sent back by *Voyager 2* in 1989.

Before *Voyager 2*'s mission, astronomers had based their calculations concerning the structure of Neptune's inner regions on the planet's known mass, size, and position. From these they deduced correctly that, like other gas giants, it probably has a central rocky core made of iron, probably with a mixture of silicates. This core is similar to that of Uranus.

Outside the core, Neptune's mantle is thought to differ from that of Uranus, being made up of ionized molecules of water and ammonia and probably hydroxyl. In fact, it is sometimes described as an "ion ocean." Above the mantle is an envelope of helium, hydrogen, and methane similar to that of Uranus.

Like other gas giants, Neptune emits more energy than it receives from the Sun, by a factor of 2.8. This is much greater than the energy emission of Uranus, which is only 0.1 percent greater than the energy received.

Earth-bound observations also led astronomers to think that Neptune possessed a magnetic field – a theory confirmed by measurements taken by Voyager 2. However, the field proved to be weaker than those of the

Blue, cold, and remote, *Neptune is the outermost of the known gas giants. This Voyager image shows the blue cloud surface and the white wispy clouds of methane above it. The high clouds rotate with the planet, rather than with the faster-moving, deeper-lying clouds.*

NEPTUNE AND PLUTO

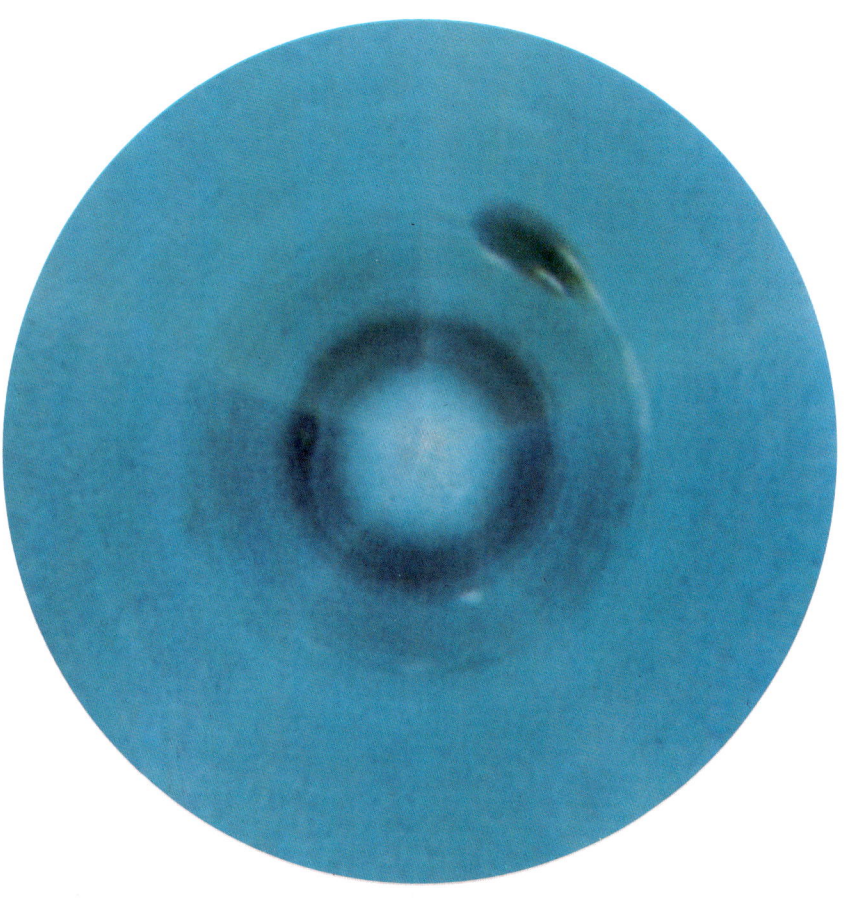

White clouds of methane crystals are striking features of Neptune, visible in this *Voyager 2* image. They lie high above the blue cloud layers and move independently of them. One, called "The Scooter," moves so fast that it catches up and overtakes the Great Dark Spot every few Neptunian days.

other three gas giants, and is inclined at an angle of 50 degrees to Neptune's axis of rotation. Surprisingly, the field does not pass through the planet's center, but lies about 6,200 miles to one side. It is thought, therefore, that the electric currents within the planet, which produce the field and its associated magnetosphere, must exist fairly close to the surface.

Because of the nature of Neptune's magnetic field, the point at which the magnetosphere meets the solar wind is more than 500,000 miles from the planet – about twice as far as that of Uranus. Neptune also transmits at some radio wavelengths, and experiences auroral displays over its equatorial regions.

Like the other gas giants, Neptune has its own ring system. It has at least four rings, one of which is rather dim, and all are considerably less substantial than those of Jupiter, Saturn, and Uranus. Before the *Voyager 2* flyover, it was thought that the rings might be in the form of fractured arcs of matter, but they are, in fact, complete. They do, however, have arcs which are brighter than other parts. These are probably due to uneven distribution and clumping of the particles of which they are composed. Some ring particles measure a few miles across, and are termed "moonlets."

The dimmer sections of Neptune's rings are of such low brightness that *Voyager 2* was only just able to detect them. A single "shepherd" satellite appears to be associated with each ring, the two brighter rings being linked with two of Neptune's newly-discovered satellites. Because the material in the rings is widely spread out, it occupies only about a tenth of the ring space. However, a disk of tiny particles accompanies the rings; this is spread throughout the entire space the rings occupy.

Of all the information transmitted back by *Voyager 2*, that concerning Neptune's cloud system was the most dramatic. The planet's cloud surface, colored a beautiful blue, is swept by strong winds, blowing in some latitudes at speeds up to 100 feet a second. *Voyager 2* recorded a dark blue spot, named the Great Dark Spot (GDS), which moved more slowly than the clouds surrounding it in the southern hemisphere.

Neptune's GDS measured 8,700 miles long and measures almost 4,143 miles in the north-south direction. The entire Earth would fit inside the east-west extension. The spot was seen by *Voyager 2* in 1989 but when the Hubble Space Telescope looked at Neptune in 1994 it had disappeared.

Suspended about 30 miles above the blue surface clouds are white, wispy, cirrus-like clouds of methane. These do not take part in the fast relative motion of the cloud layer below. Instead, like clouds above some high mountains on Earth, they seem to hang still, moving in step with the planet's rotation.

Neptune's atmosphere extends above the methane clouds. It is composed mainly of hydrogen, which makes up 85 percent of the total, with 13 percent helium and 2 percent methane. Under the influence of ultraviolet radiation from the Sun, the methane undergoes an unusual cycle of chemical change. The ultraviolet splits the methane into carbon,

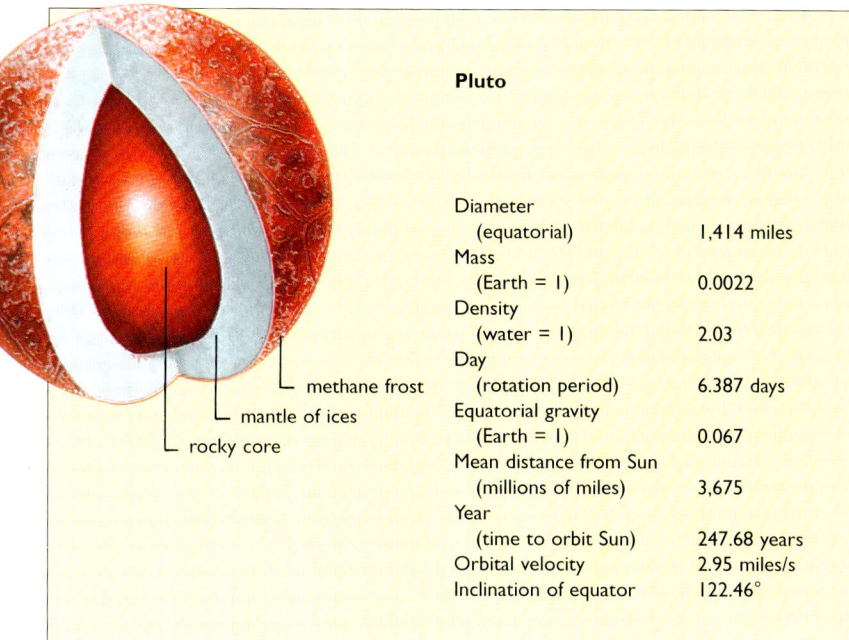

Pluto

Diameter (equatorial)	1,414 miles
Mass (Earth = 1)	0.0022
Density (water = 1)	2.03
Day (rotation period)	6.387 days
Equatorial gravity (Earth = 1)	0.067
Mean distance from Sun (millions of miles)	3,675
Year (time to orbit Sun)	247.68 years
Orbital velocity	2.95 miles/s
Inclination of equator	122.46°

hydrogen, and a mixture of hydrocarbons, including acetylene.

These hydrocarbons then sink down to cooler levels of the atmosphere where they condense into hydrocarbon ices. In turn, these ices sink still lower, to where the atmosphere is warmer. Here, they evaporate and recombine to form methane, which rises again.

Even after Neptune had been discovered, astronomers continued to observe irregularities in the motions of Uranus and of Neptune itself. So, following the example of Adams and Leverrier, Percival Lowell and William Pickering calculated the position of another planet, which they believed to be the cause of the disturbances. In 1905, a search for the planet began, but it was not until 1930, at the Lowell Observatory in Arizona, that a young astronomer, Clyde Tombaugh, photographed a tiny object about 5 degrees from the predicted position. The object possessed a planetary orbit and was named Pluto.

The planet Pluto is extremely difficult to observe from Earth because it is so distant and so small – it is only about 1,430 miles across, that is, about one and a half times the size of Triton, Neptune's largest satellite. Pluto's mass is equal to 0.22 percent of Earth's.

Pluto's orbit has been revealed as quite unlike that of any other planet in the solar system. It is extremely eccentric – considerably more so than Mercury – and although it sometimes orbits more than 1,740 million miles beyond Neptune for part of its circuit around the Sun, at other times it comes closer to the Sun than Neptune ever does. Its orbit is inclined at 17 degrees to the plane of the solar system, more than any major planet.

Its size, mass, and orbit all support the view that Pluto is not a major planet at all but a large asteroid, or even perhaps an escaped planetary satellite. Its gravitational attraction is much too small to disturb the orbits of Neptune and Uranus on its own, which means that these disturbances must have a different cause, possibly the influence of some other, currently undiscovered, planet.

Only a thin, hazy atmosphere has been detected around Pluto, but infrared observations with a spectroscope have shown that it is covered by ice. Much of this is frozen methane, turned a reddish color by sunlight, but there are also water and ammonia ices present.

Pluto has a satellite of its own. This was discovered in 1978 by James Christy at the Lowell Observatory. Named Charon, this satellite orbits Pluto at a distance of about 12,430 miles. The axial rotations of Pluto and Charon are synchronized, due to tidal forces between these closely-orbiting neighbors. Charon is too small to have retained any methane it may once have possessed, and is covered with water ice.

Pluto and Charon have been mapped by observing the satellite's occultations of the planet. From several years' observations of the varying brightness of the combined system, features such as polar caps could be mapped roughly.

SATELLITES
• *Companions of the planets*

Earth is not alone in possessing an orbiting satellite. This fact remained unknown, however, until 1610 when Galileo published observations he made using the recently-invented telescope. These included his observations of the four largest satellites of Jupiter. Since then, it has been discovered that satellites orbit every other major planet, except for Mercury and Venus.

It is no surprise that there are such satellites in a solar system that condensed out of the solar nebula (pp. 102–3). During the process a lot of debris, in the form of planetesimals, was left over, and some of this was captured by the gravitational fields of the major planets. The evidence for this comes from examination of the orbits of the various satellite systems.

As far as Earth is concerned, the Moon's orbit is inclined at 5 degrees to the ecliptic, which marks out the plane of the Earth's orbit around the Sun. But this is exceptional and the orbits of most of the larger members of other satellite systems lie closer to the plane of the equator of their parent planet.

As for the major planets, Mars has two tiny satellites that orbit in the planet's equatorial plane. The four largest satellites of Jupiter also orbit in its equatorial plane, as do its four nearest but tiny ones: Metis, Adrastea, Amalthea, and Thebe. Jupiter's outer satellites, however – which are also small and were, presumably, captured some time after the solar system formed – orbit at greater inclinations.

A group of four (Leda, Himalia, Lysithea, and Elara) orbit Jupiter at an inclination of about 27 degrees. A second group comprising the four outermost satellites (Ananke, Carme, Pasiphaë, and Sinope) orbit at an inclination of around 150 degrees, which means that they are in retrograde motion.

All of Saturn's satellites, with the exception of the outermost, Phoebe, orbit in the equatorial plane. Phoebe is thus probably a late addition since it moves in an eccentric orbit, and is about 8 million miles beyond the next nearest satellite, Iapetus – four times farther out. Uranus is similar in that every one of its 18 satellites orbits in the planet's equatorial plane.

Neptune's six inner satellites have equatorial orbits, but the two outermost do not. Of the latter, Triton, the larger with a diameter of 1,412 miles, has an orbit inclined at no less than 160 degrees, giving it a retrograde orbit. The other, Nereid, which orbits 16 times farther out, is inclined at about 27 degrees.

The number of satellites possessed by the major planets varies widely. The inner, terrestrial planets have few, with one for the Earth, two for Mars, but none for either Mercury or Venus. The gas giants are in a different league. Jupiter possesses no fewer than 16 satellites, Saturn 18, Uranus (at least) 18, and Neptune 8. This is not surprising since all four have much greater mass and, therefore, stronger gravitational fields and must be expected to have captured more planetesimal material.

The mass of the Earth is only 81 times that of the Moon, whereas the mass of Jupiter is more than 12,000 times greater than the most massive of its satellites, and Saturn is more than 4,000 times greater than its giant satellite, Titan.

In the case of Uranus, measurements show that it is 4,000 times greater in mass than its densest satellite, Oberon, and Neptune is 800 times more massive than its comparatively immense satellite Triton. Clearly, the Earth-Moon system is significantly unusual.

Jupiter's largest satellite, Europa, *is a little smaller than Earth's Moon. This image of it was captured by the probe Galileo, the first probe to orbit Jupiter and its moons. A liquid ocean may lie below Europa's smooth ice surface.*

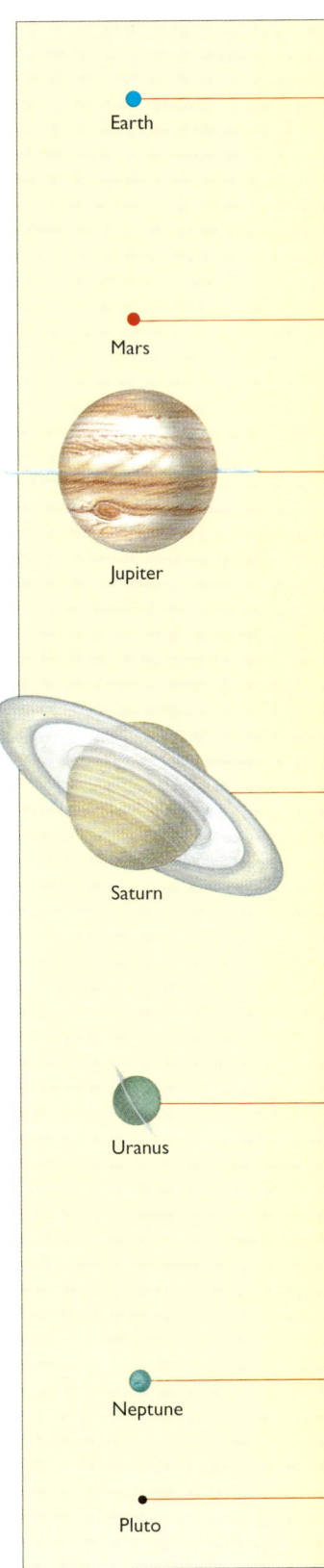

THE GRAND DESIGN

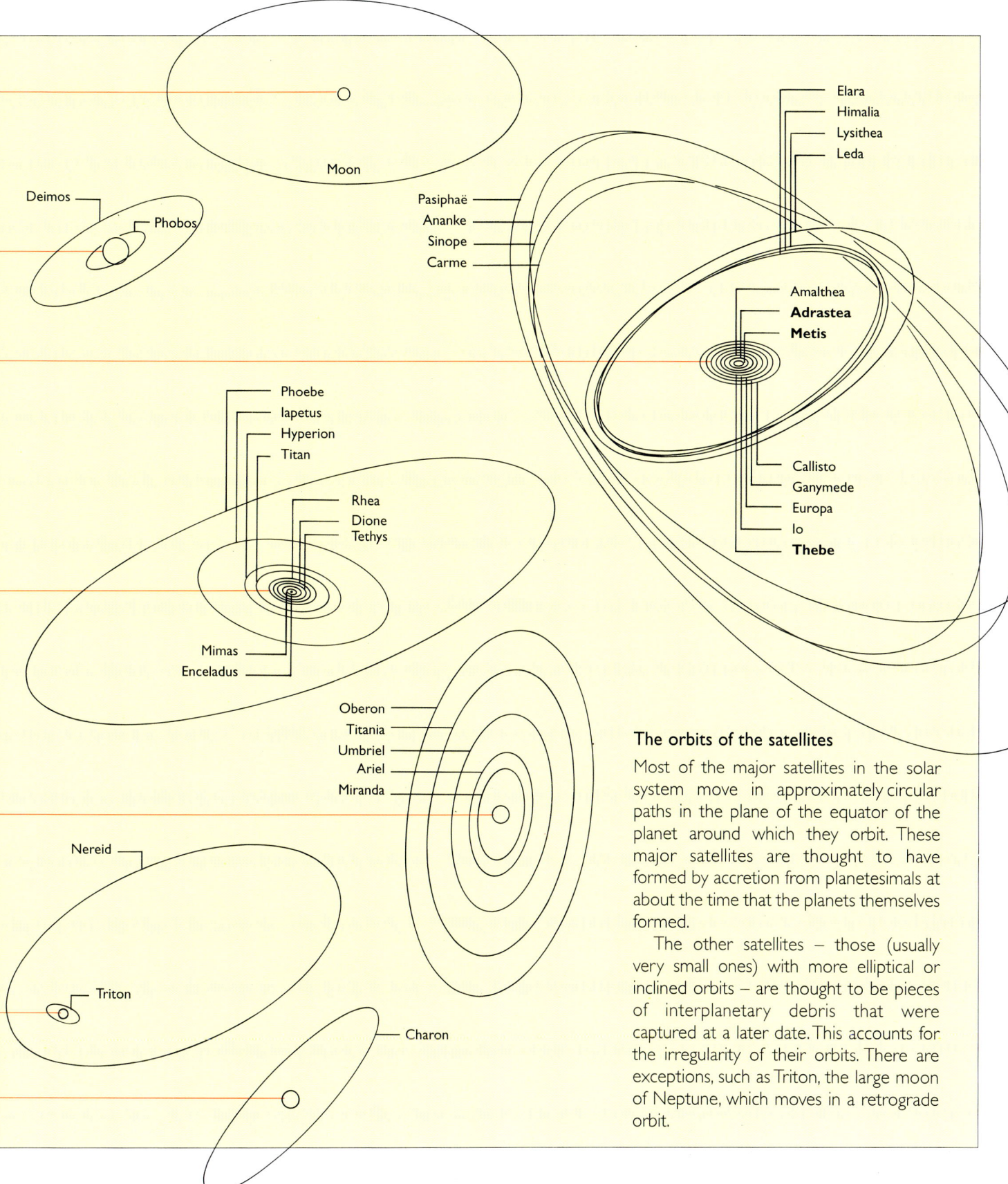

The orbits of the satellites

Most of the major satellites in the solar system move in approximately circular paths in the plane of the equator of the planet around which they orbit. These major satellites are thought to have formed by accretion from planetesimals at about the time that the planets themselves formed.

The other satellites – those (usually very small ones) with more elliptical or inclined orbits – are thought to be pieces of interplanetary debris that were captured at a later date. This accounts for the irregularity of their orbits. There are exceptions, such as Triton, the large moon of Neptune, which moves in a retrograde orbit.

The Earth's Moon shines by light reflected from the Sun. It displays phases, from a thin crescent to a full Moon, because different proportions of its illuminated side are presented to us at different times. Compared with other satellites, the Moon is large in relation to the Earth. Thus, the Moon and the Earth may be regarded as a binary planetary system.

You do not need a powerful telescope to observe the Moon, it is there for all to see. It has fascinated astronomers over the years and there have been various theories about the Moon's formation and its internal structure.

Earlier, in the 20th century, it was generally believed that the Moon broke off from the Earth when the Earth was rotating at high speed while still in a "plastic" state. However, opinion has now changed and today it is becoming accepted that the Moon was formed when a Mars-sized body crashed into the young Earth. It is therefore a planetesimal, but a large one, which has allied itself to the Earth to make a binary planetary system in which both components are in orbit around each other.

Shining in the sky by reflecting sunlight – a characteristic of all the solar system's planets and satellites – the Moon thus shows phases to Earth-based observers as it orbits our planet. The length of time for a complete cycle of phases, called a synodic month, is 29.5 days. But if we measure a month by the time taken for the Moon to complete a circuit across the stars in the night sky, it is only 27.3 days, because the motion of the Earth around the Sun also has to be taken into account.

Revealed even to the unaided eye during the changing phases is the Moon's mottled appearance; it displays large dark areas, once thought to be oceans and seas. These are still referred to by their Latin names of *oceanus* and *mare*. Yet we now know, from optical observations, data from orbiting spacecraft, and manned landings on the Moon, that these are not seas at all, but vast flat plains formed by outflows of molten lava from beneath the crust. These plains, which are often bounded by mountain ranges and pitted here and there with craters, also show geological faulting.

Although there are craters on the plains they are more numerous elsewhere on the lunar surface. These craters have an enormous range of sizes. Many large ones display a central peak, and the majority, at least, are generally agreed to be the result of bombardment of the surface by planetesimals.

The plains and the faulting, though, give evidence of geological change from within the Moon. More evidence comes from seismic equipment set up by astronauts who have landed on the Moon, and this has provided independent confirmation that the Moon is still geologically active. Together with results of analysis of samples of lunar rock brought back from the manned landings, it is now possible to obtain an insight into the Moon's internal structure, as well as its likely early history.

It appears that the Moon probably has a central core, perhaps 375 miles in diameter, with a partially-molten rocky outer core stretching about 220 miles above this. On top of this outer core lie a mantle and a crust – the lithosphere – together stretching upward about 665 miles. Deep moonquakes have been recorded in the molten (outer) core, and there have been others nearer to the surface.

Samples of lunar rock give an age of 4.5

billion years; this is more than the oldest known terrestrial rocks, which have an age of 3.8 billion years. The discrepancy may be due to still older terrestrial rocks having been destroyed during erosion, mountain building, and volcanic eruptions. The lunar age does fit in well with ages of meteorites (pp. 146-47), and goes back to around the time when the solar system was forming. Probably, then, the Earth and Moon were both born at about the same epoch.

Certainly, there is evidence that during the first few hundred million years of the Moon's history, the outer layers of the lunar surface were completely melted to a depth of several hundred miles. This was either because of violent bombardment by other planetesimals or because of heating due to the quick decay of the radioactive metal, aluminum 26.

While the Moon was cooling, continued bombardment by planetesimals, perhaps as large as 155 miles in diameter, created huge basins like the Mare Imbrium (Sea of Showers) and Mare Orientale (Eastern Sea). About four billion years ago, this prodigious bombardment slowed down, leaving highlands peppered with craters and a deep covering of broken rock. Heat produced below the crust by radioactive materials such as uranium then led to melting about 125 miles below the surface. This produced the upwelling of great lava flows over the surface for perhaps as long as 500 million years, and resulted in those vast dark plains, now so clearly visible.

After that active age, about 3.1 billion years ago, things quieted down and there has only been sporadic bombardment by much smaller material. Although the Moon is much more quiescent today, it is clear that seismic events and probably a little bombardment continue. Our satellite is not as dead as was once thought.

The lunar mountains in the Hadley-Apennine area were explored by the Apollo 15 mission. Covering about 14 miles, the Lunar Roving Vehicle was used to help collect rock samples. These showed a maximum age of about 4.5 billion years.

The Moon's changing face

When the Moon had formed it was initially molten, with no permanent surface features. But when it cooled, it began to retain the scars of its bombardment by debris, consisting of planetesimals up to 155 miles across. This left it, about 4 billion years ago, heavily cratered. After the era of bombardment, layers 125 miles below the crust melted, probably due to radioactive heating. The maria were formed when lava welled up and flowed over the lower areas of the Moon. This happened about 3.3 to 3.1 billion years ago, and obliterated many of the craters. Little has happened between the formation of the maria and the present, except for the occasional impact of interplanetary debris to form new craters, such as Copernicus, on the relatively smooth surface of the maria.

4 billion years ago

3.1 billion years ago

The present

Io is the most volcanically active body in the solar system. The volcanism is caused by the tidal effects of Jupiter's powerful gravity generating heat inside Io. The moon's colorful surface is made of

The discovery of the satellites of the other planets in our solar system had to wait until the invention of the telescope. More recently a lot of new information has come from the revelations provided by spacecraft such as Voyager and Galileo. The available evidence shows that the satellites of Jupiter, Saturn, Uranus, Neptune, and Mars vary widely in terms of shape, size, structure, and surface features.

Beginning with Jupiter, the four large satellites that were discovered at the start of the 17th century (Io, Europa, Ganymede, and Callisto) display some astounding features. Of these, Io, orbiting at an average distance of 262,000 miles from the center of the planet, is perhaps the most fascinating of all.

With a diameter of 2,263 miles – a little larger than our Moon – Io orbits in Jupiter's equatorial plane. This orbit carries it within Jupiter's magnetosphere, causing Io to generate a billion watts of electrical power from one side to the other. This is a greater output of electrical energy than can be generated by all the power stations in the entire United States. It is the reason why Earth-based radio observations of signals from Jupiter were found to be linked with Io's orbital motion.

At 1.77 days, the orbital period of Io is half that of the next satellite out, Europa, which orbits Jupiter every 3.55 days. This causes Europa to periodically modulate the tidal effects of Jupiter on Io, so that Io undergoes a regular series of strong gravitational effects, which squeeze it out of shape and, in doing so, heat up the satellite's interior. The outcome is that Io experiences a great deal of volcanic activity, as Voyager observed.

Io's surface is composed of sulphur, which at the very low temperature there – about 120 K (–153°C/–243°F) – should be white, not yellow as we are accustomed to seeing at Earth surface temperatures. However, volcanic activity and hot spots on Io's surface melt the sulphur, which then becomes orange or red as it flows over the exterior. When it cools and solidifies, the orange and red colors remain. This is why Io's surface is such a spectacular reddish color, and so smooth. It does, however, display some dark spots where sulphur has been heated to 300°C (572°F) or more, in eruptions.

As for Europa, it is a little smaller than our Moon and is unusual in displaying very few impact craters. Instead, it is covered by ice crisscrossed with veins. This veined appearance is caused by muddy ice that welled up and flooded the surface after a period of bombardment by meteorites (pp.146–47) about four billion years ago. Liquid water may still lurk under this frozen surface.

Farther out, at an average distance more than 2.5 times that of Io, is Jupiter's giant satellite Ganymede. Just a little larger than the planet Mercury, although less massive, it is probably composed of ice and silicates in approximately equal proportions. Its icy surface is pitted with craters and seems to be divided into two types of terrain.

The first is made up of dark patches with craters and large furrows; between these furrows, the second type of terrain, made up of light-colored streaks, predominates. The streaks contain depressions several miles wide and hundreds of miles long. The evidence suggests that Ganymede is an ancient satellite, with a surface once heavily bombarded, but which later iced over.

The other large Jovian satellite is Callisto. Almost as large as Ganymede, and of similar density, it has a highly-cratered surface. Of note is Callisto's so-called Valhalla basin, a circular area about 375 miles across. This feature displays about 15 concentric ridges and it appears to have been caused by the impact of a large meteorite or asteroid. Callisto's surface may even be even older than Ganymede's.

The satellites of Saturn also show many interesting features. For instance, the "shepherd satellites" have the effect of retaining ring material in place, rather than letting it spread out into space. But more remarkable are two other smaller satellites, Janus and Epimetheus, which orbit just outside the rings.

They travel in almost identical orbits, with only 30 miles separating their two paths. In consequence, their orbital speeds are similar, although one catches up with the other every four years. Small and oblong, they could have

Jupiter's second largest satellite, Callisto, has a prominent "bull's-eye" feature, believed to be a large impact basin. This image, taken by Voyager 1 *from a distance of 220 miles, shows features down to 4½ miles across. Callisto, composed of rock and ice, with a dirty ice surface on which the temperature is –153°C (–243°F), has more craters than any of the other large Jovian moons, leading astronomers to believe that it has the oldest surface.*

collided long ago, but are now prevented from doing so because, when they approach each other, mutual gravitational effects cause them to exchange orbits. The faster-moving one becomes the slower-moving one, and vice versa.

The shapes of these two satellites lead to the supposition that both once formed a single satellite, from which Epimetheus broke away. Perhaps they were originally like the next more distant satellite Mimas, which is a shade more substantial, having a body with a diameter of 247 miles. A satellite composed of ices, its cratered surface is dominated by one enormous crater, indicating that something extremely large once struck Mimas, and probably very nearly tore it apart.

Enceladus, Saturn's next most distant satellite, is still something of a mystery. Part of it is cratered, the other part smooth, and it also appears to have suffered severe bombardment. This seems to have ceased at least four billion years ago, after which something occurred to make material well up from inside the satellite, covering part of the surface which now looks like one large flat plain.

Still farther out are some larger satellites – Tethys, twice the diameter of Enceladus; Dione, which is a little larger at 700 miles; and Rhea, with a diameter of 950 miles. All of them are cratered, with Dione showing some strange wispy markings.

Probably the most intriguing of Saturn's satellite family is the giant Titan, which has a diameter of 3,200 miles, making it almost 1.5 times the size of our Moon. Titan is massive enough to retain an atmosphere, but this is an orange-colored smog thick enough to obscure the surface completely.

Infrared observations from *Voyager 2* show clearly that although the atmosphere is 90 percent nitrogen, methane is also present – as expected from earlier Earth-based observations. A small probe, named *Huygens*, will test the atmosphere as it descends to the moon's surface in 2004. *Huygens* will be released from the larger probe, *Cassini*, in November of that year. Instruments on board will help determine the density, temperature, and composition of the atmosphere. No one knows what *Huygens* will find once it has traveled through the smog to Titan's surface. There could be ocean-like lakes of liquid methane, or a dry, rock surface. Whatever the case, there are scientific instruments on board to give us the results.

Beyond Titan lies Iapetus, with a diameter of 892 miles; it is also heavily cratered. The outermost satellite is Phoebe, whose odd orbit suggests an origin separate from that of the other Saturnian satellites (p. 134).

Most of the 18 satellites of Uranus are small, not more than 100 miles in diameter, but the largest five – Miranda (293 miles), Ariel (720 miles), Umbriel (726 miles), Titania (981 miles), and Oberon (946 miles) – are all of interest. Pictures of all five taken by *Voyager 2* show that they are made up of ice and rock; the rock, being heavier, has sunk to the interior. All display a series of darkish cratered surfaces.

Detailed pictures of the smallest of them, Miranda, revealed an intriguing surface. Mostly cratered, it possesses two almost rectangular areas which look as if they have been smoothed by the flat side of a giant knife. One of these is covered with lines, while the other displays a chevron of lighter material.

Planetary geologists suggest that Miranda, the closest of the larger satellites of Uranus, was once broken up, so that for a time it turned into a number of boulders in orbit around Uranus; later, these clumped together to reform into a satellite. If this interpretation is correct, it can provide a clue to the reason why the axis of rotation of Uranus is so unusual (pp. 128–29). It could have been that large lumps of material once fell toward Uranus, material which not only tilted Uranus over, but which was large and fast-moving enough to also disrupt Miranda.

Neptune has only one large satellite, Triton,

Uranus' satellite Miranda, seen here in a composite image from Voyager, is 310 miles across, and is the smallest and closest of the planet's major moons. Its unusual surface has areas that are predominantly cratered and others with fractures, grooves, and fewer craters. The grooves and troughs are a few miles deep.

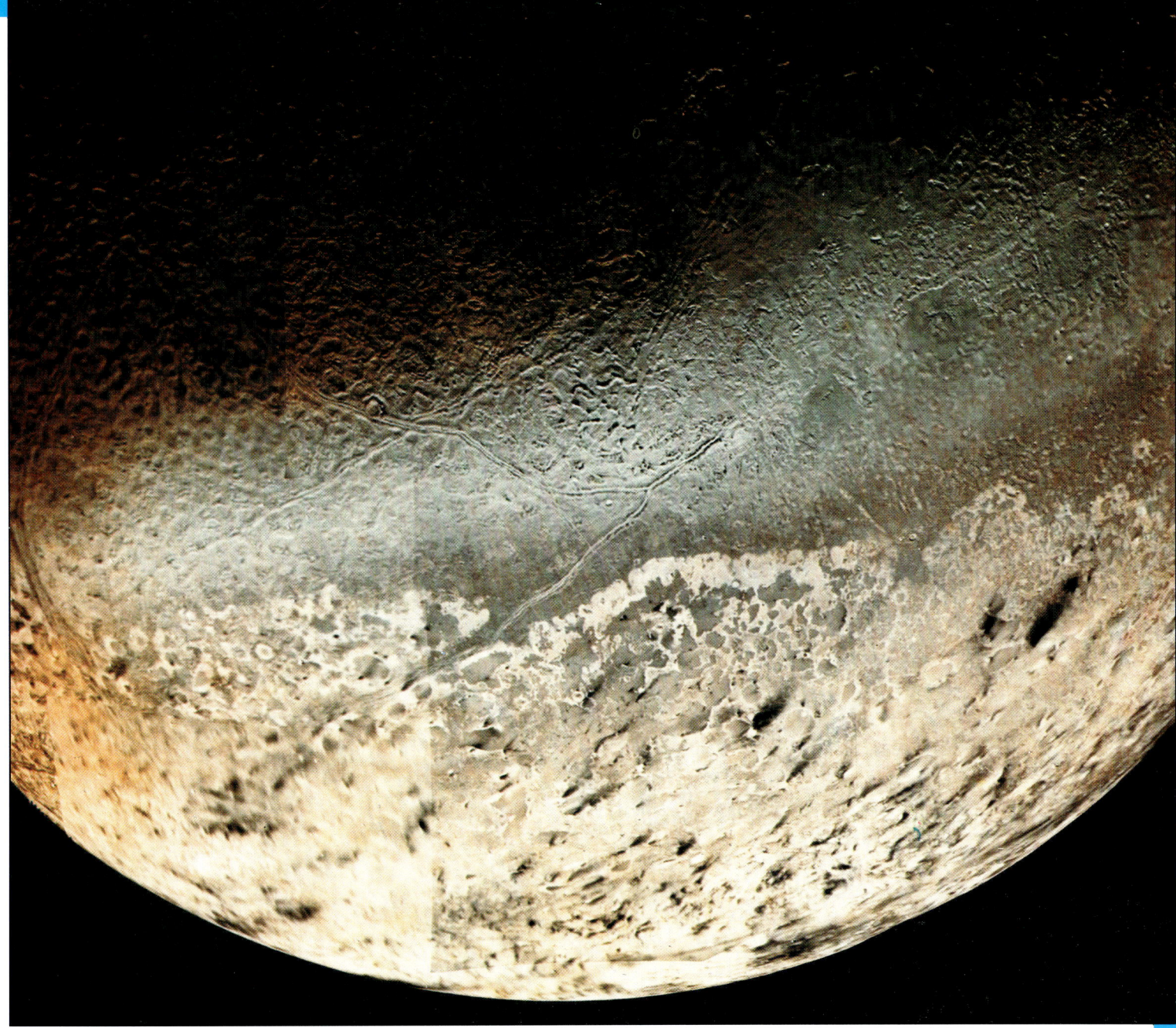

with a diameter of 1,348 miles. This is an extremely cold place, with a temperature of no more than 37 K. Even liquid nitrogen is warmer than Triton's surface. Study reveals that the surface is covered with water ice over which lie methane and nitrogen ices. The nitrogen ice forms a transparent coating about 3.3 feet thick.

The poles of Triton are inclined at almost 160 degrees to Neptune's equator, and since it orbits once every 5.9 days, one of Triton's poles will point toward the Sun for half of every Neptunian year, that is, for 82.4 terrestrial years, followed by the other for a similar period. As a result, night on Triton lasts for about a human lifetime, and during this time the ice builds up, especially around the satellite's poles, up to about 5 feet.

Coming back closer to the Sun, the two tiny potato-shaped moons orbiting Mars are both very small, with Deimos only 9½ miles and Phobos 16½ miles long. Deimos takes 1.26 days to orbit; Phobos does so in just 7 hours 39 minutes, and thus rises and sets three times a day. Very pitted from bombardment by meteors, these are not moons, but are more like small captured asteroids.

Clearly, the range and diversity of the many satellites in the solar system point to several different processes of formation and of capture by the parent planets.

The south polar cap of Neptune's large satellite Triton is highly reflective and pinkish in color. Astronomers think it is made up of nitrogen ice deposited during the very long winter. Away from the cap, the surface becomes redder and darker. This coloring may be due to the action of ultraviolet radiation and charged cosmic-ray particles, affecting methane in the atmosphere and on the surface.

ASTEROIDS
• *The minor planets*

Asteroids, sometimes referred to as "minor planets," are small pieces of planetesimal material orbiting the Sun. The first was observed in 1801, but astronomers had begun hunting for them 16 years before this, following the discovery of Johann Titius of a numerical relationship between the distances of the planets from the Sun.

Titius found that if he wrote down the numbers 0, 3, 6, 12, 24, 48, 96, and 192 (each one double the last), and then added 4 to each, the number series he ended up with was 4, 7, 10, 16, 28, 52, 100, and 196. He realized that if he then assumed the distance from Sun to Earth to be 10 units, the number 4 represented the actual distance from the Sun to Mercury, 7 that from the Sun to Venus, and so on. The distance from the Sun to Mars was 16 units on this scale, 52 represented the Sun to Jupiter, 100 the Sun to Saturn. The numbers 28 and 196 were spare.

Titius published this relationship only as a footnote to a German translation of a French book on science, but it was rescued from there by Johann Bode, a young astronomer who drew wide attention to it. It became known as Bode's law, or the Titius-Bode law.

When William Herschel discovered Uranus in 1781, it was found that its distance matched the number 196 in Bode's law. Before this, Bode had suggested there should be a planet between Mars and Jupiter, with a distance expressed by the number 28; after Herschel's discovery his suggestion was taken seriously.

In 1785, therefore, the Hungarian Baron Xavier von Zach began to search for the missing planet. Despite dogged persistence he had no success for the next 15 years, and even when he organized a cooperative effort in 1800, the "missing" planet could not be found. It was, in fact, a chance observation in 1801 by Giuseppe Piazzi, who was busy making a star catalogue, that detected a planet at the correct distance. It was named Ceres, for the goddess of Piazzi's native Sicily.

Once Ceres had been observed, other dim planets were also found orbiting between Mars and Jupiter. Each appeared as a starlike point, being too small to show a detectable disk, and in 1802 the name asteroid was coined to describe them. Over ten thousand are now known, but only 33 of them are larger than 125 miles in diameter, and not all orbit between Mars and Jupiter.

Those that do have such orbits make up the asteroid belt. There are asteroids everywhere within the belt with the exception of some gaps called Kirkwood gaps. These gaps are caused by the gravitational effects of Jupiter, which is now thought to have been responsible for the fact that no major planet coalesced from planetesimals in the gap between Mars and Jupiter. Asteroids are pushed out of the gaps and leave the belt, due to Jupiter's gravitational pull.

Calculation shows that only in orbits well separated from the major planets can planetesimal debris collect into reasonably stable orbits. The gap between Mars and Jupiter is not the only one in the solar system. Theory predicts that there is another gap inside Mercury's orbit; whether asteroids ever collected there in the early days of the solar system is uncertain.

Some asteroids have been so influenced by Jupiter that they have been pulled into the same orbit around the Sun. They have not become Jovian satellites, circulating around the planet itself, but orbit the Sun at Bode's distance, 52. These asteroids are gathered into two groups and are called the Trojans. Each

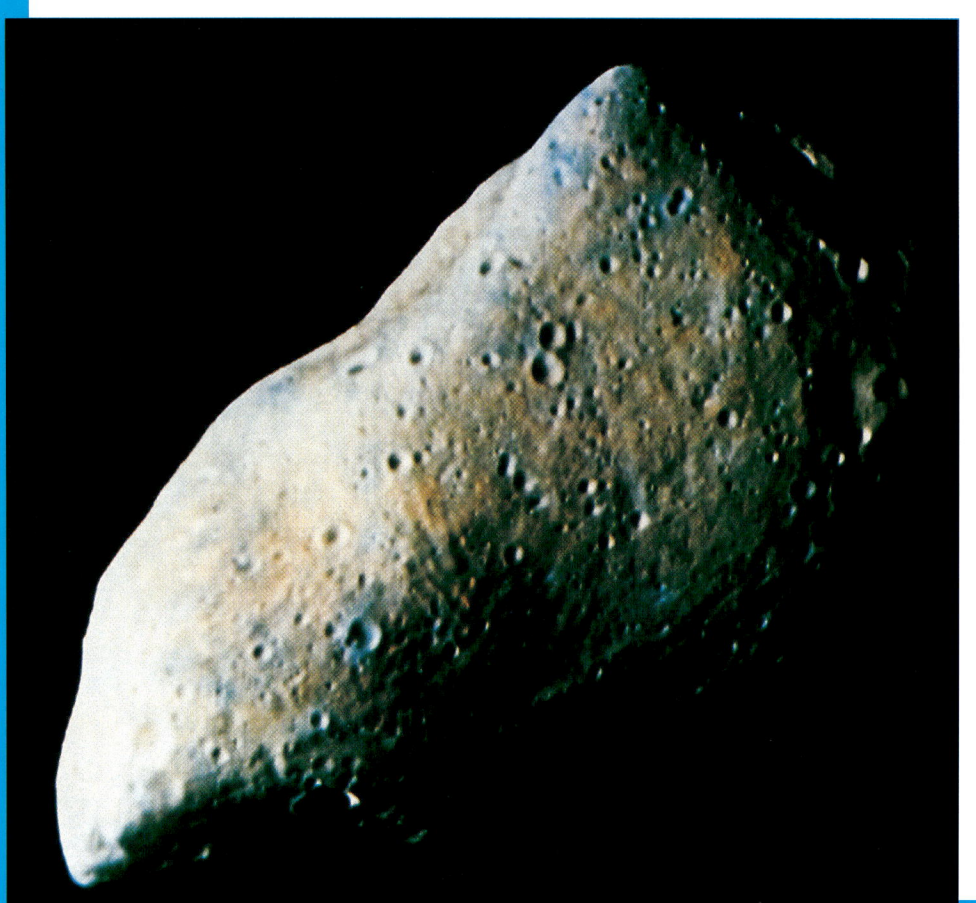

Gaspra (below) was the first asteroid to be seen in close-up. It was imaged by the Galileo space probe in October 1991 as it journeyed to Jupiter. Gaspra is a rock asteroid, about 12 miles long, and orbits the Sun every 3.3 years.

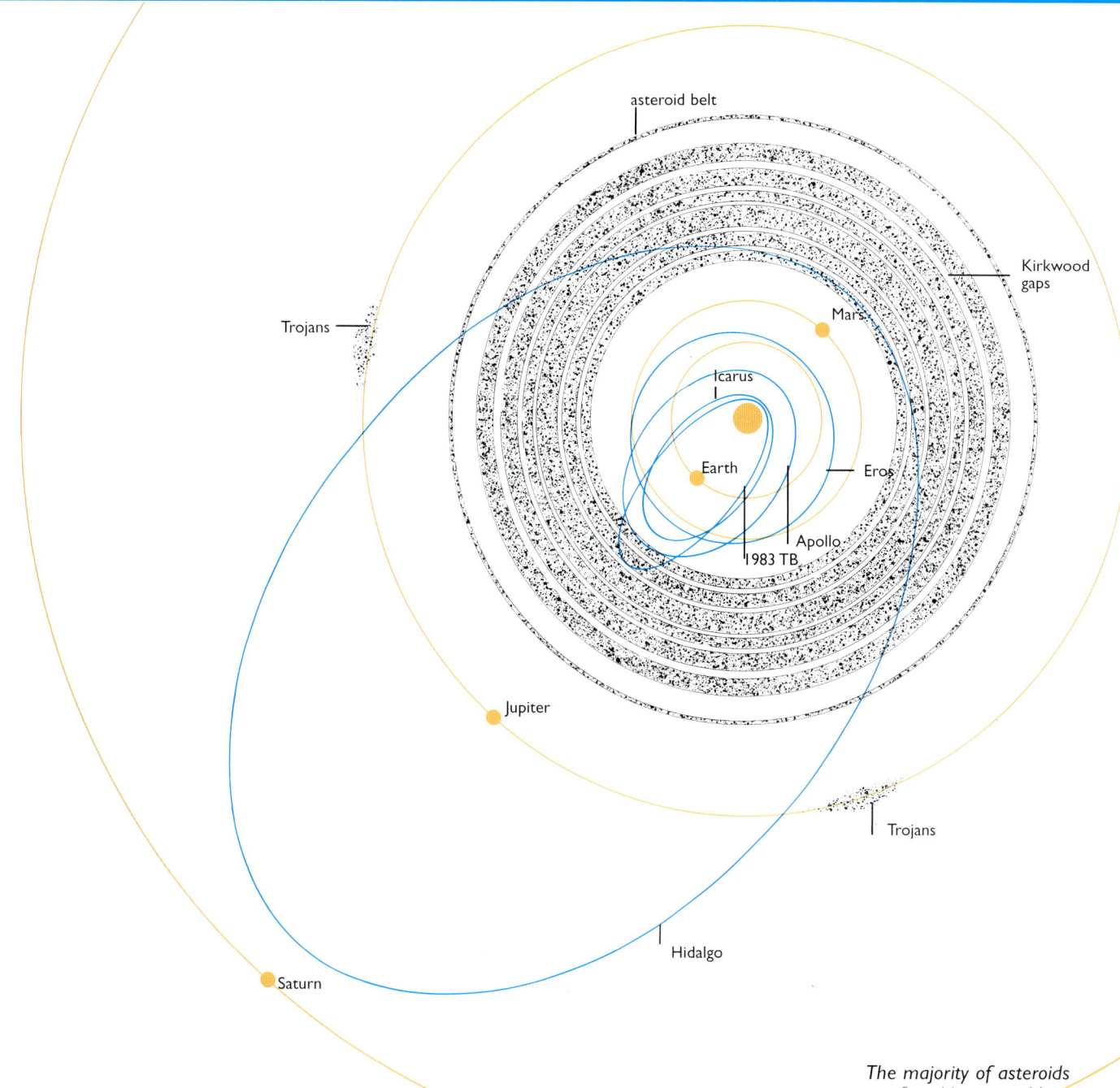

The majority of asteroids are found between Mars and Jupiter. Their orbits are grouped into bands separated by so-called Kirkwood gaps, caused by the influence of Jupiter's gravitational field. Some asteroids, however, have eccentric orbits that cross those of the inner and outer planets. Icarus, for instance, comes close to the Sun, while Hidalgo's orbit nearly reaches out to that of Saturn. The Trojans are two groups of asteroids that orbit 60 degrees ahead of and behind Jupiter.

group is separated from Jupiter, and imaginary lines drawn between Jupiter, a group, and the Sun form an equilateral triangle in space. One group lies ahead of Jupiter and one behind.

About 5 percent of the asteroids have very eccentric orbits, which cross the orbits of Earth, Mars, Jupiter, or Saturn. Other minor members of the solar system have orbits that take them further from the Sun, as far as Neptune. But these are comet-like objects, possibly from the Kuiper Belt.

Asteroids are pock-marked rocky lumps, the great majority of which are potato-shaped. Only a handful have been seen in close-up. The first was Gaspra, encountered by the *Galileo* probe as it traveled to Jupiter in October 1991. The only craft designed solely to investigate an asteroid was NEAR (Near Earth Asteroid Rendezvous). It was the first to go in orbit around an asteroid when it started orbiting Eros in February 2000.

COMETS
● *Visitors from deep space*

Cometary orbits around the Sun are often exaggerated ellipses. Comets develop tails only when they get so close to the Sun that material is blown or boiled off them. Each comet has two tails, one consisting of dust and the other of plasma, or ionized gas. The dust tail is usually yellow, because it reflects sunlight. The ionized gas tail is often bluish. Tails reach their maximum brightness at the comet's perihelion, the point of closest approach to the Sun. At aphelion (when the comet is farthest from the Sun), comets are merely lifeless lumps of dust and snow.

One of the most astonishing sights in the night sky is a large comet, with its bright head and long glowing tail. In ages when astrology held sway and celestial events were thought to presage events to come, comets were seen as signs predicting disaster; no wonder their occasional appearances struck terror into humankind. Not until the late 16th and 17th centuries were their paths studied with astronomical precision.

We now know that trillions of comets live at the outer edge of the solar system. Together they make up the Oort cloud. Each comet follows its own orbit, and these orbits can be so long that the most distant comets take ten million years to complete their paths around the Sun. Comets are visible when they leave the cloud and move on to orbits which bring them close to the Sun. Of these, about 135 are short-period comets, they orbit the Sun in periods of less than 200 years and, therefore, regularly appear in Earth's sky. Independent confirmation of the elliptical orbits of comets came in 1758 when the bright comet of 1680 returned as predicted by Edmond Halley, whose calculations had been based on Newtonian theories.

Halley's Comet, as it has come to be called, has made many appearances since. On its last passage close to the Sun in 1986, it was investigated by a number of space probes, one of which penetrated the comet itself.

Comets are composed of material that formed within the solar nebula. Each is an icy conglomerate of frozen gases and dust, just like a "dirty snowball." Halley's Comet is no exception. It is also probably fairly typical of other comets in that it has an elongated central nucleus — with the asteroidal potato shape — and is no more than $9^{1}/_{2}$ miles long, 5 miles wide, and 5 miles thick. Its surface is dark and the whole object rotates once every 52 hours. This is a slow rotation and when the comet is close to the Sun, the sunward side becomes strongly heated and dust and snow evaporate from the nucleus. This happens to every comet as it approaches perihelion (the closest point in its orbit to the Sun).

The darkness of the Halley nucleus — it is darker than coal — was unexpected. Moreover, material does not evaporate from it as rapidly as would be expected if it were just an icy conglomerate. The comet loses material at a

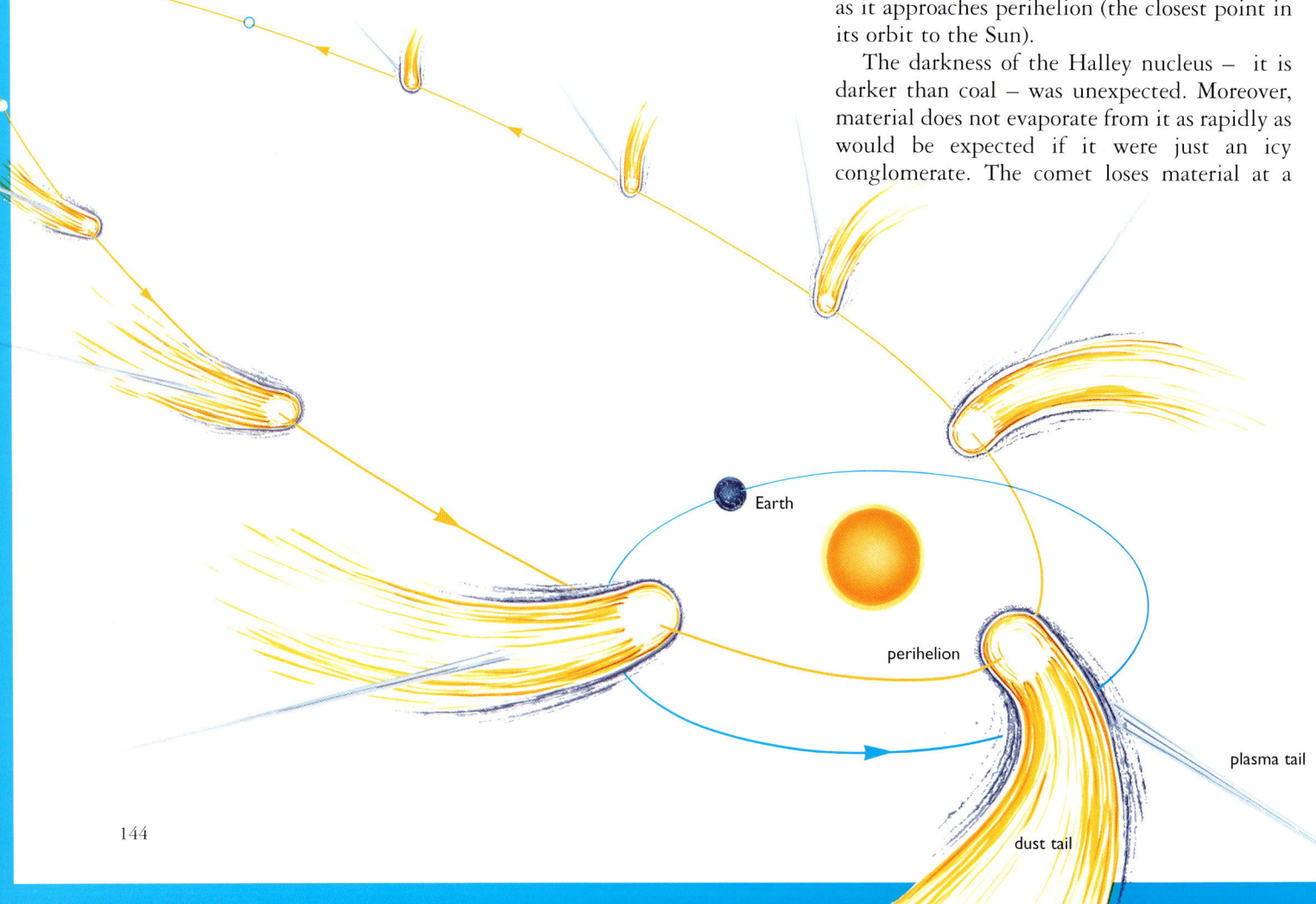

144

The two tails of Comet Hale-Bopp are distinct in this image. The bluer, narrower tail is made of gas pushed away from the Sun by charged particles. The broader, whiter tale (below) is made of tiny dust particles pushed by the pressure of the Sun's light. Hale-Bopp was discovered in 1995 as it journeyed into the solar system. During 1997 it was visible to the naked eye.

rate of only 16.5 tons per second, whereas ice would evaporate from 10 to 100 times faster.

From this evidence, astronomers have concluded that the nucleus is covered with a mantle of dark porous material several inches thick. Such material might be too heavy to blow off from the comet, or at least too heavy to reach escape velocity, so that it drops back before it gets very far. In fact, only some parts of the surface of the nucleus poured material into space; other parts remained inert.

Nevertheless, Halley's Comet, like every other, loses some of its substance every time it passes perihelion, and some estimates suggest that Halley's ejected about 330 million tons at its 1986 apparition. This still leaves the 11 billion or more tons of the nucleus with enough material to sustain many more reappearances.

The materials that compose the nucleus of Halley's Comet consist of the expected mineral elements such as carbon, calcium, iron, magnesium, oxygen, potassium, and silicon. Yet the spacecraft *Giotto* also showed that there were many light elements present, notably hydrogen and nitrogen. But the most prevalent material was a combination of both light and heavier elements to give organic molecules. The presence of these has led a few astronomers to suggest that Halley's Comet – and others – did not originate in the Oort cloud but came from interstellar space.

The tails of comets are generated by two mechanisms, both of which cause them to always point away from the Sun. In the first mechanism, the pressure of radiation from the Sun pushes the dust outward from the comet's body to form a tail. Dust particles from the comet form a large bright tail, which is usually slightly curved because the material first ejected lags behind material thrown out later as the comet increases speed when it moves closer and closer to perihelion. Such dust tails reflect sunlight and appear yellowish in color.

In the second mechanism, ionized gases emitted from the comet are driven away in a straight tail by the solar wind. As the principal emission of radiation is due to ionized carbon monoxide, such gas tails appear blue in color.

A cometary nucleus was seen for the first time in 1986. In March of that year, the Giotto space probe flew inside the coma of Halley's Comet and photographed the nucleus from 375 miles. The nucleus has a crust of dark dust. As it faces the Sun, the surface is heated and gas and dust are released.

METEORS
• Cosmic dust

Barringer Crater, Arizona, is 52,000 years old. The crater is well preserved because the climate in northeastern Arizona is extremely dry and there is therefore little erosion. The crater is 2,640 ft across and 660 ft deep. It is thought to have been made by the impact of an iron-rich meteorite, weighing thousands of tons, which largely vaporized because of the violence of the impact. The largest fragment that has been found weighs 1,400 lbs.

Falling or "shooting" stars are familiar to all civilizations but they are not stars in the true sense – they are meteors. Most meteors are, in fact, produced by bits left behind by a comet. During a comet's orbit around the Sun, it emits gas and dust and leaves a stream of material around its orbital path. As Earth moves along its orbit it cuts through such streams of material.

When this happens, the trail of mineral and rock dust and larger debris is swept into the Earth's atmosphere where it is burned up by friction with air molecules. As a result, the molecules become ionized and emit radiation. Thus, a short-lived trail of light – a meteor – is produced by each piece of incoming material.

When the Earth crosses a cometary orbit, the result is not just one or two meteors but often thousands, and such meteor "showers" can be spectacular events. The meteors all appear to come from one point in the sky called the radiant and the constellation in which this point happens to lie gives its name to the shower. Thus, there are the Perseids with their radiant point in Perseus, the Leonids with a radiant in Leo, and so on. However, radiants are no more than an effect of perspective; they merely indicate the direction in the sky from which the meteor shower comes.

Not all meteors are the debris of comets. There are also sporadic meteors, which come from all directions in space and are produced by dust and rocks, left over from the solar nebula, which are pursuing their own orbits around the Sun. During an average night, a visual observer can see about 10 per hour.

These bits of debris enter the terrestrial atmosphere at velocities of anything between about 7 and 46 miles per second. The trails they leave depend upon the size of the incoming material, which may be no more than a dust particle or could be a large lump of rock. The lengths of the trails vary between about $4^{1}/_{2}$ and $12^{1}/_{2}$ miles.

In rare cases, when an extremely large lump of rocky material enters the Earth's atmosphere, the trail can be sufficiently bright to light up the countryside, and the meteor looks just like a glowing ball of fire.

Most meteoric material is completely burned up in the Earth's atmosphere, but on occasions the ablation is incomplete, and a lump of rocky material actually lands on the ground. Examination of such meteorites provides a clue to the formation and composition of the asteroids from which they are probably derived.

Meteorites can be classified into three main types: metallic, lithosiderite, and chondrite. There is also another minor group, the achondrite. Metallic meteorites are composed mostly of iron and nickel. The proportion of nickel varies; some contain less than 6 percent nickel, while others have between 8 and 20 percent. The metals are crystallized, and show that the asteroids from which they came underwent either slow cooling or sudden solidification.

The lithosiderites are made up of metallic material and silicates, which overlap each other. Such meteorites can be further classified into two types: mesosiderites and pallasites. In mesosiderites, the silicates consist mainly of the minerals feldspar (an aluminum silicate) and pyroxenes (silicates which include iron, magnesium, and calcium). This indicates that the asteroids that mesosiderites came from received their metallic components after they had solidified. Pallasites contain a lot of olivine (a silicate of iron and magnesium). They were probably formed by the intrusion of a metallic liquid between the core and the mainly olivine mantle of the asteroid from which they came.

The chondrites, or stony meteorites, contain small spherical particles, or "chondrules," which give them their name. These contain iron, feldspars, olivine, and pyroxenes and are thus chemically similar to many terrestrial rocks. A subset of the chondrites, known as the carbonaceous chondrites, is a particular type composed of a mixture of carbon-rich crystals containing olivine, pyroxene, metals, glass and, uniquely, mica sheets.

Finally, there are the achondrites which, as their name implies, do not contain chondrules. They are very similar to lunar rocks.

Study of meteorites thus provides some indication as to the chemical nature of the solar nebula about 4.5 billion years ago. By that time, iron and other heavy atoms had been synthesized, presumably within those early stars that had become supernovae and therefore distributed their contents into interstellar space.

The Geminid meteor shower occurs in the second week of December every year. The short diagonal streaks in this time-exposure photograph are star trails, while the long one (at left) is a meteor. The Geminids peak at about 100 meteors per hour on the night of December 13.

THE LIVING UNIVERSE

From the moment of the Big Bang, around 13 thousand million years ago, the universe has evolved into the vast, expanding mass of galaxies we now observe, which stretches farther out into space than even the most sophisticated modern instruments can measure. In the immensity of the universe, our Earth is a mere speck of rock in orbit around a not very significant star. Yet the findings of modern science have led some astronomers and physicists to the view that the crowning glory of the whole cosmos is humankind.

In our solar system, only planet Earth supports life, and in this sense our planet certainly is unique. Only here, it seems, are conditions suitable for the evolution and existence of complex and intelligent life forms. Such life, based essentially on the versatile chemical attributes of the element carbon has, after millions of years, led to the astounding appearance of the self-reproducing, highly sophisticated, communicative species that is *Homo sapiens*.

The range of life existing on Earth is so vast and so complex that scientists have been led to speculate whether even the time span that has elapsed since the Big Bang has been sufficient to accommodate that evolution. An alternative explanation for the success of life on Earth might lie, for instance, in the arrival of complex molecules from outer space, which aided and speeded the evolutionary process. Painstaking study of meteorites and similar evidence now suggests that this is a definite possibility.

At the pinnacle of evolution, we human beings are endowed with powers that mark us out as special. Arguably, the most important of these is our ability to communicate, not merely with spoken alphabetic languages, but through the universal symbolism of mathematics. But are we humans really the only intelligent beings in the universe? If we are not, then communication with alien civilizations is desirable – and, indeed, possible. Yet even if this were achieved, the problems of dialogue seem, at present, insuperable; for example, the time taken for messages to span the immense distances involved would impose the most daunting of limitations.

In considering the universe, and our place in it, we must also speculate about the evolution of the universe as a whole. Is our view too prosaic and parochial? What are the new laws of physics we must learn in order to unravel the mysteries of the actual creation, existence, and future of the universe? And will these reveal that our ideas of the Big Bang and an expanding universe are, in fact, completely erroneous?

Evidence gleaned thus far suggests that our general picture seems to be correct. But the central role that human beings appear to play in the scenario raises some more enthralling questions, most particularly whether the ability of human beings to observe and understand the universe has an effect on its reality.

As the sciences of physics and astronomy become more sophisticated, they tell us more about the distant future of our universe. This may not be as simple or straightforward as was once thought. For example, black holes and the space they enclose present strange possibilities and provoke the speculation that the scientific outlook of today is still too limited in its concepts to allow full understanding of the universe.

The lunar landing module, **Eagle,** *ascends from the surface of the Moon after the Apollo II mission in July 1969. For the first time, the human race had established a foothold on a world beyond its own. In the view of some scientists, the appearance of intelligent life and its colonization of the universe was "designed into" the laws of the cosmos from the very beginning.*

EARTH
● Our living world

Here on Earth, our planetary spaceship, we humans live and evolve – along with a vast complex of vegetable and animal life, from microscopic bacteria to huge elephants, and from minute viruses to giant sequoia trees. Although the Earth possesses, like the other terrestrial planets, a dense core surrounded by a mantle overlaid with an outer crust, it has this unique difference – life.

During the 20th century, it became clear that life exists on Earth only because conditions are right. It is neither too hot nor too cold, and potentially deadly radiation is kept at bay by the terrestrial atmosphere. Moreover, the chemical and biochemical environments are in perfect balance, to ensure not only the existence of living things but also their proliferation. This benign equilibrium has evolved, in tandem with life itself, over geologically long periods of time. Unbalance or destroy it, and the rich variety of life risks being diminished or even threatened with extinction.

Critically, the prevailing temperatures on Earth allow water – a vital ingredient of life – to exist in liquid form. Earth's average surface temperature is 15°C (59°F), well above freezing, and about 33°C (91°F) higher than it would be if it were heated by solar radiation alone. The reason for this is that some heat radiated by the planet is trapped by the "blanket" of carbon dioxide and water vapor in the atmosphere and warms the sea and land below.

Yet if such warming increased, and continued unabated, this heating or greenhouse effect would run away with itself as it has on Venus (pp. 110–13). Consequently, the temperature would increase too much to allow life to continue. On Earth, it seems possible that this galloping effect has been prevented by modification of Earth's environment by its living inhabitants. This forms the crux of the Gaia hypothesis.

The hypothesis was proposed in the late 1960s by the British atmospheric scientist James Lovelock who, with the American biologist Lynn Margulis and other colleagues, had begun to study the Earth's natural system of checks and balances. Looking at the whole Earth as a living organism, much as the Chinese had done in past times, Lovelock and Margulis pointed out that, despite continuing climatic change, the Earth enjoys climatic stability at any particular time. Carbon dioxide and water vapor help warm the surface, but there is no runaway effect because both are recycled in various ways.

Such recycling takes many forms. Sunlight, for instance, enables green plants, through the process of photosynthesis, to use carbon dioxide and water to manufacture carbohydrates – substances such as sugars, starch, and cellulose. In exchange, oxygen is released as a waste product, and is used by animals for respiration.

When plants die and decay, the carbon in their tissues combines with atmospheric oxygen; carbon dioxide is released, therefore promoting the greenhouse effect. Some of the carbon is, however, removed permanently. It then becomes locked into the shells of marine creatures as insoluble calcium carbonate and finds its way into sediments on the seabed. This, the Gaia hypothesis argues, prevents excess build-up of carbon.

The Earth owes its warmth to the carbon dioxide in its atmosphere, which traps some of the energy of sunlight that would otherwise be reradiated into space. A balance is maintained between the crust and the atmosphere; volcanic activity releases carbon dioxide, while weathering and biological processes remove it from the air.

Mars is not volcanically active, so carbon remains in the rocks. Its carbon dioxide atmosphere is too thin to retain heat.

Venus is too close to the Sun for liquid water to exist, so there is no weathering to reduce the excess of carbon dioxide.

Photosynthesis by green plants produces the oxygen that makes Earth's atmosphere unique in the solar system.

Living things are also ultimately involved in the control of other cycles and in greenhouse effect regulation. For example, minute marine plants, or phytoplankton, are involved in the cycling of sulphur and iodine. Particles containing these elements are exuded by the plants and form the nuclei around which atmospheric water condenses to form clouds.

Clouds form as rising water vapor cools and condenses. And the higher the Earth's temperature, the greater the production of water vapor from the oceans that cover two-thirds of its surface. Being white, clouds reflect sunlight back into space, and therefore have a cooling effect. In due course, clouds bring rain, returning water to ground level.

Biological factors undoubtedly play a large part in environmental control on Earth. But the presence of humans on the planet – although we are part of the Earth's biology – introduces other factors. Two human-centered effects of current concern are the presence of "holes" in the ozone layer and global warming.

Ozone, a molecule composed of three oxygen atoms, is formed when solar ultraviolet light breaks down atmospheric oxygen. Most atmospheric ozone lies 8 to 15 miles above the Earth's surface. Crucially for life, ozone absorbs dangerous short-wavelength ultraviolet radiation which, if allowed to penetrate to the surface, causes deadly cancers in humans and other animals. It also threatens the existence of the phytoplankton, the plants crucial to element cycling. Additionally, food crops produce lower yields when exposed to ultraviolet radiation. Cooling of the upper atmosphere by ozone depletion is also likely to cause climatic change.

The destruction of ozone is caused mainly, it seems, by the commercial use of chlorofluorocarbons (CFCs), which are used in aerosols and as refrigerants. In cold air regions, for example over the Antarctic, the CFCs react with nitrous oxide in the atmosphere to form chlorine compounds. And one chlorine molecule can destroy 100,000 ozone molecules.

The use of CFCs is also thought to contribute to global warming – a long-term increase in the temperature of the Earth. Global warming appears to be

Carbon and water are recycled continually by natural processes. Water evaporates from the oceans and land, and is given off by the transpiration of plants. The water is released as rain, snow, or sleet when air rises and cools. This water eventually flows into the sea, to resume the cycle.

Carbon dioxide gas is absorbed from the atmosphere by land plants, and dissolved carbon dioxide is absorbed by water plants, to be utilized in photosynthesis. When the plants die, they decompose and return carbon dioxide to the atmosphere. Animals consume carbon in plants and release it in wastes or on death, when they decompose. Marine organisms form carbon-rich sediments when they die.

accelerating, due mainly to the widespread use of fossil fuels such as coal and oil. Burning these fuels releases carbon dioxide into the atmosphere, which in consequence traps more heat.

Unless carbon dioxide emissions are reduced, it is estimated that by the middle of the 21st century the average temperature of our planet will have increased by between 1.5° and 4.5°C (3 and 8°F). This would lead to significant melting of land-based ice, and a rise in sea level of about 3.3 feet due largely to an expansion of the warmer upper levels of the sea. Parts of New York, London, and Tokyo could be flooded, and small atolls such as the Maldives submerged totally.

The climatic changes accompanying such warming would have mixed effects. Parts of northern Canada and Scandinavia would become crop producers, while production would increase in central Australia and parts of Africa, China, India, and South America as they became wetter. On the debit side, the grain-producing areas of America and the former USSR would suffer droughts.

It remains to be seen whether the forces of Gaia will indeed "rescue" the planet from violent climatic change. What is irrefutable, however, is the fact that chemical and biological balance are crucial to the continuance of life on Earth. If chemical equilibrium were total – if there were no chemical cycling – Earth would be devoid of life.

Conditions observable on Earth also affect our opinions about the requirements for life to occur elsewhere in the universe. The *Viking* probes of 1875 made it clear, for example, that although Mars probably had surface water at one time, and may even have supported life forms in the distant past, it is barren and desiccated now. Perhaps a future generation of space explorers may find fossil remains and evidence of the processes that brought such life to an end.

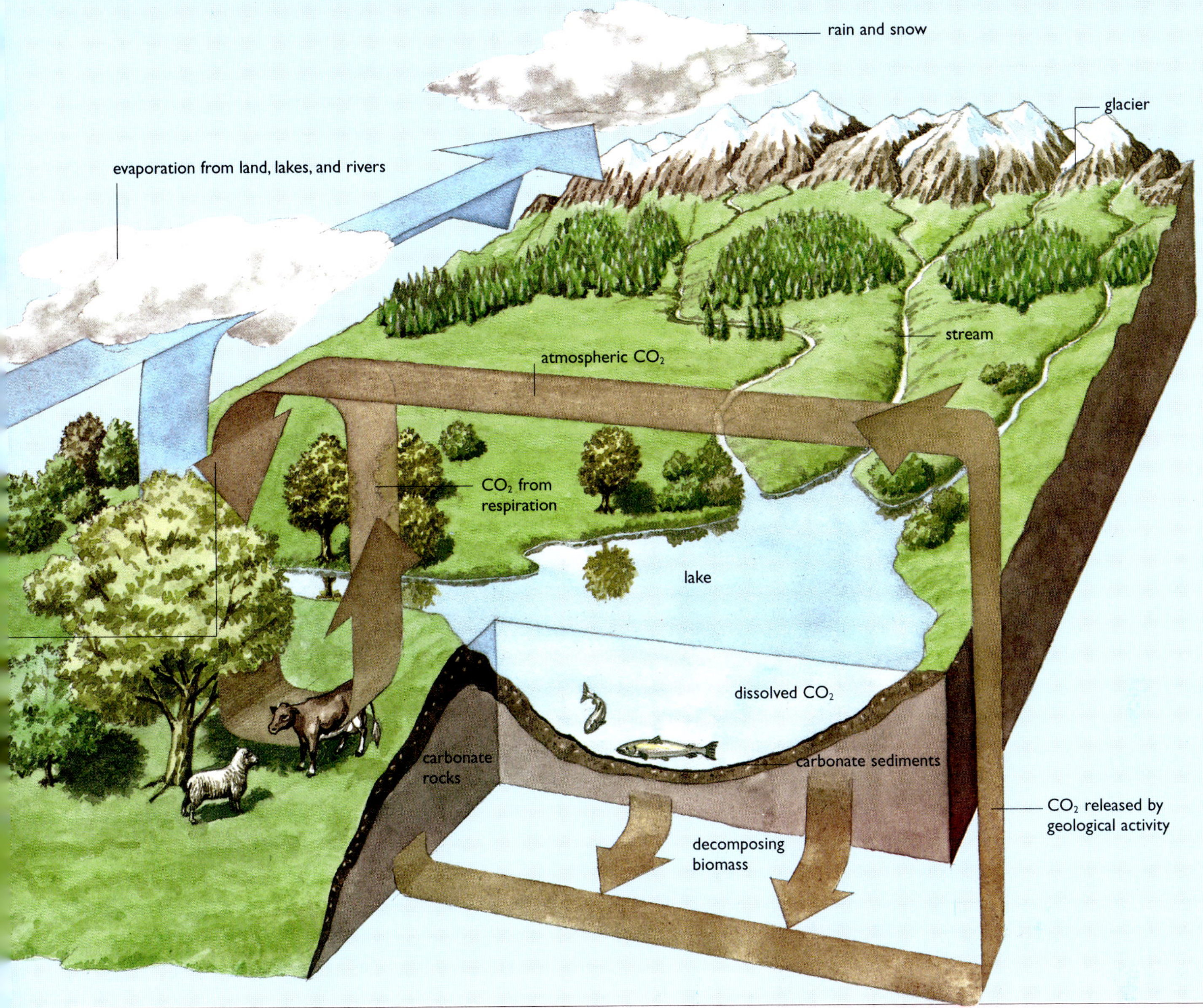

Daisy worlds

Criticism of the Gaia hypothesis centers chiefly around the fact that the organisms involved need to "know" in advance what is required to keep the planet in balance. Lovelock countered this by designing a "Daisyworld" computer program for a hypothetical Earthlike world. The daisies were either white or black and began growing at 5°C (21°F), flourished at 20°C (68°F), and stopped growing at 40°C (104°F).

At 5°C (21°F), the black daisies would thrive best because they would absorb more sunlight and so become warmer than the soil. Ambient temperatures would slowly rise, causing the demise of the black daisies but allowing the white daisies to flourish because they could cool themselves by reflecting heat from their petals. Ambient temperatures would drop and, in time, the black daisies would thrive again.

Subsequently daisies of ten different shades were studied; results showed that regulatory mechanisms continued, but in a more sensitive way.

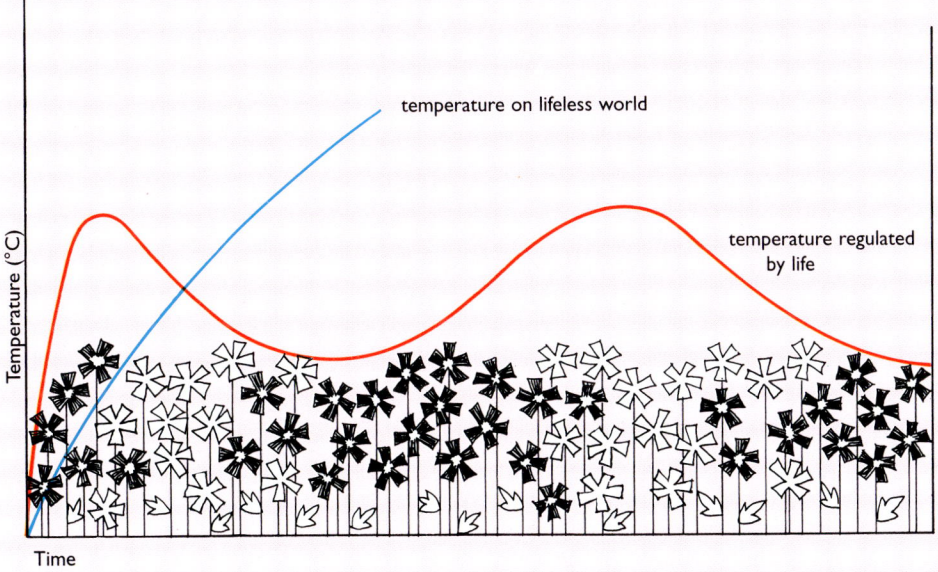

THE NATURE OF LIFE
● *Features of living organisms*

Life on Earth is all-pervading but elusive. It can be found from ocean floor to mountain peak, yet cannot be "made" in a laboratory. New life can begin in a test tube with a human egg fertilized *in vitro*, but this cannot be nurtured to viable babyhood outside the womb.

More than 2,500 years ago, the Greek philosopher Aristotle, who was also a fine biologist, stated "We mean by 'possessing life' that a thing can nourish itself and grow and decay." Aristotle was correct in that these are certainly recognizable attributes of living material, but they are not the only ones.

Crystals, for instance, are non-living materials that display the behavior he described. Suspend a piece of string in a strong solution of sugar and water and it will become covered in sugar crystals. The string will be "nourished" by the sugar and, in time, the crystals will multiply and grow until all the sugar has crystallized out.

Such growth and replication is certainly a characteristic of living material, but clearly not of living things alone. The sugar can be put back into solution, so that the crystals decay or "die," but this does not make them living. Clearly they lack other fundamental qualities.

Since nourishment, growth, and decay are not sufficient alone to define life, other attributes must be required. Living organisms are defined by their ability to reproduce and repair. They must, in addition, be able to react to their environment – yet, even here, valid comparisons can be made with the behavior of some inanimate objects.

The Hungarian-born US mathematician John von Neumann, a pioneer in the development of modern computers, devised a mathematical representation of an idealized self-replicating machine. It is composed of two parts – a constructor and a computer program containing data and instructions. The constructor's actions depend on both the computer program and on the environment.

In order to operate, the self-replicating machine is dependent on being able to pick up and use the materials it has at hand. In favorable conditions, the new "daughter" machine it produces has its own data bank and is self-reproducing. And because it uses raw materials available to it externally, the machine can be said to interact – albeit in a limited way – with its environment. The machine was not capable of self-repair, but this cannot be ruled out as impossible.

If such a machine is compared with a living organism, crucial differences emerge. First, all living matter is composed of cells which are complex constructs of simple biochemical materials. These cells can grow, divide and, in so doing, replicate themselves.

The essential difference between the processes in living organisms and non-living things lies in the way materials are used. The growth, replication and self-repair of living material occurs *inside* the body of an organism. Unlike a crystal or von Neumann's machine, which take in materials and use them as they are, a single cell can transform itself from the simplest of ingredients into a complex organism containing millions of cells. Once inside the crystal or machine, material is, by contrast, immutable. These inanimate objects cannot take in simple substances, then adapt and use them in order to replicate and multiply, in the way that cells do.

The power of self-repair in living things is remarkable. If a starfish loses an arm, it can grow a new one. A deep skin wound can heal itself unaided. The human liver can regenerate even if reduced to less than a quarter of its normal size.

The way in which living things react

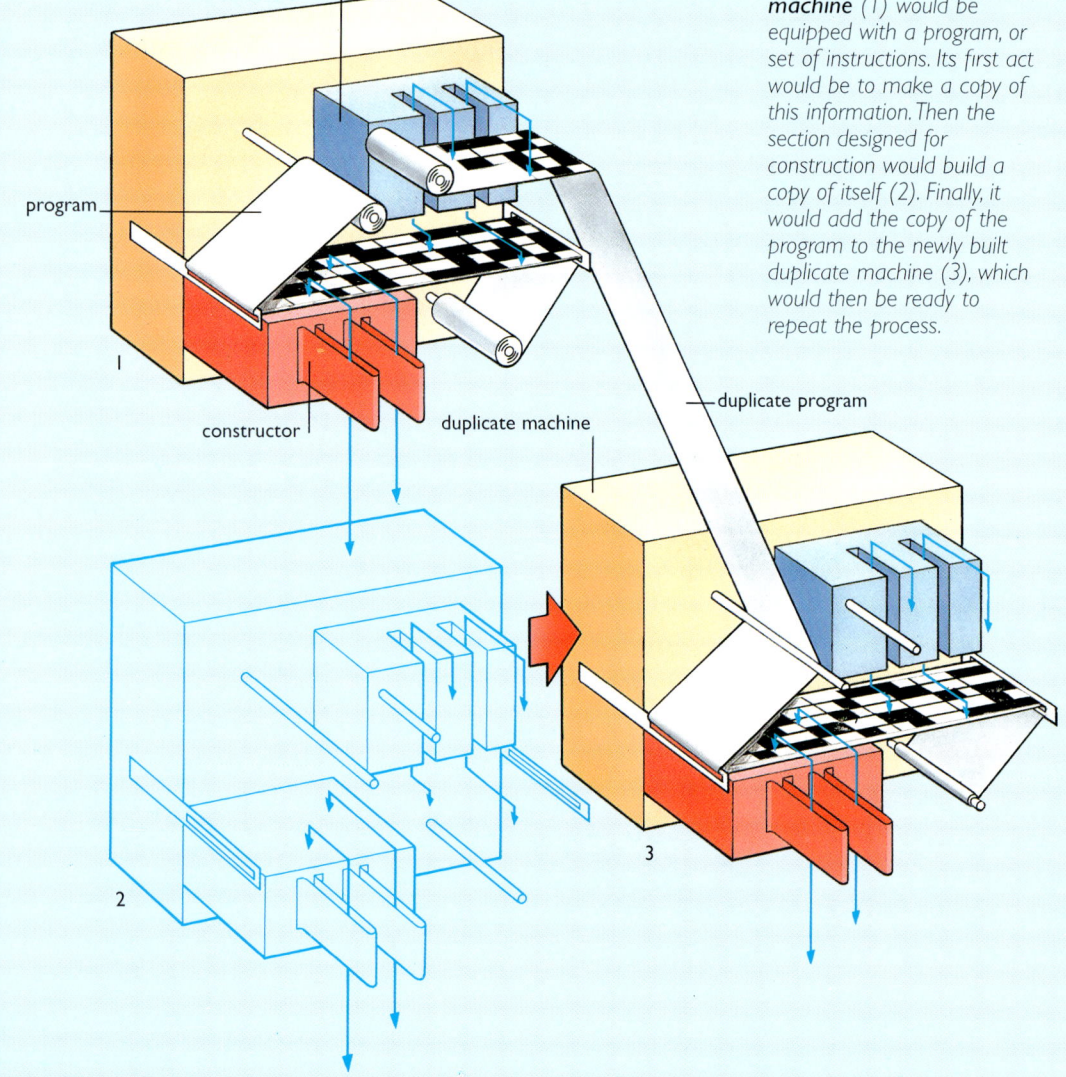

An ideal self-reproducing machine (1) would be equipped with a program, or set of instructions. Its first act would be to make a copy of this information. Then the section designed for construction would build a copy of itself (2). Finally, it would add the copy of the program to the newly built duplicate machine (3), which would then be ready to repeat the process.

Crystals appear to grow and reproduce themselves (above and left) in certain environments – usually liquids in which the substance from which the crystals are made is dissolved. But they are not able to adapt to changes in their environment or to repair themselves – both essential for living organisms.

A virus cunningly exploits a host cell to sustain and reproduce itself (below). Here, new viral bodies are budding from such a cell. The virus cannot manufacture copies of itself, but highjacks a cell for this purpose. The fact that it can arrange for – if not actually perform – its own reproduction means that it just qualifies as living.

to their environment is equally varied. Bacteria, among the simplest of organisms, respond to changes in their environment – that is, their food supply – by subtle alterations in their internal chemical processes, something inanimate things cannot do.

Differences are even more marked in highly evolved plants and animals. The Venus fly trap, for instance, is a carnivorous plant which eats insects that settle on it. By means of chemical "sense organs" it can detect the presence of food and react accordingly. In animals, the senses of hearing, sight, touch, smell, and taste all allow complex reactions to the surroundings, including the vital tasks of finding food and a suitable mating partner.

The acme of development is, arguably, the human being. Our complex reactions to the environment, our powers of memory and constructive thought, and our consciousness put us far above any other species. Certainly, it may one day be possible to create robots capable of imitating some of these functions, but they will ultimately be of biological origin since they were designed by humans in the first instance.

CHEMISTRIES OF LIFE
• *The basic building blocks*

All living things on Earth, from micro-organisms such as bacteria to the largest plants and animals, are based on the chemical element carbon. This seems to be because the carbon atom is unique in the way it links with other carbon atoms and with atoms of other elements, notably hydrogen.

Carbon atoms can join together by sharing electrons. Through such covalent bonding, a carbon atom can link itself to another carbon atom and, at the same time, join with other elements or compounds. The result is that carbon can form molecules that have a "ring" formation based on the linkage of six carbon atoms. Such molecules are essential to many life processes, including photosynthesis, and many are soluble in water, an attribute vital to earthly life.

As well as forming rings, carbon atoms can also combine with each other to make long chains, or polymers. In nature, these polymers form the chemical backbone of complex organic molecules that are vital to the structure and maintenance of life, from the supportive walls of plant cells to insulin, the hormone humans need in order to metabolize sugars.

The versatility of carbon means that the temperature range of carbon-based life is enormous by earthly, if not by celestial, standards. It covers approximately 100°C (212°F) – the range between the boiling and freezing points of water, the substance that makes up a high proportion of all living organisms. At the upper end of the range, some algae flourish at temperatures of 70° and 80°C (158° and 176°F), and some bacteria still grow actively at 95°C (203°F). In the laboratory, they will survive above boiling point, at 104°C (219°F).

Many lichens, plants in which an alga and a fungus live in close association, can exist equally well in the heat of the Sahara or the cold of the Arctic. And some fishes with special blood can survive a few degrees below freezing. Outside Earth, low-temperature life might be possible, but because there would be little energy in such systems, evolution of complex life forms would be so slow that it would be outpaced by environmental change, and thus could not survive for any length of time.

Short-wave radiation also puts limits on life. The complex molecules of living material are broken down or changed by such radiation, as, for example, when white skin becomes sun-tanned by ultraviolet light. The ultraviolet causes the production of the pigment melanin, to help prevent the radiation from penetrating the skin. But because such radiation is everywhere in space, life can exist only where it is protected by, for example, a planetary atmosphere.

On Earth, the element oxygen is an

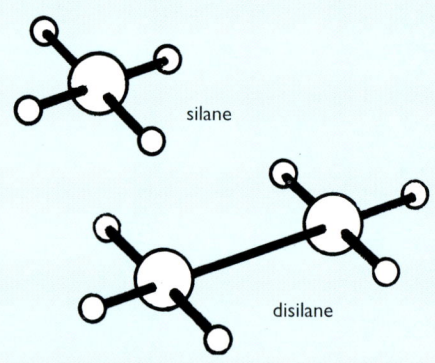

Silicon resembles carbon in its ability to form large chain molecules. The molecules of the compounds silane and disilane, for example, are analogous to those of methane and ethane. Such silicon compounds are, however, generally less stable than their carbon-containing counterparts.

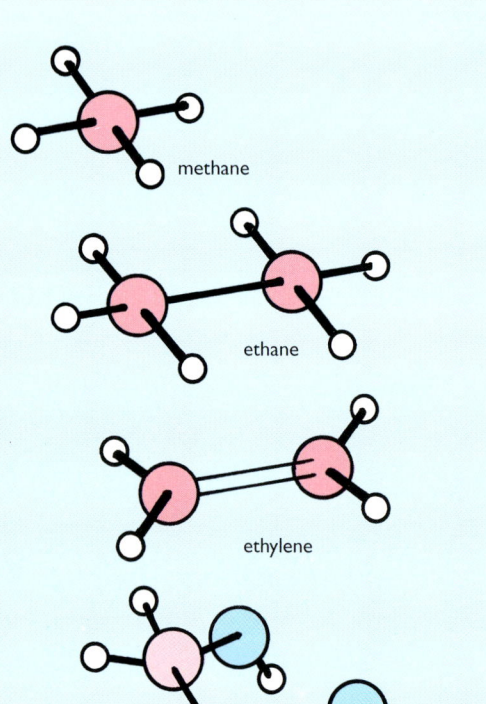

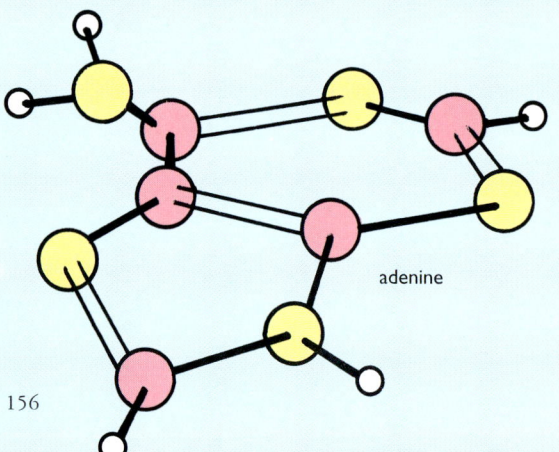

Carbon is unique in the variety of molecules it can form. They are typified by methane (one of the simplest), ethane and ethylene (in which a single or a double bond links two carbon atoms), glucose and adenine (containing carbon rings). Here, carbon atoms are pink, hydrogen white, oxygen blue, nitrogen yellow.

THE LIVING UNIVERSE

Tiny diatoms (left), the base of the food chain in the oceans, have found a way to use the element silicon in their shells to harden them.

On the sunless floors of deep ocean trenches live these tube worms, 12½ ft long. They, and other shellfish, live off bacteria that extract energy from hot water, rich in hydrogen sulphide, that pours from thermal vents. Such communities are the only ones known that do not ultimately rely on photosynthesis for energy.

important constituent of living things and a part of the atmosphere. Atmospheric oxygen is released by green plants as a waste product of photosynthesis, and is used by animals for respiration. Without oxygen, the range of life would be very restricted, but in other situations in the universe, could some other element take its place in living systems? Sulphur, for example, has been found to act in similar ways to oxygen, and provides the energy source for some bizarre deep-sea creatures.

Biochemists have thus considered the possibility of living systems based on the element silicon rather than on carbon. Although it cannot form ring-shaped or long molecules, silicon does combine readily with oxygen. Sand, or silica, is one of the commonest chemicals on Earth, and compounds of silica with calcium and aluminum are found in many terrestrial rocks.

These compounds are not, however, soluble in water, which makes them unsuitable as a basis for life. This is not to say that such a scenario is impossible. Some terrestrial bacteria can produce insoluble substances, such as sulphur, and process them perfectly adequately.

In general, silicon molecules are more resistant to heat than those based on carbon. Silicones – polymers with a central chain of alternate oxygen and silicon atoms used in lubricants and synthetic rubber – are stable up to 350°C (662°F). If silicon-based life were possible, the biochemists argue, it would probably be stable up to 250°C (482°F). If the atoms on the side of the silicone chain were more widely varied, it is conceivable that they could provide the basis for life elsewhere in the universe, where conditions prevailing on planets are hotter than on Earth.

Most recently, biochemists have considered alternatives to both carbon and silicon. Germanium, selenium, and sulphur compounds have been suggested, but the restrictions seem to be similar to those for silicon.

THE BEGINNING OF LIFE
● *Building the essential molecules*

How life began on Earth remains a mystery, but it probably happened in one of two ways. Either it started when conditions were suitable for the synthesis of its basic chemicals, or it came about through the arrival of complex organic materials from space.

In the 1930s, the Russian scientist Aleksandr Oparin, and the Americans Melvin Calvin and Harold Urey, concluded that life could have begun in an atmosphere where there was no free oxygen – at an early stage in the Earth's history, it was all combined with other chemical elements. Their supposition was upheld by the Chicago-based chemist Stanley Miller in a notable experiment that attempted to mimic conditions on Earth as they were 3.5 to 4 billion years ago.

Miller passed electric discharges through a mixture of ammonia, hydrogen, and methane, in a flask containing distilled water. The chemicals represented the components of the early atmosphere; the electrical discharges represented flashes of lightning. After only one week, the water had become deep red and, besides simple acids, contained amino acids – the organic molecules that are the building blocks of proteins, themselves a vital constituent of living things.

This chemical mixture represented the primordial soup, life's supposed starting place. It later became clear that other chemicals could also be transformed into fundamental organic molecules under similar conditions. These could have included sugars and nucleotides (components of the genetic material DNA).

Although plausible, this scenario raises a problem of time scale. The fossil record suggests that simple organisms such as bacteria and algae lived on Earth at least 3.2 billion years ago. Yet for all their biological simplicity, the chemical construction of these organisms is relatively complex. The first synthesis of the basic component chemicals here occurred no more than a billion years earlier. Was this enough time for the necessary evolutionary changes to take place?

Synthesis of amino acids may, of

The genesis of life may have occurred about 4 billion years ago when energy from electrical storms, acting on the primordial atmosphere of methane, ammonia, carbon monoxide, carbon dioxide, and water, produced amino acids such as phenylalanine, tryptophane, histidine, glycine, and valine.

course, have happened even earlier than 4.2 billion years ago. And synthesis of complex materials may have taken place more quickly than subsequent evolutionary changes – at least as detected in the fossil record. Or, maybe suitable organic materials arrived from space.

In the late 1960s, radio astronomers discovered the existence of organic molecules in dark nebulae. Since that time, there has been a very real possibility that even more complex molecules may have arrived on Earth from space. Such molecules must have come from some scientifically plausible source, and must have survived the risk of destruction by short-wave radiation on its way to our planet. There are two possible ways in which the molecules might have traveled: in meteorites, or in cometary dust.

Analysis of a meteorite that fell to Earth in 1969 has shown it to contain at least 74 amino acids. Skeptics suggested that the meteorite was contaminated after it fell, but in fact, the 74 amino acids showed both similarities to and differences from those known on Earth. The atoms in the molecules of such amino acids can link with the internal carbon chain of the molecule in either a left-handed or a right-handed arrangement. Terrestrial amino acids are left-handed, but both left- and right-handed forms were found in the meteorite.

Approximately 70 million years ago, a giant comet deposited dust into the solar system. About 20,000 years later, a meteorite from this comet landed at Stevns Klint on the coast of Denmark, an event pinpointed by evidence at the interface between rocks of the Cretaceous and Tertiary periods, which were laid down at this time. However, the Earth began to sweep up the comet's dust about 15,000 years before the meteorite impact, and also for thousands of years after the collision.

It has been proposed that any amino acids within the meteorite would have been destroyed on impact, but that they would have been retained in the fine dust. This dust could have brought amino acids to Earth and, in fact, such compounds have been detected many inches below the boundary between Cretaceous and Tertiary rocks.

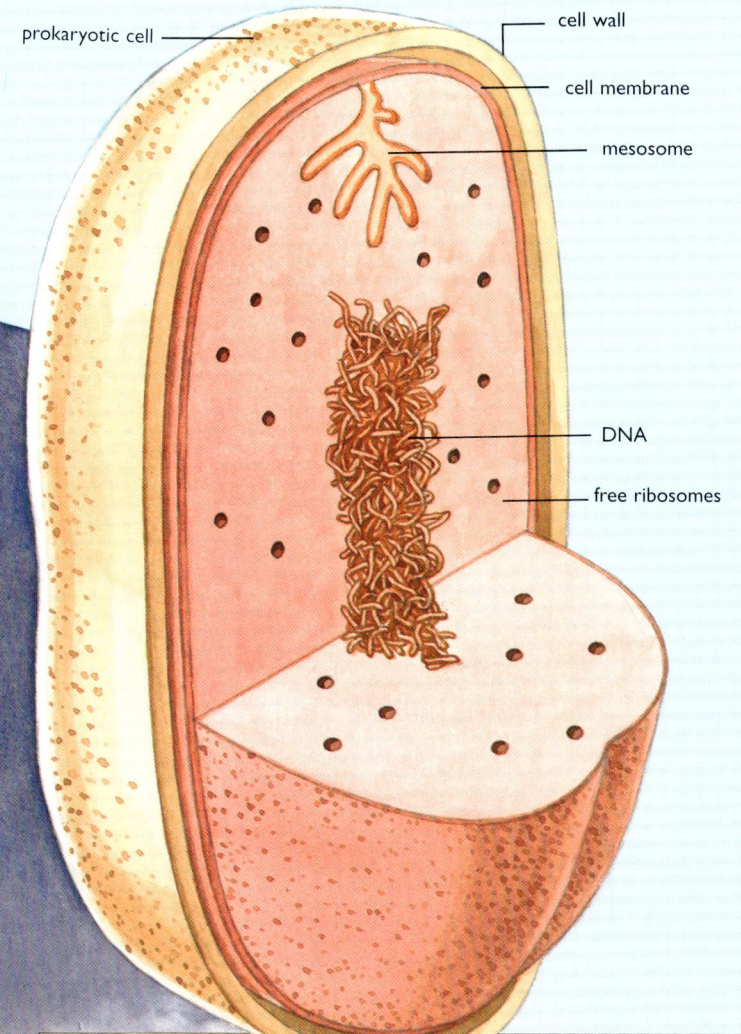

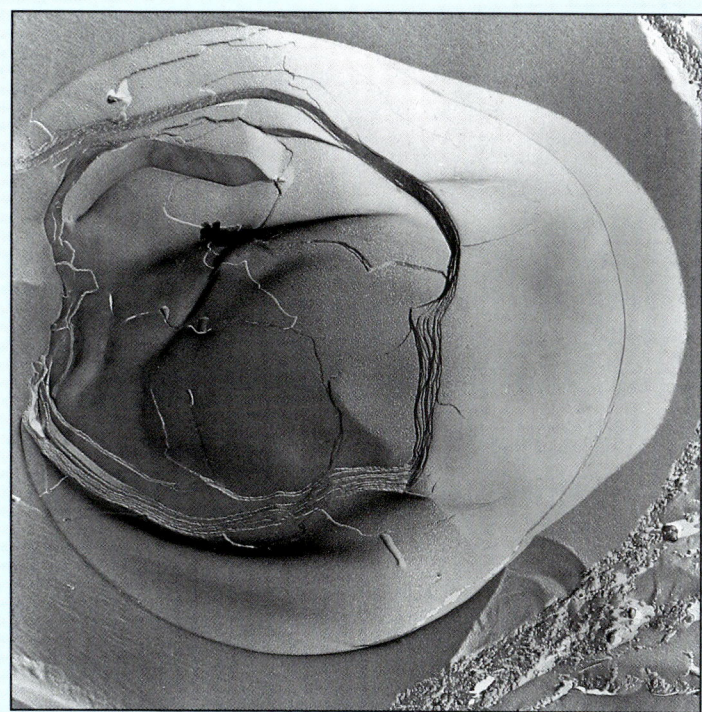

The earliest evidence of life on Earth appears in the fossil record about 3.8 billion years ago. It consisted of very simple single-celled organisms known as prokaryotic cells. This kind of cell does not have a clearly defined nucleus. Much of its chemical activity goes on at sites called mesosomes, which are highly-folded openings in the cell membrane.

In the laboratory, scientists find that phospholipids — molecules with one end that can dissolve in water and one that can dissolve in fat — spontaneously form globular, multilayered structures, called liposomes, when placed in water (above). The outer wall of a liposome is a membrane, and in the evolution of life, the appearance of cell membranes was as important as the evolution of DNA.

Molecules from space

When Halley's Comet (right) made its approach to Earth in 1986, observations from spacecraft showed that the central core of the comet's head was protected sufficiently for the substance formaldehyde to be safely tucked away within it. Such a substance could have formed the basis for the development of many other organic molecules. Evidence is gradually accumulating to show that the conditions in Halley's Comet might occur in large meteorites reaching Earth, and that organic substances might reach us in cometary "dust" swept up by the orbiting Earth.

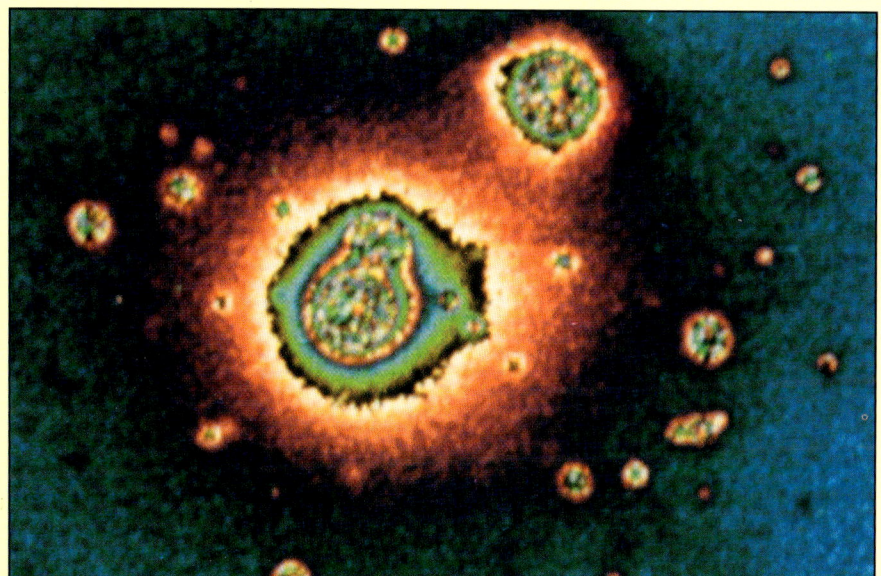

Comets might incubate the molecules of life, protected within the cometary nucleus.

MOLECULES OF LIFE
● *The vital reactions*

Life's essential attributes come about through the biochemical properties of carbon-based molecules. These exist in a vast range of forms, but among the most significant are the amino acids, which are strung together and folded in three dimensions to form proteins. These amino acids come in 20 distinct chemical configurations. Their complex arrangements of atoms include not only carbon but also hydrogen, oxygen, and nitrogen. Sulphur is another common constituent of proteins, as are iron and phosphorus.

Proteins fulfill major roles in the formation and maintenance of living materials. Structural proteins, for instance, form the building blocks of plant and animal cells. Other proteins control the chemical reactions that take place within and between cells. In this latter role, they act as catalysts, facilitating and speeding the biochemical reactions on which the maintenance of life depends.

The catalytic proteins of living systems are known as enzymes, and each specific enzyme usually controls one particular biochemical reaction. This is achieved in an extraordinary way. The amino acid chain, which forms the protein's backbone, is folded in such a way that it fits the chemical on which the enzyme is to act exactly.

Although there are only 20 amino acids, the variety of reactions they make possible is enormous because of the number of ways in which they can be arranged. A protein containing only 10 amino acid molecules would have 100 billion billion alternative forms of behavior. In fact, real proteins never have fewer than several hundred amino acids in their chain, making the scope for biochemical activity virtually infinite.

As well as proteins, all living things contain substances known as nucleic acids, including deoxyribonucleic acid, DNA. This relatively simple, large, long-chain organic molecule is composed of chemical units, each containing a phosphate and a sugar. Each sugar is linked on one side to a base – a group of atoms typified by being soluble in water and reacting with acids to form salts. DNA contains bases of four chemical patterns; these are adenine, cytosine, guanine, and thymine.

Each section of DNA – the phosphate, the sugar, and the base – is known as a nucleotide, and the arrangement of the entire DNA molecule is a double helix. Although it may contain 300 million atoms, and if stretched out could measure up to about 3 feet, it takes up little space because it is folded up on itself.

In most cells, DNA is concentrated in the chromosomes, rod-shaped bodies in the nucleus. Strung along the chromosomes are the genes, units of pure DNA in which the coded commands both for making and maintaining an entire organism – its genome – are carried.

It is the sequence of atoms in DNA that spells out the genetic code. Generally, DNA works so that one gene codes for the production of one enzyme, which then controls certain reactions within the cell. In addition, genes can act to control each other – by acting, for example, as "on" and "off" switches.

As well as making possible life's unique biochemical synthesis, DNA also allows for the vital processes of self-repair and self-replication. In the simplest form of reproduction – the creation of an exact copy of a cell – the double helix unwinds and a new copy of the complementary chain is added to it, so that two new identical chains are made before the cell divides. In sexual reproduction, the process is similar, but it is complicated by the mixing and exchange of genetic material.

While shuffling of the genetic pack brings about change in the appearance, or phenotype, of an individual, this is not due to any essential alteration in the genetic material. Genetic change, or mutation, is a permanent alteration in the nucleotide sequence in DNA.

Mutation, which can come about spontaneously, or be caused by environmental influences, such as radiation or certain drugs, can be deleterious to an organism, causing severe congenital deformity, for example. But it can also have beneficial effects, making an organism better adapted to its circumstances. Mutation allows for such adaptability and makes evolution possible.

While even single-celled bacteria can reproduce independently, this is not so for viruses, whose successful DNA replication depends critically on outside influences. A virus can reproduce only by making use of chemicals inside a suitable host – another living cell. For this, it "highjacks" the cell's ribonucleic acid or RNA, a messenger molecule used in replication and protein synthesis that is very similar to DNA.

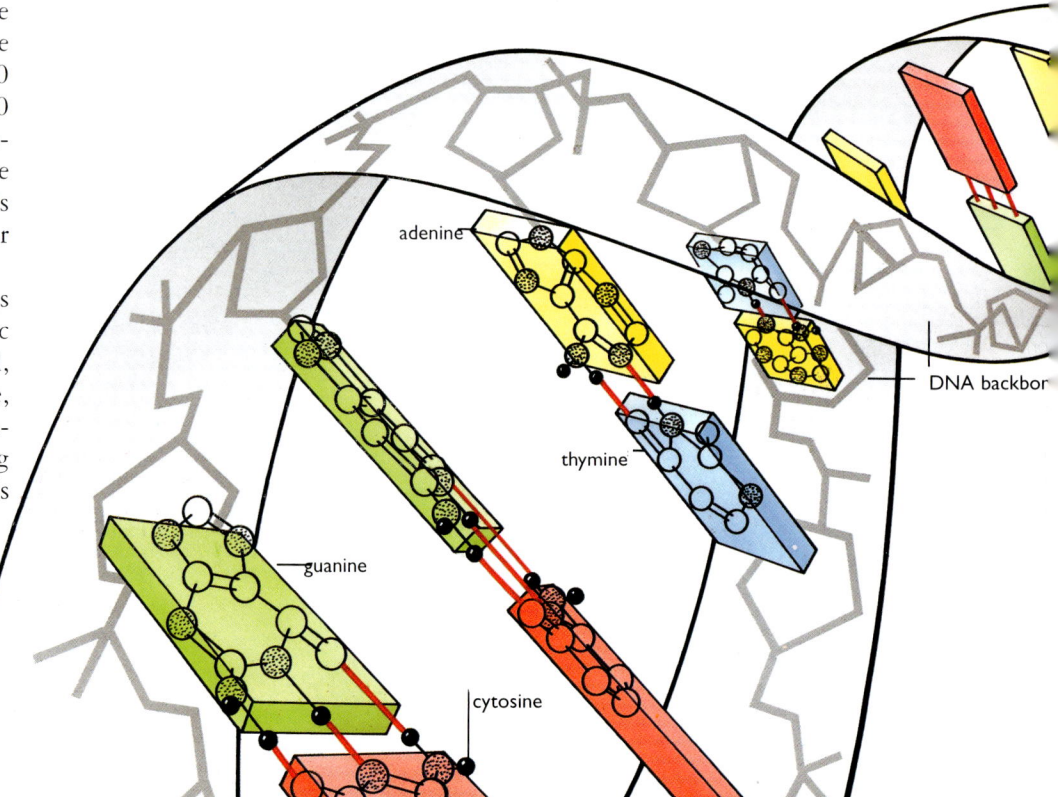

THE LIVING UNIVERSE

The virus is thus a kind of "hybrid" between the living and the non-living. Isolated viruses may cluster together, looking like clumps of inanimate crystals. Yet, within them is enough DNA, the molecule of life, to provide sufficient coding for existence in favorable circumstances. It has even been speculated that viruses arrived on Earth from space. Whether or not this is true, they graphically illustrate the thin dividing line at the edge of life.

Cell replication (right) involves passing on a precise copy of all genetic information to the offspring cells. As two nucleotide chains of the old DNA uncoil, new nucleotide chains are synthesized on their surfaces. Here, newly-replicated chromosomes separate.

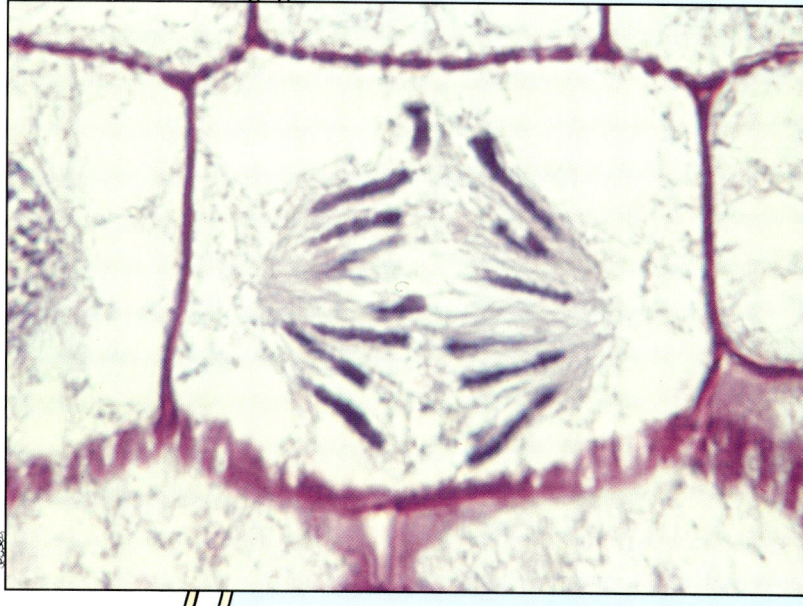

The DNA molecule is composed of two chains of nucleotides. The chains are held together by hydrogen bonds between adjacent nucleotides. The double-chained structure is coiled in the famous double helix.

chromosome pair

supercoiled DNA

To manufacture proteins, DNA strands separate and RNA polymerase molecules build up messenger RNA (mRNA) chains on them, copying the base sequence.

RNA polymerase

untwisted DNA strand

individual bases

transcription

Newly-synthesized messenger RNA leaves the nucleus of the cell.

Messenger RNA moves to a ribosome, where another molecule called transfer RNA (tRNA) reads the sequence of bases in groups of three, called codons. Each of these codons corresponds to one of the 20 amino acids in the cell which, when assembled in the correct order, make a protein.

amino acid

protein assembly

amino acids

messenger RNA

transfer RNA

ribosome

envelope of nucleus

Cross-links in the DNA molecule (left) are formed between molecules of the nucleotides thymine (T) and adenine (A), and between cytosine (C) and guanine (G). It is the order of these nucleotides along the length of the double helix that provides the code for the manufacture of proteins. The sequence visible at lower left is GGAT along one strand of the helix and thus CCTA along the other.

EVOLUTION ON EARTH
● The scope of life

The universal unit of life is the cell. Whether organized singly or in groups, cells are the substance of every living plant and animal on planet Earth. All living cells have a specific organization: essentially, variations are only matters of detail. Every cell has an outer covering, or membrane, which is, in effect, a sandwich of layers of protein and globules of fat; through it, substances pass in and out of the cell. In plants, this membrane is covered with a cellulose sheath which provides rigidity.

The cell is filled with a jelly-like substance, the cytoplasm, within which lie subcellular "particles," or organelles. These are involved in activities such as energy generation or the manufacture of fats or proteins. Many of these proteins act as catalysts, promoting other chemical reactions, either in the cell in which they are made or in adjacent cells. Still other organelles digest substances entering the cell, including noxious material.

At the heart of all cells lies a nucleus. Within it is housed the genetic material which, combined with proteins, forms the chromosomes. It is from here that the cell's activities are orchestrated; without a nucleus, a cell will die.

Unlike those of animals, the cells of plants contain organelles known as chloroplasts. Within these are tiny packets of the green pigment chlorophyll that enables plants to use solar energy and build up carbohydrates in the process of photosynthesis.

The first cells to emerge from the primordial soup and to exist independently on Earth about 4 billion years ago were probably bacteria, which "fed" on organic molecules within their chemical-rich environment. From these probably evolved the first plants, similar to the single-celled algae of today.

As they photosynthesized, these primitive plants increased the oxygen content of the water in which they lived (oxygen is a waste product of photosynthesis). Only then was it possible for animal life, which depends on oxygen for existence, to come into being.

More elaborate forms of life took many millions of years to evolve. Gradually, multicellular organisms developed, with cells specialized for particular tasks such as reproduction, and sensing and responding to the environment. By the Cambrian period, which began about 560 million years ago, complicated organisms such as large marine plants and animals had evolved.

As evolution progressed to a new era, the corals and the many-limbed trilobites flourished and proliferated, followed by the first fishes. Yet the seas were still the only theaters of life. The land masses of the Earth, which were joined together in one megacontinent, remained barren.

The first plants appeared on land about 400 million years ago and ushered in the Devonian period. They evolved into a range of types, from small mosses to giant tree ferns, and caused a change in conditions for life on Earth. Free oxygen was pumped into the atmosphere and land animals – insects, air-breathing lungfish, and the first amphibians – could now populate the planet.

The Carboniferous period, which dawned about 345 million years ago, witnessed the appearance of the first reptiles and the first winged insects. The latter evolved in tandem with the flowering plants, whose fertilization the insects made possible.

Successive periods of prehistoric life witnessed the gradual colonization by plants and animals of all the niches that the Earth has to offer. Significant events included, for example, the appearance of the first freshwater fishes, the rise of the

Evolutionary history is revealed by a study of DNA. A gene found across a range of species controls the production of the enzyme cytochrome oxidase. This molecule, involved in using the oxygen in cells, is different in every creature, and its DNA blueprint is therefore different. The degree of similarity between the versions of the gene in different organisms gives a good idea of how closely related they are. This chart of the evolutionary history of certain species is based on the similarity of their genes for cytochrome oxidase.

THE LIVING UNIVERSE

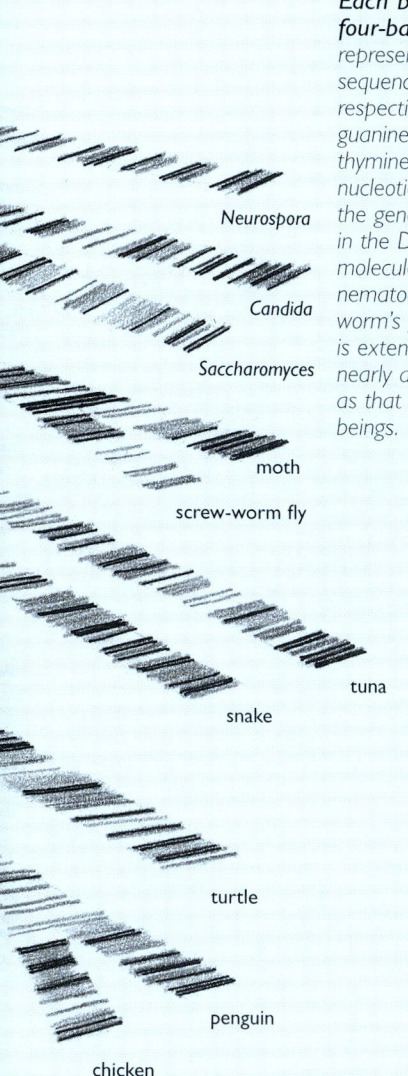

Neurospora
Candida
Saccharomyces
moth
screw-worm fly
tuna
snake
turtle
penguin
chicken

Each band in these four-band groups represents the sequence of, respectively, adenine, guanine, cytosine and thymine, the four nucleotides that form the genetic blueprint in the DNA molecules of a nematode worm. The worm's genetic code is extensive, yet not nearly as complicated as that of human beings.

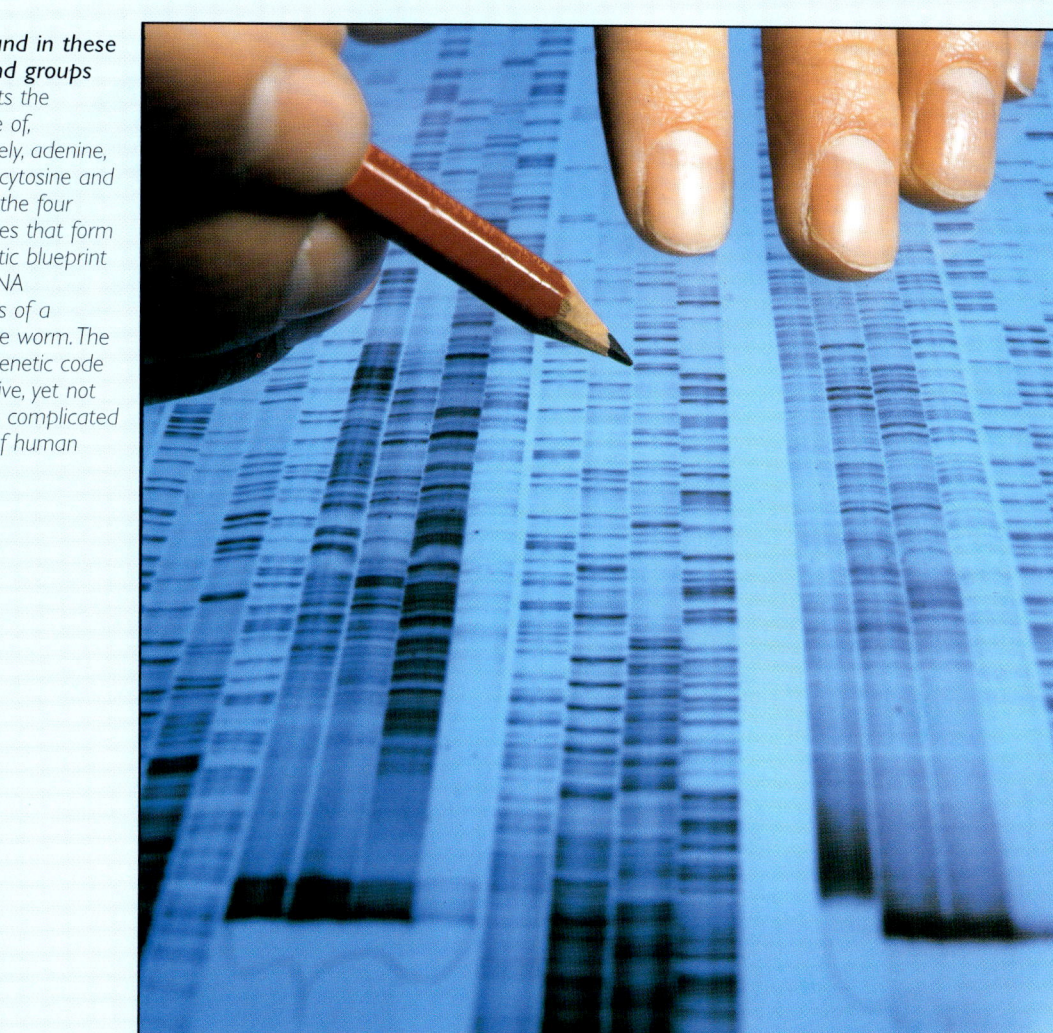

reptiles, the age of the dinosaurs, the flight of the first birds and – a turning point in evolution – the appearance of the mammals.

Only when the rule of the reptiles was ended about 65 million years ago – possibly by the collision of a huge asteroid or comet with the Earth, which would have caused great climatic change – did the mammals gradually come into the ascendant. The early mammals were small, shrew-like creatures; but in time, the group diversified to include the animals we know today, including hoofed mammals and the primates. Only about 4 million years ago did the first ancestors of humans walk on Earth.

Although comparatively small in size compared, for example, with the massive dinosaurs whose world they took over, the mammals have proved to be the great success story of evolution. This is because of the mammal's high degree of encephalization; that is, its high ratio of brain weight to body weight. This resulted in a level of intelligence far greater than that of any species yet seen.

Among mammals, humans show the most marked degree of encephalization, and this is reflected in every facet of human behavior. But the phenomenon is not confined to primates. Whales and dolphins, for example, have a comparatively high degree of encephalization, but further development seems to have ceased about 20 to 30 million years ago. As a result, these animals have an intelligence probably similar to that of a dog.

Why this brain growth stopped among the large sea mammals is uncertain, but it may be linked to the fact that encephalization seems to demand a bodily system capable of producing a great deal of energy, that is, a very high metabolic rate. This is something which the dolphins and their allies seem incapable of achieving. And the same limitation was almost certainly a significant factor in the demise of the dinosaurs.

Encephalization brings with it a need for a long gestation period for offspring. Born helpless, they have a great deal of learning to do to "train" their large brains to function fully. This might be considered an evolutionary disadvantage – helpless young are easy prey – but against it is set the versatility and ability

163

Ants appear to show great intelligence when they cooperate in the work of the colony. But individual ants are creatures of strictly limited abilities. These leafcutters, for example, are "programmed" for their tasks and are not capable of individual problem-solving. They show their limitations in the way they respond to errors: if, for example, one drops a leaf, it does not work out how to pick it up but starts again, going all the way back to the plant to chew off another.

that a large brain has to offer. And humans have used their brains to manipulate their environment to their own advantage. Science, technology, and medicine make it possible for babies to have maximum protection from accident, disease, and other threats to life.

Intelligence, as developed in *Homo sapiens*, has led to vast differences between the most advanced of the other mammals and the least developed of human societies. Apes may be adept at imitation, or even be able to "think through" the logic of placing a chair in the right place to reach an inaccessible banana, but this is the acme of their problem-solving abilities.

In addition, humans excel at communication. Certainly animals can and do communicate with each other, but in a much more restricted way than humankind. Some insects that live in communities — the social bees, for instance — are now well known for their ability to impart information. They do this by dancing around in particular patterns to tell other members of the hive where food is to be found. Although reasonably complex, this is all that the bees can "say" to each other. Ants also have certain elaborate communication systems, but again, they are used only for specific and limited purposes.

Even among birds and mammals, communication is strictly limited and centered largely around the essentials of survival: avoiding predators, finding a mate and rearing young. But among many creatures, this communication does involve smell, body language and, mainly in primates, facial expression.

These forms of communication are also important to humans, but what sets us apart from our fellow inhabitants of Earth — and possibly from all other inhabitants of the Galaxy and the universe — are our most remarkable and sophisticated powers of language, not only of speech, but of abstract and technical language. Indeed, the invention of specialized vocabularies for technical applications — of which mathematics is arguably the most elaborate and the most precise — has enabled humans to advance in a way denied to less intelligent creatures. And information to be communicated can be recorded permanently in writing, on audio tape, and so on.

As with other human attributes, the use of language depends on a brain that has a large capacity for memory, as well as highly developed powers of learning, thinking and reasoning. About 15 billion nerve cells, or neurons, in the brain alone are devoted to the tasks of processing information from the eyes, ears, and other sense organs, and from the muscles and other internal parts that make the body "work."

Yet the brain cannot act alone. To operate as the control center of the body, it needs elaborate support mechanisms to help monitor the environment and to carry out the tasks it dictates. It is thus no accident that the human brain is a central processing unit, with connecting "wires" to every part of the body to receive and transmit information.

This wiring is the nervous system and its constituent neurons. Each neuron acts as a receiver, conductor, and transmitter of nerve signals, items of electrical and chemical information which together provide all the necessary data for monitoring and controlling human activity, from walking and talking to designing a spacecraft.

The main nerve pathway out of the brain is the spinal cord, which has connecting nerves to all body parts, the nerves becoming finer and finer as they branch out to the various tissues and organs. The brain is fed by sensory messages coming in from the sense organs; motor messages are those carried outward, instructing tissues and organs to behave in a certain way.

The nervous system is also divided into two subsystems, one at the level of consciousness, the other involuntary, or autonomic. The advantage of this is that the autonomic system takes care of essential life-support activities, such as breathing and digestion, which we perform without thinking about them.

THE LIVING UNIVERSE

The ability to solve problems is one of the most astonishing of human faculties. The process often calls on complex concepts, diverse abilities, and a great deal of background knowledge. Most mysterious of all, human beings often create new and original solutions to the problems they encounter.

Nicolaus Copernicus solved one of the greatest puzzles of his time (1) when he studied the mathematics of the movements of stars and planets. He sought a better account of them than the prevailing geocentric theory.

Solving such a complex problem required deep knowledge of the known facts and previous theories, and the quest was strongly influenced by the philosophical and religious worldviews of Copernicus and the society of his day (2). The successful solution of the problem demanded abstract reasoning and mathematical calculation.

The solution arrived at by Copernicus was that the Sun, and not the Earth, is the center of the solar system (3). The theory was published in 1543, the year in which Copernicus died, but like any new idea, it could not become the common property of humanity until it had survived the crucial process of critical discussion by his peers.

165

UNIVERSAL COMMUNICATION
● *Is there anyone out there?*

The ability to communicate in complex and sophisticated ways is one of the special attributes which sets human beings apart from other creatures on our planet. As well as making us unique, our powers of communication also provide us with the possibility of making contact with other living organisms in the universe, should there be any.

The prospects for such communication have been enhanced by the development, in the latter part of the 20th century, of the technology that allows communication across vast distances of space by means of radio. Only radio waves will suffice for this; even the most powerful light beams that could be produced would be too faint. The seriousness of scientists in their approach to interstellar communication is epitomized by the fact that when, in 1967, they discovered radio pulses emanating from space, radio astronomers conjectured that these might be coded messages from an alien civilization.

One of the chief protagonists of interstellar communication, the American astronomer Frank Drake, has devised two basic ways of making contact with civilizations whose languages may be totally different from those used on Earth by humans. The first is to send a message in the form of drawings or diagrams, and to include within it an astronomical explanation of the source from which it is being sent. The second, more sophisticated, technique involves the use of mathematics – the one language that may be universal.

The pictorial approach to communication has parallels with pictographic forms of writing, such as those used by the ancient Egyptians, and the style developed by the Chinese into non-alphabetic writing. This method was used in the message sent into space with Pioneer 10 in 1973, designed by Drake and fellow American Carl Sagan.

A gold anodized plaque in the spacecraft contained drawings of a man and a woman, the man with his hand raised in a sign of peace and welcome. The figures stand against a silhouette of the spacecraft to provide a sense of scale. The plaque contained a diagram of atomic hydrogen (the simplest element), and a plan of the planets of the solar system, showing the path of the craft. Also on the plan was the position of the Sun in relation to the 14 pulsars then known. Symbols were provided representing the pulsation periods of each one. By comparing these with their values at the time the spacecraft is encountered, beings of another civilization should be able to calculate when it was launched.

For the Voyager probes, launched four years later, Drake and Sagan produced a video disc which contained both scientific information and material conveying the sights and sounds of Earth, including messages from the Secretary General of the United Nations and the American President, spoken information in a host of languages, and music from Bach to rock and roll.

The second type of message uses binary arithmetic – the system used in all digital equipment, from computers to compact disc players, and even for relaying pictures back from space. The binary system "counts" using ones and zeros, and by convention 01 is number one, 10 represents the number two, 11 represents the number 3, 100 the number four, and so on. This may seem cumbersome, but it is just what the computer needs, for 1 and 0 can represent circuits that are switched on or off, respectively.

As an exercise, Frank Drake used the code to create a message which, if translated, could provide an intelligent alien with crude symbolic pictures of a human being, the solar system, and the atomic structures of carbon and oxygen. All this was done with only 551 zeros and ones.

While we may be able to devise a message to send to an alien civilization, or to receive one from beings elsewhere in space, the possibility of interstellar dialogue seems remote, not the least because of the vast distances involved. Even if, for example, there is a highly developed alien civilization on one of the extrasolar planets discovered so far, any message sent will take upwards of 15 years to get there. Any reply will take an equal time to get back.

Yet, we cannot expect that there is necessarily an intelligent civilization at that distance. If there were such a civilization within 100 light-years, any response to a message sent by one generation of humans would not be received until seven generations later. And to receive and understand the message, the civilization must not only be intelligent, but have reached at least the technological level of Earth.

The language of modern telecommunications is binary code, which employs only the two digits 0 and 1. For example, the Voyager 2 spacecraft (right) converted its images of Neptune's satellite Triton into binary numbers for transmission to Earth. The process is shown by the sequence of pictures below. The image (right) was divided into tiny squares called pixels. The brightness of each was represented by a number from 0, for white, to 255, for black (center). The numbers were then converted into their binary equivalents, sequences of zeros and ones. These were transmitted as a stream of radio pulses (left); a white square is a "1" pulse, a black square is a "0" pulse. On Earth, these pulses were used to reconstruct the original image.

Message to the stars

In 1973, this image (below) was transmitted into space by the giant radio telescope at Arecibo in Puerto Rico. It was aimed at M13, a globular cluster comprised of thousands of stars. The message contained 1,679 "on" and "off" pulses, representing a sequence of white and black picture elements. These could be arranged in only two ways: as 23 rows of 73 elements, or as 73 rows of 23, but only the latter arrangement gives an intelligible picture.

The message begins with a demonstration of the way in which binary-code numbers are represented. There follows a list of the atomic numbers of the elements hydrogen, carbon, nitrogen, oxygen, and phosphorus, which are essential to life on Earth. Other numbers give a rough idea of the composition of nucleotides, the building blocks of DNA and RNA (pp. 160–61), and a picture of the DNA molecule is included. There are also pictures of a human being and of the Arecibo telescope, together with their sizes expressed in terms of the radio wavelength used.

An encyclopedia in code: Binary numbers from 10 to 1 (1). Atomic numbers of certain elements (2). Formulae for nucleotides (3). Number of nucleotides in human DNA (4). DNA molecule (5). Human figure (6). Height of human (7). World population (8). Solar system (9). Arecibo telescope (10). Size of telescope (11).

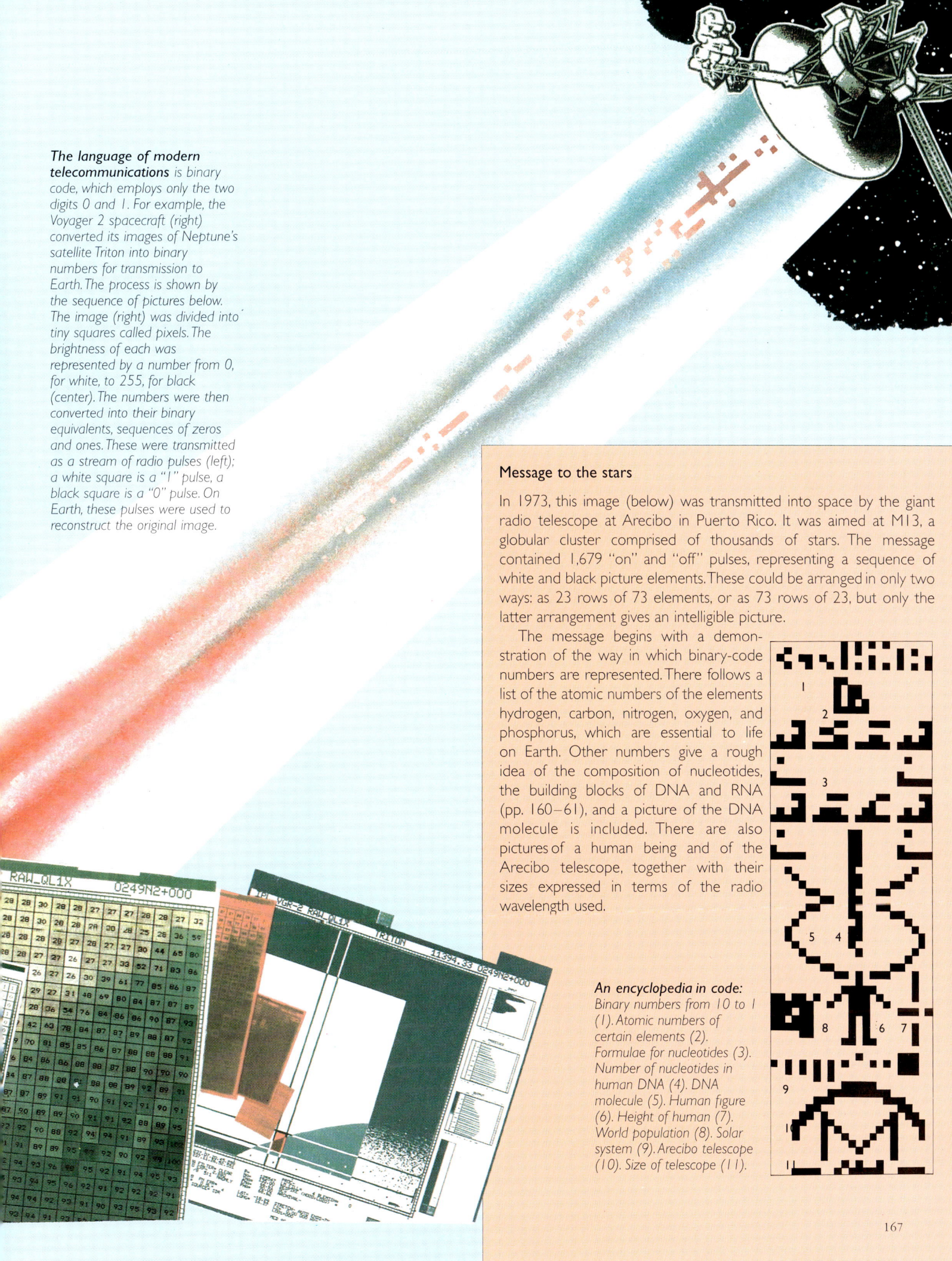

THE FUTURE OF THE UNIVERSE
● *The influence of missing matter*

In every direction astronomers look they see galaxies and quasars moving away from us. It would be natural to assume that this motion will continue forever, with the galaxies becoming more and more thinly spread. But the theory of relativity and the cosmology derived from it show that the universe can follow one of three possible courses in the distant future.

The general theory of relativity describes space-time as having a certain curvature (pp. 18–19). The first possibility is that this curvature is greater than zero, which is a mathematical way of saying that space is closed – it curves around on itself and is finite, although unbounded. The universe will expand to a maximum, before converging again.

The second possibility is that the curvature is less than zero. In this instance, space curves in a different way: it has a hyperbolic geometry (p. 18), and it is infinite. The galaxies spread apart, into infinity. The same thing happens if, thirdly, the curvature is zero, when space is Euclidean and, again, is infinite.

Astronomers would like to observe how expansion is changing with time; only then could they predict with confidence what will happen to the universe. They would like to be able to measure how the scale factor of the universe – the relative separation of two arbitrary points – changes with time. The scale factor is related to the red shift, so if we measure the red shift of a galaxy or a quasar, we can in principle obtain the scale factor of the universe at the time the light was emitted. But with the extent of present knowledge, this cannot be done precisely.

Cosmologists can approach the problem from another direction, however. The universe of today is dominated by matter. The density of matter is 1,000 times that of radiation of all kinds, of which the chief component is the microwave background. This density has a profound effect on the future state of things. Calculations show that if the density of the universe is greater than a certain critical value, expansion will gradually slow down under the influence of gravity and then reverse. The universe will contract, pulling everything together, leading to a Big Crunch.

If, on the other hand, the density of the universe is less than the critical value, gravity will not be strong enough to bring the galaxies to a halt, and the universe will expand forever.

If the density should be exactly equal to the critical value, then the galaxies will, in theory, come to a stop, but at an infinite time in the future. In effect, this is also a case of unending expansion.

The question is, therefore, how much matter the universe contains, and this is not easy to answer. Until recently, it seemed there was not enough matter to reach the critical density, but recent research is altering the picture.

Studies of our own and other spiral galaxies show that their velocities of rotation can be accounted for only if they possess perhaps twice as much material as that which appears on even the longest-exposed photographs. The recent discovery of dim, low-mass stars in our own galaxy indicates where some, at least, of such so-called missing matter may be found.

Again, it is now clear that galaxies in clusters are embedded in intergalactic material, much of which can be detected only at non-visual wavelengths. This provides a substantial amount of the mass required to close the universe; probably five items as much material is present in a cluster as was originally believed. Furthermore, unknown numbers of black holes, of unknown mass, may exist in galaxies.

In addition, we are uncertain as to whether exotic undiscovered particles, such as WIMPs (weakly interacting massive particles), permeate space. And there is still the possibility that neutrinos, which are present throughout space, have a minute mass.

Thus, within a few years, we may discover that the density of matter in the universe is above the critical value. In this event, the future of the universe will be a return to a concentrated state similar to that in which it started – even possibly to another Big Bang.

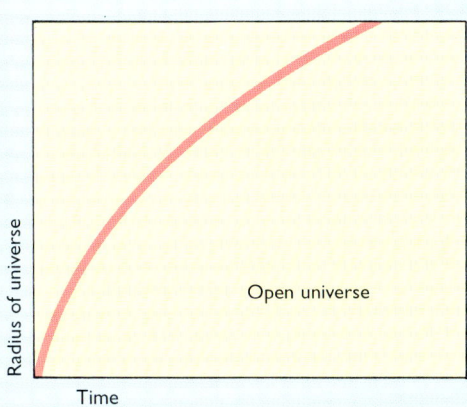

In an open universe, the amount of matter present is not sufficient to stop expansion, since its gravitational self-attraction is not strong enough. If the matter we can directly observe, or detect by its gravitational effects, is the bulk of what exists, then the universe is, in fact, open, and will expand forever.

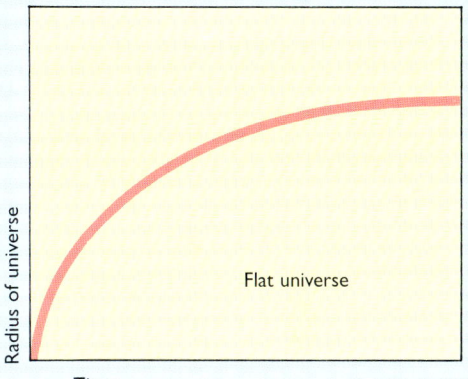

In a flat universe, the density is at its critical value, exactly on the borderline between expansion and contraction. Such a universe is infinite, and Euclidean on the large scale; its expansion will never quite come to a halt. The modern Big Bang theory gives strong reasons for believing the actual universe may be flat.

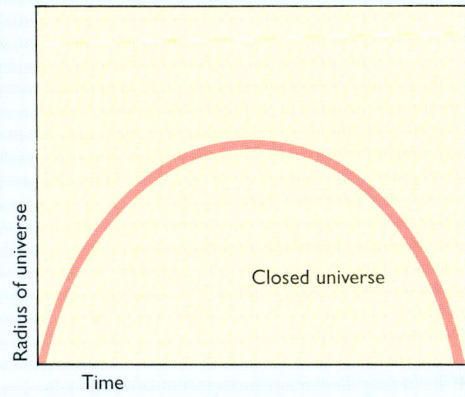

A closed universe is doomed to collapse on itself because its density is too high.

Neutrino detectors may throw light on the future of the universe. Neutrinos are ghost-like particles that swarm throughout space. They are so unreactive that they can travel through an entire planet without, in the vast majority of cases, interacting with any atoms on the way. Yet the number of neutrinos passing through us at each moment is so enormous that the tiny proportion of reactions adds up to a significant number of cases. This detector is at CERN, the European nuclear research center in Geneva. Each neutrino reaction releases a tiny flash of light which is detected by one of the thousands of photomultiplier tubes on the exterior of the device. Experiments suggest neutrinos have a mass. It is tiny – 1/100,000 of an electron. Nevertheless, it could be enough to help overcome the expansion of the universe and eventually turn it into a contraction.

OBITUARY FOR THE UNIVERSE

● *Death or rebirth*

If the universe is destined to expand eternally, the lines of the poet T.S. Eliot graphically echo the situation:

> This is the way the world ends,
> Not with a bang but a whimper.

Present knowledge of fundamental physics can enable us to foresee the history of matter over enormous intervals of time. But to discuss these remote times, it is necessary to use numbers many orders of magnitude larger than any that we have encountered so far. The lifetime of a star like the Sun is of the order of 10^{10} (10 billion) years. But the faintest, slowest-burning stars have lifetimes perhaps 10,000 times longer. So it will be in something like 10^{14} (100,000 billion) years that all stellar activity will be over, and there will be no stars left in the universe. The galaxies will consist of cold, dark matter.

In a further billion billion (10^{18}) years, the galaxies will be collapsing, because relativity theory predicts that in any system of orbiting bodies, energy will be radiated away in the form of gravitational waves. A proportion of the matter in the galaxies could be swallowed up in ever-growing black holes at their centers.

If, as some physicists believe, the proton is not stable, but decays after an extremely long lifetime, then matter itself will break down in the remote future. Protons will begin to vanish after not less that 10^{32} years from now, turning into lighter particles such as sitrons or muons. All the atoms in the universe that have not already been swallowed by black holes will disappear, to be replaced by a sea of lighter particles and radiation.

If protons do decay in this way, the final end of the universe will be marked by the evaporation of black holes (p. 65). This will occur over a great range of time scales, for the rate at which a black hole vanishes depends on its mass. A black hole with 10 times the mass of the Sun will evaporate 10^{68} years from now. One that is 10 times more massive will last 1,000 times longer – until 10^{71} years from now. The giant black holes will take longer still; they will last for approximately 10^{90}–10^{100} years.

However, if protons do not decay, the situation will be different. After the vast time of $10^{1,600}$ years, white dwarfs will all collapse to become neutron stars, and a very long time after that – too great to be able to describe conveniently, even in the powers-of-10 notation we have been using – all neutron stars will coalesce to form black holes. The end will come when these eventually evaporate, yielding a featureless universe of radiation and particles.

Most people will regard this protracted fading of matter as a less attractive prospect than the alternative of a closed universe, in which gravity will have the final word and the end will be violent. With the continuing discoveries of missing mass in the universe, such an alternative seems a definite possibility. In this case, the scenario is vastly different and the time-scale far shorter.

The actual time when expansion ceases and turns into collapse depends on the precise value of the Hubble constant, which is not known for certain. But after contraction has proceeded for billions of years, clusters of galaxies will begin to mingle, something like a billion years before the Big Crunch. Hundreds of millions of years will then elapse before the galaxies themselves begin to merge.

The merging of galaxies will result in a single super-hypergalaxy, exerting an immense gravitational pull on its constituent stars. Within another million years, the stars will approach each other so closely that the night sky will become as bright as the Sun. The temperature of space will rise until eventually it becomes hotter than the stars, which will explode. Black holes will grow rapidly in the hot, dense collapsing matter of the universe, and 100,000 years before the Big Crunch, they will be forming at a catastrophic rate, sucking up everything around them.

The end of everything could be the collapse of the universe into a singularity – a single point of space and time, where density and temperature become infinite, and theories of physics become invalid. But there might, instead, be a sequel in which the conditions prevailing during the Big Bang are re-created, the four fundamental forces are reunited, and the universe returns to its original state, ready to expand again.

If such a "Big Bounce" occurs, the universe will gain a new lease on life, expanding until gravity again takes over and brings about another contraction. The life cycle of expansion and contraction will be repeated continually, in a bouncing universe that lasts forever.

1 If the universe should begin to contract again at some remote future date, its evolution could resemble a film of its past history played in reverse. When the Big Crunch is only a few billion years away, the galaxies will be closer together than they are now, the sky will be brighter, and the temperature of the universe will have risen.

5 The universe could conceivably be reborn in another Big Bang following its collapse. If this occurs, then the universe would repeat the same cycle of expansion and collapse over and over again. In some theories, each cycle would be longer than the previous one.

4 In the last decade of the collapse, the black holes will begin to coalesce, until the entire universe is swallowed up inside a single supermassive black hole. This could mark the end of the universe and of time itself.

3 When the final collapse is perhaps only a few centuries away, the temperature of the background radiation filling space rises to the point where it tears the stars apart. At the same time, black holes will begin to swallow matter and radiation.

2 Eventually, the galaxies will be separated by distances comparable with their own diameters. They will interact strongly with each other and merge more frequently than they do now. But the increasing numbers of black holes within the galaxies will begin to make the process of collapse different from that of expansion.

HIDDEN DIMENSIONS
● *The superstring theory of fundamental particles*

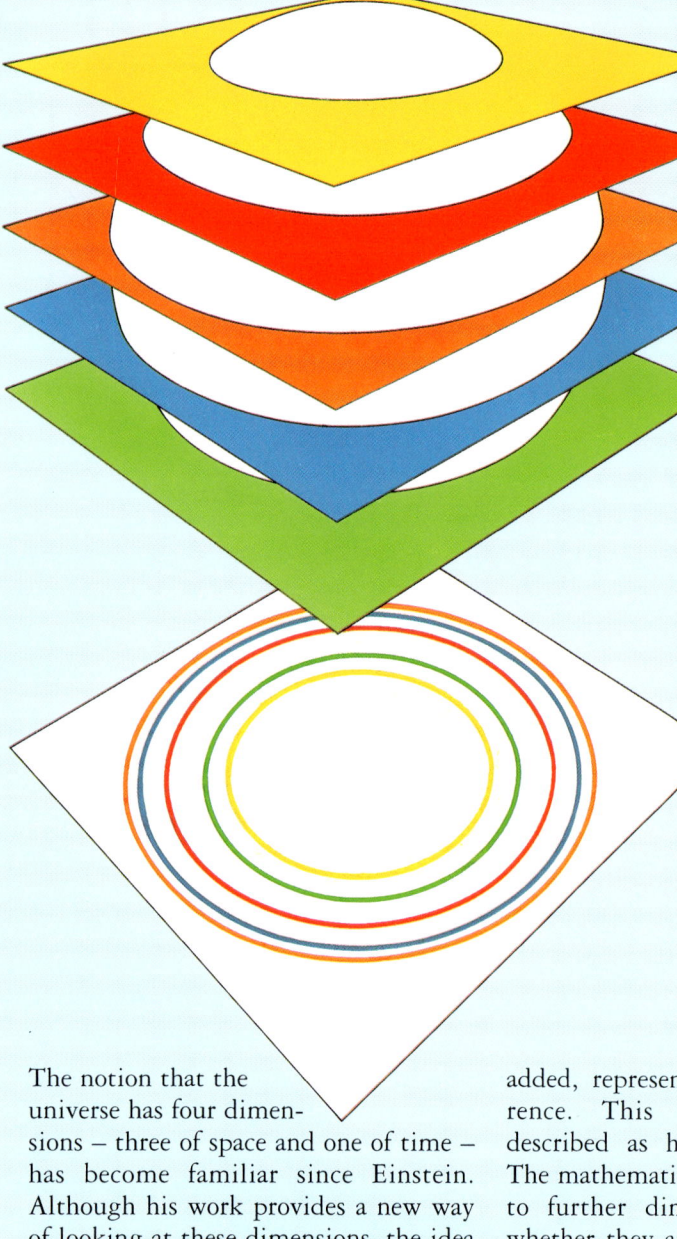

If the two-dimensional world of Flatland intersected a three-dimensional object such as a sphere (upper diagram), the latter would be perceived only as a series of two-dimensional cross-sections. To the Flatlanders, the sphere would appear as a disk, appearing from nowhere, expanding and then contracting (lower diagram). Creatures like us, living in three spatial dimensions, would have just as much difficulty in perceiving four-dimensional objects.

The notion that the universe has four dimensions – three of space and one of time – has become familiar since Einstein. Although his work provides a new way of looking at these dimensions, the idea seems to fit in well with our experience. Modern cosmologists suggest that the universe may possess more dimensions than this. Space-times with 10, 11, or more dimensions are seriously discussed.

Ordinary space is described as three-dimensional because three, and only three, numbers, or coordinates, are needed to specify any position in it. For example, the location of an aircraft is uniquely defined by specifying its longitude, latitude, and altitude.

To specify the location in space-time of an event, another figure needs to be added, representing its time of occurrence. This is why space-time is described as having four dimensions. The mathematician can extend this idea to further dimensions, regardless of whether they are considered to be real. For five dimensions, five coordinates are required; for six dimensions, six coordinates are required, and so on.

Yet while this may be legitimate for mathematicians, does it represent anything of physical significance? If there were a fourth dimension of space, for example, in which direction would it lie? Are extra dimensions just figments of the mathematical imagination?

How other dimensions could reveal themselves to us is best seen by considering the experience of a Flatlander – a being who lives on a flat surface and knows only two dimensions.

Suppose the Flatlander is sitting quietly on the beach, watching the sea (which is always smooth and calm in Flatland). Then something unprecedented happens: a ball passes right through the two-dimensional world. What will the Flatlander observe? Certainly not a ball, because he has no conception of a sphere.

The Flatlander first sees a point on the sea, where the ball just touches it. As the ball passes through the sea, a circular region of the sea is displaced. So the Flatlander sees the rim of a circle growing in the sea. When the ball is exactly halfway through Flatland, the circle reaches its maximum size. Then it begins to diminish, and becomes a point once more, before vanishing without trace.

The Flatlander has seen something very strange; the apparition of an object swelling up, shrinking, and vanishing again. But the occurrence has a perfectly simple explanation to three-dimensional creatures whose imagination is not restricted as is the Flatlander's. A Flatland mathematician, however, would in principle be able to work out by abstract reasoning something of the three-dimensional form of the ball.

Are we, as three-dimensional creatures, similarly restricted in our experience? There is reason to think that we are. Almost 80 years ago, the Polish physicist Theodor Kaluza wanted to extend relativity so that Einstein's geometrical approach embraced not only gravitation, but also electromagnetism. He achieved this without altering Maxwell's equations (pp. 14–15).

He showed that whereas in a four-dimensional world consisting of time coupled with the three dimensions of space, electromagnetism and gravity are separate entities, in a five-dimensional world consisting of time coupled with a four-dimensional space, they are aspects of the same entity. In short, he was able to unify gravity and electromagnetism by adding a new spatial dimension. But where is this extra dimension?

Five years after Kaluza's suggestion, the Swedish physicist Oscar Klein pro-

THE LIVING UNIVERSE

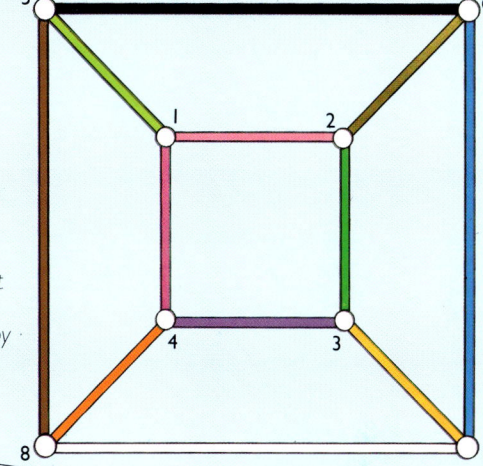

posed that the extra dimension is "rolled up" in such a way that we do not perceive it. In geometry, a point is supposed to have no length or breadth whatever. But now suppose that each point in Flatland is replaced by a tiny circle at right angles to the plane of the two-dimensional world. The circle would represent a third dimension – but one so tightly rolled up that it would be imperceptible to the Flatlanders. Just so, according to Klein, each point of our four-dimensional space-time should be replaced by a loop representing a tiny distance in the fourth spatial dimension.

The concept of a fourth dimension of space is not very difficult to imagine if it is done by stages, just as the imaginary Flatlander encountered the sphere in stages. Mathematically it is easy, and it is possible to be precise about the properties it would have. But imagination fails us when we come to picture 5 or, worse, 10 or 11 dimensions.

A strong motive for considering the possibility of extra dimensions of space is the development of the new superstring theory of fundamental particles. In the late 1960s, just before quarks were accepted as the basic units of matter, the Italian physicist Gabriele Veneziano suggested a new way of accounting for

The three-dimensional nature of a cube (left) is indicated by the fact that from each corner there extend three edges, all at right angles.

Representing the cube in two dimensions cannot be done satisfactorily by even the most skillful Flatland artist. The lattice (right) correctly reproduces certain relationships among the corners and edges – for example, it shows the cube's eight corners and the fact that each is linked directly to three neighbors by edges. But in squeezing three dimensions into two, it necessarily distorts some angles and lengths.

A four-dimensional hypercube can be represented in three dimensions by a lattice (left) in which each corner is connected by edges to four neighbors, as in the hypercube. However, in the hypercube, the edges are all at right angles, whereas in the lattice, as in any three-dimensional form, it is impossible to draw four such lines.

The meaning of coordinates

An example of a two-dimensional space is the kind of space described by a map. We can specify any point on the map's surface by two coordinates. The numbering is started at a given point – our own position on the map, say – and two distances are counted off – x miles east and y miles north. (Negative values of x or y would represent a distance west or south, respectively.)

Now suppose that information about height or depth is required. We need for our third dimension a third coordinate z, measured upward from sea level (with a negative value representing distances below sea level).

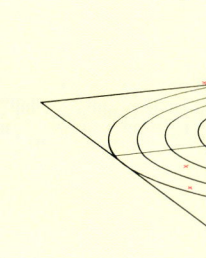

Two dimensions

In a two-dimensional map (above), each position can be specified by two coordinates, x and y. The map can picture flat objects, but not three-dimensional ones such as the Local Group of galaxies.

When a third dimension is added, represented by the coordinate z, the real spatial positions of the galaxies become apparent.

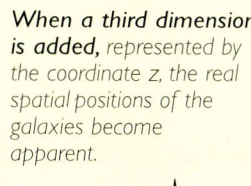

Three dimensions

173

the behavior of subatomic particles. His results were expressed in equations that turned out to be just like those of a vibrating string. It seemed, therefore, as if there were entities like elastic strings binding the nucleus together.

When the quark concept was developed, Veneziano's theory ran into difficulties and was shelved. However, it did successfully describe gravitons, the messenger particles of gravity, so perhaps string theory was really a theory of gravity.

In the early 1970s, other theories arose. One explained how quarks were bound together by messenger particles called gluons, which were later found to exist. In these theories, symmetry (p. 28) played an important part. One type of symmetry that has been proposed is called supersymmetry. It unites the two great families of particles: bosons (particles that have whole-number spins, such as protons) and fermions (particles that have fractional spins, such as photons and electrons).

But supersymmetry involves many more dimensions than the four of Einstein's space-time. One theory, supergravity, calls for 11 dimensions — 10 of space and 1 of time. Another theory is a development of string theory taking supersymmetry into account. Called the theory of superstrings, it is the result of the work of the physicists John Schwarz, an American, and Michael Green, an Englishman.

In superstring theory, the mathematics shows that particles can be described as vibrations of open strings, or of closed, loop-shaped, ones. The size of open strings is roughly equal to the Planck length, a distance of only 10^{-32} millimeters, equivalent to a hundred-billion-billionth of the diameter of the atomic nucleus. The vibrations of open strings produce massless particles of spin 1, such as the photon. Open strings can close up to form loops, which produce other kinds of particles, including the massless spin-2 gravitons, which are yet to be observed.

Open strings and closed loops are combined in the heterotic, or "cross-bred," superstring theory. According to this theory, vibrations moving around a loop in a clockwise direction are 10-dimensional, while those moving counterclockwise are 26-dimensional.

In relativity, the trajectories of particles through space-time are called world lines (pp. 62–65). In superstring theory, the fundamental strings and loops sweep out a two-dimensional surface in space-time, known as a world sheet, which is analogous to the film of a soap bubble. Interactions between strings and a shimmering motion of the world sheet account for the quantum behavior of subatomic particles, and of messenger particles as well.

Since superstring theory requires 10 space-time dimensions and Einstein's space-time occupies 4, the remaining 6 must be rolled up, in the manner suggested by Klein. However, the theory does not yet explain why this should be so. Possibly, in the earliest moments of the Big Bang, all dimensions were rolled up and were equally important. For some reason, only three space dimensions have since unrolled to the colossal size of the present universe.

The rolled-up dimensions have extremely strong curvature, which must be measured in terms of the size of the strings, 10^{-32} millimeters. If a particle could travel at the speed of light through one of the extra dimensions and return to its starting point, it would be gone for a time less than the Planck time of 10^{-43} seconds. Its absence would never be noticed on the macroscopic scale.

Furthermore, according to the Heisenberg uncertainty principle (p. 25), such small distances can be probed only with extremely high energies. In fact, to explore distances of the order of the Planck length would require giving a particle energy of a magnitude that has not existed since the Big Bang.

An interesting consequence of superstring theory is that there may be a previously unsuspected type of matter, which can be detected solely by its gravitational effects. It has been called shadow matter, and could contribute to the missing mass that is believed to exist and that could, just possibly, turn the expansion of the universe into a contraction.

Superstring theory awaits confirmation. It is too soon to make sweeping judgments about it, but it holds the possibility that it is a step toward the most fundamental basis of physics: a TOE, or theory of everything.

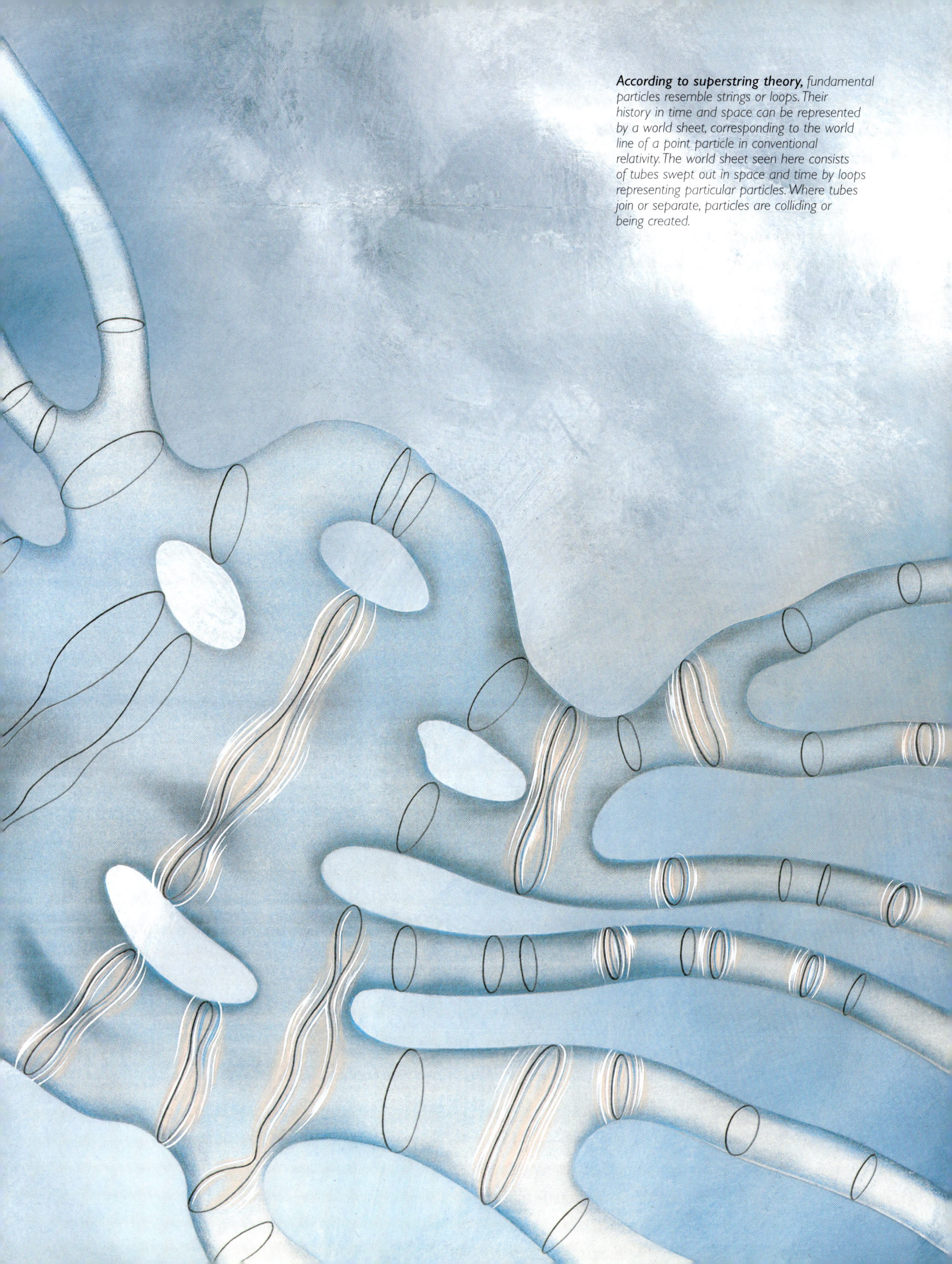

According to superstring theory, fundamental particles resemble strings or loops. Their history in time and space can be represented by a world sheet, corresponding to the world line of a point particle in conventional relativity. The world sheet seen here consists of tubes swept out in space and time by loops representing particular particles. Where tubes join or separate, particles are colliding or being created.

QUANTUM SPACE
Fabric of the microscopic world

The science of today confronts a barrier in its attempts to get back to the very start of the Big Bang, before the crucial Planck time of 10^{-43} seconds. The known laws of physics break down and become unusable under the extreme conditions of space and time then prevailing. Theorists are struggling to extend those laws or develop new ones, gaining new insights as they do so.

Another line of attack is the study of gravitational radiation. Relativity predicts the existence of gravity waves, and quantum theory predicts that, like all other kinds of wave, these should in some circumstances appear in the guise of particles. These conjectured messenger particles are called gravitons. The experimental detection of gravity waves or gravitons would be a crucial step in the unification of gravity and the other fundamental forces.

Since the 1960s, a number of attempts to observe such waves have been made. The most famous are those of the American physicist Joseph Weber, who set up a huge bar of pure aluminum, weighing 4.4 tons (US). Such a lump of metal would be too massive to respond significantly to ordinary, local disturbances, such as traffic vibrations, or seismic tremors. But it would be squeezed and stretched by passing gravitational waves, which would make it ring like a bell, even though its distortion would amount to less than the size of an atomic nucleus.

Weber fitted the bar with extraordinarily sensitive detectors. Early on, he did get what seemed to be a positive result, but it was later ruled out as a false alarm, and the instrument seems to have recorded no gravity wave events since.

Those physicists who have not been discouraged by the immense difficulties of detection have been devising new methods. In one of these, the detector is comprised of two beams from a laser. Each beam is directed along a stainless steel pipe 2 miles long, at the end of which it is reflected by a mirror mounted on a massive lump of metal to keep it highly stable. To magnify the expected effects, the laser beams are reflected to and fro 50 times, making the length of each arm equivalent to 100 miles. Eventually, the beams meet and their waves interfere with each other.

A gravity wave will stretch and squeeze space as it passes, briefly altering the distances traversed by the light beams. The interference pattern, consisting of a pattern of bright and dark lines, will momentarily shift.

Three pairs of spacecraft, collectively known as LISA (Laser Interferometry Space Antenna) are to measure gravitational waves in space. Once in orbit around the sun, the craft will use lasers to determine if a craft has shifted position because of a gravitational wave.

But observations relevant to the existence of gravity waves have already been made. The pulsar PSR 1913 + 16 is part of a binary system – that is, it is orbiting another star, taking about $7^3/_4$ hours to do so. It was calculated that it should be losing energy by the emission of gravity waves, with the two stars falling together. In 1974, a lengthening of the pulsar's orbital period of 7.5 millionths of a second per annum was detected, and it seems extremely likely that this energy loss is due to the predicted gravitational radiation.

If the messenger particles of gravity, the gravitons, exist, they will be subject to the same uncertainty relations as all other particles (p. 25). The time and place of their absorption or emission can be specified only to a certain degree of precision. As gravity distorts space, so the presence of gravitons will cause space to curve around them. The quantum uncertainty will make the curvature fluctuate: space-time can be thought of as rippling because of the presence of the gravitons. But such rippling will be minute; compared with electromagnetism or the nuclear forces, gravity is an extremely weak force. Even a body as massive as the Sun deflects starlight passing close to it by only a tiny amount.

The uncertainty principle also means that for a split second, particles can borrow quantities of energy. The amount to be borrowed and the duration of the loan depend on the strength of the force carried by the particles. In the case of gravitons, which carry the very weak force of gravity, the loan can only be for a very short time indeed – no more than the Planck time of about 10^{-43} seconds.

In that short time, a particle, traveling even at the speed of light, can travel only about 10^{-32} millimeters, the Planck length. So it is on a scale smaller than this that

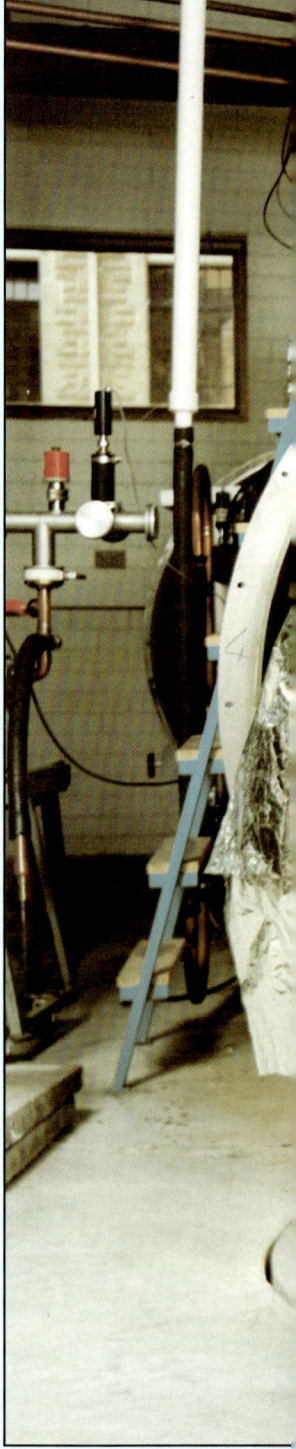

A gravity wave detector designed by the pioneer in this field, Joseph Weber. Located at the University of Western Australia, it consists of a niobium bar that would "ring" like a bell for a long time after a gravitational wave had passed through it. Although such vibrations would be smaller than the nucleus of an atom, they could be distinguished from the background noise of mechanical and thermal vibrations. It is claimed that bar antennae of this type detected gravitational waves from the supernova observed in the Large Magellanic Cloud in February 1987.

we encounter ripples in space-time. And such ripples make our present theories invalid for the very earliest moments of the Big Bang, when the universe was still of this microscopic size.

All this indicates that, when it proves possible to extend physics into this so far forbidden area, we shall find that space itself is no longer continuous, but quantized – divided into elementary units. It may be like a sponge, displaying a discrete structure of ultramicroscopic dimensions. According to the American physicist John Wheeler, we shall even find "wormholes" leading from one part of space to another. Perhaps even past and future will no longer be so clearly distinguishable in this new microcosm.

BIG BANG IN QUESTION
● *Challenges to orthodoxy*

The problems encountered in trying to probe the earliest moments of the universe would become irrelevant if the Big Bang theory were to be abandoned. Although it appears to fit in extremely well with nuclear physics and with the universe as observed from Earth, a few scientists have questioned its validity.

In the late 1940s, the British astrophysicist Fred Hoyle and his associates Hermann Bondi and Thomas Gold proposed their famous steady-state theory of the universe. This suggested that the universe always looks the same, from any viewpoint and at any time. Thus, although galaxies are born, evolve, and move away from each other, they are continually replaced by newly-created matter, in the form of hydrogen gas, which evolves into galaxies and stars in due course. Such a universe has no beginning and no end.

After many years in which the steady-state theory lay fallow, Hoyle once again proposed it as a serious contender in cosmology. He claimed the theory could now explain the observed abundances of deuterium, hydrogen, and helium, which are so successfully explained by the Big Bang theory.

One of the strongest arguments in favor of the Big Bang view of the universe is the presence of the microwave background. This is regarded as being left over from the Big Bang itself, a cooled remnant of the hot fireball that gave birth to the universe. Yet, Hoyle made a new suggestion: that the background radiation is due to comparatively recent events – supernovae. Astronomers agree that it is in such explosions that the heavier atoms in the universe – particularly iron – are formed. In Hoyle's new suggestion, the iron atoms went on to form long, narrow "whiskers."

If the vapors of metals are slowly cooled, most crystallize into metal whiskers. Typically, they have a thickness of two millionths of a millimeter and a length of no more than 1 millimeter. Hoyle argued that iron whiskers of this size, floating in interstellar space, would absorb infrared and short-wavelength radio waves and re-emit them with the spectrum of wavelengths of the background radiation.

The quasar Markarian 205, which appears green in this image, seems to be physically linked with the galaxy NGC 4319 above it; faint signs of a bridge of gas can be seen here. Yet the spectrum of Markarian 205 shows a red shift 10 times greater than that of the galaxy, which, according to most astronomers, would indicate a correspondingly greater distance. Opponents of the Big Bang theory suggest that in such cases, the active object is being ejected from the associated galaxy at high speed, causing the high red shift.

What is more, observations of the pulsar lying at the heart of the Crab Nebula – the scene of the supernova explosion in AD 1054 – show a drop in the radiation in the bands that metallic filaments would be expected to absorb. This could be due, Hoyle said, to the presence of such iron needles.

One difficulty for the steady-state theory is posed by quasars. When astronomers observe objects in the remote

The steady-state theory suggests that the universe always looks the same. As the galaxies move farther apart, new matter is continually created, so the universe does not thin out with the passage of time. The diagram shows part of the universe expanding and new material forming within it, thus keeping the average density constant.

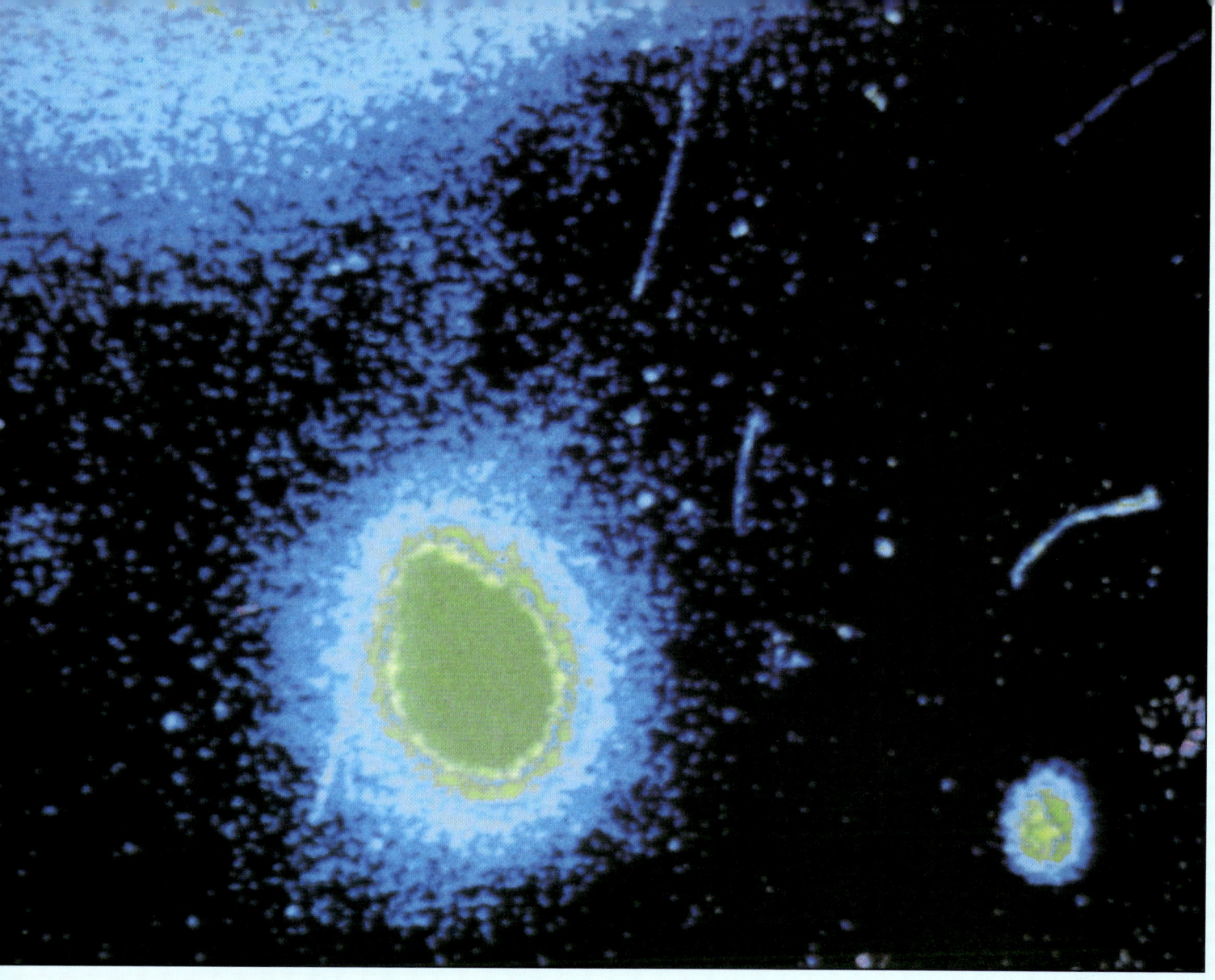

universe, they look backwards to the time when light left them. According to the steady-state theory, such distant regions should look the same as nearby ones. But it seems that quasars are more numerous at extreme ("cosmological") distances, and hence, earlier epochs.

One line of attack on this problem questions the nature of quasars. Because Hubble's law tells us that distant objects move away from us at velocities that increase with distance, quasar red shifts are normally interpreted in the same way. They are so large, it is assumed, because the objects emitting them are so distant. This view has been attacked by the American astronomer Halton Arp, among others.

Arp claimed that many photographs show quasars that are close to, and apparently connected with, ordinary galaxies. Yet, the companion galaxies do not show large red shifts, as they would if they were at the distances claimed for the quasars. Arp interpreted the quasars as lumps of matter ejected by the galaxies at very high speeds, thus causing large red shifts.

Most astronomers believe that such apparently close associations are due to quasars happening to lie almost in the same line of sight as the galaxies; the alignment is purely fortuitous, they say. Not so, say the opposers of the Big Bang, who claim there are too many alignments to have occurred by chance.

It is also said that many alignments are optical illusions, caused by gravitational lensing. Those few taking Arp's side retort that in many cases, the quasars are too far from the associated galaxies to be caused this way. They believe, too, that quasars are associated with most galaxies (though most are unobserved), whereas lensing must be very rare.

In short, they believe the quasars we observe are comparatively nearby. Hence, on the large scale, they may be distributed uniformly through the universe, as the steady-state theory requires.

Arp's alternative explanation removes the necessity of believing in a vast output of energy from quasars. If quasars are relatively close, their apparent brightness would not indicate the emission of prodigious amounts of energy. Thus, they would not require the special conditions of galaxy formation that are an integral part of the Big Bang theory.

However, the views of Arp and his sympathizers do not seem to explain why all quasars display only red shifts. Blue shifts are not observed, yet if quasars consist of material ejected from galaxies, one would expect to see as many moving toward Earth as away from it, and hence as many blue shifts as red shifts.

With the difficulties in Arp's position, and the successes of the Big Bang theory, most astronomers remain unconvinced by the latter theory's rivals.

THE ANTHROPIC PRINCIPLE
● *A universe designed for human beings?*

For most of humankind's history, the Earth was thought of as the center of the universe. This was part of a general outlook shared by all civilizations, each of which put itself at the center of things. To the ancient Egyptians, their country was the center of the world, and their universe was long and narrow, like Egypt itself. To the people of ancient Mesopotamia, whose country spread over an area that was more nearly circular, the heavens were a dome.

With the coming of the great civilization of ancient Greece, various rival cosmologies were proposed. But the dominant conception of the universe was a sphere, with the Earth, also spherical, immovable at its center. This view, coupled with an Earth-centered mathematical analysis of planetary motions, seemed so incontrovertible that the geocentric concept was accepted by all scholars for well over 1,800 years. So for a very long time, people were brought up in the belief that the Earth was the center of the universe, with humankind as its crowning glory. *Homo sapiens* was the centerpiece of creation, the lord of an anthropocentric universe.

In 1543, Nicolaus Copernicus, a Polish ecclesiastical administrator, proposed a new mathematical picture of the universe, with the Sun, not the Earth, at its center. Later studies by Kepler, Galileo, and Newton supported this view, so that within a little over a century humankind was dethroned from its prime position to that of the inhabitant of a not very large planet, in orbit around what later turned out to be a not very significant star.

This new view came at a time of intellectual revolution. Modern science was becoming established, with its insistence on observational and experimental evidence, backed up by mathematical analysis, as the touchstone for any theory. Humanity's demotion to an insignificant place seemed to fit in perfectly with this mechanistic universe governed by cold, impersonal laws.

During the 20th century, our concepts of the laws of physics suffered a severe shock with the advent of quantum theory. Now, particles are no longer to be regarded as solid, with definite positions and motions, but are best described as waves of probability. What is more, it has become clear that in observing such particles, the very act of observation has its effect on the particle itself. So now the mechanistic universe of the previous centuries of modern science has broken down – at least in the microscopic world of subatomic particles. We live in a universe that has elements of randomness.

But underlying the new universe of

Our place in the universe

It was natural for early astronomers to imagine a universe with our viewpoint, the Earth, at its center. Religious authorities, too, felt that humanity, having been created by God, must be at the center. The Greek astronomer Aristarchus thought the planets might orbit the Sun, but this idea attracted little support in his day.

Then, in the early 1500s, the Polish astronomer Nicholas Copernicus demonstrated that a system with the Sun at its

An 18th-century orrery, or mechanical model of the solar system, epitomized the new world picture in which the Earth had been dethroned from its central position.

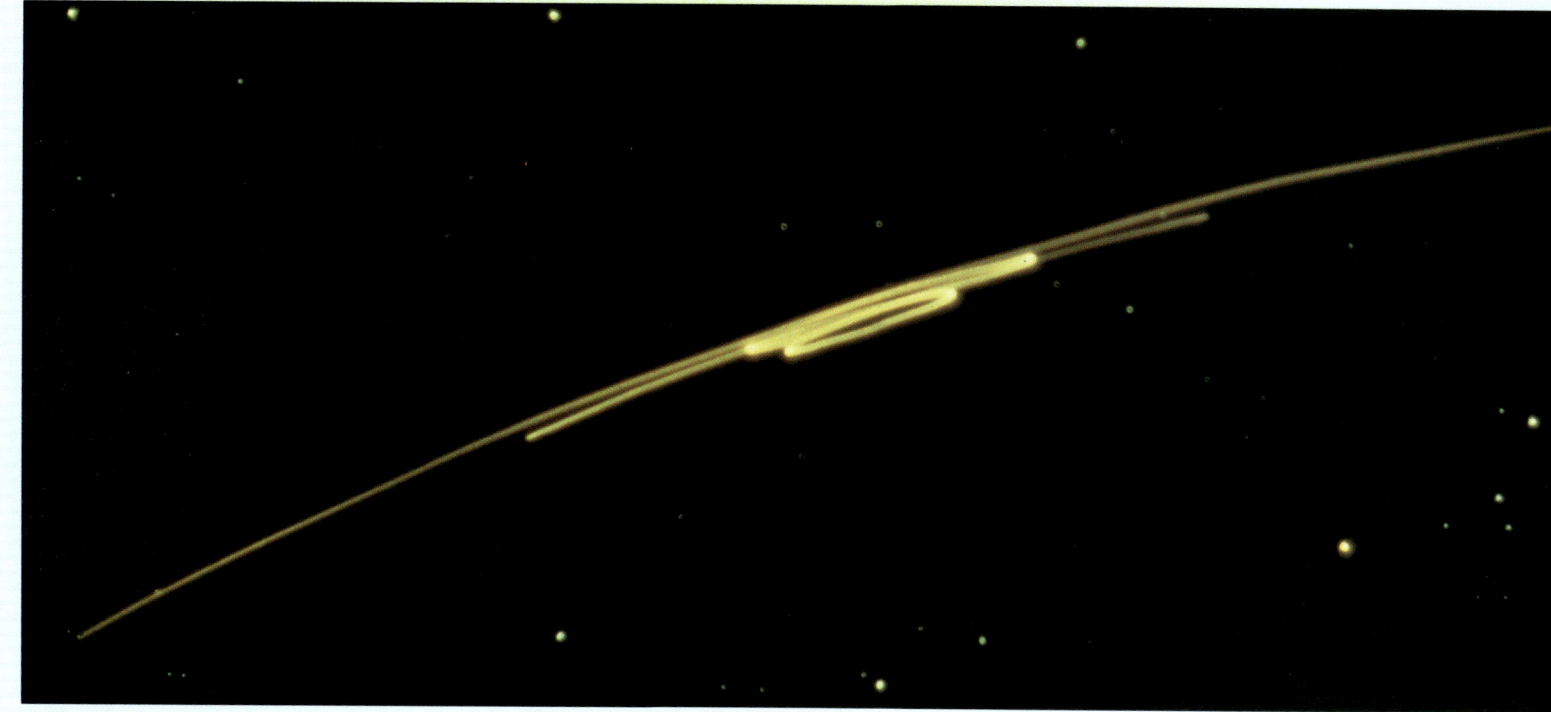

Mars sometimes seems to exhibit retrograde motion when viewed from the Earth – it loops backwards. Copernicus realized that real motion like this is unlikely, and that the effect is caused by the Earth passing Mars, both planets being in relatively simple orbits around the Sun.

21st-century physics and astronomy is a strange pattern linked with the four fundamental forces and some basic quantities associated with them. These basic quantities include the masses of particles such as the proton and electron, their electric charges, and the strengths of gravitation, electromagnetism, and the strong and weak nuclear forces. They also include Planck's constant, which sets the scale of quantum effects, and the so-called fine-structure constant, which describes how an electron behaves in an electric field.

The numerical values of these depend on our choices of units and therefore have no intrinsic significance, but pure numbers not dependent on choices of units can be obtained by considering various combinations of these quantities. The results are surprising.

They include, for example, the so-called large-number coincidences. The electrical force between the proton and electron in a hydrogen atom is 10^{39} times as strong as the gravitational force between them. This enormous ratio is almost the same as that between the size of the observable universe and the size of an electron – about 10^{40}. Furthermore, 10^{40} multiplied by itself equals 10^{80}, which is the order of magnitude of the number of atoms in the observable universe.

Furthermore, as the British scientist Martin Rees has pointed out, the lifetime of a star is related to the time it takes a photon to struggle out from the central regions to the surface. This is because the lifetime depends on the star's mass, and a link is found between the gravitational attraction of the star (and hence its mass) and the time that a photon takes to cross an atom. This link can be expressed by a pure number – which once again comes out to be 10^{39}, whatever the size of the star.

Since these numbers are so vast, there is a great range of values that they could have taken. That they should have turned out so close to each other can scarcely be a coincidence. These relationships seem to be a fundamental characteristic of the universe. They suggest a basic order to the universe that is poorly understood as yet.

There are other sorts of coincidence among the fundamental characteristics of the universe. They involve the relative "tuning" of the fundamental forces and are crucial to our own existence. For example, we live in a universe that is expanding at a rate that was set in the Big Bang. If gravity had been only a little stronger, it would have taken over early on and made the new universe collapse; if

center would provide a simpler and more acceptable explanation for the relative motions of the planets. We now know that his scheme is in principle correct and that the Earth is not at the center of our planetary system. The solar system is also nowhere near the center of our galaxy.

But the idea that the Earth has a special importance is once again gaining favor; some scientists now believe that the universe is of a kind that favors the development of living things, and the only place known to support life is the Earth.

it had been weaker, the expansion would have become a runaway process, and galaxies and stars would not have had time to form. If either of these alternatives had occurred, there would never have been an Earth on which life could evolve.

Again, if other basic ratios and constants of nature were different, the universe would not be favorable to our existence. If, for instance, the strong and weak forces were slightly stronger in relation to electromagnetism than they are, hydrogen would not exist in its ordinary form. This would mean that heavier elements, such as carbon and oxygen, would never have come into being, and there could have been no living things.

Yet again, if the weak nuclear force were of slightly different strength, supernova explosions would not occur. So some sources of heavy chemical elements would be denied to the universe, for it is in supernova explosions that the heavier elements are scattered through interstellar space, to be incorporated later in planetary systems.

The strength of gravity is crucial in another way to the existence of life in the universe. If it were weaker, then it could not crush the material in a star the size of the Sun strongly enough to ignite thermonuclear reactions. Only very massive stars could shine by nuclear processes, and such stars would probably have lifetimes too short for any evolution of life to occur.

These hints that the universe has been arranged to favor the appearance and survival of life have been taken very seriously by certain scientists. Prominent among them is the astronomer Brandon Carter. He pointed out that the time it has taken for the evolution of *Homo sapiens* from the first appearance of life on Earth is about four billion years. He also claimed that the average period for any evolution of this kind should be much longer – longer, in fact, than the 10 billion years that is the lifetime of a Sun-sized star, and longer still than the period during which conditions favorable to life could exist on the Earth.

If this is so, then it would seem that intelligent life appeared on the Earth despite its being a highly improbable event. In 1974, this led Carter to propose what he called the anthropic principle: namely, that our universe reflects the particular viewpoint of our own species. His bald statement has been the subject of intense study, and several related theses have been distinguished.

The weak anthropic principle states that observed features of the universe must be restricted by the requirement that carbon-based life can evolve and that there must have been sufficient time for it to do so.

Some scientists would go further and say that the universe was somehow arranged to bring humankind into existence. Carter named this idea the strong anthropic principle. It states that the universe must have those properties that *allow* intelligent life to develop. And some go further still and say that the universe must be such that intelligent life *will* appear.

The word "must" has triggered fierce controversy because it introduces a principle that appears to lie outside science: that the universe was designed with such a purpose in mind. Some claim that this is a metaphysical concept, and science eschews metaphysics. Those who favor it claim otherwise, because they say that the presence of humankind really is exceptional.

Certainly, the universe has very specific properties that have allowed humankind to evolve. Given that it has these properties, the evolution of human beings appears to be a foregone conclusion.

What is more, humankind has unique abilities. Other animals have evolved on the Earth, but humans are the only species that can formulate the laws of physics and thus understand the nature of the universe.

The fact that this can be done is astounding. Why should the universe be comprehensible to us? Is it just a fantastic coincidence, or is there some deep reason why it has turned out that way? If there is such a reason, then we do occupy a very special place in the universe.

If the human species does have a privileged position, is it alone in its pre-eminence? Are there other civilizations out in space, of beings that can also understand the universe? We do not know. But as discussions on communication with alien intelligences show (pp. 166–67), there is no evidence that our messages have reached anyone, nor have any been received on Earth. After all, radio and television signals have been traveling outward from the Earth for many years now, yet we have received no reaction from space.

Possibly this is because not enough

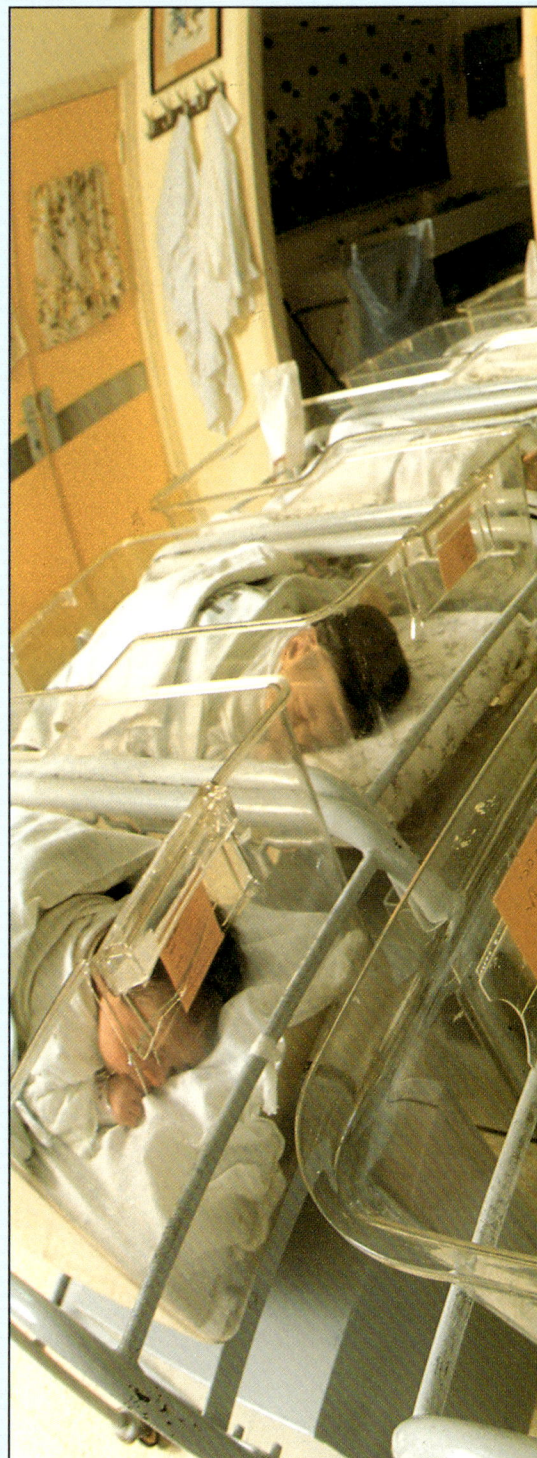

time has elapsed; it may require thousands of years before such signals can reach the nearest star system that is the home of intelligent beings that could respond in kind. However, the American scientist Frank Tipler has suggested otherwise. He has argued that within a billion years of its appearance, any intelligent species would be capable of colonizing the Galaxy. If humankind were not unique, these others should have reached here by now.

If he is correct, then we return to a belief in the unique position of human beings in the universe. It is different from older views in that humankind is not now taken to be at the physical center of the universe. But it resembles them in that our species is restored to a pre-eminent place in creation.

These children are more than six billion years ahead of their time, according to Brandon Carter, a proponent of the anthropic principle. That is how much longer he has said it should have taken life to reach such a stage of development. He has concluded that the universe has been deliberately arranged to favor the development of life. Some supporters of this theory think the universe has been arranged specifically to favor one species – human beings.

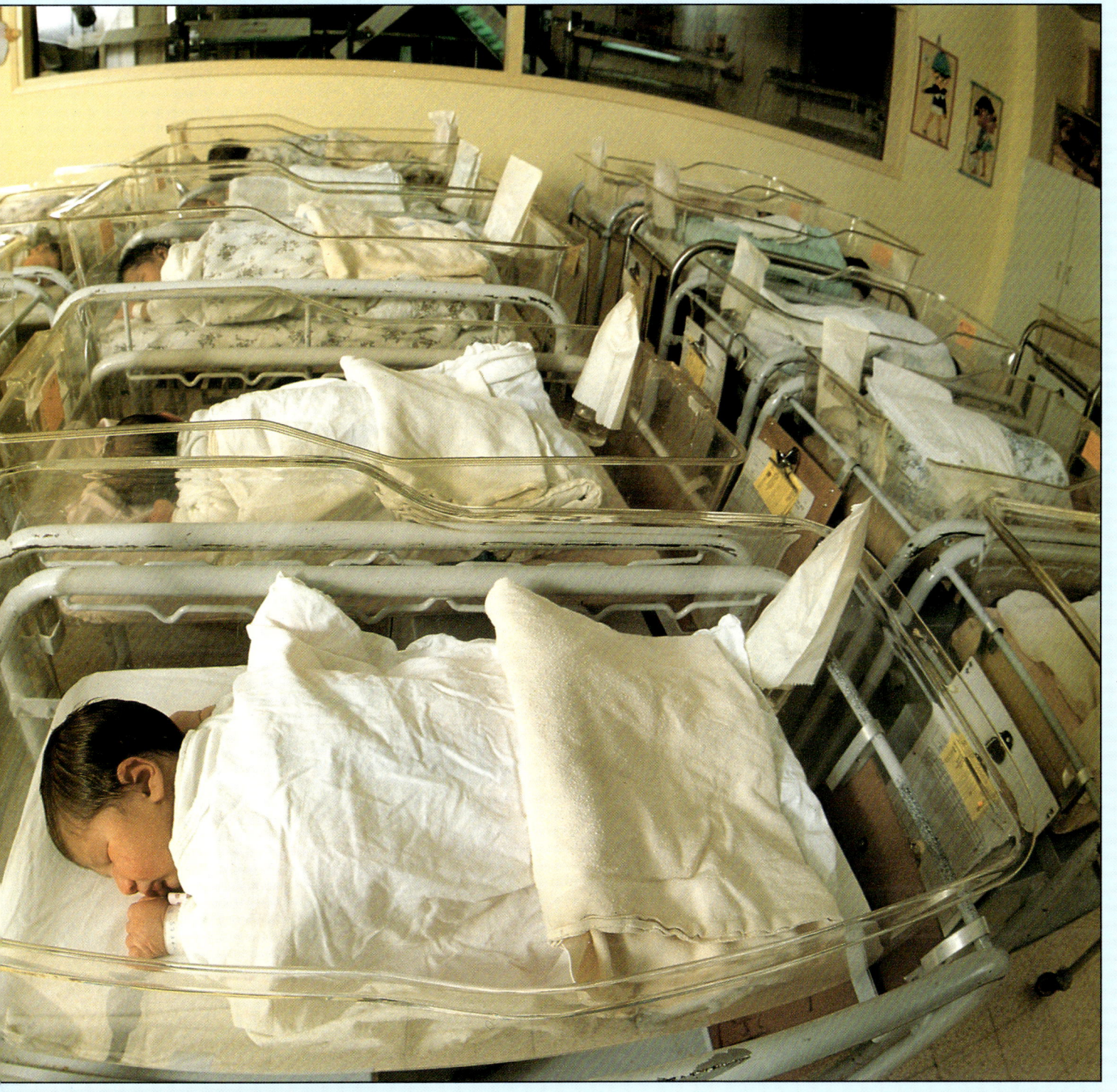

CREATING THE UNIVERSE
- *Is reality in the eye of the beholder?*

The anthropic principle brings us face to face with the problem of the nature of the universe and our place in it. The universe now has to be viewed not as a clockwork mechanism, as philosophers once believed, but as containing an element of chance. It has to be visualized, furthermore, as a blurred picture, with its particles resembling waves or even ripples rather than points. But more than this, human beings, as observers, seem to play a more important role in the universe than was formerly imagined.

A scientist can be sure that an electron is at a certain place only if it produces some appropriate event there – if, say, it hits a television screen and produces a bright spot. Until the intervention of the observer, it cannot be said even to have a definite position.

This realization has led some physicists to revive a view once expressed by Bishop George Berkeley, a younger contemporary of Isaac Newton, who said that " . . . all those bodies which compose the mighty frame of the world . . . have not any subsistence without a mind."

The physicist can pursue this train of thought further by an experiment in which a beam of photons is split into two overlapping beams that interfere with each other. Letting these beams fall on a screen produces an interference pattern, displaying alternate light and dark bands. The bright areas are caused by photon waves reinforcing each other, the dark ones by waves canceling each other out.

Suppose the original beam is now dimmed so that the number of photons is drastically reduced until, finally, only one photon passes through the apparatus at a time. Even in these conditions, interference still takes place. Although no one can say where a given photon will go, it is more likely to go to a part of the screen where a light band falls, and to avoid one where a dark band falls. If a large number of such events occur, the interference pattern will build up. Although there is only one photon in the apparatus at a time, each seems to interfere with itself – as if it were present in both beams.

If physicists probe further by adding detectors, they can discover which of the two possible paths a given photon has taken, but at the price of destroying the interference pattern. Their intrusion changes things. By compelling the photon to behave in a particle-like way, they prevent it from exhibiting wave-like behavior. Or to put it another way, only when the particle is detected – observed – does it fully become a particle, or fully real. This is why some physicists say that the universe depends on our observations. And since everything we observe, from molecules at one end of the scale to stars and galaxies at the other, is composed of subatomic particles, this argument must embrace the entire universe.

In the 1930s, the British astrophysicist Arthur Eddington remarked that when we examine the universe, we find a pattern of footprints, and when we study these, we find they are the footprints of humankind; our theories of the universe have characteristics that stem from the fact that they are human constructs.

But some quantum scientists are now saying more: that the universe exists only because we are observing it. The strong anthropic principle states that the universe is such that it must produce intelligent life, but according to the American physicist John Wheeler, quantum physics shows that without participating observers, there would be no universe. Observers are necessary to bring the universe into being.

If this participatory anthropic principle, as it is termed, is true, then our

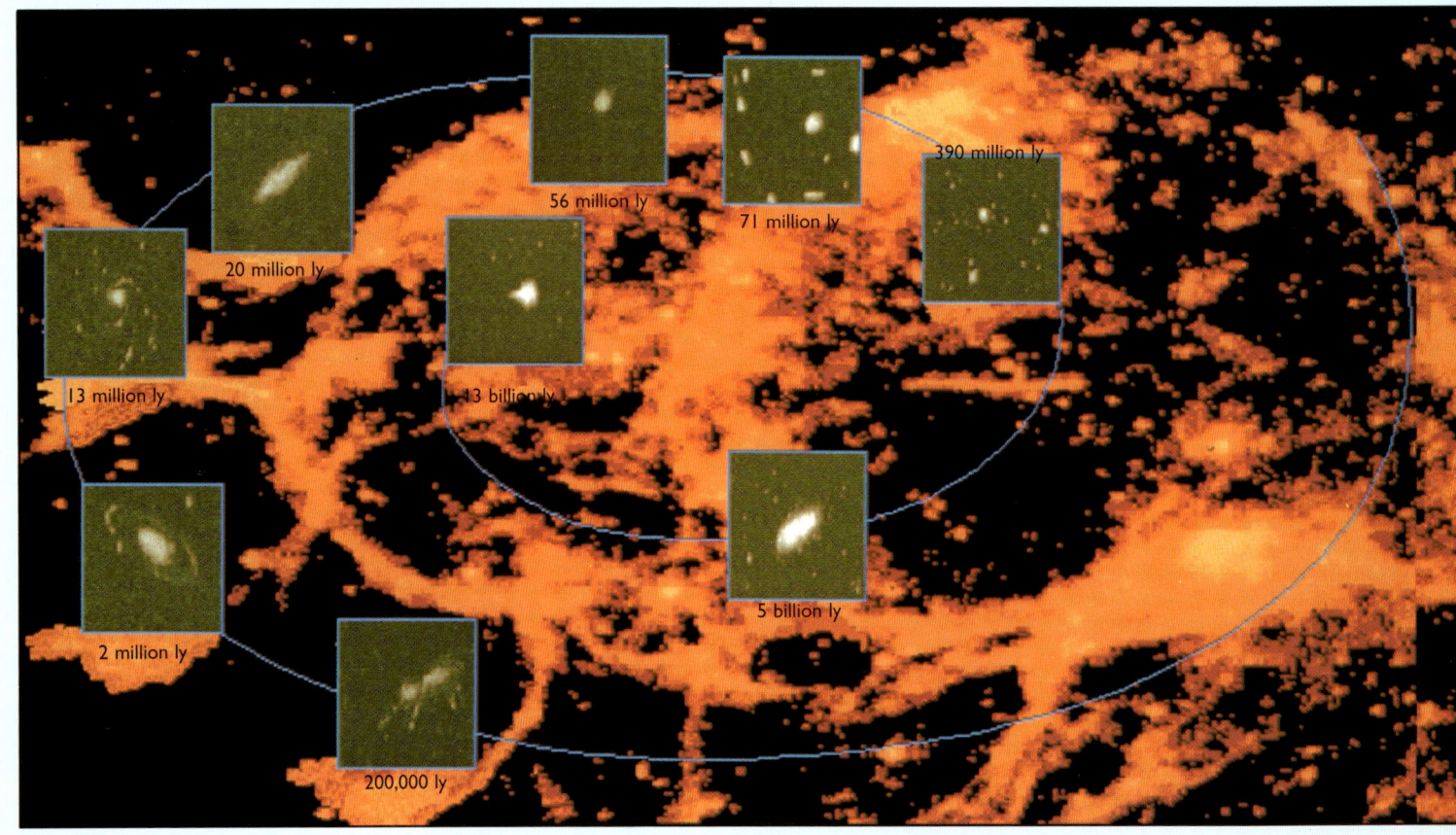

observations affect things in the past as well as in the present. When, say, a distant quasar is observed, its photons are projected into full reality in the moment that they are seen. In the preceding billions of years, they did not fully exist, and neither did the quasar – in respect of some of its properties, at least. Everything in the universe depends on intelligence comprehending it; without that, there is nothing.

With this view, the future of the universe poses a grave problem, for it seems that humankind must always be present to guarantee its existence. The American physicist Frank Tipler has tackled this question. He proposed that no other intelligent beings exist beyond the Earth, but he pointed out that the presence of humankind is not essential to satisfy the participatory anthropic principle; what is necessary is the continued existence of a high level of intelligence. His solution is therefore to distribute copies of human intelligence throughout the universe.

Tipler suggested that the mind is essentially a computer program. At present, it is housed in a particular kind of computer, the human body, but it is the program that is important. He looked forward to the creation of computers that are copies of ourselves. Then, even if humanity is swallowed up when the Sun becomes a red giant, copies of human intelligence will contemplate the universe forever – or until the Big Crunch. Thus, the participatory anthropic principle will continue to be satisfied.

As you might expect, this principle does not commend itself to many scientists. It seems to most to be extravagantly anthropocentric, because it makes humankind not merely the center of all things, but the measure, and even the creator, of the universe.

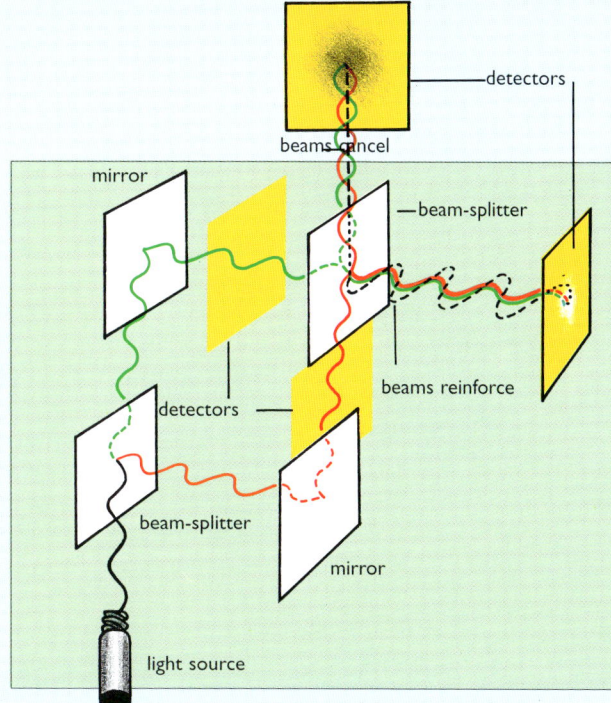

An observer can impose particle or wave behavior on light. Here, a beam that is split, recombined and only then observed, interferes with itself, so no light reaches the top detector. But if detectors are inserted before the interference occurs, they show which path each photon takes. This forces the light to behave as particles, and destroys the interference effects.

To look into space (below left) is to look back in time (below). We see the Small Magellanic Cloud as it was 200,000 years ago, when Homo sapiens was emerging; the Andromeda galaxy as it was when earlier types of human lived, 2 million years ago; galaxies beyond the Local Group as they were 13–20 million years ago, when early apes flourished. The light from galaxy M77 set out 56 million years ago, when dinosaurs were recently extinct, and from the Virgo cluster 71 million years ago, when they ruled the Earth. We see the Corna cluster as it was when animals had not yet left the sea, 390 million years ago. Light from some galaxies set out at the solar system's birth, 5 billion years ago; and from some quasars 12 billion years ago, as our galaxy formed.

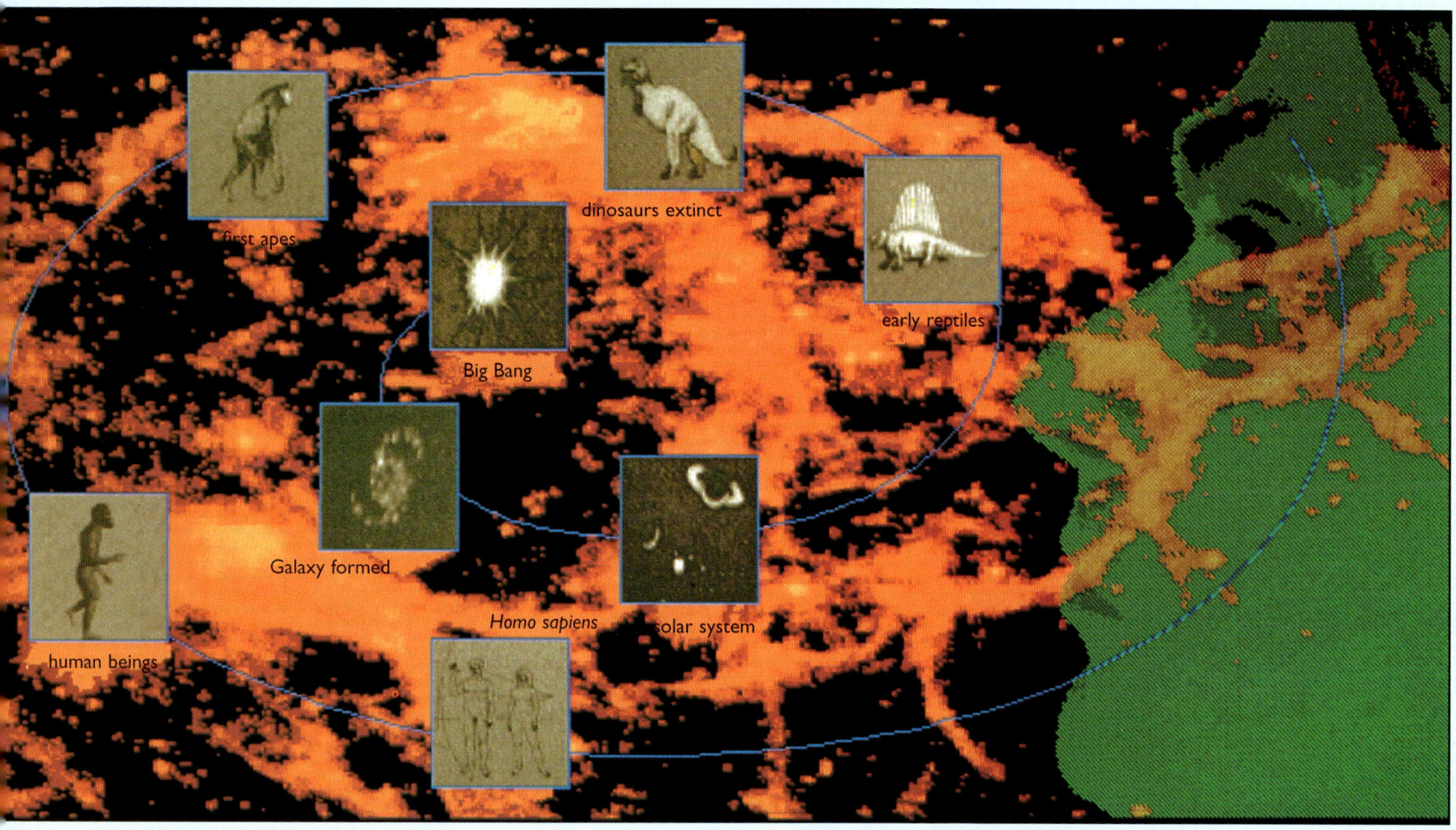

WORMHOLES
● New universes rise from the ashes of the old

The history of humanity's awareness of the universe is essentially one of widening horizons. Today, we survey a universe that extends enormous distances into space and eons into the past and future, and is of a complexity far exceeding that imagined by even the most advanced of the Greek philosophers. Yet, could our picture of the world be still as parochial in its way as those of earlier civilizations? We have now amassed impressive evidence to show that we live in a universe expanding from a Big Bang, but is that the sum total of what exists?

The Indian scientist Jayant Narlikar suggested otherwise. He proposed that the universe we observe may be only one of many such expanding universes within a vastly larger space. The Narlikar hyper-universe can be likened to a gigantic container of bubbling liquid, with our own universe as one of the bubbles, and the other bubbles being universes in their own right.

This is not the only kind of grand scheme, embracing universes wider than the one we observe, that has been proposed. Others make use of the idea that through a black hole it might be possible to reach another, totally different, region of space. Mathematics shows that, in theory at least, a black hole could be connected to another region of spacetime by a "wormhole," an extremely thin, convoluted neck or tube passing through other dimensions. The other end of the wormhole would be a white hole, spewing out material rather than sucking it in.

Such a wormhole would collapse as soon as it was formed unless a special quantum condition prevailed – the creation within it of negative energy. Negative energy, first predicted by the English physicist Paul Dirac in the late 1920s, may sound strange, but the notion is realistic enough – it led to the discovery of the positron in 1932 and is related to the existence of antiparticles in general. Wormholes, if they exist, would possess some strange properties, including that in some circumstances they would permit objects falling into them to travel backwards in time.

The American physicist Lee Smolin extended ideas about black holes and wormholes to the question of the birth and death of the universe. The interior of a black hole can be imagined as "pinched off" from the main body of space-time by a tiny neck, akin to a wormhole. But from the inside, a black hole would appear to be an expanding space – in fact, an expanding universe.

A black hole will not always contain a long-lived expanding universe; many black holes "evaporate" in a comparatively short time (p. 65). But Smolin showed that, if the hole exists for long enough, it will produce an expanding universe.

The formation of such a new universe will be violent, and the result may be that some physical processes and fundamental "constants" will be modified. As each new generation of living things varies randomly within narrow limits because of sudden genetic changes in DNA, so the physics of newborn universes will vary randomly. For example, electrons could have slightly different masses in different universes.

Some universes would be evolutionarily successful, lasting long enough for many black holes to form – by the death of massive stars, or the coalescence of stars in galactic nuclei, and so on. These black holes would each produce a further universe. Smolin calculated that in any long-lived universe, the masses of protons and neutrons, for instance, will be almost equal to each other, as is the case in our universe. So such universes will not be very different from our own.

The outcome of all this is that the successful universes would be those with conditions which would allow full development, including the production of life and the evolution of intelligence. So we should not be surprised that the anthropic principle prevails in our own universe, which seems adapted to the requirements of life; this is merely the effect of selection among a vast number of universes, of which the greater number by far were unsuccessful.

In this picture, we do not have to be

A wormhole passing through higher dimensions could, in theory, link different regions of space-time. One mouth of the wormhole would be a black hole, into which matter and energy are drawn, while the other would be a white hole, a hypothetical entity from which matter and energy would pour out. The connecting region, the wormhole, would be unobservable from the outside universe.

disturbed by the problem of a first cause to explain the Big Bang – the latter was merely a birth from a previously existing universe. Smolin's universe will last forever, propagating itself by giving birth to new universes.

Only further research can show how much truth there is in this highly speculative proposal, and, indeed, in the numerous others that will no doubt be proposed. Perhaps, with Smolin's prospect of self-propagating universes born from the ashes of the old ones, we have come full circle, returning to some very early ideas that these are now backed by scientific calculations. Only time will tell.

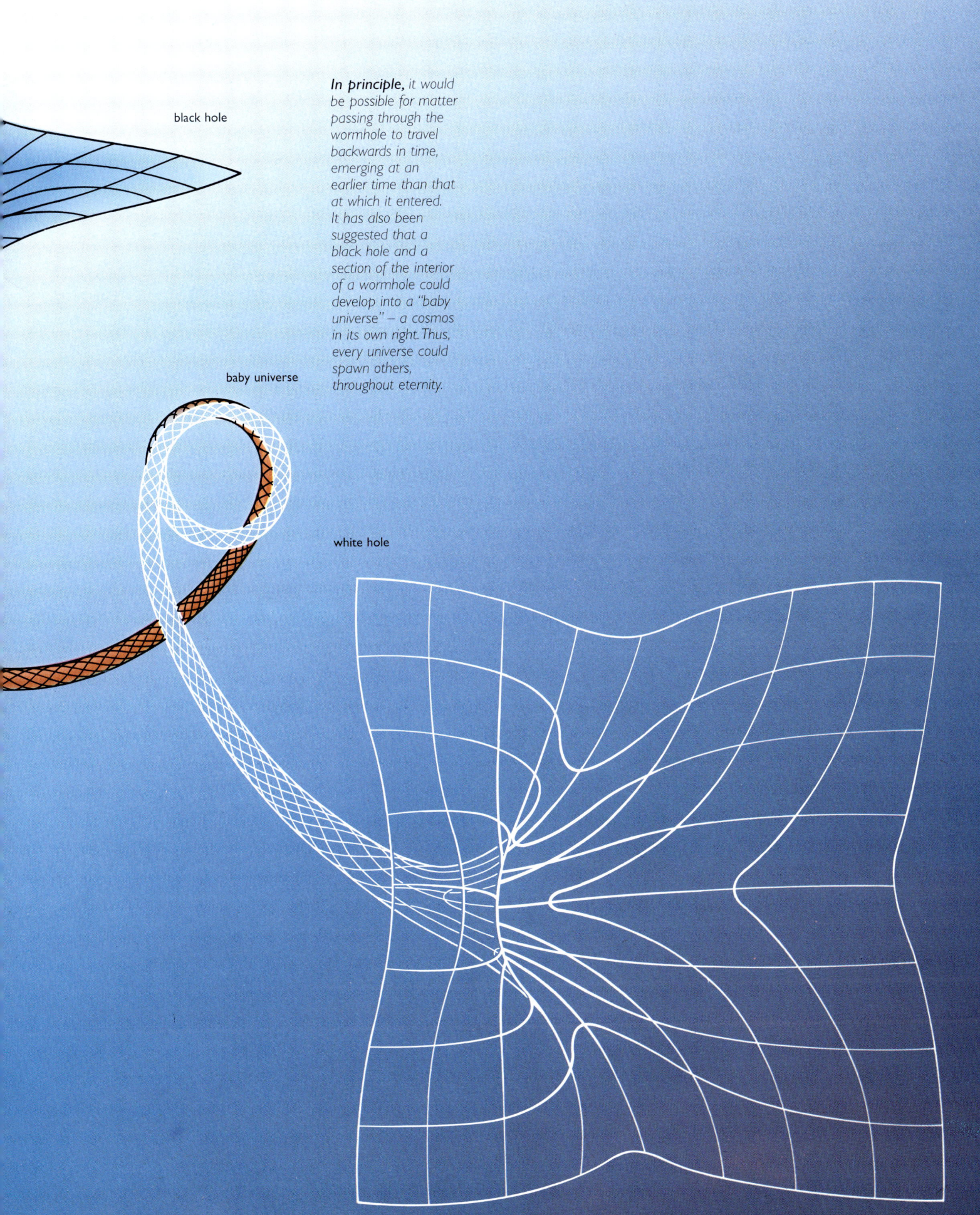

In principle, it would be possible for matter passing through the wormhole to travel backwards in time, emerging at an earlier time than that at which it entered. It has also been suggested that a black hole and a section of the interior of a wormhole could develop into a "baby universe" – a cosmos in its own right. Thus, every universe could spawn others, throughout eternity.

STAR CHARTS

The charts on this and the following pages show fixed stars of the fifth magnitude or brighter. Maps 1 and 2 show the polar regions, extending up to 40 degrees of declination (the celestial equivalent of terrestrial latitude) from the poles. Maps 3 to 8 show the remainder of the sky. Distances along the celestial equator are measured in terms of hours, minutes, and seconds of right ascension, RA (the celestial equivalent of terrestrial longitude). Each of the six equatorial maps therefore covers four hours of RA.

The names of constellations appear in capital letters. Their boundaries, as assigned by the International Astronomical Union, are shown by straight broken lines. The names of the most important stars are shown in capitals and small letters.

The only moving object whose path is shown here is the Sun. In the course of a year, it appears to move eastward around the heavens in relation to the background of fixed stars, along a path called the ecliptic. It crosses the celestial equator in a northward direction in March, at one of the equinoxes; this point is defined as the zero of right ascension. The Sun reaches its northernmost point, the solstice, in June. It crosses the celestial equator in a southward direction, at the other equinox, in September (at 12h RA), and reaches its southernmost point in December.

From this, it is possible to estimate that during May, for example, the Sun is at RA 4h (in Taurus). The stars visible during this month will be in the opposite half of the sky, centered on RA 16h.

These maps cover the entire sky, but the area actually visible is, of course, limited by the observer's latitude. An observer in, say, Paris (latitude 49 degrees north) cannot see the sky south of a line 49 degrees from the celestial south pole (that is, 41 degrees south of the equator); one in Melbourne (latitude 38 degrees south) can see only as far as 52 degrees north of the equator.

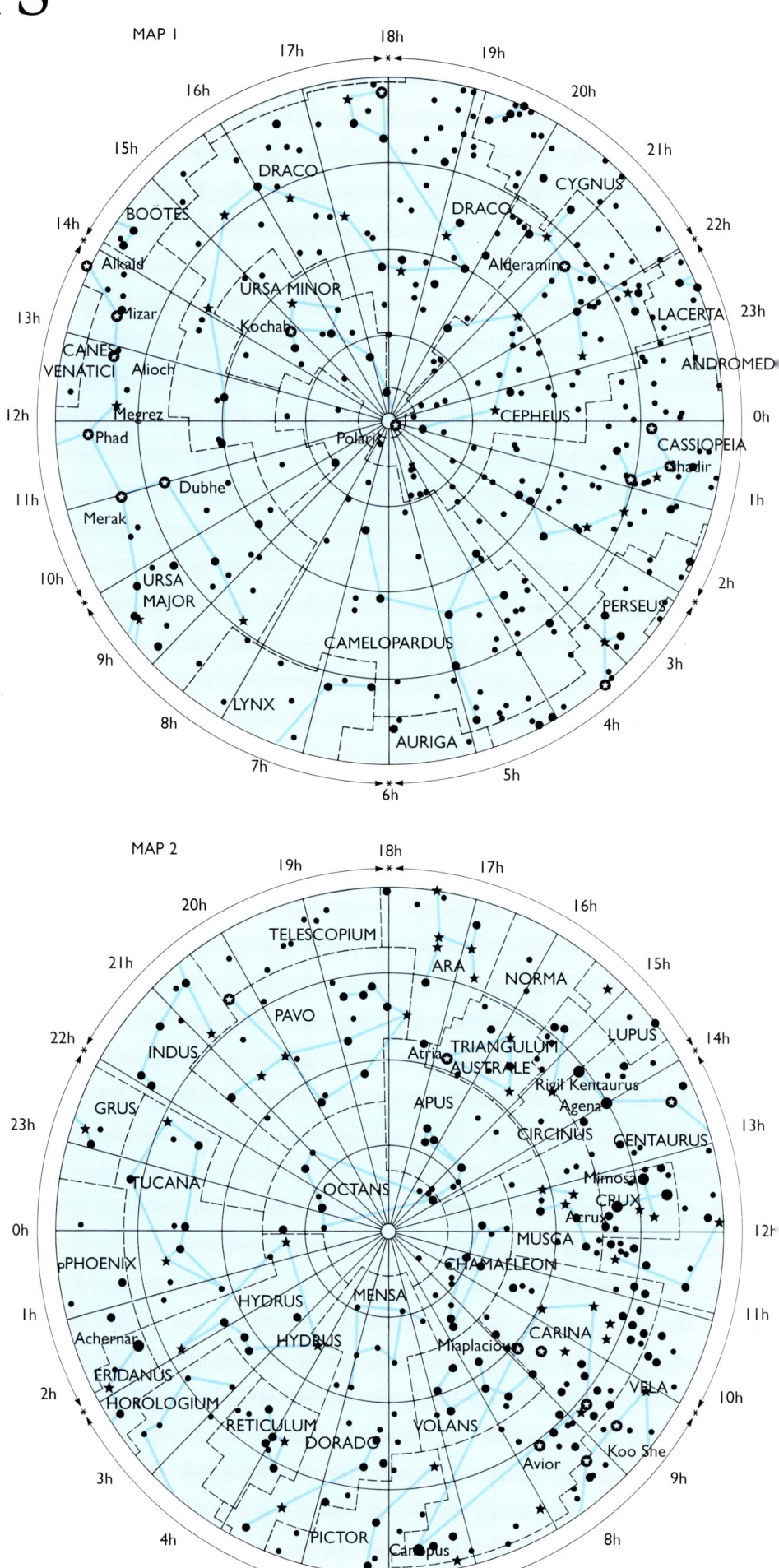

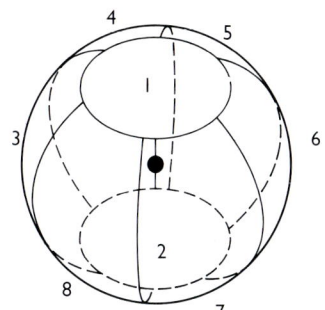

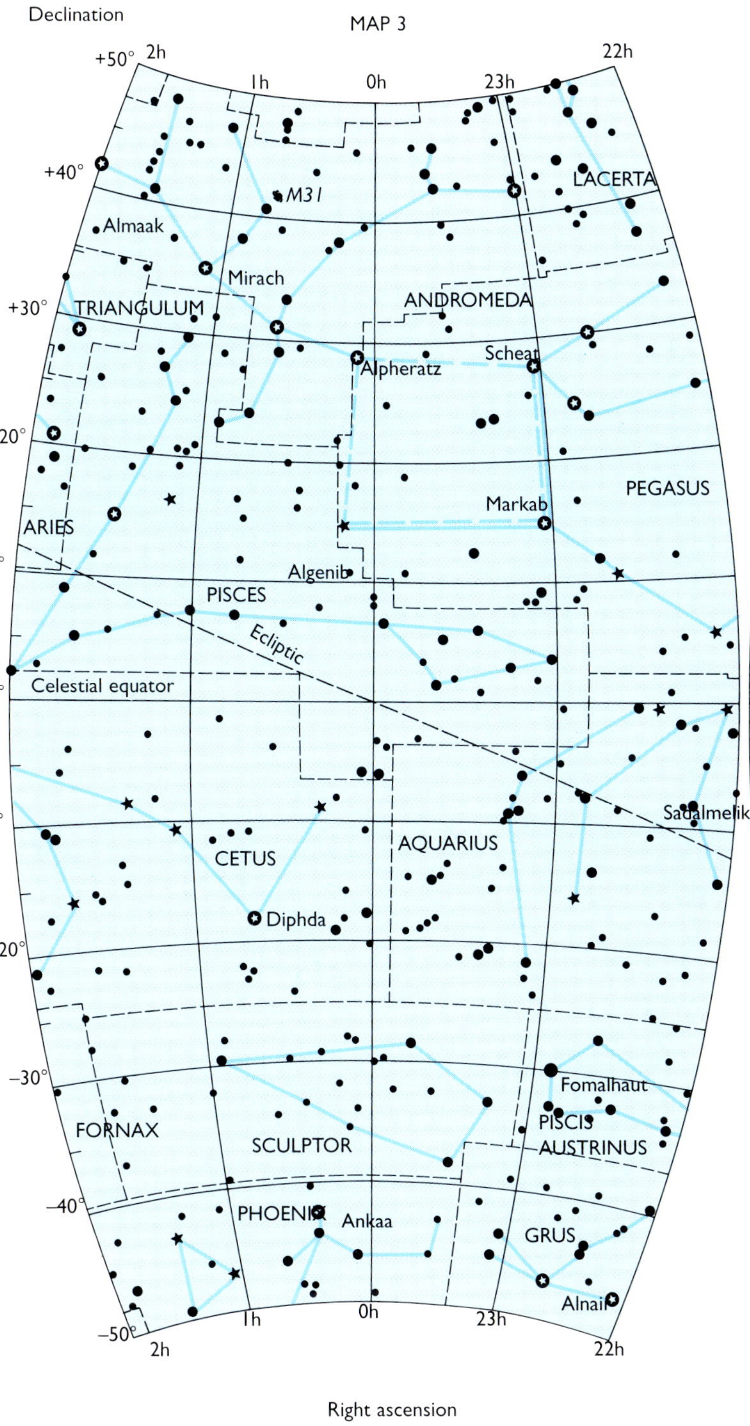

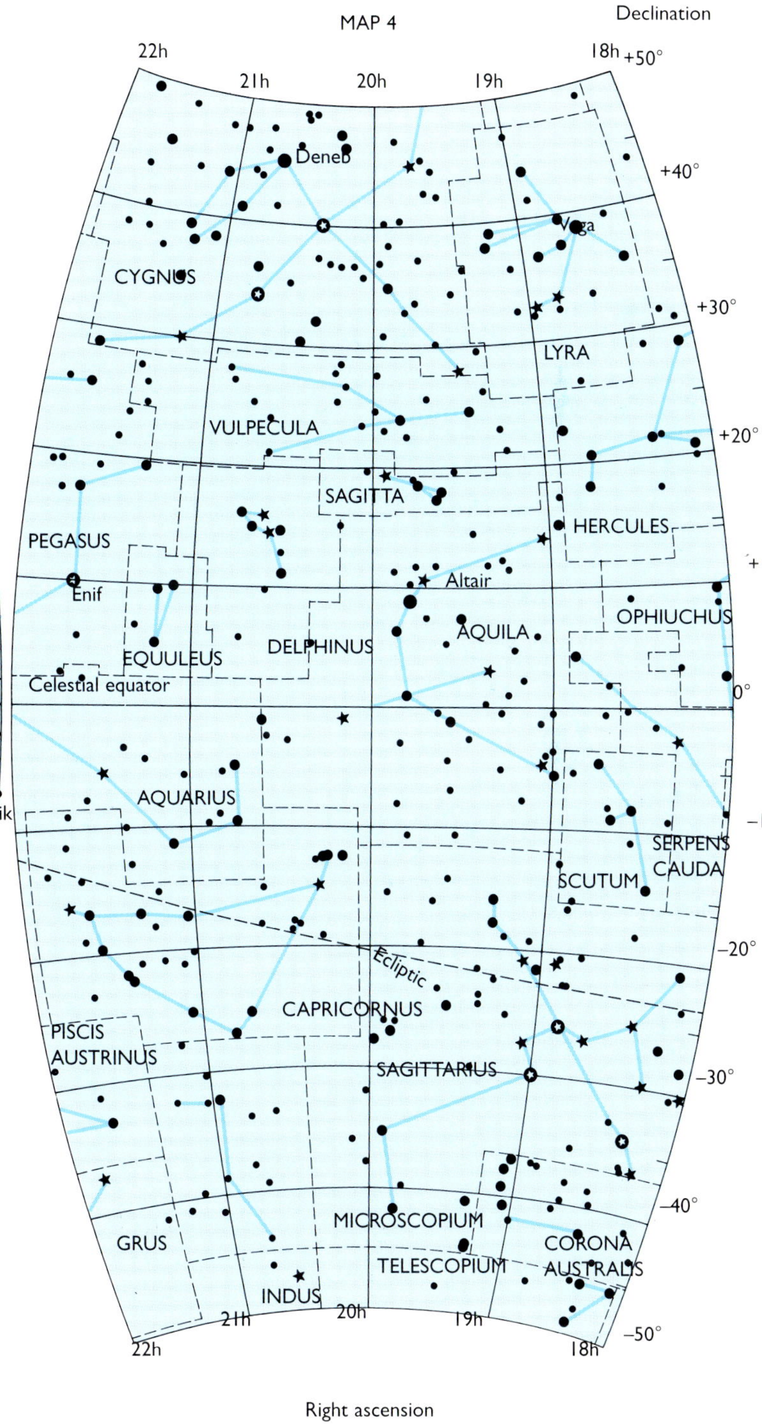

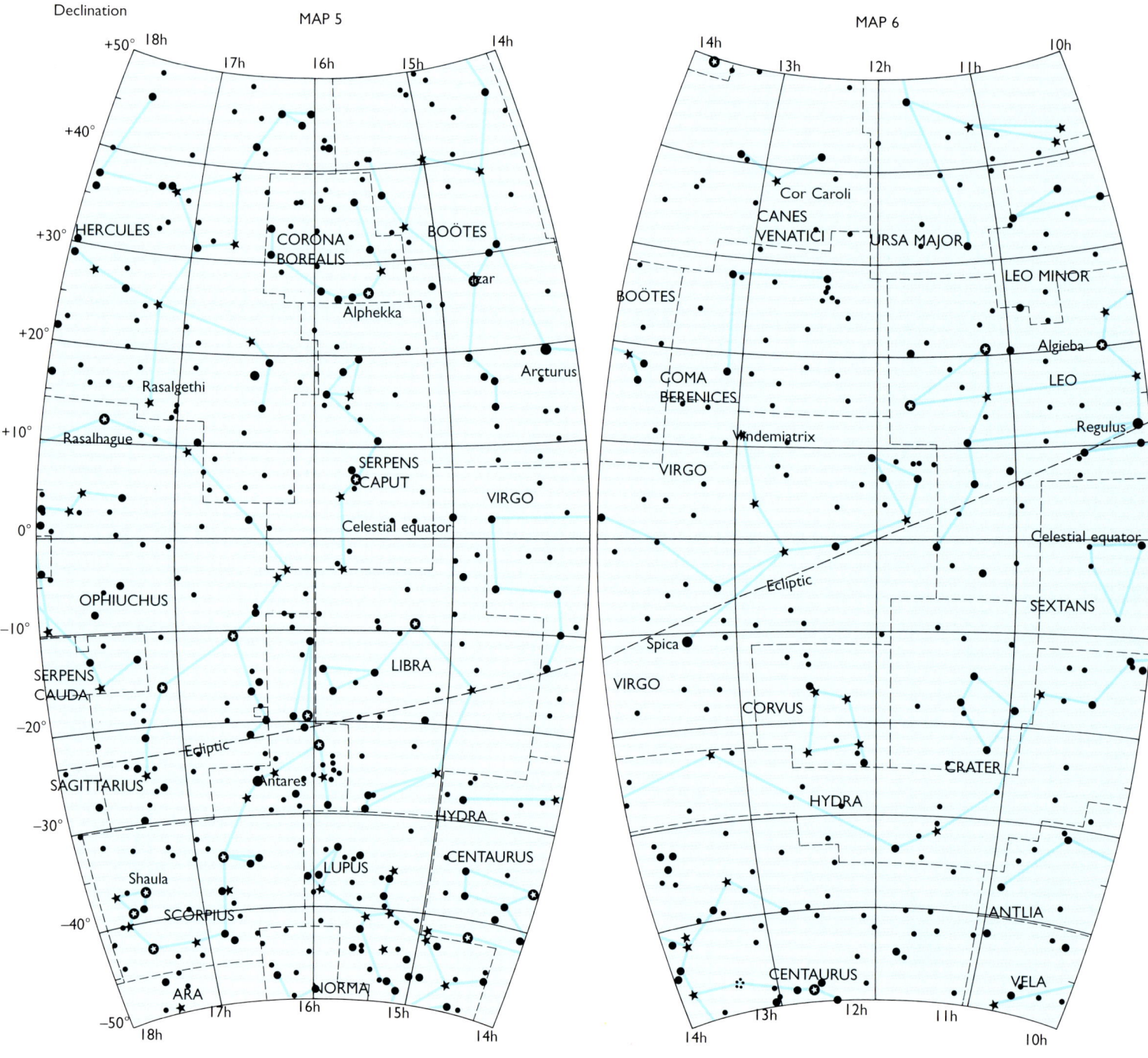

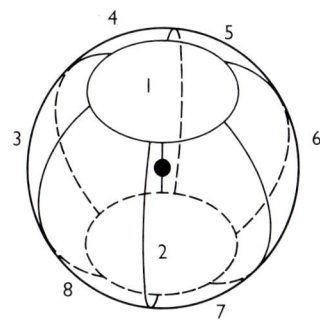

MAP 7 / MAP 8

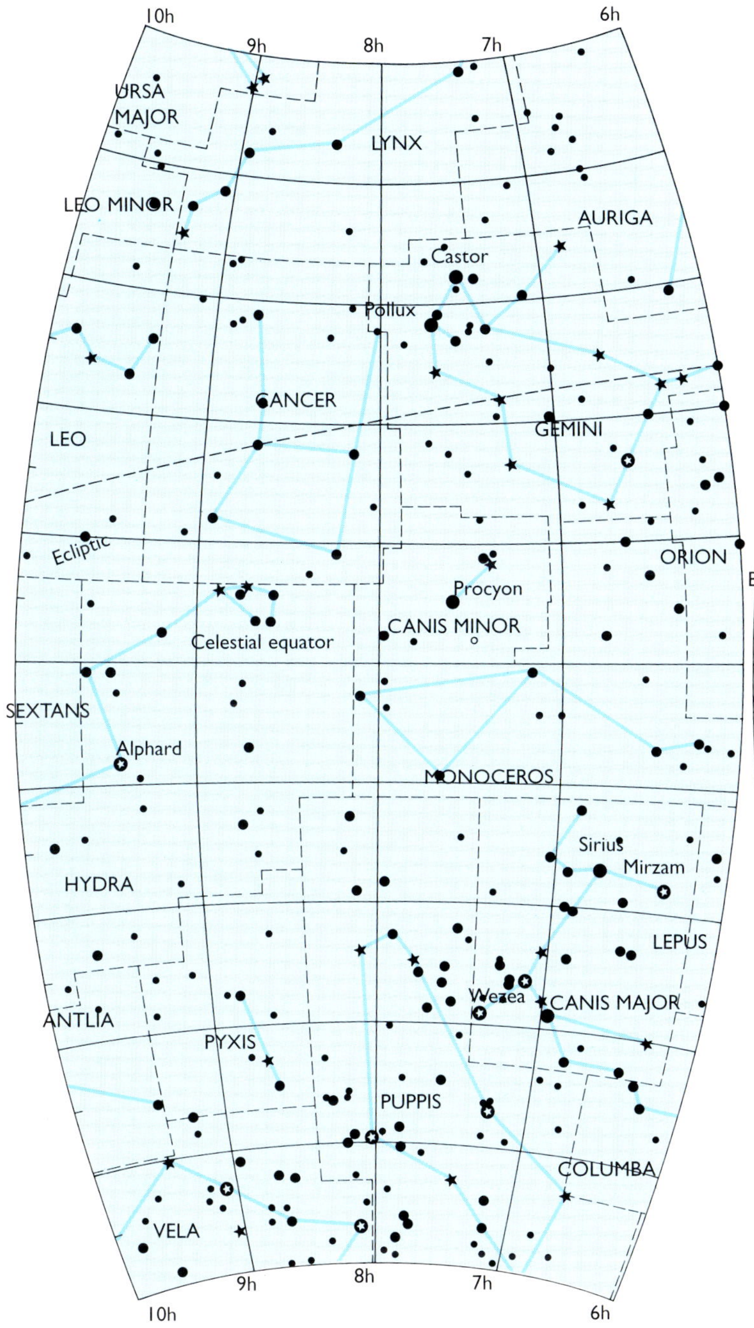

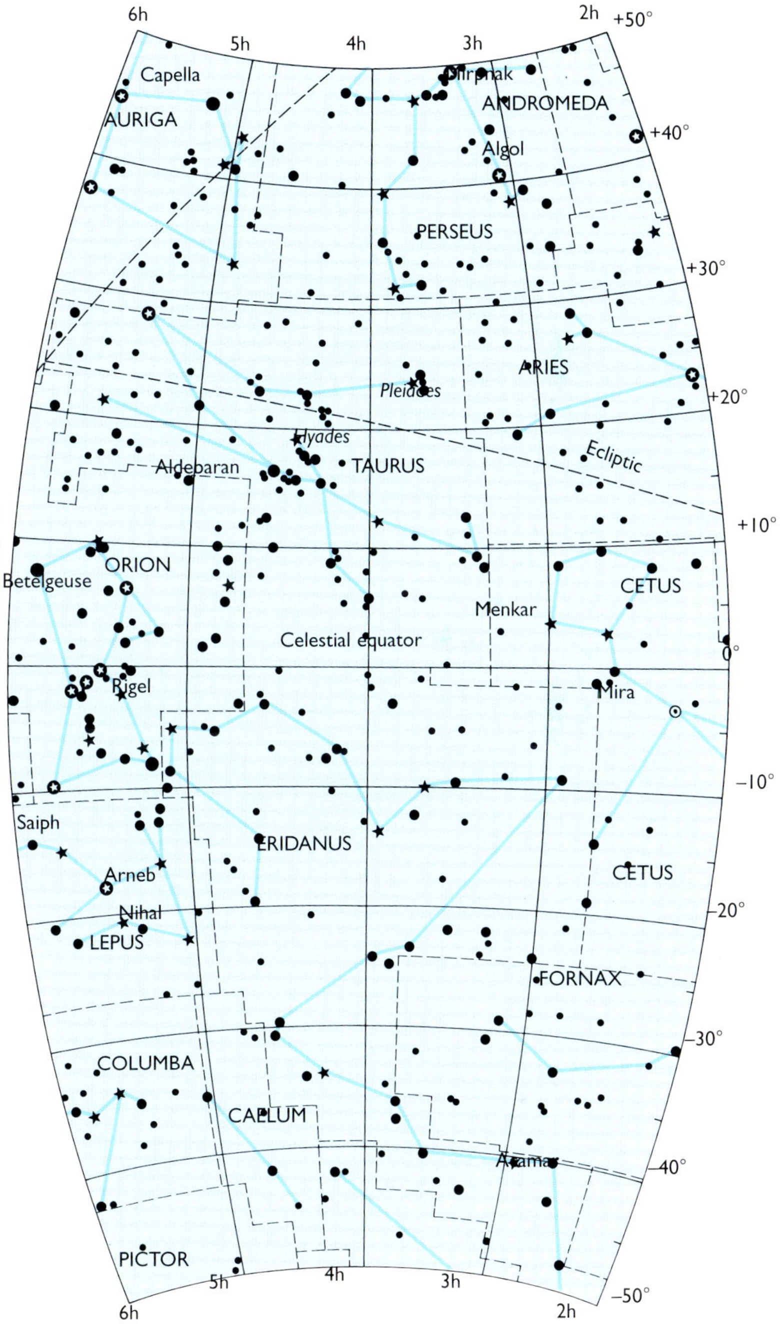

Glossary

aberration of light
An annual apparent shift in the positions of stars against the background of more distant stars, caused by the Earth's orbital motion. The shift appears as a circle, an ellipse, or even a short line, depending on the distance of the star above or below the Earth's orbital plane.

absorption nebula
A cloud of dust and gas which absorbs and therefore blocks out light from more distant objects. Observed optically as dark patches in the sky.

absorption spectrum
A bright spectrum crossed by dark lines due to the presence, between the observer and the hot source giving the bright spectrum, of cooler vapors of specific chemical elements. From it, the chemical elements in the outer layers of stars can be determined.

accretion disk
A disk of material in orbit around a celestial body such as a black hole, and onto which material is continually falling, later to be attracted from the disk to that body, or into the hole.

albedo
The reflecting power of a non-radiating celestial body. Albedo is calculated as the ratio of the total amount of light reflected away in every direction to that received. The albedo of the Moon is only 0.07, but that of Jupiter is 0.43, while that of bright, cloudy Venus is 0.76.

altitude
The angle between a celestial body and the horizon.

antimatter
Elementary nuclear and atomic particles having the same mass but opposite electric charge to ordinary matter. On meeting, matter and antimatter annihilate each other.

aperture
The total diameter of a lens, mirror, or other radiation-gathering surface of a telescope, or the separation between antennae of a radio telescope. The power of a telescope to observe detail or to separate objects close to one another is greater the larger the aperture.

aphelion
The most distant point from the Sun in the orbit of any solar system body, whether a planet, asteroid, comet, piece of space rock or dust, or artificial satellite.

apogee
The most distant point from the center of the Earth in the orbit of the Moon or of an Earth-orbiting artificial satellite.

asteroid
A minor planet or planetoid that orbits the Sun. Most travel between the orbits of the major planets Mars and Jupiter in paths that lie close to the plane of the Earth's orbit. Asteroids measure from several feet across to about 580 miles for Ceres, the largest.

astronomical unit
The average distance between the Earth and the Sun, and a unit useful in studies of the solar system. Its accepted value is 92,960,117 miles (149,597,870 km).

atom
The smallest part of a chemical element which can take part in any chemical reaction and still retain its identity.

All atoms consist of a nucleus and one or more orbiting electrons, which may be in groups, or shells. Reactions in the nucleus can cause the transformation of one kind of atom into another. In the normal state of an atom, the number of electrons equals the number of protons in the nucleus, thus keeping the atom electrically neutral. Most of the mass of an atom resides in the nucleus.

aurora
A display of glowing light of various colors in the terrestrial atmosphere at heights of about 62 miles, in the shape of curtains, arcs, or shells. The aurorae on Earth usually occur within about 20 degrees of the magnetic poles, and appear simultaneously in both northern and southern hemispheres.

azimuth
The angle between the point on the horizon directly below a celestial object and the north point.

barred spiral
A spiral galaxy in which the central region appears as a bar of material rather than as a round bulge or sphere.

binary star
A pair of stars in orbit around each other.

black body
A theoretical body that absorbs all radiation falling on it, whatever its wavelength. Such a body emits light with a characteristic **spectrum**. Stars behave in some ways like black bodies; this means that astrophysicists can explain connections between their color, temperature, and radiation.

black hole
A region of space-time in which there is such an immense concentration of material within a small volume that space-time curves over on itself, and matter and energy cannot escape.

blazar
Immensely compact objects with a vast but variable output of radiation. They are believed to be the core of remote, active galaxies. BL Lacertae is a blazar. About 100 such objects are known.

blue giants
Very hot, highly luminous stars which radiate more intensely at shorter than at longer wavelengths. Such stars are extremely energetic and therefore very bright; they have comparatively short lives, measured in millions, rather than thousands of millions, of years. Their surface temperatures are about 20,000 to 30,000 times that of the Sun, and their luminosities about 100,000 times greater.

blue shift
A shift of the lines toward the blue end of the spectrum. Such a shift indicates that the radiating object is moving toward us.

Bok globule
A small, dark lump or globule of material observed against a bright nebula or a bright background of stars. Bok globules are thought to be protostar material contracting into stars. Their diameters are estimated to be between 10,000 and 25,000 times the distance of the Earth to the Sun.

boson
Subatomic particles which have integral units of spin (0, 1, 2, etc.). Photons, mesons, and other messenger particles are bosons, and so are certain atomic nuclei in which the neutrons are equal in number to the protons.

brown dwarf
A star with such a low mass – less than 0.08 times that of the Sun – that thermonuclear reactions cannot take place inside it. Such stars will shine dimly because the force of gravity causes them to contract, with an accompanying release of energy.

butterfly diagram
A diagram due originally to the English astronomer E. Walter Maunder. First drawn in 1904, the diagram shows the variations in solar latitude of sunspots during the Sun's 11-year sunspot cycle. Sunspots first appear at latitudes of about 35 degrees north and south of the solar equator, and then move to lower latitudes, while increasing in number. As the cycle progresses, they become fewer, while moving still closer to the equator. The diagram displays a pattern which resembles a butterfly's wings.

captured (synchronous) rotation
This occurs when the axial rotation of a natural satellite equals its orbital period around a planet. This can happen to satellites close to their primaries, and occurs in the case of the Moon.

Cassegrain focus
The reflection of incoming light back through a hole in the primary reflector (mirror) of a telescope, behind which it is brought to a focus.

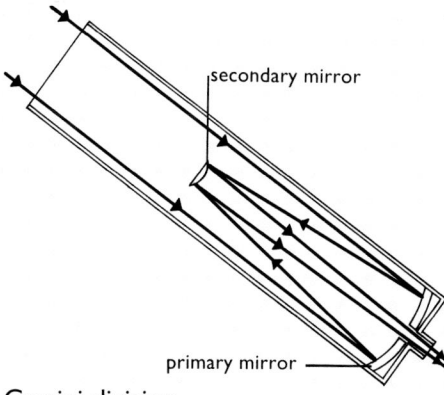

Cassini division
The chief division in Saturn's rings which can be seen in Earth-based telescopes.

celestial sphere
An imaginary sphere on which the celestial bodies are depicted, and a convenient method of defining their positions. The celestial equivalent of terrestrial longitude and latitude is **right ascension** and **declination**, and is based on the celestial equator. **Celestial latitude** and **longitude** are based not on the celestial equator, but on the ecliptic, the Sun's apparent path in the sky. **Altitude** and **azimuth** are Earth-based coordinates. All coordinates are measured in degrees, except for right ascension (RA), which is measured in hours, minutes, and seconds. Twenty-four hours of right ascension are equivalent to 360 degrees, so one hour equals 15 degrees. One minute of RA equals 15 minutes of arc; one second of RA equals 15 seconds of arc.

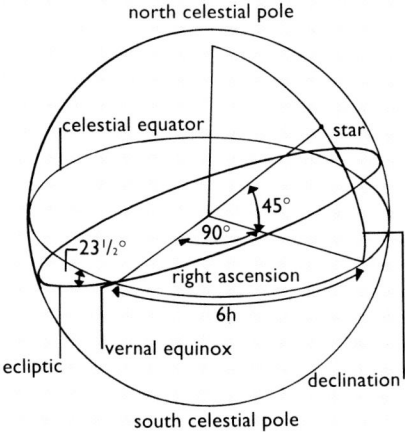

center of mass
That point in a material system at which its total mass may be considered to be concentrated.

Cepheid variable
A class of variable star of which Delta Cephei, whose variability was discovered in 1784, is the prototype. They are pulsating stars which vary their brightness in anything from 1 to 50 days.

Chandrasekhar limit
The limiting mass for a star if it is to become a white dwarf. This mass is 1.44 times that of the Sun. Stars still more massive will become neutron stars or black holes.

Cherenkov radiation
Radiation caused by electrically-charged particles of great energy passing through an electrically non-conducting material. The velocity of the particles has to be greater than the velocity which light would have when passing through the same material.

chromosphere
A layer of the Sun's atmosphere lying directly above the disk or photosphere.

circumpolar stars
Stars which never set because they are permanently above the observer's horizon.

closed universe
A solution to the equations expressing the nature of space-time which gives a spherically-shaped universe. Such a universe will collapse after its initial expansion.

comet
A body of snow and ice mixed with rocky, metallic, and carbonaceous material. About 10 trillion comets form the Oort cloud at the edge of the solar system.

conjunction and opposition
Points in the orbits of planets, asteroids, and comets. Conjunction occurs when the orbiting body is in line with the Sun, as viewed from the Earth. The body appears in the morning sky when approaching conjunction (1), and in the evening sky after conjunction (2). Opposition occurs when a body whose orbit is larger than that of the Earth lies opposite to the Sun in the sky.

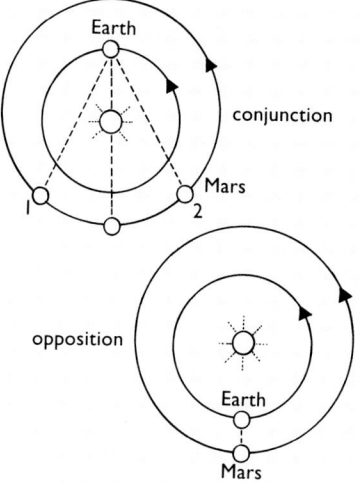

constellation
Originally, a grouping of stars to form a shape or pattern, although constellation boundaries are now delineated by coordinates agreed by the International Astronomical Union. The grouping of stars into constellations does not imply any physical connection among them.

continuous spectrum
Visually, a continuous band of colors from red to violet emitted by an incandescent solid body, or one in which the material is closely packed, such as a star. The spectrum extends into non-visual wavelengths at each end. (*See* **absorption spectrum** and **emission spectrum**.)

coordinates
Mathematical ways of delineating positions. **Cartesian** coordinates refer to mutually perpendicular directions and are denoted by x, y, z, etc. **Polar** coordinates utilize angles and a radial distance r. On the **celestial sphere** the radial distance is taken to be 1, so the position of any object is uniquely specified by two angles.

Coriolis force

The deflection of a body moving above the surface of the Earth, caused by its rotation. In the northern hemisphere, it causes a clockwise circulation of air around an area of high pressure, and a counterclockwise circulation around an area of low pressure. It has the opposite effect in the southern hemisphere.

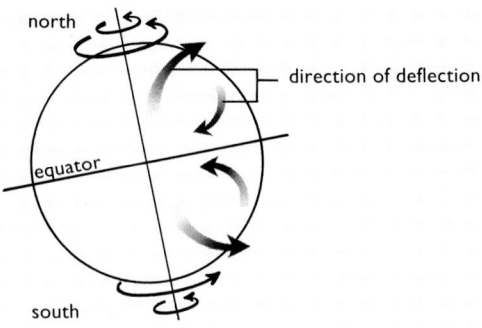

corona

The outer atmosphere of the Sun. Extending about 6,200 miles above the photosphere, it is extremely tenuous, less than one thousand-billionth of the density of the Earth's atmosphere. Composed of very energetic electrons, it reaches a temperature of about 2 million degrees C (3.6 million degrees F) at a height of about 46,600 miles. Visible only at total solar eclipses or with special instruments, it also emits extreme ultraviolet and X-radiation.

cosmic background radiation

Microwave radiation peaking at a wavelength of 1 mm, which is visible at the same intensity all over the sky. It is assumed to be the cooled remnant of the primeval fireball of the hot Big Bang that started the universe.

cosmic rays

Atomic particles, mostly protons, of very high energy moving through space. When they impinge on the Earth's atmosphere, they break up air molecules and atoms, and cause showers of other atomic particles.

cosmic string

A thin string of trapped energy left over from the earliest moments of the Big Bang, with immense mass per unit length. Cosmic strings may have acted as seeds for the formation of galaxies, clusters, and superclusters.

coudé (elbow) focus

The focus of a telescope in which light is brought out in such a way that focus always remains stationary.

crater

Shallow, circular basin found on many bodies of the solar system, thought to have been caused by impact.

dark matter

Matter not visible either optically or at other wavelengths. If present, it will add to the total mass of the universe, and may be sufficient to change its expansion into a contraction.

declination

The angle between the celestial equator and a celestial body, measured northward or southward. (*See* **celestial sphere**.)

decoupling era

A period about 300,000 years after the Big Bang when radiation ceased to be scattered by matter, and became independent of it. This occurred because the temperature had dropped to about 3,000 K, allowing protons and electrons to form hydrogen atoms, which are transparent to radiation.

degeneracy

An abnormal state of matter in which, due to great pressure and temperature, electrons are stripped from atoms to form a mass of nuclei surrounded by an electron gas. Its density can be many tons per cubic inch. The central regions of white dwarfs and neutron stars are composed of such material.

diamond-ring effect

A visual effect seen at the close of a total solar eclipse when the Sun's disk begins to reappear behind the Moon, giving the appearance of a ring surmounted by a bright diamond.

diffraction

The apparent bending of light and radiation of other wavelengths around the edges of an object. This causes bright and dark bands to appear because of the wave nature of the radiation.

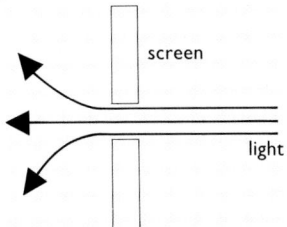

diffuse nebula

A bright nebula in our own galaxy which spreads irregularly across the sky, unlike a **planetary nebula**.

Doppler effect

In astronomy, the shift of the lines in a spectrum toward either the red or the blue, caused by a motion of the source emitting the spectrum. If the source is moving toward us, the frequency with which the crests of its waves will reach us will increase; this will give an apparent reduction in wavelength and a shift of the spectral lines toward the blue. If the object is moving away, the effect is that we receive a longer wavelength, and the shift is toward the red.

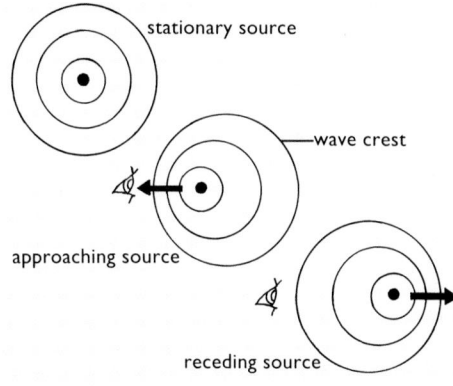

double star

Stars appearing as a pair because they are close to each other in the line of sight.

eccentricity

A measure of the amount by which an elliptical orbit differs from a circle. For a circular orbit the eccentricity, e, is zero; its maximum value is always less than 1.0.

eclipses and occultations

Eclipses occur when one celestial body passes into the shadow of another, as in the case of the Moon passing into the Earth's shadow. Eclipses of Jupiter's satellites are also observed from Earth.

An eclipse of the Sun is, in fact, occultation (one body passing in front of another), although in astronomy, the latter term is usually reserved for the obscuration of a star or planet by an asteroid or satellite such as the Moon.

electromagnetic radiation

The total radiation band from radio waves to X-rays and gamma-rays, and including ultraviolet, visible, and infrared wavelengths.

electron

A fundamental and stable atomic particle of the class known as leptons. It possesses one negative electric charge, a spin of $1/2$, and a mass of 9.1×10^{-28} gm. Its antiparticle is the positron, which possesses one positive electric charge.

ellipse

A closed elongated curve as traced by planets, asteroids, comets, and satellites.

elliptical galaxy
A galaxy, ellipsoidal in shape, composed primarily of stars with little gas and dust.

elongation
The angular distance, east or west, between the Sun and Mercury or Venus.

emission nebula
A nebula that emits visible radiation. Such nebulae may be diffuse or compact.

emission spectrum
The spectrum given by a glowing gas such as an emission nebula. It is characterized by bright lines against a dark background because the glowing gas radiates only at specific wavelengths, which depend on the chemical elements of which it is composed.

Encke division
A narrow division in Saturn's ring system. It is a "ripple" rather than a large gap.

entropy
A quantity which, together with total energy, describes the thermodynamic state of a physical system. It is a measure of the number of ways in which the positions and velocities of the molecules can be rearranged to give the same overall properties. Left to itself, a system evolves so as to maximize the number of equivalent arrangements – that is, the entropy increases as far as possible. (When no further increase can occur, equilibrium has been reached.)

Since, loosely speaking, a disorderly arrangement of objects is one that can be shuffled without making a significant change, entropy is sometimes said to be a measure of disorder.

There are more ways of arranging the molecules of a gas in a large space than in a small one; consequently, they maximize their entropy by spreading the energy evenly throughout the container, so that the initial temperature variations are smoothed out.

Variations of density or temperature can always be used to obtain useful work. Thus, for a system of given total energy, a greater amount can usefully be extracted when the entropy is lower.

equatorial mounting

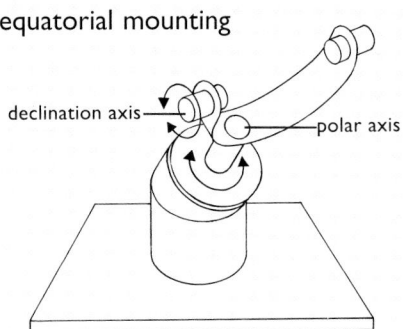

A telescope mounting in which one axis is parallel with the polar axis of the Earth. The other axis around which the telescope rotates is at right angles to the polar axis. This arrangement makes it possible for the observer to track the curved path of a celestial body across the sky by rotating the telescope around only one axis.

equivalence, principle of
The principle in Einstein's theory of relativity that forces due to gravity and to the inertia of a body (its resistance to change of velocity) are equal. As a result, objects in a box falling freely in a gravitational field will display the same phenomena as if they were in empty space, far from any gravitating body. Conversely, objects inside a box accelerating through space will experience the same phenomena as if they were stationary in a gravitational field.

escape velocity
The velocity which a body must reach if it is to escape into space from a celestial body. The escape velocity depends on the size and mass of the celestial body concerned. For the Earth, it is 6.95 miles/s, but for the Sun, it is no less than 417.15 miles/s.

event horizon
The boundary of a black hole, within which no event can be observed from the outside universe.

exclusion principle
The principle in quantum mechanics which states that no two fermions (particles with a spin that is not a whole number, such as electrons, protons, and neutrons) may occupy states which have the same quantum conditions (spin, etc.). It is because of this principle that the electrons in an atom do not all gather in the lowest orbit.

faculae
Bright active areas in the upper layers of the Sun's photosphere, often near sunspots.

false vacuum
A quantum vacuum state characterized by a vast repulsive force and active during the inflationary period of the Big Bang.

fermion
A class of atomic particles having a spin that is not a whole number. It includes protons, neutrons, and electrons.

field
A region throughout which a force operates.

flare
A sudden release of energy visible as a bright light on the Sun. Such a flare, which lasts only a few minutes, actually occurs above active regions of the photosphere, either in the **chromosphere** or the lower region of the corona. Flares are characterized not only by a visible image, but also by the emission of X-rays, and sometimes gamma and radio waves.

Fraunhofer lines
The dark absorption lines which cross the bright continuous spectrum of the Sun. So called because their positions were first carefully plotted by the German optician and astronomer Joseph Fraunhofer in 1814.

frequency
The number of times per second the crests or troughs of waves of electromagnetic radiation reach an observer. Frequency is obtained by dividing the velocity of light by the wavelength of the radiation.

galaxy
A celestial island of stars, dust, and gas.

gamma radiation
Extremely energetic radiation, shorter than 10^{-8} (one 100-millionth) of a millimeter.

giant star
A star which is very luminous and usually large in diameter compared with other stars of the same spectral class. The atmosphere of these giants are, by comparison, much more tenuous.

globular cluster
A comparatively closely-packed cluster of stars, spherical in shape. Such clusters can contain anything from a few tens of thousands of stars to over one million. Globular clusters form a major component of the **halo** of the Galaxy.

gluon
A messenger particle which holds together quarks, which are themselves the basic particles composing the atomic nucleus.

grand unified theory (GUT)
A theory that aims to unify the basic forces of nature, bringing into one scheme the strong and weak nuclear forces, electromagnetism, and gravity. At the extraordinarily high temperatures prevailing in the earliest moments of the Big Bang, these forces were in fact indistinguishable from each other.

gravitational lens
Because bodies cause space-time to curve, a massive body can deflect light and other radiation from a more distant object so that it becomes visible to an observer on Earth, though ordinarily it would remain hidden. It is as if the nearer massive body is acting as a lens. The effect can also lead to the appearance of distorted or multiple images of distant bodies.

gravitational red shift
A consequence of general relativity, resulting in the mass of a body causing a small red shift of spectral lines when light and other radiation is emitted from it.

graviton
The messenger particle of gravity in theories of quantum gravitation.

Great Red Spot
The large oval red area observed in the upper cloud layer of Jupiter's atmosphere, which rotates counterclockwise.

greenhouse effect
The absorption of outgoing infrared radiation by a planetary atmosphere and its re-radiation back to the planet's surface, thus helping to raise the average surface temperature. It is responsible for the very high temperature (737 K) on Venus.

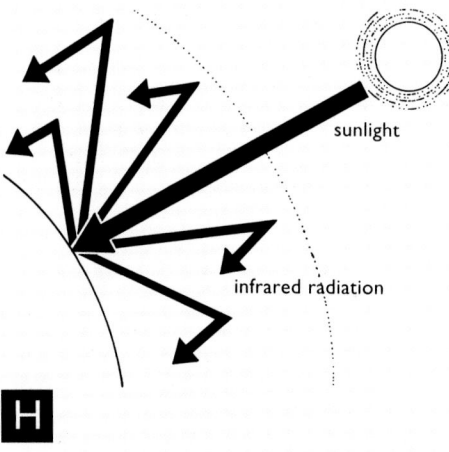

H

HI and HII regions
HI regions are those in which neutral hydrogen is present in interstellar space. Such regions are detected because they radiate at a radio wavelength of 21 cm. HII regions are those where singly-ionized hydrogen is predominant, such as in the flowing bright nebulae in Orion and other constellations.

hadron
A fundamental subatomic particle that experiences the **strong force**. Hadrons include protons, neutrons, and mesons, and are made of pairs or triplets of quarks.

halo
A glowing ring observed around a celestial body. Halos are observed around the Sun and Moon due to refraction and reflection of their light by the Earth's atmosphere. The term is also used to describe material that is spread spherically around our galaxy.

Hertzsprung-Russell diagram
A graph in which the absolute magnitudes of stars near the Sun are plotted against spectral class. It shows that there is a relationship between true brightness and spectral class.

Higgs field
A quantum mechanics field. At its lowest energy state, a Higgs field induces spontaneous symmetry-breaking (see **symmetry**) and it is important in theories that attempt to unify the fundamental forces of nature. The field is associated with Higgs particles, which are analogous to photons in electromagnetic fields.

horizon distance
The maximum distance across which light could have traveled since the origin of the universe.

hour angle
The angle on the **celestial sphere** between a celestial object and the observer's **meridian**. It is measured westward along the celestial equator in terms of hours, minutes, and seconds.

Hubble constant
The rate at which the velocity of recession of galaxies increases with distance. At present, there is some doubt about its value, which is thought to lie between 9.3 and 18.6 miles/s (15 and 30 km/s) per million light-years.

I

inertia
The tendency of a body to resist a change in velocity, whether that change is acceleration or deceleration. It is implicit in Newton's laws of motion, which state that a body continues in a state of uniform motion in a straight line unless acted upon by outside forces. The acceleration that a given force can produce in a given body depends on its mass, which is often referred to as the inertial mass.

inferior planet
A planet whose **orbit** around the Sun lies within the Earth's. The two inferior planets are Mercury and Venus.

infrared radiation
Radiation beyond the red end of the spectrum. Its wavelengths range from 1 mm to .001 mm. It is sometimes called "radiant heat" and is readily absorbed by water vapor in the Earth's atmosphere, which consequently limits the observations that can be made from ground-based observatories.

interference
(1) In radio technology, the degradation of radio signals by other unwanted signals or radio noise, which is why signals from spacecraft are transmitted digitally. (2) In physics, the superposition of one wave on another. The resulting wave caused by the waves reinforcing and weakening each other is known as the interference pattern. Optically, such a pattern is seen as alternate light and dark bands.

interferometry
Observations made using the interference of light or radio waves. In astronomy, the technique was first applied optically by the American physicist Albert Michelson in the 1920s for measuring the diameters of large nearby stars.

ion
An ion is an atom or molecule that has either lost one or more of its **electrons**, and thus has a positive electric charge, or gained one or more electrons. In the latter case, the ion will have a negative electric charge.

ionosphere
A region of the Earth's atmosphere that extends from about 37 miles to more than 311 miles above the ground. Here, most of the atoms and molecules are ionized by solar radiation. Being therefore electrified, the ionosphere acts like a mirror in the sky to long, medium, and some short radio waves and therefore permits long-range broadcasting without the aid of a satellite. However, solar magnetic storms upset the ionosphere and therefore cause radio fadeouts that disrupt such communication.

irregular galaxy
A galaxy that is too irregular in shape to be classified as a **spiral**, an **elliptical**, or a **lenticular galaxy**.

isotope

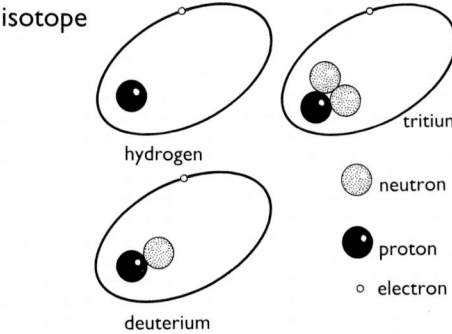

A form of an element in which the atomic nucleus has its characteristic number of **protons**, but contains a different number of **neutrons** than the element in its usual form. Chemically, isotopes do not differ from the ordinary form of the element, but their behavior may be different in other ways. For instance, tritium – a form of hydrogen with two neutrons in the nucleus as well as the single proton found in ordinary hydrogen – is radioactive.

Kelvin scale of temperature
A scale of temperature whose units (called kelvins, symbol K) are equal in size to those of the Celsius scale, and whose zero is fixed at −273.16°C (−459.67°F), often known as absolute zero.

Kepler's laws
Three laws governing the orbital motions of the planets. The laws are: (1) that planets move in elliptical **orbits**, with the Sun at one of the two "foci" of the ellipse; (2) that the radius vector (line from Sun to planet) sweeps out equal areas of space (areas 1, 2, 3 in diagram) in equal times, thus giving planets greater velocity when closer to the Sun; (3) that the orbital period squared is equal to the cube of that planet's average distance from the Sun.

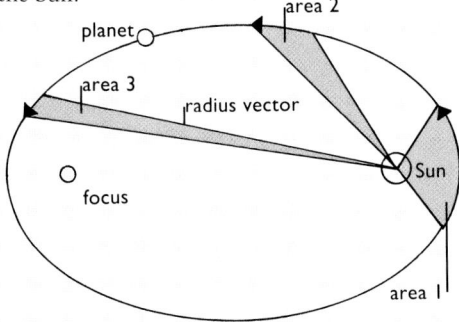

Kirkwood gaps
Regions within the asteroid belt, between Mars and Jupiter, in which few asteroid **orbits** occur, due to the gravitational effects of Jupiter. They were discovered in 1857, by the American mathematician Daniel Kirkwood.

Lagrangian point
Position in which a small body can remain in a stable **orbit** in the company of two massive bodies that are in mutual orbit around each other. There are five such points, which were discovered in 1772, by the French mathematician Joseph Louis Lagrange. The so-called Trojan asteroids occupy the Lagrangian points of Jupiter's orbit.

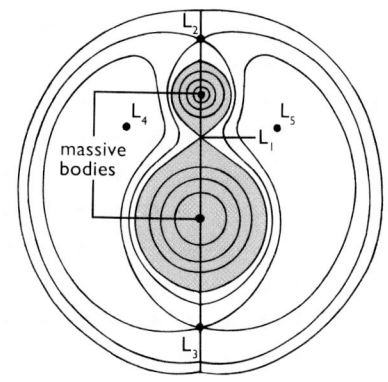

lenticular galaxy
A galaxy intermediate in shape between an elliptical galaxy and a **spiral galaxy**.

lepton
Particles not affected by a strong nuclear force. Those known to date are the **electron**, **muon**, **neutrino**, and tau-particle.

libration
Oscillations in the Moon's motions that reveal a little of its far side. For example, variations of the orbital speed, combined with the constancy of rotational speed, lead to libration "in longitude," or east-west. As a result, about 59 percent of the Moon's surface is visible from Earth.

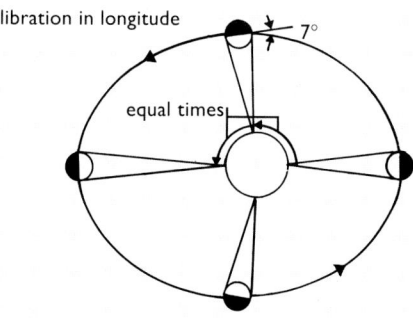

light-cone
A means of depicting events in space-time by drawing a cone to represent the space over which light from an event will spread with time.

light-year
A measure of distance, not of time. It is the distance traveled by light or other **electromagnetic radiation** in one year, and amounts to 5.88×10^{12} (trillion) miles (9.46×10^{12} km).

Local Group
The cluster of galaxies of which our own galaxy is a member. it also includes the Magellanic Clouds and covers a radius of space of about 2.5 million light-years.

magnetic monopole
A single magnetic pole on its own. Such monopoles do not occur in the everyday world because a magnet always possesses two poles, but their existence was predicted by the English physicist Paul Dirac in 1931. None has yet been observed. However, magnetic monopoles may have existed in the early stages of the Big Bang.

magnetosphere
The region around a planet in which ionized particles are under the control of the planet's magnetic field. Its edge is known as the magnetopause.

magnitude
In astronomy, a term used to specify brightness of a celestial body. Based on the way the human eye and brain estimate brightness, a difference of one magnitude represents a difference in brightness of 2.512 times. Magnitudes are measured on an ascending scale of faintness, so that a star of magnitude 2 is 2.512 times dimmer than one of magnitude 1. The scale is extended to negative values. (Magnitude zero is 2.512 times fainter than magnitude −1.)

Absolute magnitude is the magnitude a celestial body would display if set at a distance of 32.6 light-years (10 parsecs). It permits direct comparison of the brightness of the different bodies.

Apparent magnitude is the magnitude a celestial body appears to have to an observer on Earth. It depends upon not only the true brightness of the star but also its distance.

Bolometric magnitude is the magnitude taking into account radiation at all wavelengths, not merely visible ones.

Photographic magnitude is the magnitude as measured on traditional photographic plates, specially prepared for astronomical use, in which the highest sensitivity lies in the blue region of the spectrum.

Photovisual magnitude is magnitude measured from a photographic plate with which filters have been used so that the response is similar to that of the human eye.

main sequence
The area of the **Hertzsprung-Russell diagram** in which most stars lie. It extends from the lower right, where cool dim red stars are to be found, upward to the top left corner, where very hot bright stars are shown.

Markarian galaxy
A galaxy that is bright and radiates most strongly at the blue end of the spectrum. They were catalogued by by the Soviet astronomer B. E. Markarian in the 1970s.

mascon
Short for "mass concentration" and used in reference to denser concentrations of mass just below the surface of some lunar plains or "maria" ("seas"). They were located by orbiting Apollo spacecraft.

meridian
The imaginary circle on the **celestial sphere** passing through the celestial poles and the **Zenith**. It intersects the horizon at points due north and south of the observer.

meson
A particle composed of a **quark** and an antiquark.

meteor
A bright trail of light seen in Earth's night sky. They occur at between 50 and 75 miles above the Earth and are caused by tiny pieces of rock and dust, discarded parts of comets or asteroids. The tiny pieces speed through Earth's atmosphere at between 7 and 46 miles/s, and burn up to produce the short-lived meteor.

meteorite
A lump of rock or metal, or a mixture of the two, that lands on the Earth because its original size was too great for it to be consumed completely in the terrestrial atmosphere. If large enough, it will leave a crater where it falls.

Milky Way
A hazy band of light crossing the entire sky in both northern and southern hemispheres. So-named for its appearance, it is now known to be caused by myriads of **stars** as well as dust and gas lying in the central plane of our galaxy.

molecule
The smallest unit of a pure substance that will retain its composition and, therefore, its chemical properties. It may be a single **atom**, or a collection of atoms.

multimirror telescope
A design of optical telescope with a number of main mirrors, which bring their light to a common focus. The purpose is to construct a telescope with light-gathering power equivalent to that of a single much-larger mirror. The multimirror telescope on Mount Hopkins, south of Tucson, has six mirrors, each 72 inches in diameter, and is equivalent to a single mirror 15 feet in diameter. Such relatively small mirrors are less prone to distortion under their own weight.

multiple star
A star of three or more components in mutual **orbit** about their center of mass.

muon
A **lepton** having an electric charge equal to that of the **electron**, but 207 times heavier.

nadir
The point on the **celestial sphere** directly underneath the observer, opposite the **zenith**.

nebula
A cloud of gas and dust in space. Nebulae can be either dark or bright, diffuse, or compact.

neutrino
A **lepton** with no mass and no electric charge; it enters only into reactions involving the weak nuclear force.

neutron
A **fermion** that has no electric charge but a mass just a little greater than a **proton**. It is a constituent of many atomic nuclei.

neutron star
A massive **star** toward the end of its life, whose degenerate material is composed of tightly-packed **neutrons**. Their diameters are about $12\frac{1}{2}$ miles and their masses vast, because their density is 10^{15} (one thousand trillion) times that of water. When magnetized and rotating, they emit pulses of radiation and are known as pulsars. In **X-ray** binary stars, one component is a neutron star and the X-rays are emitted as material which are attracted to this massive body.

nova
Meaning "new," this is an aging **star** which suddenly flares up in brightness – perhaps by 10,000 times – and so suddenly appears noticeable in the sky. Novae seem to be associated with binary systems in which one member is a **white dwarf**. The flare-up occurs on the white dwarf as material from the companion builds up. Novae flare up within days or weeks at the most, and then sink back to about their original brightness over months or even years.

nucleosynthesis
The building up of heavier atomic nuclei from the nuclei of lighter **atoms**. Conditions for this are found in the central regions of stars, where hydrogen is changed to helium and, in the most massive stars, helium is further built up through successive elements into carbon. In explosions of **supernovae**, conditions occur that make it possible for nuclei heavier than iron to be synthesized. Nucleosynthesis was also a feature of the early Big Bang universe.

nutation
A nodding motion of the Earth's axis of rotation due to the varying distances of the Sun and Moon, and their changing gravitational effects on the Earth.

occultation
(*See* **eclipses and occultations**.)

Olbers' paradox
A so-called paradox discussed in 1826 by the German amateur astronomer Heinrich Olbers, but first recognized long before. It concerns the question why, if the universe is static and infinite, and the **stars** uniformly distributed in it, the sky appears dark at night; under such conditions it should appear bright. It is now known that the universe is not static, nor are the stars uniformly distributed, but grouped into galaxies, all of which are moving away from each other. The red shift of their radiation, coupled with the fact that the universe has a definite age and so does not extend infinitely, explains why the night sky is so dark.

Oort cloud
In 1950, the Dutch astronomer Jan Oort suggested that **comets** originate from a cloud of cometary material spread out at the edge of the solar system at a distance between about 0.47 and 1.6 light-years. When comets in this cloud were disturbed by a passing star, some of them could be deflected into orbits that pass close to the Sun. If such a comet also makes a close encounter with a massive planet such as Jupiter, it will be thrown into a short-period **orbit**.

open cluster
A cluster of **stars** within our own galaxy in which the members number less than 100 and are comparatively widely-spaced. The Hyades and Pleiades clusters are notable examples.

open universe
A universe that expands forever.

opposition
(*See* **conjunction and opposition**.)

orbit
The path of one body around another. Depending on the velocity of the orbiting body and the mass of the body about which it orbits, its path will either be an ellipse – and therefore closed – or follow the open curves of a parabola or hyperbola. In the latter two cases, an orbiting body will not return to make future circuits.

or Tietz, but only made widely known by the German astronomer Johann Bode in 1772.

If 4 is added to each number in the sequence 0, 3, 6, 12, 24, 48, 96, the sequence 4, 7, 10, 16, 28, 52, 100 is obtained. If the distance from the Sun to the Earth is taken as 10, then the other numbers closely correspond to the distances between the Sun and the other major planets out to Saturn. The continued sequence was found to be valid when Uranus was discovered, but it breaks down for Neptune and Pluto. It led to the recognition of a gap between Mars and Jupiter (i.e. at the number 28 in the sequence) and so to the discovery of the asteroids.

transit
The passage of a celestial body across the observer's **meridian**. Also used for the passage of the **inferior planets** across the Sun's disk. Such planetary transits are infrequent: those of Mercury occur at intervals of 3, 13, and 46 years; those of Venus fall in pairs 8 years apart, with successive pairs separated by a long interval of way over a century. The two transits in the 18th century were used for determining the distance from the Earth to the Sun.

T Tauri stars
Very young, rapidly rotating **stars**, probably the last stages of a **protostar** before it joins the main sequence. T Tauri is the prototype. Less massive than the Sun, they have extended and active gaseous atmospheres.

ultraviolet radiation
Radiation lying beyond the violet end of the visual spectrum. Its wavelength ranges from 380 down to 25 millionths of a millimeter. Only the longer wavelengths can penetrate the Earth's atmosphere.

umbra and penumbra
The umbra is the darker interior part of a shadow, and the penumbra the lighter outer region. The terms are also used in describing **eclipses and sunspots**.

uncertainty principle
In 1927, the German physicist Werner Heisenberg published his principle of indeterminacy or uncertainty. It arose out of his mathematical work on quantum theory and states that the position and momentum (the mass multiplied by the velocity) of a particle cannot both be specified precisely at the same time. There is an indeterminacy or uncertainty about every particle or, to put it another way, the numbers specifying its position and motion have a statistical spread.

Van Allen belts
Two regions in the Earth's magnetic field or **magnetosphere** in which electrically-charged atomic particles become trapped.

variable star
A **star** whose apparent radiation varies in intensity. This may be due to variation in output from the star, or because it is a binary system (*see* **binary star**) in which each component eclipses the other. The latter is known as an eclipsing binary.

wavelength
The distance between successive crests or troughs of a wave, particularly one of electromagnetic radiation. It is related to **frequency**.

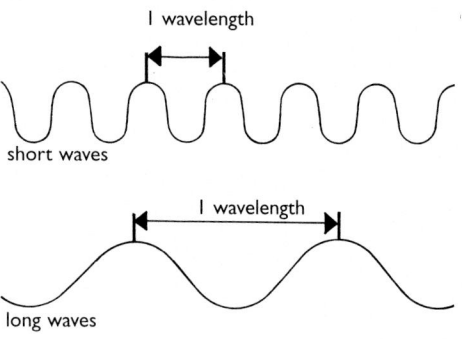

weak force
The nuclear force involved in **radioactivity** and some **neutrino** reactions. It is 100,000 times weaker than the strong nuclear force and operates over a distance of less than 10^{-14} mm.

white dwarf
A superdense state to which a star not more than 1.4 times the mass of the Sun eventually shrinks, most of its matter entering a state of **degeneracy**.

window
In astronomy, a range of wavelengths that can penetrate a planet's atmosphere. In spaceflight, the period in which a spacecraft can be launched on a particular mission.

Wolf-Rayet star
Stars whose spectral lines are all bright and therefore displays emission lines instead of the customary dark absorption lines (*see* **emission spectrum**). These are very hot bright stars with surface temperatures between 25,000 and 50,000 K; they are between 100,000 and one million times as bright as the Sun and also between 10 and 50 times more massive.

X-rays
Very short-wavelength, highly-penetrating electromagnetic radiation. The range of wavelengths runs from 10^{-5} to 10^{-8} mm (one hundred-thousandth to 10 billionths of a millimeter). X-radiation from space is evidence of highly energetic reactions on celestial bodies.

zenith
The point directly above an observer.

zodiacal light
A faint glow seen under good conditions in the west just after sunset and in the east just before sunrise. It is due to sunlight reflected from a belt of dust lying in the plane of the solar system.

Biographies

Bohr, Niels Henrik David 1885–1962
Danish physicist, noted for his model of the atom (1913) with a central nucleus and orbiting electrons. This he tied in with the then novel quantum theory, which he helped to develop further. Awarded the Nobel Prize for Physics in 1922.

Brahe, Tycho 1546–1601
Danish astronomer who built a large observatory on the island of Hven for measuring celestial positions. Brahe obtained a degree of precision unprecedented for the pre-telescope era. His observations led Kepler to his laws of planetary motion. He also devised his own planetary system.

deBroglie, Louis 1892–1987
French physicist who, in 1927, showed that not only light but also the electron and other atomic particles all possessed wavelike properties. Awarded the Nobel Prize for Physics in 1929.

Copernicus, Nicholas 1473–1543
Polish astronomer remembered for his book *De Revolutionibus Orbium Coelestium* (On the Revolutions of the Celestial Spheres), published in 1543. In this, he proposed that the Sun, not the Earth, was the center of the universe, thus completely breaking with tradition by relegating Earth to the position of a mere planet.

Eddington, Arthur Stanley 1882–1944
English astronomer and mathematician. A staunch supporter of the theory of relativity, he was also an expert on the constitution, mass, and luminosity of the stars, and was the first to suggest that spiral galaxies were similar to our Milky Way system.

Einstein, Albert 1879–1955
German physicist. Between 1905 and 1916 he developed the theory of relativity, which provided a radically new outlook on space, time, and gravitation, and for the first time proposed the equivalence of mass and energy. The theory affected the whole of astronomy and physics. Einstein also proposed the dual particle and wavelike nature of light and all electromagnetic radiation, and laid some of the foundations of quantum theory.

Faraday, Michael 1791–1867
English physicist and chemist, whose most significant research was on the relationships between electricity and magnetism. His investigations led him to the invention of the dynamo, electric motor, and transformer, while he also investigated the effects of electromagnetism on light. His results exerted a profound effect on later physics and astronomy.

Feynman, Richard Phillips 1918–1988
American physicist who developed quantum theory. With Murray Gell-Mann he formulated a theory to explain most phenomena concerned with the weak nuclear force. Feynman also developed simple diagrams to help understand the complex reactions of particles. He shared the Nobel Prize for Physics in 1965.

Galilei, Galileo 1564–1642
Italian physicist and astronomer best known for his pioneering use of the telescope, with which he discovered mountains on the Moon, sunspots and the phases of Venus, and that the Milky Way is composed of myriads of stars. His famous advocacy of the Copernican theory, which puts the Sun at the center of the universe, brought him into conflict with the Roman Catholic Church.

Gamow, George 1904–1968
Russian-American physicist. He helped develop the theory of one type of radioactive disintegration, then turned his attention to nuclear processes in stellar evolution. With others he worked on the origin of chemical elements in the stars, the hot Big Bang beginning of the universe, and suggested that there was a genetic code based on DNA.

Halley, Edmond 1656–1742
Britain's second Astronomer Royal. He was instrumental in persuading Newton to write his famous *Principia* and edited it and paid for its publication. Halley made many contributions to astronomy but is notably remembered for applying Newton's laws of planetary motion to specific comets. He predicted the return of the comet of 1682; it reappeared in 1758 as he had calculated, and is now known as Halley's Comet.

Heisenberg, Werner Karl 1901–1976
German physicist, best known for his principle of indeterminacy whereby there is an inherent uncertainty in the position and momentum of atomic particles. He received the Nobel Prize for Physics in 1932.

Herschel, Frederick William 1738–1822
Hanoverian-English astronomer and builder of astronomical telescopes of excellent quality. He was also the greatest of all observational astronomers. Herschel discovered Uranus, catalogued celestial objects, and was the first to suggest that the Milky Way marked the boundaries of an island of stars.

Hipparchos active 146–127 BC
A Greek astronomer who, among other things, discovered precession, calculated the length of the year correct to within $61\frac{1}{2}$ minutes, and was the first to complete a catalog of the stars. Hipparchos was certainly the greatest observational astronomer of the ancient world.

Hubble, Edwin Powell 1889–1953
American lawyer turned astronomer, Hubble is noted for his epoch-making work on galaxies, which in 1924 he showed were external to our own Milky Way. He also discovered that galaxies recede from us at a velocity which depends upon their distance, which implies that the universe is expanding.

Jansky, Karl Guthe 1905–1950
American radio engineer who discovered the existence of radio waves from space in 1931, but did not follow the matter any further. Nevertheless, his discovery lies at the foundation of modern radio astronomy.

Kepler, Johannes 1571–1630
German mathematician and astronomer who first made his name when he published a theory about the distances of the planets and their relationship to the regular solids of Euclid's geometry. He is noted above all for his discovery of three laws of planetary motion.

Leavitt, Henrietta Swan 1868–1921
American astronomer who, in 1908, discovered the relationship between the period of variation and intrinsic brightness of Cepheid variable stars. She also carried out fundamental work on determining stellar magnitudes.

Maxwell, James Clerk 1841–1879
Scottish mathematical physicist who discovered the nature of Saturn's rings. Most significantly, he discovered electromagnetic radiation.

Messier, Charles 1730–1817
French astronomer who catalogued all star clusters and nebulae he could detect visually. His list is the basis of the Messier numbers still used today.

Newton, Isaac 1643–1727
English mathematician and physicist. Newton discovered that white light was composed of light of all colors, devised a theory that light is composed of "corpuscles" or particles, constructed the first successful reflecting telescope, and independently laid the foundations of calculus. He is best known for his laws of motion and his theory of universal gravitation, both published in 1687. These were two of the most important contributions to the development of modern science.

Planck, Max 1858–1947
A German physicist. In order to explain the way heat is radiated by a totally absorbing body (a "black body"), Planck suggested in 1900 that its oscillating atoms receive and emit radiation only in discrete quantities or "quanta." This is the basis of quantum theory, which, with the theory of relativity, has revolutionized 20th-century physics.

Ptolemy active 161–180 AD
Greek astronomer, geographer, and mathematician, who worked at Alexandria. Ptolemy's most important work was his *Almagest*, which explained Greek mathematical astronomy and was the standard reference for the next 1,500 years.

Rutherford, Ernest 1871–1937
New Zealand physicist. He studied ionization and radioactivity, reaching the correct conclusion that atoms of one element spontaneously disintegrated to form others. This work led to his receiving the Nobel prize for Chemistry in 1908. In 1911, he made his greatest contribution with his theory that an atom is composed of a tiny nucleus accompanied by electrons.

Schrödinger, Erwin 1987–1961
German physicist who formulated a special equation in quantum mechanics which does for atoms what Newton's equations of motion do for planetary behavior.

Bibliography

WEBSITES

adc.gsfc.nasa.gov/mw/milkyway.html
NASA's Astrophysics Data Facility displays maps of the Milky Way

www.aip.org/
The homepage of the American Institute of Physics. Find out about physics today and in the past

antwrp.gsfc.nasa.gov/apod/
A different astronomical image is displayed here every day of the year

www.astronomy.com/
The homepage of *Astronomy* magazine

www.atnf.atnf.csiro.au/
Look at the universe with the Australia Telescope National Facility

cannon.sfsu.edu/~williams/planetsearch/planetsearch.html
Gives up-to-date information about planets discovered outside the solar system

www.esa.it/
The starting point for the many European Space Agency web pages

www.gamma.mpgarching.mpg.de/~hcs/history.html
A list of sites with information on the astronomical beliefs of ancient cultures

cfa-www.harvard.edu/cfa/ps/Headlines.html
The International Astronomical Union telegrams centre with details of recently discovered objects

www.hawastsoc.org/solar/
Gives a tour of the solar system

www.nrao.edu/
Look at the universe with the USA's National Radio Astronomy Observatory

nssdc.gsfc.nasa.gov/
The National Space Science Data Center of America

www.jpl.nasa.gov/
NASA's Jet Propulsion Laboratory web pages

www.nasa.gov/
The starting point for the many National Aeronautics and Space Administration web pages

www.nasm.edu/
The National Air and Space Museum, Washington, DC, main web pages

www.sciam.com/
The home page of Scientific American

www.seds.org/messier/
Information and views of the Messier deep-sky objects

www.seti-inst.edu/
The homepage of the institute searching for extra-terrestrial intelligence

www.skypub.com/
Sky and Telescope monthly magazine

www.stsci.edu/
The Space Telescope Science Institute provides information and displays pictures from the Hubble Space Telescope

BOOKS

Allen, R. H. *Star Names: Their Lore and Meaning* Dover Publications, Mineola, NY,1963

Audouze, Jean & Guy Israel (eds.) *The Cambridge Atlas of Astronomy* (3rd ed.) Cambridge University Press, Cambridge, UK & New York, 1994

Barrow, J. & J. Silk *The Left Hand of Creation: Origin and Evolution of the Universe* Oxford University Press, Oxford, 1993

Beatty, J. Kelly, Carolyn Collins Petersen, and Andrew Chaikin *The New Solar System* (4th ed.) Sky Publishing Corporation, Cambridge, MA, and Cambridge University Press, Cambridge, UK, 1998

Begelman, M. & Martin Rees *Gravity's Fatal Attraction: Black Holes in the Universe* W. H. Freeman, New York 1996

Couper, Heather & Nigel Heubest (eds.) *Space Encyclopedia* Dorling Kindersley, New York, 1999

Crovisier, Jacques & Thérèse Encrenaz *Comet Science – the study of the remnants from the birth of the solar system* Cambridge University Press, Cambridge, UK & New York, 2000

Feynman, R. P. *QED: The Strange Theory of Light and Matter* Princeton University Press, Princeton, NJ, 1985

Guth, A. *The Inflationary Universe: The Quest for a New Theory of Cosmic Origins* Perseus Press, New York, 1997

Hartquist, T. W. & D. A. Williams *The Chemically Controlled Cosmos: Astronomical Molecules from the Big Bang to Exploding Stars* Cambridge University Press, Cambridge,UK & New York, 1995

Hoskin, Michael (ed.) *The Cambridge Illustrated History of Astronomy* Cambridge University Press, Cambridge, UK & New York, 1997

Kaler, J. *Stars and their Spectra* Cambridge University Press, Cambridge, UK & New York, 1997

Kaler, James B. *Stars* W. H. Freeman, New York, 1992

Kaufmann III, W. J. & Roger A. Freedman *Universe* (5th ed.) W. H. Freeman, New York, 5th ed., 1998

Lang, K. R. *Sun, Earth and Sky* Copernicus Books, New York, 1995

Lederman, Leon M. & David N. Shramm *From Quarks to Cosmos* W. H. Freeman, New York, 1995

Levy, David H. (ed.) *The Scientific American Book of the Cosmos* St Martin's New York, 2000

Morrison, David *Exploring Planetary Worlds* W. H. Freeman, New York, 1993

Morrison, D. & T. Owen *The Planetary System* Addison-Wesley Longman, New York, 1996

Morrison, Philip and Phylis Morrison *Powers of Ten* W. H Freeman, New York, 1995

Osterbrock, D. E. (ed.) *Stars and Galaxies: Citizens of the Universe* W. H. Freeman, New York, 1990

Ridpath, Ian (ed.) *Norton's Star Atlas* (19th ed.) Longman, New York, 1998

Seeds, Michael A. *Foundations of Astronomy* Wadsworth Publishing Company, Belmont, CA, 1994

Silk, Joseph *A Short History of the Universe* W.H. Freeman, New York, 1994

Tirion, Wil & Roger W. Sinnott *Sky Atlas 2000* Sky Publishing Corporation, Cambridge, MA, and Cambridge University Press, Cambridge, UK, 1998

Trefil, J. S. *From Atoms to Quarks* Anchor Books, New York, 1994

Wheeler, John Archibald *A Journey into Gravity and Spacetime* W. H. Freeman, New York, 1990

Zuckerman, B. & M. Hart (eds.) *Extraterrestrials – Where are they?* (2nd ed.) Cambridge University Press, Cambridge, UK & New York, 1995

Index

Page numbers in *italics* refer to illustrations and their captions.

A

aberration of light 192
absorption nebula 192
absorption spectrum 192
accretion disc 68, 192
accretion theory 102, 105, 134, 135
  and the Moon 136, 137
acetylene 129, 133, 140
achondrites 147
Adams, John Couch 130
adenine 140, *156*, 160, *163*
Akna mountains, Venus 112
albedo 112, 192
algae 156, 162
Alpha Centauri A and B 94, *95*
Alpher, Ralph 36
altitude 192
aluminum 102
aluminum silicate 74
amino acids 158, 160
  chain 160
ammonia 106, 121, 128, 130
  clouds 125
  ice 120
Andromeda spiral galaxy 48, 51, 52, *56*, 58, 66, *185*
anthropic principle 180–83, 184–85, 186
  strong 182, 184
antimatter 26, 27, 34, *35*, 186, 192
ants 164, *164*
aperture 192
apes 164
aphelion 192
Aphrodite Terra, Venus 112, *112*
apogee 192
Apollo 15, Moon mission, *137*
argon 111, 115
Ariel, satellite of Uranus 140
Aristarchus 180
Aristotle, on life 154
Arp, Alton, distance of quasars 179
Arp 220 galaxy 61
asteroids 40, 102, 106, 142–43, 192
  "Trojans" 142–43, *143*
asthenosphere 114
astronomical unit 192
atmospheres 43, 110, 111, *111*, 115, 117, 120–22, 125, 128–29, *129*, 132–33
atoms 10, 22, 37, 160, 178, 192
  Bohr's model of 22, 23–24
  nuclei 22, 34, *35*, 83, 89
  Schrödinger's model of 23, 24–25
atomic bombs 14
atomic physics 22–25
aurorae 101, 123, 126, 132, 192
azimuth 192

B

Babcock, Horace 98–99
bacteria 155, 156, *157*, 160, 162
Barnard, Edward 95
Barnard's star 95
Barnes, Joshua 61
Barringer Crater, Arizona *146*
baryons (heavy particles) 26
Basri, Gibor 105
bees, social 164
Beethoven Crater, Mercury 108
Bell, Jocelyn 92
Berkeley, Bishop George 184
beryllium 34, 83
Bessel, Friedrich 94
Beta Pictoris *104*, 105
Bibliography 205
Big Bang 8, 21, 22, 64, 186
  and after 30–36, *30–37*
  conditions re-created 170
  dimensions rolled up 174
Big Bang theory 8, 178–79
Big Bounce 170
Big Crunch 168, 170, *170*
binary code 166, 167, *167*
Biographies 204
BL Lac(ertae) objects 66, 192
black body 36, 192
black dwarfs 89
black holes 8, 56, 62–65, *67*, *69*, 89, 148, 168, 170, *171*, 192
  active galaxies 66–67, 68
  at center of quasars 70, *70*, 73
  at centers of all galaxies 68–69
  connected by wormholes 186, *186*, 187
  decay of 65, *65*
  evaporation of 65, 170, 186
  blue shifts 48–49, 71, 80, 192
blue sky 74
Bode, Johann 142, 202
Bohr, Niels 23–24, 204
Bok, Bart J. 78
Bok globules 78, 84, 192
Bolyai, Janos 18
bosons 192
Brahe, Tycho 204
brown dwarfs 58, 78, *79*, 192
Bunsen, Robert 42
butterfly diagram *98*, 193

C

calcium 90, 102, 145, 146
Callisto, Jupiter satellite 138, *139*, *139*
Calvin, Melvin 158
Cambrian period 162
canals, Mars 117–18
Canis Minor 94
captured rotation 193
carbon 74, 76, 90, 97, 133, 145, 156, *156*, 160, 167
carbon compounds 156, *156*
carbon cycle in stars 83
carbon dioxide 115, 117, 150, *150*
  need to reduce emissions 152
  in Venusian atmosphere 111
carbon monoxide 111
  comet tails 145
carbon recycling 152
Carboniferous period 162
Carter, Brandon, on evolution of *Homo sapiens* 182, *183*
Cartwheel galaxy 59, *59*
Cassegrain focus 193
Cassini division 193
Cassini probe (to Saturn) 140
catalytic proteins 160, 162
Cat's Eye Nebula (in Draco) 89
celestial poles 56
  and equator 45
celestial sphere 193
cells 154
  membranes *159*, 162
  nucleus 162
  replication 160, *161*
Centaurus A galaxy *61*, 66
Centaurus arm, Milky Way 56, 57
central bulge 52, 55, 56, *60*
center of mass 193
Cepheid variable stars 46, 48, 77, 86, 193
Ceres 142
Chandrasekhar limit 89, 193
Chandrasekhar, Subrahmanyan 88
charge-coupled devices (CCDs) 45
Charon, Pluto's moon 133, *133*
chemical condensation 102
Cherenkov radiation 193
Chiron (asteroid) 143
chlorine compounds 152
chlorofluorocarbons (CFCs) 152
chlorophyll, enables use of solar energy 162
chloroplasts 162
chondrites 147
  carbonaceous 147
chondrules 147
Christy, James 133
chromosomes, and DNA 160, *161*, 162, *162*
chromosphere 97, 193
circumpolar stars 193
clouds 111, *111*, 121, 125, 132, 152
clumping 32, 38
clusters of galaxies 10, *13*, 38, *38–39*, 48, 49, 50–51, *50*, 60, 74, 167, *185*, 198
  formation 48–49
  gravitationally linked 48–49
  hierarchial 38, 40
  relaxed 51
  see also protogalaxies, superclusters
clusters of stars 80, *81*
  globular 56, 80, *80*, 195
  gravitationally linked 80
  open 80, *81*, 198
Coalsack Nebula, Southern Cross constellation 74

codons *161*
coincidences 181–82
cold dark matter (CDM) 48–49, 170
colliding galaxies 58–59, 60–61
coma, of comets 145
Coma cluster *185*
Coma supercluster 49, 50–51
Comet Hale-Bopp *145*
comet tails 144, 145, *145*
cometary orbits 106, *144*
comets 40, 102, 106, *107*, 144–45, 193
  and molecules of life 159
communication
  animal 164, *164*
  human 164, *165*, 166, *166–67*
  interstellar 166–67, *167*
complex materials, synthesis of 158
complex molecules 148, 158
computer program, intelligent 185
computer simulations, galaxy formation 38–39, *38–39*
computerization, revolution in processing information 45
computers, in self-replication 154
Cone Nebula (Monoceros) 77
conjunction and opposition 193
constellations 45, 188, 193
  Canes Venatici 49
  Carina 56, 76
  Centaurus 49, 56
  Crater 49
  Crux 56
  Hydra 49
  Leo 49, *73*
  Ophiuchus 95
  Orion 56
  Perseus 56
  Sagittarius 56
  Serpens Caput 58–59
  Virgo, galaxies in 50
continuous spectrum 193
cool matter 40
cool stars, detection of 44
co-ordinates 173, 193
Copernicus, Nicholas *165*, 180–81, 204
corals 162
cores, planetary 89, 90
  Earth 114, 150
  Jupiter 120
  Mars 116
  Mercury 109
  Neptune 130
  nickel–iron 109, 110, 150
  Saturn 124
  Uranus 128
  Venus 110
cores, stellar
  non-reactive helium 84
  of shrinking stars 89
  superdense 89
  temperature in white dwarfs 90

INDEX

Coriolis force 194
corona 99, *100*, 101, 194
coronal holes 101
coronal streamers 101
Corvus, colliding spirals 59
cosmic background radiation 32, 36, 44, *171*, 178, 194
cosmic dust see meteors
cosmic rays 194
cosmic strings 38, 194
coudé focus 194
covalent bonding 156
crab louse *11*
Crab Nebula 90, 178
   as pulsar 92, *92, 93*
craters 194
   Mars 118
   Mercury 108, *109*
   Miranda 140
   Moon 136, *137*
   Venus *112*
critical value, of density of matter 30, 168
crustal cycle, Venus *111*
crustal plates 114, *114*, 115
crystals, growth and replication 154, *155*
Curie, Marie and Pierre 22
61 Cygni 94
Cygnus A active galaxy *68*
cytochrome oxidase *162*
cytoplasm 162
cytosine 160, *163*

D

Daisyworld 153
Dalton, John 22
de Broglie, Louis-Victor 204
de Vaucouleur's law 60, 61
declination 194
decoupling era 35, 194
degeneracy 88, 89, 194
degenerate matter 62, 88
Deimos, Mars satellite 141
Delta Cephei, changes in *87*
density
   at birth of the universe 30
   critical value 30, 168, *169*
   in a shrinking star 88–89
density waves 53
deoxyribonucleic acid (DNA) 160
deuterium 8, 23, 34, *35*, 82, 178
deuterons 82
Devonian period 162
diamond ring effect 194
diatoms *157*
Dicke, Robert 36
diffraction 194
diffuse nebulae *13*, 194
dimensions
   fourth 18, 173
   rolled up 172–73, *174*
   and supersymmetry 174
   two, three, and four *173*
   of the universe 172
dinosaurs 163
Dione, Saturn satellite 140

Dirac, Paul 186
displacements current 15
distance, measurement of 46–47, 86–87
distant objects, apparent reddening of 77
DNA 160, *162, 163, 167, 167*
double helix 140, 160, *161*
   molecule, cross–links *160–61*
dolphins, encephalization in 163
Doppler effect 46, 194
double quasars 73
double stars 16, 194
Drake, Frank 166
dust 61
   interstellar 74
   see also gas and dust
dust grains 76
dust storms, Martian 117

E

$E=mc^2$ 14, *15*, 17, 73, 78
Earth 56, 114–15, 146, 150–53
   atmosphere and weather 115, *115*
   as center of universe 180
   crust 114–15
   curvature of 18, *19*
   natural system of checks and balances 150, *151*
   unique in supporting life 148
eccentricity 194
eclipses and occultations 194
eclipsing variables 87
Eddington, Arthur 184, 204
EGGs (evaporating gaseous globules) 79
Einstein, Albert 24, 172, 204
   and the bending of light 18, 20, 62
   relativity theories 16–17, 18, 19, *19*, 20–21, 62
   theory of gravity 16, 21
Einstein rings 73, *73*
electromagnetic radiation 44, 194
electromagnetism 14–15, 26, 28, *29*, 34, 172, 181
electron beams 24
electron pressure 89
electron waves 25
electrons 22, 26, 27, 34, *35*, 66, 92, 181, 186, 194
   definite energies 25
   definite orbits (Bohr) 23–24
   negatively charged 26
electroweak force 28, 32, 34
elements 8, 42, 78, 83, 97, 102, 182
   creation of 34
Elements, Euclid 8
elephant, African *11*
Eliot, T.S., quoted 170
elongation 195
emission nebula 195
emission spectrum 195
Enceladus, Saturn satellite 140
encephalization 163–64
energy *15*, 17, 36, 38, 82–83
   escaping from the Sun 95, 97,

97
   negative 186
energy density, in false vacuum 32
Encke division 195
entropy 195
enzymes 160
Epimetheus, Saturn satellite 139–40
Epsilon Eridani, wobbling 105
Epsilon ring, Uranus 129
equatorial mounting 195
equivalence, principle of 17, 195
ergosphere 64
escape velocity 195
ethane 129, 140, *156*
ethyl alcohol 74
Euclid 18
Euclidean geometry 18, *18, 19*, *19*, 172, *172–73*
Europa, Jupiter satellite *134*, 138, 139
event horizon 63, 195
evolution of life 148, 158–59, 160, 162–63
exclusion principle 25, 88, 89, 195
extraterrestrial intelligence 166–67, 182–83

F

faculae 99, 195
false vacuum 32, 28, 195
Faraday, Michael 14, *15*, 204
fat 162
feldspar 146, 147
fermions 195
Feynman, Richard Phillips 204
fibrils 97
fish 156, 162
flares 100–101, 195
flat universe 168, *169*
flowering plants 162
fluorescence 74
Formalhaut (Alpha Piscis Austrini) 105
force fields 28
formaldehyde 74, 159
formic acid 74
fossil record, simple organisms 158
Fraunhofer lines 195
frequency 195
fundamental forces 28–29, *28–29*, 170, 181

G

Gaia hypothesis 150, 153
galactic equator and poles 56
galactic nucleus, very active center 68
galaxies 8, *13*, 39, 53, 62, 168, 195
   active 66–67, *66–69*
   barred spirals 52, 54–55, *192*
   blue compact dwarf 69

   collapse of 170, *171*
   colliding 58–59, 60–61
   distances of 46–47
   dwarf 50, 69
   elliptical 50, 51, *59*, 60–61, 195
   eruptive 51
   formation of 32, 38, 52–55
   irregular 50, 196
   isolated 50
   lenticular 60, 197
   merging of 170
   Milky Way 56, *57*
   radio 66, 69
   ring 59, *59*
   Seyfert *67*, 70, 201
   spiral 48, 50–51, *52*, 52–57, *53*, *57, 67*, 168, 202
   starburst 60–61
   velocities of recession 46, 48
   velocities of rotation 53–55, 168
   see also clusters of galaxies, protogalaxies and individual galaxies
Galaxy, the *42*, 56, *57*
   center of *57*, 66
   tidal effects on the Magellanic Clouds 58
Galileo (Galileo Galilei) 42, 127, 134, 180, 204
Galileo probe to Jupiter 45, *134, 142*
gamma radiation 34, 82, 83, 95, 195
Gamow, George 36, 204
Ganymede, Jupiter satellite *123*, 138, 139
gas and dust, interstellar 40, 76–77
   becoming protostars 78
   birth of planetary system *102–3*
   clouds 40, 56, 74, 76
   density 78
   in galaxies 50
   in the Milky Way 56
   see also nebulae
gas giants 10, 106, 120–33, 134
gases, spectrum at low pressure 42
Gaspra (asteroid) *142*, 143
gauge symmetry 28, 202
Gell-Mann, Murray 26
Geminid meteor shower *147*
genes 160
genetic code, in DNA 160, *160–61, 163*
geocentric concept 165, 180
geodesics 18, *19*
germanium 157
Giotto space probe 144, *145*
Glashow, Sheldon 28
glowing gas nebulae 74
glucose *156*
gluons 26, 174, 195
Gold, Thomas 92, 178
grand unified theories (GUTs) 28, *33*, 195

INDEX

gravitation 16–17, 28, *29*, 53, 102, 181
  and solar nebula 102
gravitational energy 79
  fields 20, 21
gravitational radiation 176
  interactions of galaxies 57, 58
  lenses/lensing 20, 73, 179, 196
  gravitons 174, 176, 196
  importance of strength of 181–82
  gravity wave detectors 176–77
  unification with electromagnetism 172
grazing-incidence telescopes 44
Great Attractor 49
Great Dark Spot (GDS), Neptune 132
Great Red Spot (GRS), Jupiter *120*, 121, *121*, 122, 196
Greek, view of the universe 7
Green, Michael 174
greenhouse effect 111, 115, 150, *150*, 152, 196
guanine 160, *163*
Guth, Alan 30

H

HI and HII regions 196
hadrons 27, 34, *34*, 196
Hale-Bopp (Comet) *145*
Halley, Edmond 144, 204–5
Halley's Comet 106, 144–45, *145*, 159
halo 196
halos, of spiral galaxies 52, 56, 58
Hawking, Stephen 64, 65
Heisenberg, Werner Karl 205
  uncertainty principle 25, 174, 176–77, 203
helium 8, 34, *35*, 78, 102, 106, 109, 115, 121, 124, 125, 128, 130, 145
  conversion to carbon 83
  helium 3, 23, 24
  helium "burning" 82–83, 84, 88, *88*, *91*
  nucleus 22, *35*, 82, 83
  and steady-state theory 178
Hercules cluster 51
Hercules supercluster 49
Herman, Robert 36
Herschel, William, 56, 128, 142, 205
Hertz, Heinrich 14–15
Hertzsprung-Russell diagram 84–85, *84–85*, 196
Hewish, Anthony 92
Hidalgo (asteroid) *143*
hierarchial clustering 38, 40
Higgs field 18, 196
High-Energy Astronomical Observatories 56
Hipparchos 205
Homo sapiens see human beings
horizon distance 30, 32, *33*, 196
Horsehead Nebula, Orion 74
hour angle 196

Hoyle, Fred 178
Hubble constant 47, 170, 179, 196
Hubble, Edwin 48, 54, 205
Hubble Space Telescope 47, *73*
Huggins, William 48
human beings 10, 148, 164, 180, 186
  communication by language and symbolism 148, 164
  complexity of 155–56
  encephalization in 163, 164
  guaranteeing existence of universe 185
  special powers/abilities 148, 182
Huygens, Christiaan 127
Huygens probe (to Titan, moon of Saturn) 140
Hyades (in Taurus) 80
hydrocarbons, Neptune 133
hydrogen 78, 102, 106, 109, 121, 128, 130, 132, 133, 145, 160, 167, 182
  in interstellar space 74
  metallic 120, 124
  molecular 120, 122, 124
  proton-proton cycle 82, 97
  and steady-state theory 178
hydrogen cyanide 140
hydrogen sulphide 157
hydroxyl 76, 130
hyper-universe, Narlikar 186
hyperbolic geometry 18, *18*, 168
hyperspheres 18–19

I

Iapetus, Saturn satellite 140
Icarus (asteroid) *143*
ice coatings 76
IG 29 and IG 30, colliding spirals 59
image enhancement 45
inertia 196
inferior planet 196
inflation 28, 30, 32, *33*
inflation era 30
information, permanent recording of 164
Infrared Astronomy Satellite (IRAS) 44, 48–49, *104*, 105
infrared radiation 42, 61, 78, 196
infrared sources, M42 Orion Nebula 76
intelligence 155, 163–65, *164*, *165*
interatomic and intermolecular forces 22
interference 24, 196
interference patters 184, *185*
interferometry 196
intergalactic material, non–visible 168
interstellar communication 166–67
intracluster gas 51
Io, moon of Jupiter 123, *123*, *138*, 138–39
Jupiter's gravitational effects *138*, 139
iodine cycling 152
ionized particles, blowing out from quasars 73
ionosphere 43, 126, 196
ions 66
iron 97, 118, 120, 145, 146, 147, 160, 178
iron silicate 74
Ishtar Terra, Venus 112, *112*
island universes *see* galaxies
  isotopes 196–97
  of carbon 83
  of helium 23, 24 82
  of hydrogen 8, 23, 24, *35*, 82, 178
  of oxygen, unstable 83

J

Jansky, Karl 44, 205
Janus, Saturn satellite 139–40
jets, galactic 66, 68, *68*, *69*
Jewel Box open cluster 80
Jupiter 42, 120–23
  atmosphere 120–22
  cloud belts *123*
  gravitational field, effects of *138*, 139, *143*
  magnetosphere 123
  plasma ring (torus) *123*
  radiation belts *122*
  ring system 123
  satellites 121, 134, *134*, 38–39
jupiters 58, 78

K

Kaluza, Theodor 172
Kappa Crucis 80
Kepler, Johannes 180, 205
Kepler's laws 197
Kerr equations 64
Kirchhoff, Gustav 42
Kirkwood gaps 142, *143*, 197
Klein, Oscar 172–73
knot texture of space 38
Kuiper Belt 106

L

Lagrangian point 197
language 148, 164
Laplace, Pierre 62
large-number coincidences 181
laser beams 46, 176
Laser Interferometry Space Antenna (LISA) spacecraft 176
Leavitt, Henrietta Swan 205
Leighton, Robert 98–99
lenticular galaxies *see* galaxies, lenticular
Leonids 146
lepton-quark interchange 28
leptonic era 35
leptons 27, 34, 197
Leverrier, Urbain Jean Joseph 130
libration 197
lichens 156
life 150, *151*, 152, *158*, *159*
  beginning of 158–59
  chemistries of 156–57
  nature of 154–55
  possibly generated in space 74, 76
  reproduction and repair 154
light 14, 24
  and gravity 18, 21
  speed of 14, 16, *17*, 46, 62
light beams, interference 24, *24*
light cone 63, 197
  and black holes 63–64
light waves 14
light-year 46, 197
lines of force *15*
liposomes 159
lithium 7, 34
lithosiderites 146
lithosphere 114–15
Lobachevsky, Nikolai 18
Local Group 48–49, 50, 51, *51*, 52, *56*, *185*, 197
Local Supercluster 49
Lodge, Sir Oliver 62, 101
logarithmic notation 10
Lovelock, James 150
Lowell, Percival 118
Lunar Roving Vehicle *137*
Lynden-Bell, Donald 62

M

M 13 globular cluster 167
M 31 spiral galaxy *see* Andromeda spiral
M 33 spiral galaxy 51
M 42 Great Nebula, Orion 74, *75*, 76
M 82 spiral galaxy *55*
M 87 (Virgo A) galaxy 50, 66, *69*
machines, self-reproducing 154, *154*
Maat Mons, Venusian volcano *112*
MACHOs (massive compact halo objects) 58
Magellan spacecraft, radar pictures of Venus 112–13, *113*
Magellanic Clouds, Large and Small 51, 52, *56*, *57*, 58, *80*, 90, 92, *176*, *185*
magnesium 90, 97, 102, 145, 146
magnesium silicate 74
magnetic fields 15, 66, 99
  Jupiter 122, 123
  Mercury 109
  Neptune 130, 132
  neutron stars 92
  Saturn 125, 126
  Uranus 128
magnetic monopole 28, 197
magnetic tubes 99
magnetoids 64
magnetosphere 123, 125, 126, 197

INDEX

magnitude 197
main sequence 84–85, *84*, 88, 197
Malin, David 55
Malin I galaxy 55
mammals, gaining the ascendancy 163
mantle 114, 116, 128, 130
Mare Imbrium (Moon) 137
Mare Orientale (Moon) 137
Margulis, Lynn 150
Mariner 4, canals on Mars non-existent 118
Mariner 10, investigating Mercury 108, 109
Markarian 205 quasar, link to NGC 4319 galaxy *178*
Markarian galaxy 197
Mars 116–19, *150*, 152
   atmosphere 116, 117, 119
   polar ice caps *116*, *117*
   retrograde motion *181*
   satellites 134
   seasons 116–17
Martian surface, mapped 118
mascon 198
masers 76
mass 16–17
   center of 193
   of cosmic strings 38
   missing see matter, dark of fundamental particles 181
massless particles 174
mathematics 148, 164
   in interstellar communication 166
   precise formulation of ideas 14
   pure numbers 181
matter 14, 74, *78*
   building of 32, 34–35
   creation of *31*
   critical value of density 30, 168
   dark 56, 58, 168, 194
   decoupled from radiation 34, 35
   inside black holes 63
matter era 34
Maunder minimum 98
Maxwell, James Clerk 14–15, 16, 24, 42, 127, 205
Maxwell mountains, Venus 112
Maxwell's equations 15, 172
memory, human 155
Mercury 106, 108–9
   anomalies in motion 18, *19*, 20
   atmospheric gases 108–9
   double dawn *108*
meridian 198
mesons 26, 34, *34*, 198
Mesopotamians, ancient 180
mesosiderites 146
mesosomes *159*
mesosphere 114
messenger particles 26, 28, *29*, 174, 176
Messier, Charles 205

meteor showers 146
meteorites 146, 198
   containing amino acids 158
   iron–rich *146*
   metallic 146
meteors 40, 102, 146–47, 198
methane 102, 106, 115, 120–21, 128, 130, 133, 140–41, *156*
MG 1654 + 1346 quasar (in Hercules), gravitational lensing 73
mica sheets 147
Michelson, A.A. 16
Milky Way 51, 57, 198
   "holes" in 56
Miller, Stanley 158
Mimas, Saturn satellite 140
Miranda, Uranus satellite 140–41, *140*
missing matter see matter, dark
Mitchell, John 62
mitosis see cell replication
molecules 10, 22, 160–61, 198
   carbon–bases 74, 76, 160
   from space 148, 159
   ring formation 156
Moon 12, 42, 136, *136*, 137
   distance of 46
   formation of 102
   geological change in 136
   internal structure of 136–37
   orbit of 134
moonlets, Neptune 132
moons see satellites
Morley, E.W. 16
mountain ranges, Earth, creation of *114*
multicellular organisms 162
multimirror telescope 44, 198
muons 21, 27, 170
mutation 160

N

nadir 198
Narlikar, Jayant 186
Near Earth Asteroid Rendezvous (NESR) probe 143
nebulae 40, *42*, 48, 74–77, *76*, *78*, 85
   dark 74
   diffuse *13*, 194
   emission 195
   new, formation of 77
   planetary 10, *12*, 88, *88, 89*, 199
   reflection 74, 200
   see also gas and dust, interstellar
nematode worm, genetic code *163*
neon 115
Neptune 106, 130–33
   cloud system 132
   gaseous envelope 130
   Great Dark Spot (GDS) 132
   magnetosphere 132
   mantle 130
   satellites 134
Nereid, Neptune satellite 134

nervous system 164
   autonomic and consciousness subsystems 164
neutrino detectors 168, *169*
Neutrinos 27, 34, *35*, 82, 83, 97, 168, 198
neutron decay 26
neutron gas 89
neutron stars 65, 90, 92, 170, 198
neutrons 23, *23*, 27, 34, 37, 82, 92, 198
Newton, Isaac 42, 180, 205
   corpuscular theory of light 62
   gravitation law 16–17, 24, 130
   and planetary motion 7
NGC 1365, barred spiral galaxy *54*
NGC 2997, spiral galaxy 52
NGC 4319 galaxy, link to Markarian 205 quasar *178*
NGC 4565, spiral galaxy 53
NGC 4755, Jewel Box open cluster 80
NGC 5102, lenticular galaxy 60
NGC 5128 (Centaurus A) galaxy *61*, 66
NGC 5216 and NGC 5218 galaxies, linked *58*
NGC 5426 and NGC 5427 galaxies, linked 58
NGC 7252 galaxy (in Aquarius) 61
nickel 146
nitrogen 74, 111, 115, 140, 160, 167
nitrogen ice 141
nitrous oxide 115
non-Euclidean geometry 18–19, 172–74, *173*, *174–75*
novae 87, 198
nuclear fusion 34, 78, 82, *82*
   and life cycles of stars 84–85
nucleic acids see DNA, RNA
nucleon *29*
nucleosynthesis 198
nucleotides 158, 160, *161*, *163*, 167
nutation 198

O

Oberon, Uranus satellite 134, 140
occultation see eclipses and occultations
Olbers, Heinrich 37
Olber's paradox 37
olivine 146, 147
Olympus Mons, Mars 118, *119*
Omega Centauri 80
Oort cloud 106, *107*, 198
Oparin, Aleksandr 158
Öpik-Oort cloud *107*
opposition see conjunction and opposition
optical astronomy 44
orbital elements 199
orbits, celestial 106, *106–7*, 198
organelles 162

organic molecules 140, 145, 148, 158
   in nebulae 74, 158
Orion Nebula 74, 75, 76
oscillations, of the Sun 97
oxygen 74, 90, 97, 115, 117, 145, 157, 160, 162, 167
ozone layer 43, 199
   "holes" in 152

P

pallasites, formation of 146–47
parallax, distance measurement 46–47, 94, 199
parhelion 199
parity, non-conservation of 199
parsec 199
particle accelerators 26–27, *26*
particle-antiparticle pairs 65
particles, fundamental 22–28, *22–23, 25, 27, 28, 29*, 174, 176–77, 180
   and waves of probability 24, 25, 180–85, *185*
   See also bosons, fermions, superstring theory
Pathfinder space probe (Mars) *118*
Pauli, Wolfgang, exclusion principle 25, 88, 89, 195
Pease, Francis 48
51 Pegasi 105
Penzias, Arno 36
perigee 199
perihelion 144, *144*, 199
period-luminosity relation 46, 86–87, *87*, 199
Perseus cluster 51
Perseus supercluster 49
Perseids 146
Phobos, Mars satellite 141
Phoebe, Saturn satellite 134
phosphorous 160
   red *120*
phospholipids *159*
phosphine *120*
photoelectric effect 24, 199
photography 42
photons 23, 24, 26, 28, *29*, 32, 34, *34*, 35, 174, 184, 199
   as "wavicles" 24
photosphere 97, 99, 199
photosynthesis 150, *152*, 157, 162
phytoplankton, threatened 152
Piazzi, Giuseppe 142
Pickering, William 133
Pioneer 10, pictorial plaque 166
plages 99
Planck, Max 23, 205
Planck length 21, 174, 177
Planck time 21, 30
Planck's constant 181
planetary nebula see nebulae, planetary
planetary orbits 106
planetary systems 44, *102–3*, 104–5, *104, 105*
   see also solar system

INDEX

planetisimals 102, 134, 135, 137
planetoids, collisions between 102
planets 7, 40, 108–33, 199
    early theories of formation 104–5
    gas giants 10, 106, 120–33, 134
    minor (asteroids) 40, 102, 106, 142–43, *143*
    terrestrial 106, 108–19, 134
    see also individual planets
planisphere 199
plants, altering conditions on Earth 162
plasma 66, 199
plasma ring, Jupiter *123*
Pleiades 80, *80–81*
Pluto 106, *107*, 133
    and Charon (moon) 133, *133*
polar lights *see* aurorae
polarization 199
Populations I and II (stars) 56, 199
positrons 26, 34, 82, 83, 170, 186
potassium 145
precession 200
primates 163
prime focus 200
Procyon, multiple system 94
prokaryotic cells *159*
prominences 99, 100–101, 200
proper motion 200
proteins 160, *161*, 162
    catalytic 160, 162
    structural 160
protogalaxies 38–39, 48, 53, 60, 78
proton-antiproton collider, CERN *26*, 28
protons 23, *23*, 26, 27, 28, 34, 37, 82, 92, 200
    positively charged *10*
    possible decay of 170
protoplanets 102
protostars 77, 78, *79*, 200
protosun 102
protozoa, single-celled *11*
Proxima Centauri 46, 94
PSR 1913–16 pulsar, and gravitational radiation 176
Ptolemy 205
pulsars 92, *93*, 176, 200
pyroxenes 146, 147

Q

quantum 23
quantum space 176–77
quantum theory 7, 8, 21, 22–25, 180
    applied to black holes 64
quarks 8, 10, 23, 26, *29*, 37, 174, 200
    in early universe 28
    in mesons 34, *34*
    in nucleons *10*, 27, 34, *34*
    types of 27, *27*
quasars 20, *67*, 70–73, *71*, *71*, 168, *178*, 179, *185*, 200
    double 72, *73*

red shift problem 70
remoteness of *47*
    3C 48 70
    3C 272 70, *71*
    twin radio images *72*
quasi-stellar radio sources *see* quasars

R

R Coronae Borealis, periodic variable 86
radiation 8, 34–35, 170, 195
    electromagnetic 36, 126, 194
    from dark hydrogen 74
    from distant galaxies 40
    from protostars 78
    from quasars 70
    high-energy *83*
    synchrotron 68, 73, *122*
    see also infrared radiation, ultraviolet radiation
radiation era 34
radiative zone, Sun 95
radio astronomy 20, 44
    study of the Galaxy 56
radio galaxies 66
radio halo, Centaurus A 66
radio signals, and planetary distance 46
radio sources 66, *68*, 70
radio waves 14–15, 44, 52, *69*
    and universal communication 166
radioactive heating, Moon 137
radioactivity 22, 200
Reber, Grote 44
red shifts 46, 48, 63, 80, 196, 200
    gravitational 196
    in quasars 70, 71, *178*, 179
    and velocities of recession 46–47
Rees, Martin 181
reflection nebulae 74, 200
refraction 201
regolith 201
relativity theory 7, 8, 14, 16–21, *19*, 62, 168, 170
    and curvature of space 30
    and the false vacuum 32
    and gravitational lensing 73
    and superdense bodies 62
    combined with electromagnetism 172
reptiles 162, 163
resolving power 201
resonance 201
retrograde motion 110, 128, 134, 135, *181*, 201
Rhea, Saturn satellite 140
ribonucleic acid 160
Riemann, Bernhard 18
Riemannian geometry 18, *18*
right ascension 201
ring galaxies 59, *59*
RNA 160, *161*, 167
Roche limit 201
Rosse, fourth Earl of 42
rotation, captured

(synchronous) 193
RR Lyrae stars 86–87, 201
Rutherford, Ernest 22, 205

S

Sagan, Carl 166
Salam, Abdus 28
salts 160
satellites 40, 42, 134–41, *135*, *135*, 201
Saturn 124–27, *124*
    atmosphere 125
    cloud circulation 124–25
    magnetosphere 125, 126
    oval red spot 124–25
    ring system 124, *125*, 126–27, *126*, *127*
    satellites 126, 134, 139–40
scale factor of the universe 168
Schläfi, Ludwig 18
Schiaparelli, Giovanni 117–18
Schrödinger, Erwin 24–25, 205
Schwartz, John 174
scintillation 201
"Scooter," cloud on Neptune *132*
selenium 157
self-repair 154
    and self-replication, in DNA 160
senses, in animals 155
sexual reproduction 160
Seyfert galaxies *67*, 70, 201
shadow matter 174
shepherd satellites 127, 132, 139
shock waves 53, 59, *59*, 77
shooting stars *see* meteors
sidereal day 201
sidereal period 201
silicates 120, 146
silicon 74, 90, 102, 118, 145, *156*, 157, *157*
silicones 157
simple mounting 201
singularity 63, 170
Sirius, binary star 89, 94
sky mapping 45
Slipher, Vesto 48
Smolin, Lee 186
SN 1987A, Type II supernova 90
sodium 46
soil 112, 118
Sojourner, robotic Martian rover *118*
solar apex 201
solar constant 201
solar cycles 98–99
solar eclipses 99–100, *100*
solar nebulae, 102, 147
    detection of *104*, 105
solar radiation, Venus 112
solar system 102, 104–6, *106–7*, 108–47, *167*, *180*, 201
    birth of 102
    plan of planets, *Pioneer 10* 166
    see also planetary systems
solar wind 123, 126, 145, 201
    and Earth's magnetosphere *101*, *101*

Southern Supergalaxy 49
space 36, *185*
    and black holes 62–64
    higher dimensions of 172
    quantized 177
    stretched 20
    three-dimensional 172
space probes, Earth-orbiting 44
space technology, importance of 44
space-time 201
    curved 16, 18–19, 20–21, 62, 168
    four-dimensional 172
    higher dimensions of 172–74
    rippling caused by gravitons 176
space-time diagrams 63
speckle interferometry 201
spectral class 202
spectral line 202
spectroscope 42, 202
spectrum 193, 195, 202
spicules 97
spin 25
spinal cord 164
spinars 64, 202
star charts 188, *188–91*
starburst galaxies 60–61
stars 42, 79
    binary 80, 86, 87, 90, 94, 192
    blue 69, 70, 84
    blue giants 192
    blue supergiants 90
    death of 88–91, *88–91*
    distances of 46, *46*
    double, light from 16
    escape velocity of 62
    evolution of 78–80, *78–79*, 82–93, *84–85*, *88–92*
    formation of *13*, 32, 38–39, 44, 52–53, 59, *61*, 77, 78, 79
    giant 195
    lifetime of 181
    massive 82–83, 84
    multiple 80, 94, 198
    nuclear fusion 40, 82–83, *82–83*
    red dwarfs 84, 95, 105, 200
    red giants 84–85, 88, 200
    red supergiants 86, 88, 200
    size, determination of *85*
    triple-alpha reaction 83
    yellow 84, 94
    yellow supergiants 86
    see also novae, supernovae, variable stars *and* individual stars
steady-state theory *178*, *178*, 202
    and quasars 178–79
stellar activity, length of 170
stellar associations 80
stellar energy, sources of 82–83
Stephan's Sextet group 58–59
string theory 173–74
    see also superstring theory
strong force 26, 28, *29*, 32, 181, 202

subatomic particles 26–27, 27, 174, 176–77, 180
sugars 158
sulphur 74, 90, 115, *138*, 139, 157, 160
sulphur cycling 152
sulphur dioxide 111
sulphuric acid clouds, Venus 111, *111*
Sun 12, 42, 46, 56, 78, 94–101, 102
   as center of the universe 7, *165*, 180–81
   composition of outer layers 97
   convection/radiation 95, 97, *97*
   core temperature 95
   corona 99, *100*, 101
   flares 96
   lies in the Orion arm 56, 94
   magnetic field 98–99
   mass distorting space 19
   oscillations of 97, *97*
   proton-proton reaction 82–83, *82–83*, 97
   and sunshine 14
   see also solar system
sunlight, spectrum of 42
Sunlike star 88
   life cycle of 84–85, *85*, 170
sunspots 97–99, *99*, 202
   cycles 98, *98*
   umbra and penumbra 97–98
supergiant star 202
supergravity theory 174
super-hypergalaxy 170
superclusters 10, *13*, 38, *38–39*, 48–49, 50–51, 202
supercooling of universe 32
   and knots in space 38
superdense bodies 62
superfluid layer, in pulsars 92
superforce 202
superior planet 202
superluminal velocities, illusory 66
supernovae 8, 59, 87, 88–89, *88*, *90*, *91*, *93*, 102, 182, 202
   Types I and II 90, *90*, *91*
superstring theory 172–74, *175*, 202
   heterotic 174
supersymmetry 174, 202
Svens Klint meteor, amino acids from 158
symbols, power of 14
symmetry 28, 174, 202
symmetry-breaking 28, 202
synchrotron radiation 68, 73, *122*
synodic period 202

T

T Tauri stars *87*, 105, 203
telescopes
   infrared 44
   multi-mirror 44–45, *44*
   radio 10, 36, 44, *44, 45*, 122
   reflecting 44–45, 200
   refracting 42, 200
   X-ray
temperature 32, 36, 152
   of black holes 64
   inside a shrinking star 88–89
   Kelvin (absolute) scale 32, *32–33*, 197
   in protostars 78
   Sun's core 95
temperature equilibrium 32
terrestrial planets 106, 108–19, 134
Tethy's, Saturn satellite 140
theory of everything (TOE) 28, 174
thermodynamics 23, 32
   applied to black holes 64
thymine 160, *163*
tidal effects 58
time 16, 20
   and black holes *62–64*, 63
   dilation 20, 21, 202
Tipler, Frank 185
   human beings unique 166–67, 182–83
Titan, Saturn satellite 140
Titania, Uranus satellite 140
titanium 102
Titius, Johann 142, 202
Titius–Bode Law 142, 202
Tombaugh, Clyde 133
transit 203
transpiration *152*
Trapezium cluster 74
Triangulum 52
trilobites 162
triple-alpha reaction, giant red stars 83
tritium 34
Triton, Neptune satellite 134, 141, *141*, 167
tropopause, Jupiter 121–22
tube worms *157*
Turok, Neil 38

U

ultraviolet radiation 42, 68, 70, 74, 75, 100, 156, 203
   changing methane 132–33
   and melanin 156
   and the ozone layer 152
umbra and penumbra *100*, 203
Umbriel, Uranus satellite 140
uncertainty principle 25, *25*, 174, 176–77, 203
unified force 28, 29, 30, 33, 34, 195
United Kingdom Infrared Telescope (UKIRT) 44
universal laws 40
universe 7, 8, *38–39*, 148, 184
   birth of 30–32, *33*
   as a black body 35, 36
   closed 168, *169*, 170, 193
   cooling of 35, 36–37, *37*, 78
   death or rebirth 171, *171*
   depends on our observations 184–85
   expanding 30, 48, 186
   favoring human existence 182
   future of 168–69
   intelligent life developed 182
   measurement of 46–47
   observation of 42–43
   open 168–69, *169*, 170, 198
   oscillating 168, 170, *171*
   possible collapse into a singularity 170
   scale of 10, *10–13*
   "smoothness" 32, 38
   supercooling of 32
universes, baby 186, *186–87*
uranium 137
Uranus 128–29, *128–29*
   atmosphere 128–29, *129*
   orbital irregularities 128, 130, 140–41
   ring system 129, *129*
   satellites 134, 135, 140–41
Urey, Harold 158

V

Valles Marineris, Mars *116*, 117, 118–19
Van Allen belts 101, *101*, 203
van de Kamp, Peter 105
variable stars 46, 48, 66, 77, 86–87, *86, 87*, 201, 203
   see also novae, supernovae
vectors 15
Vega (Alpha Lyrae) 105
Venera landings, Venus 112
Veneziano, Gabriele, and subatomic particles 173–4
Venus 42, 106, 110–13, *110–13*, 150
   atmosphere 110, *110*, 111, *111*
   geological changes 112
   physical features 111–12
Venus fly trap 155
Very Large Array 44
Very Large Telescope 44
Viking 1 and *2* 118, *118*
Virgo cluster 49
Virgo supercluster 49
virtual particles 26, 65
virtual photons 26
viruses 10, 161
   exploitation of host cells *155*, 160
von Neumann, John 154
Voyager probes 45
   carried video discs 166
   data from Jupiter 121
   data from Neptune 130
   data from Uranus, 128, 129
satellites 138–41
Saturn's rings 127, *127*

W

W Virginis variable (Cepheids) 86
W and Z particles 26, 28, 29, 34
water 76, 130, 150
   ice 120, 128
   recycling *152*
   vapor 44, 111, 115, 117, 121, 150
water channels, Mars 119
wavelengths 22, 203
   and energy levels 23–24
waves 14–15
wavicles *see* photons
weak force 26, 28, *29*, 34, 181, 182, 203
   unified with electromagnetism 28
Weber, Joseph, attempt to detect gravity waves 176, *176*
Websites 205
Weinberg, Steven 28
whales encephalization in 163
Wheeler, John 184
wormholes in space 177
"whiskers," metallic 178
white dwarfs 21, 85, 87, 88, 89, *89*, 90, 94, 203
   becoming neutron stars 170
white holes 186, *186*
white light 42
Wilson, Robert 36
WIMPs (weakly interacting massive particles 58, 168
winged insects 162
Wolf-Rayet star 203
world lines 63, 174
world sheet 174, *175*
wormholes 177, 186, *186*
Wright, Gillian *et al.*, study of colliding galaxies 61

X

X particle 28
X-ray sources *42–43*, 50, 56
X-rays 62, 68, *69*, 70, 95, 100, 203

Y

Young, Thomas 24

Z

Zach, Xavier von 142
zenith 203
zodiacal light 203
Zweig, George 26
Zwicky, Fritz 48

Acknowledgments

The author is most grateful to Dr Peter Cattermole, Dr Merton Davies, Professor Michael Green, Dr David Malin, Dr Patrick Moore, and Sir Brian Pippard FRS for their advice and assistance, but must exonerate them from any errors and omissions which the book may contain.

The lower illustration on p.133 is based on the work of Mark Buie (Space Telescope Science Institute, Baltimore) and David Tholen (University of Hawaii). The illustration on pp. 162–63 is based on work of Walter M. Fish and Emanuel Margoliash (Northwestern University).

Illustration credits

Gary Thompson
pp. 30/31, 38/39, 165, 166/167, 170/171, 174/175.

David Fathers
pp. 10 (nucleon), 26/27 (nucleus), 28/29, 34/35, 82/83.

Sue Sharples
pp. 22/23, 24, 46/47, 53, 59, 64/65, 71 (graphs), 77, 78/79, 84/85, 86/87, 88, 90/91, 98, 100/101, 102/103, 110, 122 (Io), 152/153, 162/163.

Mainline Design
pp. 17, 18/19, 20/21, 32/33, 36/37, 51, 56/57, 94/95, 106/107, 108–133, 143, 144, 172/173, 178, 186/187.

David Wood
pp. 67, 68, 70/71, 92/93, 184/185.

Ed Stuart
pp. 150, 154, 160/161.

Dave Ashby
pp. 152/153, 158/159, 192–205.

Mark Iley
pp. 108–135 (planet friezes).

Technical Art Services
pp. 15, 43, 45, 105, 108, 135, 188–191.

Picture credits

t = top; c = center; b = bottom; l = left; r = right
1 David Malin/Anglo–Australian Telescope Board; 2/3 Galaxy Picture Library; 4 NASA/Science Photo Library; 5 Dr Bradford A. Smith/National Space Science Data Center; 6–7 G. Deichmann/Planet Earth Pictures; 9 Patrice Loiez, CERN/Science Photo Library; 11t Cath Ellis, Dept. of Zoology, University of Hull/Science Photo Library; 11b Earth Satellite Corporation/Science Photo Library; 12t David Malin/Anglo-Australian Telescope Board; 12c Science Photo Library; 12b John Sandford/Science Photo Library; 13t NASA/Science Photo Library; 13c–b David Malin/Royal Observatory Edinburgh & Anglo-Australian Telescope Board; 14–15 AEA Technology; 16–17t Gerolf Kalt/Zefa Picture Library; 16–17b Jean Pottier/Rapho; 25 Doris Haselhurst/The Dance Library; 26–27 David Parker/Science Photo Library; 36–37 Douglas Kirkland/Colorific!; 39 Adrian L. Melott, University of Pittsburgh; 40–41 Margaret J. Geller & John P. Huchra/Harvard-Smithsonian Center for Astrophysics; 42–43t Smithsonian Institution/Science Photo Library; 42–43b Lund Observatory; 43 Max-Planck-Institut for Radio Astronomy/Science Photo Library; 44 Galaxy Picture Library; 45 Peter Menzel/Science Photo Library; 46–47 Galaxy Picture Library; 50l Royal Greenwich Observatory/Science Photo Library; 50r David Malin/Anglo-Australian Telescope Board; 52 ESO/V.L.T.; 53 Dr Rudolph Schild/Smithsonian Astrophysical Observatory/Science Photo Library; 54 David Malin/Anglo-Australian Telescope Board; 54–55 Dr Jean Lorre/Science Photo Library; 56–57 David Malin/Royal Observatory Edinburgh; 58 Royal Greenwich Observatory/Science Photo Library; 59 Galaxy Picture Library; 60–61 David Malin/Anglo-Australian Telescope Board; 67l Royal Greenwich Observatory/Science Photo Library; 67r The Observatories of the Carnegie Institution of Washington/Science Photo Library; 68 NRAO/AUI/Science Photo Library; 69 Dr Jean Lorre/Science Photo Library; 71 X-ray Astronomy Group Leicester University/Science Photo Library; 72 Jean Arnaud/Observatoire de Midi-Pyrénées; 73 Space Telescope Science Institute/NASA/Science Photo Library; 74–77 David Malin/Royal Observatory Edinburgh & Anglo-Australian Telescope Board; 79 Space Telescope Science Institute/NASA/Science Photo Library; 80 David Malin/Anglo–Australian Telescope Board; 81 David Malin/Royal Observatory Edinburgh; 87 Yerkes Observatory/University of Chicago; 89 Space Telescope Science Institute/NASA/Science Photo Library; 90–91 David Malin/Anglo-Australian Telescope Board; 92 Harvard–Smithsonian Center for Astrophysics; 93t Palomar Observatory; © C.I.T.; 93b Harvard-Smithsonian Center for Astrophysics; 96–97 NASA; 97 National Optical Astronomy Observatories; 98t NASA; 98b Royal Greenwich Observatory; 99 NASA; 100 S. Koutchmy et al./Mission de l'Institut d'Astrophysique (CNRS); 101 Jack Finch/Science Photo Library; 104 NASA/Science Photo Library; 109 NASA; 111 NASA/Science Photo Library; 112 NASA/Science Photo Library; 112–13 Novosti/Science Photo Library; 113t NASA/Science Photo Library; 113b Novosti/Science Photo Library; 14–15 NASA/Science Photo Library; 116–17 NASA; 118–19 NASA/Science Photo Library; 120–23 NASA; 123 NRAO/AUI/Science Photo Library; 124–27 NASA; 127–29 NASA/Science Photo Library; 131–32 NASA; 134 NASA/Science Photo Library; 136 Martin Dohrn/Science Photo Library; 137t NASA; 137b Don Davis/D.W. Wilhelms/Academic Press; 138t US geological Survey/Science Photo Library; 139 NASA; 140 NASA/Science Photo Library; 141 NASA; 142 NASA/Science Photo Library; 145t Tony & Daphne Hallas/Science Photo Library; 145b ESA/Science Photo Library; 146–47 John Sanford/Science Photo Library; 149 NASA/Hansen Planetarium; 151 George I. Barnard/Oxford Scientific Films; 155t–c Peter Gould/Oxford Scientific Films; 155b CNRI/Science Photo Library; 156–57 Jan Hinsch/Science Photo Library; 157 Robert Hessler/Planet Earth Pictures; 159t Biocompatibles Ltd; 159b Royal Greenwich Observatory/Science Photo Library; 161 Carolina Biological Supply Co./Oxford Scientific Films; 163 Sinclair Stammers/Science Photo Library; 164 Dr Richard K. La Val/Animals Animals/Oxford Scientific Films; 168–69 CERN/Science Photo Library; 176–77 Dr David Blair/University of Western Australia; 178–79 Mauna Kea Observatory, University of Hawaii/S. Wykoff & P. Wehinger/Science Photo Library; 180 Science Museum/Michael Holford; 181 Science Graphics Inc, Bend, Oregon, USA; 182–3 Zefa Picture Library

For **John J. McCarthy,** selling is a demanding profession. A profession that requires its practitioners to understand and use the psychological factors present in the selling situation. His concept of teaching professionals to *manage* their customers, to be prepared for each encounter, to predict behavior patterns, has gained him international fame.

Mr. McCarthy is the author of the renowned General Electric Situation Management Program, numerous articles on selling, sales training, and sales psychology, winner of the Jesse H. Neal Award, and three time winner of the top award from the National Society of Sales Training Executives.

From his home base in Chatham, Cape Cod, Massachusetts, and from sites all over the world, he has conducted sales training seminars for the American Management Association, the Sales Executive Club of New York, McGraw-Hill's Professional Salesmanship Center, and now, his own Professional Sales Situation Management and Professional Management Institutes—institutes that have been eagerly attended by literally thousands of representatives from the *world's* leading companies.

John J. McCarthy's
Secrets of Super Selling

The Sales Strategy Bank

Pre-sale planning

- A1 Develop sources of information
- A2 Identify his needs and wants
- A3 Identify his self-image
- A4 Identify his primary buying motive
- A5 Consult your predecessor
- A6 Get help from other depts. in your company (finance, etc.)
- A7 Get boss's help
- A8 Identify your competition and its representative
- A9 Study competitors' usual strategy
- A10 Make an estimate of situation
- A11 Determine past experiences of customer
- A12 Identify his possible sources of resistance
- A13 Prepare alternative offers
- A14 Plan strategy
- A15 Plan opening tactics

Create good climate

- B1 Save his time
- B2 Help him
- B3 Offer to help
- B4 Agree with him
- B5 Give him privacy
- B6 Do him a favor
- B7 Praise, compliment him
- B8 Give him credit
- B9 Emphasize mutual interests
- B10 Welcome objections
- B11 Agree
- B12 Admit validity of his view
- B13 Find point of agreement
- B14 Admit your error
- B15 Entertain him
- B16 Be intentionally incorrect in a statement so he can correct you
- B17 Provide him an escape hatch
- B18 Listen
- B19 Ask for his help, opinion or advice
- B20 Sympathize

Compel attention

- C1 Provide him with a role
- C2 Use cause-and-effect examples
- C3 Use testimonial
- C4 Emphasize his goals
- C5 Emphasize his objectives
- C6 Omit something so he can inject his ideas
- C7 Confuse him. Or use ambiguous questions
- C8 Interest him in more ambitious goals and objectives

Overcome his objections

- D1 Give him a choice
- D2 Narrow down his choices
- D3 Provide solution
- D4 Provide alternative
- D5 Compromise
- D6 Talk benefits
- D7 Draw comparisons
- D8 Tack with his stand
- D9 Use probe for reaction
- D10 Divert him
- D11 Rephrase and overstate his objections
- D12 Get him to talk
- D13 Appeal to fairness
- D14 Fence
- D15 Use direct question
- D16 Use overhead question
- D17 Use rhetorical question
- D18 Use provocative question
- D19 Scare him
- D20 Let him do you a favor

Overcome his fears

- E1 Provide support
- E2 Provide assurance or proof
- E3 Demonstrate
- E4 Explore cost of wrong decision
- E5 Remove him from situation with third party examples
- E6 Appeal to his pride
- E7 Appeal to his prestige
- E8 Appeal to the me he wants to project
- E9 Give him recognition
- E10 Anticipate, raise and answer objections
- E11 Stimulate his recall of past experiences
- E12 Provide guarantee
- E13 Offer additional services

Force a decision (Make him take action)

- F1 Bypass him
- F2 Shock-antagonize him
- F3 Show him cost of procrastination is higher than cost of decision
- F4 Emphasize cost of not buying

Post-sale actions

- G1 Show continued interest
- G2 Make return call
- G3 Check operation and/or installation
- G4 Has his problem been solved?
- G5 Have new problems been created?
- G6 Is he satisfied in all respects?
- G7 Have all of your promises been kept by your company?

50^{00}

John J. McCarthy's
Secrets of Super Selling

BY JOHN J. McCARTHY

Boardroom® Books, Inc.
500 Fifth Avenue, New York, NY 10110

*To Martha Cavanaugh McCarthy
and
Barbara Lavoie McCarthy*

Copyright©1982 by John J. McCarthy.

All rights reserved. No part of this book may be reproduced in any form or by any means without written permission from the publisher.

Third Printing

Boardroom Books publishes the advice of expert authorities in many fields. But the use of a book is not a substitute for legal, accounting, or other professional services. Consult a competent professional for answers to your specific questions.

Library of Congress Cataloging in Publication Data

McCarthy, John Joseph
   John J. Mcarthy's Secrets of super selling
1. Selling. I. McCarthy, John Joseph.
   II. Title. III. Title: Secrets
   of super selling
HF5438.25.M395   1982   658.8'5   82-9533
ISBN 0-932648-25-8   AACR2

Printed in the United States of America

Contents

Introduction vii

Part One: Psychological Aspects of Selling

CHAPTER 1
The Needs and Attitudes of a Professional 3

CHAPTER 2
The Customer's Self-Image and Status Needs 11

CHAPTER 3
Conditioning Influences Affecting Customers 23

CHAPTER 4
How a Customer Remembers 31

CHAPTER 5
How the Customer Defends His Self-Image 39

CHAPTER 6
Making an Estimate of the Selling Situation 49

Part Two: The Anatomy of a Sale

CHAPTER 7
The Elements of a Sale 67

CHAPTER 8
Pre-Sale Preparation 75

CHAPTER 9
Establishing Contact with the Customer 93

CHAPTER 10
Gaining Customer Confidence 127

CHAPTER 11
Identifying the Customer's Need 139

CHAPTER 12
Establishing Customer Satisfactions 151

CHAPTER 13
Gaining the Customer's Agreement 167

CHAPTER 14
Gaining the Customer's Preference 177

CHAPTER 15
Establishing Priorities 191

CHAPTER 16
Asking for the Order 197

CHAPTER 17
Taking the Order 211

CHAPTER 18
Working for Continuance 217

Part Three: The Sales Strategy Bank

CHAPTER 19
How to Use the Sales Strategy Bank 225

CHAPTER 20
Pre-Sale Planning 227

CHAPTER 21
Creating a Good Sales Climate 245

CHAPTER 22
Compelling Attention 259

CHAPTER 23
Overcoming Objections 267

CHAPTER 24
Overcoming the Customer's Fears 291

CHAPTER 25
Forcing a Decision 303

CHAPTER 26
Post-Sale Actions 309

CHAPTER 27
Self-Fulfillment 311

APPENDIX: Essentials for Successful Professional Selling Performance 317

Introduction

As a salesman, you are in the most demanding profession in industry. It is upon your efforts and those of other salesmen that your nation's economic progress depends. But it is not enough to be in a profession. It is necessary today to be a professional in that profession. Yesterday's sales techniques and yesterday's sales training methods will no longer suffice ...if they ever really did suffice.

Today, the salesman must be aware, not only of his customer's product needs, but he must understand the economic limitations of his customer and the nature of his psychological needs...which often are the determining factor in the customer's decision to buy from one supplier versus all of the others who seek his business.

Indeed, as the marketplace becomes more complex, often with buying committees making the final decision, it becomes even more important for a salesman to be able to determine who is the key influence on that buying decision.

You are about to read what has been described by a visiting European team of psychologists and marketing managers as the most advanced sales training material available in the world today.

You will find that this book does not "talk down" to the salesman. Rather, it recognizes the salesman for what he is...the individual upon whose efforts companies often thrive or die.

The beginning salesman will profit immensely from the book because he will be alerted to the demands of the profession he is about to enter, and he will be sensitized to situations so that he will recognize them when he encounters them in the market. This is a far cry from...and a far less expensive method than..."letting him learn through experience."

The experienced professional, who has been successful because of his instinct and intuition and preparation, will profit as well because he will be able to add to those assets a scientific plan of action that leaves nothing to chance. He will not only be able to handle situations, but more importantly, he will be able to anticipate them and he who can anticipate is in the best position to control.

This volume consists of three parts and an appendix.

Part One concentrates on the psychological aspects of selling. It includes units on the importance of selling to the entire economy, professional needs and attitudes, and professional tools. Subsequent chapters in Part One analyze the self-image and psychic needs of the customer, the need to know the customer as thoroughly as we know our product and the importance of status symbols as clues to the self-image of the customer. Additional chapters in Part One analyze the influences that cause a customer to view a situation from a frame of reference that differs from our own; why a customer will recall experiences you wish he or she would forget, why customers forget experiences you wish they would remember. An examination of self-defense mechanisms which customers employ is presented next, followed by a chapter on planning sales calls.

Part Two, The Anatomy of a Sale, presents over 100 obstacles that can arise during the course of a sale. Part Two then presents a number of methods that can be used to prevent or overcome each of these obstacles. From these suggested methods the reader can select the one that promises the greatest opportunity for success, given the personality of the salesman, the customer's personality and the conditions under which the obstacle occurs.

Part Three, The Sales Strategy Bank presents over 80 sales strategies that can be employed with explanations of when and how each can be used.

The Appendix analyzes 32 requirements for successful selling and includes a self-examination as a tool for further self-improvement.

Included also are three charts.

The Decision Pattern graphically illustrates what happens in the mind of any individual whom we attempt to persuade. In a sense it is a table of contents of the entire volume. It illustrates all of the factors that pour into the customer's frame of reference and form his or her viewpoint. It further illustrates the forms that the customer's reactions may take which will result in either acceptance or rejection of the efforts of the persuader.

The Anatomy of a Sale chart lists the 11 facets of a sale, as distinguished from the time-worn and really nonexistent "steps" of a sale. Under each facet are listed the obstacles that can appear during that phase of the sale. The material in Part Two takes up each of these potential obstacles in the sequence in which they appear on this chart.

The Sales Strategy Bank chart lists the more than 80 strategies previously mentioned, and this sequence is followed in Part Three.

J.J.McC.

PART ONE

Psychological Aspects of Selling

CHAPTER 1

The needs and attitudes of a professional

To achieve maximum performance, the professional salesman must hold these deep-seated convictions:

1. That selling is a profession requiring extensive training.
2. That the work is important.
3. That the work can be enjoyable.
4. That the work provides a challenge.
5. That society believes this profession benefits mankind.

This book will be devoted to an analysis of the kinds of knowledge and skills a professional salesman must acquire and working habits that must be formed. In the final analysis, a professional's effectiveness will depend upon how well the strategies and tactics presented in this book are implemented. However, an individual's success will be directly proportional to the attitude that person has toward sales work—how the profession is perceived both as a means of making a living and as a contribution to the quality of life.

It is imperative that the professional salesman perceives selling as a true profession, understands its importance to society, and is proud of the contribution he is making toward progress.

For these reasons it will be of value to examine each of the five prerequisites listed above. As a starting point, it is recognized there will be individuals who will challenge the contention that sales work is a profession or even a professional kind of work. To meet that head-on, we have only to examine the dictionary definition of a profession.

"A profession is a calling requiring specialized knowledge and often long and intensive preparation including instruction in skills and methods as well as in the scientific, historical or scholarly principles underlying such skills and methods, maintaining by force of organization or concerted opinion high standards of achievement and conduct, and committing its members to continued study and to a kind of work which has for its prime purpose the rendering of a public service." (*Webster's Third New International Dictionary.*)

Selling scores high on almost every point contained in this definition. *"A calling requiring specialized knowledge."* The professional salesman has to be equipped with a veritable arsenal of knowledge. He must have extensive knowledge of the products or services he has to sell, an equally extensive knowledge of competitive products, a knowledge of the customer's needs and problems—even those of which very often the customer is unaware, a knowledge of the customer's organization, short-range goals, long-range objectives, and the psychic needs of the customer and of the image the customer wants to project as well as the corporate image desired by the customer's company. Add to that a working knowledge of the laws that govern selling and buying, and we find that the professional salesman faces a mass of specialized knowledge he must acquire.

The next segment of Webster's definition states, *"requiring... often long and intensive preparation."* No salesman who is truly a professional will enter a selling situation without exactly that type of preparation. Later in this book we have presented the technique of making an estimate of a selling situation complete with the need to identify the problem, gather and weigh facts, make assumptions, consider alternatives, and produce recommendations or take action. Add to that the need to plan on follow-up, and there is no doubt that a professional in this field of selling has to face long and intensive preparation. Obviously, this is not true in the case of every sales call. However, neither is it true in a vast number of decisions a doctor or a lawyer must make when faced with the number of problems presented to them for which there are many precedents which they can now employ with a minimum of preparation.

The definition continues, *"... instruction in skills and methods...."* Each passing year finds company managements growing more aware of the need for professional sales training. The day has long since passed when a sample case was handed to the newly hired salesman, and he was told to "Go get some orders!" Today, saleswomen in an ever-growing number of organizations are given training ranging from orientation to the company to extensive training on products, systems-selling, negotiating, feature-benefit types of selling, and including training in applied psychology. Such training makes possible an identification of the customer's psychic needs and usual modus operandi. This isn't even close to a description of the specialized sales training needs that exist in thousands of organizations that employ professional salesmen, but it is sufficient to demonstrate that selling meets this requirement of Webster's definition.

The definition contains the clause, *"instruction in...the scientific, historical, or scholarly principles underlying such skills and methods."* Here, too, selling measures up to the requirements. There is nothing capricious or unscholarly about the mass of psychological knowledge to which professional salespeople are being exposed in their employment with the more progressive firms. "Scientific?" As will be seen during your reading of subsequent chapters of this book, there is an engineering approach to solving the types of person-to-person psychological problems that occur during a selling situation or more involved selling negotiation.

Webster's definition calls for the maintenance of high standards of conduct and achievement by force of organization or concerted opinion. It is true that no major organization of salesmen exists at the moment that forces these standards of behavior or that even gives salesmen an opportunity to get together to hammer out such standards. However, the vast majority of major companies do have high standards of professional selling behavior which they require their salesmen to observe. Added to that is the fact that there are professional organizations of *sales training* men and women who do, through their training efforts, establish with their student-salesmen, the high standards required by Webster's definition. To judge selling by the performance of a relatively few order-takers who grasp the customer's money and run would be as unfair as to judge the medical profession by the occasional quack or the legal profession by the relatively few ambulance chasers that come to light from time to time. The great majority of salesmen today have high standards of behavior. Many may lack sales skills, but their principles cannot be faulted.

Webster concludes the definition of a profession with, *"...committing its members to continued study and to a kind of work which has for its prime purpose the rendering of a public service."* Again, selling comes out with a top rating when measured against this requirement.

If there ever was a field in which continuous study to keep up with the advances in that field of endeavor is required, selling is it! Customers often have no idea of the amount of pre-sale preparation that goes on before the actual selling situation takes place. In field after field, the salesman must, often on his own time, read night after night to keep up with the changes that are exploding in many areas. Equal study must be made by professional salesmen in order to keep up, and ahead of, the customer's changing requirements.

We like best of all the concluding portion of the dictionary definition of a professional when it is applied to selling; that is, *"a kind of work which has for its prime purpose the rendering of a public service."* No—repeat no—occupation or profession can lay a greater claim to this requirement than professional selling. Red Motley, former publisher of *PARADE Magazine* and widely known speaker, said it succinctly: "Nothing happens until someone sells something!"

In the free world countries, nations were not made great by engineering alone, by financing alone, or by manufacturing techniques alone. In the final analysis, someone had to make someone else drool to have the products designed, made and financed by others. Professional salesmen, through their pursuasive efforts, coupled to their product knowledge, have created the greatest mass market this world has ever known. It has enabled management to siphon off some of the resultant profits to extend research so that even better products could be produced. More has been siphoned off for automatic equipment, enabling companies to produce the same product at less cost and reflect that into lower prices so that even more people could buy. All of this starts when salesman A convinces customer B that he has a need for the product. Professional saleswomen are in the business of selling solutions to problems. They are in the business of preventing customer problems from arising by making available new products that will eliminate old problems. Consider what would occur if there were to be a national strike of salespeople. Industry would come to a grinding halt. There would be no profits upon which taxes could be levied. The impact would be felt in every area that can be imagined...education ...medicine...and every facet of the economy.

There never was a profession that was capable of making a greater contribution to the public welfare!

Yet, one of the prerequisites for maximum selling performance with which this chapter opened was the need for salesmen to be convinced that their work is important. There are thousands of salesmen who do have this awareness, but, unfortunately, there are thousands more who lack it. Many feel they are selling because there is nothing else they are capable of doing. Others are in selling because at the outset of their career they were told that if they wanted to move ahead in "marketing," they should get a few years of selling experience first. Then, locked in by the growing financial needs of a family, they found themselves unable to move. Finally, few indeed are the companies who spend any time letting salesmen know just how vitally important their profession is and how vital to the success of the organization is the performance of the sales force. All too often the absolutely false stereotype of a salesman is the basis for top management's failure to inspire their sales forces. The day of the peddler epitomized in *Death of a Salesman* has long since gone. Today, the world is being served by a new breed of young men and women, as well as seasoned veterans who are in the selling profession because they enjoy its challenge, because it satisfies their need to help others, because they are fully aware of the chain of wonderful events that is set up when a sale is made, in terms of that sale's impact upon their customer, the employees of their own company, their own suppliers, and a host of others who are affected. Many are in selling because, of all of the positions in an organization, it frequently affords the greatest opportunity to operate as an individual with confidence in one's own ability to get results alone.

A prerequisite is that the work be enjoyable. There is probably no greater enjoyment possible than the exultation that takes place when a sale is completed, a customer is satisfied, and a contribution has been made. If the need for a challenge is essential to maximum performance and a prerequisite, no problem should exist because every sale is a challenge. Every customer problem is a challenge. Every competitive counter-strategy is a challenge. Selling efforts really consist of meeting constant challenges which, in turn, give variety and excitement to the job.

Finally, it must be confessed that society in general does not have this understanding of the true professional nature of selling. The reasons for this are many. The old stereotype of a salesman who is trying to sell you something you don't need still prevails in spite of abundant evidence to the contrary. Another contributing factor to the stereotype of salesmen is the manner in which they are treated by their own management. For instance, they are usually the only people in an organization who are exposed to "contests." The inference is that of all the employees in an organization, only the sales staff has to be motivated to work harder by holding a "prize" out in front of them. In one annual sales meeting after another, salesmen are required to wear costumes, badges and the like that would be an insult to the average college freshman today. The claim is made that "this peps the boys up." Those that feel that such is the case should be privy to the remarks that the salesmen are heard making out of earshot of the manager who planned the meeting.

If we talk *down* to individuals, they will respond from that level. If we talk *up* to them, they will respond from *that* level. Professional salespeople should not allow this type of treatment or the incorrect view of the sales profession held by many sectors of society to "get them down." The professional will make certain that members of his or her own family are thoroughly indoctrinated and understand the nature, the purpose, and the tremendous importance of selling. This stereotype of selling is over 2,000 years old, and it is not going to be changed overnight. Admittedly, it would be pleasant to have society at long last understand the major role that selling plays in the economy. It would be equally pleasant to have society understand the mass of professional knowledge and the extent of preparation that is essential in selling in today's marketplace. These understandings can be developed in the future if those in selling will actively resist all attempts to deprecate the profession and will make every effort to cause each customer and potential customer to realize that the end goal of the professional salesman is to aid the customer in preventing and solving problems. If this is placed as the key objective of the job, rather than making sales, the sales *will* result because of the customer's growing confidence in the dedicated professional who is serving him.

Use of professional tools

It is not sufficient simply to believe that selling is a profession. It is essential that every salesman conduct himself like a professional and that he use the professional tools which are available. One of these is an adaptation of the military's estimate of the situation and is called "Estimate of the Selling Situation." This tool will be described later in its entirety. Suffice for now to point out that the "Estimate" was developed because of a strong conviction that one of the two major causes for lost sales is in pre-sale preparation. The second cause—failure to understand psychological factors—will be discussed shortly and is, indeed, the main focal point.

The counter salesman or the distributor salesman who works primarily "in the house" and whose contacts with the customer are hurried may maintain that he has no opportunity to make an estimate of the selling situation and to plan a strategy. It is true that he does labor with this handicap, but it only emphasizes the fact that his task may be more difficult. Maximum effectiveness can be accomplished only when he develops an ability to understand the customer and his problems in the shortest possible time and to increase and build on that understanding in future contacts.

The military estimate of the situation starts with a clear-cut understanding of the objective. Our selling estimate is more complicated and must start with an understanding of the customer's long-range objective and his immediate or short-range goals. We, too, must have both a well-defined objective for our business (as it contemplates this customer) and our immediate goal insofar as this customer is concerned in the relationship we will have with him today.

Generally, our long-range objective might be to build and maintain a lasting, favorable relationship with him. Our short-range goal might be to get today's business without endangering the long-range objective. Another long-range objective might be to increase our business with him over a specific period of time by a specific percentage or number of dollars or to obtain business from him on items he now buys elsewhere. Our short-range goal may sometimes be as mundane as to give him what he is asking for in the shortest possible time.

The military are concerned wtih accomplishing their own objectives and goals and are concerned with the enemy's *intentions*, but are not about to knowingly aid him in accomplishing *his* goals. Ours is a more difficult task. While we strive to accomplish our own objectives and goals, we must aid the customer in accomplishing his... in fact, true customer-oriented selling requires that we must first *aid* the customer in accomplishing his. The premise is that if we are successful, we shall have moved forward toward the accomplishment of our own. It is when our goals and the customer's are divergent that we have our greatest problem.

The military estimate of the situation next concentrates on what it calls "Essential Elements of Intelligence." Intelligence is more than information. *It is information made meaningful.* Information becomes meaningful when we relate each piece to other pieces and to the whole. The gathering of information must take place in the pre-sale preparation, but it continues throughout all of the future relationships we have with the customer. For maximum utilization and value, it should be cataloged and referred to frequently. Let us see what some of the elements are.

I. Product information

First is the area of product information, which we have touched upon. In too many organizations, there is a feeling that we really do a complete job in this area. This is a dangerous premise and one that will cost many sales. For example, how do managers really measure up to these challenges:

1. Do managers ever check to make certain that salesmen really know their products or merely assume they do?
2. Do managers *know* that salesmen have read, absorbed, remembered and use the product information which has been supplied them?
3. Do managers fail to tell salesmen where the product or service lags behind competition, so that they first hear of this when a customer, more knowledgeable than they are, confronts them with the information. Or do managers explain away failure to inform salesmen of product's short-comings with the rationalization that such information might find its way into the competition's hands? Isn't the danger of *not* providing them with this information so great that managers should occupy themselves with reviewing security measures? Doesn't failure to supply salesmen with such facts indicate a lack of confidence in their loyalty and/or good judgment? How do they react to this attitude?

Again, addressing the manager: Do you still think a superb job is being done in this area of product information? If so, do you feel that each salesman is aware of the specific product needs of each of his customers as they vary from the norm? Are you satisfied that your salesmen know how to convey this intelligence to the customer in terms that are meaningful and of interest to him? Are your salesmen aware of other uses to which your products can be put, other than the primary use for which they were designed? In other words, are your salesmen able to engage in creative selling... finding new uses for old products, adapting old products to new problems of their customer?

It can be agreed that most managers have recognized the need for product training, though it remains astonishing to this day that a number of firms select a likely-looking individual to sell their goods or services and then let him learn by a process of osmosis or "the hard way through experience." It is, indeed, a hard way—and a very expensive way, especially when competition is more enlightened and makes certain that product information is complete and is transmitted via modern methods of communication.

II. Economic intelligence

This is the second of three facets of sales intelligence and one that has received insufficient formal attention. Many salesmen gather information relative to the customer's economic situation and that of the competitors, but usually in unplanned fashion, merely salting away information as it happens to come to them. Effective pre-sale preparation requires that we have a clear picture of what the essential elements of economic information are in our kind of business, because they vary widely between businesses or even in the same business at the manufacturing, distributing and retail levels. This entire area will be dealt with in detail in Chapter 6, Making an Estimate of the Selling Situation. It is mentioned here for two reasons. First, because every day there is delay in acquainting salesmen with the kinds of economic information they should gather on customers and competitors may cost sales... often sales that we will never know we lost because we did not know enough about the customer's economic needs to know what new products he needs. Often he, too, never recognizes needs he has; but all too often some competitor who took a reading on his economic problems acquaints the customer with them and shows him how they can be overcome by buying the competitor's products.

Customer-oriented selling can become a glib, meaningless phrase if we do not understand its promise. It is based upon the premise that we are in business to help the customer recognize and solve his problems or to solve problems he has already recognized without our aid, and that to the extent we are able to do this, he will buy from us now and in the future. Far from being a philanthropic viewpoint, it is a pragmatic way of looking at the job.

Selling, essentially, is the exchange of need satisfactions. He needs our product or services, and we need the economic security which his money will provide. The salesman who is able to offer the best proof that such an exchange is really taking place is going to get the business of the customer. The salesman who knows the most about the customer's economic problems and status, in addition to his own product and service capabilities, has by far the best chance of convincing the customer that when he is asked to surrender his money, he is, simultaneously, going to have his own needs satisfied.

III. The psychological elements of intelligence

And so we come to the third facet of the essential elements of selling intelligence—the psychological factors. It will be recalled that we said previously that there were two areas in which more sales are being lost than in all others. The first was in lack of adequate, professional pre-sale preparation, and that included both adequate and meaningful product information and information on customers' and competitors' economic situations. The second area in which sales are being lost is in this area of failing to know the psychological factors that are present in a selling situation, usually in terms of the customer's psychological needs, but not excluding psychological influences that shape the actions and counter-actions of competitors.

Our weakness in this key area stems from many reasons. Among them is the failure to realize that although a sale is the exchange of need satisfactions, this does not mean only product needs, or even *primarily* product needs. It also means psychological needs. We may satisfy product needs more effectively and at less financial cost than all of our competitors and yet lose the sale to one of the competitors who recognized the psychological needs of the customer and moved to satisfy them.

Our failure in this area often stems, too, from the image that many of us have of "psychology"; i.e., "long-haired, interesting theories that have no application to business and especially to selling." We are more likely to think that price—in dollars—is our main problem and fail to realize that the price is too high in psychological needs, in terms of status, the customer's self-respect, his need for recognition and a lengthy list of other needs that will soon be discussed. This price we are asking may be sufficiently high not only to cost us *this* sale, but to cost us all of his future business. While the kinds of product and economic information we require may vary widely depending upon whether we sell at the manufacturing, wholesale or retail levels, the same basic human or psychological needs are present in every selling situation. The difference is only in degree and in the opportunity that we have to identify and to satisfy them.

The decision pattern chart

As you read this book, the Decision Pattern Chart will take on real meaning for you. It illustrates the path an individual's mind follows as he or she makes a decision and forms a reaction to your efforts to obtain a favorable buying attitude.

A number of factors influence the customer's attitude which we have referred to on the chart as the Frame of Reference. The milieu is the customer's total environment and is formed by the interfacing of all of the factors listed in the extreme left column, heredity, physique, etc. As a result of these influences, the customer forms an image of himself that he tries to project: This is the image individuals wish others to have of them. This image affects almost every action. Only a few of the types of self-images are listed in the second column on the chart.

This self-image creates status needs. Because customers, like all of us, do not wish our status needs to be transparent they seek to conceal them, but to the person who listens with his eyes, status symbols provide a window to the self-image. These are discussed in detail later in the book.

As a result of the self-image, each individual develops *personal* short-range goals and long-range objectives. Concurrently, the customer will develop the business goals and objectives for her company. The question for the salesman becomes: What is this customer seeking primarily, the satisfaction of personal aims or those of the company? If we guess wrong, we lose sales. If we guess correctly, we know what benefits and losses to point out in our presentation.

Psychological Aspects of Selling

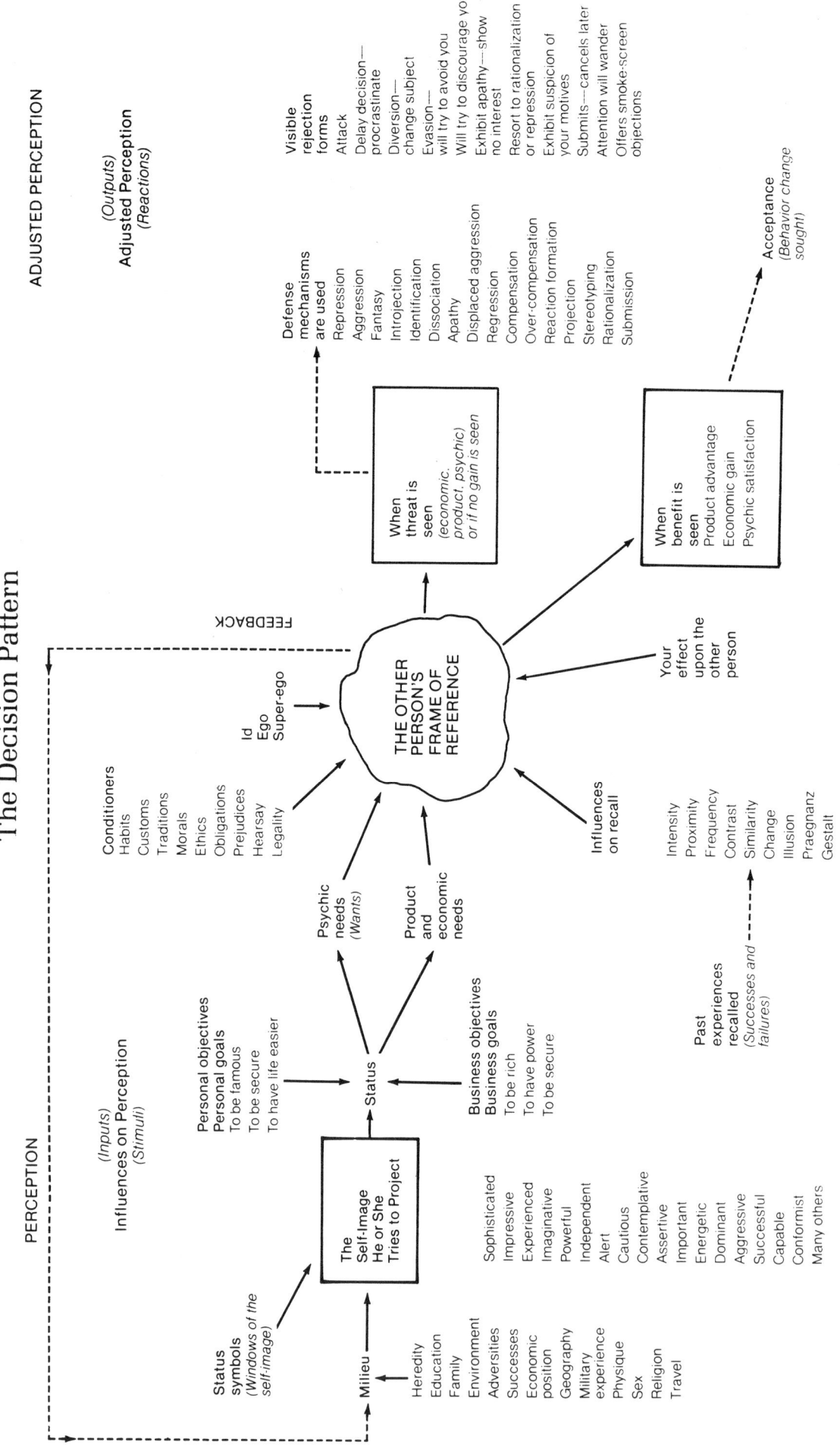

Developing a persuasive strategy includes an understanding of the path of persuasion.

Three other sets of stimuli bombard the customer and affect that individual's attitude toward you, the salesman, toward your organization and toward your product or service offerings. These are the nine conditioners listed on the chart. They are examined in detail in the text. They often cause the customer to see a situation in a manner that differs widely from your own. There are also nine influences that cause the same customer to *remember* situations in a manner far different from your own recall of the same situation. These, too, are covered in detail in the book.

As a result of all of these influences, the customer develops a viewpoint. If the customer sees a threat (impact on the self-image, an economic threat, or the feeling your product or service will not meet needs) that customer will often resort to any one of 16 defense mechanisms. It is essential for the salesman to determine what threat is seen by the customer so that it can be removed. If the salesman fails to do so, the customer's reaction will take any one of the many rejection forms listed in the right hand column of the chart. Conversely, if the saleswoman identifies the threat and can offset it by product advantages, economic gains, or enhancement of the customer's self-image, acceptance takes place and a sale results.

This is a brief explanation of the chart. As you read the book and make frequent references to the chart, it will take on added meaning. It removes the feeling that attitudes are too abstract to present in a schematic chart that shows continuity. Probably, most important of all, the chart, accompanied by the text, causes us to realize that if we had been subjected to *all* of the stimuli to which the customer has been exposed, we would probably be viewing the situation exactly as the customer views it. This enables us to have greater empathy (not sympathy, but understanding) and enables us, also, to consider, unemotionally and logically, what others would have to do to persuade us if we felt as the customer feels. With irritation, annoyance and frustration eliminated from our own attitude, we can then hammer out tactics and strategies that will convert customer suspicion, anger, and even fear, to acceptance. We can engineer agreement just as surely as any engineer can solve her problems. Engineers cannot solve any problem if there are too many unknowns. The problem in the past has been that in dealing with people there have been too many unknowns and we have not known what to look for. It is our conviction that this book identifies the specific areas in which we can seek to convert the unknowns to identifiable facts. Once we have those facts we can solve the problems that face us. The most difficult task in solving a problem is to identify the problem. Seeing customer's attitudes as a result of a definable series of reactions to specific influences should go far toward assisting us in defining human interaction problems.

The main focus, then, of this section is an explanation of the psychological factors which influence the course of a sale or the course of a long relationship with a customer. As you analyze this material, it will become apparent to you that these same factors are present in the relationships you have with your men if you are a manager or your peers and your superior if you are a salesman. This may prove to be a bonus to you as will a new awareness and understanding of your own behavior. But the next chapter will address itself to the selling situation. It will not be a "how to" piece, and no apology is made for this. There has been too much written and spoken about 76 ways to open a sale, or 50–11 ways to close a sale.

It is our contention that if we really understand the other fellow's point of view and what is forming it, our own good judgment will produce a number of alternative and successful plans of action. Future chapters will address themselves to the solving of specific problems, but they will be based on the concept presented here. To be consistently successful in choosing from available alternate courses of action, we must understand the concept of the self-image and its impact upon the sale, at every level of selling and in every field from thimbles to turbines.

CHAPTER 2

The customer's self-image and status needs

It has been said that the task of a salesman is to sell himself, to sell his company and to sell his product, in that order. A case might be made for the point of view that holds that a salesman must sell his product and then sell his company, but it is difficult to see how it can be successfully contested that the first job of the salesman is to sell himself to the customer.

Keep in mind that we are talking about the difficult-to-get sale. We may even be talking about the average sale. We are not discussing the easy-to-get sale in the seller's market. We are not discussing the situation that exists when a customer wants an item so badly that he will buy it from any source. Nor are we contemplating the problem of selling a customer whose appetite for a product has been so stimulated by advertising that services of a salesman may represent nothing more professional than that of an agent who can insure delivery at the earliest possible moment.

In the great majority of cases, we do not find the customer drooling for our product or our service, nor do we find ourselves without competition. In most instances, we find the customer not only aware that he has a need, but also aware that there are many people willing—even anxious—to serve that need. In other instances, we know that the customer has a need even though he may not be aware of it. Our job is to awaken him to the need and to convince him that we are prepared, better than any competitor, to fill that need.

Unfortunately, all too often, the salesman concentrates on identifying and supplying the product or service needs of the customer and fails to identify the psychic needs of the customer—the needs of his self-image.

Let's look at it this way. You and I often find ourselves in the role of customer, and we often find ourselves craving for a product. Unless we find that there is only a sole source for that product, we will not purchase it if we feel that the salesman is giving us shoddy treatment, is discourteous, is disdainful, disinterested or incompetent. We resent being taken for granted or being subjected to a "canned" presentation. When we feel that we are being subjected to any of these forms of "salesmanship," we will take our business elsewhere, if we have a choice. If we have no choice and our need is not urgent, we will postpone our decision to buy, hoping that a choice will present itself. Often, while we are waiting, the buying urge will die, and we will lose interest completely in the product or service which we had been prepared to purchase.

There is hardly a reader of this page who does not have certain favorite shops or firms which he likes to patronize, or who does not know of one or more shops or firms from whom he will buy only when compelled to do so by circumstances. If we were questioned as to why we will, on occasion, go out of our way to buy in a given shop, sometimes even paying a premium price, we usually reply that we "like the service" they give us. Actually, whether we are conscious of it or not, we are attracted to the shop because the treatment they accord us fills the needs of our self-image. The personnel of the shop who serve our product needs are aware of our psychic needs, and they make us feel comfortable or respected or important or whatever it is that you and I, as individuals, are seeking in the way of recognition.

In retail selling or in selling across the counter in a wholesale establishment, it is not always easy, or even possible, to identify precisely the self-image of customers. The effective salesman in this type of selling situation learns to exhibit a high degree of friendliness, helpfulness and interest. He keeps his eye on his F-H-I—he is friendly, helpful and interested—and his behavior satisfies the self-image needs of a large number of customers. We usually consider the most difficult selling job to be that of the salesman for the manufacturer or the salesman who is selling high-ticket items to an OEM (original equipment manufacturer) or to a public utility, mill, or the like. Undeniably, this type of selling demands the maximum in professionalism, but selling in such areas has an advantage that is often denied the retail salesman or the counter salesman in a distributorship; i.e., the salesman can—or at least should—learn much about the customer as an individual. This information is often denied the retail or counter salesman, except in the case of repeat customers.

Self-images

You and I and our customers all have self-images. There are actually five. First there is:

The real me

This is the "me" your customer really is, and you don't have to worry about this one because you will in all likelihood never identify it. The customer doesn't know what he is really like, and you and I don't know what we are really like. This is because we are never really willing to engage in the ruthless, agonizing self-appraisal necessary to identify our weaknesses and our strengths.

We lose sight of the fact that there are clear-cut steps to self-fulfillment, which are:

Self-analysis, which leads to *self-awareness,* and this points out what we need to do to accomplish *self-development,* which in turn leads to something closer to *self-fulfillment.* We will never achieve total self-fulfillment in our lives, and that is not all bad because it would be a boring existence if we ever achieved it—but don't worry about it because most of us probably operate at from 30% to 50% of our real potential.

The IRS forces us to take inventory of our economic situation at least once a year, but there is no one to force us to examine our growth or lack of it each year. Development must always be self-development, and to achieve it, the day may come when we will realize how much we will profit if we set aside one evening every six months to take stock of ourselves through a planned self-analysis, and not depend upon some annual review which our company may force our boss to conduct for us. If he is like most people, he will shy away from telling us about faults that we have of a personal nature because he fears this would be "unpleasant."

Your customer has probably not subjected herself to this same agonizing self-analysis, so she doesn't know what she is really like either; therefore, what chance do you have to really know what she is like?

And so we come to the second form of the self-image, and this is one we had better give much thought to if we hope to give the customer the treatment he will demand. This facet of the self-image is:

The "me-I-used-to-be"

A few real examples will make this facet clear.

Speaking to a member of a group of salesmen, we mentioned that one man he had brought with him to a course was outstanding. His reply was, "He's the best man we have." We asked, "How good is he?" The reply was, "Well, last year we gave him a territory that produced $600,000, and in one year, he brought it up to $1,050,000!" We replied, "Your management must be enthused over his performance." And he answered, "Oh, no, whenever I mention how well he is doing, they always say, 'Yes, isn't Danny doing fine for a mechanic.'"

Naturally, we asked what that meant, and the reply was, "Until last year he had been a mechanic for us for nine years, and then he asked for a chance at selling, and you've heard the results, but they'll always see him as 'Danny, the mechanic.'" In other words, they still see the "me-he-used-to-be."

In our business, clients know that thousands of salesmen go through courses we conduct and not infrequently we receive calls such as this: "You meet a good many men, and we have a spot open for a sales manager or marketing manager. Can you recommend one?" Obviously, our other clients would be furious if we did so, and we cannot comply, but often we will remember one of their own men they sent to us. So, we may say, "Say—you sent Eddie Burke here, and he was terrific! He'd make an excellent manager." We can almost see the far-away look in the caller's eyes when he replies, "Oh, yes, do you know that Eddie started out with us a stockboy ten years ago?"

They still see Eddie as a stockboy, and they have no apparent understanding of the fact that Eddie has grown intellectually and experientially...he is not the same man he used to be.

One manager, hearing this, said, "*Now* I see why I lost my biggest customer 90 days ago, and I have been wondering why!" (Why didn't he *ask* the customer, incidentally?)

He went on to say, "He and I have been fishing buddies for the past 20 years, since we were in our early twenties. Whenever I call on him, we always talk fishing, and laugh about the day he fell out of the boat, and we had to fish him back in. Now that I think of it, I should have realized that as far back as 18 months ago, there began to be a change in his attitude. When I would talk fishing, he'd act impatient, and ask about my products.

"Three months ago he gave all of the business to a new competitor, and we had enjoyed 100% of his business. What annoyed us was that we have the best product at the best price.

"Now I see what has been going on. I've been treating 'the-me-he-used-to-be,' and the competitor who never knew him in the old days, is talking to 'the-me-he-is-now,' and maybe to 'the-me-he-wants-to-be,' and he's walked away with the whole business, and I don't think I can get it back." (And we know he can't!)

While the salesman was thinking about the price

of his product, he was not thinking about the high psychic price he was asking the customer to pay. The sales manager went on to say, "I forgot that he has two expensive cars, and lives in the most expensive house in his town, and I've been talking to the guy who fell out of a boat 20 years ago!"

We have the same trouble with our children as they mature. We still see them as the kids they used to be, and we wonder why they will listen to the advice of the new neighbor next door, but won't listen to us.

This is of major importance.

How about taking an hour twice a year (once on your birthday, and once six months from that day), and promise yourself that, on those two days, you are going to stop and take a new look at your spouse, your children if you have any, your boss, your peers, and any subordinates you have, and every customer you have, and ask yourself if you are keeping up with the changes in them, or are you still seeing them "as-they-used-to-be"?

If you will do this with customers, you will get business your competitors will never get because you can be certain that they are *not* doing this.

Few, if any, of us want to be seen as we used to be. We want people to see us as we are now, and even better, as we hope, some day, to be seen. The obvious warning is: (1) Don't rely on old relationships. Friendships and faith in salespeople have to be nourished. (2) Don't take *any* customer for granted and assume that his wants and needs of yesteryear will be the same today. The test: Would you want to be treated today as you were treated five, ten, or more years ago? Obviously not. So why should we be surprised when a customer—or anyone—resents our failure to keep current with the image they have of themselves today?

We come now to the next facet of the self-image:

The "me-I-think-I-am"

This is the me we think other people see when they see us, and we're not too happy about it. We sense threats to our self-image, real and implied. Consider this example:

You are going through the revolving door of a building when it dawns on you that the fellow behind you isn't pushing his portion of the door. What's your reaction? Probably to squeeze out of the door, forcing him to push the last foot or two, as you growl, "Who does he think he is, expecting me to push the blasted door for him?"

We say, "Who does he think he is?" but we mean, "Who does he think I am, giving me that treatment?"

Or perhaps your boss calls a meeting, and he has placed namecards around the conference table. If you are like most of us, you look to see where you have been placed, and if it is at the foot of the table, your reaction may well be, "How come I'm down here... who's sitting beside the old man... who do they think I am?"

Or you attend a company sales meeting at a resort hotel, and after signing in at the desk, you ask the desk clerk if you can see which of your people have checked in. Then you scan the list, and what are you looking for? "Who got the single rooms?" And how you burn if someone at your level did get the single room. Your next step is to see who you will be rooming with, and you may be furious to find out that it is someone who is not highly regarded in the company. The reaction: "Is that the way they see me?"

It would take a book on this subject alone to list the thousand and one ways you and I see how other people see us... whether they listen to us... whether they interrupt... whether a third party breaks into a conversation you are having with the boss, while you are in the middle of a sentence. The "pecking order" is made clear to us, and we seethe with fury.

It would be possible to list the dozens of experiences you can have where you are made to realize that people don't see the image you want them to have of you. Fortunately, that isn't necessary. In dealing with customers (or anyone), just ask yourself what kinds of treatment infuriate you, and then make certain that you don't accord your customer (or others) the same treatment.

Probably the worst blow to a person's self-image is being interrupted or having someone else ignore what you are saying and start to talk while you are still talking. We keep looking for the "magic secrets" of selling, while some of them are so obvious that we overlook them. Give dignity to the other person.

Rare indeed is the individual who is satisfied with this "me-I-think-I-am." We are usually aware of many of our shortcomings, the views of Robbie Burns to the contrary notwithstanding. We know the areas in which we are not well informed. We may not be happy with some aspects of our physical appearance, our height or girth or countenance. We often make comparisons between the way people seem to regard us, as indicated by the treatment they accord us, and the way in which they regard others whose views seem to be listened to with great respect. This is not to say that we go about our day's work constantly unhappy with ourselves. But there are many times when we are less than happy with the way we seem to be regarded by others whose opinions are important to us.

The "me-I-wish-I-were"

The self-image has many facets. We can be completely happy with the manner in which we are regarded by the family at home and be most unhappy with the manner in which we seem to be regarded by the boss or by our peers or by a given customer. Except for the fellow with an extremely low threshold of satisfaction, the overwhelming majority of us have enough dissatisfaction with the "me-I-think-I-am" to cause us to create another self-image; i.e., the "me-I-wish-I-were." This is the "me" we wish others would see when they see us. It is entirely likely that every salesman and every customer has a "me" he wishes he were. It represents not only the peak of his aspirations, but it represents the degree of respect and the kind of treatment he would like to have others accord him.

If we were to assemble a dozen of our closest friends, and if it were possible to have each of them be completely frank, we would be astonished if each of them, in turn, were to tell the group what he really believes he could be if others were not so blind to his attributes. If we were to greet any of these revelations with laughter or ridicule, we could be certain that we had reduced our circle of 12 friends to 11.

The point is that each of us—you and I and your customer—are realistic enough to know that it is probably going to be impossible, at least for the present, to have others see us as we wish they would, and to treat us accordingly. We never kill this "me-I-wish-I-were." We simply put it aside for the time being. It should be emphasized that the self-image is not a static thing; it is ever-changing. Most of us have probably exceeded the "me-I-wish-I-were" that we held a number of years ago. We have set our sights higher and have newer aspirations. We are no longer content with the kind of treatment that would have satisfied us at earlier stages in our career.

This points up one of the hazards of thinking you know a man who has been your customer for years. We know that we have changed, but we frequently forget that the other fellow changes and improves, too. And with that improvement, his self-image changes. He is no longer satisfied with the kind of treatment that would have sufficed years ago. He is fair game for the new competitor who assays him as he is today and supplies him with the kind of recognition that this new self-image is demanding.

The "me-I-try-to-project"

Because we are not satisfied with the me we think others see when they see us, and because we feel it is unrealistic to expect them to see the me we firmly believe they *should* see when they see us, we compromise and we create a "me-I-try-to-project." This "me" is somewhere in between the other two. It represents all that the traffic will bear; in other words, we try to build an image that others will find it possible to accept. *Your customer does this same thing!* As the years go by, we keep pushing this "me-I-try-to-project" to new heights, subconsciously being guided in our haste by the reactions of those whom we seek to impress.

This facet of the self-image is often the very heart of a selling situation. Failure of a salesman to identify the "me" his customer is trying to project can cost a sale no matter how attractive a product or service may be or how attractive the price in dollars may be. The price the salesman is asking the customer to pay in surrendering psychic satisfactions is too high. The customer's voiced objections may be in terms of product features: price, delivery, ad infinitum. His *real*, unspoken objection is to the manner in which he believes the salesman regards him. He interprets this in terms of the treatment he is shown—the respect shown his opinions or whether he is interrupted. He may question whether he is treated as an individual or whether the salesman makes a canned presentation or presents a solution for the customer's problem that was obviously designed for and used to solve the problem of some other customer with whom today's customer has no desire to be identified or compared.

The effect of the self-image on the very start of a sale may be made clear by this basic example: If *your* self-image is active, you should have resented my use of the word "basic" in that last sentence. Who does he think I am . . . too stupid to understand unless he makes it basic?

For years, sales managers have tried to convince salesmen of the necessity to be neat, have shoes shined, ties tied, hair cut, etc. The salesman may mutter to himself, "I'm perfectly satisfied with the way I look and the way I dress. The important thing is to know my product and be able to explain it." If the salesman feels this way, the sales manager has missed the boat. Let's look at this problem from the standpoint of the self-image.

If you are I look into a mirror, what do we see? Usually, we don't see the mirror itself, we see what the mirror reflects; we see our image. When our hypothetical salesman enters the customer's place of business with an altogether unkempt appearance, what does the customer see? Essentially, he does not see the salesman; he sees through the salesman's appearance his own image as the salesman apparently sees him. The thought which immediately comes to the customer's mind is, "Who does he think *I* am, calling on me, looking like that?" Subconsciously, the customer makes a quick comparison and wonders if this salesman would call on someone higher up in the customer's organization looking this way, or would he call upon some more affluent competitor looking this shoddy. It is going to take an outstanding product to overcome this start, simply because the salesman has already asked the customer to pay part of the price in psychic needs—in this case in terms of the respect that he feels he should have been shown.

The self-image is always present in every sale. Because normal courtesy and respect satisfy self-image needs of the customer in many situations, sales are successfully made by satisfying product needs. However, many salesmen have been perplexed when a sale has been lost in spite of the fact that a superior product was offered at the best price and under conditions that were totally responsive to the customer's delivery needs. The reason for the lost sale is usually to be found in the salesman's unwitting threat to the customer's self-image.

Sales are lost through "projection"

In an age when production lines turn out hundreds of thousands of identical products, it would be in the salesman's interest to keep in mind the fact that though billions of people have inhabited the face of this earth since the beginning of time, no two of us have been exactly alike. A manufacturing foreman rebelled at being asked to take a course in human relations and was heard to say, "I don't need the course, I treat my employees fairly, I treat them all alike." The foreman was unaware of the fact that employees do not want to be treated all alike. Neither do customers. Each customer enters a selling situation with three sets of needs: product needs, economic needs and psychic needs; i.e., needs of his self-image. No salesman can afford to overlook any of these needs.

One of the most costly errors a saleswoman can make is to resort to what is called "projection." This is the tendency to think that we and the customer share the same frame of reference; that what would satisfy us if we were the customer will satisfy him. The unconscious use of projection is a substitute for market research, and it is a substitute for gathering intelligence on the customer's needs as he sees those needs. In general, it is not too difficult to identify his product needs; it requires more effort to identify his economic needs, and it requires real professional salesmanship to identify his psychic needs. We know that except in unusual cases a sale will not be made unless the customer sees, or can be made to see, that he has a product need. We know that his economic needs will have a direct effect upon whether he will choose to satisfy even his most urgent product needs. These will be discussed later when we examine the Estimate of the Selling Situation.

Next, let us examine further the psychic needs that stem from the self-image; if identifying the nature of the customer's self-image is essential to professional selling, let us see the many forms it can take and what clues are available to us to identify it.

Status needs

Every person—hence it follows, every customer—has a milieu that is unique. Milieu, in French, means the center or the middle. It is more than environment. Every customer has experienced many environments—at home, in school, perhaps in the military, geographical, religious, racial and so on. We might think of his milieu as being the total effect of all of the environments to which he has been exposed. We might think of him as standing in the center of all of these influences. This provides him with a frame of reference that will be unique to him; in *some* respect it will be different from the frame of reference of any other individual. Because of the very personalized exposures to which he has been subjected in each of these environments, he has developed certain needs for status.

Because so much has been written about status in a derogatory fashion, many of us have come to look upon the words "status" and "status needs" as unsavory; we often confuse status needs with ostentation. We should think of status as being the natural desire of a man to attain a *station* in life and to have others recognize his right to that station. As he convinces himself that he has earned his "rank" or "station," he will subconsciously examine every situation in which he finds himself to make certain that he is being accorded the treatment that a person of his rank is entitled to. Certain status needs become important to him, and he views every selling situation, for example, to make certain that his status needs are being enhanced and not threatened.

Many sales organizations recognize in a general way that customers have these status needs. For example, salesmen may be given titles that sound "more important." This is not so much to satisfy the salesmen's status needs as it is to be responsive to the need that many customers have to be served by an "important" member of the marketing team. Salesmen of large companies are not unfamiliar with the demand made by some of their larger customers that a vice president call upon them, and in some instances they are known to have demanded that the chief executive officer of the selling firm be brought in on the act. Their excuse often is that they want to be certain that the selling company will be committed to its promises, and no salesman can make commitments that will be binding (they say!). Actually, the reason for their action is that they want their importance recognized. They feel that their status is such that they should be served in the selling situation only by people of equal rank.

This impact of status needs is not confined to high-level, big business situations. Salesmen for distributors are familiar with the query by some customers, "Is the boss in?" Often the salesman is far better equipped with product knowledge and even with knowledge of the customer's economic needs than the vice president or "the boss," but this is not enough for the customer who is determined to satisfy his needs for status recognition.

The great majority of us—probably all of us if the truth were known—go through life sensitive to the treatment we are accorded by others. We try our best to disguise it. For instance, if someone cuts us out on the highway or jumps into a waiting line ahead of us, we usually mutter to ourselves or to some friend, "Who does he think he is?" Not infrequently, some of us who have a lower boiling point take out after the offending driver and cut him out to even the score; actually to regain our lost status or to demand that he respect our "rights" (to our station in life).

We are often so intent upon satisfying the customer's product needs, or so enthusiastic over our product, or so intent upon convincing the customer that buying our product will not threaten her economic needs that we overlook, totally, her status needs. Because status needs are ever-present, it will be worth our while to identify them and then learn how to identify those which are important to customers whose business we find it difficult to obtain or from whom we would like to obtain a greater share of the business they have to place.

Psychologists can identify as many as 150 separate status needs, but the saleswoman has no time to become a psychologist. And she is not interested in cataloging shadings of the basic status needs, even if she had the time. What we are attempting to do is to distill the vast amount of information made available to us by psychologists and bring to bear on selling situations the essential bits of knowledge and the strategies which they can generate that will measurably improve our sales volume. For this reason, we will concern ourselves only with six general areas of status needs which probably embrace most of the others that psychologists could enumerate.

The need to belong

Keep in mind that while all of us have these same basic status needs, each of them assumes a different degree of importance to us as individuals. The "need to belong," for instance, may be of great importance to the customer who is inclined to be a conformist or who feels insecure or who is very thin-skinned and fears criticism or ridicule. It may be of very little importance to the rugged entrepreneurial type of fellow who considers himself an individualist and seems to care little about the opinion of others. This points up the vital need of knowing your customer as well as you know your product and "your territory."

It is important, too, to realize that a given status need may assume monumental importance in the mind of your customer in one selling situation and be of very little importance to him in another. What is being said here is that we cannot afford to *type* a customer and expect that typing to hold true during the course of all of our future relationships with him.

The "need to belong" has nothing to do with "togetherness" or team work. As has been said previously, we want to be considered as individuals, as unique; but, we do not wish to be considered as strange. We want to be accepted. As the word "belong" is used here, it does not mean to be possessed by, as "he belongs to us." We think of it in this sense: After an individual has joined a group and has learned its customs, he begins to act, in a general way, as others in the group act. Other members of the group nod approvingly and say, "He belongs."

You may recommend a very new product or process to a customer, or you may recommend an old product to meet a new need, and you will find him asking you, "Who has tried this out?"...or, "Why are you so sure it will work?" Completely apart from his economic or product concerns, he is afraid that his need to belong is being threatened. If he accepts your idea, will other people in his field laugh at him later and say, "You mean to say you fell for that line?"...or, "They used you for a patsy—a guinea pig!" He doesn't want other people in his field to feel that he doesn't belong. He is looking to you for assurance, for proof, that this action you are recommending will not cause him to look badly in th eyes of these other people.

The customer to whom this need to belong is important is going to need proof from you—assurance, testimonials, even guarantees or demonstrations. In offering examples of others who have followed your recommendation, you must be sure that their identity offers him status satisfactions. In other words, don't cite as an example someone whom he regards as occupying a lower status. Try to cite someone whom you know he respects and considers as occupying a status level as high as his own or higher.

Our interpretation of the need to belong must not be restricted to the customer who is insecure, fears ridicule and so forth. The need to belong also represents the customer's insistence upon obtaining the kind of attention and treatment that would be accorded others whom he considers of equal or higher status. In other words, he feels that he "belongs" to that club and is entitled to the kind of recognition that others in the club receive.

Do not make the mistake of thinking that this need is confined to small firms or "little men on the way up." In one instance, a consultant was brought in by one of the largest corporations in the world to advise them on a problem that had to be solved. After viewing the situation, he said, "This is the same problem that the *ABC* Corporation had; let me tell you what we recommended to them because it solved their problem." The immediate reaction of the customer was, "Never mind what you did at *ABC*; we want to know what you suggest we do with *our* problem!"

The fact is that his present customer was even larger than *ABC*, even though both were giants in their fields. Therefore, being compared to *ABC* did nothing to bolster his present customer's corporate image. They were paying a high fee for his services, and they wanted personalized treatment. It apparently even bothered them that *ABC* could have the same kind of problems that they had.

Obviously, a salesman must not rule out the need to belong in any selling situation. And as has been indicated, he can often neutralize its effect by convincing the customer that she is receiving personalized attention on a unique problem. Comparisons with other companies can be made, but only after we know how the present customer views the other concern and only when such comparisons are essential as proof or assurance.

The need to dominate

The five remaining status needs are more familiar to us. We have all encountered the chap who has a "need to dominate." This need may not be constant. He may not have it at home, in the office with his peers, at the club or lodge; but, he may have the need to dominate his competitors. If we fail to recognize this need in a selling situation, it may cost us the sale. In certain selling situations, this need to dominate may be a real asset to the saleswoman. This customer may be easy to interest in a new product, service or feature if it will help him to dominate his competitors by scoring a "first."

There are customers who want to dominate the selling situation itself; that is, to dominate the salesman. Some salesmen have been told that they must always "control" the selling situation; as a result, they compete with the customer in his attempts to dominate—and lose the sale or make the sale unnecessarily difficult. Controlling a selling situation does not mean dominating it. Someone has said before us that the formula for managing a selling situation is to be dominantly persuasive while enhancing the status satisfaction of the customer. If one of the status satisfactions he requires is to feel that he is dominating the situation, we will never be persuasive unless we let him feel that he is, in fact, dominating the situation.

This kind of situation calls for professional salesmanship at its best. It requires the salesmen to develop a skill in questioning that requires the customer to talk about the things upon which you wish to focus his attention, even though he may do most of the talking. As long as he is talking, he thinks he is dominating the selling situation.

The need for prestige

The "need for prestige" is so commonplace that it requires no discussion here, but it is imperative that we identify it when it is present in the selling situation and that we satisfy it. It is even more essential that we develop such an intimate knowledge of the customer that we can anticipate this need on the part of the customer in our advance planning and determine, before the sale, how we can satisfy the need.

The need for recognition

The "need for recognition" is equally obvious and commonplace, but satisfying it is not always easy. We must avoid the temptation again to project; that is, to make statements which if they were used on us would make us feel recognized, but when viewed from the customer's frame of reference fall far short of the mark. *We do not like our behavior to be transparent.* This means that in dealing with the customer, we must avoid statements which he will regard as flattery or as insincere. All of us resent obvious flattery because the person who is employing it is essentially saying to us, "You are very transparent, and it is easy to see what you would like to be told, what kind of recognition you require, so I shall supply your need."

The safest way to supply recognition is to take a page from the manufacturing foreman who says, "Praise the work, not the worker, and be sure that the work you are praising represents real accomplishment or the worker will resent your believing that this represents his best effort." Look for some outstanding achievement of the customer and comment favorably upon it without mentioning the customer. He'll accept this as recognition.

At this point, a word of caution is imperative. Readers might believe that they are being asked to manipulate customers. Nothing could be further from the truth. Salesmen are being asked to motivate customers, and that is a vastly different thing. Manipulation is causing the other fellow to do something that is in your own selfish interest. Motivation is causing him to take actions that are in your *mutual* interests. *Remember that a sale is the exchange of need satisfactions*, and these status needs are just as real to the customer as are his product needs. To ignore them, or to ignore the roadblocks that they often place in the way of a successful sale, is to be hypocritical when we know that ignoring them will often result in a sale lost to some competitor who has done his homework in the area of gathering intelligence on your mutual customer.

The need for economic security

The "need for economic security" is ever-present if we do not interpret the phrase too narrowly. Even the individual whose bankroll is unlimited wants to be sure that the exchange of need satisfactions that takes place in a sale is an equitable exchange. Indeed, justifying price may often be more essential in dealing with the affluent customer than in dealing with the chap whose means are limited.

The need for emotional security

The "need for emotional security" is a catch-all which the psychologist could subdivide into innumerable needs. For our purpose, we think of it as containing such needs as the need for self-respect, respect of others, pride, independence and so on. From our experience, we know that all people have these needs in the area of emotional security. We know, too, that they are not static, that they do not remain constant. In selling, the task is to study the customer so well that we will be able to make accurate estimates, in advance, as to how important these and the other needs will be to him in the forthcoming selling situation.

As we study him, we will learn that he—even as you and I—has a *continuity of personality.* Under known conditions, he can be expected to act in a manner that is characteristic of him and *in a manner that is responsive to his status needs* as he sees those needs.

In dealing with criminals, the police use the term "MO," or modus operandi. Habit patterns are not restricted to criminals. Although you and I like to feel that we are free agents who size up each situation as a new experience, the fact is that certain patterns of behavior have brought us satisfaction in the past. Though we may not be aware of it, subconsciously we are ever-vigilant to threats to our self-image—to the "me-I-am-trying-to-project"—and we react almost instinctively to threats, imagined or real. Those who know us best can predict with a high degree of accuracy how we will react. Those who know us this well are able to motivate us, and unfortunately, sometime manipulate us.

If we hope to motivate customers, we must know them as individuals. We must identify the "me-they-are-trying-to-project." Once we know that, we can easily determine what status needs would be important to that kind of person, and we can set about providing the kinds of satisfaction the customer requires. If we cannot always satisfy the requirements of his self-image, at least we will not threaten them.

Continuity of personality

What key points have we emphasized up to this point? We have said that all of us, and all of our customers, have self-images. We are not satisfied with the "me" we think people see when they see us (the "me-I-think-I-am"); we have said that each of us has a me we wish people would see when they see us (the "me-I-wish-I-were"); we have said that most of us regard this as impossible to achieve for the time being. Therefore, we establish a "me-I-try-to-project" —a me we think we can cause others to see when they see us. This self-image we want others to share takes familiar forms which we will discuss in the next section, but every action of ours (and our customer) will be responsive to the needs of this self-image. We (or she) will take no actions that are inconsistent with the needs of this self-image, and we will be attracted to actions that promise a chance of enhancing it. In our efforts to protect, enhance and project this self-image, we develop a continuity of personality; a pattern of responses to stimuli that others who study us can often predict with a high degree of accuracy.

Similar study of our customer will assist us in predicting his behavior which will, in turn, assist us greatly in planning our sales strategy and aid us in providing him with satisfactions of the spirit which he requires. These are as real and vital to him as are his product and economic needs.

We have said that this self-image we try to get others to share with us provides us with a station in life—a rank, a position relative to others which we insist that society grant us and respect as rightfully ours. (Our customer is with us every step of the way, developing his own threshold of satisfaction which is *his* status.) We have said that there are six general status needs which we all share in common, although the importance of each varies greatly as do the images we try to project. We have called them status needs; i.e., needs which the self-image has to sustain itself. If society refuses to grant us these status satisfactions then, regretfully, we are forced to agree that we have not been able to project the image that is so important to us.

You and I and our customer then feed these needs into our frame of reference. This is the vantage point from which we will view every new experience. It is the vantage point from which our customer will view our selling tactics. As we shall see later, he factors other things into his frame of reference;

i.e., his interpretation of his past experiences and other elements that have conditioned his thinking historically. These will be dealt with in turn but, now, if the self-image of the customer is so important to us, let us see what forms it takes and, more to the point, how do we identify it so that we can cope with it?

Self-image forms

Aside from product problems, most of the problems we encounter in a sale will stem from the self-image needs of the customer. The selling tools and techniques we use will work only if they satisfy those needs. The following list of self-image forms is nothing more than a list of adjectives, and you can add many more adjectives to the list to describe customers you have known.

Self-image forms

Sophisticated	Important
Impressive	Ingenious
Experienced	Energetic
Imaginative	Dominant
Powerful	Aggressive
Independent	Successful
Alert	Capable
Cautious	Masculine
Contemplative	Conformist
Assertive	Submissive

Stop for a moment and consider two or three customers with whom you have had difficulty in your attempts to sell them, or consider customers you have sold to successfully although others have been unable to do so. Can you identify the adjective that best describes them in our list?

Entirely apart from the field of selling, isn't it true that as you consider a list of friends and close acquaintances, whom you have known over a long period of time, it is possible for you to select an adjective from the list that closely describes that individual as he wants to be regarded by others—or as he regards himself? Isn't it true that your chances of influencing him have been most successful when you have attempted to satisfy the demands of that self-image and when you have been careful to avoid creating the impression that you did not agree with the customer's estimate of himself?

The chicken or the egg?

What comes first, the customer's self-image or his status needs? It is not worthy of argument. An effective salesman must realize that both are present during a selling situation, and he must recognize the customer's self-image and satisfy his status needs. Our own point of view is that the customer develops a self-image—a "me-I-try-to-project"—a status or position he wants society to accept, and this position or status generates status needs. Others may contend that long before a child has a self-image, he may exhibit a determination to dominate. It could be an interesting argument, but it would not move us one step closer to the successful completion of a sale.

Others may contend that the self-image and status needs play no part in the awarding of major contracts. The contention may be that the customer is interested solely in getting the most for her cash. We have hundreds of examples to prove beyond doubt that this reasoning is fallacious, and to embrace this reasoning will cost a salesman countless sales. The examples include the customer who has no need for a computer but who buys one because he feels that it is essential if he is to maintain the image he has been seeking to build; i.e., that he is progressive.

Another example is found in an area superintendent who stops buying from a vendor because the vendor failed to throw him a party and reception when he was made vice president, as the vendor had done for his predecessors. Keep in mind that the vendor still makes the same product at the same price and offers the same service.

Still another example is the customer who refuses to buy from a vendor with a superior product because he feels a debt of obligation to another vendor who supplied his needs during a period of shortage when others were unable to do so. The new vendor has constantly pointed out to the customer that this loyalty is costing the customer money and providing him with a product that leaves much to be desired. All of these efforts have been to no avail. The customer views himself as a "fair" individual, and he will do nothing to threaten that image.

This emphasizes the fact that the self-image is not intended to impress others alone. We build self-images that are important to us even if society refuses to accept them. Self-respect is more important to some customers than the need for gaining the respect of society.

The right to know

The question is often raised as to whether the salesman has the right to pry into the customer's personality, the right to uncover his self-image or the right to identify his status needs. The answer to these questions is found in the phrase "customer-oriented selling." Customer-oriented selling is the philosophy of the firm or the salesman who sees the need for developing lasting, favorable relationships with each customer. It is the watchword of the firm or salesman who believes in long-range planning.

Conversely, those who reject customer-oriented selling as a philosophy usually regard long-range planning as "what we will do after lunch." Customer-oriented selling has its roots in the knowledge that if our first and key objective is to help the customer solve her problems and accomplish her objectives, she will turn to us and reject others—not merely out of loyalty, but because she has found that she cannot afford to take any other course of action. In this way, we satisfy our own objectives.

How can we possibly be customer-oriented unless we know the customer? This means knowing far more than his product and economic needs. It means we must understand him as an individual. One customer bought a piece of extremely expensive equipment that was far in excess of his product needs simply because he wanted international fame as the owner of the largest piece of equipment in his field. One vendor failed to understand this and offered equipment that he knew would meet product needs. The other vendor realized that the customer had psychic needs and successfully sold him what he needed to satisfy those needs. The question is sometimes raised, as it was in this case, "Was it moral to do so?" The answer to that question is clear when we remember that these needs of the spirit are just as real to the customer as are his product needs. There is also the pragmatic answer that if you don't satisfy those needs, someone else will.

There is no question but that the professional salesman not only has a right to analyze the customer for the purpose of understanding his needs and satisfying them; he has the obligation to do so. If this is so, let us see what means are available to us for identifying status needs and the "me" the customer is trying to project.

Status symbols

Consider this sequence of premises:

1. Every customer has a self-image.
2. The self-image generates status needs.
3. The self-image must be enhanced, never threatened; the status needs must be recognized and satisfied in a selling situation.
4. The customer uses status symbols to supply, at least in part, his needs for status.
5. Therefore, the status symbols often become the *windows to the self-image,* and an understanding of why the customer employs them can become the key to many sales.

In spite of this logic, a sales manager was once heard to say, in commenting upon a salesman, "Oh, he knows what kind of a car the customer drives, what schools his kids go to, what clubs he belongs to, and all the other unimportant things."

This remark gives evidence that many marketing people are unaware of the tremendous impact of the human element in a sale. If the sales manager were discussing a salesman who made it a point to know the customer thoroughly as a person but failed to know his products, then the sales manager was correct to a point; but he was not correct in identifying status symbols as "unimportant things." Nothing that has been said here should leave the impression that *only* applied psychology and an understanding of the customer is important. The need is for the well-rounded salesman who knows his product, his market, his economics *and* his customer. Unfortunately, there has been maximum concentration on product, price and market and not enough on the customer.

To understand the importance of the customer's status symbols to the salesman, we must start with the agreement that man is a rational being. Under most circumstances, he is not irrational. What he does, he does for a reason. When you or I act in a manner that appears to be irrational to others, it is usually because we have permitted emotion to rule our reason for the moment. Actually, even in our irrational moments, what is happening is that we are abandoning our long-range objectives for the sake of some attractive short-range goal. This could get us into a long discussion, but it could only serve to take our attention off our main thesis; i.e., man is rational, and there is a reason for each action he takes.

Therefore, when he utilizes a status symbol, there is a reason for it. We have heard the phrase, "He is expressing himself." In other words, he is trying to tell the world (and the salesman who serves him) what is important to him. Many of us stand in constant need of assurance. We must have proof that we are making progress toward the objectives we have established for ourselves. To some men, this means a continuous upgrading of the kind of car that they drive. To others, it may mean the kind of school to which they can send their children. Some of these actions are intended to impress others—to have them recognize status (station in life); just as often, they are means utilized to insure *self*-assurance.

In trying to assure himself of his progress, the customer betrays to the alert salesman what it is that is really important to him. The salesman who is alert to these inner needs of the customer and who can find means of helping to satisfy those needs will be a big step ahead of his less observant and solely product-oriented competitor.

You will recall that we discussed earlier the "continuity of personality," the pattern of behavior which each of us, as an individual, employs. It will be well to keep this word "pattern" in mind as we discuss status symbols and how to interpret them. There are two rules that must be kept in mind as we evaluate the significance of a status symbol which a customer may employ.

First, do not judge a customer by one action or possession—look for a pattern.

Second, do not assume that what would be a status symbol if you employed it is necessarily a status symbol when employed by the customer.

The big car that the customer drives may have been purchased at the insistence of his wife. The swank school the kids go to may have been their idea or the wife's.

For instance, a salesman attending a course looked at the instructor and said in a triumphant tone, "I see what one of your status symbols is . . . I notice you wear only Countess Mara ties!" The actual fact was that, as a result of having addressed a convention of haberdashers, the instructor had been sent a dozen of the ties.

In one firm's metropolitan office, an executive has two attractive water colors of Parisian scenes which lead a visitor to conclude that the executive has some affinity for France and things French. The truth is that the executive has never been to France and merely liked the scenes.

Another executive had a picture of hunting dogs in his office, but had absolutely no interest in dogs or in hunting—the picture had been on the wall when he inherited the office.

A third executive in the same firm had a tremendous picture of a clipper ship on the office wall, but pointed out to an inquirer that he kept it simply to cover up a bad plastering job until he could get some action out of the maintenance crew.

A pattern could not be identified in any of these cases, and to have made the assumption that these men were interested in Paris, hunting or sailing would have been in error. The chap with the photographs of Paris did not have souvenirs of France displayed in his office. There was no French-English dictionary in his book case; he didn't pepper his conversation with French phrases. Yet, a few offices away, another executive did have an interest in Germany. There were German scenes on the wall. There was the dictionary. He drank coffee, and you as his guest drank coffee, out of mugs that bore German inscriptions. A German diploma of some type adorned his wall. And when you left his office he was as likely to sing out "Auf wiedersehen!" as he was "so-long!" The pattern was there.

A speaker some years ago, before a group of sales executives, gave this advice, "Listen with your eyes." No better advice was ever given to salesmen. Customers are trying to tell you things about themselves; or perhaps to be more factual, they are trying to reassure themselves through the employment of symbols of the status they seek. In so doing, they are telling you much about themselves, their hopes, aspirations and status needs. A lapel button is worn for a reason; the "clean-desk executive" obtains some kind of inner satisfaction from this type of operation.

To a salesman, the most valuable clue to the inner needs of a customer is what topics the customer prefers to discuss when she is given complete freedom to make the choice. One of the values of golf and other moments of relaxation is that the customer unwinds and if given the opportunity will start to talk about things that are of importance to him as an individual. In doing so, he will often reveal his self-image and the things he needs to support it.

Dangers of interpreting symbols

It is often contended that the interpretation of status symbols is a dangerous business, and admittedly, it can be just that. However, electric energy is dangerous, too; but, we use it with care and discretion, and profit tremendously from its use. It is foolhardy to disregard what the customer is trying to tell you about himself via his possessions, habits, actions and words. These are far too valuable to you as keys to prediction of his behavior to be ignored or to be considered "unimportant."

We have already discussed the danger of drawing inferences from single acts, comments or possessions of a given customer. Our second warning was to be sure that we do not interpret as a symbol of desired status something that would be a status symbol if employed by ourselves but is employed unconsciously by the customer.

What may be a status symbol (a symbol of psychic aspirations) for one customer will not necessarily be such a symbol for another customer who has a totally different set of values, aspirations or milieu. We can recall standing on a European street corner and observing a large Cadillac drive past, bearing New York plates, complete with chauffeur in front and an American in the rear seat. Our instinctive reaction was that here was a "status seeker," a truly ostentatious American. Almost at once, we were annoyed with our own stupidity when we realized what was happening. True, if we, with our income and background, had been in the rear seat of that car, the only reason we would have had for using it in Europe would have been to have sought status recognition. It dawned upon us that the chap who was actually in the rear seat may have been totally unconscious of the attention he had attracted from us. He may have been so wealthy that when the need to visit Europe arose, he didn't think twice, but instinctively made arrangements to bring his car and chauffeur in order that he could have comfort and give his undivided attention to the business he had to transact.

This actual occurrence is easy for us to understand, but it is more difficult for us to be this charitable in analyzing other actions of our customer. For instance, we observed a customer in his home. After dinner he inquired if anyone wanted a cordial. A number indicated that this sounded like a good idea, and the host brought forth from the liquor cabinet about 10 or 12 different kinds of cordials. Later, one of the guests was overheard remarking that this had been ostentatious, that the host had been trying to impress people with his affluence. We knew that the host had picked up the liqueurs on trips to the Caribbean, enjoyed only one or two of them, didn't know what others liked, and merely wanted to make the selection easy for his guests.

Measure the significance of the status symbol—or what appears to be a status symbol—against what you know about the facts of his background before you interpret his words or actions as windows to his self-image. And, again, look for a pattern. If the host we have just referred to was always trying to impress, was always ostentatious, then the guest's conclusion would have been valid, and a saleswoman trying to sell the host would be well advised to find ways to show the host that she was impressed with his affluence.

Unfortunately, so much has been written in a critical vein about suburbia and status symbols that when we hear the words "status symbols," we tend to think in terms of homes that are too large, cars that are too ostentatious, mink coats for pet dogs, and so forth. This is indeed unfortunate because it can deprive us of essential intelligence on the customer. When the words "status symbols" are used, try to think of what they really mean. They mean what they actually say, not what popular writers would have us think they say. Status symbols mean indicators of the station in life which a person occupies or to which he aspires... or to which he wants you and society to agree he is entitled.

There is nothing wrong with status symbols. You couldn't avoid using them if you wanted to. Often, unconsciously, our very speech betrays our milieu. Our accents, our choice of words, our colloquialisms, the kinds of subjects which we like to discuss, the kinds of books which we prefer to read (as opposed to those we choose to display), the kinds of people with whom we associate—all of these are status symbols. They tell other people who are observant much about us. If we are equally observant, and look for patterns, we will learn more about our customer than our competitor will learn if he occupies himself *solely* with product or service considerations.

Don't play amateur psychiatrist

You are not being urged to take up psychiatry or to practice it upon your customers. You are merely being asked to keep in mind that your customer is constantly telling you things about herself, and if you are observant and look for a pattern, you will have an edge on your competition.

Three hundred years ago, Ben Jonson said, "Language springs out of the innermost parts of us. No glass renders a man's likeness so true as his speech." John Lagemann follows this up by saying, "Details that seem trivial by themselves have a way of adding up, when classified and counted, to vital information. I have found, as have many people, that certain tricks of content analysis help you to read between the lines of ordinary conversation."

Don't make a fetish of analyzing everything your customer says, trying to read some hidden or deeper meaning into his remarks. But do look for the things he repeats or for the unusual things he says. Does he refer to things which he understands as "practical" or "workable" and things he does not understand as "theoretical, impractical, cloud-nine stuff, abstract," etc.? With whom does he attempt to associate himself in his conversation? One customer had a habit of referring to certain successful people as "fraternity brothers of mine" without bothering to explain that they were 20 years and 1,000 miles apart in the same fraternity. Or, he would point out that some successful person was in his class at college, failing to point out that the "college" was a large university and there were thousands in his class, and he had never met the famous personage.

In such instances, our tendency is to silently ridicule this chap, whereas we should be alert to what he has told us. He has just told us the kinds of people with whom he likes to be identified. Therefore, lodge buttons, cars, clothes, his choice of restaurants, vacation spots, schools for his children, quality of products he buys (and makes), jewelry he wears, sometimes the kind of haircut he affects, his hobbies, and a number of other habits that will be discussed later, in the aggregate, give you commercial intelligence you cannot afford to ignore. These kinds of information present you with clues to what he wants from life in addition to a good product at the best price. Can you help supply it to him? In a number of departments in one company where salesmen have started to think this way, sales have risen as much as 25%.

the customer a chance to sound off about it and get it out of his system.

Prejudice against the product or service. Just doesn't like the appearance, the shape, design or color of the product. This prejudice may be the result of previous experience with similar products. The self-image may be involved in this type of prejudice also. For example, the customer may resent going outside of her own firm for parts she could make herself or for services that could be supplied internally. She may feel that it reflects upon her to go outside the company. She knows, of course that she will never have the time to handle this problem internally, but she hates to pay a profit on something she knows she could handle herself if she *did* have the time. As a result of this dichotomy, she procrastinates and puts off the buying decision as long as possible. Before purchasing, she wants to be sure that *you* know that she could handle this herself, and better, if she had the time. The sooner you let her know that you realize this, the sooner you will make the sale.

Hearsay

A customer will sometimes adopt the views of other people in whom he has confidence and will hold these views as tenaciously as though they were based upon experiences of his own. He will often accept the views of people with whom he likes to identify himself or the views of people who are considered authoritative on the subject. It is not always easy to determine in advance of the selling situation the identities of people whom the customer respects, and it is often impossible to *predict* that he will be prejudiced because of something he has heard about us, our product or our company. But it is possible to be *prepared* for it.

Use testimonials of people you know he will respect. If he is quoting an authority, try to present him with the favorable views of one who is regarded as a greater authority; meanwhile, don't tear down the individual he has quoted. Remember, if he has quoted another person, in some way he obtains satisfaction in being identified with that person. If you destroy the person he is quoting, to some extent you destroy him. Sales aren't made, or at least continuing business isn't built, upon that kind of a foundation.

Laws

There are occasions when either the customer or the salesman will either lack knowledge of laws or will have interpreted the laws differently. This can cause problems. The customer may ask the supplier to take an action that the supplier knows, or believes, to be illegal, or conversely, the supplier may suggest actions that the customer interprets as being illegal. This latter situation is easier to prevent than the first mentioned. If the salesman's company is large enough to have available a legal counselor, when any question relative to the legality of a proposed action is contemplated, that adviser should be consulted before the negotiating session takes place.

If the saleswoman feels that the customer may raise the question of legality of the proposal, the saleswoman should come to the session with the legal opinion in writing, or the legal adviser should be available, if called upon. In other words the strategy should be to anticipate any question of legality and be prepared should that question be raised by the customer. In any event professional salespeople should keep themselves current with the laws that affect their type of selling.

More often than not, when a customer proposes actions that the salesman knows or believes to be illegal, it is the result of the customer's ignorance of the law. An informed salesman can, in these instances, protect the customer from the consequences of violating a law. The seller in such cases should explain the law to the customer but in doing so should be certain that the explanation is phrased carefully so that the customer is not made to lose face because of her lack of familiarity with the law. The seller should also be certain that she does not indicate that the customer has knowingly asked her to violate the law. This applies even when the saleswoman knows that such intent did actually exist.

We have witnessed a number of occurrences. In one instance a salesman, in a conversation with two others, said, "I'm glad that when you lose an order, you can ask the successful bidder what price he obtained." The second salesman replied emphatically, "No, you can't!" Whereupon a lively argument took place. Finally, the third salesperson present said, "No, as of last Friday, you can't. The law was just changed." The point is that the first salesman had not kept current with the law.

In another instance a salesman had just told us of a plan of action upon which he was to embark. When he finished, we pointed out that what he intended to do was a clear violation of the law. His response was to cover both ears with his hands and run out of our office saying, "I don't want to hear about it." To put it mildly this salesman is asking for trouble, not only with the customer, with the law enforcement agency but also with his own company. Violating a law may produce short-term benefits that will exact a fearsome price in long-range losses, and some of them may be more painful than financial losses.

In a third instance a salesman offered a product to a buyer at a very attractive price. The buyer responded to the offer with, "Oh, no you don't... you can go to jail if you want to but I'm not going, too." This buyer knew that if she accepted a better price than the salesman was giving to her competitors, who were in the same geographical marketplace and were buying exactly the same product in the same quantity, that a real question of restraint of trade might be involved, and if she knowingly took advantage of the offer that she, too, might be in violation of the law. The salesman who related the incident to us, pointed out that it took him 45 minutes to demonstrate, successfully, that the product he was offering was not the same product the buyer's competitors were buying at a higher price.

Perhaps the best advice that can be given to the reader is that no salesman should accept the opinion of anyone other than a duly constituted legal adviser, and that applies, in this instance to the advice being offered above. As you read the preceding examples, don't become bogged down in determining whether or not the laws mentioned were presented correctly. Keep in mind the *principles* being stressed: (1) keep current with the law; (2) when in any doubt seek advice *only* from certified legal counselors; (3) anticipate any illegal action your customer may suggest, based upon previous contacts with that customer and be prepared to handle such situations; (4) don't argue with the customer over who is right and who is wrong—get the facts; (5) don't ever propose or take an illegal action to obtain attractive short-range benefits.

Id, ego or super-ego

For all too long, people in the field of sales training and sales management have been "talking down" to salesmen, as though they were speaking to men of limited intelligence. This book is intended to be a professional approach to professionals in the field of marketing.

Selling is largely a matter of influencing the decisions of people. This means working through the minds of people. We know far less about the operation of the human mind than we do of the human body. We have scientific equipment that lets us measure and probe the body, but we have little scientific assistance in understanding the human mind, other than the research that is being conducted throughout the world in the fields of psychiatry and psychology.

We could follow the pattern of "talking down" and omit the words "id," "ego" and "super-ego," but to do so would be a disservice to the professional salesman. They are the words that he will find in more sophisticated articles and books on the subject of psychology. Rather than change the terms, so that no reader will be "scared off," let's identify them and treat them only as lightly as the confines of a discussion of this length will permit.

We are familiar with the terms, "body," "mind" and "spirit." These terms have some identification with id, ego and super-ego, although to simply exchange the terms would be an unwarranted oversimplification. We might better think of the terms as facets of human nature. The id is the compulsive side of our nature which is completely selfish. It is usually associated with our demands for the basic needs and urges—for food, shelter, clothing, sex and the like. The ego is that side of our nature which is also self-oriented, hence selfish, but which is very aware of the presence of others and aware of the need for their approval and the cost of their disapproval. The super-ego is the *selfless* side of our nature, as the name indicates, because if we think of the word "super" as being above, we see that it means above ego; i.e., above self.

The customer who is primarily id-oriented (in the selling situation) puts himself above all else. The id is inclined to be emotional and impulsive as well as demanding. The ego is a kind of self-awareness. It injects the note of caution; call it reason, if you will. It warns us of the cost to ourselves if we follow, blindly, the urgings of the id. It urges us to do the socially acceptable thing, not because it is "high-minded," but because it tries to protect the self. We cannot oversimplify by saying that id is emotion and ego is reason, because both emotion and reason are present in the operation of either of these sides of our nature.

When the id urges action, it does so not blindly, but because our reason tells us that if we take the action, certain satisfaction will be obtained. Actions of the ego are based not only on reason, but on *fear* that the cost of disapproval will be high. The super-ego is sometimes referred to as the conscience or as consideration for the good of others above self. If the id were not restrained, we might satisfy our primitive needs regardless of the cost, and our jails might be bulging. The ego may be thought of as the guidance of the intellect which causes us to reason, albeit selfishly, and forces us to ask of the actions demanded by the id, "Will it make us look good or bad?"

The id-oriented customer may ask, "Who else has bought this?" What she is really asking is, "Will I be first?" The ego-oriented individual asking the same question, "Who else has bought this?" is really asking, "Am I taking a chance of looking bad if I buy this?" The ego-oriented customer is going to have to be supplied with references, testimonials, demonstrations and the like to enable him to be sure that he will not be criticized or lose self-esteem by making the purchase. When you are selling a *new* product or any radical change from the past, approach your id-oriented customers first. They usually care little about what others will say. They want to be first. An example of this is found in the case of a customer who said to the salesman, "I don't want you calling on anyone else in this town." The salesman replied, "Do you want me to get fired—I must call on other customers." The id-oriented customer replied, "Well, you make sure you call on me first."

After you have sold this individual on something new, you can then approach your ego-oriented customers and use the id-oriented customers as testimonials to the "ego-ites," for the assurance they require. The super-ego-oriented customers are *not* people with enlarged egos. Keep in mind that the real meaning of "super" is *not* big, it is *above*. The super-ego-oriented customer is the individual who puts others before self. He is above the demands of ego. He is the individual with whom you must be especially careful to avoid any tactics that he will regard as unethical or dishonest. Often he is the individual who will not even let you buy him a lunch and will look with disfavor upon any attempts to provide him with gifts as insignificant as an inexpensive ballpoint pen. Generally, he will put his company's welfare first. The id-oriented customer, conversely, will often buy on the basis of how it will make him look, giving less consideration of how the purchase may affect his company.

As indicated earlier, don't think that impulsiveness is the hallmark of only the id-oriented customer. Super-ego-oriented people can be impulsive as well . . . but *not* unethical. Test this out on your own past behavior. Haven't you planned on making a contribution to some worthy cause, and after listening to an effective speaker on the subject found yourself moved to give much *more* than you had intended to give, and then, later, regretted that *impulsive* action? Super-ego-oriented customers are people to whom the plea, "I need your help!" will be most effective. To the greatest extent possible, analyze your customers to see which group they fall into *in the selling situations in which you encounter them*, so you will know which strategies and tactics will be most successful.

Remember, each of these people may follow a vastly different pattern at home, on the golf course, and in any off-the-job situation, so judge them by their behavior in situations where they must make buying decisions.

In maladjusted people, there seems to be an internal state of war raging among these three combatants. Even in well-adjusted people, there seems to be a need for ceaseless adjustment, and these three sides of our nature are constantly interacting. It is probably the cyclic ascendancy of these three facets of our nature that gives rise to loose identification of "dual personalities." We can have but one personality, being but one person; but different facets of that personality can manifest themselves, depending upon the nature of the influences of the moment that are providing drive. The manner in which your customer reacts may vary radically from one selling situation to another, depending upon which of these forces within him has the strongest drive at the moment. They have a profound effect upon his behavior —therefore, upon the selling situation, and they deserve our study.

As a starting point, in order to anticipate the reactions of a customer in a forthcoming selling situation the salesman might ask himself how the customer is ordinarily oriented.

Is he id-oriented, inclined to drive ahead and satisfy the needs of his self-image, no matter what anyone else thinks or says?

Is he ego-oriented, cautious, and very concerned about what others will think of his actions and what the cost will be to the status he seeks?

Or is he a person with a deeply developed social consciousness, highly ethical, aware of his responsibilities to himself and to others; is he super-ego-oriented?

If we can loosely type our customer (remembering that typing is dangerous because it implies that he will always be the same, regardless of conditions), then we will be in a better position to estimate which of the self-defense mechanisms he will be most likely to employ in the selling situation.

How a customer remembers

The manner in which a customer perceives the selling situation is of utmost importance to the salesman. We have seen some of the factors that will influence the manner in which he perceives. First, we saw that the impact of the selling situation on his self-image will affect his perception. Closely allied to the self-image are his status needs, and he will examine the selling situation to see if it enhances or threatens those needs. We have just discussed the conditioning effects of his milieu and how they will affect the manner in which he views our efforts; i.e., the selling situation.

We can now examine the final set of influences on his perception. These are his previous experiences and the highly personal manner in which he recalls them. We know that the customer does not enter the selling situation from a vacuum. He has had many experiences, both favorable and unfavorable. He has succeeded in some efforts and failed in others. These experiences are going to play a large part in shaping his perception, in causing him to look with favor—or disfavor—upon your proposals. It is apparent that the professional salesman must learn as much as possible about the past experiences of the customer and *how the customer has interpreted those experiences.* There are influences on recall; factors that cause some experiences to have major significance to a customer, others to be of relative unimportance. Let's examine how and why customers remember experiences.

What kinds of experiences do people recall?

1. People tend to recall more experiences, both favorable and unfavorable, if they and their interests were involved. The degree of involvement, then, is the most important influence. For example, you may be with Mr. A when his superior criticizes him ruthlessly in front of you. Later, you mention to a friend B the facts of the incident. A year later Mr. A meets you, blushes, and mentions that the last time he saw you, his boss "blasted" him. Because your memory is reinforced by his remark, you are able to recall the incident. If you were to mention it again to B, he would find it most difficult, if not impossible, to recall. Mr. A was personally involved, and he recalls the incident with no difficulty. You were involved only as a bystander, so with help you can remember the incident. B, not having been involved, cannot recall the incident. Other things being equal, the degree of retention is proportional to the degree of involvement. This indicates that, although a customer may be influenced by hearsay, he will be influenced more by experiences of his own.

2. The second conclusion of psychologists is that people tend to recall more items that reflected favorably upon them than they will recall experiences that reflected unfavorably upon them. We have analgesic minds, minds that have recesses into which we store things we like to forget. We do not like to remember experiences that reflect badly upon us because these experiences threaten the image we have of ourselves. Our span of experience encompasses successes and failures. We choose to take credit for the successes and place the blame for the failures elsewhere.

This holds true even of small errors. You've seen the fellow who has just hit his finger with a hammer. His first ejaculation may be, "That damned hammer!" and the hammer is sent flying across the room. Or we may be working around the house; in striking a nail, we fail to hit it squarely, and the head flies by our ear with an ominous whine. We are a bit shaken up by the fact that it might have hit our eye, which, incidentally, was without the benefit of safety glasses. Our first words may be, "The cheap nails they sell now-a-days!" If we have an image of ourselves as rather competent handymen around the house, we can't afford to admit these errors, so we place the blame elsewhere. Later, we will have forgotten these errors, but will point with some degree of pride to the work we have accomplished.

What does this have to do with selling? Just this: Salesmen are often disagreeably surprised to find that customers will not give credit to the salesman's products for success the customer has enjoyed, but will assume the attitude that she, the customer, would have achieved success with *any* product. The salesman had been planning to build his presentation around the successful past performance of his product, and the customer's attitude upsets these plans.

Or a saleswoman has planned to capitalize on the fact that the customer has just had a bad experience with the products of a competitor. When the saleswoman brings this out in a very subtle manner, she is amazed to have the customer brush off the bad experience as having been of no consequence. The saleswoman knows that the customer actually was furious with the competitor, so her present attitude is incongruous (to the saleswoman). It not only comes as a surprise but upsets the saleswoman's planned strategy.

What the saleswoman has failed to realize is that the customer cannot admit that the experience was bad, because this puts her own judgment at stake. After all, nobody *made* her buy from the competitor. In fact, this saleswoman had urged her not to do so. The customer may sound off about the competitor to others, but she will never admit her error in judgment to this saleswoman. The saleswoman sees the incident as a reflection on her competitor's product; the customer sees it that way also, but she also sees it as a reflection on her judgment, and therefore upon the image she is trying to project of herself as a businesswoman who is equipped with sound judgment.

These are but a few of the reasons why customers will often recall experiences in a manner quite different from our own recollection, even if we were present with the customer when the experience took place. It may be irritating to encounter this in other people and in customers, but if we keep in mind that we "adjust our own recollections" in this same manner when it suits our own convenience, we will then find it easier to smother our annoyance and apply ourselves to the task of overcoming the problem presented.

3. The third conclusion of the psychologists is that people tend to recall more items that reflect unfavorably upon the other fellow than the "other fellow" will recall. In other words, if a salesman has made a mistake in a past relationship with a customer, the salesman will want to forget it because it reflects upon him and threatens his self-image. The customer will remember the experience longer than the salesman. This is why a salesman can be disagreeably surprised by having a customer bring up a past unfavorable experience which he has had with the salesman, the product or with the salesman's company, long after the salesman has successfully forgotten it.

Developing empathy

Why is all of this important to salesmen? It is important because it points up the need for thorough pre-sale preparation. It emphasizes the need for developing *empathy*—an ability to see the selling situation as the customer sees it. This is not to be confused with *sympathy* for the viewpoint of the customer. Empathy may be thought of as an ability to *understand* the customer's views and objections and the reasons for them. Past experiences which the customer has had will affect his attitude. If he sees these experiences as "good," they may assist us; if he sees them as unfavorable, they are going to create obstacles for us. He may recall experiences we wish he would forget. We can't ignore these experiences; they aren't just going "to go away."

If, by giving some thought to the customer as an individual before the sale, we can *anticipate* what experiences he will recall—and how he will recall them—we will be in a better position to plan a strategy that will build on the best of those experiences. We will use tactics that will anticipate and neutralize the experiences which he has forgotten and which will support our claims. Admittedly, it is not easy to learn all there is to know about the experiences the customer has had. It isn't easy to always predict how he will view those experiences we know he has had. Nobody has ever said that *professional* selling is easy. It's difficult, and it requires skill.

Any salesman can make a number of sales without this kind of preparation, without this skill. But even the best salesman will admit that some sales do "get away"; and he will admit that he doesn't get all of the business he would like to get from some of his established customers. He will further admit that he has given up on some customers whose business he would like to have, because he has preferred to spend his time in greener pastures.

As we study the factors that will influence the attitude of the customer, we can see even more clearly the need for keeping personal records on the customer. The maintenance of a record of experiences we know the customer has had, as well as a recording of as much personal data on the customer as we can compile, will prove to be of real value to the salesman in his preparation before he enters the sale. As stated earlier, the doctor is regarded as a "professional," and we know that he maintains complete records on each patient, and he examines those records before he prescribes treatment.

If we are determined to have selling recognized for the profession that it is, then we must use the tools of a professional. If customers' experiences influence their attitude toward us or toward our product, then we should know what other influences cause a customer to remember experiences in a unique manner.

A general premise is:

1. Customers will recall those experiences in which they and their interests were involved;
2. They will remember experiences that made them look good and will try to forget experiences

that made them look bad or caused them to be irritated or frustrated;
3. They will remember situations in which we, as salesmen, looked bad, long after we have successfully forgotten them.

It has also been stated that the customer shares with us the tendency to take credit for his successes and to blame others for his failures. There are nine additional influences that affect the manner in which he will bring past experiences into the selling situation. In our attempts to predict how he will behave or react to our efforts, we should understand these influences.

1. Intensity

The intensity of a past experience, or the intensity of a present need or want, may cause a customer to depart from what we have considered his normal behavior pattern. It is imperative in attempting to weigh the intensity of a customer's experience that we consider that experience from the customer's frame of reference, not from ours.

Take this common example: a customer buys a new car and has difficulty with the windshield wiper. He brings the car back to the dealer and complains vigorously. The dealer cannot understand why the customer makes "a big deal" out of the malperformance of a simple accessory. He tests the wiper in his dry repair shop. The customer "recalls" the experience he has been through—an extremely hot night with the car windows all closed because of a torrential downpour, with three kids and a dog in the back seat, the whole family sweltering while he has parked the car at the side of the road because he has been unable to drive with a windshield wiper that wouldn't operate. To him, this has been an intense experience, especially when he thinks of what he paid for the car and when he recalls that he never had this difficulty with the old car.

You don't sell cars, so to make this more meaningful, stop reading for a moment and transfer this premise to the kind of selling situations in which you have been engaged. Can you remember a customer who was very agitated over an experience she had, when from *your* viewpoint it was difficult to see why she should be so upset? Did your failure to anticipate her attitude make it difficult to sell her? Did it actually cost you a sale? If you had seen the experience from the customer's frame of reference, would it have been possible to have taken action to neutralize the intensity of the experience?

Sales are frequently lost because we are not truly customer-oriented. The automobile dealer may allow himself to be vexed at the customer's "excessive irritation over a small matter." This attitude won't build repeat sales. A word of sincere sympathy for the customer, an expression of regret, an attitude of understanding, a willingness to share the customer's annoyance over the product failure—these are the things of which repeat sales are made. Fixing the wiper and getting him out of the shop as soon as possible may get the customer out of the dealer's hair, but it won't erase the memory of the irritating experience at the side of the road. One experience of this kind may not lose the customer's business but it will be fed into the customer's mental computer and will have an impact on the next selling situation.

2. The law of proximity

A customer tends to group experiences. A recent experience may have a far greater effect upon him, good or bad, than an older experience. It falls into the category of "what have you done for me lately?"

A customer may have had a long, rewarding relationship with a salesman, but one recent poor experience may dwarf the impact of the good experiences he had in the past. This attitude is not confined to individuals. We see it in the behavior of nations. This nation may have given blood and money for a less favored nation, and for a period of time our image with that nation may be desirable. We may find it necessary to say "no" to that nation just once on some request which it regards as important and find that nation threatening to switch its affection to nations we regard as opponents. Over a period of time our "no" may gradually be forgotten, but at the moment the proximity of the experience causes it to take on greater importance, and it threatens the relationship we are trying to build.

This is exactly what can happen in a selling situation. It is the reason why we should make every attempt to know as much as possible about the most recent experiences of the customer and why we should try to assay the importance of that recent experience as the customer will view it. Again, to make this more than unsupported theory, stop for a moment and consider customers you have had. Has any one of them been so agitated over a recent experience that it made your selling job difficult? Was it an experience that ordinarily you would not have expected to have caused major difficulty in the selling situation? Was it an experience that with the passage of time was all but forgotten by the customer?

As these influences are being discussed, you may find yourself wondering if this isn't all rather academic, and asking, "Why does it make any difference to know about these influences on recall?" The answer is clear. Effective selling requires effective planning and this includes evaluating your customer and his probable attitude. If we know what factors have caused an experience to become very important to the customer, we are going to be able to predict with greater accuracy the nature of his probable behavior. We will then be in a better position to plan on capitalizing on the experience, if it was favorable to us, or we will be able to neutralize the experience if it was unfavorable.

Even when we fail to predict his behavior, we are going to be able to understand why a particular experience, which seems trivial to us, is important and significant to him. This means we will be able to attack the problem it presents in a rational manner and not permit our own emotions to further complicate the selling situation. We will realize that, had we lived through *his* experience, we would probably react just as he is reacting now.

3. Repetition

Unfortunately, for the salesman, the customer will tend to magnify the importance of an experience if it has been repeated frequently. This will be even more likely if the repetition has been recent and took place within a short period of time. This is why a salesman may be taken aback when a customer becomes extremely angry over some action or comment of the salesman. In this instance, the salesman may be the last link in a chain of irritating experiences to which the customer has been subjected all day, and the customer "takes out his wrath" on the salesman. We will see later in discussing defense mechanisms that this is called displaced aggression. The salesman isn't interested at this moment in knowing what it is called; it is enough for him to know that he has a problem. Many saleswomen have learned through experience, and without the benefit of any psychological training, that it's wise to try to learn "what kind of a mood is the old man in today?" before entering his office. Sometimes it is wise to leave and come back another day. At other times, this is not possible or even advisable.

The rule to keep in mind is that if a normal customer is reacting in what seems to be an abnormal manner, there is a reason. If what you have said or done is not in itself sufficient to have caused him to be irritated, then it must be something that preceded your visit, and you have triggered off a reaction. The antidote is not to fight back or to permit yourself to be irritated. You are witnessing a normal psychological reaction, and one which your wife and your friends will tell you that you, yourself, have used on many occasions. Let him talk and get it out of his system. Don't interrupt or argue. Listen and try to find clues to what is really "bugging him." Seek some point upon which you can agree, and try to exhibit understanding. Leave him escape hatches; if he is being unfair, your mood and manner will get this across to him, and try, gradually, to transfer his attention to the problem you feel your product or service will solve for him.

Customer-oriented selling must be based on understanding him. It is easy to understand the customer who permits himself to be outraged by a series of repetitive, yet insignificant, irritations, if we will but remember that we have reacted in this same manner on countless occasions. How did the people who understand you best treat you when you sounded off? The chances are that this same course of action is one you should consider taking in the presence of the irritated customer, providing your respective roles permit it.

4. Contrast

Customers tend to compare today's selling situation to previous experiences they have had, both good and bad. They compare you to other salesmen, your company to others, and your product, your terms, your delivery and your personal treatment of them to previous selling situations. Consider for a moment whether you have known customers who, in the past, have had a high regard for a given product. Have you seen them compare a new product to the old one? Turning to the automotive field, for example, you may have seen a chap who bought Packards as long as they were manufactured now compare every new car to the Packard. Actually, in the light of the product features available on today's car, he would probably reject the Packard if it were now offered to him.

We all develop value standards as a result of experiences, and often these standards are guarded by something akin to affection. An individual has an old monitor-top refrigerator in the game room which he has had for decades and refuses to part with it. For years, it gave service-free performance, but a critical analysis of that refrigerator in the light of today's products gives us an uncomfortable feeling. It held only five cubic feet of storage space, had no freezer compartment, came in one color, was high off the floor and had a dust catcher on top. You can be certain that you would not endear yourself to the owner if you were to point out these shortcomings.

The more we know about the "values" of the customer, the easier it will be to sell him. If we know what experiences have brought him satisfaction, we will try to present our product in a manner that builds on that satisfaction. We will emphasize the benefits that are obviously important to him. In this manner, we help him to recall a favorable experience.

When a lightweight motor was introduced, many customers refused to buy it because it did not "have the strength" of previous motors; it was not "rugged." The fact is that, made of a superior metal, the new motor had greater strength and was more rugged than its predecessors; yet the manufacturer, for a period of time, was forced to make a certain number of the older motors.

Customers, and you and I, don't resist change per se, but we know that certain conditions have brought us satisfaction and few problems, and we just don't want to change until we are certain that there are advantages. When there is such a thing as affection for products or even for components, it must be recognized.

Professional salesmanship means that in our pre-sale preparation, we will attempt to learn as much as possible about the customer's values... what he regards as good or bad. Then we build upon that knowledge, emphasizing product features that he will compare favorably to features that pleased him in the past. We will emphasize features that compare favorably and offer him relief from difficulties he experienced with other products he has used and which he will compare to our offering.

Keep in mind, too, that his tendency to compare will not be limited to the products you offer. He will compare you to other saleswomen. Which attitudes does he like to encounter? Which attitudes irritate him? Don't attempt to imitate the mannerisms or the familiarity which others may enjoy with the customer, but try to identify his self-image and behave in a manner that will enhance it. Stop once again and ask yourself if you have had customers who tended to compare you or your product to other salesmen or to other products that they regarded as standards. Would it have helped you to have known this before the sale? Did failure to know these things give you a few moments of difficulty in the sale?

The point being made is that customers' attitudes today will be influenced by experiences they remember, and their normal tendency to make comparisons will have much to do with their attitude today. If some of those comparisons are going to reflect unfavorably upon you as an individual or upon your product, you had better know as much about them as possible and be prepared to neutralize their effect. If you look good by comparison, so much the better; make sure he knows it by emphasizing the features that will influence him favorably. Probably no product or service is sold in a vacuum. The customer will always compare.

5. Similarity

It requires little attention here. In selling, it is much like comparison. The customer views you and your proposition in terms of similar situations, or she looks for similarities, both favorable and unfavorable.

6. Change

The customer who is accustomed to a product or to features of an offer will be quick to identify that which represents change; and the more radical the change, the more certain we can be that it will take on greater significance in his appraisal of the situation. Change attracts attention; hence the reason for the changing lights on a theater marquee or highway sign. In this chapter, we are not discussing "things to do," or we would elaborate on the value of calculated change on the part of the salesman to obtain and hold the customer's attention. Here, we are attempting to understand how and why customers can be expected to react to change. Variations in a product, a dealer policy, a discount rate, delivery schedules or the like can be expected to provoke the attention of the customer, for good or bad. If we keep this in mind, we can capitalize upon it, or avoid being surprised unpleasantly.

7. Illusion

Take two oranges of equal size, and surround one with grapefruits and the other with tangerines; and the latter will appear to be larger, as in the illustration. We are familiar with optical illusion in terms of physical objects, but we also see abstract situations in a manner that can only be described as an illusion. So does your customer. He sees one object, or one feature of an offer, in terms of its setting, and it may be difficult to change his mind. Presented with a physical illusion, such as that of the two oranges, even when we know that it is an illusion, we are tempted to take out a ruler and measure them, so convinced are we that one looks larger than the other.

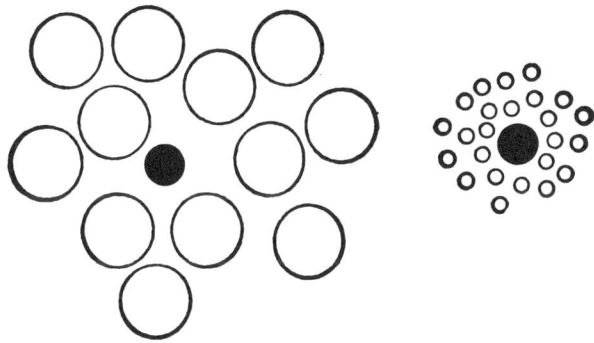

We are often annoyed when a customer tells us in advance that he is definitely going to be interested in our offer; yet when we make the offer some time later, he exhibits total disinterest. We may point out that previously he had expressed more than interest —he had assured us he was going to buy. His response may be, "Yes, but things were different then!" This is his way of saying that the setting has changed; what looked attractive then, in the light of other conditions, no longer looks attractive. This again points up the need for knowing the customer and understanding his frame of reference. It emphasizes the need to know what changes are constantly taking place in that frame of reference, so you may be prepared to cope with unfavorable changes or be prepared to capitalize on favorable changes. This also underlines the necessity for keeping current on the activity of your competitor, because his activities are part of the setting from which the customer will view your offer.

Illusion may create selling problems for you, but it can also be used as a potent selling tool. You can change the setting so that your offer becomes more attractive or his objections lose their validity. One salesman confronted by a customer with a competitor's offer to undersell him by 4% on $20,000 worth of business on items selling for 15¢ each, reduced the $800 annual saving to $15 per week, and then proceeded to prove that his service was worth $15 a week more and that the hazards of changing vendors were too great to risk for such a small sum. The fact is that the customer would continue to pay out $800 per year extra, but by reducing the increase to a weekly amount, it took on less significance in this new setting. As the thoughtful, professional salesman studies each of these influences on recall, he will see how they can be deliberately injected into the selling situation and used as selling tools.

8. The law of Praegnanz— the poor seen as good

This one can spell disaster to the salesman who thinks that the customer believes everything she says in describing her product need. A customer may have purchased a product from your competitor even though you advised him against doing so. You may not have been critical of your competitor, but you may have said, "The *XYZ* Company makes good products, but I think you should know that in the case of the particular items you say you are going to buy, they have had considerable trouble." You may have gone on to describe the difficulties *XYZ* is experiencing. The customer ignored your advice and made the purchase. Through the grapevine, you hear that he is having just the trouble you predicted. Later he calls you in to place an order with you, and it is apparent that he is replacing *XYZ*'s product with yours.

Ordinarily, you would take the new order and say nothing. But let's assume that the customer is a close friend of yours, and after he signs the order you remind him that you warned him he would have trouble. You are astonished to hear him say, "Oh, it wasn't bad at all." And he may go on to give you an entirely different reason for switching his business back to you. What is going on here? You will recall that in discussing the self-image, it was pointed out that the customer has an image of himself as an experienced person, in this case as a capable buyer. Obviously, he has made a mistake in this instance; but to admit it, even to himself, will destroy his self-image, so he *reconstructs the situation* so that he can recall the experience in a manner that will enable him to maintain the image of himself as a smart buyer. He is trying to deceive you, yes, but only because if he can get you to accept his story, it will be easier for him to deceive himself and maintain the self-image.

Salesmen are constantly amazed to see a customer have a bad experience with some competitor and then go right back to that competitor to buy the same item again. The salesman had anticipated that the customer, having been damaged once by the competitor, would now return to the salesman. The reason is that the customer cannot bring himself to admit he has been wrong and tells himself that "everyone makes a lemon now and then" and returns to the competitor, hoping that his faith in his own judgment will be vindicated by the successful purchase he hopes to make this second time.

The law of Praegnanz says, in effect, that people like things to be orderly, to be "good," complete, balanced and successful. They will distort the facts of a situation to make a bad experience look good, especially if that bad experience reflects upon their own judgment or capabilities. This not only applies to their tendency to distort their recollection of how a past experience really took place, but it applies to the customer who wants something so badly that when you try to talk him out of it, by pointing out the troubles it will cause him, he brushes off your arguments as being of little importance. It's the old story of "Don't confuse me with the facts; I have made up my mind." If he listens to you and follows your advice, he will feel the pain of frustration. Rather than risk that, he will see the poor (which you are pointing out) as good.

You and I can jeopardize the long-range relationship we seek with the customer by permitting this attitude of his to make us angry and by marking him off as "stubborn" and "pig-headed." If he is such, then he has much company, and you and I must join him, because each of us has refused to listen to advice that would have prevented us from doing something which we wanted to do quite badly. Once we view the customer's behavior with understanding (avoiding the tendency to patronize), we are in a better position to cope with the problem he presents.

If the law of Praegnanz seems nothing more than interesting psychological theory, then re-examine past selling situations in which you have been engaged. Ask yourself, honestly, if you haven't had customers who have distorted the facts of a previous experience in order to make it easier to bear, or customers who wanted something so badly that they would not listen to warnings and later, having had the predicted trouble, minimized it or even denied it.

Moral: Don't build a sales presentation on the expectation that the customer will admit he has had difficulty with a competitor's product, or that he will now come flying to your arms because he has had the difficulty you predicted. Indeed, he may go to a second competitor rather than return to you, because he is not unmindful that you did warn him, and he does not want to expose the fact that his self-image is not realistic.

9. Gestalt

"Gestalt" is a German word meaning an organized form or pattern. It means the tendency to see things at once as wholes and not as parts of the whole. For example, when you enter a room, you are conscious of the room itself but not of the various items in the room. It may be a long time before you are aware of what the ceiling looks like, whether or not there is a baseboard, where the electrical outlets and radiators are, and so forth. You see the room as a whole.

Customers see selling situations in this same manner. Unless he knows you well, the customer may tend to stereotype you as just another salesman. He sees salesmen as a group...as a whole. If you work for a firm that has more than one department, and he has suffered at the hands of another department, you may experience real difficulty in causing him to see your department in a different light. You may hear him say, "You're all part of the *ABC* Company." This can be an advantage if the other departments of your company or other products of your company have enjoyed a good reputation. He will tend to pre-judge you, your product and your department on the basis of the experience he has had with the balance of your organization.

Saleswomen often enter a selling situation behind great obstacles, or without knowing they have advantages working for them, because they do not do a thorough job of obtaining commercial intelligence on the customer; i.e., they may not even check with other saleswomen, or with other departments of their own company, to see what experiences the company, as a whole, has had with the customer.

Summary

Throughout this chapter, we have been talking about the customer's frame of reference; how he will see the selling situation. The self-image is one of the factors that will influence the frame of reference, as is the customer's need for specific kinds of status which are important to him as an individual. The conditioners, his ideas of what is moral, ethical, legal, etc., also affect his viewpoint and his analysis of you and your proposition. Now we have seen that customers do not always remember past experiences as we recall those same experiences. Each will remember the situation in a manner that provides him with certain satisfactions of psychological protections. These past experiences, as he remembers them, also are factored into the frame of reference. This is why the emphasis of this analysis of professional salesmanship has been, and will continue to be, on the twin needs of (1) really knowing the customer as an individual, and (2) pre-sale preparation, or pre-sale diagnosis, whichever you prefer.

We find then that four major areas impinge upon the customer's frame of reference: self-image, status needs, conditioning and experiences. You, the professional salesman, are the fifth influence on that frame of reference. If the preceding four influences have all been favorable to you, little remains for you to do other than to present an effective product-service story, coupled with a reasonable price. If, however, as in the case with the hard-to-get sale, or as is the case with the sales that have been getting away, the impact of one or more of these influences has been negative, then *you* must neutralize their effect. You must gather intelligence on the customer, by diagnosing the customer's needs (both product and psychological), by planning a strategy and a tactical plan, and then by combining these elements with your own personality in a manner that assures the customer of satisfaction. A job for a professional on the face of it. Next, let's see what makes it even more demanding.

CHAPTER 5

How the customer defends his self-image

It is not our purpose here to analyze each of the self-defense mechanisms in depth, as would be accomplished in a thorough course in professional salesmanship. We will identify and classify each mechanism, determine why the customer employs it in a selling situation, and hopefully awaken in the professional salesman a desire to make himself even more knowledgeable in this area so that he may develop a skill in coping with the obstacles which the employment of self-defense mechanisms by the customer can create.

In many books on psychology, the authors draw a shape something like an ink-blot and label it "the self-concept." Probably their reason for this is their opinion that the self-concept, or self-image as we choose to call it, is a formless, abstract, ever-changing thing that does not lend itself to graphic presentation. Ever-changing it may be, but we cannot agree that it is abstract or formless simply because we lack the skill to measure it. In any event, in our discussion of the self-image, the point was made that we do not like our behavior to be transparent. We don't really want to know the real me.

If we left a family we love, to volunteer for military duty, we prefer to believe that the action was a result of our love for our country and our sense of duty. We don't want to hear the inner voice that asks us if we were becoming bored with the job, anxious for adventure, possessed of itchy feet, or if there were other less laudable reasons for the psychological drive that made us volunteer. *Our real me* is hidden away where nobody can see it. In the selling situation, we might think of the customer's real me as being surrounded by a series of concentric rings, 14 in number.

Each of us is aware that the world is full of threats to us as individuals. There are threats to our pride, to the satisfactions we desire, to our economic and even physical security, and so on. We see these in the form of the chap who tries to get closer to the boss than we are, the reporter that hurts our good name, the chap that cuts us out on the highway, and so on, it seems, almost ad infinitum. This is not to say that we go through life fearful every minute of our waking hours that someone is about to take advantage of us. It *does* mean that we are mindful of these threats and are quick to recognize them when they appear; and, as we shall see, even more quick to take protective action.

Our actions often take the form of any one of the

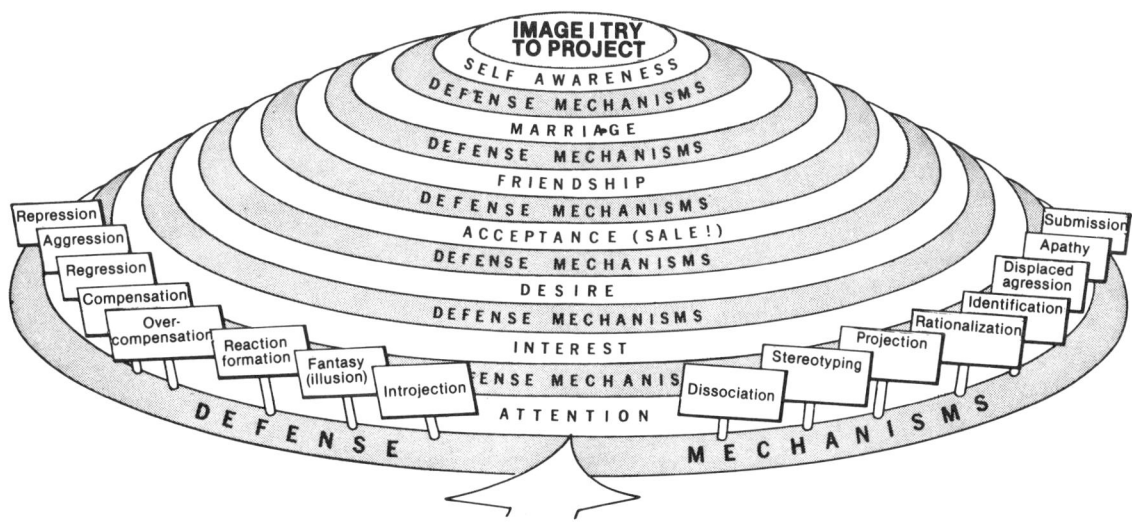

defense mechanisms. The outer ring, surrounding the self-image, is made up of these 16 mechanisms which we are ready to use against any threat, and there is but one break in the circle to allow those we trust to move in—just so far.

If the salesman's appearance and opening remarks pose no real threat, we will let him through the outside ring into the next which we can call Attention. If we give him our attention and his presentation seems to offer us something we need, we will let him through the next barrier to the ring called Interest. Again, if we see no threat we will let him through another barrier to the ring called Desire; and finally, with no threats available, we pass him through another barrier to the ring of Acceptance.

If the salesman continues to call upon us, and he enhances our self-image and poses no threat, we will let him into an area we might call Friendship. But the barriers ahead remain in place as evidenced by the question, "Don't you think that's asking a lot?" or "Aren't you imposing on our friendship?" or "Isn't that just a bit personal?" when a friend thinks we are "pushing." We let the barrier down again at Marriage, so that our spouses probably know us better than anyone else; but another barrier remains—even they are not permitted to see our self-image in its intirety.

The image we have of ourselves is not the same image others have of us, nor is it an actual representation of what we are really like. This is because the awareness we have of ourselves is based upon an incomplete self-analysis. Just as we use the defense mechanisms that follow in this text to deceive others, we use them to deceive ourselves. In fact, self-defense is primarily self-deception! Most of us are unwilling to engage in the necessary agonizing self-analysis essential for a meaningful self-awareness. Self-analysis leads to self-awareness which leads to self-development and brings us closer to self-fulfillment. To the degree that you and I and our customers avoid planned and continuous periods of helpful self-analysis, our self-image will become more unrealistic. Lacking such analysis we all tend to project self-images which we feel others will accept, but they are far short of the image we would really like others to have of ourselves. It is at best a compromise. It represents the "me we try to project" and it is the genesis of many of our reactions and those of the customer. We reject proposals that threaten that self-image and we embrace those that enhance it. This is evidenced when an individual will buy a product or service and knowingly pay more than a price available elsewhere simply because he sensed a psychic gain; i.e., enhancement of his self-image.

As we have seen, even when we are introspective and try to identify our own faults, we make excuses for ourselves to ourselves; i.e., empty self-defense mechanisms, so a final barrier of these mechanisms exists between ourselves and an honest understanding of our real motives and purposes. Our friends even help us in this self-deception with their comment, "Don't be too hard on yourself," when we make an effort to be self-critical.

The customer operates in this same manner. Although he wants product and price satisfactions, he will not pay for them at a high cost to his self-image. It is his use of the self-defense mechanisms that provides salesmen with their greatest challenge, the problem of identifying the basic fear that motivates the customer so that we can allay that fear and make a sale. For this reason, this chapter could not be complete without at least a cursory examination of each of these tools which the customer so speedily and adroitly employs.

There is no real necessity for the saleswoman to remember these mechanisms by name, although it would prove to be helpful, especially for future research. The objective is for the professional saleswoman to understand the reason why the customer employs these mechanisms, how they show up in a selling situation, and what to do about them. Great value comes from merely agreeing that you use these mechanisms daily; and if we agree to this, it will be difficult to find fault with the customer for doing so.

Once we achieve this kind of understanding, we are able to tackle the problem of overcoming the effects of the self-defense mechanisms, without allowing annoyance or anger to cloud our judgment. To understand why we use these mechanisms, we must agree that all of us will fight to preserve our self-respect, will fight to preserve our place in society (our status), and will take actions to protect ourselves from the very real pain of frustration—the pain of being denied what we seek.

Repression

We all experience painful thoughts. These often take place when we perceive a situation that is not to our liking, but about which we cannot take corrective action. A person hearing about a serious miscarriage of justice placed his hands over his ears and said, "Don't tell me about those things!" This was the self-defense mechanism of repression at work. The "self" hates to admit that it is sometimes powerless to take action, to rise to the heights. The "self" sometimes does not want to pay the price in terms of sacrifice, or work, or change in habit that is necessary to achieve an objective.

Under these conditions, the customer will resort to repression. As seen earlier, he will resort to repression (of the facts) when we advise him not to do something he is determined to do. You have seen people who torture themselves, physically, because they will not wear much-needed eyeglasses or hearing aids, because to use these devices threatens the image they have of themselves as being not much different, really, than they were at 21.

Repression shows up in the customer who has failed to follow printed instructions, and as a result fails to maintain equipment, with a resultant breakdown of the product's operation. She will swear she never received the instructions. This is not confined to individual customers. One branch of the armed services was warned that if it did not send maintenance men to a school to study the operation of certain sub-assemblies, it would have trouble. The service replied that it could not spare the funds. Later, the very disaster that was predicted occurred. The result was that the service blasted the manufacturer, refused to let him discuss the warning letter he had sent, and cut him off the list of suppliers on the project.

It is easy to become indignant as we read this until the awful truth dawns upon us that each of us has used this mechanism times without number. To cope with it when it shows up in the reactions of your customer calls first for understanding and agreement that while what he is saying may be untrue, he is not fully conscious that he is lying; he is obsessed with protecting his own self-image. After all, only a stupid person would fail to read maintenance instructions. Is he about to admit that he was stupid? (Were you the last time you failed to read instructions on something you bought?)

Some types of pay stations warn us not to deposit money until we have read the instructions, but we put in the dime and then learn that in this type of phone you don't insert the coin until the operator answers. One dime lost! Who do we blame? Naturally, the phone company.

How do we handle repression in our customers? First of all, we don't prove he is wrong in a manner that exposes the defects in the image he is trying to project. We feed him the information in a palatable form. We try to find some minor point upon which we can agree with him. We may be able to indicate that we made the same mistake ourselves and thus provide him with an escape hatch ("It was a perfectly natural mistake to make; many other smart people make the same error").

If possible, we have him discover a way out by himself, after we have subtly planted it where he can find it, so that when that moment of insight arrives (when he realizes he was wrong), we will not be a witness to his discomfiture. Above all, we don't mistake repression for suppression, which is actual suppression of facts for dishonest purposes. And we don't permit ourselves to become angry. How can we be angry with a chap who is trying to deceive himself more than he is trying to deceive us? Make this meaningful: can you recall a customer who used repression? What do you think now of the manner in which you handled the situation?

Aggression

The best defense is a good offense. We teach it in the military, and we teach it in chess. Why become annoyed with the customer who uses it in a selling situation? Recognize it for what it is, a diversionary tactic to keep your attention off something he wishes to conceal—something that may cost him psychic or economic security. He is trying to keep you so busy defending yourself, you can't attack—even if your attack were to take only the form of asking questions, questions that might embarrass him.

Too often, we become furious at the unfairness of his attack because usually the problem he has raised is a contrived and diversionary problem. Our defense often becomes in itself an attack. Then he begins to find justification for his anger. Soon he has worked himself into a state of righteous indignation, has passed the point of no return—and you are in trouble.

Too often, we spend time and effort trying to solve the problem raised by his attack, making the mistake of taking his diversionary tactic seriously. Analyze what he is saying, and if he has a legitimate complaint, recognize it and attempt to remedy the situation. If it seems to be blown up into huge proportions ("much ado about nothing"), treat it respectfully, wait him out, and don't let yourself be swayed from your original objective.

Regression

Regression is represented by an attempt to go back to old, easier ways of doing things, to habit patterns that gave us satisfaction. This is evidenced in the behavior of the customer who won't give a new idea a fair chance to prove itself. He fears failure. You may talk him into giving the idea (or the new product a try-out), but the next time you see him he will tell you that it would not work, and he has gone back to the old method or product. We are tempted to write him off as a person who resists new ideas. It's easier to do that than to look for the real reason for his behavior.

What threat did the new idea or product pose to him? Was it the threat of economic loss, actual loss of money; was it the thought of the physical work of making changes or the need for having to train men in the new method or on the new product? Was it the fear of losing face if it shouldn't work? How much help did you give him? How much proof and assurance did you give him? Was there actual physical help you could have given him or help in training his people? Did you present him with overwhelming evidence that he would make real profit or make great savings or appear dominant in his profession if he made the idea work?

In other words, did you offset his fears with a greater hunger to look good or to make money? Once again, to get the most out of this book, pause and consider your customers. Have you seen some of them use regression? Did you really learn why they did so, and were you able to offset their fears?

Compensation and over-compensation

A customer who resorts to compensation is usually an individual who has been deprived of his primary goal in life and has thrown himself into a substitute goal with tremendous energy. This type of person cannot tolerate failure in himself and needs recognition. *Over-compensation* is somewhat different in that the person who is deprived of his original objective may not only throw himself into the substitute activity, but will develop an active dislike for the object of his original affection. The boy who is poor at studies may throw himself into athletics and develop a disdain for "scholars." The customer who may have been rejected by a supplier when the customer was small and unimportant may turn to another vendor. Later, when the customer has grown and is important, he will show active dislike for the vendor who originally was his first choice.

Over-compensation is more difficult to deal with because it encompasses a real or fancied affront to the self. As children, we may have called it "getting even." When it is used against the salesman, his courses of action include: letting the customer get it out of his system by sounding off about his grievance and then indicating that you think he is a fairminded individual and won't take out on you the anger that was generated by someone else; showing recognition of his achievement; and finally, showing him the high cost he is paying for the privilege of putting your firm in its place.

Reaction formation

Reaction formation is the most difficult for many people to understand; hence, to accept. A married couple has just put their last child in college, and feeling their new freedom has all kinds of plans for their leisure time, when a new child is born. Their first action may be to resent the child since he represents a threat to their need for leisure. However, they realize that this is a socially unacceptable attitude and, feeling guilty, begin to lavish attention on the child. Others, observing them, may feel that no couple ever wanted a child as much as did these parents. At the time, nothing could be further from the truth. In time, with most situations of this type, real love follows.

Reaction formation describes that situation which exists when a customer behaves in a manner that is completely out of phase with what he really thinks and feels. Obviously, it can cause maximum trouble for the salesman who tends to judge the customer's likes, dislikes, and desires by what the customer says.

Take, for example, the chap who buys a small car, and after he buys it regrets the purchase. He now feels it was an extravagance (he doesn't need a second car), and being tall, he hates the cramped feeling he has when he is behind the wheel. He is not about to confess these feelings because to do so would reflect upon his judgment and shake his confidence in the self-image he has of himself as a man of good judgment not given to capricious or impulsive acts. If a car salesman heard this man discussing his small car, he would hear raves over its gasoline mileage (economy, to offset his feeling that it was an unnecessary extravagance) and more raves about how easy it is to park (compensating for his annoyance with its cramped interior). If the salesman were to take the customer seriously, he would assume that economy and maneuverability were the two points he must emphasize if he hoped to sell this chap a new car. This would be the worst mistake he could make.

The cure is to let the man talk. Then ask him if he could make any changes in the design of the car, what would he do. He'll point out that the car *could* be roomier, and he will go on to point out other defects. Once he has been given an escape hatch, he will talk about what he really wants. In the case of this customer, the salesman might also learn that he was wasting time trying to sell him a car because the customer is already annoyed with himself for having bought a second car. The lesson we learn from reaction formation is not to take everything that the customer says as necessarily revealing what he really thinks. Keep him talking.

Fantasy

The psychologist thinks of fantasy as the defense mechanism used by an individual who retreats from realism into an unreal world of his own where he imagines himself as something which he is not. In selling, we think of fantasy as imagination. While it can be an obstacle for the saleswoman, more frequently it may be an advantage upon which she can capitalize. We might think of it as describing the motivation and behavior of a customer who refuses to compromise between the me he wishes he were and the me he thinks he is. In other words, while most of us establish a "me-I-try-to-project" which is somewhere in between the two, he refuses to do so and tries to convince himself and others that he is, right now, the me he wishes he were.

This is represented by the executive of a company who, though not tops in its field, insisted upon buying equipment that only the top firm could really afford. There was even a real reason to suspect that he did not need the equipment. However, who is to say he was wrong, because by following this pattern stubbornly he managed eventually to project the image he desired. A salesman who discovered what this customer was trying to do managed to sell him equipment that his competitors never considered offering a firm of his size.

Housewives who may be fat and 55 see pictures of colorful appliances in advertisements. The model is trim and 21, and the customer's imagination enables her, mentally, to shed the years and the pounds, and to picture herself in that same kitchen. The salesman who appeals to this fantasy is going to make more sales.

Your business is very remote from selling dolls, but the principle involved in the following true situation should be of interest and value to you.

When a mother would come into the toy department of a large midwestern store and inquire about dolls, one saleswoman would pick up a large doll, cuddle it in her arms, then lay it across her shoulder, meanwhile caressing and "burping" it. She would then hand it across tenderly to the shopper who would find herself handling the doll in the same manner. The saleswoman was so far out in front of the other members of the sales force that there was no comparison. Respect for products begets respect. If you see beauty in things you sell, in the workmanship your products may represent, you'll be a lot closer to a completed sale if you can cause your customer to see this same beauty. He has an imagination; stimulate it.

One apparatus salesman, driving across the country, could never pass a substation without stoping his car to admire the beauty of the installation. The beauty that he saw was seen by many photographers because it is not unusual to see exhibits of substations in black and white photographs. Using your imagination can stimulate those of customers.

Introjection

When we act as though something that has happened to someone else has happened to us, introjection has taken place. It is common in selling, and it usually catches the salesman off guard or angers him. It is found in the attitude of a customer who is furious because of some bad experience another customer has had with your firm. This customer associates himself with the other customer and in effect says, "If you did it to him, you'd do it to me if you had the opportunity."

The experiences of others can be as real to the customer as if they were his own experiences, if he feels a common bond with the other customer. It is common in everyday life. We may attend a ball game at which the team we favor wins, and on the way out we can be heard saying, "We sure won that one!" We? When our vice president is shown discourtesies overseas, we become as enraged as though it had happened to us personally. She is an American, so are we; so if they did this to her, they'd do it to us.

This again points to the need for knowing with whom the customer identifies himself and knowing that when you establish a policy that affects one customer, all of your other customers who consider themselves in his category will be ready to react. If you can predict this in your pre-sale preparation, you will be able to plan strategies to neutralize probable difficulties or to capitalize on good reactions if the policy change was well-received.

Dissociation

This defense mechanism can often cause the individual exposed to it to react with inner fury. Dissociation takes place when any individual takes an action in one situation that is exactly the opposite he took in a previous situation, or which that individual criticized in others. In a current British TV series, one individual urged one of his employees to get married. A short time later, when the employee announced his engagement, the superior immediately took him to task for it and indicated that the decision was premature going into a long speech about added responsibilities, financial burdens and the like. The employee, of course, could not believe what he was hearing.

At such times, when a salesman is exposed to this mechanism, the tendency is to feel that the customer who employs it, is being inconsistent. No way! In the incident just related, the *role* of the manager had just changed. When he recommended marriage, he was acting as a friend. When the employee decided to take the advice, the manager then acted as a manager and saw a threat. He now fears the employee will not be able to devote as much time to the business so the manager changes his position completely.

You have probably never met a truly inconsistent person. It isn't people who change, it is situations that change and people adapt to the new situation as they see that situation. This is true of the wholesaler who screams about his retailers who always want a "better deal" and rages that these retailers are going to ruin any hope of price stability in the market. A few moments later, when a manufacturer's salesman submits a price, the same wholesaler will react with, "Is that the best price you can give me?" and will not see any incongruity between two viewpoints he has held in two different situations where his role changed. What color is a chameleon, blue, green, brown or red? The answer is that it is whatever color it requires to blend in with its environment for protection. We are much the same. Under one set of circumstances we will hold a view that reinforces our needs while we will completely reverse those views when faced with a different set of circumstances that pose either great threats or great gains.

One individual was completely opposed to capital punishment. Recently his wife was attacked and murdered. When asked what he thought the punishment should be, he replied without hesitation, "Execution!" He was criticized for being "inconsistent." He was perfectly consistent with his needs as he perceived those needs at the moment. A more common example of dissociation is your behavior and mine in our interchangeable roles as pedestrian and driver, and no explanation of that role reversal and attitude reversal is necessary.

Because a customer who changes his attitude to meet the needs of his changing role can infuriate us when we are selling, it is imperative that we understand why the attitude reversal has taken place and not permit it to anger or even irritate us... to do so is to endanger any hope of making a sale. Rather, the professional salesman should identify the threat the customer has sensed in the new situation and attempt to neutralize or eliminate that threat. When the role reversal has caused the customer to abandon his old views because a benefit is seen, the salesman's job is to capitalize on this and to accent the benefit without pondering over the vicissitudes of human nature.

Stereotyping

This is one of the easiest of the defense mechanisms to identify, probably because most of us use it so often ourselves. It shows up in the retail selling situation when you ask a salesman why he did not sell a customer, and he replies, "He was a looker." By this device of stereotyping customers whom he fails to sell, the salesman escapes the feeling of inadequacy, frustration and failure that he would otherwise have experienced. It is his way of saying to himself, "The self-image that I have as a good salesman is still valid because *no* salesman could have sold that 'looker' anything."

We hear stereotyping used in reference to ethnic groups. A person will say, "They are all the same; I never saw one yet that was any good." This is how he faces up to the fact that he has never been able to understand members of the group to whom he refers. Stereotyping is usually a substitute for knowledge, or is used to hide our own inadequacies from ourselves, or is a cover-up for poor preparation.

This defense mechanism is not confined to salesmen; customers use it also. A customer is shown a new device and says, "There are just too many gimmicks to go wrong." Actually, he doesn't understand the operation of the device, so he refers to its features as gimmicks. Washer-dryer customers who had difficulties with the first models that were turned out, will say, "I've never seen a good combination washer-dryer yet; they are all too complicated. I'll stick to separate washers and dryers." This is his way of saying, "They're all the same!"

Stereotyping is usually indicated by use of broad, sweeping generalities. The cure for this in the selling situation, in some instances, will be to give the customer an easy-to-understand explanation of complicated features. As his understanding increases, his fears diminish. You may find it necessary to use questions to identify the source of his antipathy. Above all, you must not cause him to lose face by pointing out that he is wrong, and you must seek an escape hatch for him in view of the fact that he has taken a strong stand. Such statements as these are sometimes effective, "I can see why you would feel that way; I felt the same way myself until I had a few of the features explained to me."...or..."I don't blame you a bit for having reservations. Let me show you how they have surmounted the difficulties which you and I experienced in earlier models."

Projection

This is a common defense mechanism and one of the most costly if a saleswoman employs it herself as a substitute for pre-sale preparation. Projection takes place when we base our sales presentation on arguments that would appeal to us if we were the customer. We are *not* the customer. A purchasing agent may say, "Our engineers won't like the feature." The purchasing agent doesn't like the feature, so he projects. In other words, he makes the assumption that if he doesn't like the product, the engineers won't.

Sometimes a customer will deliberately use projection to express views which are really his own as the views of other people. For example, a contractor may say, "My electricians won't go for that." Actually he does not like the product himself, but is afraid that if he states this as his opinion he will either be forced to defend his point of view, or he will find himself looking foolish or become involved in an argument. Therefore, he states his views as the views of "his electricians." He feels that he is now in a position where he does not have to explain the reason for their views or defend them. In other words, he can't be attacked.

The salesman cannot let the customer know that he is aware of what is happening. Rather, he must supply the customer with arguments that he can use on "his electricians." In so doing, he has not attacked the customer. If the customer is not using projection, but is actually correct in his statement that the electricians "won't like it," the salesman is still on safe ground because he has given the contractor arguments to use on the electricians.

Rationalization

This common mechanism is present when the customer supplies a false reason in place of his real reason for some action he has taken or intends to take. He uses it frequently to justify some action he has taken which he cannot defend. He is really trying to deceive himself more than he is trying to deceive you. After he has used the same rationalization a number of times, he comes to believe that the false reason is truly the reason for his actions.

We may desire a product, and when we find the price is higher than we can afford, we may say, "You're paying for the name." We resent it when conditions force us to take any action which we do not wish to take. Then we try to find some other reason why we would have taken the action anyway, even if there had been nothing forcing us to do so. In this way, we maintain the image of ourselves that we like to have of being men of free will.

Your customer rationalizes to erase disappointment or frustration or to defend a judgment he has made which produced poor results. We see it used, and use it ourselves, around the home. Someone tells us we have left a light turned on in a room we left an hour ago. Our immediate response may be, "I was just going back in there." And so we were, an hour ago, until something occurred that changed our plans, and we forgot the light. We don't relish being forced to realize that we are forgetful and careless. This conflicts with our image of ourselves as, perhaps, efficient. Therefore, we resort to rationalization.

When our customer uses the same defense mechanism, we feel that she is not telling the truth. Technically, we are correct, but she is not aware of this. She has merely re-structured the situation to make it acceptable to her self-image. Once again, if we admit that we, too, rationalize, it will make it easier for us to understand why the customer does so. Rationalization is the customer's escape hatch; don't make it impossible for him to use it, and don't let him see that you see through this artifice.

Identification

Much of the advertising we see is built around the self-defense mechanism of identification. A popular cigarette is a big seller because in the ads the people shown smoking this brand are obviously very masculine. Many men nourish the belief that they, too, are very masculine, and if our kind of people smoke this cigarette, then it is for us. Or if people in general regard smokers of this cigarette as being "masculine," then if we smoke it, we will be more successful in projecting an image of ourselves as being masculine.

In many metropolitan areas, commuters read the local tabloid at the breakfast table, but carry the local "prestige paper" with them in the commuting trains. Or they carry a nationally known paper which people associate with successful businessmen. These businessmen want to be identified with the group of successful people who are commonly seen reading prestige papers. (If we were to josh them about this, they would quickly rationalize and point out that they read the tabloid at the breakfast table because it doesn't take up much room. And they are in a hurry, so they only have time to read the brief digest that the tabloid makes available. They would add that they have more time to read the news in detail while on the commuting train.) Even though it may be unfair, many tabloids have an image with which these men do not wish to be identified.

Some years ago, Old Gold cigarettes used as a testimonial-type of ad one that said, "The only testimonial we give a hoot about is yours!" Ordinarily, we like to bridge the gap that exists between us and a degree of fame. This ad was designed to make the individual feel that he had bridged the gap on his own without reference to anyone else. It made the individual feel that he was just as important as the movie stars and athletes used by other companies in their cigarette testimonials.

The professional salesman should be quick to note the kinds of people with whom the customer identifies himself, because if the salesman uses as a testimonial some individual with whom the customer does not want to identify himself, disaster will result. One brewery experienced a serious loss in sales because ads showed its beer in a setting of white table cloths, fine linen and silverware. The tavern trade did not want to be identified with that type of atmosphere, and sales plummeted. In selling, try to learn the nature of the image your customer is trying to project, and then, in your sales presentation, identify the customer with people who have successfully projected that image. One clue is available to you if you observe any tendency the customer may have to "name drop."

Displaced aggression or scapegoating

A national magazine once ran a cover that illustrated displaced aggression. There was a series of small pictures on the cover. One showed the boss, who had had a hard day, taking it out on Jones. The next showed Jones, that evening, blasting his wife. Another picture showed Mrs. Jones taking a healthy swing at the kid. The next picture showed the child kicking the dog, and the final picture showed the dog chasing the cat down the street.

The psychologist explains this by pointing out that when an individual has failed to live up to his image, the "me-I-am-trying-to-project," he tries to bring someone else down to his level. A customer who has made an error, or who has been through a demeaning experience, perhaps reprimanded by his boss, may make wild charges about your product or your service, or charge you with having broken some promise he alleges you made to him.

You can usually identify displaced aggression because the situation about which the customer appears to be angry does not warrant the degree of anger he is exhibiting. This does not mean that we can sit back, smugly, and pay no attention to his complaint, assuming that he is simply using us as a scapegoat to vent his ire. However, we should not become extremely upset and start to spend all of our effort on rectifying the situation about which he is complaining. Scapegoating is unfair, and if we let the customer sound off long enough, he will realize he is being unfair and often will subside.

Once again, it often pays to look for some basis for the customer's complaint, express concern over it, and promise to take remedial action. Then try to switch his attention to the current sales proposition. If he is smarting under some unfair treatment he has received from his boss, or from some frustration, he is for the moment feeling insecure or unimportant. Compensate for that by making him feel important. Show respect for his opinions; ask for his opinions or advice.

Apathy

There will be times when you have completely sold your customer on your proposition, but he knows that his hands are tied by higher authority or by his belief that he cannot afford the proposition because his funds are committed to other projects. Because he wants what you have to offer, and wants it badly, he cannot admit to himself that his hands are tied or he will feel the very real pain of frustration. In such instances, he will resort to the self-defense mechanism of apathy. It usually manifests itself via apparent disinterest. He may appear to be totally indifferent to your proposals. Actually he is trying to put your proposition in a file that he maintains for things he wants to forget.

If you misinterpret his apparent apathy as real disinterest, you will lose a possible sale. Your task is to determine the basic reason for his use of apathy. If, for example, it is because he feels he cannot afford your proposition at this point in time, your job is to help him to find a way to buy. If he is apathetic because he feels he can't sell his own boss on your idea, you must supply him with arguments he can use "upstairs." Ordinarily, as is the case in so many other difficult selling situations, the task is to keep him talking and be an alert listener. If he talks long enough, you will identify the real reason for his feigned disinterest, and you can then plan your course of action.

Submission

The last of the self-defense mechanisms which we will identify is submission. When a customer is totally overwhelmed by a problem, he may submit. If his authority to make final buying decisions has been severely limited by a higher authority, he feels that he cannot sustain the image he would like to have of himself as a person of authority; therefore, he restructures himself and provides himself with a new self-image that is attainable. In other words, he will convince himself that for the time being his role is not to be that of a leader, but his role is that of a "good soldier" who follows orders. He convinces himself that this is a praiseworthy image.

This is one of the most difficult selling situations because the customer does not want you to realize that he is without authority, and you cannot permit him to realize that you know his hands are tied. You can't, for example, say, "Why don't you give this folder to your boss, it will overcome his objections." If you were to take this type of approach, he would reply, "I don't have to show your folders to anyone; I make the decisions!"

You have at least two courses of action open to you: (1) supply him with facts that he can use later to change the mind of the real decision-maker, or (2) show him that the cost of submission is greater than the cost of fighting it out. This latter course of action will be successful only if you show him how he will look good. Make him want this so badly that he will be willing to face up later to those who have restricted his actions.

One executive in a power company has had every folder he can collect on a project in his desk drawer for the past three years. He knows that the president of the utility is unsold on the potential of this project. The executive sits by waiting for the boss to resign. This has been allowed to take place simply because the salesmen know what the problem is, but none has made any attempt to help the executive with *his* selling job on his boss.

Summary of self-defense mechanisms

We have been examining the psychological techniques that customers unconsciously use to re-structure the facts of a situation or to re-structure their own self-image. If you took the time, after reading the brief treatment accorded each mechanism, to analyze your own customers, you were probably able to recall having seen each of these mechanisms used.

It is not always possible, or necessary, to identify the specific mechanism which the customer is using. We are not attempting to type or pigeon-hole customers. We *are* trying to learn enough about human behavior to understand why people act as they do. With this knowledge, we will be more understanding of our customers. It will be possible to develop empathy, an ability to understand the reasons for a customer's words and actions. We may continue to disagree with the customer's views or even to dislike his actions, but we will understand why it is psychologically necessary for the customer to act as he does.

When a customer resorts to a self-defense mechanism, it is because he sees some threat to his self-image, and our job is to identify the nature of that threat and neutralize it. This is why "selling" is more than supplying product needs; it also supplies psychic needs. It is also why pre-sale preparation must include far more than intelligence on the product, the market and the competition; it must include intelligence about the customer as an individual.

If we fail to recognize the employment of a self-defense mechanism, we may spend much time and effort barking up the wrong tree; i.e., meeting objections which the customer uses in the role of self-defense techniques instead of taking the time to identify the real problem that has forced him to defend himself with smoke-screen objections or with feigned disinterest. As we review these mechanisms, most of us will probably find that we are thinking not only of the manner in which we have seen customers behave, but we have thought also of our own past behavior. Once we understand our own tendencies, we will be able to keep them under a greater degree of control. And it goes without saying that we will be able to view our own past actions with some degree of humor and probably take ourselves less seriously. Certainly, we will not give in to the luxury (which we can't afford) of permitting ourselves to become annoyed with a customer who is probably doing nothing more than utilizing mechanisms which we, ourselves, have often used.

Perhaps in a seller's market, when demand is great and supply is short, when competition is limited or nonexistent, "anyone can sell." In the marketplace in which we customarily find ourselves, it takes a professional to sell. If we set our threshold of satisfaction too low, we can sell without giving much consideration to the psychological aspects of selling. We can content ourselves with a "fair" percentage of each customer's potential business. We can avoid the "hard-to-get" customers and keep calling on the reliable, old stalwarts. This comes perilously close to "peddling," and though some salesmen may jocularly refer to themselves as peddlers, down deep inside they feel that they are more than that.

Selling is without a doubt one of the most demanding professions; and to sell in a marketplace that is growing increasingly complex, dynamic and demanding, it is going to be essential that we use the tools of the professional. Understanding the psychological factors present in a selling situation and making them work for you instead of against you is one of those tools. We move next to a discussion of another professional tool, the Estimate of the Selling Situation...a tool that has, and will, produce measurable results.

CHAPTER 6

Making an estimate of the selling situation

Operating a business is, in a sense, engaging in a kind of warfare. You and your competitors seek what each of you feels is a fair share of the market. When the size of the market is restricted, there may not be enough business available to provide each competitor with what he feels *is* a fair share. To the degree to which you or your competitor fails to obtain a fair share of the market while making a reasonable profit, your economic welfare—even survival—and his are affected. Annually, thousands of businesses fail, and the owners of those businesses feel that they were engaged in a kind of war in which their business strategies and marketing tactics fell short of the mark.

In any war, there is the need to gather intelligence, which we will define as "information made meaningful." There is a need to identify the method of operation customarily employed by the enemy and even the "MO" usually employed by each opposing commander. Failure to gather such intelligence prior to planning strategy and tactics can spell disaster. Similarly, failure of a salesman to gather intelligence on his product, the competitive product, his competitors and on his customer can spell lost sales.

Successful salesmen have often been queried in order to determine the reason for their success, and they frequently provide these answers: "We have had good products to sell"; "I know my products and my customers"; "I developed a strong, personal relationship with my customers"; "We have a reputation for honesty, integrity and dependability"; "We don't lose interest in the customer after the sale"; "We provide good service"; and, "My company has a good image in the marketplace."

Some salesmen tell you that they work by hunch or intuition; others say they have developed a sensitivity toward their customer; yet most of them will readily admit that they have lost sales because something "unpredictable" happened during the course of the sale. If pressed, they will usually agree that the "unpredictable" could have been anticipated and alternate courses of action prepared.

Here are but a few of the conditions that can arise during a sale:

1. Unexpected competition enters the picture, or old competition strengthens.
2. Someone else in our firm makes an error and causes trouble for us.
3. A strike occurs in our firm or in the customer's.
4. The customer makes a change in personnel in his buying organization.
5. The customer's buying position suddenly changes.
6. The customer has a bad experience with our last delivery.
7. The customer raises an objection he has never raised before.
8. Our firm receives some bad publicity.
9. A new product or feature reduces the appeal of our product or service.

This is but a small segment of the list of similar problems which you could develop from your experience. Perhaps we cannot predict that any of these problems will arise in our next sale, but we can almost always consider the possibility that they will arise. We are more likely to do this when we have a personal selling plan for each of our customers. Having a plan saves moments of embarrassment. It reduces the need for snap, on-the-spot decisions. It reduces the chance that we will be caught without answers to questions that are raised. It provides us with alternative courses of action. And it improves our percentage of successful closes.

Wars are usually not lost because of a lack of courage on the part of combatants; they are usually lost because of incomplete planning, breakdowns in logistics (supply planning), and underestimating the competence of the enemy. Consequently, planning becomes a must in the military. Planning goes on in the military during periods of peace—between wars—because over-all planning is usually more thorough and of higher quality when the planner has the luxury of time and when emotional handicaps are not present. For the same reason, planning a sale before one enters the actual selling situation will produce strategies and tactics that have a better chance of success than the kind of counter-punching that becomes necessary when we enter a selling situation, depending solely on our ability to size up the situation of the moment and take action.

Because the military feel they have no choice, the armies of the Western World for years have had what is called "the estimate of the situation," or as the British and Canadians put it, "the appreciation of the situation." Here is a short paragraph out of the military manual that introduces the need for the estimate: "The situations that confront a commander in war are of infinite variety. In spite of the most careful planning and anticipation, unexpected obstacles, frictions and mistakes are common occurrences in battle. A commander must school himself to regard these events as commonplace and not permit them to frustrate him in the accomplishment of his mission."

Let us re-write that paragraph for application to professional salesmanship: "The situations that confront a salesman in a selling situation are of infinite variety. In spite of the most careful planning and anticipation, unexpected obstacles, frictions and mistakes are common occurrences. A salesman must school himself to regard these events as commonplace and not permit them to frustrate him in the successful closing of a sale."

A religious group in the 16th century listed these as steps to be followed by its advocates in their attempts to convert others: contemplate, contact, gain confidence, establish conviction, work for conversion and work for continuance. The estimate of the selling situation is the device by which we meet the demands of the first step, "contemplate"; know what you are going to say and do, and why.

Salesmen often resist this advice with the comment, "You don't always have time to plan in advance, let alone have a written plan. You often have to meet situations as they arise, and there is no chance to do any advance thinking." This is a fair complaint and carries with it the implied question, "So what do you do about that kind of situation?" The answer is found in the military manual: "The estimate should be as thorough as the time available will permit; it thus may vary from a short, almost instantaneous, mental estimate to a carefully written document, requiring hours of preparation and the collaboration of many people."

Pre-sale planning

Any good salesman spends much time thinking in advance of the selling situation which he is about to enter. He may find himself thinking about the customer while he is shaving or driving to work. Salesmen find themselves contemplating the various unpleasant turns a selling situation may take and considering what courses of action are open to them. But consider the advantages of a clear-cut pattern of thinking and planning, a pattern that is professional and scientific in its approach to problem-solving, a pattern that leaves less to chance, a pattern that provides us with a checklist of possible competitor and customer actions, and a pattern that provides us with a battery of actions that will be on tap to help us meet these situations as they arise. Such a pattern is available to you in the estimate of the selling situation.

Step I: Consider customer's objectives and goals

In a military estimate, the first step is to consider your mission, but the military planner is not interested in aiding his opponent accomplish *his* mission. If we are customer-oriented marketing people we are interested in aiding our customer in accomplishing his objectives and goals. Therefore, identifying and considering the customer's objectives is our first step. Note the use of the words "identify" and "consider"; we did not say "accept" his objectives and goals. You may find that in your opinion, based on facts which you have and which the customer lacks, that his objectives and goals are wrong for him. He may, for example, want to operate at the lowest cost possible and may want to buy equipment that will not stand up under the heavy usage to which it will be subjected. In his attempts to satisfy the needs of his self-image—his desire for recognition—he may wish to grow at a faster rate than will be healthy for his business or at a rate faster than his financial resources will support.

Generally speaking, however, we are in business to help him attain his long-range objectives and his short-range goals, and this includes his business and his personal desires. If we identify these, we will be in a better position to base our sales presentation on a basis that the customer will find possible to accept. If we ignore his objectives and base our presentation on the need to satisfy our own objectives, we will find him resistant and unresponsive, unless, luckily, our objectives and his are compatible.

By identifying the customer's objectives, we are well on the way to identifying his problems and his needs. This is a sharp departure from the old school that listed "Get his attention" as step one of a sale. This is now replaced with "Give your attention to the customer and his problems and needs, which it is your business to solve and supply."

Step II: Consider your objectives; if multiple, set priorities

Note the use of the word "objectives" rather than "goals." For purposes of making an estimate, we think of objectives as long-range, broader than goals, less flexible, and as the ultimate purpose of our work. We think of goals as being short-range and as benchmarks which we use to measure our progress toward the attainment of our long-range objective.

We might have as an objective the determination to become the major supplier for a contractor in five years. Our goal might be to get 25% of his business this year and to increase our absorption of his business at a specific rate each year. What amounts to something close to a fixed objective in selling would be to maintain a lasting, favorable relationship with the customer. It is essential that we have a clear picture in mind of what it is that we want to have accomplished in our relationship with this specific customer. This enables us to avoid very attractive, short-range goals that might bring us immediate profit and business and which might threaten or ruin our long-range relationship with him.

Often you will find that you have more than one objective. You may wish to increase your share of the customer's business in a given line of products. You may wish to become his sole supplier in some area. Or you may wish to interest him, eventually, in something close to an entirely new field of endeavor where you feel he would do well, and from which you would benefit tremendously. It becomes essential that you establish priorities for each of these objectives. In doing so, it is necessary that you consider the impact of one objective on the other. By achieving one objective, do you threaten your chances of attaining an even more important objective? Which is most important to you? Which would you be willing to sacrifice?

Step III: Compare your objectives with the customer's

This step will often point up the basic nature of the selling problem which you face. You may often find situations where your objectives and those of the customer are in conflict. Instinctively we seek ways of changing the customer's objectives, but this is not customer-oriented selling. Our first step should be to re-examine our own objectives: Can and should they be modified? If not, how can we cause the customer to change his objectives? Finally, Step III causes us to ask whether we should consider the advisability of continuing to seek this customer's business or to seek this particular order.

Step IV: Consider your goals, short-range and today's specific goal

Do your goals conflict with one another or with your objective? Do you have a clear picture of what it is that you seek to accomplish in the selling situation that you are about to enter? Will it advance you toward attainment of your long-range objective in terms of what it is that you hope to accomplish over the long pull with this specific customer? Will it affect relationships with other customers? If these relationships will be affected adversely, what is more important to you—today's customer or the others?

Step V: Gather intelligence

In the military estimate we find what are referred to as "Essential Elements of Intelligence" (EEI). The Navy's list is a long one and embraces such items as: the general situation, political factors, economic factors, psychological factors, and fixed factors such as: hydrography, terrain and topography, time and space factors, and so on and on.

You and I are not trying to launch a naval mission, but ask yourself this question: "If you were one of the men asked to take part in such an attack, would you feel secure if you knew that the essential elements of naval intelligence had *not* been listed, and if you knew that information had not been obtained on many of the items on the list?" Except for the sales that have been obtained from regular customers upon whom the saleswoman has called over a period of years, how can any saleswoman feel secure unless she has considered the essential elements of selling intelligence and has obtained information on as many of them as possible?

The major categories of sales intelligence are general in nature: (1) intelligence on your products and on the competitive offerings; (2) intelligence on your competitor and his people; (3) intelligence on economic factors which affect the customer's business; (4) intelligence on psychological factors that affect your customer and his people as individuals.

The detailed elements of essential intelligence are *not* general and can include as many as 200 questions. Faced with this statement, you may find yourself asking, "Does anyone really expect me to obtain all of these answers before I approach the customer or close the order?" or "Why have I been fairly successful in the past without checking up on all of these elements?"

These are reasonable questions and deserve to be answered. It would be preposterous to believe that a salesman could obtain answers to all of the questions that can be listed in a thorough estimate of the selling situation. It would be equally preposterous to claim that he should do so in the case of customers whom he now sells to with little difficulty. There are other times when fast action must be taken, when gambles are necessary, when even bluff may be necessary, when to pause long enough to obtain all the answers would lose the sale to some competitor who was alert to the imperative nature of the customer's needs.

Some of the information must be gathered over a long period of time. The professional salesman will find himself building up a private dossier on each major customer, not overlooking customers who, though small today, have real potential. This is to keep in mind that, generally, small companies are the growth companies. There is no doubt but that alert competition is trying to collect the same information on you, your products, the marketplace and on your customer.

Although you should have a printed list of all of the essential elements of sales intelligence available at all times, the importance of items on your checklist will vary from one situation to another. For each sale, you should challenge each item with the question, "Is this something I must know, I should know, it would be well to know, or it would be nice to know?" Consistent with the demands on your time, you will have to decide how far down that list of priorities you wish to go. You will also have to guard against the human tendency to rationalize by pushing some item down into the "nice to know" category, simply because you know it will take real work to get the answers.

This chapter will close with a typical list of essential elements of sales intelligence. It is intended only as a sample. You must eliminate factors that are not pertinent to your kind of selling and add other factors that occur to you which you regard as of prime importance. Then, beside each element, ask yourself, "Does this information I have gained represent a real advantage, a possible advantage, a real disadvantage, or a possible disadvantage to me?"

Step VI: Analyze the intelligence

No writing is necessary, though you may find it helpful, especially in complicated selling situations. This analysis points up areas of weakness where we must be prepared with answers, justifications or alternatives. Such an analysis leaves less to chance, less to memory, and we overlook fewer points. Analysis of available intelligence helps us face up to problems that we might otherwise try to repress by saying, "I'll handle that somehow if it arises." This could well be the epitaph on the tombstone above many lost sales.

Step VII: Develop a sales strategy

In this step, we consider the long-range actions open to us. It includes the following key questions:

Have I identified my major problem? What are my objectives?

What are my competitor's alternatives to my strategy?

What are my customer's alternatives?

What will be the possible reactions of my customer and my competition to each of my alternatives?

Which, then, of the alternatives open to me is most promising in terms of the accomplishment of my long-range objective?

Have I avoided attractive actions that may accomplish short-range goals, but which may threaten attainment of more important, longer-range objectives?

Step VIII: Plan opening tactics

Again, it will be necessary to consider reactions of customer and competitor to your proposed tactical opening, as well as to consider their alternatives. Your opening tactics should exploit your advantages, anticipate, meet and neutralize your disadvantages. Opening tactics will vary with each selling situation, so they will not be discussed here, but they should be planned and not left to on-the-spot impulses.

Summary

No claim is being made that analysis of the customer as an individual will be easy. No claim is made that making an estimate of the selling situation will be easy. There is nothing easy about selling. It is an exciting and challenging profession and one in which top success is assured only when the professional salesman is aware of his strengths and equally aware of the difficulties that may confront him in the selling situation. Top success is possible only when we recognize each sale as a distinctively different situation, in some respect, from all those in which we have been engaged in the past. It will be similar to other sales in some respects, but as long as we are dealing with different people, or even with the same people in different situations, each situation will require analysis.

Although the step of gathering intelligence will be recognized as the most time-consuming and of great importance, it cannot be said that it is more important than the others. It would be futile to gather all of the possible intelligence on products, customers and competitors and then have ill-chosen or poorly considered objectives. Use of the psychological principles discussed in this section, coupled with habitual use of the estimate of the selling situation, will spell the difference between selling by hunch with loss of some sales and selling professionally—and, over the long span, establishing a reputation for closing more sales with profitable results. Above all, it will inject more excitement and greater challenge into salesmanship and provide the professional salesman with a rewarding inner satisfaction that comes from the knowledge that it takes a particular kind of skill and a particular kind of person to do the kind of work he has chosen.

Essential elements of intelligence

There are many precedents to support the thesis that a salesman should be fully informed before he enters the selling situation—no matter how long he has been selling, no matter how skilled he may be. A pilot may have thousands of hours of flying time behind him and he may know his plane like the back of his hand, but he would not think of taking off or landing without running through his checklist. In naval and military combat, commanders check out the situation against their kind of checklist; i.e., the estimate of the situation. A portion of that estimate, as we have seen, contains the essential elements of intelligence.

Admittedly, no commander in combat can always have complete information at his disposal, but when, in his quest for answers, he realizes that some of the information is not available, he knows in which areas he is running at risk. Often he finds it necessary to have alternate plans of action ready. The claim is not made here that a salesman must *always* have all of the answers. Sometimes, in dealing with established customers, it may not be necessary to make a complete estimate of the situation, though we have seen the difficulties that can arise when a salesman takes the old customer for granted and his competition "cases" the customer thoroughly.

There will be occasions when it would be foolhardy to wait for full information. Situations arise that call for speedy action or risk a quick maneuver on the part of an alert competitor. These, however, are the exceptions, and salesmen must resist the tendency to rationalize themselves out of getting complete information.

Even the best salesman misses an order now and then—perhaps more often than he realizes. Some customers do "get away"; yet if they have needs and problems, someone sells them. It is rare indeed to find "out-sold" on a lost sales report. It is more common to find the lost sale attributed to price, insufficient breadth of line, a personal relationship between the customer and the competition, or the like. A thorough analysis of a lost sale will often indicate that the fortunate competitor actually did her homework and was better informed on the essential elements of intelligence regarding the selling situation than was the fellow who lost the order.

In any competitive market, you can be certain that you have some competitor who is trying to obtain full information on you and on the customer. If he is more successful in his attempts to get this information than you are, he will be that much ahead of you when you both enter the selling situation.

The checklist that follows is not intended to be complete. It is intended to be a list of "thought-starters." You should add to it the specific kinds of information that you require in dealing with your specific customers and competitors. As a starting point, you should break the list down into four basic categories. Classify each question in terms of how important it will be to you to become fully informed on that question. A suggested classification would identify questions as:

1. This information I *must* have.
2. This information I *should* have.
3. This information would be *helpful* to have.
4. This information would be *nice* to have.

Then, because time always seems to be short, concentrate on obtaining answers to all of the category (1) questions first, and work down the list as your time permits. We repeat for emphasis: Guard against the very human, and very common, tendency to classify hard-to-get information as category (4) ("nice to know") information. This is not to say that all of the "must know" information is hard to get, but when facing a difficult task, it is a normal human tendency to rationalize and to procrastinate. This is a surefire road to lost sales.

When you have the information, make a record of it—one record of the customer and another record of the competitor, so the next time you will have it available, leaving you with more time to plan your strategy. Again, as you study the following list of questions, delete those that never apply to your kind of selling—but be sure you are right. Then add additional questions. One of the best ways of accomplishing this is to consider sales that you have made and lost and identify the kind of information that you had or failed to have that brought you success or cost you the sale.

Competition

Who, specifically, represents your competitor? What do you know about his modus operandi? Is he ethical? What are his weaknesses and strengths?

What does he have in common with your customer? Does your customer sell to *his* firm; i.e., are you possibly up against reciprocal buying? Does your customer feel under an obligation to your competition, either to the representative of the competitor or to the company?

How does your competitor's image in the market compare to yours? Are his products or services in demand, or is his financial position so serious that he will reduce his prices or sell at prices that will simply meet his overhead?

How badly does your competitor want this order? Is it an order that he considers a "pilot purchase" that will lead to much more business?

Does your competitor have additional services which he can supply your customer over and above the product or service which the customer is buying and paying for? Can you match these or offer other "bonus services" which your competitor cannot match?

How does your competitor compare with you in his ability to supply effective post-sales service?

Who got this customer's last order? Why...honestly now, why?

Where has your competitor fallen down in the past?

How do you and your competitor compare in these areas:

1. Distribution facilities and ability to deliver on time?
2. Research, catalogs, bulletins and data sheets, industry position, patent edge or features, relationships with subcontractors?

Who can see the customer more often? Who can give ready answers to the customer; do either you or your competitor have to lose time by checking back with some other authority?

Does he have an edge on you because he sells across the board while you are limited to one product or one line? Do you have this edge on *him*?

Do either of you compete with the customer?

Does your competitor have other customers whom this customer regards highly; if so, will they be used as testimonials...or do *you* have this edge?

Is there any information your competitor can obtain more readily than you; if so, what kind of a problem can this create for you?

Is your competitor trying to capitalize on a past reputation or on the strength of other products and services he offers?

What family, ethnic, recreational, fraternal, social, trade or financial relationships are involved?

Customer

How does your customer want other people to see him (the me-I-try-to-project); can you help him accomplish this?

What kind of a business is she trying to run (i.e., what are her long-range business objectives and short-range goals for this year); can you help her?

What are his basic product or service needs?

What appeals will be most effective with him—appeals that are based on helping him make a fast profit, appeals based on longer-range security and growth, or appeals based on his self-image needs (need for prestige, dominance, need to be accepted, respected or imitated, etc.)?

What problems do *his* customers create for him? Can you help him avoid or solve any of these?

Do you know any of her people? How do you stand with them? Do you deal only with the key decision-maker and satisfy yourself by being only "polite" with the other people in her organization? Who are the people who influence the key decision-maker?

If your contact in the customer's organization is not the final decision-maker, what ammunition can you supply that your contact can use effectively?

Does your customer's long-range buying potential justify compromises you should make?

Does your customer indicate a dislike for your firm? Have you asked him why? (Give him a chance to get it out of his system.)

Can you use any of his products or services in your company? Have you done so?

Can your customer obtain federal financial assistance? Can you help him obtain it? Is he in a heavily regulated business; can you help him overcome problems which this regulation generates?

Can you provide him with assistance in any of these areas: storage, accounting, research, marketing, personnel, financing *his* customers, delivery, space, loan of equipment or materials, management aid? If not, what problems does he have with which you can help him, other than the basic problem your normal product or service would satisfy?

Do you really feel you know this man and know his primary psychological needs? If so, just how do you plan on satisfying those needs?

Remember, these are just a start; add to the list from your experience.

Customer's self-image

What car(s) does she drive (why?); What kind of home does he have and in what kind of a neighborhood; where does he send his children to school; what club(s) does he belong to? What is his favorite topic of conversation? Where does she have lunch (any significance to the kind of restaurants she chooses)?

What kind of clothes does he wear? What does he display in his office? Where did he go to school? Is he a self-made man with limited formal education? If so, is he proud of it? What section of the country did he come from originally? Does his wife influence his business decisions?

Does she have any weaknesses that she is trying to overcome?

Was he at any time in the armed forces? Does he talk about this?

Does he discuss his background; if so, upon what does he concentrate?

Is he satisfied in his present job, or is he impatient to move ahead?

What is his functional background (manufacturing, construction, finance, etc.)? What kind of arguments have the greatest appeal to him?

Is he technically strong or weak? What must you concentrate upon in talking with him and what must you avoid? What are his hobbies? What needs do they satisfy?

Does he have any prejudices; if so, how strongly does he feel about them? What is the basic reason for them... does it reveal something that he lacks or envies... something you can supply?

Does he feel appreciated in his own organization? If not, how can you help him gain recognition? Is he ambitious? If so, for what? What seems to drive him?

Again, these are starters; add to them from previous experiences you have had where knowledge of similar factors helped you make a sale.

Product

No attempt will be made in this section to supply the long list of questions that would be necessary to cover the diversified product fields represented by the readership. A few general, yet very important, questions will be posed, and each salesman should add many additional questions to this list:

In what respects does your competitor have a product or feature edge?

Where do you have the edge in this regard?

What hidden values does your product have that should be accented?

What features do you anticipate your competitor will push? How can you offset the effect of his arguments?

What bad features does your competitor's product possess? How can these be brought to the attention of the customer without "running down" your competitor?

Are foreign-made products involved? How does your customer look upon these? Is it an edge for you or a problem?

What product features does your competitor have that you cannot match? Where does your product excel in this regard? Which will be of greater interest to your customer?

Note: In dealing with this facet of the essential elements of intelligence, one company listed 75 questions that a salesman should consider before a sale. The list above obviously should be added to, denoting a comparison of your product to your competitor's in terms of such things as cost, service, availability, delivery, financing, transportability, compatibility with his present equipment, simplicity of operation, buying terms, size, weight, color, capacity, reliability, safety considerations, shock-proof considerations, guarantees, choice, difficulty of installation, possible early obsolescence, cost of operation, attitude of labor unions to product, etc.

A weakness in any of these areas, indeed in any *one* of these and many other areas, can spell a lost sale if the saleswoman has not considered it and is not prepared to offset that weakness in her sales presentation.

Remember, the successful individuals in any of the recognized professions don't act until they are informed, until they have analyzed the available information, considered the obstacles and planned. The professional salesman will do no less. The payoff is greater pride in being a member of the profession and in making sales that "get away" from the fellow who depends solely on hunch, intuition and luck.

The customer profile—personal and major account profiles

You are now being introduced to two valuable selling aids. As you examine them, certain objections may occur to you—just as your customers object to some parts of your proposals. So, let's examine those objections.

1. *"I can never obtain all of that information on any of the customers I know."* We agree that this may be the case. The point is, on how many of the major points included on the forms are you now uninformed? If your competitor can answer more of these questions than you can, does she have an advantage over you? The answer is an unequivocal "Yes!" Don't let her... you have the forms and she does not.

2. *"How do I obtain all of this information—I just can't ask the customer to complete the form for me."* Right again, but if you will re-read this chapter on Making an Estimate of the Selling Situation you will find much information on the subject of how to

gather commercial intelligence...which is what this is all about. Each time you visit the customer try to learn something new about him. Get him talking and listen. You are reminded again of the need to listen with your eyes as well...observe the objects she displays in her office or home and ask yourself what those possessions so obviously displayed are telling you about how she wants to be seen.

3. *"Do I have a right to seek personal information about the customer...isn't this prying into his personal life?"* The answer this time is an unequivocal "No!" As has been emphasized in previous chapters, your customer is a whole person. He not only has product and economic needs but he has psychic needs which, at times, can be the most important of all to him.

Look at it this way. You see nothing wrong with getting all of the information possible on his financial position...his credit rating, his cash flow of the moment. You see nothing immoral about seeking a Dun and Bradstreet rating or a Retail Credit Bureau rating. You would be considered foolhardy if you issued credit without such advance checking. Yet, what can be more "personal" to the customer than his financial situation?

You see nothing immoral about trying to identify the customer's operating problems, even if it means getting around him by seeking information from other people in his organization when the customer, himself, pretends that all is well and he has no problems.

You defend those actions by pointing out that you are in the business of helping the customer either solve or identify problems and you cannot possibly help him without the facts you seek. The customer has these psychic needs also, and she is usually even more determined to hide them from you and from everyone else for fear of appearing to be "socially unacceptable" as the psychologist puts it. How can you possibly identify his all-important psychic needs (what image he has of himself and is trying to project to others) unless you can obtain the kinds of facts you will need in order to make this determination?

Whether or not it is "immoral" or an invasion of the customer's privacy is determined by the motive underlying this quest for personal information. Certainly if the intent is to manipulate the customer, then the procedure is undeniably immoral. But if the motive is to help him and to motivate him to recognize and solve his problems, then the procedure is highly moral...and here is why:

Your competitor is probably focusing on the customer's product and economic needs and giving only superficial consideration to his psychic needs. In other words, as an individual, the customer is being treated as "average."

When you assemble information on the customer using the forms provided here, you are treating the customer as a unique individual. You are recognizing that he may have the same, or comparable, product needs and economic limitations as other customers but you are dignifying him by identifying and attempting to satisfy his personal needs as well. This is truly professional and moral selling.

4. *"How long does it take to assemble all of this information?"* There is no set answer to this question. In some instances it may take months as you pick up small bits of intelligence on each call. In other instances you may have customers who are prone to talk about themselves and their interests, and the information you are seeking pours forth.

It is suggested that each time you leave the customer, you note in a small pocket pad what new information you have obtained. Then, at the end of the day, fill in more blanks on the Customer Profile or the Major Account Profile. Before each successive call on the customer, look these sheets over to refresh your memory. Obviously, you do not bring the forms with you and you keep them at home in order to protect the customer's right to privacy.

Recently we called on a major company and as we walked in the main entrance a young sales engineer cornered us and asked us to come to his office. There he showed us that he had one of each of the forms made out for each of his accounts. We were delighted but as surprised as well as pleased, and we said, "You really did it, didn't you?"

His answer was, "I remembered something you said. You said you had seen more salesmen passed by for promotion for a single reason than for all other reasons combined." Then he went on to repeat an example we had given, to wit: A couple of managers get together to decide which salesman should be promoted to the job of sales manager and this is the kind of conversation that may take place (Martin and Welt are talking):

(Martin) "Abercrombie is our best man and he should get the job." (Welt) "I agree but remember he knows his area and every customer in it like a book. Our sales will take a terrible drop while we are training a replacement. We just don't have anyone to take his place." (Martin) "That's right...um...let's see now...Anderson is o.k....not as good as Abercrombie, but we could train a replacement pretty fast and not lose as much business as we would if we promoted Abercrombie."

And so Mr. Anderson gets the promotion that should have gone to Abercrombie because the latter had held close to his chest all of the information he had gathered on his customers and accounts. The young chap who was showing us his impressive file of completed forms went on to say, "I've told my boss that I never want to hear that I have been passed by for promotion because there was no one to take my place. I told him that I had gathered information that would enable him to break in a new salesman in one-tenth the time it had taken me to learn. Furthermore, I have given him copies of the information in my file.

As we have emphasized repeatedly, in using the information you have accumulated, don't judge by one or two facts... look for a pattern in the customer's behavior. Use the Major Account Profile to keep your own manager informed on the status of each account and help him to predict the future actions of that account.

You and I would be more than annoyed if we thought our dentist or our doctor did not keep complete records on our past history and that they had to start "from scratch" every time we found it necessary to see them. They conduct their business as the professionals that most of them are. They keep detailed and accurate records on each of their "accounts."

You are, in many respects, in an even more demanding business than either the doctor or the dentist. As a starter, they have a huge advantage over you. As a patient you don't dare to give them false answers to the questions they ask you. You, as a customer in this case, are a veritable fount of information, available only for the asking by the doctor or dentist.

Then they can obtain even more information by employing the battery of machines and testing devices that are available to them. Compare that to your situation. Often, you can't even get the customer to answer a question, much less get an honest and complete answer. He will often hide his economic limitations as well as his psychic needs from you... and you don't have any machines or laboratory equipment to aid you in your search for essential information if you are going to help the customer. Yours is either the only or one of the very few professions, where the person you are trying to help will often resist your efforts to do so.

This is all the more reason why you should start and maintain a file on each customer and account from which you can retrieve the facts that it took so much effort to obtain. This is another practice that separates the professional from the order taker.

A final point on this subject... it will be obvious to many readers that some items on the forms must be deleted or changed to apply to their type of selling and to their market.

In other instances you will find the need for kinds of information not listed on either of the forms. Make whatever changes are required in either form to make them apply to your business and your requirements. The forms are provided in this volume to show the kinds of information that are required in many sales activities and to indicate one type of format that may be found useful.

The Customer Profile—Personal

1. Customer's full name _____
2. Home address _____ 3. Phone _____
4. Company's name and address _____
5. Customer's nickname _____ How did he get it? _____
   How does he like it? _____
6. Birthday (and year) _____ 7. Where born? _____
8. What does he consider home? _____
9. Military service _____ How did he regard it? _____
10. Education: High School _____ College _____
    How does he regard his college? _____
    Any college fraternity? _____ College sports _____
    Other extra-curricular college activities _____
    Any collegiate honors? _____
    Graduated when? _____ Degrees _____
    If he didn't attend college, is he sensitive about it? _____
11. Previous employment:
    Company _____   Company _____
    Location _____ Location _____
    Dates _____   Dates _____
    Title _____   Title _____
    Company _____   Company _____
    Location _____ Location _____
    Dates _____   Dates _____
    Title _____   Title _____
12. Previous employment in present company:
    Locations: _____
    Dates: _____
    Titles: _____
13. Professional or trade associations _____
    Offices or honors in them _____
    How active is he in any of them? _____
14. Clubs or fraternal associations (Masons, K of C, Elks, etc.) _____
15. Service clubs (Rotary, Kiwanis, Lions, etc.) _____
16. Religion _____ Active _____
17. Highly confidential items not to be discussed with him (divorced, member of AA, etc.) _____
18. Politically active _____ Party _____ Important to him? _____
19. On what subjects (outside of business) does he have strong feelings? _____
20. Medical history (current condition of health) _____
21. Does he drink? _____ What does he drink? _____
    Does he object if others drink in his presence? _____
22. Does he smoke? _____ What does he smoke? _____
    Does he object if you smoke in his office? _____
23. Where does he like to eat? _____
    Why does he choose that place? _____
    What is his favorite item on the menu? _____
    Does he object to having anyone buy his meal? _____
24. Active in his community? _____ How? _____
25. Hobbies: Collecting? _____ What? _____
    Gardening? _____ Work shop? _____
    What kind? _____ Ham radio? _____
    Other _____
26. Recreational interests: _____
    Vacation habits: _____
27. Spectator sports interest? Sport _____
    League _____ Team _____

28. What kind of car(s) does he drive? _____
    Is it important to him? _____
29. Conversational interest: (What does he talk about when given free rein?) _____

30. With what type of people does he like to associate? _____
    Whom does he seem anxious to impress? _____
    How does he want to be seen by those people? _____
31. What adjective would you use to describe him? _____
32. What is he most proud of having achieved? _____
33. What business relationship does he have with others in your company? _____
    Who are they? _____
    Is it a good or poor relationship? _____ Why? _____
34. What other people (in your company) know him away from business? _____
    Type of connection (club, church, etc.) _____
    Nature of relationship (close, casual, clash) _____
35. Wife's name _____ Education _____
    Interests _____
36. Children's names and ages _____
    Any facts on children? (Achievement, problems, health, etc.) _____
    Does he send children to private schools? _____
    If so, why? _____
37. Any "status" symbols in his office? _____
38. What kind of home does he have? _____
    Neighborhood _____
39. Height _____ Weight _____ Age _____
40. What do you feel is his long-range personal objective? _____
41. What do you feel is his short-range (immediate) personal goal? _____
42. What do you feel is his long-range business objective? _____
43. What do you feel is his immediate business goal? _____
44. At this time which do you think is of greatest concern to him—the welfare of his company or his own? _____
    Is he thinking of the present or the future? _____
45. What moral or ethical considerations are involved when you work with him? _____
46. What recent experience has he had with you, your company, or your competitor that will be uppermost in his mind when you call upon him? _____

47. Does he feel he has any obligations to you, your company, or to your competitor? _____
    What are they? _____
48. Does the proposal you plan to make require him to change a habit or to take an action that is contrary to custom? _____
49. Is he primarily concerned about the opinion of others? _____
    Or is he self-centered? _____ Or is he highly ethical? _____
50. What are his key problems, as he sees them? _____
    Can you help him with these problems? _____ How? _____
51. Do you have equally extensive information on his economic problems, his product/service needs, and on your competition? _____
    Does your competitor have more answers to the above questions than you have? _____

Major Account Profile

(Use separate sheets where necessary to reply at length to any question.)

1. Account name _____
2. Address _____
3. Phone(s) _____
4. Key influences on the buying decision and attitude toward us: F—favorable; U—unfavorable; N—neutral

Name	Title	Attitude
_____	_____	_____
_____	_____	_____
_____	_____	_____
_____	_____	_____
_____	_____	_____

5. Primary contact: _____
6. Does the account purchase from other components of our company?   Yes _____   No _____
   If so, who in our company makes the selling contact? _____
7. Do any of the buying influences in the account have personal relationships with personnel of our company?
   Yes _____   No _____
   If so, who are those influences? _____
8. With whom, in our company, do these buying influences have these relationships? _____
9. Type of business engaged in by the account: _____
10. Their estimated annual purchases (from all sources) of the product we sell them: _____
11. Of the amount listed above, how much have we sold them, annually, for each of the past five years?
    19___   19___   19___   19___   19___
12. How much do you estimate we can sell them this year? _____
13. What do you feel will be our greatest obstacle to achieving this goal? _____
14. Have we had problems with this account?   Yes _____   No _____
15. If so, what has been the nature of the problem (or problems) and how recently have the problems occurred?

Type of Problem	Approximate Date	Resolved S—Satisfactory (to them) U—Unsatisfactory
Delivery _____	_____	_____
Billing errors _____	_____	_____
Slow payment _____	_____	_____
Quality _____	_____	_____
Shipment incorrect _____	_____	_____
Post-sales service _____	_____	_____
Price changes _____	_____	_____
Competition reduced price _____	_____	_____
Legal problems _____	_____	_____
Personality conflicts _____	_____	_____
(If so, between _____ and _____)		
Other (specify): _____	_____	_____

16. Do they use a Buying Committee?   Yes____   No____   Does the primary buying contact require approvals?
    Yes____   No____   If so, from whom? _____
17. What is their attitude toward being supplied entertainment? Check appropriate answer:
    Demand it _____ Expect it _____ Accept it _____ Opposed to it _____
18. What type of entertainment is most highly regarded by primary contact? _____
    _____
19. What is their attitude toward receipt of gifts? Check appropriate answer:
    Expect gifts _____ Accept gifts _____ Have company policy that prohibits gifts _____
    Is it enforced?   Yes____   No____   Does this include such items as pens, calendars, diaries, and the
    like?   Yes____   No____
20. Do they have problems upon which we are competent to provide assistance?   Engineering _____
    Packaging _____ Office or plant methods _____ Finance _____
    Other (specify): _____
    _____
    _____
21. What is the key determinant in their buying decision?
    Quality of product or service offer _____   Other (specify) _____
    Speedy shipment _____   _____
    Discount offered _____   _____
    Price _____
    Post-sales service _____
22. What evidence supports your answer to the previous question? _____
    _____
    _____
    _____
23. In what areas of our performance do you feel we can do a better job? _____
    _____
    _____
24. How do you regard this?   Urgent _____ Needs prompt attention _____ Would be helpful _____
25. Do you feel that this account now requires a visit or visits from executives of our company? Yes_____ No_____
    If yes, why? _____
    _____
    _____
26. If answer above was Yes, whom do you feel should call upon them? _____
27. Does the customer have plans for expansion? Yes_____ No_____ If so, how much additional business can it
    represent, potentially, for us? _____
28. If answer above was Yes, when will the impact of the expansion be felt by our company? _____
    _____
29. If the expansion is to take place outside the area you serve, have our personnel in that area been notified?
    Yes_____   No_____   When? _____ Who notified whom? _____
    _____
30. Does the customer plan on becoming part of a merger or acquiring other companies? Yes_____ No_____
    If yes, specify: _____
    _____
31. Does the customer plan on adding new products to his line? Yes_____ No_____ If yes, what are they, and do
    they represent additional business to you or other components of our company? Specify: _____
    _____
32. Do Trade Relations ("Reciprocal Buying") represent a problem to our company? I.e., does he make a practice or
    follow a policy of buying only from vendors who buy from him? Yes_____ No_____
33. Does the primary contact have specific buying days and hours; if so, what are they? _____
34. Does the primary buying contact have idiosyncrasies, unusual habits, or prejudices that must be recognized in
    calling upon him? Specify: _____
    _____
    _____
35. Have you read a Dun and Bradstreet report of recent issue on this account? Yes_____ No_____
    Any other reports? Specify: _____
36. What is account's credit rating? _____
    _____
    _____

37. What is the extent of the technical knowledge of our product possessed by the primary buying contact?
Extensive _____ Moderate _____ Limited _____ None _____

38. If less than moderate, are means available for getting technical information to people in the organization who can influence the primary buying motive? Yes _____ No _____

39. If answer is No, what is the nature of the problem? _____
_____
_____
_____

40. What should be our long-range objective for this account (include date and give justification for estimate)?
_____
_____
_____
_____
_____
_____

41. What is your goal for this year with the customer?
_____
_____
_____
_____
_____

42. Does our company provide other products or services which the account buys from other sources? Yes _____ No _____ Which should we attempt to sell? Specify: _____
_____
_____
_____
_____
_____
_____
_____
_____

43. How many salesmen from our company now call upon this account?

Company Component	Salesman's Name (if available)
_____	_____
_____	_____
_____	_____
_____	_____
_____	_____

44. Are you able to call upon this account as often as desirable? Yes _____ No _____
If answer is No, what prevents you from doing so? _____
_____
_____
_____
_____
_____

45. If answer to Question 44 was No, how much more business could be obtained if it were possible to service this account more frequently? _____
_____

46. If we are losing business, or lost the last order, what do you believe are the reasons?
_____
_____
_____
_____
_____
_____
_____
_____
_____

47. Looking to the future, what changes could occur that could affect our position with this account and what actions, if any, should we be considering now? (Probable changes in account's personnel, etc.)

_____
_____
_____
_____
_____
_____
_____
_____
_____
_____
_____
_____
_____
_____
_____

48.

COMPETITORS	% of Acct. Business	Competing Salesman's Name	Time as Supplier	Competitor's Location Serving Account
1.				
2.				
3.				
4.				
5.				
6.				

49. What are the strengths of each competitor, as you see them?
1. _____
2. _____
3. _____
4. _____
5. _____
6. _____

50. What are the weaknesses of each competitor, as you see them?
1. _____
1. _____
3. _____
4. _____
5. _____
6. _____

PART TWO

The Anatomy of a Sale

CHAPTER 7

The elements of a sale

As you progress through the material that follows, you will become aware that you are not only being presented with an analysis of obstacles that can be encountered in selling situations, but perhaps of even greater importance, this volume represents an invaluable reference book for you.

As you anticipate or encounter any of these obstacles, you can refer to the treatment given that obstacle or obstacles in this volume. You will note also that you are not being given "canned" solutions—that is, "Here is the one infallible way to handle this obstacle." Rather, you are being asked to examine the reason why customers do create obstacles (and how you and I create them for ourselves) and how to avoid them.

In Part Three of this book you are introduced to 87 sales strategies. Again, you are not told there is only one way to employ them. Instead, you are given a number of strategies that should be considered. Then you, as a professional, decide which of these strategies can be most effectively employed in the specific situation you have encountered, given the facts of that situation, your personality and the customer's personality. In some instances, the strategies presented may not be chosen by you, but they will "spark off" other methods you can use.

The contents of this book are based on the firm conviction that the most difficult task in problem-solving is to identify the problem. Otherwise we can produce a brilliant solution for the wrong problem ...and lose the sale. As most sales problems are man-made, the book endeavors to identify why people cause problems and how to prevent them from arising, or how to solve those that will arise.

You are encouraged to keep this book handy for constant reference. It is based upon the experience of countless professional salespeople. Why should you encounter the same travail they experienced when it can be avoided by recognizing the errors they made and profiting from the successes they achieved?

To solve a problem, it is most helpful to know the cause of the problem. This is especially true when we are dealing with problems involving human relationships. All too often, in dealing with people who cause problems for us, we deal with symptoms rather than causes. Is it any wonder that having cleared up one problem we find ourselves confronted with an entirely different problem created by the very same person a short time later? If we have not identified and eliminated the root cause of behavior or attitude, we can expect the *underlying cause* to generate one problem after another.

With this in mind, Part One, "Psychological Aspects of Selling," confined itself to a study of the basic reasons for human behavior as they express themselves in the selling situation. If salesmen have not identified the reasons for the customer's attitude or objections, it is most unlikely that a sale will be consummated. And it is most certain that future business will not be obtained without a struggle. For this reason, we have concentrated on identifying causes and studiously avoided, as far as possible, discussion of courses of action for overcoming sales resistance.

It is our contention that the average salesman—and certainly the experienced salesman—will generally be successful in solving selling problems once he has identified their cause. All too often sales training programs and texts on salesmanship concentrate on how to solve specific sales obstacles, blithely assuming that the true nature of the problem has been identified. In spite of this, such courses and texts have definite value because they provide salesmen with new courses of action to add to the arsenal of strategies and tactics which they have developed through on-the-job experience.

Nevertheless, such programs put the cart before the horse. Furthermore, they can create a hazard in that the salesman may come to believe that courses of action which have been used successfully by an author or teacher may be used successfully by the salesman in dealing with *his* customer. The salesman who has thoroughly studied the foregoing pages will now find it possible to make a wider selection from among the many courses of action treated in books available on selling techniques and in course material to which he may have been exposed.

In common with others who have written on the

subject, this author experiences the almost compelling urge to move rapidly on to a discussion of the "how-to-do-it" of overcoming selling obstacles. To surrender to this urge would be to do the reader a disservice. In subsequent chapters, we will present a lengthy list of selling obstacles and their genesis. And we will describe an imposing array of actions available to the salesman to deal successfully with those obstacles. This section will serve as an essential link between the discussion of how and why people react as they do and the subsequent chapters on what to do about it.

Pan American World Airways has a thought-provoking objective expressed as, "We want pilots that fly out in front of our planes!" In other words, we want men who understand the conditions that give birth to problems. We want men whose batting average is improved simply because they have become skilled in anticipating problems.

When we can anticipate a problem, two things become possible. We can take actions that avoid the problem entirely or we can prepare to meet the problem in advance. Preparation in advance is usually more effective because it can be accomplished in a less hurried fashion. Then our reasoning is not subjected to emotional strains usually present when we attempt to solve a problem we failed to anticipate. This is especially true in selling situations which are, primarily, human relations situations, fraught with the strains imposed by the conflict of the self-image demands of both customer and saleswoman.

It is our contention that if a professional salesman understands *why* people react as they do—i.e., the *cause* of the problem—he will be wise enough to know what to do and what not to do. The greater the arsenal of tactics available to a salesman, the stronger he enters the selling situation. Therefore, as specific obstacles are discussed, the salesman will be presented with many "sales prescriptions" or courses of action. The professional salesman can then consider each of the possible actions against the mosaic of the customer's behavior pattern and against the situation as he has analyzed it. She can then choose the action which promises the best chance of success.

Our objective is to develop analytical and planning skill on the part of salesmen rather than to provide them with the one best answer to each selling problem. You and I grow intellectually as we are stimulated to think. We grow very little when we attempt to imitate the successful salesman who uses tactics that have worked well for *him*, with his personality and with *his* customers, but which might be totally ineffective for you and me. The customer sees us in an entirely different light. His reactions to the stimuli we represent to him may be totally different from his reaction to the chap we have attempted to imitate.

Are there really "steps of a sale"?

When we are faced with a complex subject, one effective method for coping with it is to break it down into segments or "steps"; in other words, to make the complex appear simple and understandable.

Undoubtedly with this thought in mind, the authors of training programs in the field of salesmanship have often developed their material around what they have called "the steps of a sale." These courses had definite value, and many salesmen will state that participation in such a program marked the turning point upward in their careers.

Unfortunately, such courses tended to present formulae. Some participants failed to translate the principles taught into the very different kinds of words and actions that are needed to meet the demands of widely varying sales situations. A phrase that might be used as an example in the text would fail completely when used on a customer in a real situation. Some salesmen fell into the habit of using the same pet expression, learned in the course, over and over. Customers would detect this and feel that they were being manipulated or that they were being given a canned sales presentation.

A sale is *not* a series of clearly definable steps. Authors of sales training texts or courses know this, but presenting the sale as a series of steps has often represented to them an effective means of presenting complex situations in understandable terms. We shall see, in a moment, the hazards of presenting the sale in this light.

Today, we are engaged in a dynamic, ever-changing market. The need, greater than ever, is for thinking salesmen. We can no longer depend upon the forumla-trained salesman. We need salesmen who are as intuitive as the old-time peddler, but who are able to couple that quality with sophistication, scientific analysis, planning ability and a far deeper understanding of human motivations and reactions than are indicated by the superficial query, "What makes the customer tick?"

In spite of our rejection of the theory that there are "steps of a sale," it is readily agreed that to understand the selling situation and the problems it generates, it is of considerable help to break the sale down into elements or facets. It must be emphasized that though these elements are presented in what seems to be a logical sequence, the claim is not made that they are chronological; nor is it maintained that they always take place in a predetermined order. The ebb and flow of a sale defies any attempt to package it into nice, neat compartments.

The steps (?) of a sale

There *is* a beginning and an end to every selling situation, with a number of identifiable intermediate points. It is this sequence that has often been referred to as "the steps of a sale." There is nothing new about this approach to teaching salesmen.

Analysis of the three patterns that follow will be helpful, but before reading our analysis you will benefit more if you review them and then check your opinions against those presented below.

The so-called steps of a sale

Sequence I

O. A. Petersen of the Toronto Chamber of Commerce claims that a 16th century religious group, in instructing its "field men" in the art of making converts, used the following pattern:
1. Contemplate.
2. Contact.
3. Gain confidence.
4. Establish conviction.
5. Gain conversion.
6. Work for continuance.

Sequence II

An advertising agency once utilized this pattern in its internal teaching program:
1. Contact.
2. Arouse interest.
3. Create preference.
4. Make proposal.
5. Close the order.
6. Keep customer sold.

Sequence III

The historic pattern used in numerous sales training programs for many years followed this sequence:
1. Approach or introduction (Start).
2. Arouse attention.
3. Create interest.
4. Create desire.
5. Drive for close.
6. Sale.

Analysis of the three patterns

Comparing the three approaches, it will be noted that none of them provides any emphasis on pre-sale preparation. This is a malpractice widely followed in many sales training courses and texts today. Indeed, it is difficult to find a text that gives more than a paragraph to this element of salesmanship. Is it any wonder that selling is not widely recognized as a profession? Every profession that comes to mind has its diagnostic or "pre-operational" phase. To put it mildly, it is presumptive, if not arrogant, to think that we can move blithely into a selling situation, assured of success, without having estimated the situation.

Success, admittedly, is often attained when we are "order-taking," not selling. Without preparation, success is often achieved, but only after a struggle that would have been obviated with intelligent planning. At the very least, failure to prepare is the antithesis of customer-oriented selling. It smacks of taking the customer for granted. This is why, earlier we placed such emphasis on the need for making an estimate of the selling situation. Note that the 16th century group prefaced its sequence with the one word *contemplate*; in other words, consider in advance possible obstacles, advantages and alternatives available to both your customer and competitor.

Continuing our analysis, we find that the third sequence listed, so widely used in sales training today, makes no provision for post-sale service, product service, call-backs and so forth. Again—the antithesis of customer-oriented selling, geared to today's sale with no emphasis on developing a lasting, favorable customer relationship, with its impact upon repeat orders or continuing business.

This third approach strikes another blow at the concept of selling as a profession. It puts another arrow in the bow of those who maintain that salesmen are just "out for the dollar" and have no lasting interest in the customer. You and I know it is not true, but we should recognize it in our attempts to understand the real objectives of selling; i.e., to satisfy the wants and needs of our customers—and, in so doing, to make a significant economic contribution.

The second and third lists make no provision for gaining the confidence of the customer. It would appear that "salesmen" of the 16th century were a bit more astute in this regard. Admittedly, the customer's confidence cannot be gained in a single call. Confidence in your company, confidence in your product or service, confidence in your own integrity or ability can only be built after demonstrated performance; but there is another kind of confidence that can be gained early in the sale.

We can cause the customer to regard us as the kind of person with whom he feels comfortable, the kind of person who respects him. We can cause him to regard us as the kind of person who understands his problems. This kind of confidence may extend to approval of our appearance, our age, our deportment, our methods of expressing ourselves and even to establishing a belief in his mind that we are sufficiently competent technically to understand enough of his business to make recommendations of value. *It is agreed that this is a big order*, but if we fail we can be sure of rough going during the balance of the sale. If we fail, only the most favorable conditions can bail us out; i.e., best price, best delivery and sole-source advantages.

None of the three historical sequences includes

"asking for the order." This in the face of the fact that many sales managers regard the failure of salesmen to ask for the order as their primary problem.

Earlier, you were asked to make your own diagnosis of the three lists. In doing so, did you identify the most grievous failing of all three lists? It was not an omission of a "step" or "element." It was the orientation of all three. In other words, they are all *seller-oriented*. The emphasis is upon getting *his* attention, arousing *his* desire, awakening *his* interest. If we were to convert the sequence to a customer-oriented approach, we might see it in the following light:

1. Give him *your* attention because attention is one of the highest forms of flattery and will cause him to give you *his* attention.
2. Take a sincere interest in *his* problems, and he will show an interest in your proposal.
3. Have a sincere *desire* to help him, and he will exhibit a desire to compensate you.

There are no "steps" in a sale

The word "step" carries with it a connotation of moving forward or upward. Every experienced salesman knows a sale does not necessarily move forward from one stage of acceptance to another. Once you have the customer's attention, you cannot assume that you will hold his attention while you concentrate on the next "step" and try to arouse his desire. Distractions occur or something you say may cause the customer to allow his mind to wander, and you have lost his attention. The same is true of desire. You may have whetted his appetite for your product or service and be well on your way to a sale, when price, delivery, terms or similar factors may cause him to lose interest. You then find it necessary to rekindle his interest.

It is necessary during the course of a sale to keep a number of balls in the air simultaneously. Interest has to be maintained, attention must be held and desire constantly stimulated. Rather than a series of progressive steps, we should look upon the selling situation as having many facets or elements which demand constant attention. However, to analyze the sale, it is helpful to take these up one at a time. Within each element, we find that distinctly different obstacles can, and do, arise.

More sales will be made with less effort and embarrassment if we can anticipate the problems we are likely to encounter. To look at the sale in slow motion, we will, in this chapter, develop a logical sequence of elements. As these elements are considered, keep in mind that they do not represent a four-lane freeway to an order. Rather, they may be compared to secondary roads, complete with detours and obstacles that often cause us to reverse our direction in an effort to reach our objective with the least difficulty.

The elements of a sale

Asking hundreds of salesmen to identify the first element of a sale produced these replies: "Approach, introduction, getting by his secretary, getting an appointment, identifying the right man in the organization," and so forth. Before you agree with any of these replies, consider two problems that actually arose in two selling situations.

Jerry Cutler, salesman for a distributor, walked into a hardware store he had not noticed on previous visits to a neighborhood. He approached the proprietor and suggested the store would profit by handling Jerry's line. The owner took Jerry to an area of the store that had been hidden from his view by large displays and merchandise, and showed Jerry the very line that he sold. The owner then explained he was being serviced by another distributor and had not had time to erect his outdoor display sign.

Keep in mind that we are seeking a principle. Don't permit yourself to be lost in detail by muttering, as you consider the example we have just supplied, "This couldn't happen in our business." The fact is that it did happen in Jerry's business. The question is, where do you feel this salesman missed the boat? Was it in his approach? Was it in the way he made contact with the customer? Was it in failing to get the customer's attention? Actually, it was in none of these areas. Jerry had missed out on his pre-sale preparation.

Or, take this actual occurrence as another example: John Howell is an outside salesman for a distributor. He has just had a firm added to his list of accounts. The firm has always purchased most of its motors from Howell's firm, and specifically from another salesman, Howell's predecessor. Howell calls on the purchasing agent, and the latter, in a hurry, asks Howell for his prices and delivery on a motor. Howell supplies the desired information. The purchasing agent states that both are satisfactory to him, but he will check with his plant engineer and will phone his order in the next day.

The next morning, the p.a. calls Howell and angrily tells Howell that the motors quoted upon were "open motors," and he adds: "You should have known that in our kind of operation, we can only use dust-tight motors—it's all we ever *have* used!" Again, we can see that Howell did not miss out on his approach, getting an introduction, getting by a secretary or meeting the right man in the organization. He missed out on his pre-sale preparation.

1. Pre-sale preparation

We can agree that salesmen must sometimes skip this element. This is especially true in the case of a retail salesman or a salesman who works behind the counter in a wholesale establishment. He cannot always obtain full information in advance on the customer. However, he *can* always question.

Often, it is necessary to call upon a "cold account" or a new account and to do so before your competitor gets her foot in the door. It would be foolhardy to take the time to get full information on the customer, his past practices and the exact nature of his needs in advance. But, again, salesmen can always question.

At times, the customer may bring up some new need, in addition to the need the saleswoman intended to fill. The saleswoman may be totally uninformed on the problem or she may not be familiar with her own product as it applies to the need the customer has mentioned. In such instances, it is too late for pre-sale preparation. Most of us tend to think of these examples as unusual or unpredictable and excuse ourselves for not having done our homework. In other words, because we are unwilling to admit that we find planning distasteful and time-consuming, we miss the opportunity to improve our performance.

You may have heard a skilled lecturer called upon to make what you knew was an extemporaneous speech, and you were impressed with his performance. Yet, you would not say the speaker was *unprepared*. He has been preparing for this speech for years. Called upon unexpectedly, he must make a hasty, mental estimate of the situation. "What is the need of the audience? What past experience have I had that will satisfy their need?" Subconsciously, the speaker considers these questions as he approaches the platform. You, as a salesman, encounter unexpected demands upon your reservoir of experiences. Therefore, the more *general* information you have on the customer, her needs, her problems, and on your competition, the more effectively you will be able to draw upon that kind of knowledge to meet unexpected demands.

2. Contact

Contact takes many forms. It may take place when the customer enters your place of business or on the phone. Contact with the customer may have been previously established by some other member of your organization; it may, in effect, have been made through hearsay or when he has read one of your advertisements.

Selling is really a process of two-way communication. Contact is the beginning of that process, regardless of which form it may have taken. Impressions are formed during contact which can have a great impact upon the climate of the selling situation. Obstacles show up during this facet of the sale. We should be familiar with them so we can avoid them if possible or neutralize them when they cannot be avoided. The hazards of communication are being emphasized in countless articles and speeches, and they should not be minimized.

3. Gain customer's confidence

A contractor enters a distributor's place of business. Naturally, he seeks a salesman who will be attentive and courteous, but he looks for more than that. He wants a salesman who has the ability to grasp the customer's problem. He wants a salesman who is knowledgeable. If the customer sees an uncomprehending or uncertain look upon the face of the salesman, no amount of courtesy or attention is going to sway him from his determination to seek a salesman in whose technical competence he can have confidence. As salesmen, we know we must gain the customer's attention and hold it, but we will be able to do neither if he has no confidence in us.

You will have noticed we have not listed "Get the customer's attention" as one of the facets of a sale. This is because attention is not a "step," but it is part of the selling climate (even planned for in pre-sale preparation) and then nourished throughout the sale. If we can cause the customer to develop a confidence in our ability to understand his problem, holding his attention will be easier.

It should be repeated that this does not refer to gaining the customer's confidence in our integrity, our firm or our products. This kind of confidence, we repeat, can only be earned over the long haul by demonstrated performance—our own performance and the performance of our products.

4. Identify or establish the customer's need

Pre-sale preparation may have been neglected or it may not have been possible; hence, we may not have been able to identify the true nature of the customer's problem or need. We cannot proceed further into the sale until we have made this determination. Countless sales have been lost because salesmen have *assumed* that they understood the customer's need, but failed to see that need as the customer saw it. Salesmen sometimes see the need as a general need and offer a stock solution. The customer often resents this because he has a tendency to see his problem as unique. He wants the saleswoman to propose a solution that indicates that thought has gone into the recommendation. Obviously, if we do not

Elements in the Anatomy of a Sale—Plus Possible Obstacles

Pre-sale preparation

Too sure
Lack of time
Cold call
Customer has "new rep"
New type of account
Depends on predecessor
Deprecates competition
Overlooks new competition
No sources of information
Depends on reputation
Take customer for granted
One-sale oriented
Product oriented
Misinterprets info.—doesn't
identify a behavior pattern

Contact

Time
Priority
Recall
Competition
Mannerisms (our)
Atmosphere
Habits (customer's)
Timing (our)
Predecessor (our)
Organization structure
Personalities
Appearance (our)
Prejudice
Voice (our)
Age (our)
Interruptions
Customer has other problems
Personal worries (customer's)
Health (customer's)
Policy (both)
Trade relations
"Golf today"
Our own prejudice
Defeatism (our)
Secretaries (customer's)

Gain customer's confidence

Appearance (our)
Our own lack of confidence
Habits (our)
Mannerisms (our)
Canned presentation

Images customer has of us
of our company
of our products
Customer's self-image
Hearsay
Customer's image of our competitor
Age (our)
Voice (our)
Customer's off-beat outlook
Our reaction to customer's behavior

Identify or establish need

Unaware of need
Doesn't comprehend
Poor judgment
Ignorance
Priority
Disinterest
Doesn't want to see need
Customer is confused

Establish satisfactions

Price high
Unattractive terms
Limited use
Delivery slow
Union opposed
Installation costly
Space problem
Obsolescence
Price changes
Past performance
Service problems
Intangible benefits
Remote benefits
Customer is confused
Doesn't understand our solution

Gain agreement

Smoke screen objections
High cost in psychic needs
Won't buy your premise
Wants to avoid obligation
Has other obligations now
Influence of conditioners
Influence of experiences
Hearsay
Demands a concession
Fears
Hearsay

Gain preference

Customer has obligations
Customer's psychic needs
satisfied by competition
Competitor is local
Competitor has only one salesman
for customer to deal with
Customer has prejudice
Experience with other depts.
We threaten customer's
psychic needs
Customer's habits
Compares our product to others
Spreads business
Influence of outsiders

Establish priority

Fear of business climate
Fear of higher authority
Fear of failure
Won't plan ahead
Other needs seem greater
"Can get by with what I have"

Ask for the order

Front office jitters
Not expecting order
What to say?
Buyer's attitude
Is customer ready yet?
Size of order jitters
Impressed by customer's rank
Fear of turndown
Wants discount
Wants preference

Take the order

Hates detail (either party)
Afraid to push "luck"
Doesn't know tech. details
Afraid to lose control

Work for continuance

Is sale oriented
Forgets promises
Has no follow up plan
Doesn't see full customer potential
Geographical problems
Poor communication with
factory or office

really understand the customer's problem, we will not be able to offer a satisfactory solution.

This may sound basic, but keep in mind that all of us—when we are in the role of customers—demand what we call "service." When we say we want service, we are really saying we want individualized attention. If we have a problem that has vexed us, and a salesman gives us what appears to be an off-the-cuff solution, it does not always build up our confidence in him. It may cause us to think that he has not given our problem consideration. It may also irritate us to find he quickly solves a problem that was too much for us. This isn't good for our self-image. We don't like people who threaten our self-image. Your customers have exactly this same reaction. Hence, it is not enough to have a solution for the customer's problem, once we have identified it. It is also necessary to have him feel we have given it individualized attention and produced a unique solution.

There are times when you know that the customer has a problem or product need—or will soon have one, yet she is not aware of this. If we can't make her aware of the need, it is a sure-fire bet that we aren't going to sell her a solution. This represents, at times, a difficult selling situation if the customer has a high opinion of her ability. If we must point out her problem to her, she may feel this reflects upon her ability.

Even if we convince the customer that he has a need, and we offer him a product that will fill that need, he may go elsewhere to buy the necessary product rather than admit we knew more than he did. When a salesman suspects this may be the reaction of a customer, the salesman should lead the customer to an awareness of his problem in such a way that the customer will feel he has discovered it for himself. Customer-oriented selling requires a mutual comprehension of the nature of the customer's problem and agreement on the nature of its solution. Unless the customer agrees that there is a need, there will be no sale.

5. Establish satisfactions

A sale is primarily the exchange of need satisfactions. We have already found that this included both product and psychic (personal) satisfactions. A sale will not take place if the customer feels he is being required to pay a higher price, in cash, than he is getting back in the satisfaction of his economic needs. He will not buy if he feels he is being asked to pay too high a psychic price—if, for example, he is being asked to admit he is wrong, to lose face or to lose self-respect.

The saleswoman's job is to assess the price she is asking the customer to pay and then to prove to the customer that the satisfactions the customer will gain from the purchase are greater than the price he is being asked to pay. This calls for being constantly aware of psychic pitfalls, and it calls for stressing the benefits of the product or service in terms of what product qualities will do for the customer. Much emphasis on ways and means of accomplishing this will be covered in future chapters.

6. Gain agreement

Having pointed out the advantages of buying your product, it now becomes necessary to make certain the customer is in agreement. There is no point in proceeding further into the sale unless this agreement is present. If the customer has listened politely to your presentation, but has not indicated agreement; if you proceed without gaining that agreement, it may prove to be a waste of time. He may have listened to you but may not have comprehended. He may be giving counsel to his fears. He may have been subjected to hearsay which may not have supported your claims. This, therefore, is the point at which you must use test questions that will draw out the views of the customer. To fail to do so because you fear the customer may not agree with you will prove ultimately to be disastrous. It is far better to find out, at this point, if he disagrees with you. You still have time to provide him with a better explanation.

As will be seen by an examination of the list of obstacles that appears during this facet of the sale (see the chart), and by using the test questions, you will be able to smoke out whatever objections or fears he may have.

7. Gain preference

You must now convince the customer that not only will your product or service satisfy his needs, but will do so better than products or services he can obtain elsewhere. If you are fortunate enough to be a sole source of supply, you can skip by this element. If you are *certain* that he is considering no other supplier and that he is convinced you are offering him a good solution, you may wish to ignore this element. Just don't sell your competition short.

8. Establish priority

Your problem may not be with competition selling similar services or products. You may be up against "unseen competition" in the form of other demands upon his resources. He may actually want to buy what you have to offer, but may feel that he has other more pressing needs. He may fear the vicissitudes of the business climate and hesitate to invest at this time. He may be a short-range planner and want to put off buying until a later date. He may feel that although your product is superior, he would prefer to get by for the time being with what he has now.

Your task is to cause him to see that the problem your product will solve, or the need it will fill, is of greater importance to him than other demands upon his money. If he has listened to his fears, your job will be to provide proof and assurance. You must make him see the overriding importance of the need you wish to satisfy. Among the 87 "sales prescriptions" that will be presented later, many will provide you with additional means of persuading the customer to set a high priority on the problem your product will solve.

9. Ask for the order

It would seem that it should be unnecessary to list this element, yet failure to ask for the order is one of the most common causes of lost sales. Recently, we witnessed a movie of a salesman actually engaged in a selling situation with a customer. The salesman viewed the movie with us; yet, until he actually saw the film and heard the dialog, he would not believe he had not asked the customer to buy.

Customers usually *want to buy*. This is especially true of customers who come in to wholesale or retail outlets. They are in a buying mood when they enter the store. What salesmen do can build or bust the buying mood. After entering the establishment, the customer may begin to have worries about the cost of the product or she may be weighing her need for this product against other needs. Often, all that is required to push him beyond this stage of vacillation is encouragement in the form of asking for the order. In other words, getting him off dead center by forcing him to make a decision. This can be accomplished by *assuming he is going to buy* and making this assumption contagious by asking such questions as, "When do you want delivery?" or "Do you prefer to charge it?" or "Do you want help in installing it?" These and similar leading questions that can be used in your type of selling produce action. They should be used only if you know you have built the sale on a sound foundation and if you are sure he is convinced that your product will meet his needs.

10. Take the order

If the ninth element, "ask for the order," seemed obvious, this tenth element seems even more so. Experience shows that the manner in which an order is taken can cost sales and can seriously jeopardize future sales. Many saleswomen are intrigued by the challenge present in selling situations, and once the customer agrees to buy, the sale loses its challenge for them.

To some salesmen, the work of taking the order is looked upon as detail that can be left to order clerks. Other salesmen fear they will expose their lack of knowledge of technical details and leave details of the order to others. Still others want to get away from the customer because they are afraid to "push their luck"; i.e., they don't want him to change his mind. In some types of selling, this results in mistakes in promised delivery dates, catalog numbers, sizes, capacities, etc., and results in infuriated customers. The sale is not complete until the order is taken and product or service is supplied to the customer.

11. Work for continuance

This is not only the last element of today's sale, but it is the first element of the next sale with the same customer.

Our overriding objective is usually to build a lasting, favorable and profitable relationship with the customer. Product service does much to insure this kind of relationship, and it is the salesman's responsibility to show a continued interest in the customer. This element is all too often forgotten by salesmen who fail to see the customer's future potential. Other salesmen fail to see the true professional nature of their work: to have a continuing interest, or as Overstreet said in "The Mature Mind," "the power to sense another's need and to want to fill that need." Too often, the saleman is one-sale-oriented and lives from one day to the next, regarding each sale as an incident isolated from the next. Too often, some salesmen forget that by demonstrating a lasting interest in the customer, they will be able to accomplish the first three elements of the next sale with comparative ease.

Summary

As can be seen in the chart, the obstacles encountered in selling situations fall within these elements. The purpose of examining the anatomy of a sale in this chapter is to point up the areas in which we must be prepared and to explode the theory that a sale is really nothing more than obtaining attention, interest and desire. Even before the obstacles listed on the chart have been covered in future chapters, the listing will provide you with a catalog of possible obstacles which you may encounter in selling situations you *currently* face. If you will use the chart as a checklist, it will help you anticipate problems, consider possible solutions and plan your strategy before you find yourself in the actual selling situation, facing problems you never anticipated.

CHAPTER 8

Pre-sale preparation

If you were to compare the obstacles listed under pre-sale preparation with obstacles itemized under other elements of the selling situation, you would note one significant difference. All of the obstacles encountered before the sale are of our own making. It is true that obstacles may be created by our own action after the selling situation has started, but *all* of the obstacles listed under pre-sale preparation are either of our own creation or are within our control.

Few sales training programs or texts devote much attention to pre-sale preparation, possibly because it is felt that the professional saleman is fully aware of the need to be prepared. It is probably also believed that little assistance can be provided the salesman in this area. Yet, we find that a disproportionate percentage of sales are lost because of inadequate preparation.

Unfortunately, sales training programs all too often deal only with the challenging elements of the selling situation where the psychological struggle between salesman and customer takes place. This makes for interesting reading; it lends itself to intriguing strategies and provides material for the kind of speeches and lectures that create a demand for encores.

There is no element of the selling situation that can be categorized as unimportant, but our efforts to ride triumphantly through the balance of the selling situation may be more dependent upon the quality and quantity of our pre-sale preparation efforts than upon any other effort we make.

Salesmen readily agree that preparation is important, when it is possible. For example, one salesman was shown the list of strategies which appeared in the Sales Strategy Bank and he exclaimed, "What sales we could ring up if we could place that list in the hands of all of our salesmen." The usual reaction of sales managers is, "How soon can we have copies?" The problem was epitomized by an experienced salesman who nodded his head as he read each obstacle and each strategy. When he finished, he inquired, "Every question in the Estimate of the Selling Situation is important; why is it that we often don't even try to obtain the answers in advance of the sale?"

The answer is found, perhaps, in the fact that preparation always seems to bore us, whether it is practicing the piano, training for track events or for any form of human endeavor. The excitement of the challenge is missing. The enthusiasm engendered by competition is missing. We seem to forget that half of the fun of a vacation trip is in the planning and anticipation. It can and should be exciting to study the customer and the competition, seeking out weaknesses, strengths, their modus operandi, their reaction patterns and their psychic needs as well as their product or service needs. Then, there is the even more intriguing area of self-analysis, identifying our own strengths and weaknesses and planning how we can capitalize on the former and control the latter during the selling situation.

Essential elements of pre-sale preparation

Two things are essential. First, we must know what kinds of information should be acquired before we enter the selling situation. The list of questions appearing in the Essential Elements of Intelligence in Chapter 6 provides us with a start in this direction. These questions can be changed or added to as the nature of the specific type of selling activity dictates.

A second essential is determination, self-discipline and sweat. Unless you are fortunate enough to work for a boss who demands thorough preparation, this is an activity that you must force yourself to recognize as essential and then give it the time and attention it deserves. Productive preparation not only demands that you "size up" customer and competitors, but it requires that you develop the ability to be introspective—self-critical. It is true that much can be observed in the performance of other salesmen that will be of value, from both a study of their talents and their faults. Even more can be learned from an honest and realistic self-appraisal—always providing that you guard against the human tendency to rationalize away your own areas of weakness.

At this point, it would be helpful to make an appraisal of your own past selling efforts. Have you lost sales because of insufficient preparation? As you consider sales you have lost, do you find yourself rationalizing that things which happened during the course of the selling situation that lost you the sale "could not have been predicted"? Can you, in all honesty, say that "they could not have been *anticipated*"? Could not alternate strategies and tactics have been planned in the event that the "unpredictable" *should* occur?

Do you agree with the salesman who, with a grin, said, "I've never seen 'out-sold' on a lost sales report"? Saleswomen are frequently out-sold... even saleswomen who have built up enviable records. While we often attribute our failures to "price ...insufficient market-coverage...not enough help from the office or the boss" (and so on, ad infinitum), the fact is that more often than we care to admit, the other fellow "out-prepared us"; hence, it was easy for him to out-sell us.

We once observed a sales manager in Spain outsell his competitor. The competitor started with a great advantage. He had been in Spain longer, had better political and market contacts working for him and had an equally good product. Yet, he lost the sale. Afterwards, he approached the trimphant competitor, extended his hand and said, "You deserved it. You must have worked for months on every angle. You had an answer for every claim I made and for every point the customer raised. We thought we could come in with about 48 hours of preparation and get by on our past performance. Congratulations ...and I warn you, we've learned from this one!" You can be sure this unsuccessful competitor is going to present a real problem when the next competitive selling situation rolls around.

So, for some immediate results, analyze your own past pre-sale preparation. The beautiful aspect of this kind of effort is that you only have to be honest with yourself. You are not being asked to beat your breast and murmur "Mea culpa" to your boss or to anyone else. You *do* have to be honest with yourself if you hope to be more successful, and you must couple this honesty with a determination to be more effective in your pre-sale preparation in the future. This will call for work, but keep in mind that the more work you put into preparation, the easier the rest of the sale becomes. Just consider any other form of human activity, and you'll agree that the more we have practiced, the easier it has been when we were up against the real thing. This is so logical and self-evident that it is a mystery why more of us don't *do* something about it.

Finally, keep in mind the emphasis that we place on selling as a profession in this course. It isn't enough to think of it as a profession; we must act like professionals. And no profession lacks its diagnostic facet, which in selling is preparation before the sale.

And now let's devote some time to a consideration of the obstacles that have been listed under the pre-sale preparation element of the sale. Remember that these are obstacles which *you* can do something about today. As you consider these obstacles, resist the tendency to think of "other salesmen" whom you have seen create these obstacles for themselves. If you are to profit from this chapter, be *self*-critical and decide which of these obstacles you have created in the past, and which you need to guard against in the future.

Too sure

A saleswoman who creates this obstacle does so for a variety of reasons. She may be too confident because of past successes. She may underestimate competition. She may assume that the customer has recognized his own needs. She may be guilty of "projection"; i.e., she may think the customer shares her own keen appreciation of her product or service.

A salesman is "too sure" when he thinks that he can handle any problem that arises during the selling situation, without preparing for it in advance. He may be smart and know it. He may be a fast thinker, a fast talker, and have proven his ability to adjust rapidly to any change in the selling climate. He has demonstrated to his own satisfaction, and to his boss's satisfaction, that he can gain and hold the confidence of his customer. He has a sensitivity to his customer's reactions and is capable of developing empathy—the ability to feel as the customer feels. He knows his product, his competition; he knows where he is strong and where his competition is strong. You could add all of the attributes of a good salesman to this list, and you'll find that he has all of them—and knows it. To him, any article on pre-sale preparation is an attempt to make a science out of what he calls "the art of selling."

He is no glamour boy, perhaps, but he does have a personality that is well received by his customers. Coupling that with technical knowledge, he feels he is well equipped to meet the vagaries common to many selling situations...and he is the chap who needs the warning implicit in this chapter more than most. The salesman who is not so capable as this chap senses the need for thorough preparation, and as a result may outsell the fellow who has allowed himself to become *too sure*. Let it be admitted that he does not make a habit of ignoring preparation. As a result, he has built up an enviable sales record. The fact is that he *does* lose some sales. He may also be losing many more of which he is unaware.

Is he "too sure" because he has been dealing only with "sure accounts" while he evades new prospects? Is he "too sure" because he has been satisfied with "his share" of the business from one account while his customer places additional business elsewhere that could have been obtained by this salesman?

Too many salesmen and too many companies lull themselves into a sense of false security by patting themselves on the back because they obtain, for example, 20% of the available business. Greater sales activity can be stimulated if these same firms and salesmen interpret this as evidence that 80% of the people would rather buy from someone else. Do we become "too sure" because we lose some business by default when we consider certain potential customers as being safely tucked away in competition's order books? Would calling upon some of these tough accounts be a good antidote for the chap who is "too sure"?

We all have the very human tendency to look outside of ourselves for the cause of our failures and to take credit for our successes. When the saleswoman who has lost an order because she was "too sure" looks for the reason, she is likely to attribute it to "price," "lack of breadth or depth of product line," "location," and so on. This is a sure-fire prescription for more lost sales.

Even if all of these complaints are true—as they often are—isn't the professional salesman the fellow who can get the sale when the odds are stacked against him? Isn't the skilled salesman creating his own obstacle when he fails to analyze every selling situation in advance to decide what may have changed since his last contact with the customer? Doesn't being "too sure" lead to the possible creation of a new and more dangerous obstacle: i.e., causing the customer to feel he is being taken for granted?

What's the cure?

1. Constantly remind yourself that there are customers who prefer other suppliers. These customers represent a great challenge and a great opportunity. If you can convert them, like any convert, they develop a passion to infect everyone else with their new outlook, and they can lead you to even greater business opportunities. But, obtaining their business calls for more than verbal counter-punching ("If he says that, I'll say this"); it calls for a careful estimate of the selling situation.

2. Realize that sales situations and customers are both dynamic; they change rapidly. So does the competitor you outclassed yesterday. The chances are he learned from the experience, smarted under defeat and will be a rougher adversary today. The road to lost sales is paved with the bricks of overconfidence born of the belief that an "instinctive" ability to react quickly to new situations is all that is required. There is simply no substitute for advance knowledge, alternate strategies and tactics. The customer-oriented viewpoint insists that we enter the selling situation with facts that enable us to see the situation as the customer sees it today. Only when we understand his frame of reference can we cause our proposal to appear desirable in his eyes.

Lack of time

"I had *no time* for preparation. Something arose during the course of the selling situation that was unpredictable." It is difficult to fault the salesman who can prove that this is an accurate description of a situation. It would be unfair to charge that all sudden changes could be "anticipated" even if they could not be "predicted." Isn't it true, however, that we do have the tendency to classify the unexpected as unpredictable? If we would start using the term "anticipate," it might encourage us to be better prepared with alternate strategies and tactics. Then, too, in the midst of a selling situation when the customer does bring up a situation which we could not anticipate, we often feel impelled to react immediately. We forget that there is almost always *time to question, time to let the other fellow talk, while we think.*

Lack of time is more commonly expressed in such statements as, "I don't have time to plan each call; my territory is too large." Or, "There's no time; I have too many customers; I'm spread too thin." Or, "There's just too much to do...too much paperwork...I just don't have time." Often there is validity to these complaints, but the question remains: If the time were available, would the salesman really put it into planning, or would he simply try to call upon more customers, relying on his "instinctive ability" to meet new situations as they arise?

The problem is often not so much one of lack of time as it is of application of the time we do have available, and a general human tendency to avoid the boredom (?) of planning, except when life is at stake. All too often, when we feel a lack of time, we develop a sense of urgency, a compelling desire to *do* something. It's the same sort of drive that impelled the fictitious driver to "step on it" when he noticed his gas gauge closing in on the "E," and he wanted to reach the gas station before he ran dry.

The author is not proud of the fact that the desire to *do* something caused him to defer planning. Driving in an unfamiliar Ohio city, I was given conflicting directions on how to leave for Cleveland. I had an important meeting to attend, and I knew that my time was getting short, with 200 miles to go. On the seat beside me was a detailed map, but I was reluctant to stop the car and take the two to three minutes it would require to plot my route. "I just didn't have the time." Instead, I followed the instructions that made the most sense to me. Unfortunately, they were incorrect; I became involved in heavy day-end factory traffic that I would have avoided had I taken the time to read the map. Result: I arrived very late in Cleveland, accompanied by disapproving looks from all present.

The need to *do* something (in this instance, keep the car moving) clouded my judgment, and "no time" for planning resulted in far more time lost in the long run. The disapproving looks of the conference members were painful to take, but not half so painful as the pain of a lost sale, which is far more permanent.

A sense of urgency in a salesman is a real asset just so long as it is coupled with the realization that the more time spent in planning, the less time spent in the selling situation with its myriad of pitfalls. One salesman commented, "In my kind of selling, I only have a few minutes to make my presentation; and if it doesn't go over, I've had it. I spend far more time getting ready for the presentation than I spend in front of my customer." He then went on, with a rueful grin, that in spite of this, he realizes how inadequately prepared he often has been and how valuable the "Sales Strategy Bank" and the "Estimate of the Selling Situation" list of Essential Elements of Intelligence will be to him.

How many salesmen, like bees flitting from one blossom to another, visit one customer after another with little preparation, losing orders in some instances and being satisfied with less than should satisfy them in others? The profitable use of time requires that a salesman make an estimate of the selling situation for each customer, and this involves estimating his immediate and long-range potential. It becomes essential to determine which customers are worth cultivating and which are marginal or sub-marginal in their promise. And then to spend time in preparation... time that might otherwise have been lost in calling upon customers whose potential promises little.

Finally, most of us who complain about lack of time must, in all honesty, admit that we spend more time than we care to admit on activities that are unproductive. We, all too often, use time as the alibi to excuse ourselves *to ourselves* for our general distaste for planning. We overlook the tremendous satisfaction that is ours from making a tough sale because we acted like professionals and out-planned our competition. "No time" is a creator of obstacles that can be overcome if we are willing to be introspective and to analyze and record for a few days the way we use the time we do have. The results and the impact upon our self-respect will be worth the effort.

Cold call

Readers of this material are engaged in sales activities that vary widely. Some handle one account, others call on many accounts. Some handle the sale from start to finish, others operate as part of a team. Some locate and open up new accounts, others handle established accounts. Some have little contact with customers except across a counter, others have close and continuous contact.

As each reader explores the information here, each must from time to time guard against the tendency to examine examples presented and feel that "this material does not apply to my kind of selling." Actually, the closer the examples come to "your kind of selling," the greater the disservice the author is rendering you. You and I learn only when we are stimulated to think. When examples are drawn from "your kind of selling," there will always be a tendency for the reader to become lost in detail and to react with, "that isn't quite the way it happens in our company," or merely to agree with the author. In neither case is the reader stimulated to think.

When examples are not drawn from your kind of selling, it becomes necessary for you to search out *the principle* that is being examined and to ask yourself how does this apply to your kind of selling. When you are forced to make the transition from an example to your kind of selling, learning and insight take place. All too many effective sales training programs have been rejected by sales managers on the erroneous assumption that because they were not based on the kind of sales activity in which the men were engaged they would be of no value. Actually, if the *principles* espoused were sound, these programs would have offered maximum value.

Let us take as an example the salesman who calls upon one customer or upon a limited few. He might react to this material on the cold call with, "We never make cold calls; we always make our proposals only after thorough research and study of the customer and his needs. We couldn't possibly enter a situation where we would not be informed."

If that approximates your reaction, try looking at your selling situations from a new frame of reference. Agreed, that in almost all cases you engage in extensive preparation, you may still encounter a situation that approximates the conditions considered as cold calls. The narrow way to interpret the term cold call is to think of the tip you just received that business is available from a new company if you move fast and beat competition. But, let's look at the cold call as the situation that confronts you when you call upon a long-established customer and encounter these situations:

A new problem has arisen, and your customer feels that it requires the attention of someone in his organization with whom you've had little or no contact. When you come face to face with this individual for the first time, you are up against problems encountered by the salesman who makes a cold call on a brand-new customer.

Or take the example of the salesman who finds his former contact has been promoted, transferred, resigned or even died. The second in command, whom the salesman has also been cultivating, for some reason did not succeed the regular contact. An unknown replacement was selected. The salesman faces a new and unknown quantity.

If you find yourself in this unenviable situation, certain unresolved questions demand your attention:

1. What is he like psychologically? Will his race, religion, color, height, weight, educational background pose any psychological problems for me?

2. How about his technical competence? At what technical level should I gear my remarks? If too technical, will I confuse and irritate him; if too basic, will I insult him? If his background is impressive, how can I recognize it without being so obvious that I annoy him?

3. How will she react to "the me-I-project"... to me as she sees me? Will she have prejudices against me, my products, my firm? Are there two or more schools of thought in my business? Which does she represent? How can I neutralize any reactions that may be adverse to my objectives?

4. Just how important is he to the completion of the sale? Is he the final decision-maker? What is the basis of the influence he may have? Is it due to his position in the hierarchy, or is it due to his company's respect for his technical competence?

5. Let me scan the psychological needs which are important to him—the need for recognition, need to dominate, need for assurance etc.?

We hear much about the importance of first impressions, but we usually think of the impression we make on the other fellow. It is equally important that the impression we form of him is a *valid* impression based on facts. We must at all cost avoid making snap judgments based on one or two idiosyncrasies, habits or apparent status symbols. *We must look for a pattern.* If possible, get him talking and keep him talking. Use questions and especially note what he likes to talk about when he is given free rein to choose the subject. Remember, he will talk about those subjects which are of greatest interest to him. Remember, too, *that nobody hates a good listener.* You have never heard a person criticized by the comment, "He *listens* too much."

During a cold call, you will be guessing, but try to make educated guesses based on observations and intent listening. Don't just get him talking to flatter him; listen to what he is saying. Failure to do this will create unnecessary obstacles and may cost you a sale. Later, being human, your reaction may be, "It was the first time I ever met the guy; what did you expect me to do?" Obviously, what you are being encouraged to do is to identify this chap's self-image, the "me-he-tries-to-project." This "me" will be the basis of his continuity of personality, his behavior pattern, and it will enable you to predict, with a high degree of accuracy, how he will probably react to situations. You will find it easier to see each situation as he probably sees it, without necessarily agreeing with his viewpoint.

Can serious mistakes be made during the cold call if the foregoing procedure is not followed? Consider the following actual situations in which various salesmen testify:

1. Identifying the real "me"

"He ran a sloppy business, lengths of pipe, cable and odd-parts all over the place. He looked just as sloppy as his shop. On my first two calls, I sized him up. On my third call, I had already noticed that other salesmen called him by his nick-name, 'Jake,' and he seemed to take it well, laughing and joking with them. So, I started calling him Jake as well. Also, I received no orders.

"Later, I learned that these other salesmen only obtained minimal orders from him, and he gave most of his business to a fellow who called him by his last name with a 'Mr.' hung in front, and I noticed that this salesman used 'Sir' occasionally, always with respect and never obsequiously. I checked further and found that although Jake ran a sloppy shop, it was by design to create an atmosphere that caused his customers to think they were getting bargains. My successful competitor knew that once a year Jake took a vacation for a month and looked like something out of Brooks Brothers when he left his shop. My competitor had identified the 'me Jake was trying to project' outside his shop. I had identified only the 'me' that Jake wanted to project to his customers.

"A little homework on my part could have brought me the business that I lost on the first cold call. As a matter of act, all of my first calls on Jake were cold calls because I had learned nothing except the wrong things."

2. Ill-timed common interest

"Someone told me he loved boats, and sure enough when they brought me into his office, there were four or five pictures of sloops on his walls. I thought I'd establish a common interest and made some comment about his interest in boats and indicated it was my hobby. He turned on me with a growl and told me to never mind the blankety-blank boats, he had a problem with our equipment, and what was I going to do about it?

"I learned later that failure of our equipment had

shut down two of his production lines, and he was losing money fast. At just about that time, he was probably more concerned about selling his boats than sailing them. At any other time, I might have made some points because of our common interest. But because I didn't let him talk first, I sure ran into trouble. Later, when the trouble was rectified, we went sailing together, but I surely lost business at the start."

3. Letting him sound off

"I knew that we had the best product on the market, and I knew that this bird really needed it. I had studied his product needs in great depth and sailed into his office with complete confidence. All he did was rake our product over the coals. Nothing he said was correct, but he wouldn't let me talk. I made no progress and returned to the office feeling low. Then, I learned that my predecessor had never been able to sell him anything, even though we knew he really needed our product and was buying a product he criticized elsewhere.

"I returned again to his office, and he started in where he left off the last time. I just let him go on and I stopped defending our products and just listened. I began to suspect that my predecessor had done something to irritate him, so I brought his name up casually. As soon as I mentioned his name, the customer jumped up and said, 'Let's go to lunch, and I'll tell you something about that company of yours.' When he reached the restaurant, he started to talk about Metcalf, my predecessor, saying, 'Well, Metcalf finally told that outfit of yours where to go and went out and got himself a good job.' (This wasn't true at all; Metcalf had left our company on excellent terms, for an opportunity which our concern would have been glad to give him if there had been any opportunities.)

"Then he went on: 'You know, when I got out of school I applied to your company, and they told me I wasn't up to the standards they wanted; well, I showed them! I'm making ten times what that pipsqueak of a recruiter will ever make.' He went on at length and then asked me to supply him with equipment that represented a big order, and we've been his chief supplier ever since. He had been waiting for years to tell us off, and apparently Metcalf had never given him the opportunity.

"The only thing that saved me was listening, and also I had observed that he was a smart operator. If he really needed my product, as he did, and he wasn't buying it, there was some other reason for it. Giving him a chance to sound off was the cure. Also, I think that he wanted to identify himself with Metcalf. Metcalf had quit our company, something the customer would like to have been able to do. So he distorted Metcalf's reasons for leaving. This gave him some of the satisfaction that he would have obtained had he been able to work for us long enough to 'tell us off' and quit."

4. Turnabout in time

"I had seen his home and his office and both were luxurious, so when I talked product, I talked the top of the line. I subtly commented on the attention that possession of our top-line would draw. I could see him freeze noticeably. Fortunately, I managed to get him to talk. I began to see that he had a fear of being ostentatious; the one thing he did not want was to attract attention. He enjoyed the luxury of the interior of his home and office and the comfort they represented. He, above all, was not looking for a status symbol, as people think of these things.

"When I started to talk again, I kept emphasizing durability, high quality, low maintenance and gave figures to prove that the price he was being asked to pay was really an economy. I said no more about appearance and made the sale. Today, I feel that I jumped to conclusions and thought about his home and office as status symbols which they would have been if they were mine. I saved the day by watching his reactions and getting him to talk, but I nearly lost the order by opening my mouth before I knew what was important to him."

Handling the cold call

We can agree that what we call pre-sale preparation is almost impossible to achieve in the instance of a cold call. What can replace it? The sale doesn't really start until you, as the saleswoman, start to talk. You have a few precious minutes in which you can prepare. Consider the following tactics:

1. Use questions that will cause him to talk about his problem and his needs as he sees them. Use questions that will cause him to talk about himself and which may reveal the personal satisfactions which he desires. Use questions that will cause him to comment upon his past experiences. You may learn what he thinks about your predecessor, your company or your competitor. You may even learn what people he respects and whose opinions will be important to him, which may help or hurt your chances of making a sale.

Try to design questions, if the situation permits, which will cause him to speak of the things that are of personal interest to him, even though they have nothing to do with the selling situation. His answers may reveal the nature of the drives within him that affect his attitudes and opinions.

 2. Use the uncommon commodity, common sense. Tailor your questions to the man and the situation. If his product problem is uppermost in his mind, don't try to ask questions intended to develop common interests. Use rhetorical questions, and be alert to his replies. To such a question, his reply may be a nod, a frown, or hesitation . . . all revealing.

 3. Listen . . . don't plan your next question while he is replying to the last one. *Really* listen.

 4. Observe. Is he under pressure? Is he relaxed? Does he feel strongly about something at the moment? Does he create a flurry of activity to impress you? If so, what image is he trying to project? Can you enhance it? Observe his attitude toward you. Does he try to put you on the defensive at once? If so, is he trying to divert you and keep your attention off something else? Is he using your technique on you and asking you the questions? Is he too cordial; if so, is this a device to put you in an expansive mood so that you will make promises you will regret later? Observe the kinds of people with whom he surrounds himself as well as the furnishings of his office. Does he exhibit a tendency to "show off" or to economize? (See full list of questions in the "Essential Elements of Intelligence.")

 5. Above all, look for the pattern. Don't make superficial judgments based on any one or two of the foregoing observations. Calling on a cold account is no different from the commander in combat who must often make a "hasty, mental estimate of the situation." He still has his mental list of "essential elements of intelligence," and he obtains as much information as possible on each element and considers the impact of each bit of intelligence upon the entire situation. You, too, must often work with a paucity of facts at your command, but, like the combat commander, you will know that the perils lie in those areas where you lack complete information or where the facts you have just don't add up.

 Keep in mind that it is always possible to rationalize when we fail to collect meaningful information, whether it be on a cold call or when time for preparation is available. The hard fact is that *someone is going to sell this customer*, and it will be that person who has the greatest sensitivity to what the customer says and what he does in the few brief moments that he offers you and possible competitors to observe him and to get him talking. *You can be that person.*

Customer has a new representative

 What happens to pre-sale preparation when you call and learn that "your man" has been replaced by someone new? (The same situation exists if you are selling behind the counter, and the customer sends a new man to make his purchases.)

 Probably in common with most effective salesmen, you have spent time working assiduously to develop a lasting, favorable, and productive relationship with specific people in the customer's organization. They have either been the decision-makers or you have identified them as the key influences on the decision-maker. You have found that it has been important to sell yourself.

 This is in keeping with the premise that an effective salesman first sells himself, then the product, and finally sells his company. You have learned that if the customer is not sold on you, she will hear little of what you have to say about your product. Sell your product next because it is in your product that she will see the solution of her problems and/or the fulfillment of her needs. Sell your organization next because she will not want your product unless she feels that yours is the kind of an organization that can deliver the product, stand behind it and be fair and honest in its dealings with her.

 And now, after all of this effort to cultivate the decision-maker, you find that he has been replaced by a new man, an unknown quantity. What is equally important is that you are an unknown quantity to him. What to do?

 The situation takes on the aspects of the "cold call" (dealt with earlier) and must be met with some of the same tactics recommended there.

 You cannot prevent your customer from making changes in his buying organization, but you can reduce the impact of those changes on you and on your sales record. Here are some of the preventive measures you can take:

 1. Cultivate the buying influences, but remain aware of the fact that they can, and almost inevitably over the course of time, *will* change.

 2. Know the customer's organization so well that you will have a good idea of the identity of the probable replacement when the time comes for "your man" to retire, to be promoted or reassigned. (Even in dealing with military organizations where reassignments are regular and frequent, you will find the key influences on the decision-maker usually remain or are reassigned on a staggered basis.)

 3. Constantly give recognition to others in the customer's organization while you are developing a solid relationship with "your man." Don't breeze by the second-in-command while you satisfy yourself

with a "Hi, there," as you make a bee-line for today's decision-maker. The recipient of your careless greeting may be tomorrow's decision-maker, and your ever-eager competitor may be cultivating him today. To the degree that it is possible to do so, give him recognition. Show him sincere respect. Begin to gather the same type of information on him that you have gathered on the present decision-maker.

Failure to do so may cause "tomorrow's decision-maker" to feel that you and your company are inimical to his interests, and he may develop close relationships of his own with one of your competitors. The best way to handle this obstacle area is to consider the ultimate replacement of "your man" and to conduct yourself in such a manner with probable replacements that they will hold you in the same high regard you presently enjoy from "your man."

New type of account

You may be assigned a new customer who is engaged in a business with which you are unfamiliar.

You may be assigned a customer who uses your products for purposes with which you have had no experience.

One of your current customers may expand his business and will now use your product in applications that are new or unfamiliar to you.

Pre-sale preparation is most difficult, and at times impossible, under such conditions. Hence, it is of value to develop general patterns of behavior for meeting such situations.

Saleswomen would have to be walking encyclopedias to be thoroughly versed in all of the fields in which applications of some of their products may be made. No sales manager should expect this of the sales force. The problem, therefore, becomes one of how to familiarize yourself with the new problems of your customer without losing his confidence.

There is no pat answer to this problem. Consider, however, the following possible courses of action:

Don't bluff. You'll lose ground you may never regain if he detects bluffing.

If you know him well, if you know he has confidence in your integrity, if you have served him well, and if you enjoy a strong personal relationship, admit your ignorance of this field and ask for an opportunity to study the new problems.

Get him talking, keep him talking and listen. The problems may reveal themselves.

Don't be a loner. Call upon the resources of your own company. Seek advice from others in the company who may be aware of the problems associated with the new type of business.

Depends on predecessor

Someone before us has said, "Every generation must earn its own freedom. Freedom is not inherited; it is not entirely transferable."

Selling is much the same in that every salesman must develop his own market. Like freedom, customers may be passed from the hands of one salesman to another, from one man to you, his successor. But as his successor, you must be aware of your need to protect your inheritance from competitors. These competitors are always alert to the opportunities that we make available to them when we make organizational changes in our sales force.

Your predecessor had to sell himself, then your products, and finally your company to his customer ... now *your* customer. Because the customer learned to respect your predecessor, he may be willing to extend a kind of "polite acknowledgement" to you. He may feel something akin to a suspicion that nobody can really measure up to "Old Jim," your predecessor.

Just as "Old Jim" before you had to do, you now must sell yourself to the customer. Jim has accomplished part of the selling job for you. He has sold them on your products, on your manufacturing or service capabilities, and on the integrity of your company. The balance of the selling job remains to be done. You must now convince them that you have their interests at heart, and that your interest will not wane once the order has been taken; but, like "Old Jim," your interest in them will continue and will include post-sales service. You must convince them that you have the technical competence enjoyed by your predecessor.

Their posture may be a "wait and see" attitude. "Will this new man understand our problems... will he recognize our corporate objectives and immediate goals? Will he recognize my self-image, my personal aspirations?"

If your customer feels that "Old Jim" has been summarily or capriciously transferred, your customer may be re-evaluating your company, too. "Does that company really have our interests at heart if they would transfer "Old Jim" that way? (You can imagine, without difficulty, what their reaction may be if they feel Jim as been given a bad deal by your company.)

So what do you do?

Don't rely upon the relationships established by your predecessor. They may be transferrable only in part. To fail to do your homework (to study the customer's organization, its history, its personnel, its objectives) is to court disaster. Use the help of your predecessor if possible.

Don't imitate your predecessor or his tactics. They may have been "natural" for him and completely unnatural for you. Old Jim may have gone deep sea fishing with your customer and could always establish rapport by launching a conversation on this activity. You may only encounter a cool hostility if you attempt to do the same. You have known individuals who could use profanity and be rewarded by amused chuckles from the customer. Your efforts or mine in this area might bring open hostility.

One salesman had been handling an account for years. He and the customer were quite close. The customer would shout and pound the desk when he had problems. The salesman would sit by quietly and listen and gradually a contagious grin would appear. Suddenly the customer would cease his shouting and pounding, and his face would break into a grin while he said, "O.K. . . . so what are you going to do about the problem?"

His replacement had seen the salesman use this technique successfully over and over while he was in training. When he eventually replaced the salesman, he tried the same "grin technique" and was shocked when the customer became enraged and shouted, "Don't you sit there and laugh at me; you'd better get out of here and get some action if you want to keep this business." (The wording of this reaction has been slightly expurgated.)

Be yourself. Keep the following in mind. Your customer has a self-image. Your predecessor had an impact upon that image, and the customer found it acceptable. . .even pleasant. If your customer wanted to appear dominant, your predecessor made him feel that way. If he wanted to appear knowledgeable, your predecessor made him feel *that* way. Whatever the self-image of the customer was, your predecessor had identified it and had learned how to enhance it *in his way*, and had learned how to avoid threatening it. Through the eyes of this customer, you have an entirely different impact upon his self-image. Maybe "Old Jim" made him feel comfortable; perhaps you made him feel uneasy or inferior in some respect. Until you have identified the customer's self-image (the "me" he wants others to see when they see him), you have to proceed with caution and then find *your* best means of enhancing that self-image. You don't do that merely by emulating "Old Jim" or even "Young Jim," if your predecessor was younger than you are.

To what extent do you depend upon your predecessor?

If the facts support the assumption, you depend upon the acceptance he has gained for your products, your service, your company. But. . . *caution*. . . depend only to a limited degree upon the acceptance he has gained for your company. . . this may not be readily transferrable. Remember, to the customer he *was* the company, and now *you* will be the company, and it may appear to him to be a very different company indeed.

You depend upon your predecessor to give you such information as she may have on the needs of the customer, her company's objectives (long-range), their goals (short-range); the organization structure, economic facts about the customer, and her experiences (good and bad) with you and with competition.

How about accepting your predecessor's evaluation of the self-images, traits, habit patterns, and status needs of members of the customer's organization? Answer: *For temporary guidance only.* Let's see why.

We have been talking about the self-image of the customer and the impact we have upon it. We must keep in mind that he has an impact upon *our* self-image. He had an impact upon the self-image of your predecessor. The customer may have made your predecessor feel uncomfortable, or important, appreciated, needed or inferior. Therefore, your predecessor will see the customer in terms of the kind of impact the customer had on your predecessor's self-image.

How many times have you been in a group, and when one person left and others started to discuss him, you were amazed at some of the reactions; they were so completely different from your own. The comments may have been critical, whereas you liked the individual. Or, the opposite may have been the case. We tend to see others in terms of how they seem to view us.

Your predecessor, for example, may warn you, "You'll have trouble with this guy; he's always complaining." Forewarned, you approach the customer, not in a hostile manner, but in a manner that is more guarded than warm. As a result, he sees in you the same threat he saw in your predecessor and goes on the attack. You leave the customer feeling that "Old Jim" was right, this guy is a crank. If you had listened to your predecessor, but had approached the customer looking for the best in him, you might have brought about a friendly response.

Don't disregard the views of your predecessor. Use them as guides during that period of time which you must devote to gaining your own impressions. Even if your predecessor has been very successful with this customer, keep in mind that his advice is not a substitute for pre-sale preparation and the slow, laborious, but eventually rewarding, process of customer analysis. Your predecessor's success is no insurance policy against future failure. The identification of customer needs and problems, and the identification of the customer's self-image needs, coupled with the enhancement of those needs, are essential ingredients for a rewarding series of sales.

Deprecates competition

We often forego pre-sale preparation because of another self-created obstacle; i.e., we underestimate competition. A distributor seeking an extremely large order had analyzed each of his competitors, even to the extent of rating them mathematically on each of the points the distributor knew the customer would consider. And, as a result, was certain he had the order clinched. The order was lost to a competitor who was not even on the list. The competitor was regarded as so small and inexperienced that he would not be considered.

The fact is that the competitor almost lived on the site of the job. He developed such close relationships with the customer that the customer was always aware of him. Conscious of his small size and inexperience, he convinced the customer that he could obtain the services of specialists and could subcontract portions of the order to a *larger* company.

"Never underestimate the enemy." No pun intended. This military axiom holds true in selling.

If you, and others, have been beating the competitor to the punch and getting orders at his expense, you may be sure that with every successive defeat he becomes a more dangerous and more determined adversary. He is now running hungry and will reduce his price even if it means only meeting his operating costs. He will find methods of providing extra services. He may even sell below cost to get a foothold, to get an opportunity to prove himself to the customer. He may see the situation as one that requires him to innovate or perish. Your customers have shown him, by refusing him the orders, that they are not impressed. Therefore, he and the men in his organization have probably been working at a feverish pitch to come up with something new.

Recently, we stopped at a cigar counter in the lobby of a large office building and witnessed an incident that demonstrates the validity of the advice given above. A salesman was standing in front of the counter with his order pad "at the ready." He had already taken orders for some of the merchandise and then asked the proprietor, "How many boxes of (brand) cleaning tissues do you want?" The proprietor replied that he had already purchased the cleaning tissues elsewhere.

Something akin to pain, anguish and incredulity spread over the face of the salesman, and he blurted, "What did you do *that* for?" The owner replied, "To save money—*that's* why!" It was obvious that they were old acquaintances, probably friends. The salesman had apparently been the sole supplier for a long period of time, and he could not bring himself to believe that the owner would ever give *any* of his business to competition. There was little doubt but that he had deprecated his competition. The thought to keep in mind is that as salesmen we can blame no one but ourselves for the existence of this obstacle. *We create it.* But did you ever read "I underestimated competition" on a lost-sales report?

Overlooks new competition

A salesman may be dedicated to the principle of pre-sale preparation and yet lose orders because he fails to use a checklist of the essential elements of intelligence—matters upon which he should be informed. Any complete list of elements of intelligence will include "new competition."

It is only a few years ago since many companies ridiculed the idea of foreign competition. Many have lived to rue the day that they discounted the importance of this element.

Salesmen are often so oriented to their older, established competition that they tend to ignore the emergence of new competitors. Often this is due to our failure to gather "commercial intelligence." We sometimes just don't know that these new competitors exist.

At other times, we can trace the reason for our failure to the defense mechanism called *repression*. (See Chapter 1, The Decision Pattern Chart.) You will recall that repression is that defense mechanism which we employ when we are confronted...or are afraid we will be confronted...with distasteful information. We use it to avoid problems, and often create even worse problems for ourselves. It's usually a case of, "Just keep quiet, and they'll go away." In part, this is correct: they *will* go away, and they'll go away with the order.

It's relatively impossible to be a *great* saleswoman in this era of complex and rapidly changing markets without being a *well-read* saleswoman. Shakespeare's complete works won't be of much use to you in pre-sale preparation, but the *Wall Street Journal*, the business pages of your local newspaper (or the customer's and competitor's local newspaper), trade magazines, industrial newsletters, the business sections of some of the nation's best known metropolitan Sunday papers, publications of the Department of Defense and other government departments... these will all prove fruitful and interesting reading.

But it isn't enough to read them; it's important to know what to look for, and the following is by no means a complete list:

Mergers. Will new combines give greater strength to competitors who previously have been "too small to count"? Will they now inherit research facilities and stronger personnel who will give them the edge they have lacked? It isn't enough to consider mergers of possible competitors. How about mergers of customers? Will new alliances give you an easier entree to companies that have denied you access in the past? Do you have acquaintances in the new combine that can assist you in getting in?

Personnel changes. Watch for announcements of new appointments and/or retirements (voluntary or otherwise) of people employed by your competitor (or customer). Do such changes indicate a dissatisfaction with current levels of business or methods of operation? Do they presage changes in policy that may affect you favorably? What is the reputation of the new appointees? Can you expect your competitor to operate with new vigor? Can you anticipate that he will employ new strategies or tactics?

Changing product scope or emphasis. Is your competitor divesting himself of product lines that have not been profitable? If so, you can be sure he is going to give greater attention to those product or service offerings he retains. What form will this attention take and how can you prevent it from taking business from you?

Advertising plans. Has your competitor changed her advertising agency or her mode of advertising or advertising media? Does this portend a more effective approach? Does it indicate that a firm which heretofore has *not* been your competitor is seeking to break into your bailiwick? Is your competitor seeking to establish a new image? How may this affect you?

TV programs. To many of us, TV commercials are the bane of our existence. To a professional salesman, they may be a source of invaluable information. They indicate the features your competitor seeks to emphasize. They tell you the kind of market he seeks to exploit, they reveal the image he seeks to establish. Listeners often turn away from the TV screen or switch on the "blab-off" when commercials appear on the screen. Is this what happens when a competitor's commercials appear, or do you find that viewers seem to enjoy, even appreciate, them? Is he being successful in capturing his audience? If so, does this call for new strategies on your part?

Contract awards. Has your competitor received large awards of contracts that will now make it possible for him to enlarge his scope of operation? Will he now be able to afford additional advertising or provide improved service to customers he seeks to wean away from you? How can you meet this new strength?

Sales training. The news media often carries accounts of new sales training endeavors. This is becoming more frequent of late as more concerns recognize that salesmen must be provided with every possible aid in today's complex market. Do these articles indicate that your competitor is taking advantage of new sales training programs? Can you expect an acceleration of efforts or a change in approach to customers you seek or customers you hope to retain? And how about your customer? If he is a retailer who sells your product or service, if he has launched a sales training program, how can you take advantage of it? If some competitor is supplying him with the program, what can you do to make sure your customer's sales force does not become oversold on your competitor's products in the process?

Competitor's schools. Is your competitor offering free schooling or plant visits to key buying personnel (or salesmen) of your customer? Will they be flattered or impressed by this treatment or affected by their close contact with his personnel? What strategies must you develop to offset these influences?

Patent announcements. Do such announcements indicate that a firm you have not heretofore taken seriously now becomes a serious contender for the business of your customer?

New money available. Do financial columns indicate that some small competitor has had new money injected into the business? You can rest assured they aren't going to invest it in the fourth race at Belmont. Will the availability of this new money transform this concern into a competitor to be reckoned with?

Keep in mind that such changes as those listed above (and many others) can make an old competitor more potent as a foe. But you and I are somewhat sensitive to such changes when they affect our established competition. We are not always so alert to spot these changes when they affect a firm we have not had to consider as serious competition in the past. The very name of an established competitor in a paper or magazine acts like a red flag. The same is not true of the small fellow who is anxious to grow.

Foreign competition. This now rates a separate classification in many of the businesses represented by readers of this book. The primary need is to be eternally vigilant to mergers or new working relationships between domestic and off-shore companies. If you are a distributor, don't sell this short by thinking it doesn't, or can't, affect you. It will!

No sources of information

Checklists included in this book indicate the *kinds* of information a professional salesman requires. The need then exists to develop ways and means of gathering information and converting it into commercial intelligence. Pre-sale preparation cannot be better than the adequacy and reliability of your sources of information.

In military combat, the business of gathering intelligence is considered so important that it is given a General Staff rating, one of the four facets of a military organization. (Training is another, incidentally.) Repeated for the sake of emphasis: the parallel between combat and marketing is so revealing that it should not be discounted because of any personal prejudice that one may feel toward military activity. It is as essential to gather intelligence on the customer and competitor as it is to safeguard the same type of information about one's own operation. The competitor, and at times even the customer, will regard such information as valuable bits of intelligence to be used against you for bargaining purposes.

A future chapter will be devoted to sources of information. For the moment, consider the sources listed earlier, and do not overlook the fantastic amount of information contained in stock-advisory services, Standard & Poor's, Dun and Bradstreet reports, and the like.

Remember, you *are* a professional, and keep in mind that a professional has ethics. The need exists, therefore, to distinguish between legitimate, ethical, moral, and legal methods of gaining commercial intelligence on the one hand, and the employment of collusion and/or espionage on the other. To do otherwise is to merit the scorn that has all too often been unfairly leveled against the selling profession!

Depends on reputation

Only recently, a national sales manager complained to us about the "smugness" of his sales force. He felt that too many salesmen, representing a nationally known company, thought customers should be so impressed with the firm's name and reputation that they should be half-sold (or more) before the salesman walked in.

Only a short time ago, another nationally known company, the acknowledged leader in its field, had a survey conducted to learn what the public thought of them. Let's call the company X. Questioned by survey teams, customers replied, "Company X is a good company and makes good products." The team then asked the same people what they thought of companies Y and Z—two successful companies, somewhat smaller in size. In each case, those questioned replied, "Company Y (or Z) is a good company and makes good products."

Saleswomen, therefore, often fail to take sufficient time for pre-sale preparation because they depend upon the established reputation of the company. This applies to retail or wholesale establishments as well when they are looked upon as the "quality" organization in their locality.

Salesmen who make this grievous error rely on the infallibility (?) of the Gestalt principle; i.e., the tendency of people to see the part in terms of the whole. In other words, their tendency to equate reliability, quality and performance with the company name. Salesmen assume that the customer will transfer this image to the products or services he seeks to sell... and even transfer it to him.

Without question, the precious good will of a company has enabled many salesmen to overcome severe competition. The name of a well-regarded concern may prove to be an effective door-opener, and at times it may clinch a sale when all other things are equal. However, today's marketplace introduces new considerations that make reliance on a company name a dangerous strategy and a poor substitute for professional salesmanship or marketing.

For example, let's say that your company does 20% of the business in your area. Not bad. But what does it mean? It means that apparently 80% of the people who buy your kind of products prefer to do business elsewhere. Or perhaps you may gloat, "Our outfit has 50% of the business." The fact remains that one out of two people prefer to do business elsewhere. In other words, your customers may be impressed with your company and its products or services, but apparently your competitors, alone or in combination, have impressed even more of those who should be your potential customers. Perhaps your company's reputation has impressed all of these potential customers. But at least some of them have been even more impressed with a product, delivery, buying terms, or personalized service which has been offered to them by a competitor who knew he had to out-perform you because of the very reputation your company enjoys. These are the very customers you hope to add to your portfolio.

This is not an argument against having a fierce pride in your company. You'll sell best when you have that pride, but don't expect that pride is a substitute for preparation...your competition hopes you'll make that error.

Perhaps, even if yours is a small company, you have excellent public relations personnel, though they may not be so designated on the organization chart. In spite of their best efforts, *they* have not swayed the customers you are losing. By what kind of logic, then, can it be expected that a salesman, working *alone*, can sway customers primarily on the basis of company reputation? Isn't it true that our efforts will be more richly rewarded if we forego the temptation to speak of "our company" and choose, instead, to speak to the customer (or let him speak) of *his* own greatness, *his* goals, *his* own achievements and aspirations; and then to show him how our products and our services are designed to enhance *his* reputation and aid *him* in the accomplishment of goals which are so important to him?

This brings to mind the dilemma of a company that had long enjoyed first place in one of its product lines. Then it began to lose substantial business to an alert competitor. Searching for the reason, company managers inquired of a disinterested observer of the market, and he replied, "I can always tell one of your salesmen. He walks into the dealer with briefcase in hand and proceeds to tell the dealer the kind of a dealer his company wants him to be. In other words, he wants the dealer to conform to a pattern that enhances *your* company's image.

"But I notice that when the salesmen for your competitor approach the same dealer, they enter his store, without briefcase, and get the dealer to tell *them* what kind of a dealer *he* wants to be. Then they say, 'We've got just the kind of products and just the kind of proposition to help you be just that kind of a dealer.'" The message is clear.

This same company, a major multi-billion dollar corporation, is enjoying success while obviously unaware of even greater potential that is being lost. It is being lost because its sales force is being regarded as arrogant.

The author, who meets a large number of this corporation's potential customers, finds many of them outspoken about the attitude of the corporation's sales force...a reaction that is costing the seller untold millions of dollars of business.

In summary, if the customer has not been buying from you, or if he has been buying only a portion of his requirements from you, then your company's image has not been enough to satisfy his needs. Therefore, more of the same will leave him unsold. It will be more productive to identify what *his* needs are as *he* sees them, and then seek to satisfy them via the personalized, customer-oriented route. If you accomplish this, on future calls, you will find that you have validated your company's claims for itself, and future business will be easier to obtain from this customer.

Takes customer for granted

Time is always in short supply. Perhaps this is why salesmen often create the obstacle represented by taking the customer for granted. With only so many hours available, the salesman has the tendency to assume that his "old" customer is already sold, and that time can be more profitably spent in prospecting for "new" customers and in gathering information on them and on their problems.

A faulty premise underlies this concept. Few customers are ever sold permanently. This applies to the chap who buys from you out of a sense of loyalty or because of obligations stemming from past favors you have accorded him. These are tenuous bonds, and we sometimes stretch them successfully for years. There remain countless customers who still buy from the same suppliers because these suppliers "took care of them" during periods of shortages. Others buy from suppliers because the latter stood by them in times of adversity; i.e., strikes, periods of cash shortage, and so forth.

One company refused to buy steel from foreign suppliers during a steel strike years ago. This earned the gratitude of its steel company customers. Other things being equal, it's a fairly sound bet that the supplier will enjoy the business of these steel company customers for some time to come. But read that last sentence again, and note the phrase "*other things being equal.*" No salesman for this supplier who sells to the steel companies can ever relax and take them for granted.

If you have a customer who feels indebted or under obligation to you, it's pretty certain that you have competitors who are pointing out to this customer that he has other—and greater—obligations to himself, his family, or his stockholders. This alert competitor is certain to emphasize in dollars and cents, or in product performance, what price the customer is paying for that sense of obligation. Just how long can he resist these overtures before he resorts to the defense mechanism of *rationalization*? Won't he eventually conclude that the debt has been paid and resort to the "what-have-you-done-for-me-lately" theme?

Customers can't be taken for granted at any point of sale. It's just as true in selling gasoline at the neighborhood gas station. You trade with one dealer, perhaps, because you like his station, its location, his brand and the manner in which you have always been treated. Then one day you drive in, and before he comes out the station door, another car follows you in and drives up to the adjacent pump. Even though the dealer gives you a big smile, a wave of his hand, and a hasty, "I'll be right with you," you are seething. The significance of what is occurring is not lost on you. You are being kept waiting because he takes you for granted, and he has seen an opportunity to impress a new, potential customer.

Just how many times will you take that kind of treatment?

The point is that there is no such thing as an "old customer." The customer you have been serving for an extended period is in some respect different today. He may be subjected to new pressures and anxieties. He may have new problems, new status needs. And he may have new experiences with your company or with competition. A long list of new influences on his behavior pattern may have been injected into the situation since last you saw him, even if it was only yesterday.

Taking the customer for granted is to assume that he is static, never changing. The intention here is not to portray the selling situation as one that is hopelessly complex. The intent is to emphasize that the customer is *not* static; he is dynamic, ever-changing. He does not relish routine treatment. He does not intend to be stereotyped. He does not intend to permit anyone to take him for granted. He wants to be appreciated; he wants to feel that you are interested in him as an individual and interested in his problems and needs. He wants custom-built attention.

To see what takes place when we fail to heed this warning, one has but to observe the attitude of some nations who should be our fast friends, but who turn from us because they feel we take them for granted. We don't know their boundaries, their language or their customs, but alert competitors from across the seas make it their business to learn all of these things, and we see business lost daily as a result. This lesson should not be lost on us when we deal with individual customers.

The final test, of course, to make sure that we have not been feeding you unsupported theory is merely to ask yourself, what is your reaction when you feel that some company from whom you buy indicates by their actions that they take you for granted? Seen them lately?

One-sale oriented

It is difficult to interest a saleswoman in pre-sale preparation when she is intent upon making "one big sale" and she feels she has sufficient information to wrap up this sale. She sometimes thinks that this is the one big sale she can expect from this customer, so she unlimbers her artillery, concentrates on making this eventful sale and succeeds in wrapping it up. Then she will say to you, "You see, this time-consuming preparation stuff is bunk. If you have everything going for you, you just move in, concentrate on customer benefits and you've got it made."

True enough, single sales *can* be made that way, even important big sales; but the continued progress of a company has never been due to "one big sale" or even to a number of big sales, exclusively. The big sales are undeniably important and any salesman gets a thrill of satisfaction out of completing one of them; but it is the continuing business represented by customers who come back again and again, or who persuade other customers to come in, that is primarily responsible for the growth and success of a company.

Throughout the nation today there are innumerable companies who are concentrating on "management training." The object is to train "managers" in the techniques of managing situations which they encounter. Unfortunately, too many companies fail to realize that most of the elements and problems which are present in a "management" situation are also present when a salesman attempts to manage a selling situation. It would be of benefit to us if we were to apply some of the principles of business management to the business of managing selling situations.

We could, for example, consider the concept of objectives and goals. A business manager is taught that he must have well-defined long-range objectives (two to ten years or more) and he must also have equally well-defined short-range goals (from today's goal to those of the next two years). He learns, too, that he must avoid the temptation to grasp at attractive short-range goals which may seriously threaten the achievement of his more important long-range objectives. The professional salesman must not only have well-defined goals and objectives to guide him, but he must be aware of the customer's long-range objectives and short-range goals. At times, the salesman may find that his job encompasses the necessity for keeping customer's attention focused on *his* (the customer's) long-range objectives and for aiding the customer in resisting the allure of the short-range goals.

To avoid the siren call of today's big sale which may threaten tomorrow's business, why not make the following sequence of questions your first step in the approach to *any* and every selling situation?

1. What are the customer's long-range *business* objectives?
2. What are his long-range *personal* objectives? (What image is he trying to project for himself?)

3. Which of these is going to have the greatest impact upon the selling situation which I am about to enter? (Which needs are going to be most important for me to satisfy—her product needs or her psychic needs?)
4. What are his short-range *business* goals? (What is his primary product or service need that he wants to satisfy today?)
5. What are his short-range *personal* goals? (What personal needs must I satisfy if I am to complete this sale; i.e., how does he want people to see him as a result of this selling situation and how can I help him to accomplish this?)
6. Which of these (business or personal) goals is going to be the more important for me to satisfy if I am to make this sale?
7. What is *my* long-range objective in terms of obtaining (or holding) this customer's business? (Have I assessed accurately his long-range potential?)
8. What is my short-range (or immediate) goal; is it solely to *get this sale* or is it to strengthen and maintain a lasting, favorable relationship with this customer?

As salesmen, you and I may make evaluations of the customer's buying potential that fall far short of the mark. We analyze his present buying power and assume that this sale is probably the one big purchase he will ever be able to make. This may be especially true in our analysis of the "small customer." We often forget that small customers are growth customers. Today's pygmy may be tomorrow's giant, and he will remember most favorably the salesman who shared with him his image of eventual greatness.

I recall one customer who purchased a TV station in a remote area of the West. The salesman was convinced that this was his first and last sale to this customer and set his sights accordingly. He sold the customer a station which was (at that time) in excess of his needs from a product standpoint and gave no thought to future relationships, assuming there would be none. The customer, however, had other ideas and twice, within five years, he has bought repeater stations that have cost him as much as the original station...from another supplier.

Short-sighted or goal-oriented thinking is dangerous. Any company that has been in business for any length of time has customers who were so small as to be insignificant, compared to the great bulk of customers. Today, a number of them are in first place in their respective industries. Remember, the little fellow has only one way he can go and that's *up*. If a salesman evaluates the little fellow in terms of his *capacity to buy* instead of by his *capacity to grow*, these customers will cease to be *his* customers and will rally around the salesman who shares their dream and their faith in themselves.

Another serious problem generated by the saleswoman who is "one-sale oriented" is that this thinking almost precludes consideration of post-sale service. Assuming that the customer has "shot his bolt" in making this one purchase, the tendency is to write him off and forget his post-sale needs. When the next sale does show promise, the selling task has been made unnecessarily difficult because tomorrow's business is largely dependent upon the service we provide on today's sales. It also means—and keep this in mind—more work for the saleswoman who must now obtain her sales volume by handling a greater number of "one-sale customers."

The cause of this obstacle is two-fold; it is caused by a form of myopia that blinds us to the promise of long-range relationships, and by inadequate research on our part as salesmen; i.e., failure to identify growth potentials of businesses and customers. We said earlier that growth companies are the small companies (with exceptions). If this sounds like unsupported theory, examine the list of the 100 largest companies that existed at the turn of the century. Then examine the list of the 100 largest companies two decades ago and of today. The casualties will amaze you. Then keep in mind that these former giants were the companies for whose business many salesmen fought while they ignored the "smaller firms," many of whom you will find on today's list of "new giants." As a matter of fact, you will be surprised if you examine today's 500 top firms and note how many were either not in existence or were small firms only 20 years ago.

Don't think only in terms of today's sale. *Keep your eye on your customer's dream. Grow with it.* Help him to achieve it earlier than he would without your help and he'll love you forever.

Product oriented

In the incomparable Broadway smash hit, *The Music Man*, the salesman had a fixation expressed as "You gotta' know the territory." Other salesmen firmly believe that the one essential is "You gotta' know your product."

There is no question about the necessity for knowing your territory. In a retail or wholesale operation this means knowing the needs of your own area. There is also no question but that it is essential that you know your products. More than that, you must not only know the strong features of *your* product, but you must know its limitations and vulnerabilities. (If you don't your competitor will acquaint your potential customer with those weaknesses.)

You must also know your competitor's product and its strengths...not merely its weak points. But this alone is not enough.

While you and I might not agree with the customer, she will often view our product as "about the same" as that offered by one or more competitors. At other times she may not be able to identify any benefit which our product offers that is not also offered by competitive products. Or the real benefits which our product offers may be "hidden benefits" or benefits that will become apparent only through use. Undeniably, in such instances, knowledge of the product is an essential ingredient for a successful sale.

Because of this preoccupation with product, there are too many salesmen who are convinced that the real "guts" of pre-sale preparation is a knowledge of product features, product performance and product applications. These salesmen feel that, fortified with this knowledge and with a pleasant personality and a good appearance, they are adequately equipped to move into the marketplace and to take on competition.

An analysis of the Estimate of the Selling Situation quickly reveals how much more there is to pre-sale preparation. Consider these needs for adequate preparation:

1. A knowledge of the economic factors affecting the customer's business and his readiness to buy.

2. A knowledge of the customer's past experiences as they affect his view of your offerings and those of your competition.

3. A knowledge of those factors which have conditioned the customer's reaction to you and your offerings: his habit patterns, his idea of what may be ethical, legal, and moral.

4. A knowledge of the me-he-seeks-to-project. What image does he seek to enhance and how may you recognize this need and assist him?

The professional salesman must understand the customer's frame of reference and the elements which have formed it. Then he must develop a personalized approach to the customer. In this age of similar products and services, this is often the *only* advantage you may offer him over and above that which your competitor may provide. When products are similar or equal in the eyes of the customer, you must offer him something extra, and that something extra may be made possible by your understanding of his economic and psychic needs. Know your product and that of your competitor, but not to the exclusion of the other essential elements of intelligence. Preparation cannot be considered complete until you have touched all the bases—product, economic, competitive and psychological.

Misinterprets information

We come now to the last of the obstacles which we shall discuss in this facet of pre-sale preparation. A salesman can prepare himself thoroughly for the sale and in the process amass much information and yet lose the sale although competition has no edge in terms of product features, performance, service or price.

The reason is often found in the salesman's misinterpretation of the significance of information he has gathered. Using the term "intelligence" as the military use it, we think of commercial intelligence as *information made meaningful.* Bits of information we gather on the customer in our pre-sale preparation must be analyzed to identify a pattern. Conclusions must not be drawn from separate bits of information.

For instance, a salesman may feel that a given selling situation is strictly a psychological problem ...and this may be a correct assumption. He feels that, in his case, there is not much to choose from between his offerings and those of competition. He is convinced that the salesman who unearths the customer's basic buying motive will make the sale, and again he is right. Then, however, he judges the customer by one or two things he has said or by one or two superficial possessions which the salesman identifies as "status symbols." In failing to *look for a pattern,* he makes a serious error, draws erroneous conclusions and appeals to the wrong psychic need. The kind of car he drives, the pictures on his office wall, the kinds of clothes he wears *may* all be clues to inner needs he has or to the kind of image he tries to project (and which he expects you to feed). *But*—they may be clues which if followed will lead you to the corpse...the corpse of your lost sale.

His wife may have threatened to leave him if he didn't buy the expensive car, she may have furnished his office and selected his clothing. If you identify *that* image, you'll make his wife happy, but you'll find him unresponsive and his pocketbook closed. On the other hand, the car, pictures and clothes may well be valuable clues, but draw no final conclusions until you have a substantial amount of information.

Unless you can identify a pattern in his behavior, it is unwise to attach much importance to isolated words, actions or possessions of the customer. It is even possible to identify his self-image and psychological needs as an independent individual yet be far off base because we fail to remember that *this* chap operates in an entirely different manner when he represents his company in a buying situation. Complex, eh? Well, whoever said that professional selling was easy? That's why the profession needs you and your ability to think.

Let's clear up that last point about the chap who has a self-image but suppresses it when acting for his company. You may, for example, observe that in his private life the customer is parsimonious beyond belief. He lives in a Spartan manner and stretches every dollar to the breaking point. It is understandable that a salesman could assume what appeals to this chap, to be effective, must be based on economy. Yet, this same man, when buying for his company, is determined to project his company's image which is one of top-drawer quality. Every purchase of his will be made with the determination to enhance that corporate image. The pattern to look for *is the pattern that is found in his actions on the job*. If his actions in his private life follow the same pattern, so much the better. If he follows two patterns... one in the office and another when you meet him outside, you must identify the two images he seeks to project and enhance each of these images at the proper time. In other words, your homework has to be complete, and it's worth the effort.

Another manner in which information may be misinterpreted, and hence not become intelligence, leads us to what may well be the single most important technique in the science of professional selling. For example: a salesman gathers full information on the behavior pattern of the customer and then makes the mistake of projecting himself into the customer's frame of reference (the psychological defense mechanism of *projection*). In effect, the salesman says, "If I did what the customer is doing, it would be for such and such a reason; therefore that's why *he* is doing it, so now I see what drives him and what I must do to supply fuel for that drive."

That's a sure-fire road to lost sales.

Here is the cure and the technique referred to above. We must think in this fashion: *If I were the customer* (not, if I were in his shoes, but I *were* the customer), *and if I had his milieu* (the total of all the environments to which he has been exposed), *I would feel just as the customer feels now. What could another salesman do to change my mind or make me act? That's what I must do to him now.*

The problem is to identify why a person with the customer's total background (milieu), with his goals, objectives and past experiences, takes the actions he takes and holds the views he holds. The question is *not* why you or I might take the very same stand the customer is taking. We might well take the same stand he is taking for different reasons. Let's step outside of the selling situation for a moment to study an example that makes this point clear.

In a company meeting, the subject of giving the junior executives in a retail store a day off each week was discussed. An executive expressed his opposition. A listener asked, "Do *you* take a day off?" The reply was "Yes." The listener persisted, "And you don't think these people should get the day off?" And the answer was, "That's right."

The listener was shocked. He couldn't believe that anyone would be that selfish, that autocratic. The listener knew that if *he* had made those same statements, it would have been because he believed rank has its privileges and that he considered his rank entitled him to the day off. The listener knew that these thoughts would be repugnant to him and that he couldn't possibly take such an attitude... therefore anyone who did take this attitude must believe in privileges of rank and was arrogant, selfish and autocratic.

What were the real facts?

The executive who opposed the free day was popular with his people. He was regarded as an excellent trainer of men. He felt that many of his people had the talent necessary to improve themselves. He felt there was no substitute for hard work and time on-the-job. He attributed his own considerable success to just this sort of arduous training. Also, he felt the store would lose business and everyone would be hurt if these junior executives had to be replaced on their day off with less experienced people. He was oblivious to any other interpretation that might be given to his remarks and would have been amazed to learn that they were regarded as evidence of arrogance.

The listener would not have erred in judgment had he looked for a pattern. If he sought such a pattern, he would have found no other evidence of autocracy or selfishness. He would, then, have been forced to seek another pattern that *did* exist. Seeking it, he would have found it, and drawn entirely different conclusions. Lack of basic charity leads to erroneous conclusions, and *the road to lost sales is paved with hasty conclusions.*

There is no short cut to professional selling excellence. You may obtain inspiration from books written and speeches made by successful salesmen. Some of their techniques may start you thinking about new strategies and tactics; however, selling excellence is achieved by hard work and by the awareness that selling is *not an art*, but a science. Selling excellence requires an awareness that the hardest work takes place during the period which we have called pre-sale preparation. It is at this time that you gather, sift and interrelate bits of information on customer, competitor, product and economics of the market until they add up to essential elements of intelligence.

CHAPTER 9

Establishing contact with the customer

In earlier chapters, we have emphasized that there are no "steps of a sale" with each step following the other in a predictable pattern. For example, under the step concept we might consider "getting his confidence" as a step. Logically, it would appear that this would chronologically follow "contact" or "approach." Yet many times, the salesman may have gained the confidence of the customer long before he ever contacted the customer in person.

No, there are no "steps of a sale"; but there *are* facets of a sale, and the first facet treated was pre-sale preparation. Now we deal with another facet of the sale: "contact." We think of this as that moment in time when you first attempt to meet with the customer for the purpose of making him aware of the fact that you and the product or service you sell, can be of service and value to him. As one purchasing agent put it, "You try to show the customer how you can make his job easier."

Planning for the contact (if planning is possible), or being aware of the importance of this facet of the sale, calls for keeping in mind what some unknown wit has said: *"You never get a second chance to make a first impression."*

Time

You will recall that most of the obstacles that can arise during the pre-sale preparatory facet are usually those of our own making. The obstacles that emerge during the contact facet are ordinarily caused by the customer, and we create a few of our own for good measure. In a given selling situation, you may not encounter *any* of the obstacles we are about to discuss, but it makes good sense to be aware of them so that if they *do* arise, you will be in a good position to overcome them.

Perhaps in reading these chapters, you may glance at the title or read a few paragraphs and then think, "This discussion doesn't concern my kind of selling: I'll pass it by and see what's covered later on." This may be the reaction of some salesmen when "Time" is listed as an obstacle. The chap who sells behind the counter may feel that time is not an obstacle in *his* kind of selling—the customer has a need or a problem and has taken the *time* to come up to the counter; so, where's the problem?

The problem may be that although he has come to the counter, he has other things on his mind that don't leave him enough time to really listen to what you are saying. He may not have had enough time *before* he came in to see you to identify his problem or to determine what product or service will provide him with the relief he seeks. You are going to have to capture his attention and hold it.

Perhaps in your kind of selling, the kind of equipment or service you sell is of such value to the customer that he readily sets aside the time for more than one discussion or meeting with you. But—does he have a tendency to rush by certain aspects of your proposition, leaving you with the feeling that you could do a more convincing selling job if he would stop "rushing"?

Do you have other customers who have real problems that you could solve but who never seem to be willing to provide you with the time you need, simply because they consider other problems as having a higher priority on the limited time they have available?

Time is often the key obstacle to making contact even when we, as salesmen, are not aware that it is a factor. A customer may avoid seeing a salesman, and the salesman may chalk this up to "evasion for unknown reasons" or to "lack of interest." The real reason for the customer's elusiveness is a feeling on his part that if he commits himself to seeing the salesman, the interview is going to take more time than he can allot or more time than the problem justifies. Again, the real problem is capturing his attention and causing him to see how important it is to him to solve the problem or fill the need. One customer habitually told salesmen from a supplier's firm, "O.K., if you can give the story in five minutes, go ahead; otherwise, I'm not interested." How do you handle these types of situations? There are numerous methods, and this section will concern itself with a number of them.

Method no. 1—face up to it

1. Study his problem.
2. Prepare a solution.

3. Identify the features of your offer in terms of benefit to the customer.
4. Determine which of these advantages will have the greatest impact upon him, and list them in the order of their importance as he will see their importance.
5. Prepare a presentation that will be streamlined, that will accent the attention-compelling advantages and that fits the time you think he will make available to you.
6. Make intelligent use of visual aids.
7. To the extent possible, involve him in the presentation, whether it be by adroit questioning that provides him with a chance to talk, or by having him take part in any demonstration which may be included in your presentation. Get the merchandise in his hands, if possible. Remember, a piece of merchandise in the customer's hands is worth two on the shelf.

One of these steps deserves further consideration; i.e., "use visual aids." We have all been exposed to the bromide, "one picture is worth 10,000 words." Properly used, visual aids *can* save you words and get your story across fast. They can capture attention, but visual aids must be chosen and/or used intelligently. They can bore the customer, or they can irritate him if he feels he is being exposed to a "canned presentation," and they can anger him if he is made to feel they talk down to him.

How to use visual aids in selling. Have you ever been bored by a visual aid? If you will recall, it probably was because the visual aid stressed the obvious, or perhaps too many were used, or because you knew, from experience, that this salesman *always* used visual aids. In speaking to the representative of a company, a customer said, "I can always be sure when one of your people comes in here to make a presentation: he will come with a truckload of charts. In the beginning it was O.K., but now it seems to be a standing operating procedure for one of you fellows to stand in front of us, reading one chart after another. You'd think we couldn't read."

In one instance, a salesman was selling to a general and his staff. The general sat watching the salesman as the latter read the text of each page of a flip-over chart until he could stand it no longer and burst out with, "Look, young fellow, I can read, and I don't have to be read to; so, you just stand there and I'll read the charts. When I say 'Flip', you just flip the page." Imagine the discomfiture of the salesman who dutifully stood there while the general emitted a series of grunt-like 'flip...flip...flip...' With each flip, the salesman felt smaller and smaller and smaller. It has been said that to manage a selling situation, a salesman must be dominantly persuasive while enhancing the status needs of the customer. Just how dominant did this salesman feel?

Many customers have fast-acting minds. Perhaps this is why they have reached decision-making levels of authority in their companies. They read quickly; they comprehend quickly. Nothing bores them more than being required to follow charts at the slow reading pace that a person who reads aloud must follow. They also become annoyed when a saleswoman turns a complicated chart too quickly, before they have had a chance to comprehend the chart. Visual aids should be just that—visual and aids. They must be designed to get across, in a brief period of time, that which would otherwise require many minutes of reading or explanation.

Your visual aids must be understandable and readable. They should not be used just because "it is the thing to do." They should not be used when they are below the level of the customer's current knowledge. They must not be too complex because this type of visual aid is no aid at all. You may find yourself spending more time explaining the chart than you would have used in explaining the point around which the chart was built.

If the charts are too small for the room or for the size of your audience, the audience will become irritated, and they will charge it up to "poor preparation" on the part of the salesman. No amount of explanation that you realize the charts are too small and that you will read the small type is going to erase the impression that you came poorly prepared. The suspicion will be sown that perhaps the balance of what you have to say will reveal the same inability to cope with their problem...the problem you came to solve.

Visual aids can anger the customer. Too often, they reveal a tendency on our part, as salesmen or vendors, to concentrate on *our* importance. He is usually well aware of our size, our achievements and our capacity. What he wants now is to be sure you are aware of *his importance and his problem.* In Part One, we discussed the needs of the self-image. We pointed out that all of us, customers included, think of ourselves as being somewhat unique, and we often think that our problems are unique. We don't like to be robbed of that feeling, whether we are describing an operation we have been through in the local hospital, or some business problem we have. The scar the doctor left us with must be bigger, or more beautifully stitched, or in some way unique. You customer wants custom-built treatment, and sometimes the use of a visual aid leaves him feeling that you are giving him a treatment designed for someone else. This does not mean we should not use visual aids; it means they must be used with care.

The preparation of visual aids presents two problems. First, they must not be carelessly put together, leaving the impression that the salesman doesn't consider the customer important enough to spend sufficient time in preparation. Second, they should not be so attractive that the customer pays more attention to the aids than to what the salesman is saying. Color has definite advantages in some situations, but color may also be a disadvantage. In a factory, employees were more impressed with the skin tones and texture of the skin of a girl's hands in the film than they were in the safety practice being taught. When the same film was produced and shown in black and white, accidents dropped. Price of visual aids and their effectiveness do not go hand in hand, necessarily. Visuals should not be so attractive that the message gets lost.

The advantages. Reading the foregoing, one would think that this is intended as an argument against the use of visual aids. Such is not the case, but their disadvantages should be considered and overcome or avoided by considering their use from the standpoint of *this* customer; and pre-sale planning should include planning on how to design and/or use the aids.

Visual aids *are* one of your most effective selling tools in combatting the "Time" obstacle; i.e., when the customer feels that he hasn't the time to listen to a really comprehensive selling presentation. Consider for a moment the basic cause of the time obstacle. The customer uses time as a valid excuse when he does not have a high degree of interest in what you have to offer, or when he is not fully aware of the nature or severity of his problem. In other words, he does not give this problem a high priority, and he conserves time for other problems and activities which he does consider important. Obviously, then, the time obstacle must be combatted with interest-arousing tactics.

If you are selling small items, remember that the product itself may be your best visual aid. Get it into his hands. Have you had someone hold a device at a tantalizing distance while you felt a mounting desire to handle it, manipulate it . . . just feel it. By the time you finally had it in your hand, your irritation was probably so pronounced that you had lost interest in it, or you had become annoyed at the salesman.

Make it *easy* for your customer to get salient facts in the shortest possible time. Use words and devices that whet her interest to the point where she gives you additional time. Try such techniques as letting *her* run the slide projector, if you are using one, so that *she* can decide when she has looked at a slide long enough.

Above all, *don't—don't—don't* talk when you have handed him something to read. He simply cannot read and comprehend while you are talking. He either reads and hears nothing you are saying, or he listens to you and doesn't read a word—and the chances are great that he has become annoyed.

Hand him the material to read, and when he has finished, you'll have your chance to talk. This fault is not confined to salesmen. Time after time, subordinates bring documents in for the boss to read. Then, as he begins to concentrate on reading, they start to talk, and they wonder why he becomes irritated.

Anticipating and preventing visual aid problems. There are many voluminous publications dealing in depth with the use of visual aids, hence the following is not intended to represent complete coverage of the subject. Furthermore, the more detailed volumes cover situations rarely, if ever, faced by the professional salesman. The following represent only a checklist of considerations which must be kept in mind by salesmen when they intend to use the type of visual aids customarily employed in many sales situations.

Before you definitely decide to resort to visual aids, consider these factors:

Is the room in which they are going to be used large enough to accommodate the group comfortably, and will all present be afforded seating that makes visibility possible?

Is ventilation going to be adequate?

Will lighting conditions be such as to make for poor visibility?

Is printed material used on slides, films, or charts going to be too small for everyone present to read? If so, don't use them.

If audio systems are to be used (sound films, etc.), can sound level be controlled so that those in adjoining areas will not be disturbed or antagonized?

Will ventilators be blowing on any member of the audience?

Will anyone be subjected to glare from windows or light sources?

If any of these conditions present problems, you can be sure that much of your message will be lost, and the effectiveness of even expensive visual aids will be diminished.

Before you decide to use the visual aids, consider these factors also:

Are you *really* totally familiar with the films, tapes, or slides?

Have you anticipated questions they may create in the minds of the audience and are you prepared to answer them satisfactorily?

Will your audience be able to relate to the aids?

If slides are to be used, have you made absolutely certain that they have been arranged in sequence and are properly positioned in the projector, including some that may have to be positioned vertically? Have you given them a dry run in advance make sure that no slide has bent corners that may stick?

Do you have a spare bulb on hand, or are you gambling that the bulb is good for "just one more showing"?

Do you know the location of the power outlets? Do you have an extension cord with you of sufficient length, or must you annoy the customer by asking her people to obtain one?

Is the voltage and current supply right for your equipment? (In a major New York hotel, a sales trainer encountered only direct current supply and the visual aid showing had to be cancelled...unusual, but it happened.)

Have you made provision for an adequate screen?

Do you have a flashlight available in the event that you do have difficulties with projection equipment?

If you are using sound film, did the last user of the film rewind it on the original reel, or will you be embarrassed to find the need to do so while the customer waits?

Did you check the film itself to be sure that someone has not put the wrong film in the canister?

Has the lens been cleaned?

Did you check the threading of the film in advance of the showing to make sure sound and film are synchronized? Lack of synchronization can really annoy an audience.

If you are using charts or a blackboard:

Is there an easel on hand? Do you know how to erect it without appearing to be unfamiliar with it?

When you must turn pages of your chart, have you made provision for turning each page separately... either by using tabs or cementing small pieces of sandpaper on the bottom corner of each sheet...or will you annoy the audience by fumbling while trying to separate the sheets?

Is there anything your audience will find offensive ...put in there by someone with a unique sense of humor?

Do you have an adequate supply of crayons...are they the right color to be visible?

If using a blackboard, have you checked to see that chalk is available in adequate supply, and that an eraser is available? Will an eraser really clean the board? (Recently in one of the nation's finest conference centers we were unable to wipe the boards clean using a normal eraser, even though the boards were brand-new. Use of the boards had to be abandoned while we resorted to flip-charts.)

If using one of the white "blackboards" owned by the customer, are you using water-soluble crayons, or is there a chance that you will scratch the surface of his board or find it almost impossible to erase what you have written when it must be replaced by additional information? Know in advance what the customer's conference room has to offer. Inspect the room in advance to cover all of the points above which are germane to the situation you face.

These are all basic, and saleswomen often find them boring, but a slip-up on one or more of these precautions not only can destroy the effectiveness of the most expensive visual aids, but they project an image of something less than professionalism on the part of the saleswoman who is using the visual aids.

Method no. 2—prevent it

The second method for eliminating the time obstacle is to recognize it when it exists and attempt to eliminate it before contact is made. Again, this means that ways and means must be found for convincing the customer that the need you wish to fill, or the problem you hope to help her solve, are so important that they are worth her time and attention. This section will discuss a number of ways and means available to you, but only you can decide which will work best with *your* customer in the specific selling situation in which you meet her today.

In this book you have been provided with a "Sales Strategy Bank." It will be taken up in detail later in Part Three, but for the present it can serve as a checklist of tactics to help you arouse the customer's interest and overcome the time obstacle. But, again, it is you who must decide which tactics serve best in dealing with this customer. This is why we—and you, too, we suspect—are annoyed with superficial sales training programs that attempt to furnish salesmen with the one best way of meeting a given obstacle. Only the salesman, knowing his customer and the situation in which he will meet him, is the judge of which tactic will work. The Sales Strategy Bank provides the salesman with reminders of tactics that are available to him.

The material that follows is not a complete list of attention- and interest-arousing tactics that can be used. It *is* a helpful list. In choosing your tactic from the list, bear in mind that the reasons for the customer's actions will often stem from her status or psychic needs.

Exclusive of the product or service need which you hope to help him solve, what is the customer's most pressing need? What appeals will arouse his interest? Is he ego-centered and can't stand the idea of any competitor surpassing him? Is he hungry for recognition? Is he motivated most strongly by economic needs? Is he a short-range planner oriented to a quick return on investment, or is he a long-range planner oriented to conservative operating procedures? Who does he want most to impress: his boss, his peers, his customers...or perhaps you? How can you help him in this regard? Can your behavior at the moment of contact cause him to see you as a person who can aid him with these psychic needs as well as with his product needs? If so, time will cease

The Sales Strategy Bank

Pre-sale planning
- A1 Develop sources of information
- A2 Identify his needs and wants
- A3 Identify his self-image
- A4 Identify his primary buying motive
- A5 Consult your predecessor
- A6 Get help from other depts. in your company (finance, etc.)
- A7 Get boss's help
- A8 Identify your competition and its representative
- A9 Study competitors' usual strategy
- A10 Make an estimate of situation
- A11 Determine past experiences of customer
- A12 Identify his possible sources of resistance
- A13 Prepare alternative offers
- A14 Plan strategy
- A15 Plan opening tactics

Create good climate
- B1 Save his time
- B2 Help him
- B3 Offer to help
- B4 Agree with him
- B5 Give him privacy
- B6 Do him a favor
- B7 Praise, compliment him
- B8 Give him credit
- B9 Emphasize mutual interests
- B10 Welcome objections
- B11 Agree
- B12 Admit validity of his view
- B13 Find point of agreement
- B14 Admit your error
- B15 Entertain him
- B16 Be intentionally incorrect in a statement so he can correct you
- B17 Provide him an escape hatch
- B18 Listen
- B19 Ask for his help, opinion or advice
- B20 Sympathize

Compel attention
- C1 Provide him with a role
- C2 Use cause-and-effect examples
- C3 Use testimonial
- C4 Emphasize his goals
- C5 Emphasize his objectives
- C6 Omit something so he can inject his ideas
- C7 Confuse him. Or use ambiguous questions
- C8 Interest him in more ambitious goals and objectives

Overcome his objections
- D1 Give him a choice
- D2 Narrow down his choices
- D3 Provide solution
- D4 Provide alternative
- D5 Compromise
- D6 Talk benefits
- D7 Draw comparisons
- D8 Tack with his stand
- D9 Use probe for reaction
- D10 Divert him
- D11 Rephrase and overstate his objections
- D12 Get him to talk
- D13 Appeal to fairness
- D14 Fence
- D15 Use direct question
- D16 Use overhead question
- D17 Use rhetorical question
- D18 Use provocative question
- D19 Scare him
- D20 Let him do you a favor

Overcome his fears
- E1 Provide support
- E2 Provide assurance or proof
- E3 Demonstrate
- E4 Explore cost of wrong decision
- E5 Remove him from situation with third party examples
- E6 Appeal to his pride
- E7 Appeal to his prestige
- E8 Appeal to the me he wants to project
- E9 Give him recognition
- E10 Anticipate, raise and answer objections
- E11 Stimulate his recall of past experiences
- E12 Provide guarantee
- E13 Offer additional services

Force a decision (Make him take action)
- F1 Bypass him
- F2 Shock-antagonize him
- F3 Show him cost of procrastination is higher than cost of decision
- F4 Emphasize cost of not buying

Post-sale actions
- G1 Show continued interest
- G2 Make return call
- G3 Check operation and/or installation
- G4 Has his problem been solved?
- G5 Have new problems been created?
- G6 Is he satisfied in all respects?
- G7 Have all of your promises been kept by your company?

---

its role as an obstacle for you.

There are other needs with which a salesman must be concerned. Does the customer have a myopic view of his market; i.e., does he, for example, fail to see his needs of one or two years hence? Does he wait until problems are upon him before searching for a solution? If this is the case, are there other customers who have had the same problem he now has but fails to recognize? He may feel that you are more concerned with making a sale and as a result resist you, but if his attention can be focused on another customer who had the same problem, you will find his interest quickening and the time obstacle diminishing.

Can you use in an effective manner, mailing pieces that are so well designed they will alert the customer to his needs so that the time obstacle will be neutralized when you call upon him? Is there anyone else in your company who calls upon him and in whom he has confidence, who can alert him to the problem which he now under-estimates? Is he fully aware of the cost of *not* buying from you . . . the cost of not giving you some of his time?

Are there examples that you can supply the customer, or have supplied to him by others whom he trusts, that will shock or scare him into an awareness of what will happen if he disregards the subject? Remember, again, we are not talking about manipulating the customer, causing him to do something in our selfish interest. We are talking about motivating him, causing him to do something in our mutual interest. If we use "scare" examples, they must be valid, believable, sincere, and must represent a threat to his economic needs or his emotional needs . . . whichever, at the moment, are most important to him.

You are urged to refer to the Sales Strategy Bank in Part Three to determine which strategies you can best use to arouse interest and compel attention to the point where the customer will see the situation as one which is worthy of *his time*, thus overcoming the obstacle. To save *you* some time, we have examined the Sales Strategy Bank, and the following are some of the strategies which we recommend you consider, always keeping in mind the unique nature of your customer and the relationship between you.

Actions to consider

1. Provide him with a role. This includes the use of "fantasy"; i.e., cause him to see how you can help him look well to others, in keeping with the needs of the "me" he wishes others would see when they see him.

2. Save his time. Through thorough preparation and a streamlined presentation, make him aware that you are trying to conserve his time. This also indicates to him that you consider him important. A note of caution: Do not carry the streamlining to the point where he regards your proposition as of minor importance; don't compromise too much.

3. Do him a favor. If possible find a way of aiding him in some other area so that as repayment he will give you the few precious moments you need to concentrate his attention on the problem or need you hope to solve or fill. Obviously, you don't mention the past favor, and your proposition must be able to stand on its own feet. The problem you want to help him solve, or the need you want to fill, must be real.

4. Talk advantages (benefits to her) always. Mention the advantage first and then back it up with product or service features, and, if possible, with proof or demonstration. Select those advantages which are most important to her as your kick-off ammunition.

5. Demonstrate. Save words; give credibility to your claims; demonstrations produce action, and action holds interest.

6. Use cause and effect examples. This means using case histories of other customers or similar situations. Be sure you know how he regards the other customer. Don't exaggerate. Be believable. Don't cause him to lose face by using examples that make *his* past actions seem something less than smart.

7. Use testimonials. Don't overwhelm with a long list. Be selective in those you use. Again, determine what he thinks of the people you intend to use as testimonials. You may regard them as important; he may not share your view. If he finds that you regard someone highly whom he holds in contempt, it will reflect upon your total judgment in his eyes if he feels strongly about the person you have used as a testimonial.

8. Emphasize the cost of not giving time; i.e., of not buying. Alert her to the financial cost of not buying, *and let her see* the psychological benefits (needs of her self-image) she may be losing by not buying.

9. Stimulate his recall of past experiences. Don't make him lose face by doing so; if this is a danger, couple it with "give him an escape hatch" (also from the Sales Strategy Bank). Remember, he—as you and I—has an analgesic memory that erases memories of mistakes and past examples of poor judgment. We, and he, need to have someone remind us of the cost of our past mistakes. When you and I, as the salesman, find it wise to remind the customer of a past mistake, we must find ways of reminding him without his being aware of the fact that we remember his past error. This calls for the best in professional selling—but, again, who ever said that professional selling is easy?

10. Raise his threshold of satisfaction. This is a long-haired way of saying that you must make him realize that he can be better than the "me he wishes to project" (the me he wants others to see when they see him). He may have reached that stage already, and the job is to elevate his sights, to make him set new and more demanding goals and objectives. Whet his appetite for things he has heretofore considered beyond his grasp. This is where a professional salesman makes his greatest contribution, causing people to want to *be* something better than they ever thought they could be.

11. Get him to talk. Probably the most effective strategy to make a customer forget the time obstacle. Get him to talk and you'll find that the customer who didn't "have five minutes to spare" will say good-bye an hour later. And you'll either have an order in your pocket, or you'll be a giant step ahead in the long trek to the sale of the kind of equipment that requires many calls.

12. Ask for help or advice. The greatest of all human appeals is, "I need your help." This is not to be confused with the infantile "I've just *got* to get an order today"; or "I need help in making my quota in the contest." As a reader of this book and a professional salesman, you have a right to be annoyed that we would ever sound that note of caution. No professional would ever resort to these types of appeals, but because many non-pros *have* done so, the note of caution to newcomers is injected.

Here is an example of the right way to ask for help or advice. A saleswoman asked a customer, "I'd like your opinion of the marketability of (a device)." She then went on to describe its features in terms of advantages to this customer and finished with, "How do you think such an item would sell?" The customer not only gave his opinion but added, "Incidentally, keep me in mind. I'd like the first one you produce." The customer called two days later to remind the saleswoman.

This, incidentally, was a customer who had been too "busy" to give the saleswoman much time; but the satisfaction of his "need to be needed" was too great to resist. It might be added that the device was already in production and the saleswoman's approach represented excellent pre-sale preparation. Nor was the customer "manipulated"; he *did* have a need that the device would fill, and the saleswoman knew it. Asking for advice *motivated* the customer. Such appeals to pride and to the other needs of the self-image not only produce sales, they produce valuable market research because if the customer is negative in his comments about the product, he may show you what is wrong with your product planning.

13. Restate, and overstate, his objections. If he has indicated that he does not regard the problem which you wish to discuss as being important enough to warrant the expenditure of his time, overstate his objections or comments. This often causes him to say, "Wait one minute. I didn't put it that strongly." Then, in an effort to show you what he *did* say, he dilutes his objection considerably. He may begin to sell himself or to give you an easier objection with which you can cope. You can then turn to your Sales Strategy Bank and use appropriate tactics such as: "Admit your error," "Find point of agreement," or "Tack with his new stand."

14. Shock or antagonize him. This is last resort selling and should be used only when everything else has been tried and you have nothing to lose. Even then, consider its effect upon future business, upon your company's image and upon other salesmen for your company who may sell her other lines. This strategy will be discussed in detail in a future chapter with examples of its use.

15. Bypass him. This strategy is always there to be considered, but it is fraught with obvious dangers and all too often used as the refuge of the witless. In spite of that, there are times when it is warranted and essential; so, it must be included in any list of strategies to overcome the time obstacle.

Priority

Your customer may be aware of the needs or the problems which your product or service will fill or solve. He may not share your feeling that these are necessarily his most pressing needs or problems. He may have placed a higher priority on other needs and problems. Worse, he may have completed his budgeting for the year and may not have allocated funds for your type of service or product simply because he did not consider it as important as other demands upon his resources. To put it mildly, this type of situation makes it most difficult to make contact with him or to ever get beyond the "contact" facet of the selling situation.

What to do?

One person may view an object through a telescope and see that object magnified. Another person viewing the same object through the other end of the telescope sees the object reduced. If you want the other chap to see the object as you do, you must make him view it through your end of the telescope—and so it is when the obstacle of "priority" enters the selling situation.

One way to solve the problem is to depend upon luck, and that's not the way of a professional. A better method of solving a problem is to isolate its cause. If a customer sees his other problems as being of greater importance, it is our job to determine why; in other words, we must be able to see the situation from his viewpoint. This is *empathy* and is not to be confused with *sympathy*. When we develop the ability to empathize, we then find it possible to ask ourselves the most helpful question in professional selling; i.e., "If I were the customer, with his background, milieu, and problems, I would see the situation exactly as he sees it today. If that were the case, what could somebody else do to change my mind—*that* is what I must do to this customer now."

Keep in mind that it is not enough to say, "If I were in the customer's shoes." It would still be *you* in his shoes. Rather, we must say, "If I were truly the customer with all of *his* fears, aspirations, needs and problems, I would feel just as he feels." This is a far cry from bland sympathy which can result in lost sales because it causes us to say, "I see what is bothering the customer, and I sympathize with him and don't blame him for not buying. This is not the time to try to sell him." That kind of thinking not only loses the sale, but it represents a disservice to the customer. He doesn't require sympathy; he requires *understanding*.

If the customer feels that she has other problems that are more pressing than the problem your product or service will solve, your task is to determine why. She sees certain needs threatened by her other problems which are greater than any needs threatened by the problem you are trying to emphasize. What is the nature of those needs? Are they product needs or economic needs, or are they status needs? How can you cause her to see these needs as being less important than the needs you are trying to help her fill?

Selling is a highly personal situation, and each customer has unique reasons for placing priorities on the various problems he encounters. Often, as you well know, he will not willingly divulge his reasons, especially if they are based on status or if they will reveal him as being status-hungry. He will often use smoke-screen objections to hide his real objections, or he will resort to simulated apathy and/or disinterest.

The best method for determining why the customer places a greater priority on other problems is to get him talking and keep him talking. Remember, too, that we are still in the "contact" phase of the sale, where we are trying to make a good impression or trying to reinforce a good impression we have made in previous contacts with him. Customers, as you and I, like good listeners, and listening is a sign of respect. If it is *sincere* listening, it avoids the pitfalls of empty flattery. It also aids us in identifying his self-image (see Part One) and provides us with valuable intelligence.

From the Strategy Bank, in Part Three we get a number of strategies that merit your consideration. You, as a professional, must decide which of these fits *this* situation and *this* customer.

D15—Use direct question.
E4 —Explore cost of wrong decision.
D16—Use overhead questions.
A2 —Identify his needs and wants.
A3 —Identify his self-image.
D9 —Use probe for reaction.
B9 —Emphasize mutual interests.
B18—Listen—really listen!
C4 —Emphasize his short-range goals.
C5 —Emphasize his long-range objectives.
D11—Rephrase or overstate his objections.

This list is not in any "approved sequence." Your immediate goal is to get the customer talking to identify the problems to which he has assigned a higher priority than that which he has assigned to the problem you wish to help him solve. During the Contact facet of the sale, you are only trying to compel his attention. Therefore, you may wish to save some of the strategies listed in this chapter for later facets of the sale.

Once you have identified the problem that he sees as more important, you have the task of making him see that his "top priority problem" and the problem you wish to solve have something in common. You must cause him to understand that they have the same inherent threats to his goals, be they economic, product or status (psychological) goals. Finally, if possible, you must make him see that these threats are greater in the problem you are trying to center his attention upon, and you must excite his interest by showing him that this problem is immediately solvable. It can be seen that E4 *(Explore cost of wrong decision)* would not be one of your first actions.

We have used the word "threats" rather freely. It is not always some threat to his goals that moves him in a direction which leaves him disinterested in you and your proposition. It may be that he sees greater benefits or advantages in giving higher priority to other needs. You must whet his appetite for the advantages or benefits which your product or service offers and cause him to regard these as more desirable and *attainable* than other benefits he had in mind.

We have placed emphasis on the need to identify her long-range objectives and short-range goals. Why? Because your customer is a rational, thinking person. She will do only those things which, in her opinion, lead to the achievement of these goals and objectives. Her failure to show interest when you contact her, in your product or service, is the result of her conviction that the problem you are discussing does not have as great an influence on the attainment of her goals and objectives as does the solution of other problems. She therefore sets a higher priority on the solution of other problems and the assignment of her funds and her personal interest to those other problems. If you are to be successful in arousing her interest (which is what must be accomplished during the contact facet of the sale), then you *must* identify her objectives and goals, and your presentation must emphasize how you can help her achieve them.

Before leaving this problem area, let us be realistic and face up to the fact that some customers have problems which, to them, are so important that they will never give our presentation their attention because it is focused on problems which they consider of less importance. In such instances, the best a salesman can do is to cause the customer to see the problem which the salesman's product solves as being second in importance. The salesman should work for a commitment to discuss this problem on a future, specific, date. Failure to accomplish this will probably mean that some new problem has been given top priority and another call will be wasted. If you can't get by the Contact facet today, make the next contact certain.

Recall—influence of past experiences

Every action of yours serves as a stimulus to the customer. You hope it stimulates him to move in the direction you desire. Your words, your actions, the name of your company, your product or service, the very problem you are attempting to help him solve, may spark off memories of past experiences in the mind of the customer. No sale is every made in a vacuum. The influences on recall were discussed in detail in Part One, and review of those influences is urgently recommended.

Recall is not always an obstacle. Your presentation may cause the customer to recall satisfying experiences that will make the sale easier. In such instances it will be to your advantage to use recall as a tactic. Try to draw a parallel between today's offering and some previous experience which you know satisfied him. Don't depend upon the customer to remember unaided. He may not see the parallel without assistance.

Too many superficial sales training programs encourage the salesmen to "know the customer." The question is, "What should we know about him?" One of the things we should know is as much as possible about his past experiences. Which did he find profitable, which did he find threatening, disagreeable or costly?

We begin now to see the value of the general treatment given the psychological aspects of selling in Part One. What may have been regarded, in part, as "interesting theories" begins to take on new meanings. If a customer has had a previous experience, and she was unhappy about that experience, and if she sees a parallel between that experience and your proposition today—you've got a problem.

It is quite obvious that it is in your interest to know as much as possible about his previous experiences as you can. The experiences may not have been with you. They may have been with another person in your company or with another department, if you are a multi-department concern. They may even have been with a competitor. In any event, these previous experiences represent an obstacle, and the best way to handle an obstacle is to recognize it in advance of the sale, and then plan tactics and strategies that will neutralize it or prevent it from arising.

The psychologist says, "Restructure the situation" or "Change the setting" so that today's presentation will not excite his recall of disagreeable experiences. If this is not possible, then the next best strategy is to be prepared. It may even be wise if you *know* that the obstacle will arise, to bring it up yourself and answer it before the customer does. If he raises the obstacle without your assistance, and you answer it, he may consider you to have given him an off-the-cuff reply and he will be unimpressed. If you raise an obstacle that he was prepared to raise and you make a reasoned presentation, he will feel that you have given thought to his problem and have not dismissed it lightly *(E10—Anticipate, raise and answer objections)*.

Turning to our Strategy Bank, it becomes evident that A1 *(Develop sources of information)* is a must. A10 *(Make an estimate of the selling situation)* is *always* in order. If the situation that he remembers was his own fault and he knows it, consider using B17 *(Provide an escape hatch)*. If the previous sad experience was of your creation—and he knows it, consider using B12 *(Admit validity of his views)*; D13 *(Appeal to fairness)*; B4 *(Agree with him)*; D8 *(Tack with his stand)*; or B13 *(Find a point of agreement and develop it)*.

Examine the block of strategies listed under "Force a decision" and consider two of them: F3, slightly paraphrased *(Show him that the cost of his anger over an earlier experience can cheat him out of a profitable experience now)*; F4 *(Emphasize cost of not buying)* which is another way of saying much the same thing. Make the Sales Strategy Bank work for you by examining it for strategies we have not mentioned.

Competition

Little need be said at this time about this problem area. Often, your chances of making contact with the customer will be impaired by your competitor's efforts to achieve the same objective. Future chapters will consider specific problems created by competition; and for now, we will consider general strategies, but you are urged to consider them.

First of all, overlook no competitor. The chap you beat easily the last time may be much smarter and more determined as a result. If you have ever lost a sale to a smart competitor, you didn't retire to lick your wounds. If you are a professional, your pride was hurt and you determined to "even the score—and then some." Why should you or I feel that our competitor is less human? Keep up to date on new competition, and don't overlook the little fellow.

What is that slogan? He tries harder? The little guy often has greater flexibility and can make fast decisions on the spot. Avoid smugness. You and your company did not become successful by being smug, and it's a sure bet that you won't *stay* successful by being smug now.

Know your competitor's representative. During World War II, we tried desperately to hide from the enemy the identity of our command generals and admirals. Why was this so important? Simply stated, because each of them had a modus operandi, or as the psychologist states it, "a continuity of personality"—even as you and I. For months, the Army had General Patton hidden in Great Britain while the Germans tried to determine on what front he was fighting. They knew, for example, that he did not "buy" the philosophy of protecting his flanks. They knew that he would drive ahead with force. Knowing the identity of the opposing commander, the enemy would adopt its strategies to most successfully meet his established technique.

For the same reason, it is not enough to know the company which you consider your competition. It is a great advantage to know who the individual(s) is who represents them in your area. What does he do when you have a better price? What does he do when you have a feature he lacks? What is his rapport with your customer? How does your customer look upon him? What advantages does he have which you lack? These may be advantages of product, price, reputation, or even his own personality. How can you neutralize these or even surpass them?

Consider recall again. What experiences has your customer had with your competition? Do these experiences represent advantages or problems for you? If they will be advantages, how can you stimulate the customer's recall without being in the position of "knocking your competitor"? How can you present your proposition at the moment of contact so that your customer, on his own, will recall less satisfactory experiences he has had with competition?

Be a fast second. You can't always be first in making contact with the customer, but you *can* be a fast second. Don't leave the field of battle uncontested because the other fellow moved in first.

Identify the needs the other fellow supplies. If a competitor has had an inside track for a long time and is a favored supplier, your course of action is well defined. Determine what needs he is supplying. Why is he favored? Is it because he has price or product advantages, or is it because he supplies some status need which the customer must satisfy?

We are not referring solely to such things as the competitor's ability to take the customer to a swank club to which the customer might otherwise not be admitted. We refer also to the fact that something about the competitor makes the customer feel *comfortable*. The competitor may make the customer feel secure or superior or needed.

Don't fall into the fatal trap of believing that these feelings have no place in the sale of "high-ticket items." Some salesmen feel that customers who must buy expensive items will not let their personal feelings enter the situation. This is a fallacious belief. An intelligent competitor can create a setting which a customer will like so well that he will pay a high price for it. He is often willing to pay a price in economic security to insure continued emotional security, a sense of well-being, a sense of feeling important, or being recognized, or of belonging.

You may have to answer the question: "What status satisfaction has been supplied by the competitor which we have not recognized, or if we have recognized it, we have not filled it?" If we can't supply the status satisfaction, can we show the customer how much he is paying for that satisfaction? When you adopt this strategy, be sure that the customer learns it himself, and *by himself*. Do not let him become aware of the fact that you know it is status satisfactions that have drawn him to your competitor. Your task is to create a situation wherein this will become clear to him, without a witness being present during this moment of discomfiture. Remember, the psychologist warns us, "We (and your customer) do not like our behavior to be transparent."

Examine the Sales Strategy Bank again and decide which strategies can be adopted to meet competition. You cannot remove an entrenched competitor by depending on "luck" or by sheer "force of personality." Replacing established competition can be accomplished by (1) an analysis of the real and fancied needs of the customer, and by (2) planning your moment of contact so that his interest and attention will be aroused, maintained, and developed to the degree that his appetite will be whetted for more. This means making a serious, well-planned Estimate of the Selling Situation with tactics and strategies complete with alternatives based upon the alternatives available to your competitor.

Cause the customer to wonder what price he has been paying for the satisfaction of status needs and to wonder if the full cost has been on the price ticket. If he has been buying elsewhere out of habit, he has probably been doing so because habit relieves him of the necessity to make decisions. The name of the game is often "avoidance of risk." Convince him there is no risk and that his habit has cost him economic security. If all else fails, don't overlook *F1 (Bypass him)* and *F2 (Shock or antagonize him)*, but remember they are last-resort methods.

More and more, we see that selling is a profession. Our goal: to be a professional in that profession.

Mannerisms (our)

The easiest selling situation that a salesman must cope with is created by the customer who has the urge to buy. In other words, he is in a buying mood. In other selling situations, it is the salesman who must develop and sustain the customer's buying mood. The buying mood thrives on such "mood builders" as friendliness, sincerity, attention, courtesy and similar attributes. Mannerisms which the customer finds objectionable are devastating in their effect upon the buying mood and may be considered as "mood busters."

At the start of the selling situation, the customer often has an opinion of the company represented by the saleswoman. This opinion is frequently based upon the image that has been created through advertising, or by the customer's past experiences with its products or services, or upon its reputation for research or for post-sale service. This initial impression or opinion is important, but the customer's buying mood will be affected by his opinion of the company's representative with whom he is doing business at this moment.

There are many experienced salesmen who have been successful and have enviable selling records. If you speak to them about the adverse effect of mannerisms, they will often "shrug off advice" and point to their record of success as evidence that whatever their mannerisms may be, they cannot have had too adverse an effect on their customers. They seem to be oblivious to the fact that their records might have been even more impressive had they discarded mannerisms which affronted or annoyed some of their potential customers. They forget, too, that often a customer will buy in spite of the fact that a salesman irritates him, simply because he is completely sold on the product; but, at some future date when he is given a choice by a competitor of that salesman, he will transfer his business.

All of us have some mannerisms which annoy even our closest friends and strongest supporters, yet we are usually unaware of their nature. Some of these mannerisms are probably unique to each of us, while others are more common. In any event, little would be gained by listing a litany of irritating mannerisms. We shall discuss a few, but more is to be gained by identifying the root cause of these mannerisms. If they work to our disadvantage, why is it that we don't take corrective action? Let's examine a hypothetical situation to obtain one answer.

A specific salesman whom we have in mind would call upon a customer, sit on the edge of his desk, use profane language, smoke strong cigars in the customer's office and, worst of all, tell dialect jokes that reflected upon the ethnic group to which the customer belonged. He would wear his hat in the customer's office and call the customer names which were semiprofane, and all the while he would wear a big, disarming grin. And—he usually got the order.

You may now be working for a company where this kind of salesman would never be employed, or if employed, wouldn't last a month. If this is true, don't make the error of assuming that we do not have salesmen who follow one or more of the procedures listed in the previous paragraph in some of their calls.

We once asked this salesman why he behaved in this manner. He replied that the customer "liked" this kind of treatment. This salesman was exceptionally well-versed in the technical aspects of his product line, and he did not realize that many of his customers would have left him long ago had it not been for their dependence upon his technical assistance and the desirability of the products and service of the company represented by the salesman.

We pointed this out to the salesman, and he said again that the customer enjoyed this treatment because the customer always smiled when he was with him and laughed at the ethnic jokes. The salesman never realized that the customer acts amused, while internally seething, because this is the only defense mechanism available to him that will enable him to save face. This reaction is his way of saying that he understands that the salesman "really does mean it."

After reading of this salesman, you may feel repelled and surmise that the description was overdrawn and, therefore, that this material will not apply to you.

There *are* salesmen exactly like the chap we have described; but all of us have mannerisms which are not so blatant or as rude or crude as those we have just described, yet our mannerisms may irritate and alienate customers, weaken the buying mood and never permit the sale to get beyond "contact."

The actions of the salesman we have discussed are recognized as being in bad taste and certainly evidence of poor judgment. Other mannerisms may not be as obvious, and they are not always considered to be particularly offensive. However, it will be valuable to see some of these actions through the eyes of the customer.

1. Profanity. The customer may use profanity without being aware of the fact that he does so. Public speakers are sometimes told, "That was an excellent speech, but you should cut down on the profanity." Often such a speaker is incredulous and refuses to believe that he sprinkled his speech with "hells," "damns" and assorted profanities until he is confronted with a tape recording of the speech. The customer may use profanity, but this does not represent his sanction of our use of the same language. At its best, it is evidence of a limited vocabulary, and evidence that when we feel strongly about something, we lack the capacity to express ourselves; hence, we choose profanity as the easy way out. At its worst, it ranges from disrespect to a display of vulgarity toward our customers. Today, four-letter words are common, but some salesmen would be surprised to know how many orders their use has cost them.

2. Poor listening. Salesmen often give the impression of not listening when the customer is talking. The glazed eyes, the far-away look, the mechanical nodding of the head in agreement when there is nothing requiring agreement being said; these are the tip-offs to the customer that she does not have the salesman's attention. Imitation may be the highest form of flattery, but certainly attention runs a close second. You may well be miles or minutes ahead of your customer and be fully aware of the point she is making, but it certainly does not enhance her prestige to be made to feel that what she is saying is unimportant or is boring to you. She becomes convinced that we spend our "listening time" thinking of what *we* are going to say when she stops talking.

3. Reading his mail. Even though you have no intention of reading his mail, your eyes may flit across his correspondence in such a manner that you give the impression you are reading it or attempting to do so. You may not have seen or comprehended a word that was exposed, but sometimes you will see the customer slowly move his mail into a pile, or slide it into a drawer, or even turn the top letter on the pile upside down. When this happens to us, as salesmen, we cannot help but redden and show our embarrassment. The customer, now aware that his maneuver has been detected, is also embarrassed, and a gap is created during the contact phase when the entire strategy is to bring the customer and ourselves close together.

The cure is to keep as far away from his desk as possible. If your work makes it essential that you hover over his desk and a pile of unopened mail meets your gaze, you might try turning the top page upside down yourself, or sliding the pile away from you. A good impression is created and the customer may then move the pile of mail out of the way.

4. Using his first name. Europeans are often amazed at the speedy use of first names in America. All Americans are not "first-name fellows." Some men actively dislike this practice. You and I may regard them as stuffy, but the fact is that they are the customers. You enhance his self-image by using the "Mister." If he finds your use of his last name and the formal "Mister" cumbersome and impersonal, he will tell you to call him by his first name. Let him make the first move in that direction. Above all, never—but never—use the customer's last name without the prefix Mister. We have seen this malpractice create a gap that literally took years to bridge. The customer usually reacts with the thought, "Who does he think he is?" What the customer is really thinking is, "Who does he think I am that he can talk to me as though I were inferior to him?" It's the same old story of the customer subconsciously asking, "Would he call my boss by his last name that way?"

5. Smoking. Great though the "sacrifice" may be, smoking is a pleasure we should forego in a customer's office until we know how he feels about it. One key executive in a large New York store said, "I don't smoke and I don't like smoking by exposure." There is little point in asking the customer if he minds if we smoke, because he feels trapped and thinks the only permissible reply for him is, "No, go ahead and smoke." But, if we were to come back a few minutes after our departure, we would find that his ash trays had been emptied and his office aired out. The mere presence of ash trays in his office means little. If, however, the customer is smoking, you can probably smoke without any danger of offending him.

6. Respect his office. There are men who take considerable pride in their office furnishings and will resent any actions of ours that indicate a lack of consideration for the manner in which we may use his chairs or desk.

7. Finishing his sentence for him. The customer may think more slowly than you do. Or perhaps he just speaks more slowly, but thinks even quicker than you. Some salesmen who are impatient and can't wait for the slower speaking customer to complete his sentence do it for him. At the very least, the customer will see this as evidence that the salesman is impatient with him. It is a form of human behavior that does not enhance the self-image of the customer, and it is a form of aid that he can do without.

8. Familiarity. Another obstacle is created when the salesman becomes too familiar with the office staff of the customer. His secretary may be worthy of a second glance, or even a third and fourth. The customer isn't blind; he, too, is aware of this and with commendable restraint has confined his observations to one glance a day. He won't take lightly your efforts "to make a federal case out of it."

9. Using his desk for your material or briefcase. Don't—until she suggests that you do so. If she sees you struggling with material in your lap, she'll make a space on her desk and invite you to use it.

Countless sales have been lost because of irritating mannerisms, and many of these mannerisms are seen by the customer at the moment of contact. They include the way in which a salesman expresses himself, how he conducts himself and how he appears; i.e., what he wears and how he wears it. Unfortunately, we are often unaware of many of these mannerisms. They may be little things, but they irritate the customer who has "a low patience tolerance." The salesman drumming his fingers on the table, the fellow holding his hand in front of his mouth when he is talking so that you can understand what he is saying only with difficulty, the chap tapping his feet incessantly on the floor or the lad picking up items off the customer's desk to handle while he is talking . . . these men are probably employing outlets for excess energy or they are displaying evidence of not being at ease.

One of the most annoying mannerisms that has held back many capable men and has infuriated countless customers is the habit of crowding up on a customer and talking with your face only a few inches from his—or while at lunch, talking directly across his plate. Someone afflicted with this habit will actually follow you around if you shift your body to avoid his breath in your face. In his enthusiasm for his proposition, even in his keen desire to help a customer, he makes himself so objectionable that all he arouses in the customer is an overwhelming desire to bring the contact to a close.

The salesman with years of experience may read this material on mannerisms and react with an "anyone knows that you shouldn't do these things—how basic can you get?" The point is that in spite of the fact that we "know we shouldn't do these things," we still commit many of these errors. Usually, we are so sold on our product or proposition that we forget.

The cure is large and frequent doses of introspection, of self-analysis. Try to be mindful of the way you use your hands and feet, the percentage of time you listen, the respect you show your customer by your behavior in his presence. Consider whether or not you have known a customer so long and are on such friendly terms with him that you no longer seize opportunities to show him respect and enhance his feeling of importance. Ask yourself this challenging question: "Do I treat him as I would my most important customer?"

Don't assuage your conscience after one of these self-analysis sessions by arguing that you get most of his business. The habit that we develop in our relationships with our friends can carry over into situations involving our most status-hungry customer and can cost us sales.

In the last few years we have been, relatively speaking, in a buyer's market. Throughout the country, salesmen have been seeking sales training opportunities. Most of them are looking for "something new," something that won't be superficial. Yet, sales are being jeopardized and sales are being lost by the mannerisms discussed briefly here. As salesmen, we *should* seek new ideas and new knowledge, but at the same time we should use the abilities we now have to the maximum advantage and not let their force and effectiveness be dissipated by thoughtless and unnecessary mannerisms.

These obstacles, or mannerisms, fall within our own control once we know what they are. This time, therefore, we have not referred you to the Sales Strategy Bank. No strategies are necessary to neutralize obstacle-creating mannerisms. All that is required is self-analysis and a determination to eliminate the faults within ourselves.

Atmosphere

Salesmen frequently find it necessary to make contact with their customer in an atmosphere that is less than desirable. A successful contact requires that the salesman must gain and hold the attention of the customer, but there are often influences that distract the customer, the salesman, or both. Here is a partial list of the distracting influences which we have in mind.

1. The customer works in an open office with many other people in close proximity. There may be a high noise level, and the proximity of other people —even other customers or competitors—makes the discussion of confidential matters extremely difficult.

2. Salesmen from competitive firms are sitting within hearing distance or waiting outside the customer's office, with the customer waiting to see what their proposals will be.

3. The customer has an office that is furnished in an extremely expensive or impressive fashion. In spite of yourself, you feel somewhat awed by this fellow and find it difficult to concentrate on the presentation as you had planned to make it.

4. The customer accepts all incoming calls while you are with her and may even find it essential to make a few outgoing calls. You may be able to follow your presentation sequence as you planned it, but the interruptions may make it difficult for the customer to follow.

5. The customer has a pile of mail on his desk, and it is obvious that this is causing his attention to be diverted... a major roadblock during the Contact facet of the sale. While you are talking, he may even pick up a letter, glance at it and then seem to lay it down reluctantly. Your story may be important to him, and he may find it interesting—but something in the pile of mail is distracting him.

6. Other visitors drop in on him during your presentation.

7. Her secretary seems to delight in interrupting, and she continually brings in papers for her to sign. Each time this happens, you lose precious time—and she loses the trend of your presentation.

8. In retail sales situations, customers constantly come in and require—and obtain—the attention of your customer, the store owner or the buyer. Or subordinates constantly ask for instructions or make other demands upon his attention, such as credit approvals, etc.

You can undoubtedly add to this list, but compiling a complete list is of no real value. The effect of each of these obstacles is the same: It is difficult to hold the customer's attention long enough to feel that you have really established contact, let alone get an order. In some instances, an order is obtained only after months of preparation and many contacts. Obstacles of the type we have just listed often make it difficult to feel assured that today's contact has moved the sale one step closer to completion.

What to do? How do you overcome these obstacles?

The choice of actions varies widely. If, from previous experience with this customer in his native habitat, you know the conditions under which you must serve him, perhaps the only thing you can do is to anticipate the interruptions and distractions, and plan more time to make your presentation. Obviously, your main effort must be to plan a presentation that gives tremendous emphasis to *benefits* or to showing him how he can avoid loss. You must in a covert manner show him what the interruptions are costing him. This means thorough advance study of the customer and of his current situation so that you will know exactly what appeals will obtain—and hold—his attention.

In some instances, the best course of action is to remove the customer from the situation if you cannot change it. Consider this, always, as the most effective means of meeting the obstacles listed. But let's be realistic. It is often difficult, even impossible to take the customer away from the distractions. How could you, for example, in trying to sell the proprietor of a small store, expect him to agree with you that it would be better if you left the store and went out for a cup of coffee where you will both have privacy. He isn't about to leave his store

The same type of situation exists when you are selling to men who work for large companies or for military organizations. Often their freedom to come and go is restricted. Frequently, you cannot even plan your call just before noon so that you can take them to lunch because some will avoid actions that tend to place them under an obligation to you; some are prevented by regulations from letting you buy them lunch. If your customer is working for an organization that gives him little freedom to leave the plant or office, and you suggest going to some spot outside that offers greater freedom from interruptions, you succeed only in embarrassing him. This is obviously a dangerous thing to do and impairs the effect you wished to create at contact.

This book is intended, primarily to aid you in overcoming obstacles. Why, then, have we spent so much time in the foregoing paragraphs on a discussion of "what it is not always possible to do"? We have done so simply because these are the first seemingly logical actions that come to our mind when we are out there, face to face with a customer, in a situation that is making the going difficult. We should therefore be aware of the hazards involved in taking what seem to be "logical" actions with *some* customers. With other customers, attempting to remove them from distracting situations is the right thing to do.

You must always weigh your actions against the conditions. Once again we see that professional selling requires that we consider each customer as an individual, often different today than he has ever been before. Professional selling requires that we avoid inflexible methods of operating and that we adapt our modus operandi to the conditions of the moment, no matter how satisfactory a given line of action has been in the past. A choice of actions might include the following (letter-numeral combinations refer to actions listed in the Sales Strategy Bank in Part Three).

Consider D12 *(Get him to talk)*. He won't be quite so willing to permit interruptions if he is holding the floor. You may be able to get him to talk by using D15 through D18 *(Use provocative questions)*; applying B19 *(Ask for his opinion, help or advice)*; using C1 *(Provide him with a role)*, or C6 *(Omit something* so he can become involved and inject his ideas), or B16 *(Be intentionally incorrect in a minor statement* so that he will correct you and start talking).

Another method is to use B20 *(Sympathize with him)*. Indicate that you can see the difficult conditions under which he has to do business and suggest there may be an empty conference room that you can use where he can get some relief from "these demands upon his time." If appropriate—considering the situations discussed earlier—suggest you leave the office for some place in town where you can both relax and talk. "Entertainment" expense is not always spent wisely, but this is a legitimate and effective use of such funds.

Consider E9 *(Give him recognition)*. In this instance, give her recognition—and we refer to the secretary who seems to delight in interrupting. Learn her name, call her by name when you come in and find ways of letting her know that you don't regard her as just a fixture. Before you ask to see her boss, ask her how she is, and find ways to compliment her subtly and sincerely.

There are really only two ways to combat this obstacle of "atmosphere." Try to eliminate it by removing both your customer and yourself from the situation—having a specific place in mind and suggesting it; or, face up to the situation and find ways of compelling his attention in spite of distractions.

Making the best of it can be accomplished in many ways. If you know in advance that you are going to encounter the situation, take these preparatory steps:

1. Have any confidential aspects of your offer in writing so that you can hand them to him to read in the event that competitors are seated nearby. If the area is noisy, he can read better than he can hear. Note: When you give him something to read, keep quiet while he reads it. There is nothing more irritating to a customer than to be given something to read and then to have the salesman keep up a running line of chatter—or explanations—while the customer is trying to concentrate on the material before him. Hold your explanations or comments until he has finished. If the material needs a running explanation, it is probably poorly prepared; it should be able to stand on its own feet.

2. Consider the use of visual aids, especially charts that are not complex. These enable you to refocus his attention on the point you were making when you were interrupted. This device also helps him follow the sequence of your presentation in spite of interruptions. Keep in mind the admonitions of the discussion dealing with the use of visual aids. Don't irritate him by reading the text of charts when he is perfectly capable of reading the same text. Avoid creating the impression that you are giving him a "canned" sales presentation. Prepared charts *can* give this impression. Take the time to prepare charts for *this* presentation or, at the very least, constantly relate the charts to what he will certainly regard as his unique situation.

3. Don't underestimate the time your presentation will require. Consider the probable impact of interruptions and factor these into the estimate of time you will need when you ask for the appointment.

4. Years ago, Borden and Bussey emphasized the need for what they called "The ho-hum crasher." Have a "ho-hum crasher" ready. Know what he will regard as the most important advantage of your offer and hit him with it *hard* at the outset. Identify what he considers his most vexing problem and show him, early in your presentation, how you can help him solve it. If this is really important to him, he may suspend the incoming calls, close his office door and give you privacy. Keep in mind that the other demands upon his time, which *you* naturally consider interruptions, must offer him some kind of satisfaction or he would disregard many of them. You must find a way of offering him greater rewards, and you must cause him to feel that interruptions will threaten these rewards.

5. In some cases, it is possible to arrange meetings after hours when privacy is assured. This brings up a point. Why should you put in extra hours if you have not been doing so?

The other day we were told about a young man who had a fixed determination to become a millionaire. He is putting in 16-hour days and has now passed the $450,000 mark in terms of cash assets and owns a fleet of busses, completely paid for. Some will look at this fellow and think to themselves, "Lucky!" You may have no desire to amass a million dollars, but the fact is that your material success may be in direct proportion to the effort you are willing to give the job. Furthermore, we know that selling is a profession. In spite of the jokes about how difficult it is to get a doctor, the fact is that thousands of them are on call day and night, and the same applies to lawyers.

There is no need for you to I to put in 16-hour days, but if a customer represents an account we value, the sacrifice of an evening may pay off handsomely. In the bargain, it is the highest compliment you can pay the customer—that you are willing to give of your free time to the solution of his problems. It is customer-oriented selling at its apex.

Habits (customer's)

No one can list 101 ways in which you can meet the "habit" obstacle and even begin to hope that he has covered all of the situations which you will encounter. Each situation must be considered as a separate and new one. You, and you alone on-the-scene, must develop your own solution. One thought bears repeating: Customers form habits for a reason, and you can't change their habits unless you identify the reason and either neutralize it or cause it to appear too costly to the customer.

You will find that the Sales Strategy Bank will offer you helpful clues, especially if, from a thorough presale analysis, you know you are going to encounter the problem. Consider *E10* (*Anticipate, raise and answer objections*). This is often the best way to avoid placing the customer in a position where she feels she will lose face if she changes. *She* has not brought up the objection. . .you have.

For example, how might we handle the executive who will not interfere with the decisions of his new purchasing agent?

If we are going to follow the advice just given, we must determine why the executive (Mr. Spalding) takes this attitude. Let's assume that Spalding was formerly the purchasing agent and has been promoted and replaced by "John."

What is the relationship between Spalding and John? Was John the choice of Spalding? If not, does Spalding hope that John is going to "mess up" the job? If John was selected by Spalding, does the latter think that the best developmental route for John is to leave him on his own and learn through his mistakes?

What is Spalding's business objective (as part of the company's over-all business objective)? Will Spalding's concept of a development path for John cost his company heavily? If John is about to buy from your competitor, will this action have an ill effect upon the company objectives which Spalding is committed to achieve?

What does Spalding want most; to let John develop or to protect the immediate interests of the company? If John makes an error and the company knows Spalding could have prevented it, what will be the effect on the image Spalding is trying to create in the company (his me-I-want-to-project)?

How can you make Spalding see that John *and* the company need protecting at this time—and that by appropriate action, Spalding will accomplish both of these objectives and enhance his own self-image? Or, if one of these three objectives must be sacrificed, which will he sacrifice first?

If he really wants to help John, how can you cause Spalding to see that, here, he should *break his habit* of letting subordinates learn through their errors? The point being made is this: You can't treat habit as *one* obstacle. It appears in many forms, and we must understand the basis for the habit in *this* case. Habit patterns are established for many reasons: to save us time, to avoid unpleasant situations, to avoid decision-making, to give us a sense of security, and to provide us with smoke-screen objections so that we do not have to reveal our true objections (which might not be socially acceptable or might be hard to defend). These are but a few of the reasons for habit formation.

What satisfaction is your customer seeking—or has he obtained—through the habit route? How can you supply him with a greater satisfaction through another route? How can you cause him to see the high price he is paying for the satisfaction the habit provides him?

Habits are always formed for a reason. Your customer is not stupid; he is a rational being who does things for a reason even when he is not willing to reveal the reason and gives you some other excuse for his action. Often the reason upon which the habit was originally based has ceased to exist, but the habit persists. The customer may not realize this, and it is your job to show him that such is the case. Failure to do so may cost you a sale.

Timing (our)

The potential problem area of "timing" shows us how fruitless it is to develop pat solutions for sales problems. Timing, as a problem, should be anticipated and considered when the sale is being planned and long before contact is made. In spite of your best planning, you may encounter sales situations where timing as an obstacle presents you with an unexpected problem. New salesmen will often ask, "How do you handle this problem?" The experienced salesman knows that there is no one answer, unless you can call "Use judgment" a pat answer.

Let's be sure semantics don't get in the way. What are some examples of what we have called timing? Here are but a few:

1. Just as you walk into the customer's office, she is having trouble with one of your products or your delivery.

2. You come in after he has had a series of annoying experiences. He chooses to use the self-defense mechanism of "displaced aggression" and takes out all of his pent-up emotions on you, the innocent bystander.

3. Perhaps you represent a company that has many departments and separate sales organizations. You may just happen to be the third individual from your firm who has called upon him today, unknown to the others. The customer sounds off about companies that can't have salesmen who can represent them "across the board."

4. The customer has other problems on his mind which he considers of greater importance than those on which you are attempting to concentrate his attention. This is not uncommon in all types of selling. We find it in the retail store where retail customers are demanding the attention of your customer, the proprietor. You can find it in the situation that exists when you call on the purchasing agent of a public utility who has just been told by his management to get some figures together to help support an increase in rates the utility is about to seek. You can find it in the military where you find an entire staff working on a presentation that must be made before an appropriations committee of Congress. These situations all spell trouble for the salesman who is trying to establish contact—which means getting the *attention* of the customer.

The perennial question is, "What do you do about it?" Here are some general rules for proceeding:

1. Anticipate when possible and plan your own timing. This means gathering intelligence on your customer's operation so that you will be familiar with peak work loads. A little homework here will go a long way. There *are* peak periods in almost any business; there are times when budgets must be prepared. Knowing what these are can help in avoiding bad timing. But we have seen that it is not always possible to anticipate events that add up to poor timing.

2. If you are unable to anticipate the problem, try to determine the reason for it. This is often easy because bad timing is frequently obvious.

3. Once you see that timing is poor, you have a choice:

(A) Get out of the situation as gracefully as possible, leaving the climate in such a condition that the customer will be willing to see you at a more propitious time.

(B) Neutralize the situation and focus the customer's attention on your proposition so he will forget the cause of his initial irritation. Consider the following from your Sales Strategy Bank and adapt those which are applicable to the situation of the moment:

A12—*Identify the source of his resistance.*

B20—*Sympathize with him* (not necessarily agreeing).

D10—*Divert him.*

B19—*Ask for his opinion, help or advice* (the strongest of all appeals).

B18—*Listen!* (let him get it out of his system). If he is emotionally upset, don't try to use logic. The more compelling your logic, the more angry he will become.

D8—*Tack with his stand.* Be flexible, don't be a slave to some planned opening that won't work under these conditions.

C5—*Emphasize his objectives—carefully.* Let him see for himself—and by himself—that though his anger and irritation may be justified, surrendering himself to either at this time will be costly to him. Add to your Sales Strategy Bank: Emphasize *new* goals or objectives. Make these attractive in order to divert his attention from the cause of his rancor.

D6—*Talk benefits!* Hit hard on those you know have the greatest appeal to her. Don't scatter your shot by listing all benefits. Remember, when she is upset, you are trying to re-focus her attention; so hit the most attention-compelling benefits. Don't talk product features or qualities; talk in terms of what these features and qualities will do for her.

B2—*Offer to help him;* if possible use D3—*Provide a solution for the problem that has vexed him.*

B12—*Admit validity of his view if a complaint is the source of bad timing.* Look for something in his expressed views with which you can agree.

D13—Appeal to his sense of fairness. We all like to be considered fair ("unfairness is un-American"). Use this with caution: "I've always considered you to be a fair person, and I think of you that way now. . . . Let's see what we can do together to remedy the situation."

There are many other sales strategies that should be considered. . . appeals to his pride, prestige, his self-image and his key interests. Above all give him recognition, and *give him the right to feel as he does.* Your approach should be: His attitude has given me a timing problem, but if I were truly the customer, I would feel exactly as he does. Now, what could somebody do to change my mind. . . that's what I must do for the customer.

By taking this approach, you eliminate any emotion from your own analysis of the situation; your mind is unencumbered by self-pity, by anger or by contempt. You are then in a position to study the problem as he sees it and to produce solutions that are the result of reason unhampered by emotion. Often, the customer only wishes to blow off steam, and he may be seeking sympathy, or at least, an expressed understanding of his problems.

If, on the other hand, the customer is beset with problems that are more important to him than your sales presentation—at this point in time—you must recognize this and not attempt to soften him up by assuming all he wants is a sympathetic ear. He may want just one thing right now: to be left alone.

The task is always the same—to remember that your customer is a rational being. He doesn't wake up in the morning with one all-consuming thought, i.e., "How can I make life miserable for some salesman today?" He does things for a reason. He acts in a manner that fills a need that is uppermost in his mind at the moment. If he has received what he considers a "bad deal" from us, he feels we have not recognized his importance. He is going to make sure that we recognize it now and from now on.

To a customer who is under attack by *his* customers, by a public utility commission, by a legislative body, by his own superiors or by anyone who has threatened his economic or emotional security, protecting that security is going to be uppermost in his mind. Calling upon him now will be bad timing. We either must show him that he has other needs that require attention, or we must learn how to bow out with grace, leaving the door open for another and better day.

It is not bad selling to know when to retreat. . . it's just exercising common sense. Every other profession recognizes the need for timing: in surgery, in law and in combat. Why should you in the profession of selling regard strategic withdrawal as anything less than professional behavior?

Predecessor (our)

Salesmen frequently change territories or, because of organizational changes, assume responsibilities for an entire product-line in a given geographical area. Or, conversely, the salesman may now be assigned a more specialized task—that of selling a single product or a limited line of products in a much greater geographical area. In either event, he probably assumed responsibility for serving some customers who were formerly sold—or not sold—by another salesman.

But our use of the term "predecessor" is not confined solely to such situations. If you are selling in a distributor's organization or a retail store, your customer may, on past visits, have been served by another salesman, whom you should consider your predecessor. This predecessor may represent a source of advice or help to you. Or, because of a sorry relationship that existed between your predecessor and your new customer, he may represent an obstacle to you, especially when you are trying to establish a meaningful and helpful contact with this new customer. Someone has said before us that you never get a second chance to make a first impression. Your predecessor may influence greatly the impression you now make. Let's consider how he may help you or hurt you and what you can do about it, in very specific terms.

Following a disliked predecessor

Let's start with the situation that exists when you follow hard on the heels of a salesman for your company whom your customer did *not* regard favorably. Often, the customer will sound off and make his opinion of your predecessor very clear. As a matter of fact, when he does so, you are fortunate because there can be no doubt in your mind about the customer's feelings or opinions.

The temptation presents itself to agree with the customer's evaluation of your predecessor as an immediate means of getting "in" with him. At best, this can represent nothing more than an orientation to short-range goals. Over the long haul, it will not necessarily result in gaining the customer's respect. This can be gained only through your own subsequent performance. To say nothing of the questionable morality of such a procedure, the fact remains that your predecessor *did* represent your company, and agreeing with the customer's low evaluation of your predecessor does not enhance the prestige of the company you represent.

At the same time, it must be agreed that your best course of action does not consist of remaining silent—hence, by your silence implying agreement.

Your silence can also be interpreted as disagreement, and this will not strengthen your initial contact.

Step one is to listen and let the customer give vent to his feelings. Often, if he talks long enough, he will begin to feel a bit guilty and may even compensate by saying something good about your predecessor.

Step two may be to make a neutral comment, such as "I can see how you feel." Then as rapidly as you can make the transition, focus the attention of your customer on the advantages that will accrue to him by discussing the product or service you are prepared to offer instead of discussing your predecessor. Neutral statements tend to imply *respect* for his views without necessarily agreeing with them. The transition can sometimes be made by saying, "I can see how you feel, and I am sorry you have had a problem, but today I can help you with another problem you have. I've noticed that. . ." Continue on to emphasize the problem or need that you feel your product can remedy.

Another approach is: "I know that situations like this do arise, often through natural misunderstandings, and I am sorry to hear about this problem you have had. However, I am looking forward to working with you, and I'm very glad that I have been assigned to your account because I am sure it is going to be pleasant working with you. Right now, I am sure we can make some progress on . . . (another problem) (this project)" or whatever phrase is applicable.

The effort should be to focus his attention on the immediate problem. It is sound psychological practice, when an individual is concerned about one problem, to make it appear less important if you face him with a more pressing problem—especially one that is current! This is often referred to as "changing the setting." You have undoubtedly used this device when dealing with a customer who is concerned about what he regards as the high cost of an item. If shown a higher priced (more expensive) item, the "high cost" of the lower priced item takes on less significance.

By using the neutral approach, followed rapidly with the focus on his current problem, you have not damaged your predecessor (who may be selling elsewhere for your company). You have also given the customer respect, and a sympathetic audience. And now you have turned his attention to a problem that he wants to solve—and upon which you can help him.

Following a highly regarded predecessor

Customers often resent organizational changes that deprive them of the services of a salesman with whom they had enjoyed rapport, salesmen whom they liked, or for whose competence they had respect. Often, if you follow such an individual, you will sense that the customer inwardly questions your competence and she may show actual resentment or hostility. What to do?

We will do well to consider what *not* to do as well as what course of action to pursue. First, then, do *not* try to imitate your predecessor. Imitation may consist of dressing as he did, adopting his mannerisms, using figures of speech or profanity that he may have successfully employed, or using selling techniques that were unique to him. These devices may have fit him like a tailor-made suit, but you may look "overdressed" or "badly fitted" if you employ the same devices, with a resulting loss in effectiveness.

The question naturally arises, "Why not adopt the well-liked fellow's modus operandi—obviously the customer admired his behavior, speech or dress?" The answer is so obvious that it hardly warrants discussion. These devices were natural to the chap who preceded you. If they are not natural to you, the customer will resent them; he will regard your actions as proof that you, too, consider your predecessor as worthy of imitation. Hence, he will feel his resentment is justified. Worse yet, your actions serve to *remind* the customer of the chap he liked, whereas your object is to cause the customer to regard you as his source of help and to *forget* your predecessor. This does the previous salesman no disservice and is a realistic course of action. Adopt your own techniques, use your own personality, your own strengths and be friendly, helpful and interested.

So much for what *not* to do. Addressing ourselves to a positive course of action, we find there are at least three avenues to follow:

First, try to determine the reason for your customer's high regard for your predecessor. We like people, usually, because they fulfill some need we have . . . not necessarily a product or service . . . which you can supply as well as the previous salesman. For example, here are a few of the reasons why you and I may have liked salesmen who have served us:

1. He may have made us feel important.
2. He may have made us feel respected.
3. He may have made us feel "comfortable," at ease, safe, not at risk.
4. He may have made us feel "understood" (he saw the "me" we want others to see when they see us).
5. He may have had full confidence in his technical capacity.
6. He may have had the habit of giving us assistance "above and beyond the call of duty" (more than we felt we had a reasonable right to expect).
7. He may simply have liked the things that we like

(hobbies, sports... you name it). We tend to like people who share our likes, other things being equal.

8. He may have been a good listener (this is one area in which you cannot go wrong in imitating a highly regarded predecessor).

From your experience, *as a customer*, you can add to the foregoing list. It is by no means complete. The problem now is to identify the need that the other chap filled for the customer. You may find that you cannot supply this same need. For example, you may not share his enthusiasm for boating, poker, or a dozen other "likes" on which your predecessor and the customer are both "tuned in."

What to do? Find other needs that the customer has which you *can* fill. The rule, then, is simply stated: "If you identify the need the other fellow filled, and you really can fill it, do so. If not, seek out another satisfaction the customer requires and cause it to become so important that he will look to you to fill it."

Second, in your first contacts with your new customer, make it possible for him to do most of the talking, even if you must find means for doing so. This places you in the role of a listener.

Ask yourself: When did you dislike someone who listened to you appreciatively and sincerely? The listening course of action pays many dividends. You begin to identify his self-image, the "me" he wants others to see. You begin to understand his problems and their priority as he sees them. You begin to earn his respect. And you disarm him by exhibiting a humility that indicates you don't regard yourself as the new broom that is going to sweep away all of the mistakes made by your predecessor, whom he liked.

Organization structure

In our efforts to purchase a product, we phoned the distributor for a nationally known company. Three times, on the phone, three different young ladies let us make our request, and three times we were shunted to another individual. Our own fury at this treatment is unimportant, but the lesson to be gained is important. We had no way of knowing whom we should contact and precious time was lost.

As saleswomen, we should know the proper persons to contact, or time will be lost—and often sales lost as well. If, in error, we contacted the wrong person, and the person we *should have contacted* hears of it, he may be so incensed that he won't even let us make contact with him. Sometimes suspicious that we were trying to undermine him, he will be determined to stop us cold.

If we make our presentation to the wrong person, we may use the wrong product benefits and eventually lose the sale. For example, you may concentrate on low cost or economy. This could appeal to a purchasing agent, but *might* fall flat on the ears of an engineer or manufacturing man who responds only to quality or trouble-free benefits. Even if we avoid this danger, we lose time—probably the salesman's most precious commodity.

What is the course of action to pursue? Nothing dramatic... just do our homework. If we want the moment of contact to be fruitful, we had best answer these questions:

1. Where can I obtain information on the customer's organization?

2. Am I familiar with the nomenclature used to describe functions, levels or responsibilities that are used by the customer's company? If not, is there a chance I may unwittingly contact the wrong department—hence, the wrong individual?

3. The purchasing power may nominally rest in the hands of one individual, but is he really the decision-maker? Who *is* the key influence in the buying decision?

Additional questions that should be asked are:

Is the customer's organization centralized or decentralized? If centralized, am I in error in approaching a component of the company? If, in this instance, the component does buy, will I have alienated the purchasing authority at the centralized corporate level—hence, have sold "a car" and lost "the fleet"?

Is this a sale that is going to require a number of exploratory calls before I identify the key influence; if so, should I make my first calls appear to be "researching the need" and avoid asking for the order prematurely?

Sources of information were given in the pages on Pre-Sale Preparation and will not be repeated here. Tools available to you from the Sales Strategy Bank are:

A1—*Develop sources of information.*
A6—*Get help from other departments in your company (finance, etc.).*
A5—*Consult your predecessor (guidance only).*
A7—*Get your boss's help.*

If you find, during your first call, that the man you have contacted does not have decision-making authority, try to identify the proper person and the best means of contacting him.

Personalities

A group of people in a room, polled for their opinions of a stranger who had spent an hour with them, might run the gamut from keen dislike to liking, while others would plead "guilty" to having no opinions because the stranger failed to impress them at all.

This is often the problem the salesman faces at the moment of contact. The customer may instinctively dislike the kind of person the salesman appears to him to be. The salesman may immediately dislike the customer (which, because it affects his emotions, may affect his behavior). Worse, it may be a case of mutual dislike. Unless the salesman is a sole source of a product or service the customer needs badly or keenly desires to have, this attitude can end the selling situation before it gets off the ground—unless the salesman does something about it.

Over and over we have said that we cannot cure a problem unless we have identified the cause. Treating *effects* of the problem brings no lasting solution. Therefore, if we are going to overcome the problems that arise from personality clashes (even if they are only one-way clashes; i.e., unilateral dislike), we must consider the reasons why one person dislikes another, even at the moment he has met him for the first time.

Such reasons would include:

1. Saleswoman reminds him of someone he knew and disliked.

2. Feels that the salesman does not like him, even though this may be pure fantasy.

3. Thinks that the salesman feels superior to him.

4. Correctly or incorrectly identifies the salesman as a member of some ethnic or geographical group which she does not like; therefore, dislikes the stranger (stereotyping).

5. Jealousy: His self-image is threatened because he feels the salesman has been more successful, is more intelligent, better educated (there *is* a difference), technically superior or wealthier.

6. Fear: This can assume the form of imagined threats to his economic security, his emotional security, or both—"Can't trust this chap... probably will try to take advantage of me... selfish type, won't worry about what's good for me," etc.

7. Physiological stereotyping: Has preconceived ideas about people with red hair, no ear lobes, jutting jaws, close-set eyes... the list is endless.

8. Age prejudices: May not like young salesmen to call on him, or, conversely, wants *only* young fellows "with fresh ideas" to call on him.

9. Prejudices against habits: "He's a pipe smoker" (this meaning, "He's a loafer, full of theory"). Some dislike those who drink; others only want people around who can match them drink for drink and resent those who don't.

10. Resents the fellow who thinks more quickly than he does... considers quick answers as evidence that the salesman with the fast mind is superficial and gives off-the-cuff answers. Or he sees in this fellow a threat to his own self-image. The same thing applies in the case of the very confident salesman. This kind of a customer interprets the salesman's self-confidence as an act of superiority.

11. Resents the saleswoman's cultural background or advanced formal education, which makes the customer feel inferior to the saleswoman.

These are only a few of the reasons why one person may dislike another. To add to the list, be honest with yourself—completely honest. Avoiding rationalizations, ask yourself why you have disliked some people you have known. In some instances, you have had valid reasons for the dislike. But in others, honesty will force you and me to admit that our reasons have been spawned from the need for self-defense. In some way, we felt the other fellow threatened us ...our own aspirations, our beliefs, our set of values, our judgment.

Doing business with a customer you dislike

Unfortunately, the customer can, with a certain amount of impunity, hold such prejudices. But they are a luxury no salesman can afford. Realistically, we know that there will always be certain customers whom we dislike, and we must find ways of neutralizing or eliminating these feelings. Identifying the source, as we said, is the first step; you have read 11 sources and probably added a few of your own.

We dislike people because they do not conform to what we consider acceptable behavior in one form or another. Our judgment tells us that our form of behavior (or belief) is correct. Therefore, when others do not subscribe to it or conform to it, they are, in essence saying that our judgment is wrong. We do not accept that verdict happily. To preserve our own judgments, we reject those who do not conform as individuals along with their ideas and/or behavior.

For example, we may dislike a person because she holds beliefs which are contary to our own. We may have gained these beliefs from experience or they may have been passed on to us by people we respected. We now meet another person whom we consider intelligent and find that he does not subscribe to our set of values. We are forced to re-examine our own values, and this we have found to be an upsetting experience. Sometimes, rather than subject ourselves to this analysis, we reject the other fellow.

On other occasions, instead of rejecting him, we choose another form of self-defense; i.e., we attack. We choose some point or some subject on which we *know* we are right. If we can prove him to be wrong on *this* point, then, we can with greater conviction convince ourselves that he is wrong on the more important points upon which we find ourselves in disagreement with him. Keep in mind that your customer uses these same methods.

Even as you and I, your customer uses another device to protect his self-image—he evades. This may come out in the form of an instruction to his secretary, "Tell him I'm not in... or I'm in a meeting." Thus, he does not have to face us and subject himself to the threat we pose. And he does not have to re-examine his own opinions to check their validity... no need to ever utter those terrible words, "I was wrong."

Before venturing to consider what actions are available to us to "reform" our customer, we should examine our own behavior because *contact* with the customer (new or old) is often made difficult by our own dislike of him. First, let's agree that forming opinions based on any of the 11 reasons listed is against our own long-range interests. We may instinctively (and unfairly) dislike a customer because of his ethnic background or because of "feelings" we have about people who hail from his part of the country. Conditions may force us to do business with him, and we may learn to like him very much. Later, observing our relationship with him, a third party may say to us, "I thought you didn't like those people?" We find ourselves replying, "Oh, he's different." What has really happened is that we have found he is not a threat and we have accepted... even sought him.

In our arm chairs, reading this, with no threats in sight, with no emotions on the razor's edge, we can agree that it is foolish to make snap judgments based on anything as immoral as prejudice. When our work brings us into contact with the other fellow and our emotions are awakened, it is more difficult to reason objectively. The other fellow is almost certain to sense our reactions though he may not understand their basis. Our reaction may not be as obvious as the grimace of a child who is forced to eat something he doesn't like or to obey an order he resents. We, as adults, may be more subtle, but if our target is a customer, he will sense our dislike.

If you have a customer whom you dislike, identify the *real* reason for your feelings. Often, you won't like what you discover and you will then find it possible to identify something about him that you like. Once you have accomplished this, you'll find it easier—even enjoyable—doing business with him.

Doing business with a customer who dislikes you

More common is the situation where you sense that the customer dislikes or avoids you. What then? The first course of action is to ask yourself when you last went out of your way to do a favor for him... to do anything to *cause* him to like you... to cause him to see in you an enhancement of his goals rather than a threat?

Unhappy with the answer you gave to that question? Don't be. Most of us would be unhappy with our answers, too, because we usually "do favors" for people we like—and, resenting those who we feel dislike us, we leave them high and dry in their search for emotional satisfactions.

The customer has the same basis for his dislikes of people that you and I have... fear in a mild or acute form. The commodity which we must supply to overcome that fear is *assurance*. One of the best forms of assurance that we are not threats is to give him your sincere *attention*, display real *interest* in him as an individual and seek ways to provide him with *meaningful help*.

Consider the following from your Sales Strategy Bank:

B18—*Listen!* (We always come back to it!)

B19—*Ask for his help, opinion or advice.* (Not only does this approach lack threats, but your customer sees it as your recognition of the validity of his "me he wants others to see when they see him."... capable, important, well-informed, etc.)

B8—*Give him credit* or E9—*Give him recognition.* Keep in mind, "Praise the work, not the worker." Seek out something laudable he has done and comment on *that*; then he knows you are sincere. Praise only him, and he will suspect flattery—and you are back where you started.

A12—*Identify his source of resistance and do it by getting him to talk.*

A3—*Identify his self-image.* Do this by listening with your eyes (keeping in mind that everything she surrounds herself with was placed there by her or with her consent and was intended to tell a story). Note her key interests by the subjects she chooses when she talks. Then seek out ways to enhance the image that her possessions, surroundings and conversation have revealed. As we have discussed in Part One, seek out a pattern... don't judge by isolated deeds or possessions that may really not be symbols of the status (station in life) she seeks to have others grant is rightfully hers.

We constantly ask audiences of marketing men to supply us with adjectives that describe the way one of their customers wants others to see him. We have yet to find an audience that doesn't deluge us with adjectives ranging from "cautious," "conservative" and "thoughtful" to "impressive," "powerful" and "important." Recently, one saleswoman responded, "Immortal!"

There's not much that can be done with that last image. But in the case of all the others, these salesmen should ask themselves what they have been doing to cause the customer to feel that the salesman will help the customer to be seen in the desired light. If your customer seems to reject, evade or dislike you, try to identify the ways in which you have threatened this image of himself that is so important to him. Then, determine how you can change this concept he has of you.

Example: the customer who has little formal education and resents "college guys." Obviously, to remove the threat that your education seems to pose, you must find ways of showing him that there is nothing so uncommon as common sense and that he possesses plenty of it... something "no school can give a fellow." Or find subtle ways of showing your appreciation for "the school of hard knocks and hard experiences." In doing so, have no fear that you are manipulating him. If you are successful in your efforts, you have not only erased the basis of his resentment or dislike, but you have made it possible for him to render judgments that are not obscured by the blurring effect of prejudice and fear.

Summarizing: Personality "clashes," even unilateral dislikes, are usually the result of fears that "the other fellow" threatens some basic need we have. Remove the threat, and you create a climate within which both men can operate harmoniously and effectively to their *mutual* advantage.

If *you* are the one who does the disliking, large doses of introspection are recommended. If *he* does the disliking, large doses of listening, providing attention, obvious interest and sincere praise are called for.

Appearance (our)

The appearance of salesmen has become a subject upon which most sales managers are loath to talk. At best it is difficult, if not embarrassing, to comment upon the other fellow's appearance, and we are usually afraid of creating resentment. Unfortunately, there are still some saleswomen who feel that they are paid for selling a product, not for how they dress or appear. The trouble with that premise is that few products sell themselves, and customers want advice, information or answers to their queries about product characteristics and/or applications. It is not easy for them to accept the judgments of men who show (in their opinion) poor judgment in their attire or appearance. This is the psychological phenomenon of Gestalt at work (see Part One). This is the tendency we have to see the part as the whole... in this instance to equate the judgment of a salesman on *all* things in terms of his judgment on *one* thing (dress, appearance), which we feel we are in a position to evaluate.

A salesman is a walking mirror, as we have said and as we must say again for emphasis. The customer sees what he, the customer, looks like in the eyes of the salesman—by the way the salesman looks and acts.

The salesman, to be effective, does not have to be a fashion plate. If he is, this may cause other problems. We know of one public speaker from whom we always learn much; yet, large segments of intelligent audiences reject him. They are heard to say, "He's too Madison Avenuish." They are referring to the cut of his clothes, haircut and general appearance, not to the tone or content of his remarks. It is clear, then, that a salesman should avoid either extreme. He should know what colors become him, which colors compliment others and which clash.

We will not discuss such points as clean nails, neat haircuts, and the like. If we were to comment on these facets of appearance, the average salesman would say with annoyance, "Everyone knows that!" The question is not whether we know it or not, but how long is it since we checked our judgment on clothing, for example, with the evaluation of someone whose own appearance indicates his judgment in such matters is above average? Our own appearance can change so gradually we are often not aware that a change has taken place... especially if we become so enamored of our work that we forget appearance.

The customer does not create appearance as an obstacle. We create it unassisted. And if we create it, we can eliminate it.

Prejudice

We have maintained consistently that selling is not only a profession but one of the most demanding of professions. A fact that supports this contention is the tendency some customers exhibit to thwart our attempts to help them. Usually, other professions do not share this handicap. Prejudice can be a real obstacle when we attempt to make progress during the "contact" facet of the sale.

If we are unfortunate enough to hold racial or religious prejudices, we will often not permit them to get in the way of our good judgment when we need a good doctor or a good lawyer. At that moment, we are primarily concerned with the doctor's ability to return us to a condition of good health. Or, we are concerned with the lawyer's professional ability and his capacity for helping us keep out of trouble.

Unfortunately, the same stimuli are not being brought to bear on our customer. Some customers will not keep their prejudices in check. As a result, we have problems in establishing the rapport essential to making a satisfactory contact.

Our first reaction, when we can clearly identify prejudice as the obstacle, may be to react to it with anger, even though we may be able to keep that anger concealed. We are outraged at the unfairness of prejudice. Prejudice is prejudging, and we do not like to be prejudged or stereotyped. When we permit anger to take over, our judgment is seriously impaired, whether we realize it or not.

When we permit ourselves to become angry, we are guilty of a stupidity which is just as bad as the stupidity that causes the other fellow to be prejudiced. Fortunately, there is a remedy. We should keep in mind that if we had been exposed to every environment to which our customer has been exposed; i.e., family, geographic, educational, groups, military, and so on, we would have shared his milieu . . . we would indeed have been the other fellow. We would, therefore, share his prejudices. This does not make his prejudices acceptable to us as we see them from our mileu, but it does make them *understandable*.

Prejudices are an impairment of the thinking process . . . a disease, and the holder of prejudice merits our sympathy, not our anger. If the saleswoman will develop this frame of mind, she will be able to keep her anger in check, and she will attack the obstacle of bias and use wisdom in doing so.

In reading the foregoing, even though you may agree with this analysis, you may feel that prejudice is just *so* unfair that you cannot look upon it in the clinical fashion recommended. Then try this as an alternative: Ask yourself, honestly, are *you* completely free of prejudice?

Are you sure you have never prejudged someone because he came from a certain racial stock or a certain religious group, or because he was an atheist or an agnostic?

Have you ever prejudged someone because he came from a company which you have never liked, or because he was very pro-military or anti-military, or harbored and spoke freely of other views which were diametrically opposed to views which you held and about which you felt equally strongly?

If you can answer these questions in the negative, then you are a rare bird. On the other hand, if you acknowledge that, perhaps occasionally, you have harbored a pet bias of your own, how can you really become angry with the customer who holds prejudices that get in *your* way and make it difficult to establish a successful contact?

Make no mistake about it, the foregoing is not an apologia for prejudice. Prejudice is an affliction upon mankind, or an affliction which mankind has placed upon itself. It is an excuse for thinking, and evidence that the holder of prejudice has not been able to understand the nature of the stimuli to which the object of his bias has been exposed. It is often a defense mechanism which you and I may use when we experience what is really a fear of the group toward which we are prejudiced. We may see members of a group make progress in business, and because we may not have been as successful, we find ourselves thinking that they are unethical, dishonest or immoral. This may well be true of the members of any group, but it just doesn't make good sense to judge all of the members of that group by the worst examples.

Keep in mind that the members of any group to whom prejudice has been shown are not unaware of that bias. Sometimes what seems to be prejudice on their part is nothing more than their use of a defense mechanism, an equally juvenile attitude, an attempt to "get even." No, prejudice is a sorry commentary on all of us, but it can't be solved with anger; in fact, anger can only aggravate an already unfortunate situation. Most of the reasons that underlie "personality clashes" form the foundation for group prejudices.

Over and above the remedies previously discussed, consider the following as especially applicable to the problem of group bias. Prejudice is a fact, and it won't go away. At the same time, a salesman does not go out of his way to create prejudice or to aggravate it. He does his homework on the customer. If he knows the customer has strong feeling on a subject, he should do his best to avoid discussion of that subject.

If you know that a customer has a built-in bias against the group to which you belong, seek to sell him on you as an individual. Prejudice is usually a result of ignorance or of hearsay. If you sell the customer on yourself, you may hear later that in his attempts to prove he is "broadminded," he will use his friendship with you as proof. He may resort to the old bromide, "Some of my best friends are (whatever group you belong to)." When you hear this, you may experience a feeling of anger. Again, this is not an intelligent reaction; it is an emotional reaction. Consider your success with this fellow as a new benchmark. You are perhaps the first experience he has had with members of your group, and you have impressed him favorably. He will meet others of your group, and as he grows to like them, his prejudice begins to wither and may die. In any event, other members of your group will have less difficulty with prejudice as an obstacle when they attempt to contact him in a selling situation.

Frequently, salesmen ask, "What do you think of wearing lodge pins?" It isn't what I think of wearing lodge pins, it's what your customer thinks of the pin in question. You should know how he feels. If you don't, then why create unnecessary obstacles? What do you seek—to make a sale or make a point? Are you concerned that taking the pin off is sacrificing a principle? Is a principle really involved? Let's take a real "for instance":

We met an individual who had made a spirited attack on the American Legion. If we were selling this man, or attempting to establish a productive contact with him, we would have to make a decision if we belonged to the Legion and wore its lapel pin. We would certainly have to ask ourselves these questions:

1. There are people in my company whose existence depends upon the sales I make. Do I have the right to jeopardize their welfare while I "stand on principle"?

2. Since the customer dislikes the American Legion, won't he have a tendency to question the judgment of anyone who belongs to it? (Stereotyping.) Can this cause him to distrust my judgment on other matters and specifically on the matter I wish to discuss with him?

3. How important is this sale to me and to those who depend on me? Am I letting the Legion down by taking the button off when I call on this chap, or am I letting my company down by wearing it and jeopardizing the sale? In this instance, to whom do I owe my first loyalty?

4. Do I know my customer so well that I know he is "big" enough to separate his judgments on my Legion membership and my professional competence in my field. Is he the kind of fellow who may joke with me about my Legion membership, yet buy my product?

As the human race becomes older, we lose some of our old prejudices. But new ones grow in their place. You can flaunt the banner of some of your beliefs in the face of the biased customer, but this luxury may carry a high price tag. Only *you* can decide if it's a bargain at that price.

The best way to overcome prejudice is to show the customer he is paying a high price for it. To accomplish this, *concentrate on his problems, and show him that your product or service will relieve him of those problems.* In other words, accent a need that is greater to him than the one he satisfies by harboring a prejudice.

Keep in mind that when you and the customer differ, it is not only a difference of opinion (as she sees it), but it is also a reflection on her judgment. She will fight to defend her judgment. When she cannot do so with facts, she will resort to the self-defense mechanism of prejudice! Help her find an escape hatch.

One last point will be of value: Ask yourself when you last went out of your way to make the prejudiced customer *like* you. We usually lavish our attention, and entertainment, on the fellow we like to be with. We avoid the biased chap, calling upon him only because the sales manager insists upon it or because of the magnetic attraction of the business which he represents for us. You are not being asked to *like* prejudice. You are being asked to understand its roots, and to attack it professionally and not with the anger which most of us use to combat it, thereby destroying our effectiveness.

Voice (our)

Voice can be an obstacle, and it can be controlled. You have met people who literally fatigued you because it was difficult to hear them, and others who tired you because their voices were strident. Still others caused your attention to wander because of the monotony of their voices. These examples are apart from the chap who talks with his hands across his mouth or while he has a cigarette, cigar or pipe in the corner of his mouth. Customers experience the same annoyance under these conditions; only they don't have to put up with it. They can bring the selling situation to an abrupt close before contact has really been established.

We have seen Paul Mills of Word Power actually change the voices of his students in a matter of a few hours, and we are sure he will agree that this change will be transitory if the salesman does not concentrate on his newfound power. There are records available to help the salesman develop voice control. Use of a tape recorder can provide you with assistance at home in your attempts to develop a new voice or to improve the quality of your present voice. If you feel that voice is a handicap, it is one which, in most cases, can be eliminated—if you are willing to work at it.

We might also keep in mind that another way to avoid the voice obstacle is to let the customer do most of the talking. This, alone, is obviously not enough, and concentration on our voice defects will improve our professional image.

Age (our)

Age might well have been considered as one of the prejudices we have discussed. Some customers stereotype age groups. One customer may feel that only a salesman with years of experience can possibly cope with "his" problems, or that only the older salesman will have judgment upon which the customer can rely. This customer makes it very difficult for the younger salesman to establish a meaningful contact; little credence will be given to his advice or opinions until he has proven himself. Unfortunately, it is not possible to "prove one's self" in the early stages of making contact.

Prejudice is not limited to the customer who resents having young salesmen call upon him. Other customers regard anyone who is over 45 as being out of touch with progressive ideas. Such customers feel that older salesmen will not be able to provide them with "new" solutions for old problems. Obviously, both of these forms of prejudice are based on fallacious precepts. As one chap in the Southwest said, "We have people who, when they die, will have placed on their gravestones, the epitaph, "Died at 30, buried at 70!" Such people are not limited to any section. There are old men who think young and young men who think old. Unfortunately, the salesmen against whom the prejudice is directed cannot use this argument on his customer. Whether the problem is one of being "too old" or "too young," the selling strategy is much the same.

When the customer thinks you are "too young"

When the customer has a prejudice toward you because you are young, consider which of the following strategies will be most effective in combatting this prejudice (from Sales Strategy Bank):

E10—Anticipate, raise and answer an objection you feel he may have toward the product or service you wish to provide him. Stay in "Low key" so that he will not be reinforced in his opinion that being young you must also be inexperienced and brash. Be sure the objection you raise is one that can be answered only by a salesman who has considerable "savvy." This approach, alone, will not convince him, but it will usually obtain for you "polite acknowledgement," a grudging awareness on his part that perhaps, though young, you are not without experience.

B19—Ask for his opinion, advice or help. Rare is the older man who does not relish the opportunity to give advice to the young. Usually, the basis for his prejudice toward the younger man is not because he really feels the younger man lacks competence. The very opposite is true. The older man is afraid that the younger man may be totally competent. This is usually a blow to the self-image of the older man and raises the question, "How could this young fellow get so smart so fast, when it has taken me so many years to get as smart as I am?" This also shakes the assurance of the older man. Sometime, to avoid such situations, he will try to "scare off" the younger fellow. When you seek his advice, help, or opinion, you provide him with the assurance he needs.

In asking for advice, you must not downgrade yourself or ask for help on such a basic point that he will only be reinforced in his opinion that you are inexperienced. This can be avoided by such phrases as, "I have an idea on this application, but I value your experience and would appreciate your opinion." Often this small bit of recognition of his experience is all that he needs. "I need your help" is the greatest of all human appeals, so don't abuse it. No professional salesman ever uses it with the juvenile approach, "We are having a contest this month, and I need your help."

D12—Get him to talk. People like audiences, and sincere attention is one of the highest compliments you can pay. If you really listen, her attitude will soon change from, "This kid they sent to call on me" to "This is a bright young man; he knows when to listen."

E6-7-8—Appeal to his pride, prestige or needs of the "me he tries to project." Do this with care, and do not permit your remarks to appear as flattery. Rather than praise him, praise something he has done; i.e., praise the work and not the worker. Be sure that whatever you praise represents a real accomplishment in his eyes. Don't leave him annoyed and thinking, "Is that the best he thinks I can do?" This can be avoided by selecting some accomplishment of the customer to praise and being certain that he is proud of it.

D6—Talk benefits! Focus his attention on what your product will do for him. This tends to remove the focus from you, as an individual. Use voice control and control your enthusiasm (he regards enthusiasm as the hallmark of youth). Rely upon your *facts* to speak in a loud voice. Make him so interested in his other problems and your proposed solutions that he will forget your youth. Above all, *don't use canned sales presentations.* He associates such presentations with youth. "Those birds at his home office hire a bunch of kids and put words in their mouths... Get 'em beyond that canned 'pitch' of theirs, and they are lost." This epitomizes the reaction of the customers who hold this prejudice.

C6—Omit something so she can inject her ideas. Leave some relatively minor, yet obvious, point out of your presentation. She will call your attention to it. This permits her to feel that "this young fellow didn't have the whole answer." All of these strategies seek to accomplish two ends: (1) transfer the focus of her attention to her problem and away from your youth; (2) enhance her importance. Examine the Sales Strategy Bank for other means which may be employed to accomplish these two objectives.

Finally, don't be annoyed or irritated at her prejudice toward your age. These reactions on your part will only cloud your judgment and as a matter of fact, give her justification for the views she holds of youth.

When the customer thinks you are "too old"

How about the customer who looks upon the older salesmen as "has-beens" and out of touch with new ideas? Actually, the remedy is not unlike those proposed in the foregoing. Focus his attention on *his* problem—talk product features and customer benefits, let him talk, find a point of agreement. We admire people who agree with our views.

Emphasize mutual interests. Avoid bromides.

Use current business terms. *Don't, don't don't* talk about your past experiences and accomplishments . . . avoid discussion of the past; he already regards you as being wedded to it, so don't give him justification for this feeling.

The younger chap, against whom age prejudice is directed, is advised to hold some rein on enthusiasm; the prescription for the older man is just the opposite. Show enthusiasm and especially relative to *new* features of your product. Concentrate on how your products will help him accomplish his goals and long-range objectives, and emphasize losses he will sustain if he does not buy. Don't be paternalistic or condescending. Don't be too conservative in dress or manner when calling on the customer who is prejudiced toward older salesmen. But avoid going too far in the other direction and trying to look and act like a youngster.

Interruptions

This potential obstacle during the "contact" facet of the selling situation was covered previously during the discussions on "Atmosphere." The solutions recommended then apply here.

Customer has other problems

In considering the barriers to be surmounted at the moment of contact, we must anticipate the situation that arises when the customer has other problems on his mind. This is not unlike the situation that exists when the customer gives other demands on his time a higher priority than he is welling to give you. These were covered in the discussion of "Priority."

Essentially, the remedy consists of focusing his attention on the problem you wish to help him solve and making it appear more important than his other problems. Your opening remarks, especially, need to be well-considered, and they must be planned for their attention-getting effect.

Personal worries (customer's)

At first glance, one might not see much difference in this obstacle and the foregoing, but the customer who has personal worries presents a more difficult problem. Frequently, he will not reveal the fact that he has such anxieties. Yet they hold his attention, almost completely, and it becomes most difficult to establish rapport during contact.

The nature of personal worries ranges widely from those connected with work-related problems to money problems at home, marital problems, children problems, and ad infinitum. To the customer, these anxieties sometimes become so accute that it is only with great difficulty that he can transfer his attention to work-related problems, which your presence represents.

You may be fortunate enough to have him bring his personal worry out in the open. At least, you will know what you are up against, and you won't mistake his lack of interest or lack of enthusiasm for apathy. When he does not mention his problem, but merely exhibits lack of interest, we can easily misinterpret this situation and write him off as a poor prospect.

A saleswoman may encounter a customer when he is in one of these moods produced by a personal problem, and the saleswoman, not recognizing the reason, may leave and return another day. Unfortunately, when the saleswoman returns later—not having identified the cause of the customer's lack of interest on the first call—she may expect the same lack of interest on the this second call. Anticipation of apathy may affect the saleswoman's approach.

To be effective, the saleswoman must proceed as though she had never talked to the customer about this proposition before. It is a safe bet that the customer heard little of the original presentation, submerged as he was in his personal problem. Nonetheless, on this second call, avoid irritating him by repetition, because there is always the chance that he did hear much of your original presentation but just didn't feel like taking action because of the frame of mind produced by his personal problem.

Therefore, to avoid irritating the customer, you may use the following strategies from the Sales Strategy Bank: *D9—Probe for reaction; D15 through D18—Use questions, with heavy emphasis on the rhetorical question.* This type of question makes it possible for you to present information, ostensibly through questioning even though you are not seeking a spoken reply but are watching for a reaction. Consider the following example:

You called upon a customer and because of his attitude, you decided that his concern over personal problems was so great that it would be impossible to attract and hold his attention, much less arouse interest. Therefore, you bowed out after setting up a date for a return call. You now return at what you regard as a more propitious time. You do not know how much attention the customer paid to your previous presentation. How much will he recall? How far did you progress into the sale? One way to find out, as stated above, is to probe for reaction. You ask this rhetorical question, "As I recall, on my last visit, we agreed that inventory was not a problem, but the problem was one of space...?" The inflection you placed on the last few words makes this statement a question. You pause only momentarily long enough to listen for—and look for—a reaction.

The customer may react with, "No, that isn't so; I think inventory is a real problem." If you obtain this reaction, you may be sure that he heard little, if anything, of your original arguments on the matter of inventory. This means you must go over this ground again. If, on the other hand, the customer merely nodded agreement to your rhetorical question or maintained a placid silence, you may have some assurance that you had sunk inventory as an obstacle, and you may proceed with the balance of your presentation. You may decide to use additional rhetorical questions to determine just where you were in your original presentation when you lost out in the battle with his personal worries for his attention.

Let's agree that all of this is fine when it is possible to back gracefully out of the situation and set up a new appointment when his personal problems will not create an obstacle. The fact remains that this is not always easy to do. You may not want to leave and thus give competition a chance to step into even this temporary vacuum. Geographical considerations and other demands upon your time may not make it easy for you to return...also, you may not want to lose the time you have already invested in this customer today. What do you do under these conditions?

At least two courses of action are available to you. They are not presented in any order of priority. *You* must decide which fits the situation the better.

One strategy is to get him to talk about what is bothering him. If it is a personal problem, it may be difficult to cause him to talk, and it may even be undesirable to do so. He won't want to talk about money problems, and he may avoid discussing marital problems with you... and you should not encourage him to do so. If his problems are of a nature that do not inhibit him from bringing them up, let him do so; this may be just what he needs.

Let him unburden himself, but *never* surrender to the temptation to give him advice, even when he asks you for it. Be sure that if you give him advice, he is going to blame you later if it doesn't work out. Furthermore, even though he asks you for advice, he does not want it.

If he asks you, "What would you do if you were in my place?" respond with such statements as, "What have you considered doing?" You will usually be surprised when he tells you *exactly* what he is going to do. Your final comment should never be more than "That's an interesting approach." Don't indicate your approval or disapproval. Often, following this, you'll find he experiences a certain sense of relief just to get the matter off his chest, and he will be able to concentrate on the subject which you came in to discuss with him.

The second approach is one which we have mentioned over and over...try to capture his attention by causing the problem you wish to discuss to appear to be far more important than the problem which is now holding his attention. Holding out benefits will not be enough; you may find it essential to emphasize the losses he will suffer if he makes the wrong decision or if he procrastinates. In other words, you attempt to cause the prospects of *economic loss* to appear more important than the *psychic loss* about which he is presently concerned.

Health (customer's)

Unfortunately, there are still some few sales managers and salesmen who, in effect say, "These psychological aspects of selling are so much nonsense; all you need is a good product, a good price and an effective product story." It would be interesting to hear how these individuals would solve the problem presented by the customer who is in poor health and is concerned about it. Our customer does not live in a germ-free cell awaiting our call and our happy solutions to his business problems. He is being bombarded constantly by stimuli that affect his viewpoint and his immediate area of interest, and health represents some of those stimuli. Nor does the customer have to be afflicted with some serious malady...his interest can be completely dissipated by virtue of the fact that he has a headache or an annoying attack of hay fever.

Under normal circumstances, the customer would be eager to solve her business problems and to discuss them with you. Consider her frame of mind, however, if it's late August and she is a perennial hay fever victim just opening her third box of tissues when you walk in the door. She is not going to be in a very good mood to make the decisions that a completed sale is going to necessitate. At least if it is something as obvious as hay fever, we can identify it and take appropriate action. If we have also been unfortunate enough to have experienced sieges of hay fever we can establish rapport and offer understanding sympathy. This will remove the mild embarrassment she experiences each time she has to reach for a fresh handkerchief.

Customers have health problems which lack the visible effects of hay fever. Such problems are difficult to overcome because the salesman may be unaware that the customer is not feeling well or is concerned about his health. If you cannot bow out of the situation after setting up another date, just what actions can you take now? Consider these from the Strategy Bank:

D2—Narrow down his choice. Simplify his decision-making if there are more than two choices of action from which he must choose. None of us likes to be confronted with difficult decisions, and this is even more true when we are not feeling well. *B1—Save his time...* boil your presentation down to its essential factors using appropriate visual aids if they are available. *B5—Give her privacy.* Present your arguments; then, if possible, find something else to do... someone else to see. Use any legitimate excuse to let her think over your arguments in privacy at her own pace, which will probably be slower than usual.

E5—Use a variation of the "Remove him from the situation" strategy. Suggest that he might like to have you take up the matter with some subordinate. If the matter is one that he will trust to a subordinate, he may leap at the chance, whereas ordinarily he might not be willing to relinquish control. If he does not respond to this offer, at least, in a subtle manner, you have indicated that he has not been interested in meeting the problem, and he may start to give you his attention.

With a customer who is not feeling well, the key objective should be to make decision-making *easy* for him. The greatest service you can render him is to focus his attention on a problem he *can* do something about, and take his mind off of his health problem. Therefore except in the case of minor, yet annoying afflictions, such as headaches and hay fever, do *not* discuss his health. Discussing it only serves to remind him of it, and this will make it increasingly difficult for him to listen to your presentation.

Policy (both)

Cartoonists and humorists have beaten to death the phrase, "There's no *reason* for it, it's just our policy!" Sometimes this humor is nearer to fact than fiction. Policies are not established capriciously, and there is usually a sound reason for them... or there was a good reason when the policies were established. Often the original reason no longer exists, and the policy is permitted to continue. Policy presents many of the aspects found in our discussion of "Habits." In addition to the various strategies which may be used to combat habit as an obstacle, consider the "policy" obstacle from these viewpoints:

1. Does the customer sincerely believe there is a valid reason for the policy that presents an obstacle to you?

2. Is the "policy" really not a policy at all but simply (a) a device that makes it easy for the customer to say "No" and avoid decision-making; (b) a device that protects incompetence in the customer's organization; or (c) a device that has political roots; i.e., a means of preserving certain organizational relationships?

Don't just assume that the policy makes no sense and permit it to arouse your anger. Policies are usually established to prevent undesirable situations from arising or to stop the re-occurrence of undesirable situations that have arisen in the past before formation of the policy. Admittedly, some policies are established by the customer, or one of his people, as protective devices. We may not entertain the same fears or concerns held by the customer and we will, therefore, consider the policy unnecessary. He will not. So long as he considers the policy as one which is valid and with purpose, we are not going to change his mind by attacking its validity.

Here are a few objectives of many customers: (a) uniformity of action and thought within the company on important matters and procedures; (b) maintaining a posture in the eyes of suppliers; (c) safeguarding the company against capricious actions of people within the company who may not have all the facts or are inclined to act impetuously, or to act in their own selfish interest without concern for the effect of their actions on the company. To help achieve this type of objective (and many more), policies are established.

If you are going to neutralize the effect of the policy, you must identify the reason for that policy. What need is it intended to satisfy? Is it psychological, economic, or is it an organizational operating need? If the policy is presenting you with an obstacle in your selling efforts, then you must appear to be a threat to the need that the policy seeks to fill. If you don't identify the basis of the policy, you can't determine the course of action to follow. Essentially, you must identify the need it protects, then prove to your customer that your product or service will not work at cross-purposes with their policy.

If you cannot accomplish this, then you must show the customer the high price she is paying to maintain the policy. In other words, she is making a greater sacrifice than the value of that which she seeks to protect through "policy." To get down to brass tacks, you must show her that you can fill a need that is more important to her than the need her policy protects.

Once again, we see the need for pre-sale preparation. In this case, learning as much as possible, in advance, about the customer's policies, so that we can prepare strategies in advance and not find it necessary to rely upon off-the-cuff actions. Once again, we are reinforced in our opinion that professional selling requires a professional competence. Selling is not merely a stepping stone to another job, it is an attractive end-objective unto itself.

Trade relations

Trade relations are sometimes referred to as "reciprocal buying." It may be described as a quid pro quo arrangement; i.e., "You buy from us and we'll buy from you." This sounds like a cozy and helpful arrangement if it really *is* your company that buys from the customer and not your competitor who buys from him. Unfortunately for the salesman who would like to take advantage of this situation, the law regards such "arrangements" as restraint of trade, evidence of monopolistic practices or as any one of a number of violations of anti-trust legislation. Not only does the law take a dim view of such agreements, but it has some painful punitive measures which it will take if the salesman is foolish enough to indulge in this practice which, incidentally, is at best short-range planning.

Therefore, let it be clear that trade relations are not being presented as a selling tool which you may consider using. We are discussing trade relations as an obstacle which you may encounter when you meet the customer in a selling situation. He may not want to go beyond contact for one of two reasons:

1. He may want a commitment from you that if he buys from you, your company will buy from him.
2. He may not wish to give you the opportunity of an interview because he is wedded to a policy of buying only from suppliers who already buy from him, whenever he finds this possible.

If we were to beat around the bush on this point, some readers might think we were being coy and were really, in some subtle manner, indicating that reciprocal buying is a great tool to use if you can "cover your tracks" and do it with impunity. To disabuse anyone of this idea, let's make it clear, at once, that reciprocal buying should *never* be employed as a selling tool. If you feel that such a situation may be implied by the customer's behavior, there is one course of action to take. Discuss the situation *fully* with your company's legal counsel. Together, you should consider the full legal, moral and ethical implications of the proposed transaction and its effects. This is a matter where competent legal advice is essential.

It is quite possible that there may be situations where reciprocal buying arrangements are legal. In some business climates, however, such activities may be seized upon and blown up out of all proportions to their true significance and intent. There are forces at work that seek to pillory "business" at every opportunity. Even from a pragmatic standpoint, it doesn't make sense to engage in reciprocal buying arrangements. The profit which you might make on an order could be wiped out by legal costs, to say nothing of the damage to the image of your company which would result from publicity, even in instances where no *intent* to violate the law was present.

We once heard a sales manager say, "Don't tell me what the lawyer covered at the meeting. I missed it, and I don't want to know what he says I can't do." This manager is laboring under the delusion that so long as he doesn't *know* what is legal and illegal, he won't be held responsible if, in his ignorance, he breaks the law. This attitude can only lead to personal and corporate loss. Consider how long this manager would be kept on his job if his management, which pays lawyers to insure that only legal practices are followed, were to have heard his comments.

You are undoubtedly wondering just what one does when the customer's attitude, or her comments, expressed or implied, make it unmistakably clear that she will buy only on a reciprocal basis. Consider first what you *don't* do:

1. You don't give her legal advice.
2. You don't give her a lecture on ethics.
3. You don't threaten either directly or by implication.

What, then, can you do?

First, you recognize this as a challenge—an economic challenge and possibly a customer-relations challenge. You marshal your facts and present them so convincingly that your customer becomes aware, possibly for the first time, of the full price she is paying for this maintenance of the quid pro quo with a competitor. You drive home forcefully the other benefits of which she is being deprived because she is not doing business with you. When possible, present these benefits in such a manner that they transcend the immediate benefits she now receives from doing business with the competitor who "buys" the customer's business by purchasing from him.

Analyze your product or service and determine whether either of the following may be presented as benefits that surpass the supposed advantage of buying only from those who buy from him:

Research: Can you provide benefits of research to this customer in a greater measure than your competition?

Fast action on service: Can you move more rapidly than your competitor in aiding the customer when he is in difficulty. Will he lose more money in down-time or delay than he now feels he makes through his arrangement with the competitor? You probably will be hard-pressed to use price as a benefit because he has already mentally deducted from the price your competitor is asking, the profit that he, your customer, is making on the items your competitor is buying from him. It is most unlikely that you can bring your price down to that level and still make a profit.

Look at the problem represented by reciprocal buying this way:

1. Your customer sees an advantage in buying from those who promise to buy from him in return.
2. Your customer pays a price when he buys from those who buy from him, and that price is not in money alone. It *may* be in deficiencies in the products he buys, in service, in delivery, in appearance, prestige, etc.
3. The key question is this: Does the *total price* that he pays cost your prospective customer as much or more than what he makes on the product he sells to your competitor with whom he has a reciprocal buying arrangement?
4. If this trade relationship adds up to a total profit to the customer, as compared to how he would fare buying from you, then you have a problem.
5. If the net result of reciprocal buying is not a profit to your prospective customer, then you must alert him to the fact.

Another obstacle sometimes arises when you attempt to sell to a customer who makes products which your company *could* use, but which you buy elsewhere. Your customer develops a suspicion that while your company finds him desirable as a customer, it doesn't regard him as "good enough" to be a supplier. Does he have grounds for this suspicion? If he does, then your problem with this customer is not a "price" problem; it is a psychological problem. He may resent losing the profit he would make by selling to your firm, but he resents—even more—the reason why your firm rejects him as a supplier. This is an affront to his self-image... to the "me-he-tries-to-project" and to the "me-he-wishes-he-were." Into your company's refusal to buy from him, he reads an evaluation of his company and of himself that he does not like; i.e., "They don't see me or my company as we want others to see us."

This may, in some instances, be the case—that your company may not regard him or his products highly. In other cases, and far more common, the rejection of this man as a supplier may be the result of:

1. Ingrown buying habits of your company;
2. Lack of awareness on the part of your purchasing people of the true cost they are paying when they favor certain suppliers to the exclusion of others who could be customers;
3. Unawareness on the part of your company that this potential customer makes products which you *could* use—without any reciprocal selling-buying arrangement;
4. The quality of his product or service fails to meet standards you have set, even though you may have respect for the man and his company.

If your problem *is* the result of an imagined affront to his self-image, you must remove this feeling, or your chances of making a sale are slim indeed. You must convince him that you do respect him as a supplier. Admittedly, a problem does exist and reciprocal agreements must not be entered into or implied. They must not be held out as a lure. If you decide to buy from this customer, it should not be done in connection with the sale you are trying to make, or one which you are arranging for the future, without obtaining competent legal advice.

Purchases which are made from this customer should be made because he does make products which meet your buying standards as well as those you could buy from *his* competitors. These purchases should not be a result of a quid pro quo arrangement. They must be the result of a need to satisfy buying needs of your company or to eliminate any feeling in the customer's mind that you have no respect for his products or his organization. In other words, if the basis of his resistance to your selling efforts is resentment because he feels your company doesn't consider him "good enough" as a supplier, you must either give him sound reasons why you buy elsewhere, or your company should be spreading its purchasing and including him as a supplier—but not as a part of a "deal."

It is repeated for emphasis that the foregoing paragraphs are not written with tongue-in-cheek. Business obtained by illegal or unethical means almost inevitably carries a high price tag in damage to your company's posture or in costly litigation.

As a professional saleswoman, even if you are working for a small company, you are in the major leagues. The type of buyer-seller relationships in which you engage are far more complicated than those that may take place at the local tradesman's level. The grocer decides to have his house painted and selects a painting contractor who buys groceries in his store. We don't find it hard to understand this arrangement because, as individuals, in our personal buying arrangements, we are affected by similar considerations. No power on earth can make a local barber buy from a hardware store if the barber sees the owner of the hardware store get his hair cut in the shop of a competitor. But the law takes a dim view of this same psychological reaction when it appears outside the neighborhood level. Here, there are many more claimants for your business, and many more people who will sense that they are being wronged by such buyer-seller relationships. As a result, there is some hard-hitting legislation on the subject.

The legislation is in place just as much to *protect*

you from such arrangements as to prohibit you from engaging in them. When you do encounter a selling situation where reciprocal buying is the problem, you will need to exercise real professional salesmanship. These situations, more than almost any others, call for thorough product knowledge, an understanding of your customer's business and his economic problems. And they call for the application of sound psychological principles. When you can obtain business in the face of the obstacle represented by the potential customer who does business only with those who buy from him, then yours will be a great satisfaction... unless the sale was obtained by the same blunderbuss methods of "buy from us and we'll buy from you" which your competitor may have been using.

"Golf today"

"Golf today" is an umbrella title that covers the many situations that cause a customer to avoid talking and/or doing business when you try to make contact. It may be the call of golf, the lure of a fishing rod or a lakeside cottage, or literally dozens of non-business lures. Your appearance on the scene constitutes a threat. He wants something else and you are in the way... it is as simple as that.

You have a decision to make. Can your proposition be made so attractive, in a few short minutes, that it can completely crowd these other desires out of his mind or at least enough to capture his attention? Can your proposition be speedily stated so that he will realize what his golf game is really going to *cost* him? This is a last resort method, because if you show him he will lose money and he goes to the golf course in spite of this, you will probably lose the game for him or create such conscience qualms that he will blame you.

You have another choice. Can you postpone your presentation? If you can, he may be so relieved that he will gladly give you another appointment. If, however, you feel that your proposition is *so* important to him that he will be willing to forego the pleasures of golf, then be sure you are not falling into the trap of using "projection" (see Part One); i.e., that you are not really saying, "If *I* were this chap, *I'd* be willing to forego a round of golf for the chance to make some money (or to solve this problem) that is possible through the proposition I want him to hear."

You are *not* this chap. His values may be quite different from yours. At *this point in time*, the siren-call of profit or the solution of an operating problem will not interest him compared to the "Id satisfaction" represented by the outside attraction you are suggesting he forego. Neither can you say he is *wrong* in making this choice. He is not wrong, he is *different* (from you). If you are sure you are right, hit hard on the advantages of your proposal, stressing such timing features as may exist. Go for his *attention* and let proof of your claims wait until you are certain you have his interest and "golf today" is no longer a problem.

When this approach fails, try for an alternative date and take advantage of the delay to be even more completely prepared. He will be so relieved to be on his way to golf that he will be more willing than usual to give you the time you need for a full presentation later.

Our own prejudice

This obstacle is listed only to emphasize that prejudice is not always on the customer's side. We sometimes create this obstacle for ourselves and it threatens our chances of ever getting beyond the contact facet of the sale. Handling of prejudice has been thoroughly covered, and you are referred to those comments on prejudice and coverage of "personalities" as an obstacle.

Defeatism (our)

This is still another obstacle we usually create for ourselves. It is just as costly as over-confidence. You hear it expressed as: "We don't have a chance, he has always had it in for our company"..."he thinks we 'did him wrong' when he was a 'little guy'"..."he is buddy-buddy with our competition"..."he wants us to give it away"..."he wants features we just can't supply." Such statements as these, and a long list of similar expressions of gloom and doom, precede the inevitable, "Well, you can't win 'em all!"

We can't win *any* of them when we think we are beaten before we start. We can list many companies of long vintage who are Number One in their field who have never had a price advantage since their companies were formed. Some of them not only are first but also are now largest in their field.

Defeatism is conveyed to the customer even though we may try to conceal it. If she *is* inclined to favor competition of ours, she reads our defeatist attitude as lack of confidence—not in ourselves but in our product or service. She seizes upon this as proof of her belief that we are not able to measure up to our competitor. This reinforces her in her decision.

When we suffer from defeatism, more often than not we never even ask for the order. It's almost as though we *want* to be able to report back to the boss, triumphantly saying, "See, I told you so...he wants us to give it away" (or some similar excuse for our own failure to ask for the order).

Defeatism is usually a result of being short-range oriented in our thinking. The customer may well be wedded to his present supplier, but that supplier didn't get into this favored spot overnight, unless he married into the customer's family, which is hardly likely. It took our competitor a long time, and it will take us a long time to push him over on the perch and give us some room. In the meantime, we will lose some orders we would like to have received. If, each time we lose an order, we have learned a bit more about the customer as a person, about his problems, his fears, his aspirations, and his needs, and, if we have learned a bit more about our competitor and his pet strategies, we can be certain we will eventually "cop an order" from this customer.

Admittedly, in some businesses, the loss of an order from a potential customer means we "have had it." He is a one-time proposition; so all of the arguments above about short-range thinking fall on deaf ears. The fact remains that defeatism not only communicates itself to the customer, but it kills creative selling; it stifles initiative and imagination; it dulls our thinking capacity and it robs us of the essential enthusiasm. Salesmen are often called incorrigible optimists, and it is well that this is the case. The only practical antidote for defeatism is not starry-eyed optimism, which is based on little more than unsupported hopes, but it is in doing our homework. Obtain as much information as possible on your product, the competitor's and on the needs of the customer (product *and* psychological). *Have—and show that you have—a driving desire to help the customer.*

This means making a new survey of your proposition, comparing it to the offers being made by your competition, comparing it to the customer's needs and then determining how you can better express and demonstrate its advantages. This calls for creative professional selling of the highest order.

What *new* aspect can I create? How can I present my proposition in a new setting that will cause the customer to see it as an enhancement of his needs? This approach calls for humility to the extent of asking others to review your strategies and tactics up to this time. It may mean trying out your strategies and tactics on others in order to obtain their reactions and advice. Out of this may come confidence born of good preparation and the assurance that you are right.

If you are about to attempt to establish contact for the first time with a customer—and you experience a sense of defeatism—avoid such negative openings as: "I realize this is a particularly bad time to try to see you."...."I can see that you are busy, and I'll try to make this short."...."Would you prefer I come back at some other time?"

Defeatism reveals itself, but so does confidence. The time for confidence is when you know you are up against a difficult proposition. This is the time you must exhibit confidence. Don't apologize for taking his time...what you have to offer him will more than repay him for his lost time or you wouldn't be there to start with.

Address yourself as soon as possible to his problem. Enhance what you believe to be his self-image. Get him talking—then *he* is the one that is using the time, not you. If you think your chances are poor, it is obviously because you have not done the best job of convincing the customer that you can satisfy his need better than anyone else. Perhaps this is because you have not identified his *real* need. You may have been concentrating on meeting the obstacles (objections) he has voiced instead of identifying his real or hidden objections. By getting him talking as soon as possible in the contact facet of the sale, you may identify new areas to explore. These may give you the foothold you need in what is, undeniably, a difficult selling situation *(the type of situation that breeds professional salesmen).*

When defeatism sweeps over us, we must guard against making concessions we will live to regret. Keep in mind that it is often the customer who is most willing to compromise. Accent the features of your offer which you know she finds acceptable. Accent the features of your competitor's product or service which you know the customer regards as less than desirable.

When you are the victim of defeatism, *don't cut the price* in order to get the order and restore self-confidence. It takes no professionalism to cut prices. There is something worse than losing an order. The mortal sin of professional selling is to cut the price and then lose the order. By doing so, you have established your new price in that area for months to come.

Secretaries (customer's)

"Secretaries" is used as a classification of obstacle in this discussion of potential problem areas. But the term is used loosely and is meant to include receptionists, office personnel and others who "bar the way" to contact with the customer. As the word "secretary" is used, then, keep in mind that what is written here applies to many others who can represent trouble for you.

Secretaries can be one of your most difficult hurdles, but they can also become one of your most reliable sources of help, information and advice.

When a problem exists, it usually stems from the secretary's concept of herself and of her job. It may also stem from her own basic and normal human desire for recognition.

We are going to *consider* some of the approaches that can be taken to this problem. There is no one approach because there is no one secretary. Each of these persons is an individual with his or her own unique aspirations and concepts; you can no more "treat them all the same way" than you can treat all customers the same way. You may not have the time to "research" the office personnel. It takes much of your time gathering and analyzing the intelligence you need on the customer, let alone his staff. Therefore, it is essential that we try to understand the most common reasons for the blocks thrown by secretaries. Then, the on-the-spot estimates of the situation, which you must make, will be more effective.

Here are some of the basic causes why *some* secretaries block your progress.

1. The individual who resents the fact that she is a woman, in what she regards as the man's business world. In this customer's organization, there is a definite ceiling for her aspirations, and she knows she cannot go beyond that, no matter how well-endowed she may be with intelligence and ability. She feels that she has "more on the ball" than many of the men she sees in executive positions. Therefore, she seizes any opportunity to exert her authority over any male that crosses her business path.

2. The individual who doesn't resent the fact that her destiny seems to be the working-life of a secretary, but she isn't about to let you—or anyone else—treat her like an office fixture. She is going to insist upon recognition as an individual with capabilities, intelligence and judgment...which she is.

3. The secretary who has been taken over the coals so often for letting "time-wasters" get into the boss's office. She has long since learned that she has to put on the "officious act" if she is going to prevent visitors from catching her with her guard down.

4. The secretary who thinks her boss is the greatest person in the organization, and she isn't going to permit his importance to be minimized by letting "just anybody" get in to see him or to get in without an appointment...or worst of all to just "walk in on him" without being announced. Furthermore, she is going to make sure that you don't minimize his importance by seeing him on matters that someone "less important" could have handled.

5. The secretary who feels that the image she is trying to project of her self (as someone of importance) will be adversely affected if the impression is created that she works for a woman who is not important in the hierarchy of the company. In building up her boss, she feels she is enhancing her own prestige. Many organizations, unfortunately and illogically, base secretaries' salaries on the levels occupied by their bosses. There is, therefore, valid reason for the opinion held by this gal...if her boss isn't considered important, she won't be, either.

6. The secretary who plays favorites or who just doesn't like something about you or your company. Insofar as he can, he will make it difficult for you to see his boss if it means that the competitor, whom he prefers, is going to be at a disadvantage as a result of your seeing his boss.

Luckily, the vast majority of receptionists, secretaries and office personnel are not only agreeable and pleasant, but also helpful. However, we are discussing the problems that arise at the moment you are attempting to establish contact. It would be unrealistic to ignore the fact that there *are* people similar to those just described. They won't "just go away" if we ignore the problem they create...so, what does one do when he encounters one of these roadblocks?

First, remember that each of us, to a considerable degree, is the product of our experiences and our milieu. Our reactions are based on the thirst we have, individually, for one or more psychological needs; i.e., for recognition, to belong, for prestige, emotional security, economic security, and other similar needs. This is equally true of the secretary who bars her boss's door.

If *you* were that woman, with all of her experiences, *you would react just as she does.* This doesn't make you or me *like* her actions, but it helps us *understand* them. With understanding comes the elimination of emotion which clouds our judgment and causes us to take ill-advised actions. Hence, your first actions must be to "give her the right to feel as she does," to consider what "threat" you represent to her goals, and then to determine how to eliminate that threat.

Regardless of whatever course of action you decide to pursue, *it is certain that recognition is always in order*. Don't just ask, "Is Mr. Baldwin in?"...or..."May I see Mr. Baldwin?"...or..."Would you mind letting Mr. Baldwin know that I am here?"

All of these openings are commonplace; and while they are not antagonistic, they have one failing in common. They do not recognize the secretary as an individual. How much longer does it take to ask, "How are you today?"...or to call the secretary by her name and make some nice comment about *her* before you ask if the boss is in. It may only be, "It's nice to see you again, Miss Lavoie...how have you been?" As a matter of fact, it may indeed be nice to see Miss Lavoie...so, let her know that it is.

Then, depending upon how well you know the secretary, you can decide whether or not you should make further comments of a more personal nature, ranging all the way from a comment about "the attractive dress" to "isn't that a new hairdo since I was here last...I like it." Obviously, this last technique is fraught with hazards. You'd better be sure you are right about the new hairdo, and you'd better be sincere in your praise. It might be emphasized that sincerity is essential...and it is very easy to be interested in the people who surround the customer.

Techniques for supplying recognition will be as varied as the kinds of secretaries you encounter. We repeat it is *essential* that it be supplied. Find a method that comes naturally and sincerely. Keep in mind your obligation to speak to him as you leave. This paves the way for a good reception the next time, but this is not the main reason for doing so. The key reason is that he is an individual, and as an individual merits your respect.

Any attempt to deal with this subject in a few paragraphs, and to consider such hasty treatment as complete coverage, would be ridiculous. "How to deal with secretaries" is almost as complex a subject as "how to deal with customers." Secretaries follow the same general pattern the rest of us follow. The task is always to determine what satisfaction the individual is seeking to obtain through his—or her—behavior pattern. Then we must either provide that satisfaction or switch their attention to another satisfaction that is even more appealing, which we *can* supply. This is exactly the nature of your task in eliminating the problem posed by the difficult secretary. Any attention you give this problem will be rewarded.

Earlier it was said that you may represent a threat to the secretary. Obviously, then, the task is to identify the nature of the threat and eliminate it. If you want to get the most out of this book, don't let it do your thinking for you...*think your way through some of these situations.*

For example, let's take the secretary who wants to be sure that what you wish to discuss with her boss is important. We agreed that if he isn't regarded as important, it is clear to her that her importance is threatened—and you are that threat. If you have analyzed her past actions, you should know that you'd better make the subject of your call seem important. After greeting her by name and having an exchange of sincere pleasantries, you might *in her case* say, "I certainly hope Mr. Elliot is in today because the problem I have to discuss is so important that I don't want to discuss it with anyone else until I get his opinion."

Now, that is just one way of causing her to see that you feel her boss is just as important as she does; you may be able to think of dozens of better openings that all accomplish the purpose. How to do it is easy, once we understand what the problem is.

Many companies are using this book as a vehicle around which they are conducting regular training meetings. So, here is an opportunity for such groups. Why not go over the list of "secretaries" just presented and determine how you would overcome each of the "types of obstacles" which they represent. Why not start out with the secretary who resents the fact she is a woman and wants to exert her dominance over you. How would you handle that one?

Later in the book "strategies for secretaries" will be included in the discussion of the Sales Strategy Bank. Meantime, why not analyze each of the obstacles we are discussing, consider solutions which will be presented here, and then try to produce better solutions of your own. In this way, the book can become the vehicle around which you can build a tremendous—and financially rewarding—learning experience.

Summary

This completes our discussion of the obstacles listed under the Contact facet of the sale. You may have others which we have overlooked or with which you are more familiar than we are. So much the better.

By sharing our knowledge and experience, we can come up with a body of hard knowledge. We can, together, rapidly advance the prestige of selling as a professional kind of work, which you and I know it to be. Then, perhaps, college polls will find selling as a career popping up high on the list of the 10 most coveted careers. Society, as a whole, will have a better appreciation of the tremendous contribution the professional salesman has played and continues to play in our success as a nation. After all, Gross National Product is the accepted barometer of the nation's progress, and GNP is not what we made...it's what we sold.

CHAPTER 10

Gaining customer confidence

This book commenced with discussion of the Psychological Aspects of Selling. Since then, it has been devoted to examination of the Anatomy of a Sale with special emphasis on identifying obstacles that can, and do, arise during selling situations. At this point, we have discussed the obstacles that may arise during the facets of a selling situation which we have identified as Pre-Sale Preparation and Contact. In this chapter, we initiate our examination of the obstacles that may arise during the third facet of the sale in which we attempt to Gain the Customer's Confidence.

It is a bit redundant, but necessary for the purpose of emphasis, to state once again that these facets of a selling situation must not be considered as a chronological "steps-of-a-sale" arranged neatly in a convenient order. Anyone who has been selling for any period of time knows that a sale can take unexpected turns and twists. We have all been exposed to the "start-attention-interest" approach to the development of selling skill. We know, for instance, that once we obtain the customer's "interest," we can't forget about it and devote ourselves to arousing his "desire." Interruptions occur; we say something that starts him thinking about something else, and we find the need to recapture his interest. The same may be said for "confidence." Not only must we earn his confidence as soon as possible, but we must hold it throughout the sale. The sooner we forget about steps-of-a-sale, the sooner we will view selling as the professional kind of work that it is.

When we speak of gaining the confidence of the customer in the early phases of a selling situation, we have in mind the need to cause the customer to feel that you are the kind of person he is willing to meet. Except in unusual cases, it is difficult to have the customer recognize your integrity or to have him realize you are customer-oriented, i.e., that you have his welfare at heart or to have him accept your technical competence without question in the first few minutes of a selling situation. This kind of confidence takes time to build.

In the early phases of the selling situation, there is a need to gain the customer's confidence so that the two following questions may be answered in the affirmative:

1. Will the customer feel that you are sufficiently well-versed in your field to understand the general nature of his problems and needs?

2. Will your appearance, your mode of speech, your deportment and your attitude help to build his confidence in you, and thereby make it easier for you to gain and hold his attention and interest?

Consider your own frame of mind in many of the selling situations in which you have participated as a customer. Haven't you found it necessary to make on-the-spot decisions relative to the salesman's ability to handle your needs intelligently? On more than one occasion, haven't you thought, "This fellow doesn't even understand my problem, much less be able to help me solve it...I wonder if they have anyone here who knows anything?"

In other selling situations, have you been approached by a salesman whose careless appearance caused you to raise a question in your mind about his capability? In other words, wasn't your confidence in his ability shaken a bit? You've probably been sold by such an individual and found him to be extremely competent technically, yet you may have experienced a feeling something akin to relief when the sale was concluded and he had left. It is difficult to reconcile a salesman's competence with his very poor judgment in appearance or behavior.

There is an old bromide with which you are familiar, "First impressions are lasting." This may be exaggerated, but certainly first impressions are not quickly dispelled. They *are* important. They are stimuli, and the customer reacts to any stimuli he encounters. We never get a second chance to make a first impression, and that first impression has an effect on the customer's confidence.

It avails us little to be extremely competent in a technical sense, in our product knowledge and even in our knowledge of the customer's problems if something else about us shakes his confidence so badly that he will not listen to our advice. It helps little to have all the right answers if something about our appearance or behavior irritates the customer to such an extent that his mind is on the cause of the irritation rather than on the facts we are attempting to present to him.

Let's make it clear that we are not talking about having clean hands, pressed suit, and that sort of thing. Every saleswoman in this day and age would, we hope, be fully aware of the need to be well-groomed. We are discussing the need for adapting ourselves to the customer's moods—even, to a degree, to his likes and dislikes. We are talking about an action that many of us find distasteful—to compromise.

The average salesman is inclined to be an independent sort of person, indeed this is often the reason he chose selling as a career... the opportunity to be largely on his own. As a result, instinctively, he may resent the need to compromise. Yet there is probably no form of human endeavor in which you and I are not called upon to compromise. To a degree, we sacrifice our independence everyday... we find it necessary to conform.

We conform to plane, bus and train schedules. We conform to working hours and sometimes to vacation schedules. We conform to countless laws and ordinances which we often find most irritating. We may consider it stupid to sit in front of a red traffic light on a deserted highway at midnight, but we usually conform. Ordinarily we understand the reasons for these restrictions on our independence, and we find ourselves conforming or obeying, almost unaware of the fact that we are bending ourselves to the will of others.

In spite of this, occasionally a salesman will be heard to say, "I'm not going to change myself to satisfy anyone—I'm going to be myself." But this salesman is not completely himself any hour of the day. He is constantly making compromises between what he would *like* to do and what he knows society will *force* him to do if he rebels. It is difficult to see why some salesmen will insist upon "standing on principle" when it is a question of wearing what they feel like wearing, "cussing" if they feel like it, and talking rather than listening when they are so moved.

In some areas of human behavior, if we disregard the desires of others, the penalty is jail. In other areas, if we follow the course we plot, oblivious to the reactions of others, we find ourselves ostracized. We have this same freedom of choice in a selling situation, but if we insist upon behaving without reference to the desires of our customer, the penalty is lost sales.

The customer's confidence in us, as people with whom he is willing to discuss business problems, is essential if the balance of our efforts to persuade him is to have any favorable impact upon his buying decision. It will pay dividends if we will take the time to consider the nature of the obstacles that can undermine his confidence in us.

Appearance (our)

This obstacle also appears during the contact facet of the sale and has been discussed previously and, therefore, will not be repeated here. Suffice it to say that appearance affects confidence.

Our own lack of confidence

"Defeatism" as an aspect of our behavior affects the customer's attitude toward us. Defeatism was discussed as an obstacle under contact previously. It is suggested that you review that portion of Chapter 9. There is, however, a difference between defeatism, as it is used here, and lack of confidence in ourselves. We think of defeatism as describing an attitude of, "We're beaten before we start in attempting to get this sale. Even if we have the best solution, he won't buy from us." Lack of confidence is quite different. It embraces those attitudes which are conveyed to our customer by our attitude or our speech, leaving him with any of the following impressions:

That you have no confidence in your product.

That you don't feel you are equipped to understand his problem.

That you have no confidence in the people in "the factory" (or back at headquarters) or in others in the company who must follow through on commitments you make.

That you have no confidence in your ability to get necessary approvals of others in the company.

Creating any of these impressions in the mind of the customer can lose a sale. If *we* don't exhibit confidence in ourselves, our company or in our product, there is little reason to believe that he will—or should.

These impressions of our own lack of confidence are revealed to the customer by:

Openly expressed doubts.

Facial expressions that reveal worry, concern or bewilderment.

A passive or negative attitude... a lack of enthusiasm.

Statements that deprecate others in our company.

Attempts to appear to be so completely a "customer's man" that we make disparaging remarks about the competence or lack of imagination of others in our company. Instead of causing the customer to think he has a loyal friend in us, his confidence in our company is undermined. He may end up thinking the salesman is the finest fellow he has ever met, but that the salesman's products aren't for him.

Another way to lose the customer's confidence is by agreeing too readily with him and with his complaints. There are times when concurring with his complaints is sound strategy, but there are other times when being too ready to agree reinforces his belief that your company lacks some quality necessary to give him the service he wants. Before concurring, consider the implications he may draw from your ready agreement.

You may pose this question: There are times

when a salesman really does lack confidence because he feels he does not understand the customer's problem. What do you do under *those* conditions? The antidote for this type of lack of confidence is to get the customer talking and keep him talking. And listen and absorb.

It isn't always possible to possess all of the knowledge we would like to have regarding every possible application of our service or product. When we are face-to-face with an application we do not fully understand, the remedy certainly is not to act confused or bewildered or to plead total ignorance. The proper action is to indicate that this is "an interesting application" and one that merits more study. Indicate that you will obtain more information for her immediately. Then set a specific date for a return visit. This will provide you with an opportunity to do some homework... or at the very least to make a call to your own people to get the information you lack. Bluffing your way through is almost a sure road to disaster. If the customer detects bluffing, you may be sure that her confidence is destroyed.

The action recommended here indicates that you consider his problem as almost "unique" and one that deserves more time and attention. The customer feels you are giving him personalized service and not providing him with off-the-cuff answers. Obviously, it is essential that you follow up at once.

There are times when this approach will not work because of time limitations. He needs action, and he needs it now. He has no time for you to go back and get information. Assuming that you have contacted the customer previously and he has confidence in your general ability to understand most of his problems, it may be the best policy to admit that you need more information before you make recommendations or make a commitment on price, delivery, or whatever the issue is that is at stake.

In your first contacts with a given customer, avoid such phrases as these:
"I'm not sure..."
"I'm not sure our people will go along with that."
"I'm not sure we can meet these requirements."
"I'll have to find out."
"I doubt that we can..."
"I don't know..."
"This is the first time I've encountered anything like this."

Your words should convey your confidence and should be positive, not passive or negative:
"I can certainly get that information for you."
"Our people will certainly be interested."
"This is important; I intend to look into it further, and I'll do so at once."
"I'm certain that we can help you in this matter."
"This is an interesting application, and I am sure our people will move on it."
"I'm sure!" "I know..." "I will..."

Confidence begets confidence: Show it... express it... feel it.

Habits (our)

This obstacle appears in many facets of the sale; hence, it is one that deserves your continuing attention. The subject has been covered, and your attention is called to earlier discussions. There *are* certain habits that are contributory to loss of the customer's confidence, and these have not been discussed previously.

Habits that undermine confidence

Constant note-taking. If you must take notes, tell the customer that it is because of your desire to render him optimum service and that you intend to take action on each of the points that he is raising.

Letting your gaze wander while the customer is talking...failing to listen with your eyes... failing to give him what he considers full attention. This robs him of confidence that you are trying to understand his problem, and this grows into a suspicion that you are either not interested or that you don't—and can't—understand it. He begins to feel that you have no appreciation of the importance of the problem or that you are waiting until he has finished and that you will then give him a "canned" solution. Unfortunately, you may be listening intently to every word, but you may have developed a habit of not looking at the person talking. Strangely enough, if you look at the customer when he talking, you'll find your own interest in what he is saying increases. Try it.

Treating him with too much familiarity... joking...making light of his concerns. Probably you are only trying to allay his fears and reassure him or calm him, but he begins to suspect that you don't understand the problem or are not interested in it.

Supplying him with quick answers to a problem that may have kept him awake nights. You may have a faster mind than he has, or you may have encountered his problem in previous selling situations with other customers. You are, therefore, able to supply him with fast answers. He sometimes resents your demonstrated superiority in the "quick-think department." The answers were not obvious to him.

He cannot believe that you possibly could come up with valid solutions or see through the problem so quickly. He will suspect you are bluffing, and he will question your competence.

Don't fall into the habit of comparing her problems to similar problems in other companies whom you serve. She doesn't want to believe that she is "just like" your other customer, so avoid phrases like: "This is just like the problem *ABC* company had." Such phrases lead her to the belief that you are jumping to the conclusion that her problems are standard problems requiring standard treatment. She begins to suspect you are taking the easy way out. All of us like to feel that we are unique—as we are. We also like to think that our problems are unique—which they are not. We don't like having this proven to us.

Mannerisms (our)

The Anatomy of a Sale chart which appeared earlier shows this obstacle appearing in both the Contact and Gaining Confidence facets of a sale. Though you may make contact successfully, mannerisms may rob you of the customer's confidence. It is suggested that the material offered earlier be reviewed for its impact on confidence.

The canned presentation

Admittedly, there are times when a canned presentation is just what is needed to give the customer assurance that his problem is *not* new or difficult to solve, but that it is one you have handled successfully many times. There are other occasions, however, when use of the canned presentation affronts the customer or causes him—as stated earlier—to feel that the salesman doesn't really understand the nature of the problem.

Unfortunately, for the salesman, the customer may regard what is really a unique solution as a canned presentation, simply because the salesman is so experienced he has sized up the problem and has made a very fluent and comprehensive presentation. It is so letter-perfect that the customer erroneously regards it as canned.

There are a number of tactical solutions to this problem:

1. *Adopt a thoughtful mien.*

2. *Present facts at a slower pace than is your custom* (in other words, compromise with what you'd *like* to do).

3. *Give reasons for your conclusions and solutions.*

4. *Above all, get your customer in the act* by presenting questions he must answer and use the probe for reaction (see the Sales Strategy Bank).

5. *Omit something of minor importance* but something sufficiently obvious to the customer to give her the opportunity to interrupt and to contribute to your presentation. Keep in mind the point you omit must be minor. Reason: So that its absence will not cause her to lose confidence in you, yet its importance is enough for the customer to feel she must inject it. The omission of the point causes your presentation to look more natural and less canned.

6. Ask a number of questions sufficient to convince him that you are not treating the problem lightly and that you developed your solution *for him* only after you obtained information from him.

Images customer has of us

The customer has an image of you, of your company and of the products and/or services you provide. Don't feel that because you have faith in yourself and confidence in your company and its products or services, the customer will necessarily have the same confidence. Avoid making groundless assumptions of this type; it represents the self-defense mechanism of projection in operation; i.e., assuming the "other fellow" will see the situation as we do.

Just as many young college graduates depend too much on their diploma when they seek their first job, so, too, do many salesmen depend too much on the image their company has projected. An attractive company image is a great asset, but it offers you the most help when you are making contact. It will often open the customer's door for you, at a time when some salesmen representing a company less favorably known is having real difficulty in even seeing the customer.

Once contact has been established, the customer is going to judge your company *by what it does for him* rather than by the reputation it has with other people. He is going to be far less impressed with your company's reputation and far more impressed by the image he has of you, as well as by the "nuts and bolts" of your product or service and the price you are asking for it.

A favorable company image can be a mixed blessing (but always an asset). The customer considers the benefits of doing business with a company that backs up its products. He considers the advantages of buying from a supplier whose products or trademarks in his place of business...or as parts of his finished product...will enhance his own prestige. At the same time he is considering what this will cost him in terms of cold cash, as represented by your price. He is also wondering whether he will receive personalized attention or will he, as one of many customers of a large firm, be considered unimportant when he is having difficulties and needs service.

Out of this measurement the customer makes will emerge the final image he has of our company. His decision to buy and to place confidence in our promises will be predicated, to some degree, on that measurement. So, once again we see the need for doing our homework—the need for making an estimate of the selling situation in advance. We need as much information, in advance, as it is possible to obtain, about the customer's views, his aspirations and his past experiences with us as individuals as well as with our company. What psychic rewards does doing business with us offer him? What conclusions has he drawn from his past experiences with us or with other representatives or departments of our company? Do these experiences leave him with a feeling of confidence?

Admittedly, it is not always possible to obtain all of this information prior to entering the selling situation. So, we must resort to use of the types of questions that appear in the Sales Strategy Bank.

If the customer has an image of you or of your company which is not favorable, *identify the source of his resistance (A12)* by *asking direct questions (D15); use the overhead question (D16); employ the rhetorical question (D17)* and, as a last resort, use the *provocative question to make him talk (D18).*

Once having identified the reasons for her lack of confidence, consider using the following strategies (the question *you* must decide is, which of these will be most effective in *this* situation with *this* customer, considering your personality, her personality and the effect you have on one another):

B17—Provide him with an escape hatch. If he has expressed himself and exhibited lack of confidence in you or your company, he is not going to "change his mind," i.e., reverse himself, unless you make it easy for him to do so. Before he will reverse himself, he must find some valid reason for the opinion he held previously, so that he will not appear to have been something less than "bright." You might help him find this "way out."

B9—Emphasize mutual interests. Show him that both of you have much to gain *only* if a successful result follows the sale. If his past experiences have been unsatisfactory, show him that you have much to gain now by providing him with satisfactory service and, as a result, *he* will solve a problem. Both of you gain. What you are attempting to do is to supplant emotion with reason, though you cannot say this.

C4-5—Emphasize her goals and objectives. Emphasize what she has to gain and *D10—Divert her* by directing her attention to her immediate problem.

If he has expressed his negative opinion of either you or your company, employ *D11—Rephrase and overstate his objections.* His tendency will be to deny that he said anything "that strong" and he will tend to restate his objections, toning them down as he does so and removing much emotion, thus giving you something you can tackle.

D8—Tack with his stand. Listen, show respect. This does not mean you agree with him.

B10—Welcome his objections. Steer clear of vapid remarks such as, "We always welcome objections because it is only by knowing where you think we have slipped, that we can improve." Rather, express your regrets for his past difficulty, which created the image he now has of you, and express your conviction that this new situation will afford you an opportunity to satisfy him.

A4—Unearth his primary buying motive and concentrate on satisfying it.

A12—Identify the source of his resistance and *(A7)* consider getting help from higher echelons in your own company or *(A6)* from other departments of your company. If someone else in your company was responsible for his loss of confidence, consider how they may be helpful in restoring his confidence.

In summary, the task facing you breaks down into these elements:

1. Identify the image he holds of you or your firm.
2. Isolate the reasons for this image.
3. Neutralize his reaction by showing respect for his views (without necessarily agreeing with them). In other words, give him the right to feel as he feels...thus eliminating emotion from your own frame of reference, permitting you to think more clearly.
4. Concentrate his attention on his current problem.
5. Evidence a keen desire to be of assistance to him now.

The courses of action selected from the Sales Strategy Bank and presented in the foregoing paragraphs represent some possible courses of action. Examine the Strategy Bank carefully to see if we have overlooked any strategies that you feel will be more effective with your customer. Consider possible problem areas you may encounter (from the Anatomy of a Sale Chart) and then contemplate the various strategies that may overcome those obstacles.

This is a far cry from the kind of "sales training" that provides you with the "one best way" to meet an obstacle. You, as a professional salesman, know better than anyone else what actions will be most effective with *this* customer in *this* situation. The charts supplied recognize your professionalism and merely supply you with checklists that are helpful in enabling you to consider potential problems and possible strategies which you might overlook.

Customer's self-image

The point has been driven home repeatedly that a customer's self-image (especially the "me-he-tries-to-project," i.e., the way he wants to be seen by others) is often more important to him than the satisfaction of an operating need. For instance, a customer may *want* to buy a very expensive car but may not do so because people will ascribe motives to him which conflict with the image he is trying to create in their eyes. He may, therefore, buy a car which falls short of his "product needs" because it threatens psychic needs. To amplify this point, we might consider the customer who drives thousands of miles each month and needs a big car for comfort. He may buy a less expensive and less comfortable car because of his fear of appearing ostentatious.

A customer may avoid buying from you if you represent a "big" company because your very bigness makes him feel small. He will rationalize his way into this decision by forcing himself to believe that your "big company" won't really have an interest in his "small company" once you have rung up his sale in your cash register. Remember, self-defense is really self-deception.

Conversely, a customer who considers himself important (and wants to be seen that way) may be difficult to sell, if you represent a small company. He feels more important when "big" companies are competing for his business. You may be asking, "Just what has this to do with 'Gaining his confidence'?" Just this—he may already have confidence in you as an individual, in your company and your products, your experience and your integrity, but he does not have confidence in your ability to make him look as he wants to look in the eyes of those he seeks to impress. So your next question may be, "What can you do about it?"

The antidote is to talk, "You...you...you!" Play down references to your company (if you are large), its greatness, its size, its experience, its reputation. He is already aware of these and it is this very awareness that is causing your problem, i.e., his lack of confidence. *A2—Identify his needs and wants* and *A3—Identify his self-image.* Present your proposition with heavy emphasis on how it will make him appear to others (subtly) and with only essential emphasis on how it will help him solve his operating problem. Make him feel he is dominating the selling situation by making him do most of the talking. This can be accomplished by remembering the power of questions and using them.

By using questions, you steer the conversation into channels that suit your strategy. He is flattered because you are asking his opinion. He is doing the talking; hence, he feels he is dominating the selling situation and is not being "high-pressured." He begins to reinforce his confidence in you because you are showing respect for him and his opinions, and because *everyone loves a good listener.* (Listeners show such good taste.)

Hearsay

In Part One on the psychological aspects of selling, we saw that the opinions of others can influence our opinions and actions almost as much as experiences of our own. If we have confidence in the chap who gave us his opinion (or recited his experience to us), we tend to accept it as valid. What's more, we will fight as hard to support that opinion as though it had been generated from our own experience. It becomes a part of your judgment. We have confidence in our own judgment, and if a salesman attacks opinions we have formed as a result of hearsay, he is attacking our judgment. We, in turn, begin to lose confidence in the salesman because if we believe what he has to say, we develop a sense of insecurity...we must question our own judgment, and this is asking too much of us. This, then, is what you are up against when the situation is reversed and *you* are the saleswoman up against a customer who has been exposed to and is vulnerable to hearsay.

Again and again and again, we must repeat the old refrain...gather intelligence on your customer before the sale. Know as much as possible about the people with whom he associates, the people whose advice he respects or will accept. What kinds of people does he view as having interests similar to his? What experiences have they had with your product or service and how do they feel about it? What have they probably told this customer about you or your product?

A customer may mention a situation in which it appears that another customer has fared badly as a result of an experience with you or your company. If he identifies himself with the other customer (similar interests), he will feel as strongly as though he had personally lived through the same experience.

(This is *introjection*, one of the 17 defense mechanisms mentioned in Part One.)

We are often surprised to find a new customer has already formed negative opinions of us, as a result of hearsay, before he has ever had actual contact with us. We should *not* be surprised because honesty forces us to admit that we, too, have rather firm opinions of people whom we have never met. We form these opinions sometimes because our favorite columnist (in whom we, for some reason, have "confidence") has told us about this fellow. Why, then, should we be surprised to find the customer reacting to hearsay just as we have reacted thousands of times?

You are fortunate when the customer brings the problem out into the open... you know what you are up against... you know the nature of your problem. It is far more difficult when the customer does not mention the incident but quietly harbors a lack of confidence in you as a result of the hearsay. Examining the Sales Strategy Bank, we find these effective tools:

Welcome objections (B10). If he raises an objection you recognize as being a result of a problem suffered by another customer, meet it head on. "I'm glad you raised that point; we *did* have that problem in the case of one customer, but here is why it occurred..." Then present your customer with facts which his friend may not have supplied. Or you admit that the situation did exist and show the steps taken to prevent a recurrence.

Probe for reaction (D9). As you offer your explanation, it is important to determine whether or not the customer is responding favorably to it. You may decide to use the direct question or to use the rhetorical questioning technique and watch for her reactions. In a future chapter, examples of the "Probe for reaction" will be given. At this point, the key thing to keep in mind is the need for determining whether your words are having the desired effect upon your customer. It will avail you little to keep talking if she hasn't agreed with the basic premise upon which your subsequent remarks are based.

Anticipate, raise and answer objections (E10). Some salesmen consider this a dangerous policy. They fear that you may raise an objection which the customer might not have considered had it not been raised by the salesman. This is a valid viewpoint; hence, you use this device *only* when you have real reason to believe that the customer has been exposed to and affected by hearsay. If you do not raise the objection... and if he is the kind of person who will nurse his objection privately without giving voice to it, that objection will still color his reaction and may result in a lost sale. If it is not brought out into the open, it cannot be disposed of.

You may wish to present facts (without mentioning the objection), which will show him that the objection is groundless. This has the advantage that it does not create a situation in which he will lose face as he has never been obliged to express himself.

If you *do* bring up the objection, this course of action has the advantage of demonstrating your honesty by raising a point which he will feel you might have avoided by remaining silent. Recently we purchased a machine because the salesman told us what it would *not* do. He had first ascertained our needs and was on safe grounds, but at the moment we were so impressed with his forthright attitude that it clinched the sale for him.

Appeal for fairness (D13). The greatest of all appeals is "I need your help." Close behind it in effectiveness is the appeal to fairness. We all like to think we are fair. Ask for a chance to present your side of the story when his opinion has been shaped by hearsay. For example: "I can understand how you feel, sir, and I would feel much the same way, had I heard of the situation which you encountered. I know you are the kind of person who wants *all* the facts, and because you are fairminded you'll give me the chance to give you the complete story."

In giving "the complete story," don't tear down the source of his hearsay. Remember, he would not have believed the gossip if he didn't respect, in some way, the source. He has something in common with the person whose opinion he valued. In tearing down that person, you may, in effect, be tearing down the customer too. Stick to facts.

Hearsay is a molder of opinion

If we didn't believe that, we would cease using testimonials in advertising. Testimonials can backfire because we do not always know how potential customers consider the individual whose testimonial we use. In hearsay, the situation is quite different. Here, the customer selects his own testimonial. If it is negative, it creates a problem for us. He regards the source of the "adverse testimony" favorably or he would not be impressed by it.

Try to anticipate hearsay. Have tactical plans ready. Know *specifically* how you will handle it if it arises. Don't be content with "I'll think of something to say if and when it arises as a problem." Keep in mind the power and influence of words and have the right words planned in advance. Don't delude yourself into thinking that the problem will go away if it is not mentioned. Sheer force of your personality usually will not be enough to dissipate the corrosive effects of hearsay which is adverse to your interests. He has reacted to it already. You must neutralize this reaction at the very least if you wish to open his mind to the balance of your sales presentation.

Customer's image of our competitor

This is often the most difficult of all obstacles to overcome. It frequently shows up at the moment of contact, with the potential customer refusing to even give you a hearing because she is "perfectly satisfied" with her present supply source. In the sale of some equipment, customers pay for, and use, the services of consulting firms and may decline to "talk business with you." You may run into the "there is no point in wasting your time" routine.

Keep in mind that if you allow this to irritate you —much less anger you—you won't be in a position to do your best thinking. Someone has said that "anger blows out the lamp of the mind."

A salesman added: "I save my anger for those things worth getting angry about, and I have found that these are the very things that getting angry won't help."

We can all agree that to the degree that we become irritated, annoyed, or angry—to that degree— our thinking ability is impaired.

We are not talking about *simulated anger,* which a salesman should rarely exhibit, but which customers are free to use as bargaining tools.

To make the customer's attitude more understandable, and to remove irritation completely, examine your own buying patterns. Isn't there one store you will patronize over and over again? Wouldn't a competitor of that store encounter a real problem in trying to change *your* buying pattern? If you will honestly examine your reasons for patronizing the store you favor, you will find that it is not always price that attracts you. Sometimes it is the pleasant surroundings; perhaps the treatment you receive there appeals to the "me you wish you were" (or the me you wish people to see when they see you).

You and I have specific psychic needs (needs of the spirit) as well as product needs. When both are satisfied, we are not about to change our buying pattern unless a new buying pattern can be made to appear more attractive. It will be more attractive only if it promises to provide even greater satisfaction to the same needs, or if it will satisfy other needs which we have that are even more important to us.

If an element of gamble seems to be involved, we won't take the gamble unless the odds seem to be very much in our favor or unless the potential advantage is great.

We resent it if we are accused of being "resistant to new ideas or methods"... and rightly so. It is simply a case of finding that our present methods bring us satisfaction, and we won't abandon those methods unless someone can prove to us that satisfactions we now enjoy will not only remain unhindered but will be enhanced.

It is easy to understand this when we apply it to our own buying behavior. We must keep in mind that our customer reacts the same way to our efforts if he is already satisfied with a supplier other than ourselves. In other words, he has confidence in his present supplier, and we are attempting to cause him to have even greater confidence in us.

The task is to identify the satisfactions the competitor provides him which seem to be so important to the customer that he resists changing. We must resist the common tendency to assume that price is the *only* stimulus that affects him, or that product features or performance are the *only* problems we have to overcome.

What tactics are available to you to overcome this obstacle of the prospect being "perfectly satisfied"?

If he indicates that he has great confidence in your competitor, you may assume that he has less confidence in you. This is especially hard for us to "buy" if we happen to be the best-known or the "largest" supplier in the area. If others have confidence enough in us to make us the "most successful," why doesn't he? Perhaps those readers who are with large companies should take a new look at the situation, as has been emphasized in previous chapters. Let us say, for example, that you have 40% of the entire market. The reaction is to feel that you and your company are pretty good. The cold, stark fact is that for one reason or another, 60% of the market would rather buy from someone else and they prove it by doing so.

With the customer who is not impressed with your size or your image in the market, try the direct approach; i.e., the *direct question* (D15 from your Sales Strategy Bank). In a non-argumentative tone, you can ask him why he likes buying from your competitor... or you can ask him why he doesn't want to buy from you.

This approach is so simple that it is often overlooked. One company refused to buy from a supplier for 20 years, and it took that long before a salesman had the simple courage to ask why. He found that his company had refused to take back certain equipment the customer found he didn't need. The customer waited 20 years for an opportunity to sound off about it. Once he was asked the direct question and vented his long pent-up ire, he started to buy again from the supplier.

Another real example in which we will change the names for obvious reasons is the case of a salesman who called on a customer for the first time. The customer asked, "Where are you from?" The salesman, thinking he was being asked what city he was from replied, "Hammond."

The customer immediately acted disinterested

and the salesman left without an order. There followed a sequence of unsuccessful calls on the same customer.

Finally, the salesman asked the customer, "Would you tell me why we can't seem to get any of your business?" The customer replied, "I had plenty of bad experiences buying from you people at Hammond Steel in the past!"

The salesman said, "I'm not with Hammond Steel... I'm with Zenith Steel." Then the customer asked, "Why did you tell me a few months ago that you were with Hammond?" And the salesman said, "I though you were asking what city I was from and I'm from Hammond... the city in which both we and Hammond Steel are located."

He then received his first order. Let us emphasize that the city was not Hammond and the two companies were not steel companies. The point is that if the salesman had asked why he wasn't getting any business after the first call he would have learned the reason and would have obtained business much earlier.

Granted that this is an unusual situation even though it did happen exactly as related, the point is that many times, by simply asking, you can smoke out the real reason for the customer's refusal to buy and then you have something tangible with which you can grapple. It costs nothing to ask!

Using the direct question accomplishes two things: (1) You can determine the reason he gives himself (and it may be the *only* reason), and (2) you can obtain leads as to the nature of the underlying reason. If he states openly that it is price, you may be fairly sure you have the whole story. The underlying reason may be that he is satisfying his need for economic security or his desire to save face (he doesn't want to be taken for a "patsy" by being required to pay a higher price than that paid by his competitors).

You then have two choices: (1) Meet or beat the competitor's price, or (2) show your customer what the *real* price is that he is paying by demonstrating what goes into price; i.e., service, delivery, guarantee, long life, ease of maintenance, prestige, research, and so forth. Then show him that your price is actually lower because you pack more into your offer than your competitor does.

The reasons that a customer favors a given supplier are so varied and so great in number that it would require a separate volume to discuss ways and means of handling each of the possible situations which they create. You, through adroit questioning and observation, can determine what it is that your customer wants *most* and which your competitor seems to be supplying. Then review the Sales Strategy Bank and determine which of the strategies is the right one for *this* situation.

Age and voice (our)

These obstacles arise during the moment of contact. If not then, they may crop up when you are attempting to gain the customer's confidence. These obstacles were dealt with in earlier discussions devoted to contact, hence will not be repeated here... but don't overlook them as possible obstacles to gaining confidence even if they did not loom up when you first made contact with the customer. Age and voice may not have prevented you from gaining an audience, but they may shake the customer's confidence in you.

Customer's off-beat outlook

People are rational; they do things for a reason. If they do not, they are regarded as mental health problem cases, and society deals with them. We are all fully aware of this, yet we often insist upon judging as an "odd ball" the customer who resists standardized treatment and solutions for problems that are accepted by other customers. If the customer's viewpoint of our efforts creates a problem, we try to change *his* actions. There are all too many salesmen who try to reconstruct the customer in the image of what the salesman... or the salesman's company ...feels is the perfect customer. They try to show them how to run their store in the manner in which *they* feel a store should be run... and there are sales engineers who attempt to accomplish the same kind of results with manufacturers, utility companies, or even defense department installations.

As a case in point, Company A was losing business to Company B. An observer pointed out to Company A the reason why. He said, "When we see one of your salesmen enter a store, he does so with full briefcase in hand, and then proceeds to tell the customer the kind of a dealer Company A would like him to be. When the salesman from Company B enters the same store, he does so without briefcase, asks the dealer the kind of a store the customer would like to run. Then he listens. When the customer finishes, the salesman says, 'We've got just the kind of products to help you run that kind of a store.'" This is no hypothetical image but is an actual occurrence.

Many of us go through life trying to change people's actions. We forget that you can only change the actions of a fool or of someone who is in a position where he *must* do as we say... and few customers are in that category. People are reasoning beings. They take actions that will lead them to the satisfaction of specific short-range goals or long-range objectives. They often view the actions we suggest as leading in a direction that is not in line with either their goals or objectives. Naturally, they resist our advice and we lose their confidence. They cannot help but feel that we are either "seller-oriented" (i.e., unmindful of their aspirations and wedded only to the satisfaction of our own) or they mark us down as incompetent because we are unable to sense the nature of their objectives and goals.

Often, customers will not *tell* us what those objectives and goals are because they are "socially unacceptable." For instance, the chap whose buying motives are almost 100% psychic (wants to be seen as important or impressive or dominant, etc.) won't tell you this because these are the hallmarks of what society regards as ostentation. He wants you to satisfy these needs without any indication, on your part, that you are aware that these are his real needs.

Only recently, we encountered a classic example of this. A salesman had been trying to sell a piece of equipment that ran into the six figures. The customer was in a remote area where it was completely out of character for a machine of this type to be purchased. For weeks, the salesman hit hard on product features and performance. The machine was an excellent device, but the customer continued to point out that his present equipment was doing the job perfectly well.

Then, the salesman heard a discussion of this material and returned to the customer with a new approach. He casually mentioned that if the customer bought this product he would have a "more sophisticated machine" than anything owned by the best-known company in the customer's field, and the salesman named this nationally known competitor. The customer immediately called in members of his staff and the decision to buy was made at once.

This was no accident. The salesman had been gearing his efforts to changing the customer's *actions* without success. Then he sensed the customer's key psychic need, and the customer's interest and desire were aroused at once. Once the customer saw that the equipment being offered would meet his *real* need (to be regarded as more modern than the nationally known competitor), *he changed his own actions* because now the purchase made sense.

Incidentally, the customer has a platform built around the equipment, and all guests to his plant are brought in to see this machine which is "more sophisticated" than any owned by the nation's largest competitor in his field. *Moral: Customers will change their viewpoint and their actions if it makes sense to them.*

The task of a saleswoman is to determine the nature of the customer's real need and then show him —or better yet, let him discover—that only a purchase of the product or service the saleswoman provides, will meet that need. *Don't try to change the customer's actions...* try to get him—on his own—to change his goals or objectives. The other course of action for the saleswoman is to identify the customer's real goal or objective and then to present her product in a manner that causes the customer to see that only by purchase of this equipment or service can those goals be achieved.

Maxwell Maltz in his book "Psycho-Cybernetics" drives home the fact that few of us ever reach our real potential. In other words, we set our thresholds of satisfaction too low or we are satisfied with gaining acceptance for the "me we try to project," and we abandon attempts to achieve the "me we wish we were" (the me we wish people would see when they see us).

You may find it necessary to cause the customer to see that he *could be* the kind of person he wishes he were. Then be prepared to show him how your service or your product can help him accomplish that objective. Again—don't try to change his actions. He will change his actions voluntarily when he sees that only by changing his actions can he accomplish these new and appealing objectives. Make the new horizons so alluring to him that he will abandon his old objectives.

There is, of course, the line of least resistance. You take the customer as he is, and simply offer him products and services that supply the needs of that kind of person. There are reasons why this is not always the wise action:

1. You may not have the products or the services that "his kind of person" needs. In this case, you must cause him to awaken the "me he wants to be" (and usually *can* be). Your products may then meet a need he can recognize. Incidentally, this builds confidence in you (in his eyes) because we always admire the person who feels we could be greater than we are.

2. If you content yourself with merely trying to satisfy the needs he currently recognizes, you may find yourself up against the competitor who is meeting these needs to the customer's complete satisfaction.

3. His present objectives and goals may indeed be short-sighted and wrong for him. If this is so, to pursue a policy of merely selling him what he "wants" is something less than customer-oriented. Furthermore, it fails to recognize that if you can cause the customer to want to *be* more, he will, if successful, *need* more; hence your own selling opportunities will be increased.

The foregoing should not be construed as an attempt on our part to make salesmen a corps of reformers. We are not suggesting that you create problems where none currently exist. It is not recommended that present business be imperiled by taking a chance of alienating the customer through a process of "educating him as to what he really needs." We are talking about those selling situations in which the customer does not act favorably to our interests because it does not seem to him to be the intelligent thing to do—because he sees no enhancement of his present objectives or goals.

Summary

You have one factor working for you. The customer's present supplier is probably making no effort to change the customer's goals and objectives ...he's doing real fine as things are. Therefore, what seems to be an obstacle to you can become a real asset if you are successful in lifting the customer's sights. The first step in accomplishing this is to stop considering the customer's actions as "off-beat" because they don't conform to the action pattern *you* would embrace if you were the customer. His actions make sense to him because they enhance *his* objectives. If you want his action pattern to change, cause him to establish new objectives.

Our reaction to customer's behavior

The sales profession comes completely equipped with an extensive kit of built-in problems. That's not enough for many of us, so we create a few of our own. One of the self-generated problems is found in our failure to control or conceal our reactions to the behavior of a customer that displeases us.

A chess player or a military commander cannot permit himself the luxury of anger when his adversary initiates actions that threaten his strategies. In chess and in combat, we know that this is—of all times—*the* time for a cool head and an analytical approach to the problem. It becomes imperative to determine what is causing the other fellow to act as he does. In chess and in combat, the opponent's actions must be studied to identify his motives and to reveal his strategies. Above all, we must identify moves that are intended solely to divert us.

The same situation exists in professional selling. Customers *can* irritate. They can be unfair and even abusive. The question is, what do you gain by losing your own patience, becoming angry or exhibiting disdain for the customer's line of reasoning? The answer to that question is that we gain only the satisfaction of some basic need of our own psychological drive pattern. What we *don't* gain is an order. Nor do we neutralize or eliminate the reason for the customer's behavior.

When the customer irritates us, we might coolly examine these questions:

1. Is the customer intentionally trying to divert you or to make you so angry that you will behave in a manner that will provide her with an excuse to terminate the interview?

2. If he debates a premise you are attempting to establish, is it because he "really wants to be difficult" or is it simply because you have not made the reasons for your recommendations clear to him?

3. If he has become really abusive, is it because you just happened to come along at a time when the culmination of assorted problems has just become too much for him to take and he vents his displeasure on you so he can "pop-off"? A customer can act in what seems to be an abnormal or irrational manner, becoming excited over a trifle when that trifle happens to be the last of many he has endured all day.

Regardless of the reason for the customer's actions, it doesn't make it any easier for us to take abuse. Fortunately, we have a choice of actions. It is not a case of choosing between fighting back or just "taking it." The third choice is to keep calm, adopt a sympathetic mein, and listen. This course of action has often turned a customer's abusive attitude into an apologetic frame of mind.

Keep in mind that the customer's attitude is providing him with the satisfaction of some psychic need. Can you provide him with the same satisfaction by playing a sympathetic or listening role? Often, if you wait, the storm will abate and you can once again reason with a calmed or composed person. Don't make the error of believing you can use logic and reason with an individual who is irate. It just can't be done. He will brush aside logic which under ordinary circumstances would impress him.

Here is a list of positive actions which you can take if your customer is angry or if he has adopted an attitude of disinterest or apathy to your proposal:

C7—Use the ambiguous question. This puzzles him and slows him down. It then gives you a chance to answer some of the points that have annoyed him or it permits you to make points that will arouse his interest and curiosity.

D15—Use the direct question—with care! If your customer has been making broad, sweeping statements or charges, you can slow him down with the simple, "I wonder if you could give me a few examples?" This will brake him or stop him, *if* he has been exaggerating. It is also a dangerous tactic because it can be embarrassing to him. If he can't supply examples to substantiate his wild charges, he will lose face. It is vital that the direct question, when used for this purpose, be coupled with a pleasant expression, and that you convey the impression that you are sincerely anxious to help, revealing a deep-rooted desire to get at the cause of his problem. It is imperative that an appearance of defiance be avoided. The request that he provide you with examples should not appear as a challenge but rather as an appeal.

D13—If the occasion seems propitious, appeal to his fairness. Again, consider carefully the words and tone of voice you intend to employ. There is a world of difference between (a) "You're being unfair!" and (b) "I've always found you to be extremely fair, and I know you don't want to be unfair now. Can you give me a chance, now, to explain....?"

B12—If it will not imperil the selling situation, and if the customer really does have a point, admit the validity of his view. Couple this with your assurances that the point at issue will be resolved to his satisfaction. *B13—A variation of the foregoing is to find a point of agreement and build on that point.*

D11—Rephrase and overstate her objections. If you repeat a charge the customer has made, and in restating, you make it even stronger than she did, she will usually deny that she made such a strong charge. In repeating her previous statements, she will be inclined to tone it down a bit to prove that her original charge was not so strong. This gives you something easier to handle than the original statement. Don't attempt to use this strategy if the customer is really angry. If you rephrase and overstate his objection, he is likely—in the heat of his anger—to embrace the more serious charge.

Above all, don't permit yourself to be provoked into betraying your own emotions. This is the time for professional selling, using all of your resources.

Don't be satisfied with the selection of strategies we have just presented. Examine the Sales Strategy Bank and see if you can find other approaches that, in your opinion, would be more effective with this customer.

Admittedly, situations will arise without warning when you find yourself face-to-face with an irate customer. Far more often, however, you and I can anticipate that he will be irate. There is, then, no excuse for our failure to be prepared to meet his anger and to have our strategies well-planned in advance.

When the customer is angry, remember once again, "The power to feel another's hurt and to *want* to feel that hurt; the power to sense another's need and to want to fill that need, is the root of true professional selling." If the customer is upset, look upon this situation not as an obstacle, but as an opportunity to serve him, if only as an escape hatch for his frustrations and vexations.

CHAPTER 11

Identifying the customer's need

In previous chapters, we have been discussing the obstacles that may crop up during the first three facets of the selling situation; i.e. (1) Pre-Sale Preparation, (2) Making Contact, and (3) Gaining the Customer's Confidence. We now take up the obstacles that may occur during the fourth facet of the sale: Identifying or Establishing Need. First, let us consider just what is involved in this highly important facet of the sale.

This facet can take two forms: (1) you must identify the customer's need if you are not already aware of its nature; or (2) if *you* understand his need, but he is not yet aware of it, you must create this awareness in his mind. To the first point, if the salesman doesn't fully understand the nature of the customer's need, he will be in a poor position to recommend the *right* product, service or application. To the second point, if the customer is not aware that he *has* a need, he won't be interested in the salesman's proposition.

You will note that only eight potential problem areas are listed under "Identify or Establish Need." From your own experience, you may be able to supplement that list. The danger is that because so few obstacles are listed, the tendency may be to skip over this facet of the sale too lightly. The truth is that it may surpass all of the other facets in real importance, because if there is no customer need, there will not—indeed, *should* not be—a sale.

Only the short-range thinker will sell a customer something he does not need. This may be effective for the hit-and-run "order taker" who never expects to sell the customer again, or perhaps may not even meet the customer again. This is just the kind of activity that has caused too many people to regard selling as something less than a professional kind of work. It would be short-range thinking at its worst if we were to employ these tactics in *our* selling efforts. There are also moral and ethical overtones that should prevent us from selling products or services to customers who may be gullible enough to buy what they do not need. This always evokes the question, "What do you do if the customer 'wants' something he does not need and insists upon obtaining it?"

The complete moralist says, "Refuse to sell it to him."

The psychologist says, "No such situation could ever arise; if the customer *wants* something, he *needs* it. He 'wants' it because it satisfies some inner psychic need, as contrasted to a product need. The only exception that comes to mind is the situation that finds a customer 'wanting' something because he erroneously thinks it will solve a product need when actually it will not."

The realist says, "Sell it to him, but first determine whether it is possible to divert his attention to legitimate needs he has that you can fill. Then, if such attempts fail, try to make him realize that the decision has been his."

The viewpoint of the professional saleswoman must be a blending of the attitudes of the psychologist and the realist. If you know that the customer is in error and actually has based his decision to buy on faulty information, it is your responsibility to present him with the correct facts. Do not delude yourself into thinking if he buys—and later finds that his purchase was the result of his own ignorance—that he will blame himself. We look outside ourselves for the reasons for our failure, and so will your customer. He will recall that *you* sold him the product or service, and he will restructure the situation in his own mind so that he can blame you for his own decision. To do otherwise would be to admit that his own image of himself as a man of good judgment is not valid. He will then employ self-defense mechanisms. That, after all, is what they are for. "You should have told me!" will become his battle cry.

Whenever you face a customer who seeks to buy something you "know" will not fill a need, you should consider the following:

1. Perhaps there is no product need, but is there a psychic need that he wants to satisfy? This will be as real to him as any operating need. Pride of possession is such a need. Perhaps the owner will rarely, if ever, use the product he has purchased, but ownership provides him with real satisfaction.

If you find this hard to believe, we suggest you make a search of your cellar or attic. Identify a few items that you "needed" and bought, but rarely used. Perhaps you don't even have to go to these remote outposts to prove the point. Is there possibly a camera around the house that hasn't been used in quite some time? When you bought it, you "needed" it. What would have been your reaction to the salesman who knew how often you were probably going to use that camera, had he attempted to dissuade you from buying it?

Buying to fulfill psychic needs is not limited to cameras, video recorders and the thousands and one other items that we "need." It shows up in such a purchase as that of a computer which a certain manufacturer did not need for his operation but which he "needed" to project the image of a progressive organization. If a salesman had tried to argue him out of such a purchase, he would be doing the customer a disservice—and if the need were sufficiently strong, the customer would have purchased elsewhere.

2. Perhaps a customer may have a "pet concept" that requires a product or an application of a product which you know is unnecessary or is not her best solution. You must keep in mind that the customer wants to be identified with the solution to her own problem. The purchase and use of the product required by her pet concept provides tangible evidence that she has been part of the solution. If you fight the issue, you will lose the sale. In doing so, you will jeopardize long-range relationships, and some competitor will take the business.

Your best course of action is to present your facts in such a way that they do not seem to clash with her opinions. Do this so subtly that she develops new insights (Aha! I see!) and realizes how inappropriate was her own original solution. Above all, seek a way of making her part of the solution. Admittedly, this calls for professional selling effort of the highest order.

Following are some strategies that may be of assistance:

D15—Use the direct question, coupled with an attitude on your part that is free of antagonism, free of implied disbelief or ridicule; make it a sincere quest for information. This may accomplish two gains: (1) it will help identify the true nature of his problem, including any psychic needs that may be involved; and (2) as he answers your inquiries, he may develop insights that cause him to see he has misinterpreted his own need or its solution.

B7—Praise or compliment him. B8—Give him credit. B19—Ask for his advice, his help, his opinion. E9—Give him recognition. E6-7—Appeal to his prestige or pride. Any of these—correction—*the appropriate one,* combined with the following, may prove effective: *A3—Identify the me he seeks to project. A4—Identify his primary buying motive. E4—Explore cost of wrong decision. C5—Focus his attention on his long-range objectives* rather than the attractive short-range goals that may have pre-empted his attention and dictated his course of action.

For the sake of emphasis, it must be repeated that it is folly to believe he will not blame you later if he makes a bad decision now. Don't, therefore, use phrases like, "I'm selling you this against my better judgment." Don't think that by putting your views in writing you will protect yourself when he later realizes his buying decision was wrong. It would be the height of folly, at a later date, to call his attention to the correspondence. It would serve only to infuriate him or cause him to lose face. The result would be that he would avoid you in the future; hence, his business would go elsewhere.

The foregoing is not an armchair theorist's version of what would happen. Our files bulge with instances of customers who made wrong decisions against the strong (and written) advice of professional salesmen, and later placed the blame squarely on the shoulders of the salesmen. In one case, an organization dealing with a major government agency was appalled at the fury of the customer's attack when documentation was produced to prove that the decision had been the customer's alone... against advice that had been reinforced over and over in writing. The company concerned lost a major account as a result.

As stated previously, the eventual reaction of the customer is, "You should have insisted if you knew that I was misinformed"; "If you knew that this was wrong, why did you sell it to us?" Fortunately not all selling situations are that bad. If the salesman has developed a close rapport with the customer, he can often advise against a specific purchase or course of action, and the customer will take his advice. This emphasizes the importance of the personal relationship side of selling... something that older salesmen have long since mastered. Try to anticipate the customer's actions and move in fast with your advice before he has committed himself to the wrong action. If you can accomplish this, you avoid any need to help him save face later.

There is another possible action that always leaves managers of salesmen aghast. Pushed as they are by "top" management to produce sales, they worry when their men are given the following advice: "Keep your eye on your long-range relationships with the customer. Your objective should practically *always* be to maintain lasting, favorable relationships with the customer... and this means there are times when you should tell him that in his best interests you feel you cannot sell him the product, service or system upon which he is insisting."

Consider these two questions:

1. Will he, later, appreciate your honesty when he learns that you were right and some competitor who sold him was wrong?
2. Will he lose confidence in your judgment because you disagree with him?

There are other risks which must also be considered before taking this drastic action: Does the possibility exist that some competitor will not only get the customer's business on *this* order but will keep a firm grip on his future business? Any competitor who was shrewd enough to take advantage of the customer's psychic needs will be smart enough to recognize other psychic needs the customer has and will play up to them. The customer, in turn, will look upon the competitor as a person who really appreciates the nature of his total problems (operational *and* psychic). Furthermore, if you don't sell the customer, and later he realizes you were right, he may avoid you because he will feel that he is losing face when he meets you. Finally, there is always the possibility that the customer is right, and that our view of his needs and the proper solution for those needs is in error.

All of the foregoing tends to appear as so much rationalizing and camouflage which is really saying, "If the business is there...grab it!" No such intent exists. What we are really saying is that selling is far more than order taking, and it calls for professionalism. There are no easy answers. Anyone who poses easy answers in sales training is doing you a great disservice.

There *is* a tremendous and exciting challenge in professional selling. There *is* the need for analysis and consideration of the pros and cons of the many actions open to us, as there is in any profession. The basis for your ultimate solution of the problems just considered is to try to effect a balance between your own short-range goals and your long-range objectives, coupled with your moral responsibility to your customer and your responsibility to the people of your organization whose livelihood depends upon your ability.

No one can make the decision for you when you are face-to-face with a customer who is intent upon satisfying her needs in a manner that you view to be unwise. You must draw upon your judgment, your patience, your empathy for her and your conscience in deciding what course of action to pursue. The foregoing is intended primarily to alert you to the probable results of certain actions available to you.

Now, we shall deal with the eight obstacles that arise in many selling situations when you attempt to identify the customer's need or to awaken him to a need that he has of which he is not aware.

The customer is unaware that he has a need

Four of the many reasons for this condition are:

1. The customer is unaware that such a product as yours exists.

2. She has reached a level of satisfaction that has made her content.
3. He has a short-range outlook and does not see potential problems until they are thrust upon him.
4. He has lived with his problem so long he thinks there *is* no solution. He believes his problem is a fixed condition of his business and cannot be relieved.

Often, this poses no particular problem for the salesman. He merely explains—or demonstrates when possible—his product or service. He points up the need that it meets or the problem it solves, emphasizing the advantages to the customer. The customer, now made aware of his need and the availability of satisfaction for that need, becomes sufficiently interested, is stimulated, and a sale results.

The salesman faces a problem when the customer is so satisfied with his present situation that usual methods neither motivate nor stimulate him, and further action is necessary. The ability to solve the problem this type of situation presents separates the order-takers from the professional salesmen. The order-taker leaves the scene grumbling about the stupidity of the customer; the professional salesman considers strategies similar to those which follow, selecting the ones he believes will be most effective with *this* customer in *this* situation:

E6-8—Appeals to pride, prestige or to the self-image (the person he wants others to see when they see him). You may recall the example cited earlier about the customer who was totally disinterested in an expensive machine until the salesman pointed out that ownership would provide him with a more sophisticated machine than his nationally known competitor. Use of this strategy cannot be accomplished without effort. It requires research on the part of the salesman. The "research" may consist solely of listening with one's eyes and ears. It means determining what image the customer would like to project and who it is that he wishes to impress. The salesman must then present his product or service, emphasizing how its ownership will satisfy these psychic needs.

It should go without saying that psychic needs are never mentioned by the salesman. Remember the warning of the psychologists; i.e., "We do not like our behavior to be transparent." Neither does your customer. Rarely will he ever tell you that he wants to impress anyone. You must determine this for yourself and then concentrate on the features of your product or service that the customer will realize promise him the satisfaction he seeks.

C4-5—Couple the foregoing with an emphasis on his goals and objectives. Consider the possibility of interesting him in new and more ambitious goals and objectives. This strategy is based on the fact that most of us, including your customer, set our sights far below our true potential. Indicating to your customer (sincerely) that you believe he has great potential is a sure way of cementing a warm relationship.

D7—Draw comparisons. This is exactly what the salesman did (successfully) when he compared ownership of the "more sophisticated machine" to that possessed by the nationally known competitor in the example cited previously. When drawing comparisons, mention the advantages first. Then indicate the disadvantages the customer will suffer by not buying. Conclude by hitting hard on the advantages again.

This advantage-disadvantage-advantage sandwich has great psychological effect. Citing the advantages first awakens the customer's interest. Then indicating the losses he will sustain by following his present method awakens his sense of dissatisfaction. Mentioning the advantages again, at the conclusion, leaves the strongest stimulus for the moment when he must make a decision.

C3—Use testimonials. Alert him to what others are doing and induce him to emulate those he admires or respects, or cause him to envy those with whom he competes. Use testimonials *only* when you are absolutely certain of his attitude toward the person you are using as an example. Wisely used, the testimonial has great value; but keep in mind that most of us consider ourselves "unique" (as we are), and we consider our problems to be unique (which they usually are *not*). Using the testimonial can leave the customer with the feeling that you do not consider him unique and, therefore, its use presents some hazards.

When you select a person or firm to be used as a testimonial, be sure that your customer considers this person or firm as one that represents the "me he wants to be." If your testimonial is intended to arouse envy or the competitive spirit, be sure that the person or firm used is one which your customer would like to surpass.

C2—Use the "cause and effect" method. There are times when it is not wise to come directly to the point and show the customer that he has a problem. If he is proud, he may look upon this as a reflection upon his judgment in that he did not discover the problem or its solution without your help. The "cause and effect" technique usually means using a third person as an example. For instance, "We have a customer who has been having some difficulty with (cite problem), and he found that by using our (product or service) he was able to (cite solution of the problem). If you have had a similar problem, you may be interested in this approach."

Consider what this approach has accomplished. *You* have not said your customer has a problem. As he listens to your third party example, he becomes aware that he really does have a problem. He can now say to you (as though he had been aware of it all the time), "Yes, I've had that type of trouble." Furthermore, you have not indicated that you consider him as less than unique. You did not say, "If you have *this* problem, you may be interested in this approach." You said, "If you have had a *similar* problem...."

Words are tools, and they have tremendous power. They can produce sales, and they can destroy buying moods. Most of us *hate* preparation and we hate writing, but time spent in advance, planning exactly what kind of statements you are going to make, will pay huge dividends. It is not necessary to plan every word of your presentation so that you become stilted. But it will prove to be of value to plan every word of your key statements, especially when you are trying to make the customer see that he has a need he has not recognized... or at other vital turning points in the selling situation.

These words are addressed primarily to the experienced salesman who has been most successful. He has already demonstrated that he has great ability. He does lose *some* sales... and most of these could have been made, not by more effort, not by working harder, but by working just an ounce smarter.

The younger salesman is harder to convince. He can't believe that selling is this sophisticated or involved. He will learn as he becomes involved in more selling situations. The hope is that as he encounters the more difficult situations, some of the material he has read here, and from other sources, will ring a bell and a new insight will be created; i.e., "Aha! I see what that author was driving at." Hopefully, he will have time to rearrange his strategy before the sale is lost.

E4—Explore the cost of the wrong decision. This is much like the "sandwich" discussed above, in which you alert the customer to the losses she will sustain if she decides against your proposal. Be on guard here against "projection." In other words, don't use as "losses if you don't buy" those disadvantages that would be "losses" if *you* were the customer. Be sure that *she* regards these as losses. Incidentally, the same consideration applies when you select the advantages you decide to emphasize. Be certain that *he* regards them as advantages. Again—know as much as possible about the customer's viewpoint, his goals and objectives, and what will appear as losses or gains to a person working from *that* frame of reference, as contrasted to how *you* would respond to the same arguments.

In dealing with the customer who is unaware that he has a need or a problem, hit hard on attention-getting devices. Use opening sentences that will grip his attention. Don't be corny or dramatic. The use of the questioning technique will prove a valuable tool. Have you noticed how you are inclined to stop and read an advertisement that starts with a question, especially if the pronoun "you" is used. Your sentences should start with such phrases as, "How would you like...?"

We have just said, "Don't be dramatic." It is possible to *dramatize* your presentation without being dramatic. The late Jack Lacey told how he used to fare before he learned to dramatize. Walking with a friend who inquired, "Jack, what are you doing these days?" Jack says he replied, "I'm selling life insurance." A few minutes later, the friend said, "Jack, I forgot I have an appointment...I'll have to leave you, but let's get together again soon." Jack said he learned a lesson from this experience. Whenever he was asked the same question after that incident, he would reply, "I'm in the business of finding money for people!" The friend would always beam and ask, "Is that so...do you think you could find any for me?"

Hit *hard* in your opening remarks on advantages to the customer, and back them up later, after he has evinced interest. At this point, don't dwell on the features of your product or service. Exhibit enthusiasm ...and this does not mean being sensational, loud or flamboyant. Keep in tune with your prospect's interests. Don't talk about technical advantages if he is primarily interested in money; don't talk money if he is primarily interested in technical facts.

The customer doesn't comprehend

Saleswomen often encounter situations where the buying influence in the customer's organization —an employee or the customer himself—is not informed in the area which the saleswoman services. A saleswoman may be trying to influence the decision of a purchasing agent in an area in which the purchaser is not informed, or she may be attempting to sell an executive whose background is an unrelated facet of the business. There are times when the saleswoman is up against the problem of selling an otherwise competent customer in an area that is new or so far advanced that the customer is not well-informed. In all of these instances, the advantages of the saleswoman's product or service may be lost on the customer.

Communication is going to present real problems, yet it is imperative that the advantages be made apparent to the customer. At the very least, the salesman must find a way of arousing sufficient interest to cause the customer to call in someone from his organization who *will* understand the benefits to be derived from the service offered.

This situation generates these types of problems:

1. How can you present your information so that it will spark your prospect's interest?

2. How can you educate your prospect without indicating that you know he is not informed?

It is suggested that the following strategies be considered:

1. Identify the advantages that are so general in nature that he will see them as desirable, even if—at this point—he does not see how your product will overcome the problem.

2. Avoid technical language or terms and initialized abbreviations that will be foreign to his ear. (On this last point, it is always well to avoid initialized abbreviations, acronyms, unless the first time you use them you use the entire phrase. Use of this type of abbreviation can annoy the customer or make him feel less secure if he is not conversant with them.)

3. Avoid creating the impression that you are talking down to him. Be careful using the phrase "as you know" (when you know he didn't know until you told him) too often.

4. Use visual aids when possible.

5. Boil down your presentation to its essentials. Avoid detail.

6. Use probing questions with great care to determine whether or not you have lost him or if he is following your explanation. Don't use questions that will embarrass him.

7. If it is impossible to avoid exposing his lack of understanding, provide him with an escape hatch; i.e., "I imagine that with all of the pressures on your time, you may not have had an opportunity to look into this sufficiently to keep up with all the details."

8. Consider offering to help, when this appears safe to do.

9. Ask for his help, opinion or advice. "Whom would you suggest I see in your organization?" "I'd like your advice; we have a new product, and I just don't know who should see it in your company... can you help me?" (Admittedly this won't always work, especially with the type of purchasing agent who lets *nobody* get by him. But "I need your help" is still the most powerful appeal that can be made to man.)

If your customer's organization is small, the most effective approach may be to offer him your help.

If you must make a technical explanation, beware of the little "nods of comprehension" the customer may use to indicate that he is following your explanation. All too often, we all use this device as a means of hiding our embarrassment. The tip-off comes when he fails to buy, and this is a little on the late side if we want to make a sale. Be patient, check your progress, present your points slowly, and gear your pace to his attitude.

This can be a source of real trouble to those readers who sell outside of their country. For instance, we ordered a hamburger in another country and emphasized to the waiter that we did not want ketchup by saying, "No ketchup." The waiter beamed and replied, "No ketchup." and left.

Of course, the hamburger came back smothered with ketchup. Without boring you with the hilarious events that then took place, suffice it to say that each time we gave the waiter instructions, he would nod, smile and repeat the instructions, and then proceed to ignore them.

All of us present were mystified until we realized what was happening. The waiter would repeat the instructions each time without having the slightest idea of what they meant.

Face-saving was involved. The waiter did not want to admit that he did not have sufficient command of English.

This phenomenon is not limited to those who are unfamiliar with English. English-speaking people are often observed overseas, nodding their heads in agreement to comments being made by an individual speaking in his native tongue, when the English-speaking person doesn't have even a vague idea of what is being said but does not wish to expose a lack of knowledge of the other language.

This is just one of the problems that makes international selling even more difficult. However, even in their own country and language, customers will often nod their heads in agreement to points made by a salesman, when these customers actually don't understand a word being said about the product, service, or problem.

Poor judgment

It will be well to define "poor judgment" as it is used here. In a selling situation, the customer may show poor judgment when:

1. He makes assumptions or draws conclusions before he has all of the facts.
2. He exhibits desires, plans or intentions or takes actions which are based upon:

The desire to achieve short-range goals which appear very attractive, but which are in direct conflict with his long-range objectives.

His becoming emotional and abandoning reason.

His concentration on satisfying status needs (be they his own or those generated by the image his company strives to project) to the detriment of product or business needs.

His treatment of you personally is such that it arouses your own emotions and causes you to react in such a manner that he is deprived of the help or advice you could ordinarily provide. His attitude and/or actions may take the form of coolness, hostility, abuse or disdain. Or, he may ignore you or indicate a lack of respect for your judgment.

There are two *basic* methods for overcoming poor judgment when it is the customer's poor judgment that is involved:

1. Control yourself and don't complicate the situation by adding your own poor judgment to it.
2. Redirect the thinking of the customer.

Your problem is to cause him to see his product or service need. If his judgment is poor and prevents him from doing so, you can't afford to exhibit the slightest bit of irritation, impatience or rudeness. If he senses the presence of any of these attitudes, his mind will be closed to any appeals to reason. The solution to this problem is to be found, almost entirely, in your own mood and manner; in the things you say and do and how you say and do them.

Be patient. Don't make this appear to be tolerance. Don't create the impression that you are listening only until she stops talking so you can "straighten out her thinking."

Be understanding. Try to determine why he feels and acts as he does. Remember, he is probably successful thus far in business; so, his judgment must ordinarily be good, and he is a rational person who does things only for reasons that appear sound to him. His actions are a reaction to certain stimuli to which he has been exposed and to his interpretation of those stimuli. Your task is to determine the nature of the stimuli. He sees the present situation (including you, your service and your company) because of his interpretation of past experience he has had or because he wants to satisfy some psychic (status) need of which you may be unaware.

Consider the use of the following strategies: First...identify the real problem. Get him talking and keep him talking...even if he is abusive...until you obtain clues for his behavior and/or reasoning. *Use rhetorical questions (D17). Seek a point of agreement (B13).* Then, if he is trying to satisfy some short-range goal, try to redirect his attention to his long-range objectives and *explore the cost of the wrong decision (F4).* If he has already placed himself firmly on record, and to back down now will cause him to lose face, *seek an escape hatch for him (B17).*

More difficult to accomplish is to seek a solution that will enable him to meet the product or service need you know he has, and at the same time *supply the status or psychic needs he feels he has. Talk benefits (D6).* Make the goal that your products or service satisfy appear far more alluring than the psychic need he seems to feel is so important. In other words, try to substitute a new goal for the old one. If you can concentrate his attention on the problem your service will solve, he will listen.

It is urged that in dealing with this very difficult problem you review the Sales Strategy Bank. There are numerous other strategies that you may find more applicable to the situation represented by the customer's attitude. In order to assist you, the following summary of "Poor Judgment" is provided:

Why does the customer who in other activities uses good judgment use poor judgment now?

1. This customer has long-range objectives and goals. She might have difficulty in defining them, but they are there and they will play a vital part in her reactions. Under normal circumstances, she will take actions that promise her attainment of these goals and objectives.

2. If an attractive short-range goal captures his attention, he may grasp for it, abandoning for the moment his interest in his more important long-range objective. In his emotional state, he does not realize the price he may later be required to pay. It is your task to bring this to his attention, sometimes subtly, sometimes directly.

If his judgment is based on insufficient information, it is your job to supply the full facts now and at the same time provide him with the escape hatch. "I can see how you feel and I think I'd feel the same way if it weren't for information I recently picked up"... and then provide him with the facts that "he would not be expected to have."

The customer will let his judgment be influenced if he feels that status satisfactions or his self-image are at stake. If he feels that your proposition threatens them, or if he feels that your competitor enhances them, you can expect a problem. Don't consider his actions "irrational"—they're not. Your task is to show him that your proposition is in his greater interest or that it meets other needs that are more important to him.

Ignorance

The dictionary reveals that the word "ignorance" has many meanings. For our purpose, we mean that the customer is *uninformed* on certain facts. He may not know that a specific product or service will do a job for him, or he may not know that your competitor's product or service will create new problems for him.

This problem of customer ignorance is compounded because it appears in at least three forms, and these cannot be handled in the same manner.

1. The customer is uninformed but does not reveal this to you.
2. The customer is uninformed and doesn't hesitate to say so.
3. The customer is uninformed and deliberately tries to conceal the fact.

The problem is more serious when the customer is not only uninformed, but now has opinions which are based on erroneous or incomplete information, and he does not know that this is the situation. How do you overcome the problems generated by these attitudes?

In the case of the customer who does not realize he is uninformed and has no fixed opinions, you can ask questions and avoid embarrassing him by having escape hatches ready for him. For example: "I know that things are moving so fast that it is difficult to keep up with every new development. I know *I* have this difficulty, and probably you do, too. I'd like to save your time by boiling this down to the essentials."

At a glance, you may see that you have some customers with whom you would not use the sample sentence above. Other customers would grasp at this "out" and readily admit that they are misinformed and reply, "As you say, I *have* been swamped with work." You must develop the right sentence for each customer... and only *you* can do it, because you know him better than any author or sales training specialist. Any approach you use must not aggravate the situation by causing him to lose face. Your tone of voice must be casual, and you should not make a major issue of the matter. All you are attempting to do is to pave the way for the explanation you realize the customer requires if his interest is going to be stimulated.

Use the *probe for reaction (D9)* (usually rhetorical questions) or use statements that call for a conclusion and watch for his reactions. *Omit something (C6)* so he can have an opportunity to fill the gap. If this sale requires a number of calls, leave reading material with him if you have it. If such literature provides the information he requires, it skirts the necessity for letting him know that you realize he is uninformed.

Eliminate the entire problem at its source by doing a good job of pre-sale preparation, especially in gathering information on your customer's background and previous experience. If you find that it is likely he will be ignorant on the subject of your call, you can come prepared with a presentation that concentrates on essential facts presented in easy-to-follow language and visual aids if possible.

In the case of the customer who does not hesitate

to let you know that he is uninformed, the problem is easier to solve because "face-saving" will probably not be a problem. All that is called for is a straightforward presentation of the facts, using the techniques just outlined. You may consider using the strategy of removing him from the situation or *bypassing him (F1)* by asking him if he would like to bring in someone who is technically informed in the area, or if he prefers that you approach the engineers directly and that he, then, obtain their reactions.

The customer who is uninformed and tries to conceal it is the most difficult problem because his key purpose in concealing his lack of knowledge is probably a result of his desire to save face.

1. Avoid any form of presentation that reveals your awareness of his ignorance on the subject. Use the cause and effect method by citing the example of another customer who had a similar problem and the action he took to relieve the problem (using your product).

2. Explore the cost of the wrong decision (E4). He is hiding his ignorance to protect one psychic need: show him that a wrong decision will cost him something he may value as highly...economic security ...money. Confronted with this choice, he may prove willing to listen and learn to understand your proposal. Consider this strategy in the "last resort" category.

3. Bypass him if this can be done safely (F1).

4. Live with the situation and make your presentation so clear that he will ultimately understand your proposal.

Priority

The customer sometimes gives a low priority to the problem which the saleswoman's product will solve. She places a higher priority on other problems and will give the saleswoman neither time nor interest. This is so common and so important that we have given this problem area the rank of a major facet of the selling situation, and it will be discussed in future chapters.

Disinterest

It is sometimes difficult to distinguish between priority and lack of interest because disinterest exists when the customer places a low priority on the problem we wish him to consider and which we are prepared to help him solve. In other instances, it is not a matter of relative priority; it is simply that he does not see a need or he does not see in the salesman's proposal an opportunity to enhance his position in any way. In still other instances, the customer may sense that there is a need, but he is not sufficiently stimulated to do anything about it.

For example, a salesman may call on *you* to try to sell you a fire alarm system for the home. You may be fully aware that there *is* a need for fire protection, but you are not sufficiently stimulated to do anything about it. If the cost of the system is quite high, you exhibit a total lack of interest to this salesman because you don't want to sell yourself on his system and then realize you can't afford to buy it. This would only produce frustration and you and I (and your customer) know that the pain of frustration can be great. We resort to apathy, which to the observer appears to be lack of interest.

Lack of interest, as is the case with so many of the problem areas, is caused by different stimuli (or the absence of them). Hence, the problem always requires an identification of the cause in *each specific instance*. Only when this has been determined can we take appropriate steps to eliminate it. Some of the causes of lack of interest follow, with some general solutions you may consider using.

1. The customer has a low threshold of satisfaction or has reached the objectives and goals he set for himself. This fellow feels that the status quo has much to offer. His motto is "don't rock the boat" and "most problems will go away if you ignore them." His economic and status needs (as he now sees them) are satisfied. He, therefore, lacks interest in your proposal because it seems to offer him no reward in which he is interested. Change to him is the antithesis of status quo and can only spell trouble... and your name to this fellow is "trouble." You are lucky that you meet only with lack of interest; he could have used aggression.

One antidote is fear. It may be necessary for you to awaken the customer to a realization that his status quo may be in jeopardy if he neglects the problem for which you are offering him a solution. If this action of yours is going to be effectively and ethically applied, you must *know* what is important to him and you must be able to determine how neglect of the problem can deprive him of the satisfactions he now enjoys.

For instance, a retail store owner might be unwilling to expand his space or purchase advertising because his present income and prospects for the future satisfy his needs, as he sees them. The only way to shake him out of his apathy may be to awaken him to the cost to him if another dealer is franchised in the area, or if his competitors decide to expand or engage in a heavy advertising program.

A customer may enjoy the esteem and respect of people who know her; she may be well-satisfied with her string of accomplishments and the position of her company. Sometimes the fear that she will miss out on "something new" and be made to appear as a "has-been" who is not in tune with new trends will shake her out of her indifferent attitude and stimulate interest to the point where she will recognize a need.

There is a more desirable cure for lack of interest than fear. A customer may resent your intrusion on what has, until now, been his satisfied frame of mind. You may, by injecting fear, awaken him to a need... but because you have disturbed him he may go elsewhere to satisfy the need. In this instance, you have thrown business to your competitor; or the customer may exhibit resentment in his future relationships with you.

To avoid these complications, you may find it more profitable... and safer... to concentrate on another solution; i.e., awaken the customer to satisfactions greater than those he now enjoys and make these so appealing, even alluring, that his interest will be quickened. If you can set his sights on *greater business objectives than he has envisioned (C8)*, or on greater psychic rewards (the me he wants to be), he will do those things that will lead him to the new objectives.

2. "Assumed cost" is another reason for lack of interest that causes the customer to refuse to recognize needs he has. He assumes the cost will be great if he tries to solve the problem you have raised. When he believes the cost of solution is greater than the cost of living with the problem, he will rationalize and regard the problem as one of the necessary tribulations with which "you have to live if you are in this business."

Here are some alternative solutions available to you:

F4—Emphasize cost of not acting (buying now!).

D7—Draw comparisons.

F3—Show him the cost of procrastination is higher than the cost of making a decision.

E10—Anticipate, raise and answer his objections.

D6—Hit hard on the advantages and make them seem to dwarf the disadvantages.

A4—Above all, identify the basic source of his resistance; unearth what should be his primary buying motive and concentrate on that.

B12—Don't argue or disagree, but admit the validity of part of his views. Then proceed to show him you have considered those views in your preparation and are prepared to show him how to overcome the obstacles he has raised.

When we refer to "cost of solution," we are not thinking solely of the financial costs which the solution of the customer's problem may involve. Other expenses may be in his mind, such as:

Cost of additional work necessitated by the implementation of your proposal.

Cost of human relations problems it will possibly create, and which *he* will have to solve. (Getting his people to accept a change.)

Added work burden it may cause (financing, taxes, etc.).

It is obvious, then, that a clear identification of the *cause* of his lack of interest must be made before you can decide which strategy you will choose.

Customer doesn't want to see need

There are times when a customer does not want to see a problem or a need because:

1. He visualizes no solution and wants to avoid frustration.

2. He doesn't want to pay the cost of rectifying the situation (in time, money, effort, argument), and this gives him a guilt complex every time he thinks of the problem or need. Therefore, he denies, to himself, the existence of the problem or he minimizes its importance and pushes it out of his mind. (Ever do this on some maintenance job around the house? How did you regard the person who brought it to your attention after you had successfully forgotten it?)

3. He sees the problem, but doesn't believe the saleswoman has the answer... so he tries to avoid further discussion of the subject.

4. He hopes that, left alone, the problem will go away. This often means that he hopes conditions will change or that someone else will assume responsibility for the problem.

The question is, what action can you, as the salesman, take?

As always, first identify the cause. Which of these is the reason for the customer's disinclination to see the problem? There is, however, something working in our favor in this problem area. There is a thread of similarity running through each of the four basic causes just outlined; i.e., an assumption that the cost of solution is difficult or costly.

Therefore, you must be prepared with a strategy that recognizes this and you must show the customer that his fears are groundless. Heavy emphasis must be placed on providing proof (past experiences of others), using testimonials, demonstrations, building his confidence in you and in providing assurance. As has been stated, in essence, you have two general courses open to you; i.e., show him the cost of solving the problem is low or that making an adverse decision will result in a greater cost.

In choosing your action, consider these points:

1. Does he really understand there is a need but is refusing to face up to it, or is he honestly unaware of the need?

2. What is his basic underlying reason (perhaps even a subconscious reason) for not wanting to see the need? What is he afraid of? More work? High financial cost of solving the problem? Afraid of failure? Afraid of criticism? Afraid of being criticized by his superiors for not having acted sooner? Afraid that spending time and/or money on this proposal will make it impossible for other projects upon which he had his heart set?

It becomes readily apparent why we keep emphasizing the need to see the *real* reason for the customer's unwillingness to act. It becomes equally clear why so much emphasis is placed on *really* knowing the customer as a person and not regarding him solely as a source of business opportunities. Professional selling is far more than representing a company with a fine image, and it's more than getting people to like you. It's providing service to people, and the professional salesman knows that he can provide that service only when he knows what it is that the customer wants most.

Professional selling *is* a profession. That is why you, as a professional salesman, should reject canned solutions. Each solution must be custom-built for the customer.

A doctor can have three patients with the same problem... an ulcer.

Patient A is strong, but can't be off the job long because he has a family to support, so the doctor operates.

Patient B is not able to stand an operation, but he, too, must remain at work. The doctor may prescribe medications and a diet, knowing it will take Patient B longer to recover, but knowing that the treatment given Patient A would just not work with him.

Patient C is wealthy, can be off the job—but can't stand an operation. Again, the doctor considers him as an individual. He may prescribe relaxation in a rest area and put the chap on medication and a diet.

Three men with the same problem, but the solution for each is different. So must it be in professional selling. What will work for one customer won't necessarily work for another.

You and I would be outraged if a doctor didn't take the time for a thorough diagnosis. We are even suspicious when he prescribes some "ready-made" medicine... our problem is unique and we want to feel that he has considered us unique. For the same reason, customers will not react favorably to "standard treatment."

The professional salesman must recognize the source of the "disinterested customer's concern" so that the customer will realize the salesman is not unmindful of the facts that disturb the customer. The customer must be assured that the salesman is not ignoring, discounting or minimizing those concerns. This is not to imply that the salesman should agree with the customer, but the salesman's approach should include consideration and mention of those concerns about the "high cost" of taking action. Then, the salesman must show that the cost of *not* taking action is even higher.

Customer is confused

A customer may not realize he has a problem or that a need exists because he is confused. This confusion may have been the result of his having been exposed to many people, and acting as advisors, they have offered him sharply contrasting opinions. (Try telling a number of your friends that you are considering the purchase of a hi-fi audio system for your home. If you were really serious about it, after hearing all of their opinions, you'd probably decide to settle for what you already have at home.)

The customer's confusion may have resulted from inability to analyze statistics that have been made available to him. He may not have been able to gain insights or to draw conclusions from his study. In either case, the customer needs assistance. If his confusion stems from conflicting opinions to which he has been exposed, your course of action may embrace some of the following tactics:

D2—*Narrow down his choice.*
D1—*Give him a choice.*
D3—*Provide him with a solution.*
C2—*Use cause and effect examples (explain and demonstrate).*
E11—*Stimulate his recall of past experiences.*
D7—*Draw comparisons; couple with* E4—*Explore cost of wrong decision.*
F3—*Show him the cost of procrastination is higher than cost of making a decision.*

Narrow down the choices

A wild animal trainer uses a chair because it has

four legs and confuses the belligerent animal who doesn't know which leg to attack, hence doesn't attack at all.

Humans, too, when confused by too many choices will exhibit apathy or will reject all of the choices. The task is to make it easy for the confused customer by narrowing down the choices.

Any experienced retail salesman knows that he has an excellent chance of losing a customer if the customer is presented with so many choices that he cannot make a decision. The salesman makes it easier for the customer to decide by removing from the scene all but those items in which the customer has evinced the greatest interest. If the customer does not seem to favor any of the many items in front of him, but seems interested in all of them, the salesman will, nevertheless, remove some of the items. He knows the customer will ask that an item be brought back if he is really interested in it.

Whether you are dealing in electrical supplies, turbines, jet engines, toasters or insurance policies, the *principle* is the same. If the customer is confused because of conflicting viewpoints to which he has been exposed, the course of action is clear: Try to help him concentrate on a comparison of the smallest number of conflicting views. In other words, give him a choice—but make the choice easy.

Couple the foregoing with *providing him with a solution (D3)*. Your solution should emphasize all of the key advantages to him, but it should also include any disadvantages of your idea and how they will be overcome. If this is not done, your competitor will undoubtedly bring up the disadvantages of your proposal and the customer will either think you were hiding them or were unaware of them. He will either lose confidence in you or become more confused than ever.

If he has been confused by a multiplicity of choices open to him, you must buttress your proposal with facts, proofs, testimonials, or cause and effect examples drawn from the experience of other customers. Your entire presentation to a *confused* customer must carry with it *conviction* and *confidence*.

More than anything else, this customer needs assurance. Only complete assurance must show in your presentation. If his confusion is the result of lack of knowledge, present your facts slowly and clearly, avoiding the hazards of appearing to be patronizing. Avoid the use of expressions or jargon to which he may not have been previously exposed and which would only add to his confusion.

Stimulate recall

We know that our current opinions are heavily influenced by our past experiences. We may listen to a saleswoman and *want* to believe that she is right, but the impact of our own past experiences will be greater than the arguments of the saleswoman. On other occasions, we are quick to see a saleswoman's point because our past experience supports her contentions.

When it is possible to do so, refresh the customer's memory of situations that you know he has experienced which will prove your point *(E11)*. Some of these may be experiences that will remind him how his faith in your advice or in your company has been rewarded in the past. Some may bolster his confidence in his own decision-making ability as demonstrated by his past decisions. You may remind him of experiences he has had which will enable him to draw parallels between his problem of the moment and problems which he has met, understood and solved in the past. This helps to dispel any confusion that currently exists. You may introduce analogies, based on his past experiences, but avoid making the analogies so complex that they create further confusion.

When you stimulate his recall of his past experiences, rely heavily on the cause and effect method; i.e., remind him of a past problem he has had and its cause. Then, remind him of the successful solution of that problem (the effect) and conclude by drawing a parallel between that experience and his present situation. By moving in this way from the known to the unknown, you create understanding and eliminate the confusion that has prevented him from making a decision to buy.

Draw comparisons

Draw comparisons and emphasize advantages versus the cost of a wrong decision—or no decision. Visual aids can be helpful here. First, list the facets of the problem to which you are attempting to alert him. Then compare each of the conflicting methods of solution or proposals to which he has been exposed. Stress the advantages of your proposal in terms of the points he seeks to solve. Total up the advantages versus the disadvantages and show him how, on balance, your proposal offers the best solution to his problem.

When it is *you* who is confused

We have been discussing the problem that is represented by the customer who is confused. There are times when we, as salesmen, are confused in that we are unaware of the nature of the customer's need. He may be a new customer or we are new to this business, or the technology has moved faster than we have. What to do? Your choice of actions follows:

1. Admit that you do not understand and ask for his help. This is most effective when in your past relationships with this customer you have demonstrated an awareness of his problems and an ability to be of assistance. It is most hazardous when you are new and have not yet developed within your customer a confidence in your knowledge or ability. Still, even in these cases, it is a disarming approach and it is based on the most powerful of human appeals... "I need your help."

It is also a subtle way of satisfying the customer's need for recognition—even his need to feel needed. Although it seems to permit the customer to dominate the selling situation, in reality *you* still control it because it is you who keeps the customer talking on the facet of the selling situation that is of interest to you. A request for the customer's help can be couched in language that reassures him rather than causes him to lose confidence, i.e., "I wonder if you would mind helping me by explaining the nature of the problem... I want to be *certain* that I understand it fully so that I can make available to you the best service we can provide."

2. Use a variation of the strategy *D14—Fence*; i.e., stall for time. If the problem is one that does not call for immediate action, get time to do your homework. Get him talking, keep him talking and *really* listen.

relationships with this customer you have demonstrated an awareness of his problems and an ability to be of assistance. It is most hazardous when you are new and have not yet developed within your customer a confidence in your knowledge or ability. Still, even in these cases, it is a disarming approach and it is based on the most powerful of human appeals... "I need your help."

It is also a subtle way of satisfying the customer's need for recognition—even his need to feel needed. Although it seems to permit the customer to dominate the selling situation, in reality *you* still control it because it is you who keeps the customer talking on the facet of the selling situation that is of interest to you. A request for the customer's help can be couched in language that reassures him rather than causes him to lose confidence, i.e., "I wonder if you would mind helping me by explaining the nature of the problem... I want to be *certain* that I understand it fully so that I can make available to you the best service we can provide."

2. Use a variation of the strategy D14—Fence; i.e., stall for time. If the problem is one that does not call for immediate action, get time to do your homework. Get him talking, keep him talking and *really* listen.

four legs and confuses the belligerent animal who doesn't know which leg to attack, hence doesn't attack at all.

Humans, too, when confused by too many choices will exhibit apathy or will reject all of the choices. The task is to make it easy for the confused customer by narrowing down the choices.

Any experienced retail salesman knows that he has an excellent chance of losing a customer if the customer is presented with so many choices that he cannot make a decision. The salesman makes it easier for the customer to decide by removing from the scene all but those items in which the customer has evinced the greatest interest. If the customer does not seem to favor any of the many items in front of him, but seems interested in all of them, the salesman will, nevertheless, remove some of the items. He knows the customer will ask that an item be brought back if he is really interested in it.

Whether you are dealing in electrical supplies, turbines, jet engines, toasters or insurance policies, the *principle* is the same. If the customer is confused because of conflicting viewpoints to which he has been exposed, the course of action is clear: Try to help him concentrate on a comparison of the smallest number of conflicting views. In other words, give him a choice—but make the choice easy.

Couple the foregoing with *providing him with a solution (D3)*. Your solution should emphasize all of the key advantages to him, but it should also include any disadvantages of your idea and how they will be overcome. If this is not done, your competitor will undoubtedly bring up the disadvantages of your proposal and the customer will either think you were hiding them or were unaware of them. He will either lose confidence in you or become more confused than ever.

If he has been confused by a multiplicity of choices open to him, you must buttress your proposal with facts, proofs, testimonials, or cause and effect examples drawn from the experience of other customers. Your entire presentation to a *confused* customer must carry with it *conviction* and *confidence*.

More than anything else, this customer needs assurance. Only complete assurance must show in your presentation. If his confusion is the result of lack of knowledge, present your facts slowly and clearly, avoiding the hazards of appearing to be patronizing. Avoid the use of expressions or jargon to which he may not have been previously exposed and which would only add to his confusion.

Stimulate recall

We know that our current opinions are heavily influenced by our past experiences. We may listen to a saleswoman and *want* to believe that she is right, but the impact of our own past experiences will be greater than the arguments of the saleswoman. On other occasions, we are quick to see a saleswoman's point because our past experience supports her contentions.

When it is possible to do so, refresh the customer's memory of situations that you know he has experienced which will prove your point *(E11)*. Some of these may be experiences that will remind him how his faith in your advice or in your company has been rewarded in the past. Some may bolster his confidence in his own decision-making ability as demonstrated by his past decisions. You may remind him of experiences he has had which will enable him to draw parallels between his problem of the moment and problems which he has met, understood and solved in the past. This helps to dispel any confusion that currently exists. You may introduce analogies, based on his past experiences, but avoid making the analogies so complex that they create further confusion.

When you stimulate his recall of his past experiences, rely heavily on the cause and effect method; i.e., remind him of a past problem he has had and its cause. Then, remind him of the successful solution of that problem (the effect) and conclude by drawing a parallel between that experience and his present situation. By moving in this way from the known to the unknown, you create understanding and eliminate the confusion that has prevented him from making a decision to buy.

Draw comparisons

Draw comparisons and emphasize advantages versus the cost of a wrong decision—or no decision. Visual aids can be helpful here. First, list the facets of the problem to which you are attempting to alert him. Then compare each of the conflicting methods of solution or proposals to which he has been exposed. Stress the advantages of your proposal in terms of the points he seeks to solve. Total up the advantages versus the disadvantages and show him how, on balance, your proposal offers the best solution to his problem.

When it is *you* who is confused

We have been discussing the problem that is represented by the customer who is confused. There are times when we, as salesmen, are confused in that we are unaware of the nature of the customer's need. He may be a new customer or we are new to this business, or the technology has moved faster than we have. What to do? Your choice of actions follows:

1. Admit that you do not understand and ask for his help. This is most effective when in your past

CHAPTER 12

Establishing customer satisfactions

In a selling situation, you, as a salesman, seek to establish a buying situation or bring about *buying action*. The buying action will not take place unless the customer is convinced that a problem or need does exist and that your proposal will solve the problem or fill the need. It is obvious, then, that you face the task of establishing satisfactions; i.e., whether your "sales presentation" is formal or informal, rehearsed or extemporaneous, it must be built upon a foundation of advantages to the customer.

The Anatomy of a Sale chart indicates there are at least 15 possible impediments in the path of a saleswoman who attempts to establish satisfaction in the mind of the customer. You, from your sales experience, may be able to add many more. In some sales situations, you will encounter none of these obstacles, but this chapter does not address that kind of "happy" situation. We are concerned with ways and means of identifying the existence of problems and considering what means are available to cope successfully with them.

Price high

The customer may be willing to believe that your solution offers relief from the problem he seeks to solve, but he may feel the price is too high. Therefore he feels that you have not really established a satisfactory solution. The "price" may not be solely in terms of money; he may feel that your solution will result in one of the following conditions:

1. Cause him to lose face for not having thought of the solution without your aid or for having gone on record, originally, as being opposed to this solution or as having favored another solution. He is unwilling to pay that "price."

2. The customer may feel that you solved one problem but your solution creates other problems which he regards as too high a "price" to pay. Your solution, for example, may require extensive layout alterations or changes in procedures in his organization that may create problems for him in human relations with his people, or may necessitate that he invest time or money in training. Your solution may require him to revise some of his goals and objectives of long standing.

Your task in any of these instances is to help the customer see the whole picture. This includes those situations in which he feels that the actual monetary cost of your product or service is too high. You must make the advantages of your proposal so meaningful to him, and so attractive to him, that these other "prices" will lose their impact.

This means that *product or service features* must be expressed in terms of *customer advantages*.

The customer advantage which you decide to stress must be carefully selected to meet the *specific needs* of this customer as he sees, or can be made to see, *those needs*. This is customer-oriented selling in action!

It is conceivable that we can "miss the boat" in our treatment of other facets of the sale, but it is difficult to see how we can fail to establish satisfactions and still complete the sale successfully. Remember, a sale is an exchange of need satisfactions. If the customer does not feel that such an exchange is taking place, in which he feels that he is at the very least getting an even break, the result will be no sale.

Product analysis sheet

It is imperative, therefore, that we make certain we are fully aware of the advantages of our product or service; that we know the nature of the customer's need, as *he* sees that need, and that we have a means at our disposal for making the advantages clear to him in terms of the satisfaction of the need. A simple work sheet similar to that shown can help you.

This example may not fit the kind of product selling in which you are engaged. It will—believe it or not—be more valuable to you than to the reader who sees that he can use this claim exactly as it is. That type of reader may be satisfied with merely using the example in his sales presentation. You, on the other hand, must seek out the principle we are establishing and then decide how to apply it to your kind of selling.

An examination of this example which represents a brief excerpt from a Product Analysis Sheet will be revealing and profitable. It will be seen that

---

Product Analysis Sheet

(What you claim it will do)
Customer advantage

"You won't lose time because of breakdowns due to overload."

(Basis for your claim)
Product feature

"Our ratings are conservative and provide for line voltage fluctuations and normal overload."

(Significance of your claim)
Explanation or contrast

"Foreign manufacturers' ratings are maximum ratings with no provision for operating above rated capacity.

"This often results in lost production time."

---

the four statements can be assembled into one continuous statement, reading from left to right. First, the salesman has made a claim for his product based on his understanding of the needs of this specific customer (and based on the fact that he has learned that this customer is contemplating the purchase of lower-priced equipment of foreign manufacture). The salesman has stressed *time* because he knows this customer is very concerned about lost production time resulting from the failure of electrical equipment. If the salesmen knew this would not be a significant advantage, he would have concentrated his fire on an issue that was important to the customer.

For example, let us assume that the salesman knows this customer is primarily motivated by money. The salesman might have made his claim in this manner: "You won't run into *heavy expense* caused by breakdowns due to overload." Or he might have said, "You won't run into *expensive* idle time caused by breakdowns due to overload."

This again points up the hazards of canned presentations. Our product claims must be geared to the customer's greatest area of interest.

In the center column, the salesman proves his opening claim by stressing the product feature that makes the advantage possible. If practicable, he would back up his statement with a demonstration, a sample or a visual aid. He would thereby appeal to sight as well as to the sense of hearing, knowing that this will have an even greater impact on the customer and be more certain to establish creditability... an answer to the customer who is thinking, "Prove it!"

Note that he takes care to make his claim believable. He doesn't rashly claim that he has a product that will stand up under all conditions of overload. He stresses that it will withstand normal overloads. He knows this is all the customer hopes for. The salesman intimates that the customer wouldn't be likely to use the product improperly under conditions of heavy load far beyond the use for which it was designed.

The salesman continues in column three, to make his point meaningful to the customer. He also realizes that eventually he may face the problems represented in the "Gain Preference" facet of a sales situation. He begins now to make the customer aware of the superiority of his product over competition's offerings. Otherwise, it is conceivable that his claim would make little impact upon the customer who might be thinking, "What's so unusual about a provision for overload? Don't all of my potential suppliers provide that?"

Notice, also, that the last thought the salesman leaves in the customer's mind (on *this* point) is the implied loss or disadvantage the customer would suffer if he were deprived of this feature; i.e., "This often results in lost production time." This is one way of using the strategy, "Explore the cost of a wrong decision." Again, if the salesman were dealing with a customer to whom *money* was more significant than time, his final remark concerning this product feature might have been, "This often results in *costly downtime*."

The Product Analysis Sheet is especially helpful as a pre-sale preparation tool. It helps to avoid the tendency to assure yourself that "you'll know what to say when you face the customer"... only to find at the time of actual contact that you have forgotten some of the key points you had intended to make. Writing down these points helps to reinforce our memory, and the Product Analysis Sheet can be scanned at the last moment before we meet the customer.

Sales managers—or more properly, managers of salesmen—will profit by having such sheets drawn up, with the sales staff playing a big part in this preparation... so that each sheet will represent the best thinking of the salesmen. The sheets then will be more effective because the salesmen thought the problem through, and they will use them because they are a product of *their* thinking and experience.

Even when we do not use the actual sheets, we

should keep in mind the following principles on which the sheet is based:

1. In talking to a customer, mention the advantages to her first.

These advantages are *not* product features. They are *not* facts about us or our company.

They are statements that promise the customer some form of meaningful satisfaction.

They are custom built; i.e., selected on the basis of what we know will be most important to *this* customer. In other words, they are designed to satisfy her primary buying motive or to eradicate any fears she may have had.

They must be believable and fact supported.

They must be designed to arouse her attention and interest... their prime purpose.

They will help to gain or reinforce her confidence in us because they indicate to her that we understand her problem and chief concerns.

They must be couched in language that does not talk down to her. The phrase "as you know" may be used with discretion, i.e., "As you know, foreign manufacturers' ratings do not provide for operating above rated capacity."

2. Having mentioned the advantage of your product or service, follow this up with the feature you feel insures the advantage.

Be specific.

Suport your statement, when possible, with demonstrations or visual aids, and try to get the "merchandise" into the customer's hands... get him involved.

Talk the customer's language; don't use terms to impress him if you are not certain he will understand them.

3. When it is possible to do so, provide him with an explanation or contrast so that you can be certain he understands why this feature of your proposal is important to him.

You may wish to draw comparisons with competition's offerings when they lack the feature you are describing.

You may wish to show him the difference between a previous model you have offered and the product you now offer him.

In the foregoing, we recommended the use of the statement "as you know." Probably nothing irritates us more than being told, in detail, something we already know and understand. The cause of this irritation is the implication that we have been ignorant on the point being explained. Your customer may also be irritated if he senses this implication when you furnish him the explanation contained in column 3 of the Product Analysis Sheet.

If you preface your explanation with the words "as you know," you may escape this obstacle. If he did *not* know, he knows now, and is flattered at your assumption that he did know before you presented your explanation. If he *did* know, he does not become irritated because you have indicated your awareness that he possesses the knowledge, and he considers your remark to be nothing more than a helpful reminder.

Caution: Use this approach with judgment.

Don't say "as you know" when you know that he does not have this information you are about to impart *and he knows that you know.* To use the phrase under those conditions will cause him to feel you are resorting to flattery. Flattery, to many people, is an indication that we think we know what will motivate them. Psychologists remind us that we do not like our behavior to be transparent. We dislike people who think they can "read our motives and status yearnings."

When we openly flatter a man and he realizes what is taking place, he feels that essentially we are saying, "I am fully aware of your status needs and I'll feed them with flattery." The customer will then be on guard from that moment on because he will suspect your approach will be to his emotion rather than to reason, and to his status needs rather than to his product or service needs.

Use positive and specific statements

You can't rely upon "general" claims to overcome the type of resistance being discussed in this section. General claims lack punch and carry little conviction. They are not customer oriented. They are more likely to be oriented to the general market. In other words, most customers might be interested in the subject of your general claim, but will *this* customer be interested? Remember, too, that the customer likes to feel that he and his problems are unique (except when he feels he can profit by the experience of others).

A weak statement might be, "This cable will give you reliable service"; or, "This cable will give you extra years of service." These are general statements and general claims. They are not always attention-getters. If you fail to attract the customer's attention with your claims, he won't be listening when you present your proof and explanation. Make strong statements.

"You can be certain this cable will give you uninterrupted service, and you won't find yourself worrying about cable problems no matter how much the temperature changes or how roughly the men handle it when they are pulling or bending it because ...(state product features and explanation)."

Or: "Once you install this cable, it will be a load

off your mind because it provides years of trouble-free service because..."

Or: "You're going to avoid maintenance expense for years when you install this cable because..."

In each case, you select the focal point of *this* customer's key interest (or concern). If you know he has had difficulties with cable because of temperature, weather or rough handling by his men, you will build your first statement around that problem. If you know maintenance expense is his worry, you use that as your kick-off. You are *stimulating his recall of experiences* (E11) which he may have tried to forget because previous purchases may have been a reflection on his judgment. At the same time, you do not cite these experiences...let *him* bridge the gap between your statement and actual experiences he has had.

Notice, too, how you can weave *assurance* into your remarks with such phrases as "You can be sure"..."You can be certain." Notice, also, how you begin to ask for the order this early in the sale by implying there is no doubt but that he is going to buy when you use such phrases as "You are going to avoid maintenance expense *when* you install this cable." Notice the difference in impact when a salesman makes an unfortunate choice of words by saying, "You will avoid maintenance expense for years *if* you install this cable."

Some salesmen will avoid this confident approach. They will say, "If I use that approach, I can hear him now saying, 'Hold on! I didn't say I was going to buy your cable...yet'." It is true that many customers will adopt that attitude and may even sense you are rushing them into a buying decision. This can be handled if you adopt the right posture and follow up with such comments as, "No, sir. I know you haven't, but I'm sure that when you have seen all of the features of this cable, you won't be satisfied with anything less, and you'll be as sold on it as I am."

So his objection is turned into an advantage. You have not argued with him. You have agreed with him and you have indicated, once again, your assurance that he is going to buy. You have grasped the opportunity to parade in front of him your confidence in your own product. Confidence begets confidence.

The price ticket

There will be times when *price*—the figure on the price ticket—is his key objection and is preventing you from establishing satisfactions. He raises the question of price and can't hear anything else you have to say; and the fact is that your price *is* higher than that of your competitors. What then?

A successful marketing executive with years of experience as a successful salesman answers this question: "When a customer concentrates on price, I say, 'O.K., let's talk about price. Now, in order to do that, we've got to see what goes into price. First, there's service...freedom from costly delays and shutdowns; quick replacement or servicing of parts. Now, how well equipped are we to provide you with the kind of service to which you are entitled? Well, we have a service operation right here in Middletown. As you know, our competitor doesn't have a service operation for this equipment within a hundred miles. So you see, when they offer you what appears to be a lower price, they are actually charging you more, because you must add in the cost of delay and lost production into their price. Now let's see what else goes into price. Life of the product is factored into price...not just what you pay for it now, but what you pay for it per year. I'd like to show you some comparative figures on length of useful operating life, etc.'

"I continue to go down the list of values that constitute elements of our price. I hit hard—and first—on the factors that I know are of key interest to *this* customer. I cause him to see price from an aspect that may never have been presented to him before. You will notice that although I say, 'O.K., let's talk about price,' I never actually talk about price again.

"You may say that this smacks of being a bit dishonest. It isn't. Face the facts. You can usually manufacture a product or provide a service as inexpensively as the other fellow. If your price on the invoice is higher than his, there is little doubt but that you are offering something better than he is. Often these are hidden—but real—advantages to the customer, and it is your job to point them out."

The general premise is that you play to your strengths and not to your weaknesses. You strive to convert what appears to be a weakness into a strength by changing the customer's frame of reference. Remember the discussion of the Gestalt concept? This approach causes the customer to see an object in a different setting. When he does not like what he sees, you change the setting. In this instance, you cause him to see price in terms of its relationship to the whole picture.

Summary

If this chapter is to be of maximum value to you, it is essential to refer again to the Product Analysis Sheet and make out similar forms for the products or services you sell. Concentrate first on those which you sell with the greatest difficulty. The chart presented represents only a small portion of each sheet. Additional customer advantages should be listed in the column at the left and would be supported in the other two columns. Every advantage and supporting feature should be listed, but in making a presentation you should select those which will have the greatest impact on the specific customer. Avoid the tendency to snow him under by reciting all of the advantages when some of them may be of little interest to him... although possibly of *key* interest to another customer with different product or psychic needs.

You will have no difficulty in distinguishing between "customer advantages" and "product features" if you keep in mind that a customer advantage is *what it does*, i.e. results it produces in terms of customer needs. Product features are: *what it is or what it has.*

The third column is used, when necessary, to explain why the product features will provide the advantages you claim *or to explain why these are advantages* to *him*. This is especially valuable in the instance of highly complex products or services where the customer may not understand why he should seek this advantage.

You may also use the third column, as in the example, to draw comparisons with other offerings he may be considering or to sell him on the need to buy a new and improved model. It makes your claims significant and more meaningful to the customer.

Unattractive terms

This is a problem area that can be anticipated if we have done our homework on the customer. If we have gathered significant information on his economic position, we may anticipate that this consideration will or will not be a problem. If it is, it often shows itself for the first time at this point in the sale; i.e., when we are attempting to prove to him that our product will satisfy his needs.

The customer may agree that your product will meet his need and that the price you ask is an equitable one, but he considers the basis upon which he must pay for it to be objectionable. An effective strategy is to use the psychological tool of illusion as one means of meeting this obstacle. The proposition must be presented to the customer in a manner that will cause it to appear to be more favorable.

You may recall, from a previous chapter, the example of the salesman who caused a customer to see a $780 annual savings that would accrue to him if he bought from another supplier as actually being $15 a week. Then the salesman demonstrated that $15 a week was a small price to pay for the top-flight service the customer had been receiving. The bold fact is that the salesman merely made the $780 per year appear to be less attractive when viewed as part of the whole picture.

This practice is not being recommended. It is cited solely as an example of how illusion affects our viewpoint. To be effective, the illusion must be believable, attractive and honest. If it meets the first two criteria and fails to meet the third, the salesman may obtain this order, but lose the customer's future business.

One method, then, for meeting the objection of "unattractive terms" is to find a means of presenting them so that they will appear attractive. Another method is to cause your offer to become so attractive that the customer's desire will become intense to the point where she will find ways to rationalize herself out of her feeling that the terms, as originally viewed by her, are unattractive.

A third way to meet this problem is to consider the resources of your company. Can people from other sections of your company provide other free services that will cause the customer to regard the offer, in its entirety, in a favorable light.

When terms appear unattractive to the customer, consider these courses of action which appear in the Sales Strategy Bank:

D1—Give him a choice; i.e., if possible, provide him with an alternate solution that appears less distasteful than the proposition which he finds distasteful.

E13—Offer additional services which you may be in a position to provide without cost to him.

D5—Compromise, if possible. Keep in mind that no matter how forcefully a customer may have expressed himself, he will usually compromise if you provide an escape hatch so that he will not lose face.

E2—Provide assurance and/or proof.

D6—Talk benefits so that desirable features or advantages overcome his fears.

E3—Demonstrate. Show him how savings will offset the disadvantage which he feels your buying terms represent.

C2—Use cause and effect examples (this is similar to *E5—Remove him from the situation with third party examples*). In other words, show how other customers, who held viewpoints similar to his, made the decision to buy under the terms you have offered; explain why they did so and show favorable results. Be sure you know how he feels regarding the customers whom you choose to use as examples.

E11—Stimulate his recall. In other words, remind him of his previous experiences that will cause him to see your proposal in a favorable light. For example: occasions where he had viewed terms as unfavorable but which he later found to be satisfactory. Or, an occasion where he did not buy because he regarded terms as unfavorable and then suffered a loss because he did not buy. Obviously, this latter course, though effective, must be handled delicately so that he will not feel his past actions are being criticized; and don't cause him to lose face.

C4-5—Emphasize his goals and objectives, and

show him how failure to buy will threaten these. Make new goals and/or objectives so alluring that they will create a setting in which he will regard your present terms as not unattracitve. Arouse his "me-I-wish-I-were" so he will want to project this image so badly that the terms will appear to be a minor obstacle.

E4—Explore the cost of a wrong decision or of procrastinating. Make these losses appear so great that the objectionable terms seem minor by comparison.

D7—Draw comparisons with other expenses she has. (These are all strategies that change the setting in which he sees your terms as unattractive.)

D11—Rephrase and overstate his objections so that he will tend to restate them and, in doing so, will usually bend a bit backward as a way of denying that he went as far as you have gone in stating his objections. This gives you easier objections with which to cope.

Limited use

Customers will sometimes be disinclined to buy because, although they recognize a need for a product, they consider the cost of the product too high when measured against the limited application it has or the infrequency of occasions upon which it will be used.

You have at least two courses of action open to you. You may emphasize the savings or other advantages as one course of action. You and I have purchased many items for which we suspected we would have limited use, but the value...the satisfactions they promised...were so great that the cost was of little concern to us. For instance, you hope you *never* use a fire-extinguisher, a car jack, a spare tire, and any number of comparable products or services—but you wouldn't be without them.

If your product or service can be presented in such a manner that the customer can be made to view the purchase as insurance, emphasize the cost of the wrong decision and hit hard on the fact that he has good judgment.

The second course of action is to be creative! Before you enter the selling situation, be prepared by *anticipating* that limited use as an obstacle may arise and be ready to provide other uses to which your product or service can be put. Present your proposal in a manner that causes the customer to hunger for the advantages and to regard infrequency of use as a minor point. Among other appeals, hit hard on pride of ownership and availability if he *wants* to use it. Emphasize advantages of convenience and explore cost of inconvenience when the product or service *is* needed. Know your customer and know which appeals will register most favorably and forcefully with *him.*

You and I have often talked ourselves into buying a product on this basis, and because our judgment is at stake, we rarely regret the purchase. We may have purchased a film developing kit or photographic darkroom equipment, assuring ourselves that we would henceforth do all of our own developing. Once purchased, the equipment may have been used extensively for a period of time and then used rarely, if at all. How many cans of car wax are resting unopened on shelves in private garages throughout the nation? How many of us have ever computed the cost per photograph of the pictures we have taken with an expensive imported camera we purchased? Yet, few of us regret having purchased the camera ...or even cameras. Take a look in your attic or basement for examples that will strike closer to home.

The question here: Is it ethical or moral to use this type of argument on a customer (i.e., alert him to other uses to which he *can* put equipment) when you suspect that he actually never will? The answer is found in your own reaction to past experiences of your own. You and I obtained value for the money we spent when we purchased the items that fill our attics and cellars. These items supplied us with certain satisfactions, if only the pride of ownership, the pleasure of possessing quality items, or the satisfaction of knowing, "*if we want to use it, we can.*"

It's "nice" to own a pair of expensive binoculars even though they are used only on rare occasions in most households. Consider, however, the real pleasure we obtain, when, on those rare occasions, they are available. We don't stop to compute how much they cost us per occasion of use. We measure their value in terms of how much satisfaction they bring us *when* they *are* used. The envious glance of our neighbor at a football game or on a mountain top more than compensates us for "our foresight" in having bought a pair of *good* binoculars.

Therefore, if you *know* the primary use of the product you are trying to sell the customer will solve a problem (even though the product will be used but rarely), you are quite justified in alerting him to other uses to which the product may be put, even though you suspect he may never make these applications.

To the rigid moralist who says, "You should never sell anything to a man unless he needs it," we would issue this reminder: *you can't!* If the customer does not want a product, he won't buy it. If he wants it, he needs it. He needs it not to satisfy product needs necessarily, but to satisfy psychic needs, such as pride of ownership, or to appear progressive or affluent...or to satisfy any of over 100 such needs.

You and I may know the customer's product needs, but we are a bit presumptuous to decide what psychic needs he *should* have, or to decide it would be immoral of him to want to own something he "doesn't need from a product standpoint." The rigid moralist may label this type of argument as specious and as arch rationalizing. Yet the rigid moralist —along with all the rest of us—spends more time than he (or you or I) is willing to admit in trying to get people to do the things he would do if he were in their place... trying to get them to conform to his standards.

A professional saleswoman's job is to identify needs—product, service and psychic, wherever she finds them, and to fill those needs, leaving contentment and satisfaction in her wake.

Delivery slow

Just so long as we have customers, it is certain that we will have complaints about late delivery or slow delivery. Granted, customers often establish delivery dates that are either unrealistic or unnecessary, or both. But isn't it just possible that you, too, in your frequent capacity as a customer, have raised a rumpus with some firm for being late in its delivery of items you had ordered? Isn't it just possible, too, that once the article arrived, it stood in a corner for days—even weeks—until you could find the time to assemble or install it? If you won't make such an admission, this writer will.

You and I may take months trying to decide if we will buy a new car, a color TV set, a boat or any one of a dozen other items. Once, however, that we have made the decision to buy, the red hot question becomes: "How soon can you deliver it?" And woe unto the dealer who indicates there may be a short delay.

It's almost as though we were afraid that if we didn't obtain immediate possession—hence, commitment—we might change our mind. When the decision to buy has taken a long time, it has usually been accompanied by concern that the cost might outweigh the benefits. Once we decide to buy, we undergo a certain amount of concern until we begin to realize the benefits.

The point is that it is a bit presumptuous of us to become annoyed with a customer who makes "quick delivery" demands upon us, when, in all honesty, most of us admit we have been guilty of the same fault. There is another and more important reason why we should not permit ourselves the luxury of irritation. When we are emotionally aroused, our capacity to think clearly is impaired. Indeed, it might be said that our reasoning ability will be in direct proportion to the degree to which we can control emotion.

We have seen in the foregoing that there are desires for fast delivery based on purely psychological drives. There is also the problem of fast delivery requirements of the customer that are based on his actual operating needs. These are more difficult to handle because the customer equates late delivery with the expense he is incurring as a direct consequence.

The fact is that often the only solution is to *get the goods delivered*, and no amount of talk, assurances, apologies or "buttering up" is going to mollify the irate customer.

From our observations of current problems of many firms, we are compelled to inject this note: The reason for late delivery is that people in the organization—outside of marketing—often are not customer-oriented. They tend to put "production convenience" first. Anyone who has worked in manufacturing can appreciate the problems of that hard-pressed group, but sympathy for manufacturing problems is absent from the customer's frame of reference—especially when late delivery is costing him money.

It would be helpful if someone outside of sales and marketing, especially someone from general management, were to impress upon the other functions just how essential it is for *everyone* to think in terms of the customer. This doesn't imply that "The Customer Is King" and must be idolized, but everyone in your firm must understand that the customer's money fills the payroll and keeps the business operating.

While the prime objective is getting the goods to the customer fast, even when the order is incomplete or partially in error, the opportunity to soothe an irate customer is available. Unfortunately this opportunity for psychological methods is completely muffed at times; for example:

A customer of a retail firm purchased a TV antenna that came in kit form and required assembling. During the assembly process, the customer found he had not been supplied with a paste needed to make good electrical contact between the elements of the antenna that telescoped together. He called the store just before five o'clock and asked if they would send him some paste, special delivery, at once.

The salesman who answered the phone replied that the last mail had gone out. The customer became quite indignant and stated, "There ought to be someone with sufficient interest in my problem to pick up a can of paste, wrap some paper around it and mail it from the post office which is directly across the street." In the face of of this attack, the salesman retreated and said he would "talk to the sales manager."

Moments later, he returned to the phone to announce that the package would be put in the "next morning's mail." The package arrived 40 hours later. To complete the picture, the customer had spent well over $4,000 in the current year with this store—a figure far above average in the type of business involved. The paste, which, incidentally, he was willing to pay for, was priced at 50 cents, and lack of it held up the job.

The question then: Is this good selling? What is going to happen to the relationship between the customer and the retail store? Will this customer ever really believe again that this company is truly concerned with his interests?

When delivery is late, you must recognize your responsibility to get some fast answers for the customer. Even though you know that delivery will be delayed, show your customer that *you share his concern*, and this effort will dilute the problem, and in some cases eliminate it. These unwavering efforts are especially helpful with the customer whose creed is, "now that I've decided I want it, I want it now."

The customer will almost always compromise—a principle to keep in mind—and this applies to the problem of handling most late deliveries. If you are successful in expediting delivery, you may be able to convert what was originally a problem into a strengthening of your relationship with your customer.

There are times when closing a sale hangs on your ability to deliver as quickly as your competitor. In such instances, consider these courses of action:

Know your competitor—know his tactics. Is he one who has the reputation of promising but not coming through? If so, his number is legion. Have facts available on his past performance and bring these to your customer's attention with irrefutable proof. Try to avoid the necessity for presenting these facts directly to the customer and attempt to set up a situation where the customer will "discover" these facts for himself. If this is not possible, don't hesitate to prove to the customer that your competitor's claim will not be backed up with performance. This seems to fly in the face of the adage "don't knock your competition." The disadvantages of "knocking" can be overcome by the manner in which you present the information to the customer.

Take a lesson from an automobile dealer in Europe. He faced an outraged customer because a car, promised for immediate delivery upon the customer's arrival in Europe, was not ready. The customer was depending on the car for his tour and faced a delay. His attitude was almost apoplectic.

The dealer chose this course of action: "Sir, this car is for export, and we want nothing but the best models returned to the U.S.A. We want you to have a trouble-free vacation in Europe. Your car was especially selected and ear-marked for you. Yesterday (Friday) it was found to have one minor defect. It is now being fixed and will be ready for you Monday morning. In the meantime, I am sure you will enjoy our city. Our chauffeur will pick you up Monday morning and take you to our factory. There you can see your vehicle as it comes through final test."

The cutomer was still irritated at the delay and suspicious that the car would not be ready Monday. When all of the dealer's promises materialized Monday, the customer was elated and drove off only too willing to forget the incident. Since that time he has bought a second car of the same make. The actions, incidentally, cost the dealer probably not more than a very few dollars.

Why does this kind of treatment work well? To understand this, we must look upon late delivery from the viewpoint of the customer. He often interprets it as our failure to recognize his importance (the "me-he-tries-to-project"). When we fail to show that we take his problem seriously, he is 100% correct in this analysis.

At other times, his annoyance is based on disappointment...being deprived for a bit longer of the satisfactions he had hoped to receive from immediate ownership or use of your product.

Aggression is one of the self-defense mechanisms—an escape from frustration. The customer will usually direct his aggression at the most convenient object, i.e., you. When a customer feels he is being deprived of *one* need (be it enjoyment, ownership or whatever), and there is nothing we can do to satisfy *that* need through immediate delivery, we must satisfy some *other* need that can be made to appear just as important—or even more important—to him. Usually, we will find it necessary to "over gratify" the other need.

For instance, ordinarily the car dealer would not have had the customer driven to the factory in a chauffeured limousine. Now that the customer feels his self-image has been kicked around and because his need for enjoyment (immediately take-off on his trip) has been denied him, the dealer realizes that he must fill to overflowing the customer's need for recognition.

The cure for slow delivery will be found in variations of the following actions:

1. Be sure your competitor is not "out-promising you" rather than out-performing you. Know, and make known to the customer, the competitor's past record for keeping such promises, if it is not a good record.

2. Demonstrate your concern for the customer's problem. Go all out to do this and do it where she will see you. Develop some empathy for her position.

3. Find another need you can supply that will neutralize the customer's irritation. Don't put the blame on "the plant," on "manufacturing," the "shipping people" or on any other segment of your organization. Build his confidence in your entire company; you may need it later. Blaming others in the organization may get you, personally, off the hook, but it will only reinforce his present belief that your company does not have his interests at heart.

Union opposed

You may encounter a problem caused by the union that represents your customer's employees. The union may object to the use of your product for a variety of reasons. These may range all the way from the union's antipathy to your company to its feeling that the product or service you are selling will reduce the number of jobs in your customer's company.

The first step should be to obtain sufficient information about your customer's business, and about union attitudes and the reasons for them. Use sources in your own organization. If your company is large enough to have a union relations section, consult these people. They may have encountered and solved the problem before.

You may find that union opposition is based on lack of facts, on a misunderstanding, or it arises because the union may be oriented to a short-range viewpoint. For example, a product or service may cost some jobs temporarily but may result in more jobs ultimately. It may, for instance, enable a user to incorporate your product into his end product. This, in turn, may make possible lower cost and price reductions which increase sales and jobs. Such has been the regular pattern in the history of industrial America and is not far-fetched. Don't approach the union. Give your customer facts so that he may do so.

The customer may be emotionally involved in the situation, and because of this be incapable of thinking of logical arguments to present to the union. You can provide a valuable service by giving him this information. The union may represent the employees of the end user of your customer's product. Determine how important this customer is to your customer. You may be able to point out that although your product may impair this relationship, it may improve more profitable relationships with other customers where the union problem does not exist.

Stress the advantages of your products and what they will mean to your customer so strongly that he will be willing to take on the problem represented by the union attitude. In other words, make the losses he will suffer from *not* buying appear much greater than the trouble he thinks he will encounter if he *does* buy.

The obstacle of union opposition is rather narrow in concept, and we should translate it broadly. The customer may not be concerned about a union, per se, but he may be worried about the attitude of his employees or about the attitude of his customers. Your task remains much the same.

1. Give him arguments he can use on those who he believes will object.
2. Convince him there is no problem by revealing experiences other customers have had. Select customers with whom he can compare himself. Use testimonials.
3. Divert him by concentrating his attention on the advantages.
4. Explore the cost of the wrong decision.
5. As a last resort method, shock or antagonize him. "You don't mean to say, with your long experience, you can't handle that type of situation—I don't believe it!" The foregoing actions may be found in the Sales Strategy Bank; explore it for other strategies that you feel fit your situation.

Installation costly

The customer may resist buying because of what he regards as the high cost of installing your product or service. In his eyes, this can take many forms: It may be represented by the cost of training his employees, loss of time while he is selling employees or customers on the change, loss of productivity while changes are being made, or the actual financial cost of making necessary changes.

It is going to be important to focus his attention on his long-range goals. Concentrate on what you have identified as his ultimate objective; i.e., the kind of an organization he is trying to build, the kind of a reputation he is attempting to establish, or the long-range profit realizations which he seeks. You cannot change his actions; you must direct his attention away from short-range losses and make him consider the cost of delay in meeting longer-range objectives. If these can be made to appear important to him (as they truly are), he—being a rational being—will do the things necessary to attain those objectives, even if it means a short-term cost.

From the Sales Strategy Bank, consider these actions: *(B4) Agree with him* that the costs he mentions are present. Then *(D10) divert him* by comparing these transitory expenses to long-range benefits. *(E2) Provide assurance or proof* by mentioning the experience of others... especially others whom he hopes to surpass. Offer him layout help and draw upon others in your company who may be able to provide him with assistance.

Space problem

This last is especially important if the problem is one of space utilization, which is often the case with retailers or with a manufacturer to whom you are attempting to sell equipment that requires extensive floor space for effective operation. Do not hesitate to use other members of your team; i.e., others in the company whose forte is space planning.

Many saleswomen are primarily entrepreneurs and think it is a reflection upon their ability to ask for help from other facets of their organization—or from their manager. It is, of course, no such thing. It is a reflection on their judgment when they *fail* to do so. This is especially true when they fail to realize that providing help is the primary responsibility of a sales manager. He may not be as effective in actual selling as some of his men, but he does have more sources of information. And, being away from the scene of the actual sale, the sales manager is not emotionally involved and may be able to reason more effectively.

The positive approach to this problem is to enter the selling situation armed with answers that are well planned. Show the customer that you have considered thje problem in advance and are not, therefore, giving him spontaneous answers. Prove this by coming with visual aids, layouts, pictures and plans when possible.

Another approach that is often effective is the use of the question. When he raises the space problem, instead of answering him directly, ask: "You've a point there... how do you feel that problem could be overcome?" Often he is more competent in this area than the salesman. By asking his opinion, you are accomplishing at least two gains. You have transferred the subject from its negative aspect to a positive facet and concentrated his attention on possible remedies. You have also appealed to his pride and prestige. In a way, you have supplied him with a challenge, and you have given him a chance to obtain the satisfaction that comes from solving a problem. If he offers you only a partial solution, try to tie in your prepared answer with his so that he will regard the combined solution as primarily his idea.

Don't be a hog—don't look for both the sale and credit. Give him the credit, and walk away with the sale.

Obsolescence

Economic satisfaction threatened: "It will be out of date and I'll have to spend more money to buy a new model."

Self image at stake; need for recognition threatened: "I'll look foolish if I buy now, and you'll come out in a short time with a new product." (This is rarely stated—but he thinks this way.)

Status threatened: "I won't have the latest. This will be out-of-date."

These are typical of the reactions of customers who fear obsolescence. In other words, such persons are afraid they are being asked to pay a high price in psychic satisfactions. They sometimes are fretful that they will look foolish in the eyes of their contemporaries for having made a purchase just before products of superior quality or at a lower price are introduced. The fear that "I won't have the latest" is a status-oriented concern of the customer who wants to appear to be progressive and alert to new trends.

Fear of obsolescence can also have its roots in concern over economic loss; i.e., that the price will drop when you come out with new products with better features. It is obvious that all of these worries cannot be handled in the same manner. They do have one facet in common: They stem from the fear that although your product may answer the immediate product or service problem, purchase of it at this time may deprive the customer of other satisfactions—the need to appear "smart," the need to appear "up-to-date."

Essentially, you have three courses of action, each of which must be tailored to the needs of *this* customer:

E2—*Provide assurance.* You must be able, in conscience, to give him what amounts to your word.

If you cannot make such a commitment, you should emphasize the cost he will be paying if he has to wait for a new model.

E12—*Provide guarantee.* Do this only if it is approved by your manager and, if necessary, by your legal counsel.

There will be times when you cannot provide such assurance because you feel there is a reasonable chance that products with new features will replace what you now have to offer. If so, concentrate on *immediate gains.* Focus his attention on present problems and make him feel keenly the cost of leaving those problems unattended.

Stress any flexibility your product may have and point out the uses it can be put to if a new product does come out. If possible, emphasize possibilities of "turn-in" allowances or re-sale possibilities. Show him—mathematically if possible—exactly what the cost of waiting will amount to, in terms of lost production, high cost of scrap, or in such terms as are applicable to his situation.

Above all, don't exaggerate to make this sale and lose his future business as a result.

Price changes

Customers often hesitate to buy because they

anticipate that you will soon change your prices — downward. They fear losing money and/or they fear the impact on the image they try to project of themselves as shrewd businesswomen who never come out on the "wrong side of a deal." Solutions to this are much the same as those presented in our discussion of obsolescence. Assurance, when possible; advantage of immediate relief; rebates—if authorized.

This is another problem you should *anticipate* if there has been a recent history of price fluctuations or if "the trade" believes you have new models almost ready for release. Don't wait until you are with your customer to come up with the right action — plan it in advance when the emotional setting of the selling situation does not impair your ability to think.

Past performance

It is suggested that you again read that portion of Part One that deals with the "Influences on Recall"; i.e., how and why a customer remembers as she does. You can be certain that a customer's experiences are factored into her frame of reference... how she sees you, your present offering and your company. You can also be certain there is much better than a 50-50 chance that if she has had trouble with your products, through her own fault or failures, she has successfully "restructured" the facts. In her mind, the fault is now that of your company or of your products. This restructuring enables her to continue to harbor an image of herself as a capable person.

You can be sure, too, if his troubles with your product have been relatively insignificant but *repeated*, he will have distorted them into an importance far beyond what they deserve. Further, if the experiences have been *recent*, they will take on a magnified importance in his eyes. If the difficulty he has had was with other products made by your firm, he will be affected by the Gestalt principle; i.e., the tendency to see the part as the whole. In other words, to blame you for the transgressions of other people in your company or to lump all of your products into one stereotype... "I've had nothing but trouble with your products."

One old-time, professional salesman encountered this type of situation and offered *"to become completely responsible for the results of his recommendations."*

Actually, the salesman made a rather empty "promise" when he agreed to take responsibility. He *knew* his product would do the job. There is something about the phrase "I'll be completely responsible" that is persuasive. It relieves the customer of any fear that *his* judgment will be exposed to ridicule if the idea proposed doesn't produce the desired results... after all, *he* wasn't the one who proposed the idea. At the same time, if results are good, he will assume pride of authorship become wasn't he the one who made the investment and "took the gamble"?

When a customer's past experience causes him to doubt that your proposal will satisfy his needs, you will find it is often due to the human tendency to generalize. His past experience may have been with a product that has since been improved, or it may have been with your product when it was employed to solve an entirely different problem than the one you are now trying to relieve. At the source of his attitude is the feeling that his own judgment was faulty when he "listened to you the last time." He doesn't want a repetition of this wound to his self-image.

Step one is to get him to talk so that you can determine the exact nature of his complaint. If he is blaming you or your product because of experiences he has had with other elements of your company or with other products, you can resort to the *"appeal to fairness" (D13)*. For example: "Joe, I know you have always been fair, and I am sure you don't want to be unfair now. Consider, then, Joe, that I am being blamed for...(point out how neither you nor your product were involved, without casting any reflections on the person or product that was the source of his difficulty)."

You may find it wise to resort to B19—*ask for his advice*. For instance: "Joe, we never have had any difficulty of that kind with this product before. I wonder if you can help me find out what went wrong?" He will usually respond to this type of appeal. Often he *knows* what went wrong, or the conversation that follows will reveal "what went wrong." You may be able to convert this obstacle into a gain. After getting his opinion, you can respond with, "Joe, I think you're right, and I'm sure that when we implement your idea, we'll get the performance you have a right to expect. What do you think?" Your chances are good that he will calm down because he is part of the solution.

Recently, we encountered a situation where a customer was annoyed at a number of problems he had with a line of products he was selling at retail. The difficulty was not with product performance but with his ability to sell as many as he should have been able to move. The manufacturer's manager of sales invited the retailer to be guest speaker at the annual sales meeting, asking him to explain how their product could be sold. It doesn't take much imagination to guess at how his own sales skyrocketed when he returned from the sales meeting.

At one time a smart department head at Macy's used this technique to motivate his sales force and move a product. Each Saturday morning he would select a bottle of slow-moving wine. Then at the weekly sales meeting, he would ask, "Who can make this wine move?"

Each time there would be more than one volunteer, and he would ask one or more of them to demonstrate how they sell the wine. The result would be that a week later that brand of wine had been sold out.

He had made them part of the solution. He had appealed to their pride, and he had recognized the fact that perhaps some of them had better ideas than his own. This same technique can often be used with customers who have problems that could be solved if they would give those problems some thought.

This problem area takes on so many different aspects that each solution must be searched out by studying the Sales Strategy Bank. Your actions will range from asking for the customer's help to making him part of the solution. It should go without saying that when he is right, the right action may be to admit our previous shortcomings and show him what has been done to remedy them. He is going to need proof and reassurance if the fault was that of your product or your company. If the fault was his own, he is going to need an escape hatch.

Recently, in handling a "case" at a Professional Sales Situation Management Institute, a young salesman pointed out that the fault (for which the customer was blaming the supplier) was the customer's own. This young salesman insisted the only course of action was the direct action... "He's going to have to be told, point blank, that it is his own fault, and we are not going to pay for the damage he caused."

This approach is not limited to "young salesmen." It is all too often used, when the indirect approach would be far more effective and less abrasive in its effect upon future business and customer relationships. The right answer to the "case" was to set up a situation in which the customer would find out, by himself, that he had been at fault. Hence, there would be no witnesses at that "horrible" moment when the customer would have to say those most difficult words that come to the tongue of man; i.e., "I was wrong."

Too often, we are not satisfied with having a customer learn that our alleged "poor performance" was his fault... we want him to realize that we know he has been at fault. When we ask him to pay *that* kind of price, we can be certain it will prove a greater barrier than any other "price" obstacle we could encounter.

Save his face and make the sale.

Service problems

Obstacles that arise from the customer's dissatisfaction with service are similar to those we have just discussed under "Past Performance." the same general considerations apply here also. The customer who is sold on your product but feels that your post-sales service capabilities give him difficulty is going to require proof, assurance and action.

Once again, we should look upon dissatisfaction with service as an opportunity to actually *strengthen* the bond between customer and salesman. This is not to say, of course, that we should *create* service problems so that we can then move in and play the part of the hero who gets the customer out of difficult situations. However, if the problem with service does exist, it provides the salesman with a real opportunity to demonstrate his personal interest in the customer.

Service problems sometimes originate with the salesman when he fails to understand the real nature of his job. The responsibility is not to obtain *one* order. It is to develop and maintain a favorable, lasting relationship with the customer so that subsequent orders become even easier to obtain.

"Peddlers" and "order takers" inherit service problems because their interest in the customer is not real—it is feigned. The professional salesman understands that he is in a profession. He will not create the type of service problem that occurs when the order taker grabs an order and runs off in pursuit of other customers.

More often, service problems originate because other members of the organization are not customer-oriented. This can happen most frequently in companies that are manufacturing-, finance- or engineering-oriented. Service problems are likely to arise when the man at the top, or men in other functions (engineering, manufacturing and finance), fail to realize that essentially their work, though vitally important, is really the means to an end—producing products that will bring in the revenue to keep the company solvent.

We have no ready-made solution for this type of problem. Asking sales and marketing people to "educate" men in other functions to the importance of the customer brings to mind the Aesop fable that raised the query, "Which mouse is going to tie the bell on the cat's neck?" It can only be hoped that as men with other responsibilities attend more seminars and post-graduate schools in which problems of general management and inter-functional relationships are studied, they will become aware of the need to be marketing conscious; i.e., *customer conscious*, regardless of function.

Service problems are caused by the isolation of other non-management personnel from the customer. This applies to men and women on assembly lines, in shipping rooms, in order-service offices and in every area of the company where mistakes and/or thoughtlessness can cause problems for the customer (and for *you*, the salesman). Speaking as one who worked for years in manufacturing, the author is all too aware of the unbelievable pressures that are placed on men and women in these other functions. It is understandable that the people subjected to these internal and constantly visible pressures will seek *first* to remove those pressures. This can result in production at the cost of quality. It can result in a host of other actions that all spell poor service from the customer's viewpoint. Isolated from the customer, these employees are not always aware of the ultimate cost of their attempts to "keep out of trouble" in the plant or stockroom.

The cure is essentially in the hands of management, but managers of salesmen and managers of marketing can do far more than they are doing to alleviate this situation and to develop a customer consciousness throughout the organization. Certainly, we can do more to bring salesman in direct contact with employees who often create service problems.

Most progressive companies see the absolute need for annual sales meetings. How many consider the necessity for at least annual meetings between salesman and non-selling employees? The content of such meetings is so obvious that it should not require discussion here.

How many companies ever bring the actual customer in to meet some of the employees? It is not unusual to have customers visit the organization for the purpose of seeing products made, but how often is dialogue between the customer and non-selling employees made possible?

This entire area is of such vital importance that it is hoped that managers of salesmen and marketing managers will do more than agree that "something should be done" and will see to it that something *gets done*.

Intangible benefits

Customers sometimes are reluctant to buy because the advantages of buying are too remote, or the customer does not see how the purchase will help her *personally!* This is very common with prospects encountered by insurance salesmen.

It is not uncommon to find, for example, a contractor who may not want to buy equipment at the top of the line. He feels that equipment costing less will get the job done... even though *his* customer may not be obtaining the best material available.

The same obstacle presents itself when a dealer who is being asked to buy in large quantities in order to obtain a better unit price tends to measure the potential increase in profit against the trouble of keeping inventories, moving stock insurance, danger of obsolescence, etc.

This is the self-defense mechanism of Praegnanz at work. Praegnanz is the normal human tendency to "See the bad as good." In other words, it is actually "bad" not to take advantage of the lower cost available through quantity purchase; but the customer will want to see it as "good," because by not buying he will avoid the possibility of the problems mentioned above. As a result of this psychological obstacle, what appears to you to be very tangible advantages will seem far less tangible to him. He will tend to see long-range advantages as "pie in the sky."

How can these obstacles be overcome?

The need here is to personalize the advantages as much as possible. It may be necessary to look beyond product or economic needs and alert the customer to the possibility of satisfying psychic or status needs! If you are really a professional, you have identified the "me he would like to project" (the image he wants others to have of him). In other words, you will know what adjective best describes the way he wants others to see him.

The next question, therefore, is: What psychic needs will this kind of man have uppermost in his mind? In other words, if he wants to appear important, he will probably have a desire to dominate his field—whether he is a small contractor who wants to look important in a small town, or whether he is a large manufacturer who wants to be highly regarded by other manufacturers.

The final question—the $64 one—is: How can I show him that following my proposal will satisfy these psychic needs he has? In other words, how can I create a setting in which he will see my proposal as essential to the satisfaction of his status ambitions?

In the foregoing paragraph, we have picked—only as an example—the chap who wants to look "important." It is your job to substitute for the word "important" the adjective that best describes the ambition of *your* customer; then, decide how you can present your proposal in a manner that will cause him to see that its adoption is essential to his unique status needs.

Another technique is to drive home hard the losses he will suffer if he does not buy your proposal. If he tends to procrastinate because he regards the advantages as "intangible," remember he is following a course of action that you and I often have followed when we were presented with a distasteful decision. We tend to "put it off" or "change the subject." So does he. It is your job to show him that the cost of procrastinating may be much higher than the cost of making a favorable decision. If the advantages of your proposal, though very real, do appear to him to be intangible, then you must accent the losses he will sustain by not buying and make *them* as tangible as possible.

He may sense your proposal has advantages but that they are clouded by other problems that your proposal will create. The solution is found in the Sales Strategy Bank; i.e., *ask for his advice and make him part of the solution (B19).*

For example, "Joe, you have a point there. Do you have any idea of how we could overcome that disadvantage so you can obtain these other great advantages that are available in this proposal for you?" The advantages of this strategy are many. First, it gets him talking and gives you time to think. Second, it poses a problem for him and a challenge. Third, he will, in all likelihood, be as capable as you are of solving the problem. Because the solution is *his* idea, he will work harder to make it succeed. Finally, it will cause him to feel that he is not being pressured... that *he* controls the situation.

Remote benefits

In some selling situations, the real advantages will not be obtained by the customer for some time to come. Such is the case when you are asking him to invest in the future. This becomes a real obstacle if your customer is a person who normally is inclined to think in terms of his immediate or short-range goals rather than of his long-range objectives.

The first antidote is to compress time. Make him see how fast time really passes. Make the future seem just around the corner—as it is.

For example: "Joe, I can see how you feel. Right now you would rather invest in things that pay off immediately. But Joe, time goes by quickly. It seems only yesterday when you installed (name product or service) and your foresight paid off—yet that was three years ago."

Or this: "The success of your business is a tribute to your foresight. I know it has not grown as it has without real concentration on long-range objectives. What we are talking about will pay off in less time than other long-range investments you've made, and you are going to be far out in front of your competitors who think only of the present."

The best way to make the future seem imminent is to focus the customer's attention on the past. The past always seems like yesterday, and it makes next year seem like tomorrow. This is one way to use the phenomenon of illusion. It is *not* manipulative because you are doing this in the real interests of the customer, as well as your own.

Hit hard on the customer's long-range objectives and make their realization seem important. Emphasize prestige (subtly), the image she will project, the reputation for foresight she will achieve.

Don't make the costly error of assuming because your customer may be "old" that he will not have long-range objectives. That is stereotyping of the worst sort. All "old" men do not feel the same way. It is also a good example of "projection"—thinking, "If *I* were that old I wouldn't be thinking about the long-range future, so *he* won't be thinking that way." He has no intention of passing from the scene if he can help it. He may be most anxious to create an image by which he will be remembered. He may be one of those wonderful men who keep young by thinking young.

At the same time, some older customers as they approach retirement, admittedly, will respond more slowly to appeals that are based on rewards available only in the future. In such instances, you have at least two courses of action open to you:

1. Dust off their past reputations for being foresighted.
2. Shine up the desire of most people to make decisions that will affect operations after they have retired from the scene.

Most of us, as we leave middle age, will be gratified if we find that our contributions have not been forgotten. Most of these people who are considered "old" are especially gratified to find that we still think of them as being alert to new ideas and new trends. They want their reputations to outlast them. Consider these points in planning your strategy and your tactics before entering the selling situation. Above all, avoid insincere flattery; the older man—because of his experience—can spot it more rapidly than others.

It is always impossible to over-emphasize the need for professional pre-sale preparation and planning. Don't wait until you are in the middle of the situation and then try to think of some example where the customer has profited from foresight or by seeing "remote benefits" and later cashing in on them. Gather information on the customer in advance. *Know* of instances where he has been foresighted. Even the most capricious of us has, at one time or another, been aware of long-range benefits and has acted on that knowledge. So has your customer in all likelihood. By selecting such examples, you follow the principle of "praise the work, not the worker." The customer does not regard your use of such examples as flattery. You may actually be doing him a favor by causing him to begin to think in terms of long-range objectives, not just in this instance, but as a regular habit.

Customer is confused and doesn't understand our solution

The next-to-last obstacle on our list that may

arise when you are attempting to prove that your product or service can satisfy a need concerns the customer who is confused by the multiplicity of proposals to which he has been subjected. This was covered in a previous chapter when it occurred as an obstacle at the time you try to convince the customer he has a need. Its treatment will not, therefore, be repeated here.

The final obstacle relates to the customer who doesn't understand your solution. You provide him with the same treatment covered in the same previous chapter under "confused." Briefly, some of the choices for action include: narrow down his choice; provide him with a solution; stimulate his recall of past experiences; or show her that cost of procrastination is higher than cost of making a decision.

CHAPTER 13

Gaining the customer's agreement

In the preceding chapters we have discussed the obstacles that can arise during those facets of the selling situation which we have identified as: (1) Pre-Sale Preparation, (2) Making Contact, (3) Gaining the Customer's Confidence, (4) Identifying or Establishing the Customer's Need, and (5) Establishing Satisfactions.

There still remains the task of getting the customer's agreement that your product or service *will* satisfy his need. You may have, in your opinion, thoroughly established that your product is just what he needs. But *you need to know* if he agrees that this proof has really been furnished. Obviously, if he is not yet satisfied, you are going to find it necessary to hit harder on the product features that will appeal to him or to accent losses which we wish to avoid. Therefore, a verbal commitment—when it is possible to obtain one—is your only real proof that he has been impressed by the arguments you have thus far advanced.

The tip-off

We have constantly emphasized the value of asking questions. It is just as important to listen to the questions the customer asks of you. Listen not only to his actual questions, but look for the *reason* behind them. His questions often provide you with the "tip-off" to worries or concerns he may entertain, and they may reveal his belief that although your product or service will fill his needs it may also threaten other needs he has. Here are some examples of tip-offs which are tucked away in questions customers ask.

The cue to needs for emotional security he may have is present in the question, "Where have you made an installation of this kind?"...or..."Who has used your service?" and similar questions. Here, the customer is probing to determine whether or not you are making a guinea pig of him; i.e., getting him to be first, or among the first, to buy a new method or product. He may be concerned that your design is to have him pioneer some heretofore untried application. If it flops, he doesn't want to be ridiculed by others in his company or in his industry for having been "taken in." His judgment is at stake, and he wants reassurance and proof.

This same type of question may indicate the customer is seeking assurance that you are technically qualified or that you really understand his problem. You may have successfully left the impression that you are qualified, but he may harbor some lingering doubts...he is not yet completely sold.

The kind of question represented in the preceding example may indicate the customer is looking for prestige by association, especially when he asks what other firms have followed your proposal. How does he identify himself with those firms? Are they firms that will impress him? Are they companies with which he likes to think he is identified, or are they organizations that will not satisfy his need for ego satisfaction?

The customer may on the other hand be looking for the prestige of being *first*. In this case, he's completely sold on your proposal and hopes to be the first to adopt it...but he will not want to reveal this to you because it is an open bid for prestige, and he does not want his motives to be transparent. Therefore, he simply asks, "Who else has used this?" He may hope your answer will be, "You're the first to have the opportunity."

In other words, two different customers may ask the same question with reasons that are directly the opposite. *You* must know the customer well enough to fathom the kind of person he really is, so that you will be able to understand his reason for the question. Obviously, it can be fatal to the sale to make a mistake here...so, once again, we see the need for knowing the customer as well as you know your product.

The same questions listed above may be the tip-off to the customer's desire or need for *economic security*. For example, he may be probing to determine if organizations of his size (be it large or small) have adopted your company's proposal. Have they felt it was a wise investment or something they could afford? He may not be totally sold on the claims you have made and may wish to check with other customers you have served.

At this point, with justification, you may ask, "It would seem that this simple question, 'Who has used this?' could be based on a number of fears which the customer has, and each is quite different from the others; how do you know which of these possible reasons is *this* customer's reason?" The answer is that you must have sought out a pattern in his behavior to identify the kind of person he is in this selling situation. He will usually ask a number of questions; you should seek a pattern in those questions and observe his reaction to your replies. If you find that a number of his queries have a common base, you can usually tell what is really bothering him.

All too often we are so busy talking or thinking about what we are going to say that we don't really listen to the customer. Usually, when she asks a question, we are intent upon providing her with satisfactory answers and we don't stop to wonder what prompted the query. We cannot afford to do this in a highly competitive market. Nor can we afford to forget that as long as she has questions, she has doubts—and we have not gained her full agreement that our product will provide complete satisfaction.

Sometimes a casual question voiced by a customer provides us with a clue to his need for recognition or assurance. For example, he may describe some action he has taken or some procedure he has installed. Then he may ask, "What do you think of it?" If this question is prompted by a need for recognition, don't make the mistake of giving him a perfunctory reply. Don't go to the other extreme and resort to out-and-out flattery. Select something about his idea or his procedure which you really like and comment meaningfully on it. You don't have to openly praise *him*. Your comment shows you have paid him respect by listening, and when you praise what he has accomplished rather than praise him, your words have the ring of sincerity.

Actually, you are lucky when the customer at this point in the sale does ask questions. When he does not, you can't be sure he is in agreement with you. It is then that you use Sales Strategy D9 from the Sales Strategy Bank—*Probe for reaction*. Use the *direct question (D15)* or the *rhetorical question (D17)*. For example, consider the rhetorical question: "I'm sure we agree on this point...don't we?" Here, you pause for only a fleeting moment to observe the customer's reaction to this request...and long enough to permit him to disagree. We gain nothing by denying him the opportunity to disagree. If he really disagrees, this disagreement isn't just going to "go away." It will be there, unspoken, as a barrier to a completed sale. Let him get it out in the open where you can deal with it.

The direct question also has value. For example: "Is there any point you would like me to cover more thoroughly?" In using the direct question, avoid any possibility of embarrassing the customer by asking her questions she cannot answer or does not want to answer. The successful completion of the sale depends upon how well you have done from your presale preparation to the point where you have tried to establish satisfactions. It is essential that you pause to determine whether or not the customer is in agreement with you. Obstacles get in the way here, and we should consider some of them.

Smoke screen objections

There are times when the customer really believes your proposition will answer his needs—but not only will he fail to say so, he will also offer objections. Often, these are smoke screen objections which he uses to camouflage his real reason for not agreeing openly.

In the Psychological Aspects of Selling (covered in Part One), we were reminded that you and I and the customer do not like our behavior (or the reasons for our actions) to be transparent. It is another way of saying this: "I do not want others to identify my status needs. If others can identify these needs, then they will also know the image I like to project, i.e., how I want others to regard me. In view of the fact that I have not fully attained the reputation I seek, I don't want others to recognize the needs I am pursuing."

This kind of thinking causes the customer to take actions that would appear irrational or stubborn if we did not understand the psychological basis of his actions, and if we did not realize that you and I follow the same patterns of behavior when it suits our convenience—or saves us from ridicule or embarrassment.

For example, a customer may object to the amount of chrome trim on a model displayed in an automobile showroom. Actually, he may like the trim, but he hesitates to buy the car and seeks some reason for his action which he feels the salesman cannot attack successfully. His real reason may be that he is afraid to buy the model because it is so luxurious or expensive that he fears his friends and neighbors will say he bought it for "prestige."

If the salesman takes the objection to chrome trim seriously and attempts to prove there is not too much, he will only be wasting time. The customer will have been successful in dragging a red herring across the psychological trail he does not want to leave. The chrome trim is a smoke screen objection used to hide the real reason for his resistance to buying. To make a sale, the salesman must identify the underlying causes for objections and overcome them.

The first action is to ignore what appears to be a smoke screen objection. If it his real reason, the customer will come back to it again and again until he obtains a satisfactory answer. On the other hand, don't be tempted to disregard his objections by assuming that they are "all smoke screens." Often, your own good judgment will tell you whether the objection being offered is really important when it is measured against the entire proposal you have made.

**Smoke screen objections often are difficult to

identify because they have a certain validity. In the example of "chrome trim," the car in question may really have so much that it could be offensive to many people. The point is that the customer knows this, too, so he can give this as his objection without having it appear to be an irrational objection. In fact, the more valid the question, the more difficult it is to detect.

One test for the validity of the smoke screen objection is to ask yourself if it is consistent with the overall pattern of behavior of this customer. Even this is not an infallible test. The best first reaction is to try to channel the discussion away from the feature to which she objects and to concentrate her attention on features and advantages which you feel certain will be important to her. If she persists in returning to her objection, you must assume that it really is important to her, and then address yourself to it.

Another test for the validity of his objection is to resort to strategy B19 (Ask for his opinion, advice or help). If he opposes a particular feature of your proposal, ask him what he would suggest to overcome it. If his objection is of the smoke screen variety, it will be his turn to evade—and you will find him dismissing the point as not being "*that* important." You may then make progress toward uncovering the real reason for his delay or resistance.

Worthy of consideration, also, is the technique of rephrasing and overstating his objection (D11); i.e., "You mean you prefer a car that has *no* chrome trim?" The customer in his attempt to deny that he ever made that sweeping an objection will restate it in a watered-down form, handing you a negative statement that is much easier to neutralize. If, on the other hand, he replies, "Well, if you put it that way—yes!" you can be fairly sure the objection is real.

You will also find that smoke screen objections can be converted into assets. For example, "Mr. Brown, we'd be happy to remove the trim if you feel it is excessive. This would give us a chance to show you we are interested in more than selling cars—we are interested in satisfying you." The customer will often back down in such situations with, "Oh, it isn't worth all *that* trouble...I just thought there was a little too much trim."

Still another technique is to use welcome objections (B10). For example, "I'm glad you raised the point because that was my first impression, too. But then it was pointed out to me that the chrome strip gives your door complete protection against those people who open their doors in parking lots and never worry about how much they scratch the doors on *your* car. You'll save money on touch-up jobs and your car will always look new because of those strips." You have thus provided your customer with an escape hatch by indicating you also had previously held the same view which he now expresses, and it was not because of your superior intelligence that you changed your mind but because "somebody else" pointed out the advantage to you.

Two more strategies to consider are to admit the validity of his view (B12) and/or to divert him (D10). You do not have to be in actual agreement with him, but you can use such phrases as, "I can see how a person would feel that way—but, of course, Mr. Brown, these other advantages you are going to obtain more than offset that feature. May I point out, for example..." (and go on to hit hard on important advantages to him). Not only does this evade the smoke screen objection but it gives you a new chance to accent advantages.

One other method exists for determining validity of the smoke screen objection: Meet it head-on. The customer may state, for example, "All that you say is fine, but you haven't made any provision for overload." Your reply might be, "You've got a point there, but we've purposely omitted this provision for two reasons. We want to keep the price low, and we can't anticipate any situation where this apparatus will be subjected to overload...do you have any condition in mind?" If he were using this charge as a smoke screen objection, he won't have "any condition in mind"—and he might dismiss the subject with, "Well, no, I suppose what you say is true."

In the foregoing, we have used the example of chrome trim on cars. You are not selling cars...and let's admit that "removing the objectionable trim" might present some problems. We chose this example simply because it is readily understandable to all of us. The job ahead is to examine the *principle* we have been discussing. Ask yourself, what kind of smoke screen objections may I encounter in *my* kind of selling...how can I anticipate them...how can I recognize them and convert them into assets?

High cost in psychic needs

This category is a catch-all for many objections. The customer may resist or stall because he sees needs—other than product needs—threatened. In other words, the psychic price of buying from you is too high.

For example, he may feel that the size of your company "dwarfs him" and threatens his self-image. This leads him to think that when he needs help or service in the future, your firm will treat him as just another unimportant "little fellow." Or he may dislike to do business in situations which he cannot dominate; if you are obviously better informed or more competent, he may resent this and attempt to avoid you. He may dislike doing business when all of his problems have been worked out by you or by your organization because this robs him of the need to be "necessary."

These are only a few of the basic reasons for the customer's failure to agree that your product is what he needs. The solution, as it is so often, is (1) do your homework in advance of the sale; (2) learn as much as you can about the customer as an individual, his status needs, how he wants to be seen, and how he likes to operate in a selling situation; (3) gear your sales presentation to these status needs as well as to product needs. Let him do most of the talking and ask for his opinion. Consider the advisability of leaving minor loopholes in your proposal so that he can fill them, thus making him feel important to the solution.

Won't buy your premise

Any edifice made to last must have a solid foundation. This is true of building lasting customer relationships. At least one of a number of premises must be established if you are going to gain the ultimate agreement of your customer that you and your product can fill his need or solve his problem.

Let's examine some premises upon which we must, at times, base our appeal to the customer:

1. If the customer has had a poor experience with our company, we must base our presentation on the premise that we have "changed our ways" and that we will prove to be satisfactory if he will buy from us at this time.

2. We can try to establish the premise that it will pay the customer to anticipate her future needs and buy in quantity now.

3. We can try to convince him that it will pay to buy in large quantities, even if it means pooling his purchases with others.

4. We can maintain that this is the wise time to buy because prices are going to soar.

5. We can try to get his agreement on the premise that his territory is expanding and that he should increase his capacity now in order to meet the rising demands of an impending population growth.

6. We can base our claims for a newly established line of products or service on the premise that because our company's performance has been excellent in its traditional field it will be equally meritorious in this new field of endeavor.

To make this apply to your kind of selling, stop now, for a moment, and consider the many types of premise upon which you find it necessary, from time to time, to build your sales presentation. Sometimes you do not state the premise in so many words, but it is implicit in the claims you must make or in the credibility you seem to assume the customer will concede is rightfully yours.

The customer is not always ready to concede that your premise is valid. He accepts only so much on faith and on your previously demonstrated performance. If he is a new customer, he may require solid proof before he agrees with your premise. If he has had difficulty with your firm in the past, this will be especially true.

You have seen customers who have been assured by a salesman that prices were now stabilized on some product. On the basis of such an assurance (or this premise), the customer made a purchase. Then, a short time later, prices dropped. Just how much credence will the customer place in the premise that the salesman now seeks to establish that today's quoted price will remain in effect for some time to come? This customer may be completely sold on the product, but until he is convinced that the premise is valid he is going to postpone buying. And postponements can be fatal because they create a vacuum which competitors are only too happy to fill.

What is the antidote? Try to look upon the sale from the customer's frame of reference *before you enter the selling situation.*

What premise is the customer being asked to accept as valid?

How would you regard that premise if you really were the customer? What further substantiation or proof would you require?

Would you find yourself attacking the premise? If so, on what grounds?

Now ask yourself, if you really were the customer—with his views, his suspicions and his feelings—what appeals to you could neutralize adverse attitudes and gain your agreement?

Then, on the basis of your answers to those questions, plan your strategy and tactics. Determine what incontrovertible proof you are in a position to supply the customer. Decide, specifically, what words you will use, and under what conditions you will inject these arguments into the selling situation.

Finally, forget false pride and try out your strategy and tactics on someone else in your organization—perhaps another salesman or even the sales manager (after all, that is what he is there for—to provide you with advice). Explain the facts of the situation to him and tell him your plans and what you are going to say to the customer. Then ask him how *he* would react to the tactics you propose to adopt if he were the customer.

It is much better to find out in advance of the selling situation the weakness of your tactics than to have them fail in the presence of the customer. If your manager or fellow salesmen can fault your approach, you always have a chance to revise it—which is more than you may be able to hope for in the actual selling situation.

Wants to avoid obligation

Strangely enough, on occasion, some customers will withhold their agreement that your proposal meets their need, even when they are convinced that it does. They may feel you are offering them so unusually attractive a proposition that you will have a claim on them for future business if they accept the help represented by your present offer. Rarely will they say this in so many words; they will usually resort to smoke screen objections.

How do you overcome *this* obstacle?

When you are convinced that such is the reason for the customer's reluctance to buy, emphasize unequivocally that this purchase puts him under no obligation. Drive home the fact that each offer you make him stands on its own feet. Tell him you have confidence in your products or service, so that you are quite willing to have him consider each of your selling efforts on its own merit. (Actually, this technique is similar to the one used by a lawyer where information elicited from a witness is objected to successfully by the opposing lawyer, with the jury being told by the court to disregard it. Just as the jury cannot truly disregard anything it has heard or witnessed, neither can the customer forget courtesies you have extended to him or "good deals" that he has capitalized on through your efforts. At the same time, it does help to relieve him of the feeling that he is placing himself under an obligation.)

Another course of action is to neutralize the customer's fear of being obligated by asking him to render you some small favor, so that he can feel "the books are balanced." *Let him do you a favor (D20).*

Has other obligations now

This obstacle can crop up during any facet of the sale, including the Gain Agreement phase. It may even appear so early in the selling situation that it becomes difficult to establish contact with the customer.

For instance, the customer may feel he is under an obligation to one of your competitors. He will see no point in even discussing his needs or problems, or of permitting you to discuss your proposal with him. You may have encountered the customer who still feels a lasting obligation to the competitor who supplied him through lean years. Another may feel indebted to a supplier who helped him when he first started in business, or when he was in trouble with production problems or needed credit. Any of these and similar situations can create a sense of obligation that can spell trouble for you.

What strategies can be employed?

Have the customer *explore the cost of a wrong decision (E4)* in the Sales Strategy Bank. If possible, identify obligations he has to himself, to his family, to his company, to stockholders, or to others who have invested in his business. In other words, change the setting so that he sees the obligation to your competitor as insignificant when compared to other obligations he has, which can be met by buying from you. Don't push this idea by the lecture method, but try to emphasize the concrete advantages he will gain from buying your product. Compare these gains to the losses he—and others to whom he is really obligated—will suffer if he buys from the competitor to whom he has felt indebted.

On other occasions, compliment him for his loyalty and fidelity to the firm to which he feels an obligation. Then demonstrate to him—with facts and figures—how he has long since written off this obligation. Show him what it is costing him to continue to meet this "obligation."

Avoid the hazard of causing him to lose face in presenting this proof. If time permits, and there is no real danger in leaving the selling situation, be subtle and leave the facts and figures with him so that he can discover for himself—*and by himself*—what his continued loyalty to the competitor is costing him.

If you feel it is hazardous to leave the scene of the selling situation, try to find an escape hatch for the customer. Do this so that when he discovers how much he has been losing by patronizing the competitor (out of gratitude for past favors), he will not run the risk of appearing foolish in your eyes.

Influence of conditioners

In common with all men, your customer's concepts of what is legal, ethical, moral or socially acceptable are an amalgam of his many environments (business, social, family, military, educational, etc.). In a word, they are the result of his *milieu*—the total effect of the interaction of each of these many environments upon one another.

Many of his concepts are the result of his interpretation of rules of behavior which have been passed on to him. When he tries to transfer these principles or beliefs or standards from one field of endeavor to another, he can fault. And some of his concepts may be in error because of his ignorance...of law, for example.

If he envisions that to agree with your proposal and the premises upon which your proposal is based —or the conditions you have laid down—will cause him to violate one of the codes he has established, the progress of the sale will slide to a dead stop.

What can you do?

First, discern the reason for his resistance. Then determine whether or not he is correct or in error. If he *is* right, you should not attempt to sell him or change the conditions under which you are asking him to buy, so there will be no need for him to violate his code.

If "ethics" is the stumbling block, remember that in a buyer-seller relationship two people often see the situation from vantage points that are diametrically opposed. What is thoroughly ethical to the seller may appear unethical to the prospective buyer.

A seller, for instance, may see nothing unethical in suggesting that the buyer "change her mind" when the buyer finds she can obtain a better product or a better price. The buyer, to the contrary, may feel that once a commitment to another supplier is made she is morally bound to honor it regardless of cost.

As the seller, we may easily rationalize this situation by feeling that if the competitor was ethical he would not take advantage of this customer and would either counter our offer with a better product or a better price or release the potential customer from his implied commitment. This is a dangerous line of argument to follow with the customer who feels his word is involved.

Some sales will be lost because of your reluctance to urge the customer to take actions which you —or he—will regard as unethical. In other instances, the customer may not know about certain laws or regulations, and because of this ignorance will balk at taking actions which you urge upon him. In such cases, proceed with care and obtain the advice of your legal counsel. Ask him to recommend methods you can employ to "educate" your customer and to show him the actions you recommend are, in fact, legal and ethical.

Making a study of your customer will usually enable you to predict the effect of his past conditioning on his present frame of mind and on the actions he will undertake or refuse. If you can predict an unfavorable attitude, you are in position to plan in advance how to neutralize the effect of his past conditioning. It avails you nothing to become irritated at the customer's resistance. You make progress only when you isolate its cause and choose strategies that will show the customer that the action you recommend is legal, moral, ethical and in his interest.

Influence of experiences

The influence of the customer's past experiences is a factor that must be recognized in our pre-sale preparation. If we are engaged in the type of selling that provides little or no opportunity for pre-sale preparation or if we are on a "cold call," we will nevertheless find that past experiences affect the customer's attitude. When his attitude is negative, we must find ways of neutralizing the effect of those experiences which create problems for us.

A sale does not really start when we first make contact with a customer or even when he first discovers that he has a need and calls on us. The roots of the problems we may encounter—hence, the beginning of the sale—may often be found in experiences he has had weeks, months or even years ago. These experiences have been factored into the frame of reference from which he will view our proposals.

In the first portion of this book, we discussed the psychological aspects of selling in some depth. You will find in that portion, Chapter 4 dealing with the influences on recall. These influences affect the manner in which a customer will remember an experience—often in a manner that contrasts sharply with our own recollection of the event. We also discussed the kinds of experiences a customer will remember— often long after we have forgotten an event, and sometimes because we *wanted* to forget it. We learned why he will sometimes have forgotten an experience because he wanted to forget it, even though it would be in our interests now to have him remember it. This material and an understanding of the influences on recall have such a vital bearing on the course of many sales that it will be of great value to you to reread that chapter at this time.

Naturally, the best course of action when possible is to *anticipate* the kind of experiences the customer will recall and to estimate how they will affect his attitude in the forthcoming selling situation. This means doing your homework in advance of the sale. Admittedly, this is work but it adds a dimension of professionalism to your job. It also adds excitement to the work of selling, and a real sense of achievement is created when you find that your pre-sale estimates have hit the mark. The order-taker refuses to look upon selling as a profession that requires advance planning. He would rather operate by "feel," by hunch and by intuition. The professional salesman knows that selling *is* a profession—one that shares with all true professions the need for diagnosis in advance.

Learn as much as you can about previous experiences the customer has had with you, with your competitors...and possibly with other elements of your company. Don't wear rose-colored glasses in recalling his past encounters with your company—or with you. Don't regard these experiences as you would like to think of them. Try to see them from *his* viewpoint. Really, now, if you were truly the customer (not you in his shoes, but the customer—with his objectives, his goals and his milieu), how would you regard the company represented and this chap who is trying to sell you a product or service?

In estimating the effect of her experiences with your competitor, don't wear dark glasses. Avoid evaluating those experiences as you would *like* to think she will evaluate them. Keep in mind that if she has had a bad experience—or even a series of such encounters—with your competitor, there is always the chance that she will regard them as "not so bad." You see, no one made her buy from your competitor. Her decision to buy represented *her* judgment. To admit that this competitor has let her down requires her to admit that her own judgment was bad. She may not be "big" enough to do this. Your job will often be to provide an escape hatch for your customer through which she can see that her experience, bad though it has been, was not a result of her own judgment.

In other words, try to interpret his past experiences for him in a way that will provide a positive attitude. Then, consider the applicability of the following courses of action from the Sales Strategy Bank:

D13—Appeal to his sense of fair play.

B13—Find a point of agreement; then *D6—Hit hard on benefits.*

E12—Provide guarantees. This will be most helpful when the customer has had poor experience with your product or other service you provide, or when you have introduced new models about which he has some reservations. It will also be helpful when he has had trouble with your competitor's product and wonders if he will have the same trouble with yours. Most salesmen don't use this strategy enough. If you really have confidence in your company and the product you sell, there is no reason why it cannot be used.

E2—Provide assurance or proof. Use third party examples, testimonials; or *E3—Demonstrate.*

E11—Stimulate his recall of past experiences which reflect well upon your product. A customer will often resist because of one bad experience, possibly because the timing made it intense; or, he will remember such an experience because it is so recent. Conversely, he will tend to forget past good experiences. Try to get him to see this single bad experience in the setting of all of the past favorable ones he has had with you.

B14—There are times when the best course of action is to admit your error. This often takes the steam out of an irate customer. If you use this strategy, quickly follow the admission with *E2—Provide assurance and proof.*

B9—Emphasize mutual interests. Point out that it is as much in your interests to provide him with satisfaction as it is in his. You are interested in more than *this* sale... you want his continuing business. You are as much disturbed about the remembered bad experience as he is. (The seasoned salesman knows that it is easier when all of your experiences with the customer have been good. He also knows that a bad experience—if handled properly—represents a great opportunity to demonstrate your sincere interest in the customer and in his welfare.)

Don't be satisfied with just the strategies we have listed. All we have done is to try to save you time by scrutinizing the Sales Strategy Bank and extracting the ones that, to us, promise the best courses of action. In examining the Bank for yourself, you may identify other strategies that meet the unique requirements of this specific selling situation better than any we have proposed. Furthermore, examination of the Bank may remind you of tactics you have used successfully in the past.

Hearsay

Hearsay was discussed in an earlier chapter dealing with the obstacles that can and do occur at the point in the sale when we are trying to gain the customer's confidence in us as qualified salesmen. It can occur again when you are attempting to gain his agreement that your product will satisfy his needs.

Essentially, hearsay, when it provides you with a selling problem, is a *negative testimonial*. Hearsay would be ruled out in a court, but unfortunately the judge in this case is the customer—and he sets the rules.

One antidote: provide the customer with other testimonials from people whose creditability is even greater in his eyes than the opinions of the people whose remarks or experiences have influenced his attitude up to this point. Consider using the "appeal to fairness." If he indicates dissatisfaction, ask for "examples"—but be careful to do so in a manner that does not antagonize him or seem to cast aspersions on his honesty.

Hearsay can be a powerful enemy because it stems from one of 16 defense mechanisms. When we see or hear something that represented a bad experience to someone with whom we identify ourselves in some manner, we feel as though it had happened—*or could happen*—to us. This is epitomized in the thought, "If they did it to him, they would do it to me if I gave them a chance."

Another antidote is to use strategy *(D7) Draw comparisons*. Try to show the customer that he is completely unlike the other individual (if this seems appropriate). Help him identify himself with someone who thinks highly of your company and product. In doing this, use introjection as a selling tool rather than have it used against you as a selling obstacle.

Hearsay may well be based on fact...perhaps your customer's informant really did have trouble with your product or service. If so, you will need to provide assurance that the same trouble will not be experienced by your customer. As stated earlier, you might profitably use the "guarantee" strategy. If you use assurance, alone, you should be able to support it with evidence. If the hearsay is based on erroneous information, you must supply proof—and in this manner impeach the source of hearsay.

Demands a concession

In some selling situations, your customer may actually be in agreement that you can satisfy his need but will not admit it because he is seeking a concession of some kind from you. Obviously, the number and types of such concessions are so large and varied that it would be impossible to give them coverage here. Each would be unique.

If the customer has already indicated the nature of the concession he seeks, the problem is easier to handle. It is obvious, or at least probable, that he is sold on your proposal but feels his bargaining position will be weakened if he admits it. In some situations, the customer may be wondering just how he is going to ask you for a concession. And, as you may be unaware that he wants some kind of concession, you will be puzzled as to why he is so reluctant to admit your proposal will provide him with advantages that have been established beyond doubt. To make this situation clear, here is an example from a field with which we are all familiar.

Mr. Elkhorn is shopping for a home, and has located one that meets all of his requirements except for price. Barnes, the present owner, is asking more than Elkhorn (who really wants the house) is willing to pay. He is wondering just how to go about the business of gaining a concession (lower price). This may be no problem to many customers who would quite simply offer a lower amount. Elkhorn is not that kind of person. He doesn't want to create the image of being "cheap" or of being "low on funds." He knows that comparable homes in the same area *are* being sold for the asked-for amount. Therefore, he feels he will encounter resistance from Barnes if he asks the latter to reduce the price.

Elkhorn would prefer an outright reduction in price, but he is wondering if there is a less difficult way of getting more for his money. For example, can he ask the owner to install wall-to-wall carpeting, or can he ask that the house be painted or that appliances be installed? Obtaining any of these concessions would be equal to a reduction in price because Elkhorn is going to have to spend more money on them later.

For this reason, Elkhorn will not admit to Barnes that the house meets his needs. He will certainly not show any enthusiasm. He may even inject smoke screen objections; i.e., find fault with things that do not really bother him significantly. He senses that if the owner or agent suspects he is sold on the house, his bargaining position will be weakened and they will be reluctant to make any concessions.

When the salesman (in this case the owner or his agent) finds that Elkhorn does not express approval —"agreement" that the house meets his needs—he may wonder what has been wrong with the presentation up to this point. The salesman may decide to "back-track" and re-emphasize some features harder. This can be dangerous because there is always the chance the customer may hear something he missed on the original sales presentation and be diverted from his original intention. As a result, the sale may really be jeopardized. It has another drawback: in that it may cause the customer to feel the seller is worried about making the sale, and he may then ask for concessions even greater than those he had in mind originally. If the salesman does not read the signs correctly, he can make the fatal error of assuming the customer is merely a "shopper" who has no intention of buying.

What to do? There is nothing more effective at this point than *asking for the order*. If the customer has been hesitating—not knowing how to go about asking for concessions—this will smoke them out in most cases. If asking for a concession is not in his mind, then he will probably raise the objections he really has to the offer.

If, instead of getting the order, you find yourself faced with new problems to solve, nothing has been lost. The "new" problems are not really new...they were there anyway, and you couldn't deal successfully with them until they were revealed. You thus answer his objections, try again to obtain his agreement, and ask for the order once again.

But, you ask, what if the concession he seeks— and bluntly asks for—does present you with a problem; what then? The best reply is the same message we have continuously stressed...pre-sale preparation. Don't wait until you are immersed in a tricky sales situation to plan a solution. Sure, you can't always predict what problems the customer will raise, what concessions he will seek... but you can *anticipate* what they may be. For example, we can assume that some customers, at one time or another, will ask for such things as: price concessions, speedup in normal delivery time, engineering assistance, free installation, help on installing displays, advertising, special packaging, and a host of others.

These are but a few of the forms that "concession" takes: prepare in advance and know what your company's posture will be on such requests in the case of this customer. *You should have alternatives ready,* and know just how far you can go in granting concessions. Furthermore, you should be equipped with the necessary approvals from your manager so that there will not be a need to "check back with the office." Checking back—unless you can do it by phone on the spot—leaves a gap in the line wide enough to drive a truck through, and alert, high-powered competitors don't hesitate to move in and score when you leave an opportunity. If our metaphors and analogies are scrambled, our intention is not, and the message should be clear.

Fears

We received a criticism that our Anatomy of a Sale chart, in which possible objections are listed, did not denote "Fear" as an obstacle by itself. The reason is that fear cannot be classified as a single obstacle. It takes many forms, and they must be handled in different ways.

For example, we do not soothe the following customer anxieties in the same manner: fear of looking foolish (being the first to pioneer the use of your new product); fear of being criticized by his superiors; fear of losing money; fear that price will drop after he buys; fear of obsolescence; fear of service delays; fear that he may be buying something he won't need enough to justify the expense. Obviously, each of these anxieties would require different treatment, and they will be covered at various points in this book. The point is that it is the worst kind of generalization and simplification to regard fear as a single obstacle. This could lead to the erroneous conclusion that there is one (or possibly a few) best ways to handle that one obstacle.

In planning for a sale, consider the kinds of fears that *this* customer may entertain. Decide in advance —when the situation is not charged with emotion— what your best course of action will be. One of the best ways to allay fear is to include proof and assurance in your presentation so that fear never raises its head. As an obstacle to gaining your customer's agreement that your proposal meets his need, fears show up in these forms (among others):

1. He is "afraid" to agree because this puts him on record, and he may have to get approval from "higher up." If he agrees with you and then fails to obtain approval from higher authority, he has revealed to you that he is really not the key buying influence. Thus, he "loses face" (which, in this case, means he has not been able to project the image he wants you to see when you see him; i.e., *the* decision-maker). If, on the other hand, he has not agreed with you openly, and later fails to obtain approval, he can always claim that he, personally, was not sold and the decision to abstain from buying was his.

2. She is "afraid" to agree because she doesn't want to buy at this time. If she agrees with your premise that your proposal meets her need, it makes it more difficult for her when she ultimately must say "No." (This, by the way, is exactly why you should seek agreement; it does make it more difficult for her to say "No.")

3. He may be half-decided to buy from your competitor or he may want to shop around among other suppliers. Once again, if he agrees with you, it makes it more difficult for him to stall you until he has shopped around. He is "afraid" that agreement strengthens your hand and lets you control the selling situation. Once again this is just the situation you want. It is obvious why obtaining a statement of agreement from the customer, at this point in the sale, is highly important.

Some general techniques for handling several kinds of fear

Your attitude is important; make it one of confidence without showing a trace of concern or doubt. Couple the right attitude with *statements of assumption.* If you cannot get the customer to say, "This product will do the job," or "We could certainly use this," there is still a course of action open to you. Make it clear you assume he agrees that your proposal represents the answer to his problem. For example, "I know, Mr. Calhoun, that because of your experience with this problem you can see, readily, that this product (or plan or service) will meet your needs."

He won't reply, "No, it won't meet my needs" because he has seen already that it will do the job. He knows that a negative reaction on his part will simply provoke a challenge from you, such as: "Where do you feel it would fail to meet your need, Mr. Calhoun?" He would then be caught without an answer and be embarrassed.

Let's agree that he might reply, "Well, it is one solution." (This implies that there may be many other solutions available elsewhere.) Even this reply puts you ahead because it moves you into the next facet of the selling situation; i.e., "Gaining preference for your product above all others." It means you can now concentrate your fire on proving your product is best. His statement also lets you know that you have convinced him your product *does* meet his need.

Summary

1. Obtaining an expression of agreement from the customer that your proposal offers satisfaction of his needs is important because it serves as a check of the progress you have made, thus far, in the selling situation.

2. It lets you know whether you have explained features and advantages sufficiently.

3. It lets you know whether you have captured *and held* her attention and interest.

4. It makes it more difficult for the customer to say "No" later on in the sale.

5. It makes it more difficult for him to procrastinate when you ultimately ask for the order.

6. If the customer withholds his agreement, it aids you in narrowing down the nature of his resistance. Clues to this resistance are found in the words he uses when he evades your attempts to obtain a statement of agreement from him.

7. When statements of agreement are not forthcoming, they can be drawn from him by use of the direct question, the rhetorical question (*assuming* agreement) and by blunt statements of assumption.

If any selling tool can accomplish this much for the professional salesman, it should be used as the effective psychological advantage which it represents.

CHAPTER 14

Gaining the customer's preference

In Part Two we have presented the Anatomy of a Sale and listed 11 facets that *can* appear during the course of a sale. The seventh of those facets is that point in time where it may be necessary to "Gain Preference" for your product or service. This subject will now be the focus of our attention.

All 11 facets do not appear during the course of *every* sale. Gaining preference does not always create a problem. There are times, however—after you have identified the customer's needs and gained his agreement that your product or service can satisfy his need—when he will indicate that your competitors can also satisfy that need. It is at this point that you must convince him that you can offer greater satisfaction than is available to him from other sources.

Admittedly, there are instances where you have no competition or where your competition falls so far below your standards that it offers no problem. This, however, may be a temporary advantage, and it will prove beneficial over the long run to know how to gain preference when the need arises. The marketplace is becoming increasingly complex and dynamic, with rapid changes occurring both at home and abroad. As more companies in the interests of diversification embrace markets new to them, and as international competition takes on greater—and in some instances more ominous—significance, the problem of gaining preference will be a stranger to few professional salesmen.

Factors that create the problem

1. Other companies move into new fields, offering you competition in lines that may have been relatively free of it in the past.

2. Competitors who in the past were too small to offer serious problems may have grown rapidly or merged. They may now possess impressive images in the minds of our customers.

3. The rapid rise to economic strength of many nations, has introduced factors that are a force to be reckoned with.

4. The General Agreement on Trades and Tariffs and the forming of economic trading units among nations have, in many instances, drastically changed the nature of competition.

5. The rising cost of labor, always a factor, has become increasingly so, and often presents the overseas competitor with an advantage that makes it difficult to obtain preference for your product or service.

6. Programs of governmental assistance to small businessmen and to some defense contractors have provided support to many companies that might otherwise have confined themselves to more restricted fields. In some instances, because of work on federal projects, companies have gained the advantage of federally subsidized research. Often they have been able to transfer this newly acquired knowledge into the provision of service or products almost totally alien to the objectives of the original research.

7. The spectacular growth in discretionary spending power of the consumer has acted as a lodestone to attract new firms into fields which you may have occupied heretofore with a few "historic" competitors.

8. As business grows more complex, an increasing number of firms have utilized consulting firms in engineering and marketing. Old ties, old loyalties and old buying patterns have often been among the first casualties.

These are but a few of the influences that have caused the need to obtain preference to take on serious proportions. The fact that you may have a good product or service, backed by an excellent company image, continues to be an important asset. It must now contend with good products and services produced by firms that may have even greater resources, or by those who because of their youth have *drive* and are *hungry*. Still others, who, because they are small, have the advantage of *flexibility*. These last two types, in the eyes of your customer, are important because they promise all-out effort and service-plus probabilities. They also provide you with two sizable obstacles to surmount.

It would appear that time spent considering ways and means of gaining preference may be of immediate value in addition to being a wise investment in the future. With this in mind, we will turn our attention in the next few sections to the problems that may present themselves when you attempt to convince your customer that she should prefer you above all others.

Customer has obligations

Some attention was given to this obstacle, but more needs to be said. You may encounter customers who meet you in the selling situation with these thoughts—sometimes spoken; sometimes implied; sometimes, unfortunately, held but unspoken:

"They supplied us during a shortage" (or during a strike period in the suppliers' industry).

"They helped us during *our* strike."

"They gave us help (engineering, service, manufacturing aid, orders, financial help, etc.) when we were in difficulty (or when we were small and just getting started)."

"Your competitor recommended me to my present employer" (unspoken).

"They buy from us" (almost *always* unspoken).

"We think it's fair to spread our business."

"We have an obligation to buy locally."

"These people developed this—they worked on the first prototype. It's only fair that they get the first orders. Also, they have more experience."

"I have an obligation to my company to get additional bids."

"We have an obligation to help the 'little fellow.'"

"We ought to buy internally from one of our own departments."

"There are political considerations involved in placing this business" (incidentally, almost universally unspoken).

Do any of these obligations sound familiar to you? If not, do you suspect that perhaps one or another of them had something to do with a past failure to obtain preference when you *know* you had everything else going for you?

In addition to what was said earlier, it is to be emphasized that all of these objections do not stem from a feeling of indebtedness or a compulsion to be "fair." Some of them stem from fear of reprisal or from a pragmatic point of view. For this reason, it becomes evident that there is no one universal method of handling all of these objections. *Objections cannot be successfully handled unless the cause has been identified and neutralized.* Furthermore, the customer will not always give expression to his reasons for denying you preference. Often, you must smoke out the cause. Consider the solutions proposed earlier and add these to your list. Then select that solution which you feel is most appropriate to this unique customer of yours. For example:

Help her see she has already paid off the obligation.

Make point-by-point comparisons with a "price ticket" attached to each point so he will see what he is actually paying to meet his "obligation."

Subtly point out the obligations he has to others (to himself, his stockholders, etc.) which cause his original obligation to appear less important, or even short-sighted. In using this strategy, do not put him in a position where he loses face.

Again, subtly, lead him to the point where he will ask himself the question, "After all what have they done for me lately?"

Highlight the cost of a wrong decision.

Highlight the cost of further delay in making a decision. Often, torn between his obligations and the advantages you offer, he will do as we all do—i.e., procrastinate, hoping a happy solution will present itself. This can be fatal if you leave the scene and your competitor to whom your customer feels he has an obligation appears. The emotional impact of your competitor's presence will cause the obligation to appear greater, and he will make an emotional decision, denying your preference.

Keep in mind the psychic needs that are always lurking in a selling situation. In other words, build part of your presentation around what you know to be his primary *personal* objective. If you know, for instance, that he has a keen desire to appear progressive, then accent those features of your offer that will make him feel that buying from you will help him project that image; whereas, buying from your competitor will either fail to enhance his image or will threaten it. *Note:* We have selected this image (progressive) solely as an example. *You* must decide what image it is that this unique customer is trying to project. And you must have him see that giving preference to you is going to make him appear—to others as well as to himself—in the manner he considers most desirable. When all else fails this may be your most powerful strategy.

Customer's psychic needs satisfied by competition

It has been thoroughly established in previous chapters that the satisfaction of psychic needs is equally as important as the satisfaction of product, service or economic needs. Indeed, psychic needs are often paramount. In your role as a customer, you know this to be indisputably so. There are individuals, stores or other concerns from whom you *prefer* to buy. Often—indeed, more often than not—this is not because their product or service offerings are superior. It is often because of the manner in which you are treated, the respect shown you, the "way they make you feel"—and perhaps, above all, the suspicion that dealing with this chap or this firm helps you project the "me you wish people would see" when they see you.

You have probably heard a salesman say, "You'll never sell that fellow—he's been doing business with Lockhardt for *years*. He even has their calendar on his wall, and they're real buddy-buddy." Another not uncommon comment heard is, "No one but Patrucelli will ever sell to that outfit. He and Smitty, their purchasing agent, are golfing buddies." These and countless similar expressions are often heard and are given as reasons for failure to gain preference.

This is not to say that these reasons lack validity. Such close relationships *do* present obstacles to gaining preference. However, they do not constitute insurmountable barriers to the successful closing of a sale. The proof of this may be that you have enjoyed similar relationships with customers over a period of years only to find that one of them suddenly began to place some of his business elsewhere... perhaps all of it. Even the best of buyer-seller relationships prove tenuous under the impact of stronger stimuli presented by an aggressive competitor.

The problem appears to be two-fold:

1. What is the basis for the close relationship enjoyed by your competitor? Is it product- or service-oriented? Is it due only to the fact that over the years your competitor has provided unusually good service or that his products have rendered excellent performance? If so, your customer gives him preference simply because a shift in allegiance to you will present him with possible trouble. At the very least, it means he must change a habit, which none of us does capriciously if present habits bring us satisfaction and reduce the mentally fatiguing work of being called upon to make decisions.

2. Is the reason for the preference given your competitor found in the possible fact that the customer derives psychic satisfactions from doing business with him? Does he make him feel important? How? Does your competitor enhance his prestige? How? Do you really know what self-image your customer is trying to project? If so, is your competitor successful in causing your customer to feel that buying from him is the best method of looking "good"?

No one can do *this* job for you. You, and you alone, must identify the kind of satisfaction your competitor is supplying the customer. Then you must find ways and means of topping him in spades ... or of showing him what other psychic, economic or product needs you can supply that cause the satisfactions obtained elsewhere to diminish in importance. Develop your own list of questions to supplement the admittedly meagre list that follows:

Does your competitor enhance the prestige of the customer by taking him to clubs, restaurants or other points of social importance that the customer might otherwise not be able to visit?

Does your competitor cause your customer to look good in the eyes of his boss or of business associates? Does she do this by making certain the customer gets credit for some of the accomplishments achieved as a result of ideas or methods actually supplied by the competitor?

In other words, what satisfaction does your customer obtain from doing business with this competitor? How can you offer him greater satisfaction? How can you make him feel important (*if that is how he wants to be seen*)? How can you make him feel necessary or gain recognition for him? What *is* it that he wants in the psychic field and how can you surpass your competitor in supplying it?

Shortly we will supply selected strategies for your consideration, but it is imperative that we sound a note of warning at this point. We are discussing how to gain preference when your competitor seems to be solidly entrenched with the customer you seek. How about considering for a moment what hazards *you* must avoid if you are to keep the customer who now prefers you?

Remember that your customer started out with what we might call a "me he wanted to project" (how he wants others to see him). You probably aided him in projecting that image. There is always the hazard that we will continue to think of him *as he used to be*. We all know that no one is the same man he was 10 years ago. We have more experience, we have achieved greater maturity and we have gained wisdom. We often forget that our customer has grown wiser, too. We often forget that he has raised *his* threshhold of satisfaction. What appealed to him even a few short years ago may not appeal to him now. He has a whole new image he wants to project.

The tendency to forget that others change—and grow—is common. It manifests itself in strange ways. For example, we have a reunion with old buddies and come home and tell the wife, "You should see the way those fellows have let themselves go... they all looked so old!" (Everybody but us.)

In other words, we are quick to see signs of deterioration in the other fellow, but we are not quite so quick to see signs of greater wisdom. This is especially true when we are close to the customer. His growth is constant, but it so slow that we do not observe it taking place.

As a result, we often treat the customer as we did when we first knew him. When we joke with him, we tend to talk to the *me he used to be*. We "kid-down." Example: We meet him and, knowing he has been on a golfing vacation, we greet him with, "Did you break 150 this time?" Or we make good-natured fun of him in a hundred different ways. If called to task by an observing friend, we reply with, "Why, I've known Joe 10 years... we always joke that way... he understands." Yes, he understands, in a subconscious way, that you still see him as "Good old Joe!"

Your hungry competitor didn't know the customer *when*. Your competitor knows him *now*. Your competitor didn't know him when he was trying to reach that first level of satisfaction—which he has long since passed. The competitor sees what level of satisfaction—what image—your customer is seeking now and he proceeds to gratify and feed it. Then we are heard to muse, "I wonder why Joe's orders have fallen off a bit lately?" If you *must* "kid," kid-up to the new me-he-wants-to-project. It's just as much fun and far more rewarding. Don't take him for granted.

If you've been thinking as you read these last paragraphs, it may cause you to reappraise some of your attitudes toward customers who now give you preference but who won't tomorrow... when your eager competitor comes along with his fresh appraisal of your customer as he is today. At the same time, reading this may have shown you how you can capture the affections—i.e., the preference—of the customer who up until now has been solidly in the grasp of your competitor. Because this failing to see how individuals grow is so widespread, it is a fairly safe bet that your competitor may have left himself vulnerable in this area.

Some other action considerations

Your customer may have been paying a high price for that which she thinks she has been getting "for nothing" from your competitor. What economic or performance price has she been paying for the satisfaction of her psychic needs supplied by your thus-far-successful competitor? Identify this price and then cause the customer to *learn for herself, and if possible by herself,* that she has "been had."

Don't ever set up a situation where this insight is made possible with you as a witness. Present him with facts, subtly, so that he gains this insight without being forced to betray his awareness that he has been wrong. Or supply him with written information and leave it to him to discover, in privacy, that he has been wrong.

Don't ever create a situation where he must overtly admit that he has been wrong. To begin with, *he won't do it.* And if he discovers his supplier has been taking advantage of their long relationship—and he makes this discovery in such a way that he knows *you* know he was wrong—he will sever the relationship with his old supplier, but he won't buy from you. He will buy, if possible, from a third supplier who will not have been present at the embarrassing moment when he learned he had been wrong in giving preference to the old supplier.

Here are a few methods you may use in this kind of situation, always bearing the foregoing in mind:

Draw comparisons. Show him what he has been really paying and what he has been deprived of.

Hit hard on advantages to him, especially in those areas where you know he has been deprived.

Unearth his primary buying motive and gear your presentation to that desired satisfaction. Don't tear down competition... he'll make the comparison and see where competition has been lacking.

Identify his self-image. Study competitors' tactics to see how they have been satisfying the needs of that image. Take stock of what you have to offer. Plan your over-all strategy and immediate implementing tactics to increase their image-satisfying impact.

Don't flatter him. If you ask a group of people why we dislike flattery, the chances are 10 to 1 they will say, "Because it is insincere." But that is *not* why we dislike flattery. The psychologist tells us that *we do not like our behavior to be transparent*. In other words, we don't like people to be able to read our motives. When we flatter a customer, we are in essence saying, "I see how you want to be seen, so I'll be glad to say and do the things you want me to say and do." He experiences a dread that his actions have been transparent, and he hates us for being able to read through his defenses. As said before, and repeated now for emphasis, this is why the old adage "Praise the *work*, not the *worker*" applies in professional selling.

Scare him. This is a variation of the last resort method of "shock him; antagonize him." Alert him, for example, to the hazards of a single source of supply, or to the danger he runs of missing out on new features or services by sticking to your one competitor. Supply proof.

Give him an excuse for switching to you. This is what he wants. Make his past habit of buying from your competitor to appear to have been smart *under the conditions that prevailed previously*. Then point out that conditions are different (you have new features, new savings, etc.) and it will *now* be smart to buy from you. In this way, you give him an escape hatch and do not cause him to feel stupid for not having switched long ago. If you fail to do this, he may refuse to switch simply because he won't pay the high psychic price implicit in being forced to say (or even to think) he was "wrong" or "stupid" or not as smart as he would like to believe he is. In other words, make it *easy* to switch.

If your competitor does satisfy his present threshold of satisfaction, is there a way by which you can raise that threshold so that only buying from you can satisfy the needs of his new self-

image, his new objectives or goals? Can you arouse a new "me-I-wish-I-were" in him with its promises of much greater rewards? (*C8*)

Consider bypassing him (F1)...dangerous but possible at times. Will you satisfy an immediate goal (get the order) and lose the long-range objective (establish a lasting, favorable relationship with this buying influence)? On the other hand, is his present attitude really in the interests of his employer? How can this be brought out...safely?

How can you entertain him (if this is what he wants)? Can you accomplish this in a manner that will satisfy his status needs more effectively than the means now being employed by your competitor? Do you *really* use your entertainment account effectively? Do you use it ordinarily to take established customers to lunch because you feel comfortable with them or grateful to them? Do you use it to take to lunch a "tough nut to crack" so you can begin to really know him and to discover how he wants to be seen...to see what his problems are? Expense accounts are really investment accounts.

Does your competitor provide a better climate for business discussions? How? Does she provide your customer with greater assurance, freedom from worry or some other emotional relief? Your competitor seems to have tapped the roots of this customer's status needs more effectively to date. What means does she employ for "getting to know" the customer? How can you identify these same needs and aspirations?

Remember: Strong seller-buyer relationships often weaken when the answer to the question, "What has he done for me lately?" is..."Nothing!" Don't let it happen to you with the customers whose preference you now enjoy, and look for opportunities when your competitor has slipped in this area. One way to avoid this kind of trouble is to keep this checklist in mind:

What have I done for this customer lately?

What have I done to indicate that I am interested in him as a person, not solely as a source of orders and income?

What have I offered to do for him lately?

Perhaps neither my competition nor I "like" this fellow—but has my competitor disguised this feeling while I have betrayed it?

If I don't "like" him as a person, have I tried to find something about him to like?

Have I ever really made an attempt to understand him? As salesmen, we have undoubtedly lost customers over the years, or lost some of their business, because we slipped and took them for granted.

Your customer will realize your competitor is taking him for granted if you can show benefits your competitor has not been able to offer him or has not thought of offering him.

Competitor is local

There are times when a customer who may be sold on your product, service and price will withhold his business (deny preference) because your competitor is a local firm, whereas your organization is from "out of town."

This may be of minor importance when the city is a large one, but it becomes very important in a smaller community where your customer and competitor are in frequent contact in a lodge, a service club or are neighbors. It will take on a degree of importance, even in a large city, when an order is a substantial one and will be regarded as being important to the economic well-being of the community. The customer may be civic-minded, have local pride, or he may not relish the criticism he feels will come his way if he buys "foreign." He may fear retaliation by the employees of the local supplier, fear reciprocation, etc.

Frankly, the answer to this problem is the same as that given to the preceding obstacle. You must show the customer the cost of this attitude versus the gains which will be achieved from buying your product. There is no short-cut. If his actions are prompted by fear, you must counter with another emotion that will have greater effect upon him.

Only the charlatan will tell you there is "one best answer to this problem." There is *never* one best answer because the remedy must neutralize or outweigh the stimuli to which he is reacting. As has been seen, his action has resulted from one of a number of possible stimuli. The job is to find out *which one* and then to select the strategy that will offset the stimulus to which he is responding.

You have seen this well-illustrated in the strategy of many manufacturers in Europe or the Orient who seek to sell in the USA. They are often up against local pride or "Buy American" philosophies. Notice that they will often emphasize how much of their money is spent in the USA on assembly, on service, or on dealers. They show how much of the price remains in the USA. They emphasize the duties and taxes they pay...the number of Americans they employ. US firms follow suit when they sell overseas.

For instance, hard on the heels of a President's exhortation to spend vacations in the United States, one foreign airline very effectively and promptly advertised that the income from US travelers buying tickets on their line would stay in the United States.

This is an example of changing the setting. *Again, it is an example of giving the customer an excuse to buy from you...which he wants.* He doesn't *like* the idea of passing up the financial savings or the product performance which you offer him, but he fears criticism. He *wants* you to give him arguments which he can, with conviction, pass along to his critics. At other times, he is not concerned about criticism. He is civic-minded and is acting in what he feels is the best interests of his community. In spite of that, he still regrets the price he must pay and will gladly grasp any arguments you can give him that will still his conscience.

Your customer may prefer to buy from local suppliers because he feels he will obtain faster delivery, more rapid service—or because he knows the individual, personally, to whom he must go when trouble arises. What are the antidotes when you are not local? Consider these:

Have higher-ups in your company visit the customer so she will have the feeling that she knows the "people at the top" in your organization as well.

Schedule yourself to see him as often as possible to offset the feeling he may have that he is in more constant communication with the local supplier. Make yourself available to him on the phone. Always respond speedily to any communication from him. *Try, in every way, to minimize distance.*

Try to offset rapid delivery and/or service by emphasizing product features you feel may be regarded as vital to your customer. If applicable, play up advantages of the air-freight and emphasize your company's distribution system.

Highlight in as many ways as possible the price he is paying for his "buy local" policy.

The best method of overcoming the obstacle of "local competition"—like most others—is to anticipate it. Have well-thought out facts ready for the customer. Be sure they can be proven, are not farfetched and will be acceptable to him. The surest way to make certain that your arguments will have the desired effect on your customer, is to ask yourself this question:

"If I were really this customer, lived in this town, had his fears or his civic spirit, what could someone say to me that would change my mind?"

That is what I must say to *him now.*

As a matter of fact, it is always one of the most effective guides to planning sales strategy to ask a similar query of one's self: If I were really the customer (not if I were in his shoes), but if *I were* the customer, with all of the stimuli bombarding me that bombard him...his family...his religion...his customers...my competitors...his age...in a word, all of the tensions to which he is subjected, what could someone say and do to change *my* mind? That's what I must say and do when I meet him in the selling situation.

Once again we see demonstrated the advantages of pre-sale preparation—in an atmosphere devoid of emotion, with the luxury of time to think. So much better than waiting until we get into the selling situation with the need to "counter-punch," to depend upon hunch, instinct and feel, and leaving the customer with the feeling that our answers to his objections are the result of flying by the seat of our pants. Points we establish that are born of solid thinking, done in advance, have a way of carrying far more conviction. And they offer the reasons the customer needs before he can relinquish deeply felt convictions and give us and our company the preference we must obtain.

Competitor has only one salesman for customer to deal with

In our attempts to gain the customer's preference, we may encounter the obstacle that is created when the customer is required to deal with more than one representative of our company. This is aggravated when the customer is not required to see more than one representative of our competitor's organization.

The obstacle often occurs when our company's marketing effort is organized in such a manner that different sales representatives handle its various product lines. The customer often resents having a number of individuals from the same organization call upon him. He may indicate that it is too difficult to obtain prompt answers. He may feel that our organization is confusing, and he may object to giving too much time to one company. He will have these same objections when he requires product service.

This situation is frequently encountered if we work for a large multi-department concern or even a small company that fragments its marketing effort for the sake of specialization. You will recall from Psychological Aspects of Selling the discussion of the Gestalt principle; i.e., the tendency to see the part as the whole. In other words, the customer does not tend to see a multi-product or a multi-department company as several separate, autonomous, or semi-autonomous units. He sees it as one company. He does not see your organization structure as being in his interests, but rather he considers fragmentization as a technique your company has used to serve its own convenience. His annoyance should not be difficult to understand when you consider what your own reactions have been in a retail store when, upon asking a sales clerk to assist you, you have been greeted with the "That's not my department" treatment.

At the same time, the customer has no hesitancy in giving time to the individual representatives of the several one-line concerns that call upon him. Permitting this seeming incongruity to annoy you is certainly not the way to handle the problem presented by his myopic viewpoint. Indeed, annoyance can only cloud your own judgment. If that annoyance is revealed to the customer, it can be fatal to the relationships you must establish if you are to gain his preference.

Explain the values of your company's decentralized approach to him. Emphasize that this is really the height of customer-oriented service. Drive home the fact that he is being given the services of men who are specialists—men who know their products and his problems in depth. Emphasize the selling competence that this represents as contrasted to the "checklist-deep" service that is provided by the competitor who cannot possibly be equally well-versed in a wide range of products which are sold by a single individual. It should go without saying that, if it can be avoided, you do not name the competitor; you merely cite a hypothetical competitor who might service the customer with a single salesman.

Emphasize that while a number of representative of your company contact him, in the aggregate, the customer is required to spend no more time with them than he spends with a single representative of another company who will need the same amount of time to discuss his entire product line. When calling upon this type of customer be especially well prepared in advance so that you may render him efficient and speedy information. Use as little of his time as possible.

Know as much about the customer's problems—which concern your product or service—as possible and seek ways of being of greater assistance to him. Your multi-product line competitor will probably not have the time or the over-all ability to render this kind of specialized assistance. To the extent that you find it possible to do so, try to have other representatives of your company who call upon him follow the same pattern.

Once you discover that this obstacle exists, at the very least communicate it to other representatives of your company so that they, too, may successfully avoid or overcome the hazard that this customer's frame of mind represents.

Customer has prejudice

In a previous chapter, we dealt with this obstacle when it crops up during an earlier facet of a sale. The recommendations made at that time also apply here. When you are attempting to gain the customer's preference, this same obstacle may occur again.

Assume, for example, that your customer has not agreed your product meets his needs. He may, at this point, say—or think—"So you have a good product and it meets my requirements, but so do the products made by companies A, B and C. Why should I buy from you?" (Or if he is now buying elsewhere: "Why should I switch?")

What to do? The answer is found in an analysis of his question; i.e., *Tell him why he should buy from you, and do so in language that will be meaningful to him.*

1. Identify his key problem. Select your product or company advantages in an order of priority, presenting first those which offer the greatest relief for her problem or satisfaction of need. Then cite those product or service features in terms of what those features will mean to her. Mention features, yes, but *emphasize* advantages.

2. Draw contrasts when contrast will highlight the superiority of your offering.

3. Use language that will cause him to see the losses he will sustain by making the wrong decision. Select words that will avoid causing him to lose face. If he loses face but realizes that his present supplier is not giving the best available, in order to save face he may buy from a new supplier—but not from you.

4. He may be prejudiced against your company because of past experiences or because of what others have told him. If you are "big," you may encounter hostility towards "bigness," usually based on the fear that he won't be important enough to your firm to obtain prompt service or real interest. If you are small, you may encounter prejudice based on his feeling that you won't be able to supply his needs. (This will be discussed later when we examine the obstacle that is presented when you threaten the customer's psychic needs.) In either instance, as a professional salesman, you need little advice as to how these obstacles can be handled successfully. The problem, often, is to identify the cause of his prejudice so that you can cope with it.

An analysis of the Sales Strategy Bank indicates that there will be value in considering the following strategies, where applicable:

B17—Provide him with an escape hatch. When prejudice is the obstacle, convey the feeling—without agreeing with him—that you regard his feelings as understandable; but perhaps some additional facts, which you would like to supply, will change his mind. When possible, make him aware of new developments which change the situation. This causes him to feel that his reaction to the "old" conditions was justifiable, and "now that you have done something about those conditions" he is willing to "give your company another chance." The goal of your effort must be to help him surrender his old viewpoint without loss of face and doing so without creating the feeling that you have *made* him change his mind. You are after the business—not credit.

C2—Use cause and effect examples; i.e., introduce third parties by providing testimonials. Be sure to use as your examples people whom you know he regards favorably or with whom you know he likes to be associated or compared. Try to cite people who felt as he feels now but who changed their minds, and supply him with the reasons for their changes.

E11—Stimulate his recall. If you have had previous good experiences with the customer where your company has served him well, cause him to recall those experiences. Remember we all react strongly to a previous experience and so does he. If his prejudice is the result of a recent poor experience, he will magnify it out of all proportion to its real importance. The task is to "change the setting" so he will see this recent poor experience against he backdrop of previous good experiences... in other words, in perspective.

You may also find it wise to *provide a guarantee (E12), when this is possible.* He may never intend to use the "guarantee," but it provides him with an excuse for changing his attitude.

These are but a few of the methods you may employ to overcome prejudice. It is suggested that you examine the Sales Strategy Bank for others, especially those listed under the heading "Create Good Climate."

Experience with other departments

This obstacle may also be interpreted as "experiences with other representatives of your company." Once more we find the Gestalt principle capable of creating problems for the salesman... because of the customer's tendency to "lump us together" with other departments of the company or with other representatives of the company and to judge us as a single unit.

In this day of diversification, some companies escape the obstacle by operating their various departments under different names. For example, a customer might have a problem with a Buick and might buy an Oldsmobile as a result. He will not have a great tendency to "take it out" on the parent company, General Motors. If, on the other hand, your organization is not structured in this manner but sells all of its products or services under one name, Gestalt may be a frequent source of obstacles.

Nothing is to be gained by becoming annoyed or by feeling that this is unfair. It will prove more rewarding to recognize the obstacle and to take actions that will neutralize or eliminate it. It is a form of prejudice and you may find it difficult to have an angry customer sit quietly while you explain the organizational structure of your company. You may find it essential, however, to do just that—to show him that each unit of your organization must stand on its own feet and make a profit.

It has often been said that the greatest of all appeals is, "I need your help." If that is the case, then a hardy contender for the title is, "I know you are fair." Most of us abhor the idea of being considered unfair. Therefore, one approach is, "I've always known you to be very fair, and I know you want to be now; so I know you'll permit me to bring this fact to your attention..." Then you proceed to show him that he is judging you by past experiences with which you had no connection.

In some instances, it is wise to offer him your help if his complaint is a new one about which you can take action. If it is an old complaint, your best bet is to let him get his prejudice out of his system by sounding off. *Don't try to defend... just listen.* In many cases, once he has vented his spleen, he will subside. Our files are full of cases of customers who have waited for years to give voice to the cause of their prejudice, and once they have done so, they have done a right-about-face and made the purchase. To interrupt, apologize or defend can be fatal.

To agree with the customer is a sore temptation when we feel it may calm her down. Think twice before you do so. Agreeing with her may not only fail to change her mind, but it may reinforce her determination not to buy from your company. She may grow to like you, but her feelings may be epitomized in the phrase, "He's a nice 'guy'; isn't it a shame he's with the wrong outfit." The problem may be compounded if you agree because she may be heard later saying to a third party, "Even one of their 'executives' agreed with me." (You go up rapidly in stature when she uses you as a bit of proof later.)

The foregoing should not be construed as advice that one should never agree with the customer when he has a complaint. If the fault is yours, personally, agreeing is often an effective means of calming him down, especially if fast remedial action is promised—and supplied.

If your customer's prejudice is the result of a poor experience, hear him out, all the while exhibiting a respectful manner. Then respond with statements similar to this: "I can see how you feel, sir. *(note, you have not agreed)* and I'm sorry you had an unpleasant experience, but my job now is to try to help you with *(name his current problem)*, and I am convinced that the profit you can make now will more than offset any poor results of the experience you have mentioned. I'd like to bring to your attention. . ." *(then proceed to hit hard on what you believe will be of immediate interest to him. Your effort is to switch his attention from a problem about which you can do nothing to a problem with which you can help him).*

In previous examples, we used the word "profit." You may choose the word which will be most applicable to the situation in which you find yourself and which will capture the attention of the customer. What you are also attempting to do is (1) to act as a safety valve for the customer's pent-up feelings and then (2) to dilute his memories of one or more bad experiences with immediate gains you can offer him now.

We threaten customer's psychic needs

This theme, you will recognize, has threaded its way through our discussion of a majority of the obstacles covered in this volume. Heavy emphasis is being given to it because most of us have been brought up, as saleswomen, on a diet of product and price. We hear so often as the reason for lost sales, "We were beaten on price," or "Competition had a product edge." True, there *are* situations where these factors have cost all of us sales, but it is in the area of "high psychic price" that even more sales are lost.

Often "price" and "satisfactory product" are smoke screen objections to cover up the fact that we have, unconsciously, walked all over the customer's psychic needs. Here are some examples:

The customer is wrong and we prove it.

He "hears" we treated someone else badly, and that "someone" is an individual or a concern with whom he identifies himself or his firm. ("If they did it to him, they would do it to me, if I gave them a chance.")

We made him feel we regarded him as unimportant because we made him wait or paid no attention to his complaints.

If we are "big," we cause him to feel less significant.

If we are "small," we are not big enough to give him the kind of service "his kind of person" deserves.

A contractor, a retailer, an OEM, or almost any kind of customer may recognize that you know more about how to run his business than he does. This makes him feel uncomfortable in your presence, so he erects road-blocks to keep you out or to get rid of you fast. You rob him of the sense of competency that he craves.

An employee of your customer's organization may have an influence on the buying decision, and your high degree of capability may cause him to fear that he will look less competent in the eyes of other people in his company. The welcome mat will be put away for the duration.

Even things that are beyond your control, such as your size, may make him feel less significant.

The list could take pages—and from your own experiences, you could add to any list we could compile. Once again, the problem is to identify the cause of the customer's prejudice. You can be sure that he will not mention any of the reasons listed above or many more like them. As stated previously, he will either give you smoke screen objections (like price), or he will evade you, act disinterested or find some reason to attack you or your company. Here are a few suggestions for action:

1. *Never talk down to him.*

2. *Judiciously, and not too often, use phrases like,* "As you know."

3. *Avoid "I," "Me," "Us," "Our Company," and emphasize "You."*

4. *Get her talking and show interest in what she has to say.* Make comments that show her you really have been listening.

5. *Avoid any word or action that might create the impression that you regard yourself (or that you are) superior to him.*

6. *Never embarrass him in front of his superiors.* Try to find means of making him look good in their eyes. Be alert for opportunities to provide him with an escape hatch if they feel he has been at fault.

7. *Don't tell him how to run his business.* For example, if you were selling to a retailer you would avoid telling him the kind of a dealer your company would like him to be. You would ask him what kind of an enterprise he wanted to conduct and then you would point out that you "have just the kinds of products (or services) that will help him have that kind of a business."

8. *Avoid remarks about your company's size, importance, etc.* Concentrate on her success, her potential, her progress, and the like.

9. *Ask for his advice or opinions.*

10. *When you compliment him, seek some achievement of his and talk about that, rather than about him.* This avoids the appearance of flattery and conveys the impression of sincerity. Be sure that what you consider an achievement of his represents what he considers outstanding performance on his part and not something less than he feels he is capable of having accomplished.

11. *Identify what he considers his most pressing problem and rivet his attention on that.* Then emphasize how your product or service can help him solve the problem. If you can make him drool to overcome the problem, he may forget the impact that you have on his self-image.

12. *Learn as much as you can about him as a person*... his family, home, kind of car he drives, hobbies, interests, education, religion, political persuasion, etc. This does not mean that you will discuss these with him. In some instances it may mean you will avoid them when they conflict with your own likes and dislikes. In other instances, you will find a common bond and the start of rapport.

13. *Don't talk about "the old days" unless he brings them up.* He may feel he has come a long way from the old days when you first knew him. Talk of the old days causes him—or may cause him—to feel that you see him as he used to be, not as he now wants to be seen. Your competitor who may not have known him in the old days sees him as he wants to be seen—or hopes to be seen, and addresses himself to that image... and makes sales doing so.

14. *Watch the language of your body.* A sale can be lost by gestures of impatience, disagreement, disinterest or disrespect. Look at her when she's talking. Don't let your gaze stray around the room. Sit up in your chair, lean slightly forward—give the impression that you are really interested in what she has to say.

15. *Treat his office and furniture with respect.* This is not to suggest that any professional salesman would put his feet up on the customer's desk or carve his initials in the desk top. It's the little things we can do unconsciously that cause him to think, "Who does this salesman think I am?" (Or probably, "Doesn't he know who I am?") The hat left on his desk, the roving eye sweeping across papers on his desk—not reading but giving the impression that we are.

We are not about to list the 101 little things we can do in the customer's presence that don't exactly make him feel that we regard him as important. We all know what they are if we will but think of them.

The real cure is to be alert when we are in his presence and be aware of what we are doing.

It is not enough to say, "If I were the customer, I would be bigger than that... I wouldn't let that sort of thing annoy me. He's a small man." In other words, lets rewrite the Golden Rule. It's not enough to say, "Do unto others as you would have others do unto you." Maybe they don't want to be "done unto" as you would. Let's think of it this way: "Do unto others as others would have you do unto them." Or better still: *"Do unto this unique individual, your customer, as this unique individual would have you do unto him.* And to follow that advice, we had better know how he wants to be done unto. That means knowing him and the image he wants others to see when they see him.

In the foregoing, we cautioned against talking about "our company," "our facilities," etc. Keep in mind that this advice applies when we are considering the individual who feels his psychic needs are being threatened. Many other customers who sense no such threat want to have you tell them about your company, its facilities, its strengths, etc. Fashion your actions to meet the needs of the customer—be they psychic, product or economic.

Customer's habits

This obstacle, which again appears in the Anatomy of a Sale at the "Gain Preference" point, has been discussed in previous chapters and will not be repeated here.

Compare our product to others

Your customer may withhold preference because he feels, and states, that your competitor's product has features superior ot yours. In facing this problem, consider these solutions:

1. *Be certain that he is comparing apples to apples.* Make sure he is comparing the same price line and not considering the bottom of your line and comparing it to the top of your competitor's line. Don't *assume* you are talking the same language—make certain.

2. *Emphasize hidden values.*

3. *If he stresses some added feature of the competitor's offering that has caught his fancy, point out

where it has shortcomings which he has overlooked in his enthusiasm.

4. Play up any flexibility features your product or service may have; emphasize any custom-built features.

5. Stress hidden costs of the competitor's offering—lack of service, remoteness of service capabilities, unproven capabilities, lack of acceptance by end users to whom your customer may have to sell his end product, difficulty of installation, maintenance, operation or incompatibility with existing equipment he may have. Draw comparisons, letting the comparison drive home points, so that you will not be in the position of "knocking competition."

6. Use reverse testimonials. Recommend that he talk to specific people who have used the competitor's product, when you know that they have had poor experiences with it. This, again, avoids knocking the competition. In many instances, he may not bother to do so and may ask you about them. You then become a narrator of someone else's experiences.

7. Stress exclusivity of your product or some of its features. Emphasize compatibility with her present equipment or facilities.

8. Foremost—identify what features of the competitor's product he regards as superior. His concentration on those features provides you with an excellent clue to his basic buying motive—what it is that he seeks. Then hit *hard* on the features of your product that will best satisfy that buying motive.

9. Be sure he has not been misinformed about the competitor's product. If he is wrong, set up a situation in which he will find this out *by himself* and *for himself* so you will not be a witness at that moment when he must say those words which are among the most difficult that come to the tongue of man. . . "I was wrong."

Spreads business

"We believe in spreading our business." Not uncommon words for the saleswoman to hear when she tries to obtain a greater share of the customer's business. Is it really the customer's philosophy, or is it a smoke screen objection? It can be either.

Some will say they like to be fair; some say they don't want all of their eggs in one basket, and others follow this pattern in order to increase competition and the resulting lower prices for them. In each case it presents a problem when you seek to gain preference. We suggest you reread the material on smoke screen objections which appeared in Chapter 13.

When the reason which the customer gives is his *real* reason, consider the following:

If he wants to be fair, indicate the other groups to whom he also wants to be fair—himself, stockholders, employees, his customers, family, etc. Try to show him that by buying more from you, he can achieve savings that will enable him to be "fairer" to these other groups, including himself. Indicate the hidden costs he is paying for the policy.

If he doesn't want all of his eggs in one basket, it is probably because he fears the results of strikes, other interruptions in delivery or the possibility that our competitors will overlook him when they come out with improvements.

The recourse available to you is to provide assurance, back it up with facts and cause him to review your record of past performance. Stimulate his recall of favorable experiences he has had with you and your company.

Provide guarantees when you can or stockpiling for his future requirements when possible. Emphasize the losses he sustains by spreading his business.

Influence of outsiders

In this instance, "outsiders" refers to other influences on the buying decision in the customer's organization, beyond the individual from whom you must obtain the order. They may or may not favor you. When they do not, they can be real obstacles, especially if they are in a position of authority over your contact, or if he dislikes the task of converting them to your product.

One of two conditions may prevail. If you know who the "outsider influences" are and are permitted to present your arguments to them, the situation is less difficult. You are able to identify the nature of their objections to your product (or to your or your company). You then use those of the techniques which have been discussed in this book that will be most effective with the kind of person and type of objection involved.

A more difficult situation exists when the customer will not let you know that his decision is being controlled or affected by other people. He may want to keep the knowledge from you that he is not the key decision-maker. He may simply not want you to know that he is being influenced by others because he fears that you would regard this as a sign of weakness. He may not even know that the opinions of others in his organization have influenced him.

In all three cases, you may face the same problem. You certainly cannot mention these influences if he omits admitting they are having an effect on his decision. You must provide him with facts, proof, and arguments which he can use on those who are influencing him.

Hit hard on the advantages to his company rather than on the advantages to him personally (not that you ignore the latter). Give him privacy, time and the opportunity to pass along the arguments to those who influence him. Make it *easy* for him to do so by providing him, if possible and applicable, with samples, printed materials or any other visual aids you may have. Draw comparisons for him, which highlight advantages over competitors who may be favored by the people who are attempting to influence your customer contact.

Do not indicate that you know or suspect someone else is influencing her... unless she openly states that this is the case. When she makes no secret of it, attempt to obtain her agreement that both of you should approach the "influence" and, together, seek to change his mind.

If you feel for any reason that your customer contact will not carry the fight further, you have a difficult decision to make. You must consider the implications of bypassing him or of obtaining help from someone in your own organization who can contact the people who are influencing your customer contact. You may be successful in obtaining *this* sale by pursuing these methods, but you may imperil or lose all future business from this customer if he feels you have gone over his head. We have seen many instances where the customer personnel "higher up" have become angry at a salesman's efforts to have their man overruled. If, after considering these possibilities, you decide that going around or over your customer contact is the only course of action left, then be sure to base your appeals on the advantages to the customer's company if they place their business with you. Avoid emotional appeals.

Before resorting to this last-resort method discussed, always consider the possibility of having facts planted with the "influences on the decision maker" by other satisfied customers, being sure that they keep you and your name "out of the action." If there is printed material that you feel would influence them, and you know your customer contact will not bring it to their attention, seek other ways of having such material "find its way" to their desks without your being involved.

These techniques give an air of being surreptitious and underhanded. They are not. If you truly have the best product or service to meet your customer's needs, and your path is blocked (1) by an individual who is so proud that he won't admit he does not have final authority or (2) by one who lives in such fear of his superiors that he doesn't dare confront them with information that is contrary to their stated opinions, then you have a *duty* to bring this information to their attention. This duty is not only to your own employer and the people who work to build your product or service, but it is also a duty to the customer's company.

Summary

Preference must be earned. The customer withholds it when he feels that you may not be able to satisfy his needs as well as some competitor has, or could, satisfy them. He will refuse to give you preference if he feels that buying from you will cause him inconvenience, create fresh problems, upset habits or customs, or threaten his economic or psychic security in any way. He will also withhold preference as a strategic buying technique.

Emotional appeals, generally, will avail you little unless you can offer him great psychic rewards. Any time that anyone attempts to persuade you or me or a customer to take an action, subconsciously and often consciously, we are asking ourselves one of these questions:

1. Is he going to make me rich?
2. Is he going to make me famous?
3. Is he going to make life easier for me?
4. Is he going to make me more secure?
5. Is he going to give me more power or authority?

What do these mean?

Is he going to make me rich may simply mean, "Is this going to cost me money or make more money for me?"

Is he going to make me famous may mean, "How will this make me look in the eyes of people whose opinion is important to me?" (This may mean our boss, out peers, our family, our friends or others.)

Is he going to make life easier for me may only mean, "Is this going to make more work for me or is it going to relieve me of disagreeable or time-consuming tasks?"

Is he going to make me more secure could even be eliminated from this list because when we think of question (1) above we are thinking of economic security and in question (2) we are thinking of psychic security.

Is he going to give me more power or authority may mean, "If I take this course of action, how will it affect my importance in this organization... job security... promotional possibilities, etc.?"

If you address yourself to answering the wrong question, the order will be lost irretrievably. You must home-in on the customer's key area of interest.

Never, but never, make the mistake of saying, "If I were in the customer's place, this would be what is most important to me so that is what is most important to the customer."

You are *not* the customer.

If we are going to answer these questions to his satisfaction, we must come into the selling situation fully prepared. We must know his organization and know which individuals in that organization are in the position to exert influence on the buying decision. We must know as much about the customer, as a person, as is possible. We must know in which he is most interested; i.e., "being rich or being famous." In other words, will his buying decision be predicated primarily on his desire to "look good," or on his desire to profit financially or to solve a vexing operating problem.

We must not become so preoccupied with product and economic problems of the customer (as a result of a long practice of thinking along those lines) that we lose sight of the fact that *this* customer may be concerned, primarily, with looking good in the eyes of his boss or his peers, his trade association or some other group whose approbation he seeks. Considering this facet of the selling situation gives professional selling a whole new dimension.

It is not enough to assure ourselves that we are working in a profession; we must *be* professionals in that profession. The mark of a professional is preparation, analysis, diagnosis and consideration of alternatives. You and I would have a poor opinion of a doctor who would even consider operating before he had made blood tests and other essential pre-operational checks. How, then, should we classify the salesman who does not make *his* pre-sale analysis?

Gaining the customer's preference, if you are in a competitive field, is essential. The best way to overcome obstacles to gaining that preference is to consider those obstacles in advance and to consider the means available to you to overcome them. Then adapt those means to meet the attitude, fears, habits and preconceptions of this unique individual who is your customer. Predict his buying attitude, examine his buying habits of the past.

If you are in a kind of selling that does not permit this type of pre-sale analysis, get the customer talking and be alert for clues to the reasons for her attitude. Then consider the strategies that have been discussed in these pages.

This completes our consideration of the obstacles that stand between us and granting of preference by the customer to our offerings. We are sure that this treatment is by no means complete. If you are an experienced salesman, review your own past and consider sales you have lost (or successfully concluded) where obtaining the customer's preference was a key problem. What were the reasons for his attitude? If you were successful, why were you able to change his attitude? If you lost the sale because he preferred another firm, where did you miss the boat? Add the principles you can extract from those selling situations to those we have discussed; in this way, fortify yourself to meet comparable situations in the future.

CHAPTER 15

Establishing priorities

Sales training programs to which most of us have been exposed place emphasis on the need to arouse the customer's interest and to whet his appetite to buy; i.e., to create desire. Your customer may indicate interest and even possess a keen desire for that which you have to offer, yet he may resist your efforts to create the buying action. Often, the reason for his reluctance is that he places a greater priority on other needs or he entertains fears of the consequences.

This situation must not be allowed to continue because it provides your competitor an opportunity to approach your customer. His timing may be more fortunate, and when you do return you may find the order safely locked up—in your competitor's order book.

Another reason why you must not allow this situation to continue to exist is that it provides time for your customer to lose interest or to focus his attention on other demands upon his attention, or upon his funds.

We will examine some of the more common reasons why the customer will not assign top priority to the problem your service will solve.

Fear of business climate

The customer may feel that this is not the time to buy. He may think he will be well advised to hold his funds for other purposes because he fears that the business climate is going to deteriorate. He may fear that he will suffer a decline in income, and as a result either have difficulty in paying for your product or service or be short of the funds he will require for other needs to which he gives a higher priority.

The action you take under these conditions will vary with the type of product or service you are attempting to sell him. In any case, you should *provide him with assurance* (E2 from the Sales Strategy Bank) and with proof, when this is possible. As a matter of fact, you will find, in examining the Sales Strategy Bank, that any of the strategies listed in the column headed "Overcome His Fears" will be helpful in handling the obstacles discussed in this chapter.

To provide the customer with assurance or proof, you might consider the use of economic predictions and charts and graphs drawn from reputable soures which will indicate the cyclic nature of business. Provide him with information on trends that will affect his business. These may take the form of population increases and trends in the area he serves, housing starts, new industries in the area, watt-hour meter installations by the power company, new phone installations and so forth.

If the customer's prime concern is his ability to pay, consider financing arrangements that are available to him either from your company or from appropriate agencies in the area. The *principle to keep in mind* is that when the customer procrastinates or delays, and it is against your interests to have him do so, you *must* cause him to see that the cost of procrastinating is *higher* than the cost of taking immediate and favorable action. Cause him to see that he is now in a buyer's market where delivery, prices, terms and service are all in his favor. If he waits, he runs the risk of the usual conditions prevalent in a seller's market; i.e., slower delivery because of the demands made on the producer, higher prices because of the law of supply and demand, and less individual service—again because of the increased demands upon the time of the producer.

In using these arguments, you must use extreme care. Do not permit your customer to feel that any threat is being made or implied. Don't let him think that you would willingly expose him to poor service or that you would grasp the opportunity to charge him a higher price if your business were better. In presenting these arguments, you should emphasize that the conditions about which you are warning your customer are natural phenomena in a seller's market. You should emphasize that *all* producers would be up against the same problem if they were hard-pressed to fill orders.

When it is possible to do so, compute mathematically the loss he will sustain if he continues to delay in buying your product or service. This loss may be greater than he has realized, and the sudden awareness of it may be all that is needed to move him off dead center and cause him to buy.

Assuming that a customer has agreed that she *has* a need and that she has agreed that your product or service *will* fill that need, and that it will fill it better than any of your competitors can fill it, there can be only one reason for her to procrastinate...fear. Your job is to identify the nature of that fear and eliminate or neutralize it.

Fear of higher authority

In the Anatomy of a Sale Chart, this is the next obstacle listed. It will not be treated again, as it was covered in Chapter 13.

Fear of failure

Although the customer may want your product or service, if it is new or a new application, he may fear that he will be ridiculed by others "for being a guinea pig." He may fear that he or his people will lack the technical competence required to make the application work. In such cases, the customer may *deliberately* place a higher priority on some other project in order to avoid the necessity of making a decision now. In this way, he senses that he will not place his self-image in jeopardy

Regardless of the reason for the customer's failure to assign top priority to the problem or product need your product will satisfy, the basis of your problem is that somewhere along the path to the sale you failed to supply him with complete assurance. Your course of action now is clear. You must supply him with heavy doses of proof, testimonials, demonstrations, and assurances that you or your company will provide him with the help that he suspects he will require. If possible, you should heavily emphasize the advantages to him so that the thought of losing these advantages will be stronger than the fears he has entertained.

Won't plan ahead

In recent years, sales managers have placed great emphasis on the need for creative selling. One form of creative selling is *anticipatory selling*... causing the customer to anticipate his needs and to place his order now. This action on your part may take any of the following forms:

1. Anticipate price increases—save now.
2. Anticipate materials shortages.
3. Anticipate delivery problems because of imminent labor stoppages.
4. Anticipate increased demand from his customers—stock up now.
5. Anticipate wider choice of models or features that will be available to him in the near future.
6. Anticipate slower delivery later when buyer demands on you will increase.
7. Anticipate advantage to your customer of being first with your new models or your service.

You can add to this list other conditions unique to your market which you can cause the customer to anticipate.

The problem is that we often have customers who are oriented to short-range viewpoints. At times it seems that their concept of long-range planning is, "What will we do after lunch." This type of customer lives in the present and has a philosophy of letting tomorrow take care of itself. This is a costly way to conduct a business, and you must cause your customer to realize this. You must alert him to advantages that he is overlooking or to losses he will incur by not looking ahead...by failing to anticipate change. You must cause him to see that there is wisdom in managing change rather than reacting to it. His philosophy may have been to be a "damned fast second." You must cause him to see that it is far more profitable to be first!

The conditions that cause your customer to feel as he does will alter your choice of actions. If, for example, he is approaching retirement, little will be gained by offering as bait the long-range advantages that will accrue to his company if he buys or orders now. A possible exception is the case of the purchasing agent who wants to be remembered in his retirement as an astute operator. You may, in his case, make progress if you capitalize on his desire to leave something behind him or to increase the significance of the contribution he has made to his company. Few of us fail to sense a glow of satisfaction when we return to some institution with which we have been connected and see something in place for which we have been responsible. This may vary all the way from a procedure we installed to equipment for which we were the first to see the need.

Your appeals to the customer who fails to plan ahead must be to his most vital business need and interests. This includes psychic needs that stem from his business (the need for recognition, the need to dominate, etc.). He may actually be so attuned to the present that he does not plan ahead, but you may appeal, successfully, to the "me he wishes he were" by using statements that suggest he is "a sound, long-range planner, who is alert to the possibilities of the future." You need have no concern that by doing this you are being immoral; i.e., "manipulating him."

What you may actually be doing is improving his operation, and if you are successful in this, he may actually learn the value of anticipating his needs—and in time, come to be a good long-range planner. You have, in this case, done him a service.

In order to make this type of customer anticipate his needs and problems, you must present tangible evidence to cause him to feel that following your advice will make him "look good," or you must whet his appetite for financial gain. His question, remember, always is one of these: "Are you going to make me rich or famous?" "Are you going to make life easier for me?" or ". . . give me more power?" And "Can you prove it?"

Most people, inherently, are poor planners. It is not because they lack the ability to plan. It usually is that they begrudge the time it takes to plan, they find it boring; or other more exciting prospects take their time and attention. This presents you with a problem. You must cause the customer to feel that "tomorrow is right around the corner." You must bring your advantages *close* to him, and you must make them appear so desirable and so easy to achieve that he will be stimulated to take action. When you cite "tomorrow's losses if he fails to take action," you must make these appear imminent, very costly—and cause him to feel that he will "look badly" if he suffers this fate after being alerted. You must make planning look *easy*, and even *exciting*—or, at the very least, *extremely challenging and interesting*.

Finally, you should attempt to help this customer in his planning efforts. Often, a little help will be all that is needed to overcome the inertia that stands between customer and planning. Frequently, you and I don't mind doing boring tasks if someone will work with us. This applies to planning as well. Consider these questions when you intend to aid your customer in his planning efforts:

1. What statistics can you provide him that will serve as a base for his planning?

2. Can you present him with the outline of a plan, leaving provision for him to add details so that he will identify himself with the end result and come to regard the planning work as his own . . . hence take credit for it? (Give him the credit, while you take the order!)

3. What tangible proof can you supply him that he will suffer losses if he fails to plan? Are there examples you can supply him of the failure of others in this regard?

4. Can you make the plan easy to understand and can you make it appear easy to implement?

5. Can you afford to give him privacy, after presenting him with the outline of the plan, so that he may study it and then adapt it, revise it or otherwise modify it so that he can convince himself that essentially it is *his* plan?

Other needs seem greater

A customer may agree that he could use your product, but he may see other needs as being more important. In such cases, he will not be willing to buy now.

Your task is to determine what these other demands on his time or resources are and *why* he considers them more compelling. Are they product or service needs or are they psychic needs (satisfactions of one type or another)? If he openly states, "I have other needs that are more pressing," you must smoke out the nature of these needs. Only rarely are you close enough to the customer to ask bluntly, "For instance?"

If you are to overcome this obstacle, it is essential that you draw him out and identify the problem or need that he feels is more important. Your choice then is either (1) to prove to him that the need or problem to which your efforts have been directed is really more important to him, or (2) if you are unsuccessful in the foregoing, you may be able to show him how you can assist him in solving the problem to which he gives the higher priority.

Once you identify the reason for his resistance, you may have the information you need to change your tactical approach. You may now find it wise to present your offering in terms of features that are entirely different from those you have been emphasizing.

What need is he trying to satisfy when he places higher priority on the other problem?

Is it product or psychic?

How can you present your offering so that it will cause him to feel that you are offering him even greater satisfactions than those he seeks from buying the product to which he has given higher priority?

"I can get by with what I have"

If the customer has adopted this attitude, the salesman has failed to excite his interest in the advantages of his product, or the customer sees a greater gain in waiting. There are many courses of action open to you, and these will vary as the products and services sold by readers of this book vary. From the following list, you can identify the *principle* upon which you must base your action. Then, you may adapt any of these, or similar actions, to meet the situation which your customer has created with his attitude.

1. What he now has costs him more to operate than that which you are offering him. How much

more? How can you present this information to him so that the figure will disturb him? How can you present this loss so that it appears immediate?

2. Emphasize his maintenance costs can be expected to increase as his present equipment ages.

3. His equipment may already have been written off, hence no tax benefit. (Also emphasize any tax benefits associated with current laws on purchase of capital equipment.)

4. Remind him that the turn-in value of old equipment (which he now has) decreases with each month of use.

5. Rising cost spiral means new equipment can be purchased less expensively now than later.

6. Will interest rates increase on money he will eventually require when he replaces his equipment? If so, point this out...show him how delaying can be expensive.

7. Hazards of operating old equipment can be expensive in terms of accidents.

8. More down-time can be expected using old equipment. Has he considered this cost?

9. Delay in replacing equipment gives his competitors an advantage if they are already equipped.

10. Consider the cost in morale if employees must continue to use old equipment...hence reduction in employee performance and output.

Examine these, and other tactics, and be careful how you use these arguments if the "old" equipment he is currently using is also a product of your company. This especially applies to items 2, 7, 8 and 10.

If your customer sells, in turn, to others, he may argue that his customers find his present products satisfactory; hence he sees no need to purchase new items. Some of the points of view which you may bring to his attention are: Some potential customers are *not* buying from him. Is it because his present products (or service) do not appeal to them? Will they be motivated to buy if he presents your "new" products? Is he really satisfied with his present sales volume? Is he sure he hasn't been losing business to competitors who have your products?

Your arguments to this customer must not sound argumentative. Your statements must not reflect upon his judgment. You must cultivate and maintain a harmonious climate. You can easily use the "cause and effect" strategy; i.e., using examples of other customers who have increased their volume because they have invested in your product or service.

For example: "I can understand how you feel, and I would share that view were it not for the fact that I have seen what has happened when dealers have taken on this product. Let me give you an example or two..."

Or consider this type of "argument": "I can see what you mean, sir. The equipment you have is performing, and probably the purchase of new equipment at *this* time does seem like an unwarranted expense. Sometimes, though, when we are not actually required to write a check, we don't realize the hidden, yet very real, expenses that cut into our profit every hour we are in business. I believe, sir, that if you and I consider the real cost of operating the equipment you now have, we will see that the outlay for new equipment will be liquidated in a very short time from the savings you will make. For instance..." (then select the appropriate points from the list of 10 points that are pertinent to his situation and present them in figures that will be meaningful to him).

We have said much in these pages about the necessity for professional selling and the absolute necessity for pre-sale preparation. It has been readily admitted here that we *cannot predict* the customer's reactions, but we *can anticipate* his reactions. The situation we have just discussed is one that can be anticipated. It is so common that we should consider, in advance, the real possibility that the customer will raise this objection; i.e., "I can get by with what I have." We should determine in advance the kind of equipment he is using or the products he is selling (whichever is pertinent).

His argument can show up in many forms: "I've got all the insurance I need"; "All you say sounds okay, but I'll get by handling my own receivables. I can't see any point, right now, in turning them over to your credit company to collect." The words may change but the theme remains constant—"I'm doing okay." Assume, for example, that you are suggesting a change in his store layout that you know will increase traffic and sales. You may get this response: "I'm doing okay with my present arrangement, and I can't see what good would result from changing it...that just upsets thing and costs me money — money I can use for other purposes."

The basis is always much the same regardless of the type of selling in which you are engaged. The cause of his attitude will undoubtedly be found in one of the following situations:

1. He wants to maintain the status quo. ("Things are okay now—don't rock the boat.")

2. He doesn't see the problem, but he does see other problems for which he needs his money. ("We're not having any trouble.")

3. He is short-range oriented. ("We'll take care of that when it becomes a problem.")

4. He never sees hidden waste or hidden costs ...he's check-book oriented; i.e., only sees expense when he has to write a check.

5. He is uninformed. He doesn't have a familiarity with "details"...he has the "broad look" and therefore misses the significance of what your proposition can mean to him.

6. He just isn't sold, which means we have not emphasized the advantages in terms that stimulate him.

Summary

Provide him with information he now lacks. Build your presentation on comparisons of costs between the status quo and what you offer. Hit hard on comparisons. Raise his threshold of satisfaction by alerting him to profit opportunities he has not considered.

Come prepared. Impress him with your knowledge of his business and his problems. Facts and figures will help. They show him you have considered his *unique* problem...this flatters him. It convinces him that your arguments have validity, that they just didn't come "off the cuff" when he refused to buy and you had to think of something to say. Consider, again, the thesis presented in the Psychological Aspects of Selling.

As sales trainer Kelso Sutton has said, "You, as a salesman, must be a sower of internal dissension." You must cause the customer to be dissatisfied with what he has. Consider, then, the arguments that will act as stimuli powerful enough to awaken a drive in him forceful enough to make him take the action you desire. Consider his psychic (personal) needs, and determine which is paramount. Then select from the following, the best tactical approach.

1. Awaken the "me I wish I were" in your customer. Cause him to see growth possibilities or possibilities of greater prestige and recognition. How can a decision to buy from you cause him to look good? Who is it that he wants to impress, and how does he want to be seen by them? How can you help him?

2. Appeal to his desire for economic security. Don't be satisfied with the list we have included. What other losses will he sustain if he fails to buy?

3. Make appeals to pride. Aim at the pride he has in himself as a sound thinker or at his pride in his accomplishments. Which appeals to him most?

4. Make the adoption of your proposal look easy. When we hate to make decisions or to take on extra burdens, we have a tendency to procrastinate or delay. Your customer has the same "human" tendency. Determine how you can help him... how you can take on some of the burden and relieve him.

Considerable space has been given to the obstacle "I can get by with what I have" because it is so common. Often, when you and I have desired something, the financial cost or the inconvenience that its procurement will cause have found us assuring ourselves that "we can get by with what we have." The fact is we still *want* the object. If someone can show us how it will cost us more money not to buy, or if we can be shown that the inconvenience can be minimized, we will grasp the opportunity to obtain the object we desire.

We come again, then, to the best possible advice a salesman confronting an unwilling customer can follow:

1. Give him a right to feel as he feels...you'd feel that way if you were the customer.
2. Determine what someone else could do to change your mind...then try that strategy on your customer.

As a last resort, and *only* as a last resort, consider antagonizing the customer—shocking him, as it were—into a realization of the results of his apathy. This strategy will be covered at length later.

CHAPTER 16

Asking for the order

Until now, this volume has been treating those facets of the selling situation that are traversed by customer and salesman as the customer is brought to the brink of the sale. The customer's objections have been met, his appetite has been whetted, and he is at the point where the buying action must take place.

It is at this point that sales managers feel many salesmen fail. One such manager said to us, "My men are good men. They have gained the confidence of our customers. They are well-informed technically and they know their competition. In spite of these assets, they often return to the office without the order. I have taken the time to observe them on the job and, to my astonishment, I find that some of them never seem to get around to asking for the order."

This complaint has been heard so many times, from so many sales managers, that there is no doubt but that it is a major problem area and one that deserves our attention.

Ten reasons are listed on the Anatomy of a Sale chart under "Ask for the Order" and these are the most common reasons for the salesman's failure to take the necessary action at this crucial point in the selling situation.

Front office jitters

The following is a direct quote from a manager of salesmen: "My men can work their way through a customer's entire organization and they can convince every man in the company that what we have to offer will meet their need. However, purchase of our product represents the commitment of a sum of money that can be approved only at the top level of the customer's organization. Inevitably, our salesman ends up in 'Mahogany Row'—the offices of the top brass of the customer's organization.

"Then something happens. This salesman who has demonstrated the ability to convince all of the technical experts in the customer's organization that his product is 'right' now develops something that resembles stage fright. It seems that as soon as he is taken into the office of the final decision-maker he is so overawed that he forgets what he had to say—he certainly forgets to ask for the order."

The manager of salesmen added, "I can't understand what happens. I don't know if he is impressed by the appearance of the final decision-maker, or impressed by his title or by a realization of how much money the fellow probably makes, or whether he is over-impressed by the plush surroundings of the office—the rug, the big desk and the expensive furnishings. Perhaps it is a combination of all of these things. Whatever it is, it is costing us orders, or it takes a great deal of time to get orders that we should obtain much more quickly. Of one thing I am sure, we must solve this problem fast."

A marketing executive gave us this example some time ago. He said, "I have a really competent salesman—one of my very best. The two of us went to the Pentagon on a business call. Frankly, my only reason for accompanying him was to be on hand if he needed me for recent technical development information or special information on contractual arrangements that I would be able to supply. As we waited in an office, a military officer came in and he proved to be the man we had to convince. The salesman took one look at him, so help me, from that moment on he couldn't seem to place one word after another. The officer was a three-star general. I could see that the salesman was going to be totally ineffective, and I jumped into the situation *fast*. This is something I am usually most reluctant to do, but I could see the chances of our getting this sale going right out the window, and we needed the order . . . which, incidentally, we were able to get.

"Later, I asked the salesman what had happened. He said that during the war he had been a captain, and the highest ranking officer his unit had ever contacted was a colonel. He didn't ever remember seeing any general with more than one star. He said that he had been so thoroughly indoctrinated that he had never lost his sense of awe in the presence of high-ranking military men."

When the marketing man finished this recitation, I asked him why he wasn't similarly affected. He replied, "I was 4-F—and between the two of us, I'm not even sure that I knew the significance of the three stars on his shoulders, but I knew that *we needed that order*.

How can a salesman overcome these reactions which he senses are present? It would probably be most ineffective to tell the salesman that he should not be over-impressed with rank or with privileges and accoutrements that go with high rank or office in military or civilian life. The fact is that you and I are the product of our milieu, and many of our early experiences left deep impressions with us—some of them almost indelible ones. They certainly upset our emotional balance and affect our perspective.

It would be equally futile to resort to a mandate by saying to the salesman, "Don't let rank impress you." Mandates usually are ineffective. In this instance, "Do not let rank impress you" is in the nature of a mandate. It is an "order" that the salesman must not be himself. Such a mandate is the refuge of the witless—it is emotion-based. The best antidote for emotion in this type of situation, where fear is present, is reason—and it is in reason that we can hope to find a cure for "front office jitters."

"Awe," "fear," "concern"

What *are* front office jitters? Often they are the substitution of awe, fear or deep concern for what should be respect. Awe is undoubtedly the outgrowth of our realization that the customer's chief decision-maker is an individual who has achieved the level of the "me we wish we were." Knowing how difficult it has been for us to reach our comparatively "low perch," we are over-impressed with his accomplishments.

Fear is probably the result of our realization that we are facing a man who has power—*power of rank, power of authority, power of decision.* He has the power to nullify all that we have accomplished through our contacts which those who serve beneath him. He has the power to give us an order or to deny it to us. We suspect that here is a man who will not be impressed with the "psychological tactics" we can employ when we are confronting "common men." We suspect that *this* man will have greater depth of perception. This man has his eyes on the stars and the economic and/or emotional appeals that register upon and influence his subordinates will have little effect upon or significance for him. *This is what the overly-impressed salesman thinks.*

Worry is the result of concern. We are concerned that through our manner and behavior we will indicate to the customer that we feel "out of place" in the expensive surroundings of his office. If we meet him in a "neutral zone," we fear that he will regard our efforts as an intrusion upon his time. The deference shown this man by his subordinates also has an effect upon us.

Such attitudes of mind can cost us orders and the career growth within our company that depends upon our obtaining those orders. It becomes obvious that we cannot allow this negative frame of mind to impede our personal progress, to say nothing of permitting it to adversely affect the profits of our company—upon which our own future progress depends. To overcome the effects of "awe," "fear" and "concern," we must identify the causes of these self-imposed obstacles and put things in proper perspective. It will be of value, then, to analyze this problem.

First, let's clarify your position as a professional salesman in the selling situation:

1. You have a product or service to offer.
2. You have faith in that offering.
3. You are convinced that it will satisfy the need of the customer, help her solve a problem or bring her some highly desirable satisfaction.
4. You expect to be compensated for your efforts by obtaining his business, and this puts your relationship on an even basis, a basis of quid pro quo—a fair exchange of need satisfactions. You need the economic security his business will bring you, and he needs the satisfactions your product or service will bring him.

Therefore, as a starting point, there is no need for you to approach him, figuratively speaking, with your hat in your hands. He needs you as much as you need him—even if he does not recognize this fact at this point. Perhaps the favor you are doing him is greater than the favor you are seeking from him—his order. Now let's take a look at your relative importance. He may wear the "three stars" of his civilian or military rank because he has earned them. We can at least assume that he is capable. So are you. You are a professional salesman. Perhaps you could not perform his job, but could he perform yours? Probably not. Half of the people who look down their noses at salesmen do so as a defense mechanism. The very thought of trying to sell anything to anyone petrifies them.

You look at the customer's "three stars" or at his expensively decorated industrial or commercial office—or even at his expensive luxury-line car parked outside the door—and you are tempted to equate these accoutrements with greatness, with omniscience, with a great intellect. Yet, if you were selling Einstein a lawn mower, he would have had to listen to you—because you know more about lawn mowers.

Admittedly, you are sometimes confronted by a customer who is also a master of his trade. He may have come up through the ranks, is well-versed in details and may even be your superior in his knowledge of his problem. Instead of permitting this to be a cause of concern to you, you should welcome this type of situation. You are faced, remember, with a customer who will understand and appreciate the value of what you are offering him. Certainly, he will be able to ask penetrating questions, and this means you must be thoroughly prepared. But isn't this *always* the job of the professional salesman as differentiated from the order-taker?

One of the facets of this problem that causes some salesman to be slow in asking for the order is the customer's demeanor—his manner, his attitude. Some of them appear austere, even haughty. You have seen the customer who has his desk placed so that as you approach it you have a growing sense of your relative insignificance as you mentally compare it to the area in which *you* work—if, indeed, you are fortunate enough to have space assigned to you. If you permit this kind of situation to affect you, it is time that you analyzed the true significance of these "acts," because acts are indeed what they are.

To accomplish this, it is necessary for a moment that you sit behind the desk of this chief decision-maker, be he civilian or military. Now, sitting there, what do you see as your problems? Consider these thoughts that beset the individual who is really in a key position in your customer's organization:

1. Most people who try to contact you *want something*. They want you to buy their ideas, their product, their service. They seek to impress you because they feel you have it in your power to aid them in their careers. They want you to lend your name or your support to activities they sponsor. They want your money or the money of the organization that you represent. Most of all they want some of your precious *time*.

2. Generally, people do not approach you, in your high position, to provide you with help because they cannot conceive that it is possible for them to help you. They feel that in your position, you are above any need for *their* help.

3. Who can you really believe? Who can you really trust? Who among all of these claimants on your time does not have "an iron in the fire"? Who among them will have the courage to tell you something that he knows will be unpleasant for you to hear? Therefore, who among them will be telling you the truth? Which of them will choose, instead, to coat the truth with an attractive veneer?

4. Who among them aspire to your position? How many of them think you are in your high position merely because you were lucky or because you "had connections"?

5. You notice that, even at social functions, they try to impress you. Don't they ever give up? And, now, what is this latest claimant upon your time trying to get out of you?

Keeping yourself in the chair of this key executive a few minutes more, the job begins to lose some of its glamour, doesn't it? Do you begin to see the need for erecting some kind of a defense? What would some of your defensive actions be? Wouldn't you take some of these actions:

1. Above all, conserve your time. You always thought it would be nice and soft "on top" with those below you doing all of the work. Now, you find that the constant stream of decisions to make can be even more exhausting than "work." You find that the decisions at this level are more complex. You find that the battles for power are more deadly and more constant at this level than anything you have experienced before—because the stakes are greater. So how do you conserve your time so it can be more effectively utilized?

a. You make yourself difficult to reach; not impossible—but just sufficiently difficult as to discourage those without pressing problems.
b. You delegate to subordinates and require them to handle "details." When they do approach you, your time can be spent in analyzing the facts you expect them to supply so that you can make your decision quickly and with a greater certainty that you are right. You also find yourself irritated, often at what you sense is an attempt to dodge decision-making by your subordinates.

2. You insulate yourself from the many who would whittle away at your time with small talk and subtly-told tales of their own accomplishments in their attempts to impress you.

3. You adopt a demeanor intended to discourage time-wasters. You surround yourself with trappings and with an environment that are intended to impress those who approach you and to cause them to confine themselves to the broad issues of their problems. Like a chameleon you fit into... you grow up to your new job.

4. Not because you *really* feel superior, but because you seek relief, you consort with people in your own economic or executive bracket both in and out of the organization. With them, you have a common understanding of the price you are paying for "success."

The foregoing may make the behavior of "the top brass" of your customer's organization more understandable. The trappings of his office, his attitude and his demeanor are doing exactly what they were intended to do.

Don't make the error of thinking that this particular customer would never be smart enough to have thought this all out. He doesn't have to be smart... these reactions become instinctive.

With this understanding of his outlook, it becomes easier to develop a pattern of selling behavior that will make points with this kind of customer. Basically, just ask yourself what you would like to find in a salesman if you really were behind the desk of this key decision-maker. Aren't these some of the things you would seek from that salesman:

1. Be prepared, knowledgeable and concise. Save his time.

2. Paint your story with a broad brush. If he wants details he will ask for them.

3. Have proof. Don't try to bluff. He is where he is because experts have tried to bluff him and have not succeeded in their efforts.

4. Show no resentment at his attitude. You can see, by now, that it is not intended to be discourteous; it is not an indication that he feels superior. It is a form of protection he has developed because it has worked.

5. Keep in mind that his surroundings are intended to impress you. Don't rely on the wisecracking advice that you may have been given; i.e., "picture him in his shower bath without the nice rug, furnishings and office." (He may look better in the shower than you do!) Now keep in mind that these surroundings *are* symbols of his success, and they are intended to remind you that he is well paid—therefore, his time is costly to his company. Be careful of how you ask him to use that time. Show respect, and give him what he wants: brevity, information, answers and help.

6. Avoid small talk, flattery, unnecessary details, ingratiating devices.

7. Don't comment, at least not on the first call, about items you see on his office walls or desk, unless he introduces the subject. But do file these status symbols (symbols of station) away for future reference as windows to his self-image; i.e., evidence of the "me he tries to project"—even evidence of the "me he hopes someday to be."

8. Don't let him see that you are awed by his surroundings. These may be old hat to him. They no longer impress him, and if they impress you it may cause him to downgrade you.

In summary, it should be said that the mood, the demeanor, the office and the surroundings of this key decision-maker are not capricious. He is a rational being and does things for a reason. You have seen that often his reason is to make you aware of his importance in the organization and aware of the value of his time. *Respect* both his position and the demands on his time, but don't be over-awed by them. Sure, he is a success in his field and you are in yours... *or you will be if you get his order and others like it.*

Not expecting an order

Most of the obstacles that occur at the moment when we should ask for the order are created by the saleswoman, and this obstacle of "not expecting an order" is no exception.

"*I didn't ask for the order because:*
"I could see he wasn't interested."
"I could see he was only looking."
"I could see he couldn't afford to buy at this time."
"I could see he couldn't afford to buy—period!"
"I wanted him to have time to think it over, and I wanted to avoid high pressure."

These are only a few of the reasons salesmen give when they are asked why they didn't ask for the order. Then they are surprised to hear that a competitive salesman *did* ask for the order and received it...that same day.

Retail salesmen are often heard to say, "I could see he was only looking," or "I can spot a shopper a mile away." (Shoppers are employees of a service organization that check up on the honesty and/or sales approaches of the salesmen for the store's management.) These are usually poor excuses for inadequate selling.

Analyzing the reasons salesmen give for failing to ask for the order, we are forced to ask these questions:

How do you know he was not interested or only looking?

How do you know he couldn't afford to buy?

Did you consider how your company could give him the help he may have needed in order to buy?

What would have happened had you asked for the order? The worst that could happen would be for him to say "No."

"No" does not close the door on a sale. "No" can be the base upon which you build your final, and successful, presentation. For example:

The customer answers your request for an order with, "I'm not ready to buy...I'll think about it and let you know." Agreed, that if he truly wants more time to think about it, you must avoid any impression of subjecting him to "high pressure." You can, however, use this opportunity to smoke out the reason why he is not yet convinced. "Certainly, Mr. Snow, I want you to give it some thought. I'd like to make sure that I have given you all the facts. Are there any questions in your mind that I have not answered?"

This may bring out some questions, in which case you have a new chance to sell and, again, drive for a close. On the other hand, he may say, "No, you've covered everything." In this instance, you can probe further: "I know you want to be certain and want to think this through, but I cannot help but wonder if you are concerned about...(installation problems) (terms) (possible price drop) (our ability to deliver) etc."

Your choice of the question to use (similar to those listed above) will be based upon your knowledge of the customer, his problem, his general mood, the questions he has already asked, and questions other similar customers have had. When you ask direct questions such as those above, there is a good chance that he may respond with, "Yes, I *do* have a question on that."

You were not afraid to ask for the order, and as a result you have now been able to narrow down the field of his possible objections and/or concern into one that you can do something about. At the very least, you will be sure you are talking about something of interest to him because he has indicated that this is so.

Saleswomen often fail to ask for the order because they are using the self-defense mechanism of "projection." In other words, they may be selling an item that is expensive. They know that, with their income, they would give such a purchase much deep thought before they would invest the kind of money represented by the price of their product or service. For this reason, they really do not expect the customer to buy—yet. The trouble with this kind of thinking is that the saleswoman is *not* the customer. What may appear to be "big money" to the saleswoman does not appear to be big money to the customer, who may handle much larger purchases daily. The customer sees the product in terms of *need*, and that need may loom much larger in his mind than the price.

It is dangerous to avoid asking for the order when you know your presentation has been complete. The customer may interpret this as evidence that you, too, have reservations about your product. There are many buyers who deplore weakness in a salesman. From start to finish, you must be sold on your product—and your attitude must indicate that you *expect* an order. Make this expectation felt in the language you employ throughout the sale.

Avoid such statements as:

"This motor *would* solve your conveyor problems."

"I *think* you'll like this feature."

"I know you'll *want to think* about this."

Use, instead, this type of statement:

"This motor *is going to solve* your conveyor problems."

"Here is a feature that I *know* will interest you."

"When *will* you want us to ship?"

You can't go wrong in asking for the order. At the very worst, you will identify the customer's objection and be in a position to answer it effectively. If you fail to ask for the order, his buying mood may be broken, or he may buy from someone else while you are giving him time to think it over. Remember, when we leave a vacuum, some competitor is standing by ready to fill it.

What to say?

One of the reasons for the failure to ask for the order is that some salesmen actually don't know what words to use at this most crucial point in the sale. Their fear of a turndown is so pronounced that obvious courses of action open to them become clouded.

The salesman who has experienced something akin to panic when faced with the necessity to ask for the order might, with profit, consider some of the general courses of action that follow:

1. Show, by words and actions, that you expect an order.
2. Use questions; get the customer to commit himself point by point.
3. Suggest a course of action that will, ultimately, have but one possible destination—an order.
4. Seek areas of agreement and use these to launch your close.
5. Ask for a larger order than you expect to receive so the customer can "compromise" and reduce the order to the size you actually hoped to get.
6. Simply ask for the order!

Examining the possibilities of these general courses of action, we should consider *first* the course of action which is most obvious; i.e., the last mentioned—ask for the order. Essentially, when you ask for the order you are using a test question to determine how effective your presentation has been. You are about to find out whether you have gained and held the customer's attention; whether you have gained his confidence; whether you have established a need and caused him to see it as a high priority need. If the customer rejects your bid for an order at this point, it means that you must identify the area in which you have not yet been successful.

Take note that we did not say, "You must identify the area in which you have *failed.*" You have not failed until your customer has made the purchase from some other supplier.

If the customer does not agree to buy when you ask for the order, you should listen intently to the words he uses. These can provide you with clues that will suggest the course of action you should now follow. For example, he might say, "I'm not ready to buy just now." This is far from saying, "No, I am not going to buy from *you.*" He is implying that he will probably buy—later. Just how much later depends upon you. By judicious questioning you may be able to identify the reasons for his delay. We usually delay or procrastinate when we have unpleasant decisions to make, or when we are in the middle of a tug-of-war between two opposing decisions. We can be made to act in one of two or three ways:

1. If we are shown that the cost of delay is higher than the cost of taking action.

2. If we can be shown that we have more to gain by taking one action rather than the other.

3. If we are shown that taking one action will be far more costly than taking the other. Therefore, when the customer indicates that she "is not quite ready to decide," she has given you a valuable tip, and it now becomes your task to decide which of the three foregoing actions will be most valuable to pursue.

If you are, frankly, "scared stiff" when the time comes to ask for the order, keep in mind that the task will not become any easier if you delay. Once you have made your presentation, you face the *one best time* to ask for the order. To delay may be fatal to the sale.

Only a few weeks ago, a chap approached us to obtain a contribution to an institution. As we listened to his presentation, we found our interest in the project increasing. We had, mentally, decided to give a rather large contribution. Unfortunately for the institution, the chap never got around to asking point-blank for the order. His delay gave us time to think the matter over, and as we considered other similar demands on our resources, we decided to reduce the contribution by better than 70%. This 70% was lost because the "salesman" did not ask for the order when our buying mood was at its peak. All that was needed was a "little push" on his part and the sale would have been his.

So, in considering ways of asking for the order, don't overlook the most obvious, the most direct route. "May I have your order now?" The chap who failed to ask us for our order was afraid to ask for the size contribution that we—and he—had in mind. This could not help but have an effect upon us. It leads us to think, "That *is* a rather large donation—in fact, *too* large... I'll reduce it." Don't be afraid to ask for a large order—the customer always has the option to reduce it; and this, in fact, causes him to feel that he is dominating the selling situation, that he is not being "high pressured," that he is making the decision, not you.

You can also use the seemingly "indirect" approach in which you imply that you are going to get an order: "May I phone your order in now so that we can insure immediate delivery?"...."What model (color, size, etc.) do you want us to deliver?"... "Where do you want us to deliver it?" ("or install it, etc.")... How soon would you like us to provide you with drawings?"

In each of these examples, you have given the customer a choice. Certainly, he can always react with "Wait a minute—I haven't given you an order yet." "Don't rush me." If your indirect bid for an order elicits this type of response, you can always answer with, "Gosh, I'm sorry. I had no intention of pressuring you. I just thought that the advantages of (this product or service) were so desirable that you'd want to move quickly... but if there is something I haven't covered, or if there are questions I have not answered, I'll be glad to do so, because I'd like to see this in your plant" (or "in your home" or "on your shelves," as applicable to your product and your market).

So... what has been lost by asking for the order? Not a thing. You have backed off graciously; you've put him at ease, reaffirmed your desire to serve him further and given him a chance to ask questions. You've also made him feel that he is dominating the selling situation. You have followed the principle that you should act as though you expect the order.

A radio announcer concluded his commercial with, "Buy several of these suits for the summer." This approach is based on the tendency a potential purchaser has to think, "I may buy *one*, but I'm not about to buy *several*." The announcer's goal may well have been to sell us just one. His words are supported by the same solid psychological grounds that prompt us to "overstate the customer's objections." He will almost inevitably say, "I didn't say *that*. What I said was...." And he will go on to understate his previous objection—giving you something with which you can successfully cope.

There is but one obvious danger in asking for too large an order (in the hope that he will reduce it to the size of order you really had in mind). If you do not handle this with care, he may get the idea that you are not aware of his real needs. Your recommendation must be reasonable.

Use questions; get the customer to commit himself point-by-point. "I see that you feel this is a good price, Mr. Burke?" Burke replies "Yes." "And you feel that it meets your production requirements?" Burke answers, "I'm sure it will." "Well, Mr. Burke, production means money to you and delay means expense, so will you let me send in your order now so you can start benefiting from the installation as soon as possible?"

The foregoing is a simple example to indicate the nature of this technique. It can be used far more subtly in the sale of sophisticated equipment, and the questioning can be more extensive than that indicated in the example. You can ask questions on a number of points that have been covered in your presentation. They should be aspects on which you know your customer will be sold and, therefore, you can anticipate, with certainty, that his answers will be uniformly "Yes." With every "Yes" reply, he is making it more difficult to respond with a "No" when you finally, point-blank, ask him for the order.

Suggest a course of action that will, ultimately, have but one possible destination—an order. "Mr. Kramer, I'd suggest that you consider buying this basic unit *now* and then add other units as your production needs increase. If you agree, I'd suggest that we get this first unit ordered today."

"Mr. Draper, I'd suggest that you order (quantity) feet of this cable now, as a starter." In this approach, the customer has the freedom of choice. He can reject the quantity you recommend but may still give you an order.

Seek areas of agreement and use these to launch your close. "I certainly agree with your views, Mr. Rogers, on heat dissipation problems, and our men in the lab feel the same way. You're going to be delighted to find that this unit won't raise the temperature a degree in your (plant, store, home, etc.). I'd like to suggest that we get an order in now. You're going to get a lot of satisfaction out of seeing your ideas on heat dissipation proven through performance."

Once again, you have given the customer the feeling that he has dominated the selling situation. At the same time, you are *not* manipulating him. As has been said so many times in this book, you are merely moving the customer to a decision...a decision which you sincerely feel is going to be in his best interests. He will decide favorably *if* you have earned his confidence and supplied him with sufficient proof, assurance and attention. You are not working against his interests. Keep in mind that as a rational being he will not make judgments that are opposed to his vital interests.

Buyer's attitude

Salesmen have been known to fail to ask for the order because of the attitude of the customer. The salesman may shrink from a belligerent attitude, be it real or simulated; retreat, discouraged in the face of apathy—and apathy can be simulated as well. The salesman may despair in the face of seeming disinterest on the part of his customer, and he may squirm in the presence of a well-informed customer. Only this last reaction is justified—and the salesman *should* squirm if he enters the selling situation less prepared than a well-informed customer.

The customer's attitude is not accidental or capricious. Attitudes are reactions to situations. An attitude is only a visible (or audible) reaction to a set of stimuli to which the customer has been exposed. His attitude provides him with some kind of satisfaction which, at the moment, he finds desirable. Our task becomes one of causing him to see that a different attitude will promise him a satisfaction that is *more* desirable. It will be of value, at this point, to appreciate the reasons why customers adopt more common attitudes. We are not going to be able to *change* an attitude unless we understand the *reason* for it, and then neutralize or eliminate that reason.

Reaction of a customer:
Belligerent

Possible reason:

Our timing may be bad; he may have been upset by an experience that does not concern us. We may be facing displaced aggression. (This is not to say that a customer is belligerent *only* for this reason. He may be angry because of something we have done—or he thinks we have done. He is not going to change his attitude unless this situation is rectified.) In any event you can expect emotion to sway him, for the moment, and no amount of logic is going to change him. A strategic withdrawal may be in order, unless withdrawal will leave the field to your competitor. More desirable may be to wait it out and let him get the anger out of his system.

Another reason for belligerence is the customer's awareness that he has been wrong, and he is resorting to the strategy that the best defense is a good offense. Your job is to find an escape hatch for him.

A third reason is that for any one of many reasons he does not feel he can buy now. Perhaps he "has been told" by some higher authority what and where he must buy. He does not want you to know that he is not the decision-maker. Perhaps he does not have the funds to enable him to buy and does not want you to know this. Perhaps he is not technically competent, does not understand your proposal and does not want you to know that. Often, his common defense is to attack so you will not "discover his secret." The trouble is you *must* discover his secret if you are going to be in a position to aid and guide him.

Strangely enough there is a species of professional buyer who does not respect the salesman who will not fight back. Before you decide to embark upon such a tactic, you had better *know* that this is the reason for his belligerence. To "fight" a customer who is justifiably angry is about the worst course of action to pursue. So, once again, we are face to face with the rule: *Know your customer.*

Reaction of customer:
Evasive, delaying, procrastinates

Possible reason:

The reasons for these attitudes may, again, stem from some of the reasons given previously (not the decision-maker, no funds, etc.). More likely the customer is being torn between two decisions—to buy from you or not to buy; to buy from you or to buy from someone else; to buy your product or to put his cash in some other project. When faced with difficult decisions people often resort to these reactions. The task, as stated previously, is to cause them to see that delay may be more costly than making a decision.

Reaction of customer:
Confuses, discourages, "substitute act"

Possible reason:

The reasons may well be the same as those discussed in the foregoing. When the customer attempts to confuse or discourage you, it is his way of saying, "Please, just go away." He will resort to the "substitute act" (for example, talk for hours instead of buying; or go into a long, apparently angry tirade instead of buying). When we are not permitted to do that which we would like to do, we are frustrated. Frustration is *pain*, and rather than supinely submit to it, we vent our emotions through some other substitute action.

Reaction of customer:
Apathy, disinterest

Possible reason:

If the customer wants to buy but for some reason cannot (lacks authority to do so, lacks funds to do so, etc.), he will often act disinterested. If he allows himself to become interested, one of two things will occur: (1) He will become most unhappy because he cannot take the only action that will bring him satisfaction; i.e., buying. (2) If he becomes interested and shows it, and then does not buy (because he cannot do so) he has "tipped you off" to the fact that he must not have the authority or funds to do so—the very things he does not want you to discover. Therefore to protect his self-image, he will resort to apathy or seeming disinterest.

Unless the customer is so angry that he is in no mood to listen to reason, your best course of action may be to ask for the order. This may smoke out his real objection. If your efforts bring forth a "No," you have a battery of actions available to you.

Consider asking for his help—not the "I need this order, Mister" type of approach, but one of the following:

1. "Mr. Silvano, somewhere along the line we must have done a poor job today, and—because I do feel we have the answer to your need—I'd appreciate it if you'd help out by letting me know what we have failed to explain."

2. (For the antagonistic buyer) "Mr. Beauchamp, I may have unintentionally offended you in some way and I'd appreciate it if you would straighten me out on what it was—because I'd welcome an opportunity to discuss this proposal with you."

3. (When you know a personality clash—even if one-way—exists, you may find that the straightforward approach successfully used by courageous professional salesmen in such situations, will help) "Mr. Kuhn, I can't help but feel that for some reason you don't like me, and I'd sincerely appreciate knowing why." (Another version: "Mr. Kuhn, I sense that you don't like my company. We'd really appreciate knowing what we have done.")

This type of question usually takes the wind out of the sails of most men because none of us likes to be considered unfair. Even when we dislike a man, we find it possible to appreciate a frank attitude...if that attitude is reasonable and not aggressive. A straightforward request (as indicated) will often cause the customer to realize he has been guilty of displaced aggression and has been taking out his physical or mental discomfort on you.

In two recent instances, two customers who had refused to buy from a concern for 20 years changed their attitudes in five minutes when the salesman asked them directly why they didn't like his company. One instance was due to a misunderstanding; the other was due to a fancied grievance that had colored the customer's feelings for years. In each case, once the grievance was given voice and "off the chest" of the customer, his attitude changed. In one case, the customer laughed about it later.

If you are tempted to refrain from asking for the order and withdrawing from this difficult selling situation, keep these thoughts in mind: (1) The customer may be in the same frame of mind—or worse—on the occasion of your next visit; (2) A competitor may call upon him in the meantime and find him in a better mood.

Try to get him talking about his favorite subject; i.e., himself. Be a good—and sincere—listener. He may reveal the root of his difficulties, and you may be able to provide him with some relief.

We have discussed the fact that some executives surround themselves with expensive office trappings because they feel insecure, that without these external evidences of their position they would not appear to be important. Others need these tangible evidences of success to bolster their self-assurance.

By the same token, a customer will resort to belligerent or antagonistic attitudes to protect herself. She may fear that if she listens to your presentation she will be convinced that she should spend money.

She doesn't want to do this so she doesn't give you an opportunity to convince her.

Others have a stereotype of salesmen as fellows who try to take advantage of customers. Therefore, they adopt the "hard guy" attitude to let you know they are not "soft touches."

There are still others who, though they have the authority to purchase, feel they are underpaid or that they and their responsibilities are underestimated by people in general. They may suspect that your income is higher than theirs, and they resent this. They may be piqued because they are tied to a desk and that you have a greater degree of freedom. Often—and this is especially true of the chap who "starred" scholastically at school—they have had more extensive formal education and they resent the self-made man or the chap who makes greater progress without as much formal education. These attitudes produce frustration, which, in turn, gives birth to antagonistic feelings.

Such attitudes provide the buyer with a sense of power and give him a temporary feeling of dominance. But you simply cannot permit these attitudes to stop you from asking for the order. That is not to say that you plow ahead oblivious to the customer's antagonism and ask for the order. You attempt to identify the cause of his feelings and then to make him feel important, successful, powerful, or whatever need he feels impelled to satisfy. You are selling more than product satisfactions—you are selling psychic need satisfactions.

We repeat: The worst that can happen is that you get a "No"... and you may get a "Yes." If you don't ask, it is a cinch that you are not going to get the order. If he says "No," you can still seek the reason, change your strategy and end up with an order.

In his day one of the nation's leading sales trainers, the late Jack Lacey spoke throughout the nation on what he called "Find his hot button." This was Jack's catchy way of saying, "Identify his key buying motive... his *reason* for buying." Jack's advice can certainly be of value when you confront the hostile customer. You *must* identify this chap's psychic needs. His attitude is not product oriented. His is a complete psychological reaction. If he is smarting because he feels he is not recognized, recognize him. If he fears high-pressure, use low-pressure. If he fears making a wrong decision, provide him with proof and assurance. If he feels that the salesman dominates the sale, let *him* dominate it by letting him do the talking.

Some years ago, *American Salesman* carried a true story of a young salesman who, following his manager's instructions, would call upon a hostile buyer. The buyer had his desk at the top of a flight of stairs. As the salesman would mount the first step, the buyer would shout down, "What do *you* want?" (with a bit of mild profanity). The salesman would reply with, "I'm Johnny Babson from Brown & Co., and I'd like to show you our line." Whereupon the buyer would shout back... again with a profane overtone: "Well, take off. I don't want to see it."

The manager learned what was happening and gave the young salesman some advice. On the next visit, the salesman was again greeted with the roar from the top of the stairs, "What do *you* want?" This time, he shouted back, "I want to show you how to make $10,000!" To this, the buyer shouted back, "How?" By this time the salesman had reached the head of the flight of stairs and launched his presentation... successfully, it might be added.

The task with the buyer who wants dominance is to focus his attention on something he wants more than he wants to dominate you. Have your product advantages ready in language that will be meaningful to him, not general benefits. Make them crisp and clear, and hit your strongest benefits first. Don't make a litany of all of your benefits—some of them he may not see as benefits, and you'll only bore him. You can't lead up slowly to your strong points with this kind of buyer... you must hit hard with the advantages that will have the maximum impact upon him.

Is customer ready yet?

"I don't know if he is ready to place the order." This thought races through the mind of the salesman, so he hesitates and loses the sale.

There is *one* way to see if the customer is ready... Ask for the order. So he says, "No!" There are really very few *final* decisions. As we have said, the decision is not final until your competitor has the order, and we can recall some exceptions to that condition where the customer cancelled an order. If he isn't ready to buy, he'll say so. Now find out why he isn't ready. Does he need more facts? More assurance? Testimonials? Demonstrations? Proof? Supply what it is that he needs.

If you wonder whether or not he is "ready" to order, probe with direct or rhetorical questions:

"I gather our delivery time will be satisfactory?"

"I take it that I have answered all of your questions?"

"I'm sure you see this is the time to buy."

If the answer to these, or similar questions, is agreement or the nod of his head, you *know* that this is the time to ask for the order. If the reaction is silence or a "not completely," you know where you must hit harder... what advantages to stress, or what losses to emphasize that will certainly occur if he does not buy.

Naturally, you hope for a series of "Yes" answers to your questions. As has been stated previously, each "Yes" makes it increasingly difficult for him to say "No" when you finally ask for the order. Each "Yes" deprives him of a reason for procrastinating. For this reason, select questions that you are *sure* will produce an affirmative reply. Then ask questions in the areas where you feel there is a chance he may as yet be unconvinced. The first "Yes" answers create a good climate for the balance of the questions and lead you, ultimately, to the areas where your attention is needed. This strategy will help you to uncover his hidden or unvoiced objections or concerns.

Size of order jitters

"Size of Order Jitters" is, like so many other obstacles, the result of the self-defense mechanism of projection. In effect, the salesman says to himself, "If I were the customer, I'd regard the amount of money involved in this transaction as one 'whale of a lot of cash'" . . . or, "If I were the customer, I'd regard the price we are charging as high." To both of these thoughts, the salesman adds, "So, I'd resist attempts to get me to part with that kind of money, and so will my customer." Seeing "price" from this frame of reference, the salesman hesitates to ask for the order or fails to ask for it, and another bit of business is lost to competition.

Let's clear up one point that may arise in your mind if the products or service you sell are not "expensive." This "size of order" obstacle can exist if we are selling products that require the outlay of relatively small amounts of money. For instance, how many times have you walked away from a counter because you thought a store was asking too much for an item . . . whether it was a cigar, a box of chocolates or a ball-point pen. Viewing the article from your frame of reference, and comparing it to similar items you have purchased in the past, you regarded the price as unreasonable and refused to buy. It is the subconscious recall of this type of experience that causes the salesman to fear that *his* customer is going to rebel at price or at the sum of money involved.

To overcome this obstacle, it is necessary to understand the reason for it. The reason, as we have seen, is that we see "price" from *our* frame of reference, not from the customer's. You and I don't live in $250,000 homes, but there are people who do. You probably didn't pay $35,000 for your car, but there are people who have paid that much. If you should pay $400 for a suit, it is a reasonable bet that you don't stock up on a number of them at one time, but there are people who do.

Look at the sale through the eyes of your customer. Eleven million dollars is a lot of money to you and to me, but the Air Force will spend that much on spare parts for a jet engine. Five million dollars represents quite a pile of bills, but an electric utility will spend that on equipment. Five thousand dollars is not "peanuts," and neither is a thousand, but countless small businessmen spend that kind of money regularly. If your customer is considering a purchase, he has undoubtedly considered the cost and has set aside a reserve in his budget for the expense.

Keep in mind, too, that in most instances, the salesman does not need for his own use the product or service he is attempting to sell to the customer. The customer *does* need it (or must be made to see that he needs it). When a customer senses a need, it changes his entire frame of reference. What he would regard as a high price for something it would be "nice to own" becomes a reasonable price for something he feels "he must own."

Keep in mind, also, that the price of the product or service you sell was arrived at after considerable thought. This "thought" may have ranged from extensive—and expensive—market research of a product to the judgment of the individual who established the price, based on his past experience. The fact is that you do have a price, and unless you are selling a completely unique and new product, other customers have paid that price. So, why should we assume that this customer will regard it as "too high"? The fact is that it is not price that is our chief concern, it is our fear that we *have not done an effective job of justifying price or it is fear that we have not been thorough in causing the customer to recognize his need for our product.* Often, we use the excuse of "price" for our failure to have done an effective selling job.

Overcoming "big order jitters" or "high price quakes" calls for a reasonable appraisal of the situation, based on the solidity of the following:

1. Your price is established.
2. Your company feels it is a fair price for what is offered the customer.
3. Your customer is not living in a vacuum; usually he has at least a rough idea of what kind of money is involved—your price is not going to come to him as a total shock.

The simple question the saleswoman must face is this: "Have you made his product or service need so vital to his interests that the satisfaction of that need is greater to him than the economic satisfaction he obtains from holding on to his money? Have you, in a word, made the need you are attempting to help him fill appear to be more pressing than other needs he has?"

If you *must* worry, your anxiety should not be over the customer's reaction to the price; rather, it should be concerned with whether you have correctly identified the customer's problem and have stressed, adequately, the right product features in terms that will cause him to see the product as something he *must* have.

Consider for a moment the fact that when you ask the customer for an order that involves a large sum of money you are recognizing "the me he wants to be." In other words, you are saying, in effect: "I recognize in you an individual who is accustomed to dealing in sums of this size." At the same time, you avoid creating the impression that you are flippant in your regard for money. That type of attitude will provoke from the customer the irritated response, "Wait just a minute... that's *my* money you're talking about."

Ask for the order. Be prepared with solid justifications if he does rebel at price or at the total sum of money involved. Try to "change the setting" *by presenting the price in terms that appear less formidable to him; i.e., cost per day or cost per employee.* Viewing the *total* (which may actually spread out over years, even though it must be paid now) may shake up the customer. But when he views that same amount of money in terms such as those described, he sees it as an expense that he can budget and meet.

Turn to the Sales Strategy Bank and select other strategies such as *(D7) Draw comparisons, (E4) Emphasize the cost of a wrong decision* and *(F4) Cite losses he will sustain if he does not buy.* Point out what each day's delay may cost him. Make both advantages and losses seem immediate... not something he will gain or lose over a long period of time. You and I, and the customer, respond more quickly to immediate threats and gains than we do to something that "may happen" in the dim future.

Don't try to "kid" the customer by minimizing cost with such statements as, "It's only $10,000." This will get you the "What do you mean—'only'?" response. Instead, use statements of this kind: "For an investment of $10,000 you are going to make yourself a substantial profit, save yourself expensive lost time that old equipment can cost, and you'll be ready to meet the demands of the expanded market that you and I can see is ahead of us this year."

It should go without saying that each salesman should examine such recommended statements as the foregoing and revise them to meet the needs of his unique selling situations. It is the *principle* that the salesman must recognize and then choose language that will be effective. In the example given, this specific customer is interested in money, or dollar-related benefits would not have been highlighted. Notice, too, the careful selection of words: "Investment"... not "expense." "Lost time" is not mentioned solely as lost time but as "expensive lost time." The expectation that the customer is going to buy is emphasized by the phrase, "You'll be ready" ... not "you *would* be ready" ("*if* you bought").

Analyzing the statement used above, you might find fault with the phrase, "you'll be ready to meet the demands of the expanded market that you and I can see is ahead of us this year." You might well contend that the words "this year" lack immediacy... and they do. The words "just ahead" might be better—or, you may find other terms that drive home immediacy even more effectively. Just as we hope you will view selling statements recommended in these pages with a critical eye, so do we hope that you will review your own contemplated selling statements with that same critical view. Words are tools and powerful motivating devices... choose them carefully, scrutinizing their possible effectiveness from the customer's viewpoint—not from yours.

Don't try to play down price by avoiding clear statements. Recently we saw an engineer looking at some amateur radio equipment. He asked the price, and the salesman replied, "It's eighteen five." The look of perplexity on the engineer's face was obvious. Just what did "eighteen five" mean? Did it mean eighteen hundred and five dollars or eighteen thousand five hundred? It developed that $18,500 was the price, and the equipment was shipboard equipment that some millionaire radio amateur would buy in order to have the best. At $1,805, the engineer would have been interested.

There are times when buyers are *not* familiar with price structures on some kinds of equipment. Rather than betray their ignorance, they will find smoke screen objections and refuse to buy. Furthermore, they will never forgive you for subjecting them to embarrassment. Hiding the real price behind confusing terms only creates the impression that you are afraid to state the price.

Avoid sowing the seeds of doubt in the customer's mind. We witnessed an almost unbelievable example of poor selling. A couple shopping for draperies for a den saw some interesting burlap material which was cleverly designed. When they inquired about the price, the salesman said, "Oh, they are very expensive—they are $300." The customers, who had expected the drapes would be much more than that, now began to have second thoughts. Instead of being pleasantly surprised at the price, they began to think of $300 as high priced and eventually bought less costly drapes. Here, the salesman was projecting. He would never buy drapes at $300 and considered them expensive. Unfortunately, he conveyed this viewpoint to the customers. Result: lost sale.

"They are an excellent value at $300" would undoubtedly have produced a sale because when the couple left, they were heard to inquire, in reference to the $300 item, "Is that an open stock?" In other words, their interest remained and an effective salesman could easily have sold them.

Regardless of what it is your job to sell, look upon every shopping experience of your own as an educational opportunity. Watch for the errors—and the hits—scored by sales people at the retail level. Then extract the principle that caused them to make—or lose—the sale.

Ask for the order! State your price if necessary and if she rebels, justify the price. Be prepared with counter-proposals.

Impressed by customer's rank

This obstacle has been covered previously under "Front Office Jitters." Hence, there is no need to deal with this one aspect of that problem.

Fear of turndown

This is the next obstacle to appear on the Anatomy of a Sale chart. It has been discussed under other facets of the problem, but there are these points to be added:

Fear of turndown is essentially a fear of failing. To many salesmen, the selling situation is an arena of conflict. It represents to them a matching of wits. With others, it is a struggle to dominate.

All of us have self-images and most of us like to think we have powers of persuasion. We like to think that we are right in our convictions and that we can persuade others that we are right. If the customer does not buy, we may interpret his refusal as an inference that we are wrong, and we begin to feel that our self-image is less than realistic. Some salesmen tend to lose self-confidence when they lose an order.

For these reasons, some salesmen will stall throughout a sale and avoid the ultimate necessity for asking for the order. They feel that a negative response from the customer will cause them to feel inadequate. One popular writer, with whom we do not agree, states that salesmen actually *hate* their customers because the salesmen know customers have the power to prove to a salesman he is not all that he would like to believe he is. Any salesman who interprets a lost sale in this fashion is being unnecessarily rough on himself.

It has been said that "a fool can ask a question that it would take 10,000 wise men to answer." Any customer can say "No." Plagued by other problems that would take a modern-day Sherlock Holmes to detect, the customer may not hear a word the salesman has to say... no matter how eloquently he may have said it. A lost sale should not be regarded as a personal defeat or as evidence that the salesman is not capable. It *should* be looked upon as an experience from which some principle may be extracted and which, if observed in the future, may lead to far more important sales.

The key point to keep in mind is that, ultimately, you must ask for the order. Procrastinating will not make it easier when you eventually must take this action, and it may also cause the problem to become infinitely more difficult because the customer's desire may begin to wane. The real "trick" in asking for the order is "timing"... watching how the customer reacts, listening to what he has to say, to the questions he asks and deciding when the psychological moment has arrived to convert his interest into the necessary buying action.

Perhaps the true nature of a selling situation can be made very clear through this analogy. Some years ago a congressional committee produced a film called "Operation Abolition." The film turned out to be "controversial," which is the 20th century way of saying that it could be interpreted in totally different fashion by two or more viewers.

The film focuses on the activities of a crowd of thousands of people who attempted to bring the deliberations of the committee to a halt. Two viewers could sit through a showing of the film, seated side by side. One sees the chanting of the crowd as the actions of a group who, in the opinion of one viewer, were either ill-informed, subversive or dupes of Communist Party agitators. The other sees the same crowd as a group of libertarians who were determined to protect individual freedom against what they labeled the unfair tactics of the committee.

The film shows the crowd refusing to move. The assembled people joined arms, sang and shouted. The police moved in, played fire hoses and then literally slid the demonstrators out of the City Hall. One viewer of the film sees the action of the crowd as unlawful obstruction of a duly constituted arm of Congress and wonders why the police didn't get "tougher." The other viewer sees the demonstration as the outpouring of free men and the actions of the police as "brutal."

Following the film, the two viewers could argue for hours without either convincing the other. Each sees the events from the vantage point of his own milieu. Each interprets what he sees in the film in terms of his past experiences, the extent of his knowledge of the basic problems underlying the demonstration, and in terms of what he has been taught is legal, moral and ethical. Each is influenced by hearsay... by other reports he has heard of the demonstration.

As each of us reads of these incidents, even at this late date, we will agree with one or the other of the viewers. Someone is going to find it most difficult to change our minds. In fact, if we feel strongly on the subject, he will find it impossible to change our minds. Furthermore, as you have read these last paragraphs, if you have recalled the film and/or the demonstration, you have undoubtedly analyzed what has been written here to see if your side of the controversy has been presented fairly.

The same sort of phenomenon takes place in a selling situation. You are trying to convince the customer. He is going to be influenced by his own past experiences, by the impact of your offer on what he holds important (his self-image, his status needs, his economic situation, his product problems), and by what others have told him; i.e., hearsay . . . even as you and I.

For instance, you and I may hold definite views on the demonstration referred to even though neither of us may have been at the scene of the disorder. In spite of this, we will have strong feelings as to who was right. Therefore, if similar influences have a bearing on the customer's viewpoint and cause him to see your offer in a light that is unfavorable to you, it is not necessarily an indication that you have failed to act as a professional salesman. It is simply that your words, your actions, and your appeals to what you believed to be his psychic needs *have not been as powerful as the impacts of situations he has lived through.*

At best, in a selling situation, you are attempting to cause the customer to see your proposition through your eyes. He may have lived through situations which cause him to view your proposition as not being in his interests. Unless you can identify those situations, you may not sell him. We cannot always know why the customer feels as he does, but we *can* cause him to see the advantages he will gain by buying from us or the losses he will sustain if he fails to buy. This awareness may neutralize or overcome the effect of his past experiences.

For this reason, failure to consummate a sale should only cause us to be determined to be better prepared with more information on the customer the next time we contact him. Failure to sell is not a tragedy. Failure to sell should not affect our self-assurance, but failure to learn from failure merits all of the criticism we may receive.

Customer wants a discount

"I didn't ask for the order because I felt that he was going to request a discount that I could not give him."

The foregoing is certainly no valid excuse for failure to ask for the order. You can't fight a problem until it is a problem . . . until it is out on the table and you know you have a problem. If he answers your request for an order with a request for a discount, you can then consider actions open to you.

There are a number of strategies available in the Sales Strategy Bank. Among them are these:

(D4) Provide alternative: What else can you do for him, legally, that will be as valuable to him as a discount? Are there any forms of assistance that other departments of your company can provide him?

(D13) Appeal to fairness: Admittedly this will not be effective with some customers we know, but it will with others and should be considered. The tack: What you can't do for others you can't do for her.

"I know that you would resent it if I gave some competitor of yours a discount that I couldn't give you . . . you'd be pretty upset and I couldn't blame you. I have always found you to be fair (or: "you have a reputation for fairness") and I'm sure you'll see why I can't do anything by way of a discount."

Or: "I know you don't want us to lose money by selling to you and because you have a payroll of your own to meet, you'll see why we have found it necessary to adhere to our price structure."

Is it legal? If applicable, point out limitations and restrictions on your freedom to grant discounts. If he points out that this does not stop some competitor (named or unnamed) from giving him a discount, point out these facts:

"Mr. Ramsey, as you know, these laws were enacted as much for your protection and mine as they were to protect others. If we weaken these laws, you and I may be hurt by the unfair practice of others. If one of my competitors breaks these laws and gives you a discount, how can you be sure that he won't give an even better discount to one of *your* competitors? We are offering you fair and equitable treatment with the assurance that none of your competitors will be given an edge over you. Because you *are* fair, I'm sure that the fairness of this approach will appeal to you.

Customer wants preference

"I hesitated to ask for the order because I felt that he was going to counter by . . . (1) asking us to give him delivery before others who are ahead of him . . . (2) asking us to let him use his brand name on our product when we won't let others do so . . . (3) asking us for help we can't and don't give others . . . (4) asking us to provide features we don't provide to others."

These are only a few of the miscellaneous reasons why some salesmen hesitate to ask for the order. We feel that the antidotes have been spelled out in covering the foregoing obstacles. Appeals to fairness, explanation of legal restrictions, granting of alternatives are some of the choices of action open to you.

In concluding our treatment of the "customer wants preference" obstacle, it should be mentioned that the new salesman often wants hints as to words to use in asking for the order. The choice is great, and the words must be adapted to your kind of selling. As has been said, they should convey the impression that you *expect* the order, without giving an impression of high-pressure. The choice ranges from the simple, straightforward, "May I have your order today?" to:

(Giving him a choice)—"Do you want Model X or Model Y?" (Or x quantity or y quantity, etc.)

"How soon will you want us to make shipment?"

"When will you be ready for us to make installation (or delivery)?"

In many instances, actually asking for the order will be no problem. The trick is to know when to start writing. If the customer asks such questions as "When *will* you be able to deliver?" answer his question but get your order book out. When he departs from the conditional "would" and uses the assumptive "will," he is practically giving you the order . . . start writing.

CHAPTER 17

Taking the order

When do you really clinch an order? Is complete success achieved when the customer agrees to buy your product or service?

Most sales training programs that we have examined place great emphasis on all of the tactics that must be used by the salesman in his attempts to have the customer say, "O.K. I'll take it!" A few programs devote some attention to the need for follow-up after the sale. Rarely, if ever, do sales training programs give any attention to the element of the selling situation which we have chosen to label "Taking the Order."

It seems to be assumed that taking an order poses no problems and that it is automatic. Your own experiences, as a customer, indicate to you that this is not true.

For example: Have you ever emphasized to a waiter that you wanted steak without onions or without gravy and then have had it placed in front of you, smothered in both onions and gravy?

Or, have you ever had a new house built for you? If so, have you had the experience of outlining clearly to the builder or to some contractor *exactly* what you desired and then, when the job was completed, found that the builder either hadn't listened or had not comprehended?

There *are* hazards at this point in the sale, and this chapter will discuss four of them. To appreciate the importance of the "Take the Order" element of a sale, we must consider the end objective of a professional saleswoman as compared to that of an "order taker." The order-taker's objective is usually to "sell merchandise that won't come back to a customer who will." The professional's objective is to help her customers solve problems and fill needs and to establish and maintain lasting, friendly relationships.

The professional is not in business as a salesman solely to obtain today's order, no matter how big it may be. Success of the professional salesman, and his company, is represented by the continuous flow of orders that is received from customers who have found that doing business with this salesman and his company has been profitable, satisfactory and even pleasant. If we keep this objective in mind we can see that a sale cannot be considered complete when we ask for, and receive, the order. Two elements of the sale remain: (1) take the order and (2) work for continuance.

To "take the order" is to obtain *full* information on the customer's needs, to record it in written form and then to submit it...and make certain it is received and understood by those people in our company who must follow through. We share the responsibility with order service personnel to insure delivery of the right product to the right address in the right quantity at the right time.

In many of the businesses represented by readers, this is no real problem. There may be little chance for error. The order may be so simple that a mere forwarding of a catalog number, quantity or date will suffice. And on across-the-counter sales, the salesman actually completes the order on the spot.

In many other businesses represented by a large number of readers, however, the orders taken may be of such proportions that every detail is included in a contract, or so many people are involved that the opportunity for error is magnified. One member of the selling organization assumes that someone else is taking care of a detail. One or more of the "non-selling" people in the organization, who never come in contact with the customer, fail to think of him as an individual—a unique individual. As a result, they fail to note exceptions to the norm which appear on the order form. These people have yet to learn that there is no such thing as "non-selling personnel." Everyone in the organization contributes to the successful completion of the sale. It might well be said that the sale is not completed until the customer re-orders.

Even the customer can contribute to the problem. In one instance, a customer told his supplier that from now on when he bought an item, he would not require an accessory that had always accompanied the item. He pointed out that he would manufacture this accessory himself. The supplier argued against the wisdom of this, pointing out that he had enjoyed many years of experience in manufacturing this accessory and knew that it would be compatible with the equipment being purchased. The customer was adamant and insisted that the equipment be delivered without the accessory.

Result? The customer failed to follow through and notify his own people. The product was delivered without the accessory, and the customer blamed the supplier. Why? Just another customer employing the self-defense mechanism of repression.

The customer has an image of himself as a competent businessman who doesn't forget important things. Therefore, when he does make a mistake, he conveniently "forgets" his instructions to the supplier. The psychologist says, *"He restructures the situation."* You and I say he lies in his teeth. What are you going to do about it—tell him so? Better to concentrate on preventing the reoccurrence of such situations. What could this salesman have done to have precluded this situation from arising? What can *you* do to avert this sort of incident in your kind of selling?

You may have "done a slow burn" as you considered the injustice of the true incident just related. True enough, it was unfair and it was unjust on the part of the erring customer. Unfortunately, being unjust and unfair is a luxury he can afford in a buyers' market. Furthermore, it's a safe bet you will be blamed if the customer makes an error in placing his order. There is always a good chance that, being human, the customer will resort to the mechanism of repression to protect his own self-image—i.e., "forget" facts that would tend to make him appear less than competent.

You and I might have some "right" to be indignant, were it not for the haunting feeling that, as customers, we, too, on occasion may have resorted to this same mechanism. Have you ever assembled something you have purchased, failed to read the instructions, made a mistake and then complained to the supplier that the item "won't work" or arrived in bad shape?

Hates detail (either party)

Years ago, we knew a manufacturing foreman who for a while enjoyed the favors of his superiors. When a new product was turned over to him to manufacture, he enjoyed every minute of it. He would walk around the plant with drawings in his hand, select his new subordinates and ostentatiously tackle new problems every minute. Then, as soon as his production line was running, he would become bored, lose interest, take his attention off the job, and spend most of his time visiting with the engineers "where the action was." The result was that his subordinates, lacking leadership, would create problems that spawned high costs and lost production. The inevitable result was the foreman was released. Yet, for a while, because of being in the spotlight, he had been the envy of the other foreman.

There are salesmen who follow this same pattern of behavior. Many salesmen are happiest "in the heat of battle," matching wits with the customer or with a competitor. As soon as the customer has capitulated and agreed to place the order, this type of salesman might lose his effectiveness. He, like many of us, can become bored with the usual, the mundane, and with detail.

Some companies have recognized this problem and assigned another type of individual to service accounts. But this is not always possible from a cost standpoint. Nor is it always desirable. It deprives the salesman of the opportunity to exhibit a personal and continuing interest in the customer and in his operating problems. This is not to say that the idea of a team—a professional salesman and a customer service specialist—is not a good idea. It may well be an excellent idea for insuring that "the order gets taken" when you are selling to retailers. The customer service specialist can remain behind to help the retailer with displays, checking inventory and so forth.

In the great majority of cases, the salesman is not going to be given this specialized assistance and it will be *his* task to take the order. There are conditions against which he must guard. It is not always the salesman who hates detail—some customers are of a similar nature. There are customers, who, once they have given you the order, will respond to all of your inquiries for detail or specifics with, "Oh, you go ahead...you know what we need"...or, "We'll leave it to your judgment."

Beware of this approach. It is most pleasant to hear because it creates the feeling that the customer recognizes our competence and our integrity. But...make just one error in writing up the details of an order and you'll be witness to the outraged reactions of a customer who apparently has no confidence in your judgment and seems to doubt that you have any integrity whatsoever.

Just fail to point out to a customer that there is more than one way of solving his problem...more than one product available for his consideration... and take him seriously when he brushes aside your efforts to make him decide with a, "Oh, I'll leave that decision to you." Then let him learn later that there were choices open to him that you did not *insist* he listen to and consider, and again, you'll have an irate and unfair customer on your hands.

What's the antidote? If a customer leaves some aspect of an order to your judgment, always get in touch with him, preferably in writing, in person, or if necessary by phone, once you have decided how to handle that detail. Use something like the following approach: "Bill, I know that you left color to my judgment as you didn't feel this was of any particular importance to you. I do want you to know that we are specifying blue on the order. I want to be certain that you are aware of this before our people fill the order. Is blue satisfactory?"

Consider having the customer initial the specification, as rent-a-car companies ask you to initial the little square indicating that you agreed to daily insurance coverage. At the same time, don't be misguided into thinking that either of these actions is going to prevent the customer from blaming you later if, after delivery, he finds he doesn't "like" blue. Many cases in our files indicate how often the customer will still blame the salesman. So, you may ask, "Why bother, if he is going to blame us anyway?"

The answer is that in bringing the matter to his attention, he may decide, before it is too late, that "he doesn't want blue." Admittedly, in many instances, the customer's initials or signature on the specs may cover you legally, but we're not in business to win lawsuits... we are in business to keep customers.

In summary, just as some salesmen like only the exciting facets of selling but hate detail, so, too, do many buyers like the challenge of negotiating but hate detail. Attention to detail to insure proper filling of the order is going to build confidence on the part of the customer that makes the next sale easier. If the customer hates detail, make it easier for her... but don't let her get away with leaving it to you to make decisions for her unless you are absolutely certain as to what she wants.

Sales strategies to consider

From the Sales Strategy Bank, *D2—Narrow down his choices*. If he has agreed to buy, and a decision must be made on such things as color, size, style, method, model or other features, make it easy for him. Don't confuse him with too many choices or you may get from the customer the, "Oh, you decide."

Use *Save him time (B1)*. If possible, summarize advantages and disadvantages of each of the choices available to the customer. Keep her business or personal goals in mind and hit hardest on those actions (or decisions) which you believe to be in her best interests, so that she will be inclined to favor one course of action over others. Don't play up advantages which might be of real interest to other customers but are of little or no interest to her.

When you are sure you are on solid ground, *D3—Provide the solution*. In other words, recommend a course of action or a choice.

When you are working with a customer who hates detail, you have an opportunity to cash in on this situation. You can drive home the fact that you're not just looking for an order but that you are truly interested in *him*, and in helping him fill his needs or solve his problems.

If you know your customer hates detail, you must recognize the danger of annoying him by persistently directing his attention to what he will—at this point in time—consider petty details. You can, sometimes, use this approach: "Mr. French, there are a number of details upon which decisions ought to be made at this time so that we can be certain to supply you with exactly what you desire. I know you are very busy, and if you feel you can't spare the time, is there someone else who can speak for you that I might see... otherwise I would like a few minutes to square these points away?"

Or this approach has worked successfully: "Mr. Brandon, I know you are extremely busy, and I'm pleased to get your order. We are vitally interested in making certain you get exactly what you need. I'll need specific information on certain points, and I need your help. If you prefer, you may wish me to see someone in your organization who can give me the decisions I need, though, naturally, I would prefer to work with you. For example, I need some information regarding the normal load under which this equipment will be working. I wonder if you could help me obtain this information so we can give you the best service possible?"

Normal speaking time for this type of statement-question is about 32 seconds, and you'll find it a wise investment of your time. Analyze the words used and see what they accomplish (though you should use words that come naturally to you):

1. It provides Brandon recognition... *"you're extremely busy."*
2. It indicates your thoughtfulness... *"you're extremely busy."*
3. It indicates your desire to serve... *"we're vitally interested"; "...exactly what you want."*
4. Uses the strongest human appeal... *"I need your help."*
5. Provides him an escape hatch... *"someone in your organization."*
6. Appeals to prestige... *"naturally, I would prefer to work with you."*
7. It is precise... *"need some information regarding the normal load."*
8. It is customer-oriented... *"so we can give you the best service possible."*

The statement leaves the final impression you are thinking of his interests. It is offered as an example and not as the very best approach you can use. You may think of more effective words. It is presented because it drives home the power of words and illustrates how many facets of a customer's interests and psychic needs can be hit in a few well-chosen words. It drives home the wisdom of (1) making it a habit to know *exactly* what you are going to say in a selling situation, and of (2) analyzing before you enter that situation the words you are going to use, and then of (3) choosing your words for maximum impact.

As we have said, some salesmen hate detail. The questions may be raised, "Is that bad?" "Aren't salesmen basically creative people?" The answer to both questions is "Yes." It is bad, and salesmen should be creative people. Inventors are also creative people, but they know that to transform an idea into reality requires much attention to detail... to boring experimentation... to research.

You are attempting to create a favorable and lasting relationship with the customer so that your next sales will be made more easily than today's sale. To accomplish this, the ultimate results of today's sale must be satisfactory from the customer's viewpoint. These results will not be satisfactory if delivery is late or if the final product or service is not in accordance with the specifications or desires that the customer had in mind when he placed the order, *even if he did not make these clear to you because he wanted to avoid detail... or because in your efforts to avoid detail you left without getting the answers.*

Salesmen have been known to adopt the dangerous attitude epitomized by the statement, "I'm employed to deliver the business... let the people in (the office, back at HQ, the order service department, etc.) get the details." Regardless of how much help he can—and should—expect from other people in the company in processing orders, the salesman has the responsibility to make certain that the final order sent in by him represents what the customer wants.

If you feel that this does not apply to your kind of selling, try this experiment. See your product service people (or the people who service customer complaints, regardless of the title they may have in your company). Determine the nature and frequency of customers' complaints. Then make a list of those you feel might have been prevented had there been a clear understanding at the moment the order was accepted... or had the salesman sent in clearly defined details of the customer's needs. Then *use* this as a checklist.

Keep in mind that by attending to details in this selling situation, not only are you safeguarding his order, but you are making the next sale easier. You are saving time because it takes much more time to handle a complaint than to prevent one. This saved time is precious because it can be used for selling.

Afraid to push "luck"

Once the customer has given the salesman the order, there is sometimes concern on the part of the salesman that the customer really is not "completely sold"—that if given half a chance, he may change his mind and cancel on the spot. The salesman, understandably in such instances, feels that the more rapidly he can get distance between the customer and himself, the safer the sale is going to be. As a result, he may depart from the customer without having obtained details. The results are not difficult to imagine.

Is there valid reason for this concern on the part of the salesman?

It may be conceded that, in some instances, a customer may change his mind if given the slightest opportunity. This possibility does not justify leaving him in a hurry, without obtaining essential details of the order. Obviously, if the customer wants to cancel, he can do so, and more easily by a phone call later... or more easily yet, through a letter.

If you leave him before obtaining essential information, there is an excellent chance that your company will supply him with a product that fails to meet his needs, and this will be disastrous. If you leave without the information, you'll be forced to return—and he will again have the opportunity to change his mind. Or, worse yet, someone else in your organization, less skilled in meeting customers, may have to call him for the required information. The customer now has an additional opportunity to change his mind, and may find it easier to do so if he doesn't have to face you.

There are, admittedly, times when it may seem desirable to increase the distance between you and the customer. For example, you may have convinced him that spending his money on your product will bring him benefits greater than those he will receive by buying other things he has desired. You know, however, that the "tug" of these other desires are exerting a powerful pull on him and that the best strategy is to make it difficult for him to change his mind. So, you strive to insure speedy delivery.

In such instances, there is nothing wrong in leaving the customer as soon as possible, *if* there is someone else in his organization from whom you can obtain required details. Otherwise, leaving him makes no sense. You must still contact him for information, and each call gives him another chance to reconsider and cancel. As is the case with almost every selling obstacle, this one can be turned into an advantage. Ask for the necessary details in a manner that accents the advantages he is going to receive, and emphasize your vital concern over his welfare.

For example: "Mr. Boyd, I want you to start making the additional savings that this is going to produce, as soon as possible.... So, I'd appreciate it if you could take just a few short minutes to give me a few specifics we need in order to give you speedy delivery and to give you exactly what you have asked for."

Obviously, in forming this sample sentence, you will select that facet of your proposition that has been of greatest interest to him. We chose "savings" in the example; you would replace "savings" with the word that meant the most to your customer.

This problem area has been called, "Afraid to push luck." Your closing of the sale has *not* been due to "luck." It has been because of your ability to correctly diagnose the nature of his product and psychological needs. Selling him is far more difficult than keeping him sold. If he is going to change his mind, he will do so just as readily after you leave as when you are with him. In fact it is much easier, as we have implied, to cancel over the phone or via the mail, when we don't have to face the salesman. If you are afraid he is going to change his mind, capitalized on this as you ask for additional information, keeping him thinking of the advantages that will accrue to him.

At the same time, avoid over-selling or talking yourself out of a sale. Don't mention any advantages (as you seek details) that you did not mention during your original sales presentation, unless you *know* he will consider them as advantages. You made the sale without them. What may appear to you to be additional advantages may remind him of shortcomings of the product, or may appear to him as disadvantages and may cost you the sale. Repeat the advantages that you saw were so significant to him that he gave you the order, but do so only if you find it possible to tie-in the benefits to your request for detailed information.

Doesn't know technical details

Salesmen sometimes are not familiar with the technical details that are involved in the installation or operation of the product they sell. When such a salesman has just made a sale and needs more detailed information to place the order with the factory (or store), he may fear to expose his lack of knowledge of details. He feels this exposure might destroy the credibility of the arguments he has just used, and the customer may change his mind and cancel the order.

What to do in such a situation? As a general strategy, gear your action to the conditions and to the personality of the customer... in other words, don't look for the "one best way" to handle this obstacle with "all" customers. Here are some specific suggestions:

1. Leave the customer as soon as possible and try to obtain the required information from someone else in your organization.

2. Leave, but return later with a technically competent person from your organization. Make the customer view this as an extra service that starts even before delivery.

3. If you must have someone from your organization see the customer alone, be sure to brief him on the situation. Point out any hazards that exist, explain the points you made to get the sale, and what you avoided and why. Give him as much information as possible about the customer as a person and, if applicable, why you feel it would not take much to change the customer's mind and lose the order. Brief him on how to approach this specific customer. Tell him to base his request for additional information on your company's desire to give the customer exactly what he requires in the shortest possible time so that the customer may start obtaining the potential benefits with the least delay.

4. Consider what the customer's reaction will be if you point out that you are thoroughly familiar with the benefits of your product because you have seen the results, but that you need her help in terms of additional information in order to place the order.

The real stopper for this obstacle is, obviously, to know your product and not get caught in this type of situation again.

Afraid to lose control

Many saleswomen have been nurtured on the philosophy that they must *control* the selling situation at all times. This type of saleswoman worries when, after obtaining the order, she now finds it necessary to ask questions of the customer. She seems to feel, instinctively, that she is turning over control of the situation to the customer.

Nothing could be further from the truth. You are actually making the customer a part of the solution. You are also giving him some relief. He has been through the rough ordeal of making a decision and may entertain some lingering doubts about his decision. Now you provide him with an opportunity to discuss something about which he is uncertain... his needs, his problems, his operation. At the very least, you are giving him a chance to do some of the talking. You have probably said more than he has during your session together, and he will welcome the opportunity to do some of the talking.

Keep in mind the power of questions. The professional conference leader uses them to start discussion, to keep it moving, to stop discussion, to keep it on the right track, etc. He recognizes the question as one of the most effective means of *controlling* a situation, and he chooses the type of question he will use as carefully as a golfer chooses his club.

You will find, as a professional salesman, that you have a wide variety of questions at your command (to be covered subsequently in Part Three, the Sales Strategy Bank). Use them carefully and you will find that, more than ever, you "control" the selling situation. After all, when you ask the customer a question, *you* have decided what trend the conversation is to follow. You are not the one who is on the spot; it is the customer who must answer the question. When he is busy answering questions, he can't be thinking up new roadblocks to toss your way.

Directing appropriate questions to the customer keeps his mind off subjects that he might otherwise dwell upon and weaken your presentation.

Summary

The order is not complete until you have all of the information that you, and others in your organization, will require if you are to satisfy the customer's requirements. Don't shirk this responsibility and pass it off on someone else. That someone may be less skilled than you in meeting customers, and he may lose the order.

Don't fear the situation that is created when you must ask for details. Grasp it as an opportunity to drive home benefits and to show the customer that you are interested in far more than "getting the order"...you are interested in satisfying him completely.

If you know the customer is not technically qualified, don't embarrass him. Give him an escape hatch. Point out you know he is busy and suggest he steer you to someone that he feels is competent to give you the information you require.

Don't oversell. Don't bring up advantages you didn't mention when the customer gave you the order. You completed your selling when he gave you the order. Now you are starting post-sale service... stick to that.

CHAPTER 18

Working for continuance

Problems in this area of follow-up—post-sale interest—are largely self-generated. The Anatomy of a Sale chart lists only six of these obstacles. You may be able to add to the list if you will examine the reasons for customer dissatisfaction. In this chapter, we will confine ourselves to a discussion of those obstacles which we create and which are most common. It is in these examples of malpractice and nonfeasance that we find the reasons why we sometimes encounter difficulty in making successful repeat calls and in establishing a clientele that could provide us with continuous income with less effort than that which is expended in obtaining new business elsewhere.

Is sale oriented

All too often we struggle to make a sale and then, having tasted success, we relax and forget our obligation to the customer. We forget that the sale is not consummated when the customer has given us an order and has paid for it. In making the sale, we are often called upon to make promises which we, or others in our organization, are committed to keep. We have a responsibility to make certain those promises have been fulfilled. Often, we have assured the customer that buying our product will bring him the satisfaction he seeks. Do we not, then, have an obligation to assure ourselves that he has received that satisfaction?

The salesman who does not consider these obligations is, indeed, single-sale oriented, and there is a real likelihood that he will pay a high price for this selling myopia. This salesman is certainly unaware of the true nature of sales work. He certainly cannot consider himself a true professional. He does not understand the meaning or philosophy of *customer-oriented selling.*

You and I are in the business of selling solutions to problems. The sale is not complete until the problem has been solved, and you, I and the customer believe, nay *know,* it has been solved. For a moment, compare your post-sale behavior with the post-operative behavior of another professional—the doctor. What would we think of the doctor who showed no interest in us once he had removed his gown and gloves and walked out of the operating room? Would we be inclined to use his services the next time we needed medical or surgical aid? The answer is obvious. We would consider him a quack. The customer who sees little or no evidence of interest on the part of a salesman after the order has been placed cannot be blamed for looking upon the salesman as nothing more than an order-taker.

Our job is to provide a service to the customer. Our success as individuals, and therefore the success of our company, will be in direct proportion to our ability to provide that service. Your company, through its management, has established long-range business objectives. You, as a professional salesman, should have established long-range business and *personal* objectives.

Long-range objectives are not attained by hit-and-run methods. The salesman who regards his task as completed when the order is received and paid for is doing a disservice to his customer and to his profession. More than that, he is providing fuel to stoke the fires of those carping critics who seem to delight in their attempts to convince undergraduates —and the buying public—that salesmen are glib, fast-talking "con" artists who have no interest in the customer's welfare but seek only to relieve him of his money.

The salesman whose interest wanes or evaporates once the sale is made is also doing a disservice to the image of his company. He is also throwing a roadblock in the path to his own long-range progress and security.

Almost every company experiences periods of "good" and "bad" business. Often, we attribute these to the inevitable (?) gyrations of the business cycle. We experience periodic recessions which we are prone to attribute to tight credit policies imposed by government or banking interests, to over-production or to the psychology of fear that builds up when newspapers speculate on the possibility of recessions. These may, indeed, be legitimate reasons for business fluctuations; but even during the worst recessions people have needs.

It is reasonable to speculate on the possibility that some companies will weather the periods of general business decline better than others. Will you agree that the success of these companies may be due, in large part, to the excellence of the relationships they have established with their customers by virtue of their continuing interest in those customers, as evidenced by their post-sale interest in the customer's welfare?

In spite of this logic, you and I have been witness to deplorable post-sales interest on the part of salesmen in some marketplaces. The following example is not to be interpreted as a blanket indictment of automobile salesmen, but it does focus on the shortcomings that are evident in a sales field with which we are all familiar.

For instance, when did you buy your last car? When did you hear from the salesman again, after you drove that car out of the agency? It is a safe bet that your next communication from him was in the form of a Christmas card...and most likely a commercial-type card at that. Perhaps you did not hear from him again until he mailed you an announcement of the availability of new models the following season. Even then, the odds are that he didn't call upon you with one of those new cars, all wrapped up in its hypnotic new car smell. He probably sent you an announcement and stereotyped invitation to "drop in and see the new models."

Or, were you one of the lucky customers who had the salesman call you a week later to ask you how you liked the car? If you had, for example, expressed a desire for gas economy, and he had assured you during the sale that this was an economical car, did he call you a month later and inquire regarding your experience with gas mileage? Did he call you later and advise you that the guarantee period was about up and inquire if anything required attention? Did he call you some months later to remind you that it was probably time to rotate your tires if you wanted to obtain maximum possible mileage from them? Did he ever do *anything* to demonstrate a sincere and continued interest in you and in your transportation problems? Did he do anything to make you *want* to buy your next car from *him*?

To sense another's hurt and to want to *feel* that hurt; to sense another's need and to want to *fill* that need (Overstreet in "Mature Mind") is not only the root of true social justice; it is also the root of true professional selling.

Completing a sale means not only filling the customer's product and/or service needs, it also includes filling his psychic needs; i.e., his unique need for recognition, for dominance, for comfort, for prestige, for pride of ownership, or whatever provided the stimulation that caused him to buy from you. Selling is not complete until you have developed a strong bond of mutual respect and admiration between you and the customer so that, when he develops new needs, you will be the first upon whom he will call.

You have undoubtedly experienced a sense of great relief when you have finally completed a sale with some particularly obnoxious customer. If you are like most of us, you were happy to put distance between you and that customer. On the other hand you don't find it difficult to return and visit the customer whom you found "easy to like." Yet the business and the cash that represents the buying potential of the chap you didn't like is just as lucrative and redeemable at the bank of your choice as that of the pleasant customer.

Since you completed the sale with "Mr. Ugly," what have you done to make him like you? Have you gone out of your way to provide him with the satisfactions that are apparently so important to him... the desire to dominate...to be listened to...to exhibit his superiority. You didn't have to *like* his methods of seeking his particular kind of satisfaction, but you can ask yourself how you can supply that satisfaction. If you don't, you can be sure that someone else will...and he will obtain the continuing business, which could be yours if you would identify those needs and supply them after the sale is closed.

Forgets promises

Turning our attention to this self-imposed obstacle causes the author to do a bit of squirming. It is only days since a client, of whom we are especially fond, called to our attention the fact that we had not sent him some materials we had promised. This is the "stuff" of which lost sales and poor customer relationships are made. As salesmen, we certainly do not *intend* to break promises, but we do *forget* them.

The effect is the same upon the customer. He doesn't care what the *reason* is, he just knows that you failed to keep the promise. There is a good chance that he regards it as an affront to his self-image. "You apparently don't think I'm important enough to remember." He won't give voice to this thought, but it is there nevertheless. He has the impression that you made the promise to clinch the sale and had no sincere intent of keeping that promise.

Often, after the sale is made, because we sincerely like the customer, we promise him something over and above that for which he paid. When we forget the promise, and he suspects our motives, we sense a feeling of true "hurt" because we know that the promise was overlooked in the pressure of a heavy work load or other demands upon our attention. *We* know this, but *he* doesn't. We know our intentions were above reproach, but the road to hell is paved with good intentions—and the same road leads to strained customer relationships, lost friendships and lost business.

Making a promise to provide something extra after the sale is smart business. It is an honest tactic that provides you with an opportunity to contact the customer again and to demonstrate your continuing interest in him as an individual and in the solution of his business problems. When you call upon him for the sole purpose of keeping a promise and do not dilute the effect of the visit by asking him for anything, the customer cannot help but experience a feeling of appreciation.

Don't depend upon memory. Keep a small notebook in which you list *only* promises you make to customers. Couple this with a tickler file, and if you are lucky enough to have a secretary, have her remind you. Make a practice of starting the day off by referring to the book. Avoid the manana habit... take care of the promise now. It won't be one bit easier to tackle tomorrow.

During the course of the sale or when the sale was completed and you experienced that warm sense of gratification, appreciation and satisfaction, you made the promise...and you really meant to keep it. After the sale, with the order book tucked away in your pocket, the promise may be forgotten...but its impact upon the customer is not so easily forgotten; nor is the impact of your failure to make good on that promise. Don't think that you do not have this human tendency to forget. Protect yourself by keeping that book.

Avoid making gratuitous promises when the customer has not asked for anything. The effect upon the customer will be greater when you provide him with the additional service later. If you have made the promise and keep it, he will appreciate it—but he will also feel that he has a right to expect you to keep the promise, and its effect will be somewhat diluted. If, on the other hand, you have not made a promise, but now—after the sale—provide him with "something extra" that he had not been told he could expect, this continued interest in him will have a far greater effect.

Has no follow-up plan

Too many of us follow a pattern of giving the squeaking wheel some grease. We go along, from day to day, placing our attention where it seems to be needed the most. The customer who makes the most noise gets our attention. The other chap, who asks for little, gets little. We tend to give our attention to the big account, forgetting that, more often than not, small companies are growth companies. When they do become large companies, we often find that they have forgotten us because our lack of attention indicated that we had forgotten them.

Itineraries and schedules, often imposed upon us by the "system" or by the boss, prevent us from being even worse than we are. The professional salesman has a plan. He lists *all* of his accounts down one side of a matrix. Across the top of the matrix, he lists the 52 weeks of the year. Then he keeps a record of how often he has given attention to each account. Better yet, he timetables himself on that same matrix, indicating the date on which he will again contact each account. This results in a balance and protects the salesman from his human tendency to call upon "his most important accounts" and to avoid the small or the disagreeable accounts.

If we are going to obtain the maximum value from this discussion of self-imposed obstacles, we must develop a willingness to be introspective—to be self-critical. No improvement can possibly take place if the reader argues, in absentia, with the author on this point. Nothing is gained by muttering,

Matrix for Customer Contact and Follow-Up Plan

CUSTOMERS	JANUARY				FEBRUARY			
	1-8	9-15	16-22	23-1	2-8	9-15	16-22	etc.
Byron	N	✓			✓		C	
Shaw	P			✓		N	✓	
McIntyre	C	✓		P		✓		
Belotti	✓		CD		✓			
Obenshaw	✓		✓		P		✓	
Levine	✓		C		✓			
Matowski	✓		CD		P		✓	
etc.								

CODE: ✓ = Called in person
      N = Sent note   P = Phoned
      C = News or business clipping sent   CD = Card mailed

"With all I have to do...or with my big area...or with the number of accounts I have, how can I call on them as often as this fellow seems to think I should." Much can be gained if you will try to answer that question yourself.

Just what *are* some of the things you can do, *in spite* of these admittedly difficult problems, to let the customer know you are thinking about her? Is scheduling yourself to make a three-minute phone call really an insurmountable obstacle? Is it really an impossible task to drop her an occasional postal card, letting her know where you will be for the next week in case she needs you?

An effective follow-up plan, in the form of a simple matrix, includes not only the scheduling of calls on new, important and prospective accounts; it also includes provision for "call-backs" on customers who have bought from you in the past...call-backs that don't even ask for new business. These call-backs should have as their primary purpose the creation of the notion, in the mind of your customer, that you are interested in him and his problems. The "call-back" can even be a letter, so long as it does not appear to be a form letter.

To this point, some time ago, we had to purchase a copying machine. One month after we had bought the machine...and again, two months later, we received letters from the manufacturing concern, via the sales department, inquiring as to whether or not we had problems or were we experiencing satisfaction with the machine. We were so gratified with this evidence of customer-oriented selling that no competitor of that company, at this writing, could ever hope to obtain a contract for supplying our copy-paper requirements, much less sell us a replacement machine. As a customer, the gratification we received from being considered important enough to be contacted was greater than the gratification obtained from the satisfactory operation of the machine. Indeed, on occasions when we have experienced some difficulties with the machine, we found ourselves making excuses for the manufacturer.

Our own reaction is typical of the reaction your customers will experience when presented with evidence of sincere, continuing interest on your part! If you really agree that the call-back or follow-up matrix is a good tool for self-discipline, why not *do* something about it today?

Doesn't see full customer potential

It's almost amusing to hear people say, "If I had only bought that land five years ago, I could have doubled my money." It is less amusing as we occasionally study the stock market and observe the growth of a stock that "we could have bought" at a fraction of its value five years or so ago. The sad fact is that while we are bemoaning our "lack of foresight" (or courage), we will probably look at the same land or the same stock, five years hence, and give voice to the same lamentation.

So it is with customers. All too often we see them as they are today, frequently failing to visualize them as they will be three or five years from today. This fault affects salesmen in almost every field of selling endeavor, from the wiring supplies salesman who feels that this "little guy" will never have the resources to build another building to the saleswoman selling really high-ticket items. Indeed, it may happen even more often to the latter. Having made a large sale to a customer, the salesman may muse over the amount of money the customer has spent and he may assume that this customer will not be in the market for additional equipment or services for some time to come. Then the salesman learns, too late, that the customer has made additional purchases from some competitor who did not under-estimate his growth potential.

If we will but keep in mind that ours is a customer-oriented mission, this tragedy cannot easily occur. If we make regular call-backs and continue to gather economic intelligence on the customer, we will become aware of the growth potential, or merger or conglomerate potentials—as well as plans—of the customer. The one-sale oriented salesman makes the "one big sale" and then, foolishly, thinks, "We've obtained all we'll ever get from that fellow" and never calls back. There is little wrong with this myopic viewpoint that could not be cured by following these philosophies and actions:

1. The sale is not completed until you have made the second sale.

2. Yours is a profession and all professions have a follow-up procedure; i.e., visits to the doctor during convalescence, de-briefing and post-combat diagnosis in the military, etc. Why be less than a professional?

3. Have a sincere desire to help your customer and you'll find follow-up is not only essential but pleasant...and rewarding.

4. Have a plan that provides self-discipline.

5. Don't rely on memory.

6. Keep a constant watch on your customer's growth pattern, gather economic intelligence on him, set your sights high, don't underestimate his potential...even excite him to the possibilities of his own growth potential and show him how you can help him grow. He'll love you for it.

Geographical problems

Geographic obstacles are epitomized by such laments as: "My territory is too large. I *can't* give the necessary time to follow-up." "I have too many customers to see." "We must keep expense in line...I can't double back...I must follow my itinerary."

These may well be accurate descriptions of the situation you face, but as stated before, there is always the telephone. If you know the customer well enough, you can even call him at home at night with the new low rates—*if* you are calling only to ask if there is anything you can do to help him, not to sell him.

Short notes expressing your interest can always be written when the advertisements are on your TV in the motel room at night. These have the advantage of not taking the customer's time, yet they keep foremost in his mind evidence of your interest and keep him thinking of you as the most active supplier in your field. We have a client who collects stamps and we don't find it laborious or time-consuming to drop unusual or foreign stamps in an envelope and mail them to him.

Keep both your customer's personal and business interests in mind. Keep alert to items that you know will be of interest to him—news items, business opportunity items, etc., and drop them in an envelope and mail them to him. The postage can produce a big return.

Keep especially alert to devices that will help him sell *his* customers or help him solve his operating problems (even outside your sales field). As you encounter ideas, products and/or opportunities that you know will interest him, drop him a brief note... and *don't* couple these with any comments about *your* business possibilities with him. Without such comments, these items will be evidence of sincere and personal interest. Tied to requests for business, they become an artificial and transparent gimmick.

Be sure he has access to you even if it means that his calls must be relayed to you. Be sure the people who handle your calls realize that he has top priority with you, no matter how "small" he may be today.

Poor communication with factory or office

There is some validity to the complaint made by many salesmen to the effect that they can't really take care of their customers because "the factory" (or "the office"... or "the boss") never follows up on notes that are sent in. In many companies, much pride is taken in the excellence of the lines of communication that have been established "to keep the salesmen informed." But all too often, equally effective lines of communication have not been established to permit the salesmen to keep headquarters informed about the needs and problems of the customer that only headquarters can handle properly. Little wonder that many salesmen consider themselves "the loneliest men in the world."

Complaining about this situation will avail you little. Why not make constructive recommendations that, if followed, will guarantee that the voice of the customer can be heard back where some action can be taken in his behalf? For example, some companies issue a supply of bright red envelopes (addressed to headquarters with stamps affixed). The salesmen use these with great care. They are fully aware of the cost of crying wolf. They send these letters in only under two conditions:

1. When they encounter a situation where a customer has a problem that requires *immediate attention* and when, if he fails to get that attention, the account may be seriously jeopardized.

2. When the salesman identifies a need which the customer will have in the future and for which the company has made no provision. In other words, when the salesman recognizes his role as an extension of the arm of the product planner...when he realizes he is in closer touch with potential needs of his customers than any living soul. These messages can lead to new product opportunities.

Every secretary is alerted to the fact that these red envelopes are to obtain *top priority* and are to be put on the top of the boss' mail. If they must be forwarded to a second office, they remain in a red envelope. She tears a coupon off and sends it back to the salesman so he knows, *at once*, that his message has been received. Thus, in the case of customer problems, he can inform the customer that "headquarters" has been notified.

The real cure for faulty follow-up "back at the plant" lies in action by top management to cause everyone in the organization to become aware of the importance of the customer. When this customer— and enough like him—stop buying, every engineer, production worker and office worker is out of a job, because only the customer provides the money to pay salaries. This is so self-evident that one would think it requires no emphasis here. Yet, we can still remember a foreman working, by himself, in a loud speaker test booth at night. The superintendent dropped in and asked, "Freddie, why are you working so late?" The foreman replied, "It's dry in here during the cool of the evening, and these speakers will pass test. If I wait until morning, when it's humid, they will all be rejected."

Anyone with manufacturing experience can testify to the many occasions when he has witnessed arguments that almost become physical between inspectors and quality control people on the one hand and foremen on the other. If we have done a poor job in making the overworked foreman realize the impact of less-than-top-quality products on the customer, how can we expect more of the production worker who is even more divorced from the faceless customer, who will always remain unseen, miles away?

So, we in marketing agree that this can be a problem. What to do? As a starter, why not suggest to your sales manager that when you have the next sales meeting at the plant, each salesman be given an hour (or so) with the 35 or 50 men and women who are kept working because of the sales he makes. Then, show them the importance of this customer and explain how hard they make it for you to keep them supplied with orders when anything goes out of the plant that is less than "Zero Defect" performance.

And if that is too much to expect, why not discuss this idea with your boss...maybe he can do something about it.

PART THREE

The Sales Strategy Bank

CHAPTER 19

How to use the sales strategy bank

In our discussions of the obstacles that have been examined in Part Two, we have suggested from time to time certain courses of action which often have been successfully employed to overcome the obstacle being discussed. You will note, however, that at no time have we provided a "school situation," or "the one best solution."

There are a number of reasons for our reluctance to offer solutions:

1. It would be impossible to give specific answers because of the great variation that exists in the kinds of businesses in which readers are engaged.
2. We have always abhorred the tactics of those sales training specialists who offer "the one best way to overcome problem X." Usually there are *many* answers to a problem, and what may have worked for "America's number one salesman" (whoever he may be) may be the one worst possible action for you to take, given your customer, your personality and the specific conditions under which you have encountered the problem.
3. We consider it demeaning to offer "canned solutions." If canned solutions will work, then we don't need professional salesmen...we can hire just anybody.

The need today is for *thinking* salesmen. And men think best when they are confronted with problems, when they must analyze for themselves the implication of the available facts, when they must consider the relative merits of the various solutions available—and when they must then make their own choice.

In spite of this reasoning, we have never met the saleswoman—no matter how skilled she may be—who can think of *all* the strategies and tactics that may be brought to bear on a selling situation. Nor have we met the salesman who cannot benefit by hearing of strategies that others have used effectively. The Sales Strategy Bank, then, offers three kinds of assistance to the professional:

1. It provides him with a tool that he can use effectively in his pre-sale preparation.
2. It provides him with a checklist of strategies from which he may choose.
3. It gives him greater assurance that he will not overlook a strategy which he may have employed effectively on past occasions.

The task will still be yours to weigh the advantages and disadvantages of the various strategies in terms of how applicable they may be, considering the specifics of the situation in which you find yourself.

Part Three will discuss each of the strategies contained in the Bank, explaining why it is psychologically sound and under what conditions it will be found to be most effective. Letter-number combinations, such as A1, provide you with a speedy method of locating the strategy being discussed, on the Sales Strategy Bank form.

The Sales Strategy Bank

Pre-sale planning

A1 Develop sources of information
A2 Identify his needs and wants
A3 Identify his self-image
A4 Identify his primary buying motive
A5 Consult your predecessor
A6 Get help from other depts. in your company (finance, etc.)
A7 Get boss's help
A8 Identify your competition and its representative
A9 Study competitors' usual strategy
A10 Make an estimate of situation
A11 Determine past experiences of customer
A12 Identify his possible sources of resistance
A13 Prepare alternative offers
A14 Plan strategy
A15 Plan opening tactics

Create good climate

B1 Save his time
B2 Help him
B3 Offer to help
B4 Agree with him
B5 Give him privacy
B6 Do him a favor
B7 Praise, compliment him
B8 Give him credit
B9 Emphasize mutual interests
B10 Welcome objections
B11 Agree
B12 Admit validity of his view
B13 Find point of agreement
B14 Admit your error
B15 Entertain him
B16 Be intentionally incorrect in a statement so he can correct you
B17 Provide him an escape hatch
B18 Listen
B19 Ask for his help, opinion or advice
B20 Sympathize

Compel attention

C1 Provide him with a role
C2 Use cause-and-effect examples
C3 Use testimonial
C4 Emphasize his goals
C5 Emphasize his objectives
C6 Omit something so he can inject his ideas
C7 Confuse him. Or use ambiguous questions
C8 Interest him in more ambitious goals and objectives

Overcome his objections

D1 Give him a choice
D2 Narrow down his choices
D3 Provide solution
D4 Provide alternative
D5 Compromise
D6 Talk benefits
D7 Draw comparisons
D8 Tack with his stand
D9 Use probe for reaction
D10 Divert him
D11 Rephrase and overstate his objections
D12 Get him to talk
D13 Appeal to fairness
D14 Fence
D15 Use direct question
D16 Use overhead question
D17 Use rhetorical question
D18 Use provocative question
D19 Scare him
D20 Let him do you a favor

Overcome his fears

E1 Provide support
E2 Provide assurance or proof
E3 Demonstrate
E4 Explore cost of wrong decision
E5 Remove him from situation with third party examples
E6 Appeal to his pride
E7 Appeal to his prestige
E8 Appeal to the me he wants to project
E9 Give him recognition
E10 Anticipate, raise and answer objections
E11 Stimulate his recall of past experiences
E12 Provide guarantee
E13 Offer additional services

Force a decision (Make him take action)

F1 Bypass him
F2 Shock-antagonize him
F3 Show him cost of procrastination is higher than cost of decision
F4 Emphasize cost of not buying

Post-sale actions

G1 Show continued interest
G2 Make return call
G3 Check operation and/or installation
G4 Has his problem been solved?
G5 Have new problems been created?
G6 Is he satisfied in all respects?
G7 Have all of your promises been kept by your company?

CHAPTER 20

Pre-sale planning

In this chapter we discuss those strategies which the professional salesman can use during that phase of the selling situation which we have called pre-sale planning.

Because the readers of this book are engaged in a great variety of types of selling (industrial, international, retailing, banking, government, as well as consumer and original equipment manufacturers, among others), each reader should determine the most effective manner in which the strategies can be employed in his or her market. Examples given of means to imlement the strategies should be examined and the reader should determine what words can be best employed to produce the desired result in the type of situation the reader faces on the job.

A1—Develop sources of information

No strategy or tactical plan will be any better than the information upon which it is based. Computer people used to be heard to say, "Garbage in, garbage out"—indicating the need for accurate fact gathering and programming.

Taking action with a paucity of information at hand, or with important information missing, is not the mark of a professional salesman; it is the tactic of an order-taker (or order-*loser*).

There's an old gag that goes, "How would you feel if—as you were on the operating table and just going out under the influence of an anesthetic—you heard the nurse tell someone that your surgeon had cheated his way through medical school?" To put it less flippantly, how would you regard a surgeon who would operate on you without taking a blood count, pulse check, obtaining facts on your allergies, past history, etc.? You would undoubtedly, and properly, label him a "quack." He would most certainly lose you as a patient.

Yet, countless salesmen, imbued with self-confidence and assurance, plow into a sale with insufficient information. Many of them succeed in obtaining business primarily because their competitor is no more professional in his approach. But the effective professional is eager to determine new sources of information, and new methods of obtaining information. He is just as interested in making certain that he has not overlooked the kinds of information that may be helpful to him.

Strategy A1 is "Develop sources of information." But we might first ask ourselves what *kinds* of information should be sought, assembled and analyzed if we are to make an effective "estimate of the selling situation."

(1) Certainly we need to know our own product and/or service, and know it thoroughly. We need to know its shortcomings as well as its values, and to know those shortcomings *as the customer will view them*. It would be patronizing if we were to discuss how *you* go about getting information on *your* own product. (2) We need to gather information on our competition, and sources will be discussed later in this chapter. (3) We need to have information on our customer's product needs; hence, his operating problems. (4) We need information on his economic needs (restrictions on his ability to buy as well as his buying potential are just a few of those needs). (5) We need full information on the state of our market. (6) We need information on the customer's psychological needs (whom he seeks to impress and how he wants to be seen).

Before considering sources, the next question you might ask is, "How *much* should we know?" Only you can answer that question. There are some situations in which you meet regular customers where the need for new information is almost nonexistent. Your relationship and their need are such that perhaps your work with them is hardly more than making a regular call and getting a regular order. Even here, however, you run the danger of taking the customer for granted—*seeing him as he used to be and failing to see him as he now wants to be seen and as he hopes to be seen*. If an alert competitor catches you taking the customer for granted, you're in deep trouble.

The guide for determining how much you should seek to learn about the selling situation, is this:

What would it be nice to know?
What would it be well to know?
What should you know?
What must you know?

The very common and very human weakness against which the professional salesman must guard is the tendency to categorize difficult-to-get information as "nice to know stuff" and in this way rationalize himself out of the need to dig in and work to get facts—facts, which if available to and in the possession of a competitor may cost business.

You may next ask, "Assuming that I know my product, my competitor's product, the state of my market and my territory, what *kinds* of information relative to my customer could be of value to me?"

To answer that question, let us provide you with what is only a partial list. (See additional questions in "Customer Profile—Personal" and "Major Account Profile" at end of Part One.) Before you start down this list, we suggest you select a customer from whom you would like more business, or select a potential customer whom you have not, to date, been able to sell. Then ask yourself, as you read these questions, how well do you *really* know him? As you read these questions keep in mind that status symbols are the windows to the self-image (the "tip-off" to how the customer wants to be seen—or how he sees himself).

What is the *title* of his job? Does it really represent the degree of importance of the work he performs? Is it over-titled, and if so, why?

What kind of an office does she have? How much did she have to say about its selection?

How is the office furnished? What does he display on his office walls, desk and/or bookcase? What can be deduced from these items? *Is there a pattern discernible?* Don't make hasty judgments on isolated items.

Is he a "clean desk man" or is his desk and office cluttered? What does the condition signify?

How long has she been in her present assignment? Is she satisfied in it? If not, what does she want? How can you help her get it?

Does he look upon his present job as a stepping stone to something more important?

What kind of people does he seem to like in his own organization? What people does he seem to resent? (Why?—to both.)

When you give him a free rein... *completely free to choose his own subjects*... what does he talk about? How does he present himself in experiences he recounts?

Does he pass along credit to others in his organization or does he take the credit?

What kinds of clothes does he wear? Are any of the clothes he wears commonly considered "status symbols" (brand, labels, etc.)? Did *he* select them?

What kind of person is his wife? Does she have much influence on him or on his decisions? What kind of influence?

What are his hobbies? What does he seem to be seeking in those hobbies? (Are they highly competitive? Merely relaxing? Do they indicate a need to be creative? Or to keep in good physical shape?)

How competent is he on the job? If less then competent, does he attempt to *appear* competent? (If he is *running scared* and lacks assurance, will you be able to supply facts, testimonials, proof, etc.?)

Would you categorize him as a talker or as a listener?

Is he anti-big business? (If you are big, will he think you won't give him service; if you are small and he thinks he is big, will he think you can't meet his product needs—or his psychic needs?)

What kinds of experiences has he had with you, your predecessor or your company? How did those experiences reflect upon him?

What kinds of experiences has he had with your competition? Does he still remember them?

Did he ever work for your company? If so, why did he leave? Did your company ever turn him down when he applied for a position?

Does he like to talk about his previous jobs? If so, why? What did they offer him that his present job does not offer him? How can you supply it?

With what kinds of people does she like to associate? Why?

Is he a name dropper or a place dropper? How can you use this to your advantage?

Does he like to talk about his college experiences or military service? If so, what did they seem to supply him that he would like to have now?

If he did not have military experience, is he sensitive about it?

Is he a self-made man, without formal education? Is he proud of the fact or is he sensitive about the lack of formal education?

What accomplishments does he have to his credit? Books or articles written? Is he considered an authority outside his own company? On what? Military accomplishments? Graduate degrees? Travel experience? Government experience or war-time assignments? Public speaking? Any others? How can you subtly recognize these accomplishments?

What is his primary functional background (engineering, manufacturing, marketing, accounting or personnel relations)? If military, what was his arm of the service and basic branch?

What do you believe to be his most pressing psychological needs (to be needed, wanted, recognized, to dominate, need for prestige, economic security, self-respect, etc.)? How can you best satisfy them in your sales presentation?

Upon what did he place the emphasis in any speeches he has made or articles he has written? Any personal interests indicated?

What was his family background (economic level, etc.)?

How large is his family? Anything unusual about any members of his family (health, achievements, problems, etc. that either worry him or about which he is proud)?

Where does she send her children to school? Can any inference be drawn from this?

Where does he hail from originally? Do you have a common background with him in this or in any respect?

What fraternal orders or service clubs does he belong to? Is he active in them and are they important to him? What satisfactions does he obtain from membership?

Does he consider himself a member of any ethnic group? Does he feel this generates any problems for him? Does this feeling on his part generate any problems for you?

Is he active politically? How important is this to him?

Does he have any pet peeves (attitude toward government, big business, ethnic groups, political parties or on economic questions)?

What seems to be his key orientation? Is it toward: money, quality, price, service, duty, opportunity to exert authority, loyalty or something else?

What clubs does he belong to? Any inference?

When he can choose, what places does he like to patronize for lunch or dinner? Why?

What items does he like on the menu?

Does he drink? What? If not, does he object to others drinking when they are with him?

Does he smoke? What? If not, does he object to smoking "by secondary exposure"?

Anything in the way of subjects to avoid when you are with him...membership in AA, divorce, etc.?

Is physical conditioning important to him? Is he addicted to diets or to talking about diets?

Does he like to be taken to lunch or object to it? What is his "paying pattern"—let you pay, dutch treat, or share?

Are there struggles for power in her company? How do they affect her?

Does he indicate that his employer does not appreciate him? How can you make him look good?

How long has he been with the company? How much do you know about previous jobs and titles he has held there and elsewhere?

How does he feel about his predecessor?

What kind of a home does he have and what kind of a neighborhood does he live in? What implications may be drawn?

Has he had experiences with other people in your company? Good or bad?

At about this time you are probably saying, "I don't know my own *father* that well!" One answer to that is the question, "What if your competitor has these answers and you don't?"

The fact is that, with many of your customers, it is *not* necessary to have this information. But we are concerned with the ones that are getting away, the ones from whom you could obtain even more business and the customers your competitors are attempting to take from you. Obviously, we don't attempt to obtain all of this information at once. But over a period of time, you keep adding information to the dossier you should maintain on each account.

And now you ask, "Do I have a right to know this much about my customer?" This question has been answered in previous chapters. You really don't have a right to fail to know as much as possible about me if I am your customer...at least not if you are going to call yourself "customer-oriented." If I have psychic needs that are as real to me as my product needs, and if you are really interested in me, you'll want to help me fill those needs, too. You'll also know where the "quicksand subjects" are and will avoid them.

Let us not think for a moment that the many questions just posed constitute a complete list. But it will act as a thought-starter and may even convince you that you probably don't know any customer as well as you had thought.

To the individual who says that these things are not important, we pose this question: If I can get a product about as good as yours at about the same price from any number of other suppliers, why should I buy from you? I'll buy from you when you solve needs that I have, ones that your competitor has never thought about—my psychic needs. After all, when a salesman has no product, delivery, price or quality edge, what *does* he have going for him? The answer is found in his ability to demonstrate to the customer that he has a continuing and deep personal interest in the customer's welfare as an individual.

Finally, you may ask: "It isn't easy to get this information...just where do you draw the line?"

That's a fair question, and here's our answer to it: Weigh these considerations when you attempt to determine how much time and/or cash you should invest in obtaining information on the customer:

How much business is involved?

What is his long-range business potential?

How anxious is he to buy from us?

How well does our competitor know him?

Did the customer approach us, or did we approach him?

How much will it cost in time or money to get this information vs his potential?

Can we speedily obtain much of this information from our predecessor or from others in our company?

Does timing permit us to get this information or must we move fast, taking a calculated risk (even a surgeon has to operate in emergencies)—or are we using this as an excuse to avoid the work involved in obtaining the information?

Are moral, ethical or legal restrictions involved?

Will lack of information *probably* cost us a sale, *quite likely* cost us a sale, or *certainly* cost us a sale?

Those are the criteria that answer the query, "How much time and/or cash should we spend collecting information?"

Ways of developing sources of information

Journalists, police departments, foreign ministries and the military all have one thing in common. Their success, to a great degree, depends upon their sources of information, which they cultivate and protect to the greatest extent possible. The professional salesman also has the need for information sources. Now, what are they?

First, listen with your eyes. There is no such thing as an idle word or gesture. Your customer is a rational being; i.e., he does things only for reason. Therefore, the things with which he surrounds himself, buys, displays or uses, have a purpose for him. If you can identify a pattern in his behavior, you'll have a real insight into the customer's psychic needs. If it's a lodge button, he has a reason for wearing it. If it's a picture on his wall, there's a reason for it. A salesman once bitingly asked us, "What are you suggesting... that I go into a customer's office, look at his walls and then talk to him about what I see there?" The answer was, "We didn't say *talk*, we said *listen* with your eyes... he's trying to tell you something!"

Second, listen with your mind. Don't just hear words. Why does she use the particular words she employs? Why does she seem to enjoy talking about certain subjects? What needs does her conversational pattern reveal?

Get him talking and keep him talking. We so often forget that we learn very little when we talk, but the customer can be a fountainhead of information if we will just let-him-talk and if we will really listen.

So, to begin with, we find that your customer himself may be one of your best sources of information. Here are some others; you can learn much about both customer and competitor from:

1. Things they do which you can observe.
2. Things they (or their people) say publicly or write for publication.
3. Things others say and write about them.
4. Public records that indicate their goals, objectives, assets, liabilities and current state of the business.

Specific sources

1. Their operating reports. We have known salesmen who have purchased one share of their customer's (or competitor's) stock in order to obtain the annual reports and, in some cases, interim reports. This is not always helpful if your customer is so large that the department with which you deal is lost in an overall company report.

2. The various investment services such as Value Line are excellent sources of information. Most large brokerage houses turn out special reports. One investment service produces reports on 800 companies and special situation analyses that run to as many as four pages of detailed information. You may find that the financial people in your own firm have these reports available.

For example, one investment service includes such items as the following in its reports:

Recent price of the common stock.

Growth potential and the reasons.

Capital structure, debts, interest being paid, number of shares, value of grades of stock.

Approximate share of the market the company controls, extent and nature of its sales diversification, return on net worth, number of stockholders, who owns the company and the names and addresses of principal officers.

History of dividends paid over five years, sales volume, profit margin, cash earnings per share, working capital, reserves.

History of the company, its financial condition, a three-to-five-year projection.

3. Standard and Poor's is another source. A sample report on a company listed this information:

Kind of stock issues, approximate price, dividend rate, yield, and a chart comparing the stock to 425 industrials and seven other firms in the same business.

Trading volume of the stock, sales for each quarter for period of four years, capital earnings for same period and what its short- and long-term earnings will probably be. Recent developments in the company.

Income statistics and pertinent balance sheet information that included such items as: net sales, per cent of operating income to sales; operating income, earnings, dividends, price range of stock, gross properties, capital expenditures, cash items, inventories, receivables, current assets and liabilities, net working capital, current ratio of assets to liabilities, long-term debt, etc. These reports also provide per cent of business in consumer, producer and defense business when applicable. It identifies products made, plant locations, number of employees and shareholders. In one report, we found a detailed table that covered a period of 30 years, making it possible to draw deductions based on trends.

Other sources

Alumni publications often contain much information about individuals in your customer's organization, including facts about family and outside interests.

Your own accounting department—and under some conditions, your plant credit union.

Merchants' reporting services (Retail Credit Bureau).

Registrar of Deeds. In most states there is a Uniform Commercial Code, and municipal offices have various mortgages listed. Often all credit purchases must be recorded and are available for public inspection.

Other components of your company if you are a multi-department or multi-plant company.

Vital information can often be obtained by putting together small, seemingly unrelated bits of information. These can be obtained by reading, visual inspection and listening. For example:

Observation of the customer (or competitor's) freight yard or parking lot.

Reading the local press, his employee newspaper, union publications, his advertisements and reading material in his reception room. Even the presence of certain suppliers in his reception room may provide you with leads to his future plans. Local conversation spots; i.e., lunch rooms near the plant, etc.

Technical papers written by his employees often reveal future plans. Because of the universal need for recognition and credit, your customer's (and competitor's) key people often rush into print, or in unguarded speeches to local groups—reported only in local papers—they say things they would not ordinarily mention.

Trade shows...representatives in display booths in conversations often produce helpful information.

Don't overlook government records and statistical reports. The Superintendent of Documents, Washington, D.C. 20402 publishes a report (free) of its publications which, if requested, will be mailed to you. Special congressional committee reports on many concerns often reveal information which would otherwise be jealously guarded.

Dun and Bradstreet reports.

Open house and plant tours. Don't feel unethical in participating...you can be sure that your own open house tours are not missed by your alert competitor.

Trade publications and yearly directories printed for many industries provide a vast amount of information; one firm publishes a business periodical index of such publications.

Other government publications and sources: Distribution Data Guide published by the Business and Defense Services Administration; U.S. Department of Commerce: Census of Business, Census of Manufacturers, Census of Wholesale Trade by areas, Census of Retail Trade. These indices provide over-all information, not individual company information.

Congressional reports including subcommittee reports. A subcommittee on business practices report contained information on the aircraft manufacturers that certainly would not have been made available on any other basis.

Army-Navy (etc.) Officer Registers contain professional records of each regular officer on active duty.

The Congressional Record is one of the best buys in the world at a low cost. Your Congressman has a number of free copies he can also distribute. Most companies of any size have it in their libraries. You can become adept at reading the headlines in the Congressional Record and quickly identifying areas of interest.

Fortune's 500 largest companies, Moody's index of companies, D&B's "Million Dollar Directory."

The newspapers, as mentioned before, in the home towns of your customers or competitors are of value only if they are read with the question uppermost in the mind of the professional salesman: "What possible business can this item represent for me?"

Help wanted advertisements placed by customers and competitors indicate the numbers and kinds of people being sought and provide you with clues to their intentions.

All too often salesmen fail to obtain essential elements of intelligence on the customer or the competitor and then attribute this failure to the fact that "it takes too much time." If you will but read some of the publications listed in the foregoing, you will find that a wealth of information is readily available in a few minutes for the inquiring mind and the seeing eye.

How can we expect others to consider us as professionals in our field if we insist upon getting by on the basis of hunch, feel and instinct. These qualities *are* valuable and there may have been a time when they alone were sufficient armament for the salesman, but in this marketplace today we need something more reliable than instinct. We need *facts* upon which we can make quality decisions. The first step is developing sources and protecting them.

A2—Identify his needs and wants

Strategy A2 is labelled "Identify his needs and wants." At first glance it would seem that little needs to be said relative to this strategy. It would seem obvious that a customer is not going to make a purchase unless he is aware of a need or feels "a want." But it's just not that simple.

First, the salesmen must guard against "projection." This is the self-defense mechanism that causes you and me to say, "If I were Mr. X, I would want this; therefore, Mr. X will want it." This kind of thinking is a good starting point for a lost sale. You and I are *not* Mr. X. We see a want or a need from an entirely different viewpoint than Mr. X's frame of reference. If we are going to sell him, we must develop empathy, i.e., an ability to see the situation as he does —without necessarily agreeing with his viewpoint.

Preparation for the sale must start with the salesman asking himself, "If I were Mr. X, the customer— with *his* point of view, with his fears, his goals, objectives and aspirations—how would I see the situation he now faces?" Then: "What obstacles would I see and what advantages would appeal to me?" *These* obstacles must be overcome by the salesman, and these advantages must be brought to the attention of Mr. X, the customer.

The features of your product that appeal to you, the features that may actually excite you, may not so much as stir a flicker of interest in the customer. This becomes readily obvious if we consider a chap who sells automobiles. He may be young, and the "quick pick-up" or the "streamlining" of the car, or the "gadgets" with which it is equipped, may have a great appeal to this salesman. His customer may be of a far more conservative bent, and these "features" will not appear to him to be advantages; indeed, they may even be disadvantages. Suppose, for example, his teenagers are going to be using the car frequently. The "quick pick-up" may appear to him to be something to be avoided.

To make this more meaningful, select something you sell and identify the features that appeal to *you*. Now consider how some of your customers might view those same features as either disadvantages or as not being of very much importance.

Every sales training program exhorts the salesman to "emphasize the benefits." We agree...but only if you are sure that the features you intend to emphasize will be considered as benefits by *your* customer. A dangerous practice is to convince yourself that "the average" customer will see these features as benefits so it is safe to emphasize them in your presentation. You and I are not dealing with "averages." We are dealing with *unique* individuals. It is this very practice of developing selling strategies that are geared to the "average" customer that is the basis of countless lost sales. And it is one of the key reasons why so many people do not regard salesmen as professionals.

We hesitate to beat the "doctor analogy" to death, but once again, how much confidence would you have in a doctor who treated you as an average customer (client) and gave you treatment that "worked for most" (the average) of his patients? The professional salesman looks upon each of his customers as being unique—as being different in some way from any of his other customers.

Success in identifying the customer's wants and needs depends largely upon being willing to spend some time getting to know him, to know his business and to know his problems... especially the latter. It means that the chances of making a sale are greater if you get him to talk than if you do the talking.

Agreeing that the statement is a bit simplistic, it will still be helpful to think of customers' "wants" as psychological needs and to think of their "needs" as product needs. Most of us do fairly well at identifying the customer's product needs but we don't always do as well in identifying his "wants"—psychic desires. This is because we really have not tried to develop an empathy for him.

You and I may not have an insatiable desire for power, or for recognition or for supremacy. As a result, we often develop blind spots—a kind of selling myopia that leaves us unaware that our customer *has* these needs. To sell him, our product may need to come wrapped in a presentation that emphasizes power, supremacy or recognition. Unfortunately, these are only three psychic needs. Psychologists have listed as many as 154 more of them. They include such things as the need to be needed, to belong, the need for security and so forth. The task of the professional salesman is to identify the adjective that best describes the way *this* specific customer wants to be seen (and by whom he wants to be seen that way). Then, step two is to determine what features your product has that will appeal to that need; and, step three is to present your product in such a manner that the customer will, by himself, see the satisfactions he craves.

At that same time, your presentation must be so subtle that it will not indicate to your customer that "you have seen through him." Keep in mind the warning of the psychologist that "we do not like our behavior to be transparent." We resent the chap who, by his behavior, indicates that he has successfully identified our psychic needs.

There is no intention to over-emphasize the importance of psychic needs. The reason that they are mentioned frequently is that they are not given sufficient attention by all too many salesmen. They *are* selling situations where *product needs* are paramount. Product needs usually stem from operating problems, the anticipation of operating problems or from the customer's business aspirations (as contrasted to his personal aspirations). To identify these needs requires more than a superficial understanding of his business. Even being "checklist deep" in an understanding of his business will bring you *some* sales. The more thorough your knowledge of his business, the more certain it is that you will be able to predict his needs in advance—often before even *he* has been able to predict them.

You need no direction in how to become informed about your customer's business. A checklist of sources for information has been supplied. These can be supplemented by observations made at his place of business and by getting him to talk.

A3—Identify his self-image

In Part One, we discussed in some depth the concept of the self-image and how it appears in the selling situation. We might summarize that material for our present purposes by saying:

1. *Every man has a self-image*... the way in which he wants to be seen by other people.

2. *There are specific people whom she is more anxious to impress than others.*

3. *He is a rational animal*, i.e., he will not do anything that will threaten that self-image and he will embrace any idea that promises to cause others to see him as he wants to be seen... even at times if it causes him inconvenience and expense. If the product or service you offer him will enhance that image, he will be receptive; if it threatens the image he will resist.

How, then, do you identify his self-image, if it is that important?

Before answering that question, let us readily admit that there *are* customers who have become most skillful in hiding from others their self-image aspirations. These are the kind of people who display no visible outcroppings of the self-image via what are commonly called "status symbols." The office walls are barren, the desk gives no clues. This kind of customer keeps you at a distance and is unresponsive to friendly overtures on your part. He turns down invitations of all kinds and keeps his distance. Admittedly, if status needs are important, he is going to be a tough customer to analyze—and to serve. Even here, however, the obvious attempts to maintain privacy are clues to you. In approaching this kind of customer, your presentation should appear to be based solely on your product and what it will do for him business-wise.

Fortunately, the great majority of customers do give indications of "the me they want others to see when others see them." Part Two listed a number of clues to the self-image in the form of questions that the salesman should ask himself. Keep in mind that "status" is not necessarily synonymous with "ostentation." It *is* synonymous with *station*, obviously coming from the same root. Each of us has a station in life and probably another station to which we aspire. We will do things that are consistent with that station. We will respond to suggestions and buy products that we feel are in keeping with our station in life. Status symbols, then, are essentially "symbols of station," and to the alert observer they tell much about us.

The task is to "listen with your eyes" because through status symbols, the customer provides you with real clues to the type of presentation that will be most effective as well as clues to how he wishes to be seen. Previous chapters have warned about misinterpreting "status symbols"; we repeat the warnings for emphasis:

1. Don't judge the customer by single status symbols. That Eldorado may have been purchased because his wife threatened to leave him if he didn't get it. His clothes may reflect her judgment, and *her* status needs rather than his. *Look for a pattern.* Do a number of his possessions, a number of his actions and the subjects he is prone to discuss add up to an identifiable self-image aspiration?

2. Again, repeated for emphasis: Don't project. That is, don't assume that a given status symbol is employed by the customer because if *you* employed it, that would be *your* reason for doing so. Ask yourself, why a man with your customer's background, financial position, and so forth, would do things or own things that *you* would use and own only as status symbols.

A4—Identify his primary buying motive

There are salesmen who are so sold on their product that they aren't satisfied until they cram every possible benefit down the throats of their customers. There are others who use this same technique, employing it as a buckshot approach. Not knowing the customer's primary buying motive—because they haven't done their homework—they throw every benefit and advantage in the book at him, hoping one of them will have appeal. This is not the approach of the professional salesman, to put it mildly.

This buckshot approach introduces problems needlessly. By mentioning *all* possible benefits you may (1) bore the customer, (2) convince him you are guilty of exaggeration, (3) mention something that, to you, is an advantage but to him will be a disadvantage (hence, talk yourself out of a sale), and (4) convince him that you don't understand his problem and are not customer-oriented.

The remedy, obviously, is to know the customer's problems so that you will know which product features or advantages to mention. There are many reasons, possibly, why he *could* be interested in buying, but the job is to identify his primary buying motive. Above all others, what is the primary reason for his interest? When possible, before the sale, list the advantages of your product in the order in which they will seem important to the customer—not as they appeal to you—and not as they might appear to that "average customer" (whom you have never met, and will never meet).

Even with the advantages so listed, once you have awakened real interest in the customer... stop! Don't insist upon overwhelming him with the additional benefits simply because you are determined he is going to realize that he has made a good decision. Save the other advantages as extra ammunition if his interest shows signs of wavering. Always have something saved... or as those in vaudeville used to say, "Always leave them asking for more."

A5—Consult your predecessor

Earlier, we discussed at length the problems that can arise if you follow in the footsteps of a competitor your customer did not like. It was also pointed out that following a popular predecessor can present problems, and the warning was given not to try to emulate the chap who preceded you.

At this point, however, we are looking upon your predecessor as a source of meaningful information that will aid you in developing a strategy. For some reason, this is often a difficult subject to discuss with salesmen. Most salesmen who are successful are blessed with an abundance of self-assurance—or are constantly running scared. The latter presents no problem because they are only too happy to obtain help and advice from any source. Strangely enough, it is the competent, self-assured salesman who is often most reluctant to seek advice and opinions of others. It is as though he fears that seeking advice will be looked upon as a sign of weakness or insecurity. Actually he ignores the wisdom of the old bromide that "two heads are better than one."

Why reinvent the wheel? Why go to the gathering of intelligence on the customer by starting from scratch when your predecessor may already be in possession of that same information? Why run the risk of committing a faux pas when by obtaining information from your predecessor you can avoid pitfalls?

This is not to say that you should accept your predecessor's evaluation of the customer without question. First, consider how effective the predecessor was with this customer. Even if he *did* produce results, was it because of his efforts or his acceptance by the customer? Or was it because the customer wanted your product so badly that he bought it in spite of your predecessor? In other words, the mere fact that your predecessor was successful with the customer does not, ipso facto, indicate that he was personally effective with that customer.

On the other hand, if the chap preceding you did not obtain a satisfactory portion of the customer's business, there is some room to doubt that he is really an authority on the customer. In either event, successful or not, he *should* have factual information as well as impressions of the customer that you do not have.

You can be quite certain that a rookie catcher on a major league baseball team would be only too happy to learn as much as possible about the opposing batters as he possibly could from his predecessor. In so doing, he would not for a minute feel that he was showing "weakness or insecurity." Certainly, he can learn the strengths and weaknesses of the opposing batters in time... but at the cost of seeing more than one ball go over the fence in the meantime. We, too, can assemble information on the customer "in time," and in the meantime lose business that could have been ours had we put a foolish pride in our pocket and obtained advice from the chap who had already gained the knowledge.

Evaluate carefully any advice your predecessor gives you... especially if it is critical of the customer. In any case, when she gives you an opinion, ask her, "Can you give me an experience you had with the customer that brings out your point?" In the field of personnel relations, we constantly warn managers, as they rate their men, to avoid the "halo-halitosis effect." In other words, all too often, most of us are tempted to evaluate men in terms of the outstanding things they have done or in terms of troubles they have given us during those moments when they were at their worst. If a subordinate does something that gets us out of a jam, we are tempted to rate him highly instead of asking, "Is this typical of his performance?" If he gets us in a jam by some stupid action, we tend to rate him low, instead of asking, "Is this typical of his performance?"

The same tendency shows up in our evaluations of customers and may color the opinions of your predecessor. When you ask him for real examples to support his opinions of a customer it helps you to determine whether or not he has been using "the yardstick of fair play" or good judgment in rating the customer. Watch for the role that your predecessor assumes in any experiences he relates. If the experience was a poor one, was it your predecessor who was probably at fault rather than your customer?

We have been accenting the *danger* of turning to your predecessor for opinions primarily to emphasize that to base one's strategy *solely* on possibly distorted views of another is unsound. Yet, it would be unwise not to obtain those views, provided that they are subjected to the type of scrutiny discussed above. The professional military man, even in today's modern warfare, studies the campaigns of the Civil War and the Napoleonic Wars to learn from the errors and mistakes of others in planning his own strategy. This modern military man does not consider himself "weak" or anything less of a professional for having done so. The professional saleswoman will add to her own stature and success by studying the views and the strategy of her predecessor.

A6—Get help from other departments of your company

Today, to be "big" is in the eyes of many to be "bad"; yet, we never cease to be amazed at the failure of many salesmen (and others) in large companies to profit by their "bigness." The larger your organization, the greater are the possibilities of obtaining help, information and specialized assistance from other components (finance, credit, personnel, etc.). Yet, once again, we often find men who ignore the strengths available to them, preferring to go it alone, feeling totally self-sufficient.

Throughout these chapters, there is a constant theme that accents the professional in selling. This is because we firmly believe that selling is not only a profession but perhaps the most demanding—and in some respects the most important—of all professions. Without consummated sales, our entire economy would collapse. It is for this reason that we draw analogies between the sales profession and others.

For example, the military learned long ago that many men, left to their own devices, prefer to go it alone. To "go it alone" in the military profession would be to countenance the loss of life in the name of individual freedom. Accordingly the doctrine of the complete staff study has been established. This device insures that everyone concerned with a strategy or an operational plan has not only been advised but has had an opportunity to concur or to non-concur, giving his reasons for non-concurrence. When the final plan is submitted to the commander, he is assured that everyone concerned is informed and that all possibilities of hazard have been considered and explored and that the final judgment (as reflected in the recommended strategy) is the best possible action in the light of facts available.

As salesmen, we continue to have that comparatively delightful freedom of movement that we cherish and that led many of us to choose this vocation. If we continue to neglect the opportunity to avail ourselves of opinions and sources of information represented by other individuals and functions within our companies, the day will come when managers of salesmen and marketing managers will steal a leaf from the book of the professional soldier and, by edict, put into effect something resembling the Staff Study.

How much better it will be, then, for us to recognize that the professional salesman plans his tactics and strategy, based on the total amount of information and assistance available to him in the company. The salesman working for a small company who utilizes the resources and strengths of his firm to the maximum has little to fear from the salesman who works for a larger company with far greater resources, but chooses to go it alone.

Space remains any writer's mortal enemy. Regrettably there is never enough space available to list all of the types of assistance that can be provided by other elements of a company to a salesman. Perhaps this is not all to the bad because it enables the thoughtful reader to pause at this point, pick up a pen and pad and think out, for himself and by himself, the individuals and elements in *his* company that could provide him with assistance. He can then ask himself how he should rate himself in terms of the extent to which he has used this assistance in the past. Further, he can consider some of the tough selling situations that currently face him, and determine to whom he can now turn in his company for further assistance.

Many salesmen would profit by taking another look at the job of their boss, i.e., the manager of salesmen. Have you ever really considered what the nature of his job is? As management at long last comes into its own as a professional kind of work, managers of salesmen are coming to see that their primary task is the guidance of each of their salesmen in each man's self-development. They should be available for guidance and counselling. The salesman should come to view his "boss" as less of a boss and more of an advisor. True enough, many salesmen are more skilled in the techniques of selling than their managers, but the manager *does* have the advantage of being insulated from the personalities often involved at point of sale. In the military, many staff officers might have made poor combat officers but they were worth their weight in gold as planners.

Consider the help that might be available from manufacturing, finance, engineering and legal elements of your company. Consider, too, that most men are flattered when asked for advice. Asking for that advice may be one of the best strategies available to you to develop company-wide rapport and, incidentally, to cause men working in these other functions to begin to appreciate the complexities of your work and the demands made upon you. For the first time many of them will begin to realize that selling is far more than having a good product and the best price.

A7—Get boss's help

This, we feel, has been adequately covered in A6, above, at least in so far as space limitations permit.

A8—Identify your competition

As indicated in the Sales Strategy Bank, this includes identifying the *individual* who represents your competitor in your "theatre of operations." The day is long since gone when you could keep an eye on your historic competition. In this age of conglomerates and mergers, there is hardly a salesman around who doesn't have a competitor he didn't have two years ago. Further, there is always the "hidden competition" to identify. We usually keep our eye on the chap who competes directly with us, selling a similar product or a satisfactory substitute. How about the fellow who is selling an entirely different product to meet an entirely different need? What if he can convince the customer that he should place a higher priority on satisfying *that* need rather than the need your product will satisfy?

One form of insurance against this kind of competition is to have a thorough knowledge of your customer's business and especially of his problems. What problems (other than the one you hope to solve for him) are commanding his attention? How can you make him see that the problem your product meets is the one that should be met first? There is value, while waiting in his reception room, to take note of other claimants on his limited resources. What needs are *they* there to serve?

It is not enough to know your direct competition by company name. Do you know the individual who competes directly with you, representing that company? What is her usual modus operandi? Does she cut price? Does she offer better service? Has she developed a closer relationship with your customer by providing years of good post-sales service? How close is she to the individual who is now second in command to the purchasing authority? When the present buying decision-maker leaves, will your competitor be stronger with her replacement than you are?

Knowing your competition means knowing as much about him as possible. How badly does he need the order? If he needs it very badly, what are some of the alternatives open to him? Which is he most likely to choose? What tactics can you employ to neutralize each of his alternatives? Are you going to wait until they pop up to confront you or are you, as a professional, going to be ready in advance with well-considered strategies?

What degree of rapport has he established with your customer? Upon what is it based? Is it based upon common interests, school ties, ethnic considerations, entertainment opportunities available to your competitor, a rapport with the secretary or receptionist that gives him "a most favored supplier" edge? Is it based upon obligations that your customer feels that he has to the competitor for past favors? The list is a long one and the only item on the list of interest to you is the one reason why the competitor enjoys a better rapport than you—or your predecessor—have been able to establish.

The question is twofold: (1) How can you cause the customer to see what extra or hidden costs he is paying for the favors he grants the competition? (2) What is there that you can do to provide the customer with greater satisfactions than he obtains from doing business with your competitor? Is your competitor satisfying product or psychic needs of the customer? If he can satisfy one of these better than you can, your strategy should be based upon causing the customer to see the need you can satisfy as being more important to him.

A9—Study competitor's usual strategy

We are all, to a degree, creatures of habit. We follow practices that have brought us success in the past. Your customer will also favor those strategies and tactics that have proven to be successful. If the customer you are facing today is one with whom you have conducted business in the past, you will find it profitable to recall how that individual acted or reacted in similar situations during your past encounters.

Our memories are fallible and this is one reason why we continue to emphasize the need to keep written records of customer behavior patterns that have been displayed in earlier encounters. Before entering the selling situation the professional salesman will refresh his memory and consider what attitudes the customer may be expected to exhibit, judging by past performance.

Each of us has behavior patterns that can be predicted by those who are close to us. We are infuriated when someone says to us, "I knew what you were going to say" or "I knew what you were going to do." The psychologist tells us that we do not like our behavior to be transparent. This is why we are furious when even a friend indicates that our behavior was predictable. It is obvious that we should never, under *any* circumstances indicate to the customer that we knew what his or her reaction would be. This is why the experienced saleswoman will *never* say, "I was sure you would bring up that objection." Rather, the saleswoman will say, "That's a valid point you've raised, let's examine it" (and then go on to answer the objection).

Use your past knowledge of customers' behavior patterns to anticipate the kinds of reactions, objections and the like that each customer is likely to use, and have tactics ready to employ. If we can predict we can control, or at the very least, increase our chances of controlling the selling situation. One of the worst things that can happen to a salesman is to be surprised during the course of a selling situation. Study of customer's habit patterns and tactics used in the past, can greatly reduce the chances of being surprised and being caught without answers or a tactical plan.

The best method of handling a problem is to recognize its possibility in advance and have well planned strategies for preventing it from arising in the first place or to dilute or eliminate the problem when it does arise. No strategy can be any better than the intelligence (information made meaningful) upon which it is based. One of the key elements of intelligence as we have seen in previous chapters is information on the customer's habit patterns.

A10—Make an estimate of situation

This is, without doubt, the most important of all the actions we will discuss. It is more than a strategy —it is a plan for thinking, a means of insuring that the strategies you finally decide upon will have the greatest possible chance of success.

STEP 1: Identify the customer.

Who is the key influence on the buying decision? Who is the person with whom you must develop and maintain lasting, favorable relationships?

STEP 2: What is his business objective (*Long range***)?**

What does he seek to achieve for his company in the next three, five or ten years?

STEP 3: What is his business goal (*Short range***)?**

What is he seeking to obtain for his company this year, or more immediately, from the result of this selling situation?

STEP 4: What is his personal objective (*Long range***)?**

How does he want to be seen, as an individual, and by whom in three, five or ten years; i.e., whom does he seek to impress or what reputation is he striving to establish?

STEP 5: What is his personal goal (*Short range***)?**

Whom is he seeking to impress now, and how does he want to be seen as a result of the decision he makes in this selling situation?

STEP 6: (The $64,000 question you must answer correctly.)

Which of these questions (two through five) represents his primary buying motive; i.e., is he thinking in terms of short- or long-range gains? Is he interested primarily during this selling situation in the welfare of his company or in his own welfare? Short range____ Long range____ Personal____ Company____

STEP 7: What is your long-range objective with this customer?

(This must be kept in mind as you develop over-all strategy and tactics.)

STEP 8: What is your immediate goal with this customer (In addition to getting the order)?

STEP 9: Identify the problem.

What must you overcome if you are to achieve your objective and goals?
How to:

STEP 10: Gather intelligence (essential elements of intelligence—EEI).

Intelligence on identity and nature of competition.
Intelligence on product—yours and competition's.
Intelligence on the condition of your market.
Intelligence on customer's economic position.
Intelligence on customer's psychological needs.

Intelligence on customer's product needs.
Intelligence on customer's potential.
Intelligence on customer's limitations.
Intelligence on his organization—where power lies.
Intelligence on special conditions (political, environmental, competitors who buy from your customer, conglomerate implications and similar conditions which pose problems or opportunities).

STEP 11: Analyze the intelligence, identify facts, make justified assumption.

What are your assured advantages?
What are your possible advantages?
What are your assured disadvantages?
What are your possible disadvantages?

STEP 12: Develop your sales strategy.

What result is it, specifically, that you seek?
What must be done to achieve that result: Assure him, worry him, give him an escape hatch, make him look good, etc.? (See Sales Strategy Bank.)
How many *possible* courses of action are open to you? Weigh all possible alternatives!
What alternatives are available to the customer?
What alternatives are available to your competitor?
Which of your alternatives seems most promising in the light of what your customer and competitor can do?
Avoid attractive courses of action that may bring you immediate gain and cost you achievement of your more important long-range objectives!

STEP 13: Plan your opening tactics.

Consider same questions that appear in Step 12. Have alternative tactical plans ready.
Which disadvantages will you anticipate and how do you plan to overcome them? Which of the advantages you possess will have the greatest favorable impact upon the customer? How do you plan to exploit them for maximum effect?

STEP 14: After the selling situation.

If you made the sale, why were you successful? What principle can you now use again in similar situations?
If you lost the sale—why?
What essential elements of intelligence did you fail to obtain?
Which alternatives did you overlook?

What additional intelligence must you gather on the customer's business before your next call?
Organizational facts:
Psychological needs:
Economic facts:
Operating problems:
Did you underestimate your competitor? How?
Was your timing wrong? Why?
Which of your disadvantages did you fail to consider?
Did you have advantages that you failed to exploit? Why?
Do you find yourself, now, tending toward blaming your company, your boss, "other departments" in your organization, price, late delivery? Are these really the causes or just available excuses?
If they are valid reasons for the lost sale, what could you have done to have neutralized them?

Self-development starts with self-analysis or introspection. If you can identify the reasons why you lost the sale, this lost sale may be the most valuable lesson for you—if you decide how to avoid the same errors in the future. If you can identify the reasons why you made the sale, this information may enable you to obtain even more valuable sales in the future.

Now, let us analyze the technique of making an effective Estimate of the Situation. To begin with, the foregoing represents only a work sheet—a condensed version of the Estimate.

Essential elements of intelligence on your competitors

When you use this form, you should consider Step 10 (Gather Intelligence) as only the barest of outlines on the facts you should try to gather. For example, the form suggests that you consider the identity and nature of your competition, but here are some of the specific questions you should ask yourself about the competitor:

What is your competitor's financial standing? Is it so precarious that he will sell at prices that just pay his operating costs, making it difficult for you to meet those prices?

Can he afford to give better services than you? Can he really swing an order this size?

What is his bargaining position? Are his products sought after? Does his inventory put him in good shape? How badly does he need this order? Are his products glamorous but unproven?

Will she take a loss just to get "in" (or to get her product on the prototype)?

What is his usual modus operandi? What are the personal strengths and weaknesses of the man who represents your competitor? Does the customer feel he has an obligation to him (or to his company)?

Can he supply your customer with other kinds of help (engineering, financial, promotional, and the like)? Can you do as well?

Does your competitor have anything in common with your customer (personal)? Does his firm buy from your customer—are you up against the old quid pro quo? Are they in the same town? Did your customer ever work for your competitor (or for you)? Is this good or bad?

What do you know about your competitor's promotional plans? Is he having labor trouble? How would you rate his delivery potential? Does his organization structure enable him to operate more freely than you can?

What poor experiences has the customer had with the competitor—are any so recent that they will still condition the customer's attitude?

Does your competitor have better printed materials available to the customer than you have? What is his acceptance in this geographical (or market) area? What is the extent of his market penetration? Can he use his other customers as testimonials that will impress your customer?

Is her employment up or down; what kinds of people is she employing? What do these facts indicate to you?

What are his costing methods? Is he usually customer-oriented or is he of the hit-and-run variety? Does he use your customer's products in his own finished product?

What value does your competitor contribute? What percentage does she make versus buy?

Does he have a standing in professional societies or trade associations that will give him an edge? Does he have a voice in writing standards that affect the industry?

Does he have a patent edge—or do you? What other business relationships does your competitor have (absentee landlord relationships or part of a conglomerate)?

Does the fellow who competes with you sell the whole line or is he very well informed because he specializes on one product?

This list is by no means exhaustive. You can, and should, add to it the kinds of questions that pertain to your kind of selling. It does, however, serve to indicate the kinds of information that can aid you. In the answers to some of these questions, you will find the genesis of lost—and successful—sales.

An equally valuable list of questions concerning the customer appeared in Part One. These questions involved the *psychological* needs of the customer.

Essential elements of intelligence on your customer's economic situation

To the list of psychological needs should be added an equally extensive list on the customer's economic situation and his organization. Parts of this list first appeared in Part One. Consider the following as thought-starters:

What is your customer's primary buying motive (his key reason for buying)? Is he trying to satisfy a product, economic or psychic need?

What is his reputation in his market? If weak, how can you help him improve it? Is he satisfied with his current profit picture? Can you help him reduce expense? Can you help him get new business?

What has been his rate of growth over the years? What has been the recent trend of his growth? What do you believe his potential (as a customer) will be? Has he been getting insufficient attention because he is small? How can you help him grow?

Are his people technically informed? If not, can you help him in this area? Will they understand the true nature of their problems and the solutions you are offering? How can you adapt your presentation to their low threshold of experience without "talking down"?

Is cash readily available to her or does she need credit? How can you offer help in this area without betraying your knowledge of her cash position?

Does he have funds committed to other products so that he feels unable to buy the product you offer? How can you cause him to place a greater priority on his need for your product? Does your product serve needs he has not as yet recognized? How can you alert him to these needs? Can you use other customers as examples?

What is his credit rating and past performance on meeting obligations? Is he really worth the risk and the time that you must spend on him—as compared to other prospects?

If the individual upon whom you call must obtain the approval of others before she buys, what kinds of information can you supply her that she can, in turn, use to convince the real decision-makers?

Does he buy for long-range or immediate requirements? If his pattern is detrimental to your interests, how can you show him why he should change that pattern?

If top-side makes the decisions in his organization, should you call in your top-side?

Does he have a habit of spreading his business among his suppliers? If so, what arguments can you use to show him that this is costly to him?

Do you get his annual report? Have you ever seen one? What are his key liabilities and how can you help him reduce these?

Does he have any problems with which you can help him in such areas as: space utilization, delivery, storage, financing his customers, location, tough competition, inexperience in management, finance, utilization of personnel, marketing or research? Will he welcome help or will an offer embarrass him?

Is the area she serves expanding? Is she aware of the opportunities? Is she doing anything to meet them?

Is there a "silent partner" who has to be convinced? How can your story reach him?

How long does he take to write off purchases such as an order given to you will represent? Can you use this fact to cause your "price" to appear easier to absorb?

Again, this list is not complete, and, again, it should serve to stimulate you to itemize other questions that apply to your type of customer. As you read the list, how would you rate your real knowledge of the customer, his business and his problems?

Finally, in the list of Essential Elements of Intelligence, we consider product. Obviously this list will vary widely, depending upon the type of product or service you offer. The following list is presented so that you may add to it, eliminate those that do not apply in your kind of selling, and in order that you may consider those that do apply.

Essential elements of intelligence on your product

In comparing your product to your competitor's where do you have an advantage and when are you at a disadvantage? Consider these elements:

Obsolescence—appearance—original cost—cost of operation—serviceability—accessibility—security—adaptability to other uses—room required—transportability—size—weight—color—choice—safety considerations.

Guarantee—international exchange considerations (monetary)—range of patterns—dust-proof considerations—shock-proof—moisture-proof—buying terms—delivery time—margin of safety over guaranteed rating—overload possibilities—proven market acceptance—reputation in market.

Consignment policies—floor plans—quantity discounts—drop shipment possibilities—policy on replacement of defective parts—training of customer's employees—heat dissipation—breadth of product line—stability of operation.

Hidden values not readily apparent—timing (Are new models planned for early release by your company or by competition?)—adaptability to present wiring, plumbing or available space.

Is your product or competition's so revolutionary that it will encounter resistance, skepticism, doubt or antagonism?

Life span—durability—compatibility with customer's present equipment or plans—design harmonious with other items used by the customer?

Does competition use foreign-made parts that will cause servicing or delivery delays to your customer?

It is obvious that this list can be greatly expanded. The point is that hidden away in the answers to these, and similar questions, are the reasons for lost sales (and the possibility of successful sales) if the professional salesman will consider them before the selling situation is underway.

Analysis of the 14 steps

Concluding our discussion of the value of the Estimate of the Selling Situation, it will be of value to analyze the material presented at the beginning of this section.

Step 1: Identify the customer. We often make the error of considering as our customer some individual in the customer's organization who causes us difficulty. He is not the customer; *he is the problem*. True enough, he has to be neutralized, but the fellow whose favor you must seek is the man who makes the buying decisions over the long haul. We often cause ourselves unnecessary difficulty by giving too much attention to the fellow who is creating a problem and not enough to the real decision-maker.

Another error committed by many salesmen is to give their attention to some individual who can, admittedly, give us an order today because we happen to have a product that is used in his field, while we antagonize the regular buyer who will buy almost everything else we sell. It's essential that we identify the *regular* decision-maker.

Steps 2 through 5. Here we attempt to determine what type of appeal will be most effective. It is necessary to determine whether the most effective buying advantages will be those that bring immediate gain, or those that provide greater long-range advantages. There are buyers who never think beyond tomorrow, and it is clear that talking to these fellows about the advantages of investing in the future is going to waste time. Other buyers are not impressed with immediate gains but are very much concerned with the long-range implications of a purchase. It is essential that we know how the buyer thinks if we are going to present arguments that will be meaningful to him!

By the same token, there are customers who are more interested in how a purchase will affect them, personally, than in how it will affect their company. It is imperative that the salesman determines whether the most effective appeals will be those that will "make the buyer look good" or those that will benefit his organization. For these reasons we have labeled **Step 6** as the most important.

Step 7: What is your long-range objective with your customer? If we keep our long-range objective in mind, it will prevent us from chasing after attractive looking short-range goals. Our ultimate strategy and our tactical plan can be measured for effectiveness by asking ourselves whether they are consistent with our ultimate objective.

Step 8: What is your immediate goal with this customer? The professional salesman should know specifically what he hopes to achieve during each call upon the customer. This is necessary, if for no other reason than to be able to measure his performance at the end of the selling situation. Obviously, his short-range goal should be consistent with his long-range objective.

Step 9: Identify the problem. So often in handling cases in sales training, the group will respond, "The problem is how to get the order." *Not so.* The problem is how to overcome the obstacles that exist between you and getting the order. In some cases, for example, after the customer has gone on record with some flat statement, and you have the proof that he is wrong, the problem will be: "How to present the proof without causing him to lose face or to become angry." Sometimes the problem is, "How to dissolve the situation so you can return at a more opportune time"... or, "How to cause a customer who is ready to make a hasty decision (that is adverse to your interests) to see that the cost of a quick decision may be greater than the cost of delay." We can be sure that if we fail to identify the real problem, we will come up with the wrong solutions or with the right solutions for the wrong problem. Identifying the problem is often the most difficult task that is faced by the professional salesman.

Step 10: Gather intelligence. This step has been rather thoroughly discussed in the material presented to this point.

Step 11: Analyze the intelligence. "Intelligence" as used in the Estimate of the Selling Situation is *information made meaningful*. Once you have amassed all of the information available, you should break it down into advantages and disadvantages, assured and/or likely. Such an analysis begins to point up the severity of the problem, and it indicates the strengths and weaknesses that must be considered when you develop a strategy.

Step 12: Develop your sales strategy. The key point to keep in mind is that we must guard against the tendency to think there is just one solution, often epitomized by the phrase: There is just one thing to do with that kind of a guy." There are, almost always, many actions available to us. All too often we seize upon the first attractive action that occurs to us. A good policy is to *write down* every possible strategy that occurs to you and then challenge each by asking, "What would the customer (or competitor) probably do if I took that action." It is amazing the number of times that the best possible action pops to the surface when this procedure is followed. We tend to see things in blacks and whites and ignore the in-betweens.

Recently we were given an excellent tip on the stock market... a stock was going to split. We immediately decided to invest a large amount in this stock, selling some gilt-edge stock to do so. Our tipster raised his hands in horror, as did our better half, with the result that we reduced our purchase by 90%. Of course, the stock split and with it went an early retirement. Naturally we blamed you-know-who. Only later did we think of other actions that had been available to us. At the time, we saw only two actions ... buy heavy or buy light. Later, we realized that we should have bought heavy—and if during the next week the stock did not split, we could have sold it, repurchased the gilt-edge stock and taken a small loss in the form of broker's fees. But we did not stop to weigh all of the alternatives. The cost of taking precipitous action and not considering all of the possible courses of action cost us more sales than we realize... and just as we were all too ready to blame others for our own stupidity in the matter of the lamented stock purchase, so, too, we are quick to blame lost sales on "price" when it is we who missed the boat because we saw "only one thing to do."

As indicated, the same principles apply in developing tactical plans **(Step 13)** that apply in the development of our long-range strategy. The value of **Step 14** also has been discussed. Taking measurement of ourselves, after the sale, develops insights that might otherwise be missed. We will learn things about ourselves that others will never tell us—or will tell us as they conclude with the words, "Sorry, old boy, you'll never be a salesman."

The Estimate of the Selling Situation is the most professional tool in the kit of the salesman. It places him high on the ladder above the popular and erroneous view of the salesman as an "order taker." Like all tools, it is of value only when it is used. Further, it provides the manager of salesmen with a device assuring that each of his men will enter a major selling situation with a plan, aware of the pitfalls and equipped with a strategy that will avoid them and enable him to bring the advantages of his product into sharp focus in the mind of his customer.

A11—Determine past experiences of customer

In Part One, the influences on recall (how and why we remember the things we do) were discussed at length. You are urged to re-read the pages devoted to this subject.

The sale is not a vacuum into which you just stepped... or a situation which you have just now created. The customer has been around a long time. He has had literally millions of experiences. He remembers some of them vividly. He is affected consciously or subconsciously by most or all of them. In a word, he has been *conditioned* by them. His attitude toward you, your company and your proposal will be influenced by the kinds of experiences he has had.

Which of these experiences will he remember

and which will affect him? You, admittedly, will never have the time, and few if any of us will ever have the total skill necessary, to dissect all of the customer's past experiences and determine, in advance, what their effect will be upon him; *but there are some things we can do.*

The professional saleswoman can ask herself this type of question before she calls upon the customer:

1. What experiences has the customer had with me or my company?
2. What experiences has he had with my competition?
3. Were they favorable or unfavorable?
4. Which have been the most recent?
5. Which will *he* see as most intense?
6. To what degree was *he* involved personally?
7. How can I neutralize the poor experiences he has had with us and the good experiences he has had with competition?
8. How can I cause him to recall good experiences he has had with us that he has probably forgotten?

In assessing the experiences of the customer, be sure to try to see those experiences as he will remember them. Remember that you and I (and the customer) try to forget those experiences that make us look unfavorable. We try to retain the memories that made us look good. The customer may have had an experience which you did not regard as intense, but from his frame of reference it may have been quite intense. In that case, he will remember it far longer than you do.

Before you make your call, keep in mind the tactics of the truly professional doctor who checks your past record before you even enter his office. Then, at the very least, ask yourself the eight questions listed above. Referral to Chapter 4 will aid you in developing additional questions you may ask yourself relative to past experiences of the customer and how they will probably have affected his attitude.

The rule to remember is that the customer's memory (and yours and mine) is affected by these elements:

1. We will remember those experiences longer which made us look favorable.
2. We will remember longer than the other fellow those experiences that made him look unfavorable.
3. We will remember longer those experiences in which we are personally involved.

It becomes immediately apparent that you and your customer will remember different kinds of experiences because of the quite different ways in which you were both involved. To predict how he is going to act today depends to a large degree upon your ability to develop empathy; i.e., to stir up your memory and recall things which you have forgotten because (a) they made you look bad, (b) they did not seem to you to be intense, or (c) you did not feel deeply involved personally. To the contrary your customer may recall every vivid detail for exactly the opposite reasons.

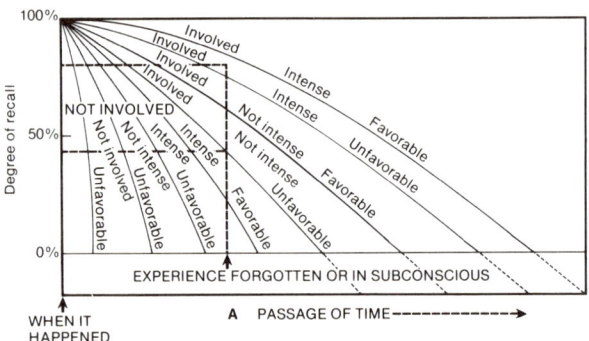

This chart points out that what your customer remembers—and how long he remembers it—is directly related to the extent of his involvement in the situation.

The chart may make this clear. It is intended only to make the point and is obviously not proportional.

It can be readily seen that the degree of involvement is the most important factor in what your customer will remember. How deeply and personally has he been involved in recent experiences with you or your competition?

The degree of intensity is the second factor, and finally, the nature of the involvement—favorable or unfavorable.

You might question how an experience could be intense to a person without that person being "involved." The answer is found in this type of situation. If you were present when a purchasing agent was being ripped apart by his boss (for something with which you had no connection), the situation would be intense (and you would wish you were not there), but you would not be personally involved.

If you were to tell a third person about the situation later, he would forget it almost at once because (a) he was not involved, (b) to him—not being present—it was not intense, and (c) it was unfavorable, so he would "want" to forget it at once.

The dotted lines on the chart show this kind of a situation and demonstrate dramatically how an experience that took place *at one moment in time* would be remembered at a later time (Point A on the bottom line of the chart) almost as though it happened yesterday by one man (the individual who had been reprimanded in front of you). You would remember it probably only if your memory was stirred up a bit. The party who knew of the incident only through hearsay would have forgotten it completely (Point A).

This points out graphically why it is essential to try to see the customer's past experiences as he sees them from the nature of his involvement... not from yours. It shows why you must not be surprised when he has forgotten the "good" things you may have done for him in the past if he has had a recent sour and—to him—intense experience. This more recent, intense experience looms up far greater in his mind than all of the "good" experiences he had in the past.

Give this matter more thought, and it will aid you in smothering your anger when you are tempted to think of the customer as irrational or as an ingrate. As you study the chart and the material in Part One, it will be brought home rather forcefully that it is you and I, not the customer, who are a bit irrational when we blame the customer for remembering things we wish he would forget. You see... you and I follow this same memory pattern. Once we understand this, it is much easier for us to control our emotions. And with emotions under control, we can use our reason to do a far better job of anticipating the customer's probable behavior pattern and plan to meet it effectively.

A12—Identify his possible sources of resistance

Many salesmen whom you and I have encountered wait until they find themselves face to face with the customer, in the selling situation, before they decide how to meet resistance. Often, because they have not anticipated resistance, they are surprised, annoyed and irritated when they meet it. None of us can operate at our peak effectiveness when we are in a an emotional state of mind. We don't do our best thinking, and we find it difficult to hide our annoyance from the customer under such conditions. We tend to "counterpunch"; i.e., take the first action that comes to mind. Or, we come up with answers to the customer's objection that are not well-conceived and, as a result, fail to neutralize his objections.

How can this tendency be overcome?

To answer that question, we should first consider the possible sources of resistance. The following are some of the more common roots:

1. The customer objects to price.
2. He feels that your product won't meet his needs.
3. She feels your product is not as good as products available elsewhere.
4. He doesn't want to change suppliers.
5. He is not the decision-maker, doesn't want you to know it; so he brings up smoke screen objections.
6. He can't afford to buy, doesn't want to admit it—hence, more smoke screen resistance.
7. She fears that she will make the wrong decision and will (a) lose face, (b) be criticized by her superiors or (c) lose money.
8. He is not technically competent in your field and raises objections that are not valid.

From your experience, you can add to the list. Essentially, as we have said before, he is asking the question, "Are you going to make me rich or famous, or are you going to make my life easier for me?" When he feels that the answer to these questions is in the negative, he is going to resist your efforts.

The first step, then, is to *anticipate* his possible resistance. Admittedly you cannot always *predict* the form of his resistance, but it is possible to *anticipate* it. We refer here to objections which he really feels are valid, not to resistance of the smoke screen variety. However, if you know—from having done your homework—that he is not the decision-maker (and resents this), or if you know that his cash position is not good, you can anticipate the type of smoke screen objections that these conditions will spawn.

How do we anticipate objections? First, by remembering that, in general, we all act pretty much the same. If we feel threatened, or if we see no gain, we raise objections, become antagonistic or we try to evade the situation. The proof of the universality of our behavior patterns is found in sociologist Erving Goffman's book, "Behavior in Public Places." The book mentions the behavior pattern of a visitor to an office when his guest answers the phone. The visitor —knowing it is not considered polite to listen—can't sit there, doing nothing, without appearing to violate the rule. So he feigns another activity... ostentatiously lighting a cigarette or looking out the window, etc. Undoubtedly you and I have been in that same position and have resorted to the same devices. Goffman, in his book, provides a multitude of actions that you, I and our customers resort to instinctively.

This is by way of saying that in order to predict what your customer probably will raise as forms of resistance, it is first necessary to *be that customer* during your pre-sale preparation. It is a pretty safe bet that he will offer the same objections that you would raise if you entered the selling situation in his milieu. In other words, if you had the same aspirations, fears, experience (or lack of experience), recall and values (opinions, standards of ethics, morals and so forth); and if you had been exposed to the same hearsay to which he has been exposed, you would probably act essentially as he is going to act in the forthcoming selling situation.

If you will examine the eight sources of resistance we have named (plus others which you have encountered), you will readily discern that each of these, having a different cause, must be treated differently. Keep in mind, too, that the name of this strategy is "*Identify* his possible sources of resistance." To handle each of those mentioned, we refer you to Part Two, the Anatomy of a Sale. At this point, we are trying to emphasize the need for spotting possible causes of resistance in advance, estimating the likelihood that they will occur, and planning to meet them *before* you walk into the selling situation. In other words, do your planning when you are unencumbered by emotion—be it irritation, anger or surprise.

A13—Prepare alternative offers

As stated previously in the Estimate of the Selling Situation, our strategy must be complete with alternative actions and offers. Often, the task may be to save face for the customer. Instead of debating with him have alternatives ready that will enable him to choose without creating the necessity for him to admit that he is wrong.

It is not always possible to identify, in advance, the customer's primary buying motive. We do our best to do so. Then we develop a plan of attack that will accentuate those advantages that best meet the needs of the buying motive we think is uppermost in the customer's mind. But even the best saleswomen often guess wrong for a variety of valid reasons. Perhaps something has occurred since her last visit that has changed the focus of the customer's attention, need or fear. In such cases, we will enter the selling situation with the right answers to the wrong problem.

In planning your alternatives, consider each of the buyer's *probable* buying motives and decide which strategy will best satisfy each of those motives. In your opening, you can concentrate on the *most probable* motive. If he responds with apathy or with more aggressive forms of resistance, switch your tactics and try the second most likely buying motive.

Another method is to get him talking and try, from the tenor and emphasis of his remarks, to identify the motive (type of gain or fear) that seems to be uppermost in his mind.

When all else fails, there is nothing wrong with asking the direct type of question..."What do you see as your key problem?" or "What is it that you are most anxious to accomplish?" Salesmen often hesitate to ask such questions because of a feeling that it makes them look weak or puts the customer "in control" of the selling situation. This is usually nonsense. A fool can ask a question it takes 10,000 wise men to answer. When you ask questions, it is the customer who is on the spot and you are in control.

A14—Plan strategy;
A15—Plan opening tactics

These two strategies have been thoroughly covered previously, and most recently in *A10—Making an estimate of situation.*

CHAPTER 21

Creating a good sales climate

The second group of strategies appearing in the Sales Strategy Bank comprises those which can be used to create a good climate. Someone has said that whenever a sale is made, an act of faith has taken place. "Acts of faith" can occur only in a good climate.

If the customer comes to you (as a customer comes to a store or to any type of vendor), he is already in a buying mood. This is, at the very least, an indication of his interest. His buying mood must be reinforced... it must be cultivated. If, on the other hand, you find it necessary to approach the potential customer, you must develop in him a buying mood. A good climate is essential in either case.

It should go without saying that such a climate can be built if the salesman evidences, by his behavior, an interest in the customer as a person and indicates a sincere desire to help the customer solve a problem. The problem may be the solution of some operating difficulty or the satisfaction of some psychic need (i.e., the satisfaction of a want).

The strategies for creating a good climate that follow will aid the salesman in creating the impression of interest in the customer if they are used sincerely. Some of them are so obvious that they need, and will receive here, little attention. Others are not the bromides that at first glance they may appear to be. It is upon the latter that we will focus key attention.

B1—Save his time

As a professional salesman, you find one of your key problems is to make the best use of *your* time, and you resent any unnecessary intrusions. Your customer is often in this same frame of mind. He will welcome anything you may be able to do to save his time and to help him arrive, more quickly, at a sound decision. This emphasizes, in some instances, the need for entering the selling situation with ready answers, upon which he can depend. Other methods of saving his time are these:

I. If, before he makes his decision, it is necessary for the customer to read what he will regard as voluminous materials, provide him with an extract of the key points (as well as making the full material available to him). Analyze the material which he must read, in advance, and prepare an *"estimate of the situation"* for him. This should include:

A. Statement of the problem. Here, you should identify, in as few words as possible, the matter he needs to solve—as you see that problem.

B. Facts bearing on the problem. Again, in as few words as possible, identify the key facts which he should consider in reaching his conclusions. Don't list unimportant facts. They can be included in your more voluminous report which he is free to read, or not, as he sees fit and his time allows. Key the facts you list to the more detailed report. Assume, for example that one item you feel he must consider is operating cost. Your fact in paragraph 2 might be: "Model 200 will cost you $650 less for maintenance yearly. (See Fact Sheet 1, Paragraph 6)." This comment in the parenthesis lets the customer know where he can find more detailed information (or proof that you are stating a fact rather than an opinion). Do this each time that you feel he may want background material.

C. Assumptions. In this paragraph, list information which you believe is factual or which you consider probable. If your final recommendation is based on any of these, it is best to identify them right now so that you can determine if he is in agreement or not. Don't list an assumption as a fact. If he can prove (or if he feels) that it is *not* a fact, and you list it as such, your credibility with the customer might be destroyed.

D. Conclusions. This paragraph should also be kept brief and it should be, essentially, a summary of the facts and assumptions—or an analysis of them. At the worst, if he does not agree, you have identified the true nature of your problem; i.e., on which points he must be convinced.

E. Recommendations. In specific terms, tell him what you feel he should do. If possible, list alternatives for him. In this case, list first the action you hope he will take. Don't try to load this paragraph with a large number of possible actions. To do so will only confuse him. At the same time, give him a choice. This gets him involved and causes him to feel that *he* is making the decision—not having it made for him. A listing of alternates also suggests to the customer that you have thought his problem through and are not giving him a "canned solution."

II. Another method of saving the customer's time is to avoid loading her up with details. If some portions of the decision can be made by her subordinates, don't spend her time unnecessarily on these details. This is especially important *after* she has given you that order. To annoy her with details may alert her to facts she had not previously considered . . . facts which may cause her to see disadvantages she had not heretofore discerned. In other words: Get out while you're still ahead.

III. You can save the customer's time by the judicious use of visual aids; i.e., photographs, models, charts and displays. Don't resort to visual aids just for the sake of using them and appearing "professional." Employ them only when their use will enable the customer to get the message more rapidly than would be possible by listening to you.

IV. Be organized. If he is like all too many people, he thinks salesmen sell by the seat of their pants. You'll make an excellent impression and contribute toward that all-important climate if you know specifically what you are going to do, how you are going to do it and in what sequence. At the same time, don't force him to be a slave to your plan.

In other words, if, during the selling situation, you see he already understands something you planned on explaining, or if you see that certain aspects of your presentation leave him bored . . . move on fast. To insist upon following a preconceived plan of action without deviation only causes him to feel he is being subjected to a canned approach.

Being organized and confining yourself to key points does *not* mean being superficial. You can always go into depth when the customer indicates he wants, or needs, more information.

B2—Help him

Use of this strategy does not mean helping the customer with the product or service you are trying to sell. It means going beyond that. It involves looking for opportunities "beyond the call of duty" to help him. It calls for being alert to other problems that he has and determining how you can assist him in areas in which he would not normally expect your help. It may require tapping other resources of your company to provide him with specialized aid. It may mean diving in to help him with some physical task in which you find him engaged. (It may even be that if he can't get the job done that is taking his attention, he won't have the time to give you.)

"Help" can range all the way from professional assistance to going out of your way to do personal favors. The latter are infinite in variety and could vary from getting his son an interview with some placement director you know at a college to such mundane items as giving the customer a lift in your car when he knows it has taken you out of your way to do so.

Only you, with your knowledge of the customer, can determine what favors or help you can give him that your less perceptive competitor has failed to give him. Personally, we recall having gone out of our way to help two individuals find housing and, years later, they haven't stopped talking about it. Offering to help shows that you have a personal interest in the customer, as an individual, that transcends the business at hand.

B3—Offer to help

See preceding strategy.

B4—Agree with him

This strategy does not mean being hypocritical. It does not require the surrender of your principles. It *does* include avoiding issues that are not important. Steer clear of debates on petty points that only threaten the climate you are trying to build. Refrain from making such statements as, "I can't agree with you there." No matter how much sugar with which you cover that kind of a remark, as far as the customer is concerned you have just walked on his self-image and reflected upon his judgment.

Agreeing with him means looking for areas of agreement and focusing upon them, so that he will not bring up other areas where you know there can be no agreement. It means avoiding flattery. Don't lay it on thick. And don't use the same phrases over and over so that they lose their value. Vary your form of agreement with, "You've got a good point there" . . . "I certainly agree with you" . . . "My experience supports your opinion," etc. When possible, and without being verbose, mention a specific situation that you have experienced which bears out some point he has made. This strategy causes the customer to feel that *you really do agree* and that you are not just attempting to manipulate him. (Incidentally, be sincere when you agree.)

Be slow to agree with premises until you are certain the customer cannot construct arguments against buying which he will develop from what appeared to be a harmless premise.

It is possible to *imply* agreement without actually agreeing: "I can see your point"..."I can see what you mean"..."There are others who will go along with you on that view"..."If I had had the pressure you've been under lately, I would probably see it that way too."

If the customer has some real complaints regarding prior purchases and/or service, and if he is right, consider the advantage of agreeing with him. An oil dealer failed to keep up the oil level in an investment house we had purchased. Result: burst radiators, etc. We called him, ready to read the riot act. The dealer immediately opened with, "It's our fault. We'll take care of it. It has never happened before!" The result was that we were so overwhelmed with this fresh and unexpected reaction that we have asked him for an estimate on a new burner and have already steered a new customer to him. In the long run he will *make* money rather than lose it, simply because he agreed.

There is nothing more calculated to deflate the anger of an irate customer than the "mea culpa"—my fault—approach.

B5—Give him privacy

When should this strategy be employed? It should be considered under the following conditions:

1. Give him privacy when you know the customer does not have final decision-making authority and does not want you to know it. Furnish information he can take to the real decision-maker, and get out of his hair until he has had a chance to convince *his* decision-maker.

2. Give him privacy when he has been furious over some error of yours or your company's and you have facts available that will show him the fault was really his. Make the facts available to him, so that he can read them... when you are not there. If you can avoid it, don't ever be present *when a customer is going to lose face* and be forced to say the hardest words that come to the tongue of man; i.e., "I was wrong."

3. Give him privacy when you know that he hates high pressure. Furnish the facts he needs, then leave him alone so that he can feel that *he* is setting the pace and controlling the selling situation (even though you know you are doing the controlling).

4. Give her privacy when your proposal is complicated, and she needs time to study it.

5. Give him privacy if you *know* that his superior has already told him he must buy from you (and he does not know that you know this). In this way you give him an opportunity to decide how he will contact you to give you the order as though it were his decision.

6. Give *them* privacy when there is disagreement among two or more people from your customer's organization. Don't stick around until your man wins or until you are maneuvered into taking sides. You and "your man" may win today, but will you have created a powerful adversary for the future?

Giving the customer privacy must be done only after you have weighed the possibility of a competitor moving into the vacuum created by your departure. As is the case with all of these strategies, you must use judgment in the light of the facts of each specific selling situation. At the same time, it is important that you avoid feeling it is *always* bad to leave before you have the order.

You will note that at no time have we ever said, "There is just one thing to do." There are usually *many* things to do. The problem is that we all fall into the habit of using the same old strategies that have worked so well for us in the past and forget there are many other strategies on tap and ready for our consideration. Don't become a "Johnny One-Note."

B6—Do him a favor

See previous comments under *B2—Help him.*

B7—Praise, compliment him

This strategy is to be used judiciously. As stated before, avoid any semblance of flattery. When we flatter a man we have, in essence, said to him, "Your behavior is transparent. I see how you want others to see you, and if that is what you want—fine—I'll tell you that's the way I see you."

Again, remember the safe way to praise is to "praise the work, not the worker." Find something the customer has done in which he takes pride and praise the achievement—not the man. Make sure it represents something worthy of him or it will backfire with the customer thinking, "Humph, is *that* the best she thinks I can do."

Praise will be most effective if you have identified that image the customer is trying to project (how he wants others to see him). From Part One on psychological aspects of selling, you may recall that each of us also has a "me we wish we were"—that is, the "me" we wish it were possible to have others see when they see us. Remember, most of us have little hope that this will be realized in our lifetime. Therefore, it is dangerous to lay it on thick and appeal to *that* "me" when you are talking to the customer. He will almost certainly feel you are not sincere. "Why should *you*," he reasons, "see me that way, when I can't get others to do so." It is better to direct your remarks to the "me he tries to project"... he feels that others may see him that way, too, so you are probably sincere.

Know your customer well before you praise him in front of other people. Some people are embarrassed at being singled out, even for praise. Rather, when possible, set up situations where people whose opinions are important to your customer see what he has accomplished and praise him. He'll remember that it was you who set up the situation that made him look good where it counted.

B8—Give him credit

There are too many salesmen who not only want the sale, but they want credit for having solved the problem. This is all right except when it is your customer who wants the credit. What is really important to the salesman is to receive credit from his *own* company. This will result from getting the order, so why insist upon getting credit from the customer's company, too? If you want repeat business, make the customer look good in the eyes of *his* boss.

B9—Emphasize mutual interests

All too often a sale is a contest between seller and buyer. Just so long as it is allowed to remain as such, the customer will be on his guard, on the defensive. This attitude does not move him one millimeter closer to granting you an order.

It is evident that a favorable climate can be assured only when suspicion is eliminated and when the gap between the customer's objectives and those of the salesman can be narrowed. For this reason, the professional salesman makes every effort (1) to concentrate on points of mutual advantage and (2) to take the focus off those areas of the selling situation where conflict exists, or seems to exist.

This is especially true when you attempt to sell a customer something that is radically new. He doesn't want to be taken for a "patsy"; i.e., to appear gullible or to have others regard him as an easy mark. He does not want to feel he is being manipulated. It is not that he is opposed to *new* ideas; it is simply that he is afraid to make a mistake. He needs assurance.

One way of supplying this assurance is to show him that your own reputation and success rest upon the satisfactory solution of his problem or the successful functioning of the product or service you wish him to buy. Show him that it is as much in your interest as his to sell him the right product.

Emphasizing mutual interests is not confined to overcoming the customer's suspicion of new ideas. She may be in difficulty with her own company because of some past decision. It becomes your task to convince her that you are in the same position with your company just so long as the problem remains unsolved. Therefore, when you recommend a course of action (a product or a service), you have as much at stake as she does. Once there is no suspicion of motives, the road to a sale is less encumbered.

No two human beings probably ever have interests that are completely identical in all respects. At some point they will be dissimilar and probably conflicting. Subconsciously, the customer is aware of this when you attempt to sell to him. He is fully aware that one of your goals is to separate him from some of his money. One of his goals is to hold on to as much of his money as he can. Your task is to cause him to realize that you have his interests at heart as well as your own and that both of you can benefit from the successful closing of this sale.

Therefore, seek to establish common interests. We tend to choose as friends those who share mutual interests with us. We feel they do not threaten our goals, and we gain a certain satisfaction from knowing that they see things as we do. We feel "comfortable" with them. The more interests we share with another individual, the closer we feel to him and the more we are inclined to listen to his opinions and advice...the more we trust him and the less we suspect his motives. *These* are the underlying reasons why it is wise to establish common interests with your customer—even if they be in the fields of hobbies or otherwise unrelated to business. It is *essential* that we have his confidence and his trust. A bridge to his trust is the interests we have in common with him.

The roots of confidence are found in rapport. Rapport starts, often, with the identification of common interests and opinions. We would be unwise, indeed, if we shared any kind of interest—be it a business goal or a personal activity—with the customer, and then failed to make him aware of it. Therefore, when you feel your progress in a selling situation has been impeded because the customer seems to lack confidence in you, try to identify an area in which you share common viewpoints and subtly steer the conversation into those channels.

B10—Welcome objections

No matter how cantankerous the customer may be, it's a safe bet that he didn't swing his legs out of bed in the morning with the waking thought, "Who can I make life miserable for today? I know...that #$%¢& salesman!"

If he raises objections, you can feel sure that he believes they are legitimate. If they are smoke screen objections, you can determine this by supplying him with facts that eliminate those objections. If he continues to raise invalid arguments as you disprove each one of them, you can be sure he really has no valid reasons but is covering up some other situation —perhaps a previous commitment to some other supplier, or following instructions handed down to him, and the like.

Objections give you an opportunity to tell your whole story. They provide you with clues to the customer's real problems. They show you where your sales presentation or your company's promotional literature has been something short of perfect. Objections show you where you must concentrate your fire. How much worse it is when he offers no objections but shows no interest. In such cases, it is indescribably difficult to ascertain what needs to be done to arouse and maintain his interest.

Objections also may show you where your product or service really does need to be improved. How much worse it is when he never lets you know why he doesn't favor your product.

If he objects, at least you know he has been interested enough to listen or to look into the wisdom of buying from your company.

Probably most important of all, the nature of his objections reveals what he feels his real problem to be. This provides you with the information you need to revise your strategy or to determine which product features you should accent. Don't greet his objections with irritation or with argument. Grant them a consideration that he rightly expects they will be given. If he makes broad sweeping objections, then dig deeper; i.e., "I'd like to look into that; can you give me the reason why you feel as you do so that I can do something about remedying (whatever is wrong)?"

Customer objections often provide us with an opportunity to present additional selling points. For example:

Customer: "I can't help but feel that when I pay your price, I'm paying for the name."

Salesman: "That's perfectly understandable, and, because you have raised the point, I'd like to discuss for a moment just why we do have a good name. For example, our name is built in part upon our service—on our demonstrated interest in the customer long after the sale is completed. Now let's see how important this can be to you in your business...."

As a salesman, you do not have to concentrate on or even mention "service" as the reason "for your good name." It is better to have identified the customer's key interest because then you can go on to point out that your company's earned good name is the result of quality, ability to deliver, service, research, or whatever will appeal most to this specific customer.

When welcoming objections, use such phrases as, 'That's perfectly understandable." This type of reply dignifies his complaint with the attribute of reasonableness without necessarily agreeing with it. Again, don't become a slave to one phrase. Vary it with such remarks as, "That's a good point, and I'd like to discuss it."..."That's a valid comment but let's see what other factors more than compensate for it."

If, for example, your customer who complains about paying for your name sells your products, in turn, to others, consider the opportunity his objection has provided you. "I agree with you, Mr. Beckwith, and I would like you to see what an advantage that is *to you*. Our name *is* important to your customers, and that is why we are in such a good position in our market. Your customers want quality, the reliability and performance that our name has come to mean, and that is why you will move more of our products than any other brand." This may cause him to see what he considered an objection is actually an advantage.

Objections, properly handled, give you a rare opportunity to recognize the views of your customer, to identify his real problem, to show him respect and to convert a seeming disadvantage into an advantage.

B11—Agree; B12—Admit validity of his view and B13—Find point of agreement

The next three strategies listed in our Sales Strategy Bank vary only in the method of application. They might well be considered as one strategic method. *B11—Agree. B12—Admit validity of his view.*

B13—Find point of agreement. In some respects, they are much like the "welcome objections" strategy we discussed.

All are based on the need for rapport or for giving creditability to the views of the customer, and for causing him to feel that you regard his views as important and his objections as worthy of consideration. All too often a salesman loses a sale because he decides to "stand on principle" when it is questionable that a "principle" really exists. In other words, he is so sold on his product or service that he will brook no criticism of it, even from the customer whose business he seeks. This type of salesman regards it as a personal defeat if he doesn't obtain 100% of the customer's agreement. Often, the customer is ready to buy but wants you to know that he is aware of some shortcoming of your product...or that he thinks it is less than perfect. To argue with him is to threaten the climate of the sale.

It should go without saying that agreeing with the customer when he raises a serious objection would be foolish. But here we are referring to minor criticisms of the customer, which if not handled well can blossom into major obstacles. For example, a customer who is ready to buy may comment, "Your competition certainly gives more attention to styling than you folks." The inept salesman will immediately argue this point with the customer.

Let's say that in selling Product X, styling is of no real consequence. Why make a federal case out of trying to convince the customer that your styling is as good or better than competition's. It is far better, when possible, to let his statement pass without comment, or even (in some instances) to agree. For example: "I suppose we *could* do something about styling, and I'm sure our people will, but of course the thing you are looking for is performance. And I'm sure you agree that our product will give you the performance you are looking for." ("Performance" can be replaced by whatever it is that your specific customer is seeking.)

In this instance, you have indicated some degree of agreement. You have not challenged him and you have put his obstacle in proper perspective by comparing it to that which you know is *really* the focal point of his need or interest.

When you find it necessary to "admit the validity of his view," it is essential that you move quickly into the presentation of additional information that will dilute the importance of the point made by the customer. For example: "If I were in your place, I would feel as you do. But in this instance you are protected against any possible price increase. So there can be no repetition of your last unfortunate experience. Furthermore, we have a new feature that is very important to you, and I'd like you to see how it can save you money...."

The preceding would be in response to a customer who has complained he lost money the last time he made a purchase because of a price change he had to pass on to his customers. There would be no point in the salesman's denying the customer was hurt. He was hurt and he *knows* it. It would also be foolish to try to minimize his loss. He thinks it was significant, or he would not have brought up the matter. It's far better to admit he has a point, give him assurance (when possible) that there will be no repetition of the unhappy experience and then concentrate his attention on that part of your offer which will be of maximum value to him.

There are two ways to tear down a stone wall. One is to use a bulldozer and try to smash it down. This may be rough on the bulldozer, and you may accomplish no more than to end up with a broken blade. The other method is to search around for one loose stone, or one key stone, and remove it. The other stones can then be removed with comparative ease. Selling is much the same. There are selling situations where you and the customer seem to be on the opposite sides of a barrier. The best way to surmount it is to build rapport by seeking one point of agreement—even a small one—and then building from that. Conversely, when no barrier really exists, it is easy to erect one by treating a small objection as something that simply must be argued until the customer agrees he is wrong. Taking that stand results in what should have been an irritating pebble becoming a solid wall of opposition.

Let's consider this example of how to handle an objection. The customer, Mr. Hale, observes that by putting in your lowest priced line of products, he will threaten the sale of the line he now sells at a higher price with a greater profit.

The salesman tackles it in this manner: "Mr. Hale, you are concerned over the possibility that if you put in this line of low priced equipment, it may cut into sales you have been making of the higher priced line. Is that a fair statement of your views?"

The customer replies, "Yes, that's just about it."

The salesman counters with this approach: "I think your concern is reasonable, and I'd like to explore it with you for a moment. Just as you have good, solid reasons for making your buying decisions, so do your customers have what they consider good reasons also. You won't buy certain items that are offered to you because you feel they don't meet your needs...you don't feel they are the kinds of items that should be sold in the kind of establishment you have built.

"Now, let's take a look at your customers. Some will buy a high priced line and others will be attracted by lower priced products. Each of these people buys for reasons that are practically unique to him. Each is looking for something different. The customers who have been buying a high priced line from you will continue to do so, while this lower priced line offers something to people who will not or cannot buy the higher priced products. What will happen is that you will attract new customers who would otherwise be lost by default. I think your point is well worth discussing, and I think you see how you can overcome the problems you had in mind."

The salesman looks for a reaction from his customer and then continues to develop his presentation. He has been able to cope with Mr. Hale's objection because he found one point upon which they could agree. But he has done more than that. He has woven into his presentation a recognition of Mr. Hale's self-image and has acknowledged Mr. Hale's success ("the kind of establishment you have built"). And he has made the customer feel his objection was a reasonable one. By taking this tack, he has opened the customer's mind so that he can focus it upon the *advantages* of the offer. To have totally ignored Hale's views or to have attempted to argue him out of them would have closed his mind to any values your offer may have had for him.

B14—Admit your error

The most difficult words that come to the tongue of most men are, "I was wrong." If you consider this subject for a moment, you will discover a surprising fact. How do you feel about a chap who, in the middle of an argument with you, suddenly says, "Hold it...hold it...I never thought of the point you just made. You're right. I'm wrong." Your opinion of him soars. You think of him as an open-minded fellow. You feel warm toward him because he provided you with a minor victory. Then, why is it that we fail to realize that the other fellow will feel the same way toward us when we are wise enough to admit that we are wrong? In spite of this logic, we will exhaust our complete repertoire of all 16 defense mechanisms before we admit our error. Yet, the most refreshing words we can hear from the other fellow are these: "You're right—I'm wrong."

Humility is always refreshing, and it has the added value of giving the other fellow a sense of well-being because it elevates him. *In effect*, it says to him: "I don't mind admitting to you that I am wrong, because I know you are so 'big' that you won't rub it in."

When we are wrong, the other fellow *knows* we are. So, just what do we gain (when we know we are wrong) by continuing to deny it or to evade the truth? Evading admission of error only builds a greater gap between the customer and ourselves and damages the climate we are striving to build. If the customer knows you have been wrong, he is probably *expecting* you to deny it or to avoid admission of your error.

Therefore, when you freely admit your error, you take the wind out of his sails and the heat out of his anger.

Most people will forgive mistakes, but they won't tolerate what they regard as a stubborn refusal to admit that a mistake has been made. We look upon such refusals as an attempt on the part of the other fellow to save face, and we can't help but feel that he has trod all over our self-image; i.e., "Who does he think he's dealing with...does he think I'm stupid?"

Incidentally, when you do admit you have been in error—or have let the customer down—don't stop with that admission. Follow it up quickly with remarks that cause her to believe that you regret what has happened and that you are taking steps to be certain that she is never affected by such errors in the future.

B15—Entertain him

This subject lends itself to chapters, and it is our intention at this time to discuss it only briefly. The whole subject of entertainment is one that should be discussed thoroughly between the manager of salesmen and his men. What are some of the considerations that should be weighed? Let's discuss a few of them.

1. Avoid projection (see Part One). In other words, in deciding *how* you are going to entertain the customer, don't decide to do something that you know you will enjoy...be sure it is something he will enjoy. This error can be avoided by finding out more about your customer and his likes and dislikes.

2. Don't put him in an embarrassing position. "A night on the town" may be your idea of pleasure, but it may be totally incompatible with what he likes to do. He may not want to be seen in places that you consider quite all right. If he is visiting your town, think twice before you invite him to your home... and especially to dinner at your home. There are men who vastly prefer the privacy of a hotel room at the end of the day, a quick dinner, a book or TV program, and enjoy the rest that often they cannot have at home. Rather than upset you, they will accept your invitation and then have a perfectly miserable evening. Others have distinct likes and dislikes when it comes to food, and your wife may serve up her most tempting platter to a chap who simply has other tastes in food. Determine his preferences in advance.

3. Don't let him be placed in a position where he will let his guard down and do something that he

will regret the next day. He may inwardly know that it was his own fault, but like all of us, he will seek to find the cause of his behavior outside of himself. You can be sure it is *you* whom he will blame—even if not openly—and goodbye to that climate you were trying to build.

4. Look upon entertaining as an investment. In our time, we have seen customers entertained when there was no profit to be made, and the men responsible did so only because it provided them with a chance to get some entertainment themselves at company expense. Consider, then, what your company is going to gain from entertaining Customer X.

Some customers react badly to expensive entertaining. They have a suspicion that they are going to pay for it in the price charged for the product. If the diversion is lavish, they may form the wrong opinion of you or your company. Others are suspicious of entertainment and get the idea that you are trying to soften them up. In other words, you had better know your customer.

Some salesmen think that entertainment *must* be expensive or it will not impress the customer. Usually, it isn't the cost of the entertainment—it's the thought that goes into it that makes the impression on the customer. We remember one chap, who, in the days when it was allowed, had a group of military customers to dinner in a rather expensive restaurant. During the course of the meal, one of the officers mentioned how long it had been since he had seen his wife who was at his military post 2,000 miles away . . . and how weary he was of being away from her so much of the time. A short time later, the waiter called the officer to the phone. When he returned to the table, he was figuratively walking on air. The salesman had gone to the phone booth, put in a call to the wife on his credit card and had the officer paged.

Later, the salesman said, "You should have seen the dinner bill . . . it was really big. But do you know, for the past six months every time I meet that officer he talks about my thoughtfulness in putting through the phone call." Cost of the call: insignificant! You can get real mileage from an expense account if some thoughtfulness goes into the spending.

Another point to consider is who you should entertain. All too often we take "Good Old Joe" to lunch with us, when "Good Old Joe" would give us an order if we never took him to lunch. It pays once in a while to take "Terrible Terry" to lunch even if he has never given you an order. You may learn more about this terror at lunch than you would learn in a dozen calls upon him in his office.

The whole principle of intelligent entertainment is that it provides an opportunity to build rapport . . . to develop a good climate . . . to learn more about the customer as an individual, and of course to meet him in a spot that is free from disturbances, interruptions and phone calls. It gives him a chance, too, to learn more about you.

And a word about the saleswoman and/or the sales manager who is inclined to be too parsimonious about the expense of entertainment. Keep in mind that, if you are in a corporation, the expense is about half of what appears on the price tab. If you don't entertain, Uncle Sam will collect it anyway. That is to say, Uncle Sam recognizes legitimate entertainment as a method that produces more revenue for him. After all, if you succeed in making a sale as a result of entertainment, the government is going to collect more in taxes from both you and the customer.

Also to the parsimonious, a word of warning: The customer will see in the methods by which you entertain him a reflection of the way in which you view him.

B16—Be intentionally incorrect

There are many customers who must buy, but who hate selling situations because they fear they will be dominated by the salesman. This feeling is intensified when the salesman has information the customer needs, and the customer therefore finds it necessary to assume a role that is primarily one of listening.

One method of easing tension is to be incorrect on purpose in a *minor* statement so that the customer has an opportunity to "get in the act" and correct you. It is apparent that care must be used in adopting this tactic. The error must be on something of relative unimportance so that it does not destroy the confidence you wish the customer to have in your mastery of your subject. It does, when wisely used, give the customer an opportunity to escape the feeling he is being dominated, and it provides you with an opportunity to show him recognition and thank him for his help.

An example of what we have in mind follows:

Salesman: ". . . And for these reasons I know you will find this model superior to the old Model 210."

Customer: "Model 210? You mean Model 213, don't you? You never had a Model 210 that operated on 220 volts."

Salesman: "You're absolutely right . . . it *was* Model 213. You do a better job of remembering model numbers than I do . . . I forget that you have been using our products for a long time."

If the salesman wants to get the customer talking and doesn't mind having the trend of his presentation broken momentarily, he can add: "I have trouble remembering some numbers. Do you have some kind of a system for remembering?"

In the foregoing example, the salesman deliberately made a minor error. If, on the other hand, model numbers were very important, he would have chosen another area for his unimportant error. In this instance, he has given the customer an opportunity to interrupt and to take the floor. He also knows the customer has been listening—so, this tactic has the same effect as a test question; i.e., to see if you have been holding the customer's attention. The added inquiry about a memory system gives the customer an opportunity to talk about himself.

B17—Provide an escape hatch for him

Never "box" yourself into a corner, and never put the customer in the same position where he will lose face. It is often necessary to avoid creating a situation where the customer must go on record. This is why retail sales people are taught to avoid asking questions that limit the customer's choice. For example, they are taught never to ask "What color do you want?" before they know what they have in stock. Once the customer has committed himself to brown, for example, he is going to be disappointed if you are unable to produce brown, and you may lose a sale.

In industrial selling, parallel situations often exist. A customer may have made a flat statement indicating that he is for something or against something, or he may have made a statement which he believes to be a fact. Then, during the course of a sale, if it becomes obvious to him that he has been in error, his position can become untenable if he is the type of individual who cannot face the fact he has been wrong.

He is going to want to get off the spot, and he won't love you for having shown him his error. He will find himself in what we have called "The Ouch Area" ("I've been wrong and Ouch—it hurts!"). Instead of thanking you for setting him straight, he may use one of the 16 defense mechanisms... in this case, "Attack!" and become antagonistic. Or, he may use another form of defense, "Evasion," and seek to bring the selling situation to a close.

People do not like to have other people present when they are confronted with information that proves they have been in error. Even otherwise intelligent people will often become defiant and fight for a point of view which they now know is incorrect and will refuse to admit they have been wrong. All of this threatens the climate you are trying to build. What to do?

1. Avoid asking questions that limit the customer's answer. (There are other times when you *want* to commit the customer, but we are discussing that phase of the sale where you are trying to build climate.)

2. Avoid asking for opinions that may leave you out on a limb or create a situation where you are going to be forced to show the customer he is in error.

Example of (1) above: "What is the latest date by which you can accept delivery?" If she sets a date that you cannot possibly meet, you've created a problem you can't avoid. Remember, the customer is almost always willing to compromise if her desire for your product or service has been made intense.

Another example: "Is it essential that the insulation be rubber?" If he answers "Yes" and your product does not utilize rubber as an insulation, where are you? A problem has been created unnecessarily, and the customer's attention has been focused on the one area in which you cannot provide satisfaction. When the selling situation opened, rubber may not have had any importance in the customer's mind. When he finds you cannot supply rubber insulation, he may exaggerate its importance or he may use this as a smoke screen objection for not buying, simply to hide his real objection, which you might have been able to overcome.

Admittedly, it is impossible to avoid situations where the customer, by his own precipitous actions and statements, may create a problem where he stands a chance of losing face. For example, a customer might say, "Nobody, but nobody, could sell 10 of those in a month." Later, he learns that someone in a situation comparable to his has sold 50 in a month.

Or, customers have been known to say, "You people never even answered my last letter" and then have the "never-written letter" pop up in the file a few minutes later. In such situations, the customer is wrong and knows he is wrong. There are some wonderful customers who will merely grin, pick up the letter and make some such comment as, "When are we going to get organized around this place—I'm sorry." But we are not talking about this kind of individual—we are discussing the frequently encountered chap who will create a problem for you by bluster or evasion. How can such situations be overcome?

One answer is found in a situation we experienced recently. In connection with a highly technical hobby, we recently made a comment that revealed our lack of deep technical knowledge in the product being discussed. It immediately became apparent to us and to the individual with whom we were speaking that we were totally wrong. Our own embarrassment was acute, and there was no way in which we could modify the comment we had made ... no escape hatch seemed to be present. The chap to whom we had made the comment merely looked at us, grinned, and said, "Join the club. I had the same opinion until I read a technical paper on the subject recently."

Our feeling of relief was intense. So—after all, he was *not* more intelligent than we were. After all, we had not had the opportunity to read the technical paper he had read . . . it was reasonable that we should have held an incorrect opinion!

The principle is obvious. When the customer suddenly encounters the letter he has just claimed, with some anger, you had never mailed, you have the perfect opportunity to use the *"Join the Club" approach*. "Join the club—I'd like a dollar for every time that the same thing has happened to me, Mr. Blake, and I'm glad you located the letter because I'd have been real upset to think that we had let you down." Then move *rapidly* into another facet of the subject under discussion and never refer to the "lost letter" again.

"Join the Club" is an escape hatch for the customer. It eases tension, minimizes the importance of his error, saves face and keeps him on a level with you. (Momentarily, he feared that he had fallen below your level.) It is also a friendly act, and people don't usually respond to friendly acts with unfriendliness.

Choose your language carefully when you supply the escape route for your customer. Certainly you would not say, "Think nothing of it; I made the same stupid mistake before I had it explained to me." We can recall an occasion when we were discussing the opportunities in the Common Market with a chap who had never been overseas. He made some flat statements about the Common Market being only a threat to us with no opportunities. Later, we presented him with evidence that he was in error. His reaction was one of enraged hostility.

Later—some days later—when he had a chance to think about it, he admitted he was wrong and that we were correct. Then added, "The only thing I didn't relish was having you tell me that you had been just as stupid as I was about the export possibilities until you had made a few trips to Europe." The fact is that we had never used the word stupid. That is not important. What *is* important is that, somehow, we had been maladroit enough to cause the other chap to think that we regarded him as stupid.

It is not even wise to say, "I made the same mistake until new information was made available to me." Don't identify her "mistake" as such. It is far better to say, "Your (statement) (feeling) is perfectly understandable. Unless, by some chance, you learned of this new process, you couldn't have felt otherwise. As a matter of fact, *most* people—and I include myself—felt just as you did until they encountered this new development."

This last approach has the advantage of being based on the fact that "misery loves company"—that most of us don't mind having erred when we find that the same mistake has been made by many intelligent people before us. As you saw in a previous section dealing with *strategy B5 ("Give him privacy")*, this tactic may also be employed as an escape hatch. In other words, if you must cause the customer to see that his view is in error, try to find a way of leaving information with him in written form that he can digest later—in privacy—so that you will not be present when he must face up to the fact that he has been wrong.

If he brings up the subject later, admits he was in error, and—in effect—apologizes, don't dismiss it in a cavalier or patronizing manner as though it were of no consequence. If he was quite intense in his previous feeling, the matter *was* of importance to him, and he doesn't want you, even now, to consider it lightly. It is better to make such comments as, "Well, Mr. Eldredge, I knew how strongly you felt about this, and I could understand your feelings; but, I'm glad that we are in agreement now." (Notice, we do *not* say, "I'm glad you agree with me now.")

B18—Listen

Listening means far more than hearing your customer's voice as he speaks. It means these things as well:

Giving your full attention to him when he speaks.

Giving him the impression that you are listening.

Comprehending what she is saying.

Understanding the *significance* of what he is saying (why he is saying it).

Seeking the reasons for words she uses—hidden tips to her product, economic or psychic needs.

Listening with your eyes. Seeking more tipoffs to his real need by observing what he wears, what he uses, and with what possessions he surrounds himself. What are these things "saying" to you?

Looking for a pattern in his remarks. Do they keep centering on some specific need? Don't judge him by isolated comments.

Giving him free rein and avoiding interrupting his train of thought. Use questions, when appropriate, to keep him talking so you can learn more.

Showing respect when he is talking. Don't let your gaze wander as he speaks; don't look at printed materials or equipment while he is talking . . . look at him. Attention is the highest form of praise; give him the feeling that you consider what he is saying to be important.

Avoiding mannerisms that may distract him.

Avoiding mannerisms that indicate your disapproval, disagreement or impatience.

When you must comment later, weave observations into your remarks that prove to him that you really were listening to what he had to say.

Listening is a skill that must be developed. Analysis of the points just listed will be of value.

Giving your full attention when the customer speaks. This means we must avoid the tendency to occupy ourselves with planning our rebuttal while he is still talking. It means we must not permit his remarks to spark off other chains of thought in our minds so that we suddenly become aware we have missed much of what he has been saying. It means concentrating. This is not easy to do because most of us have grasshopper minds; i.e., "Good on distance but hell on direction." We concentrate only with great difficulty—and the sharper our minds, the more this may be true.

Giving the impression you are providing the customer with your full attention. It is not enough to actually give him your complete attention; you must also create the impression that you are doing so. Paradoxical? Not so. For example, you may be listening to the customer and be impressed with his comments. You may not be looking at him while he is talking. You may be gazing thoughtfully at the ceiling to avoid distractions, or you may appear to be gazing out into space. He has no way of reading your mind, and he may assume you are not listening.

Blanche Marion, an expert in the field of "the language of the body," points out that the best way of showing a customer you are really listening to him and taking his views seriously is to sit upright in your chair, leaning slightly forward toward your customer, preventing your hands from distracting him by having your fingers loosely interlaced, and looking at the customer while he is speaking.

This may sound a bit basic but try it—or better yet, have someone else listen to you while *you* are speaking. Have your listener slouch into his chair and gaze off into space while you talk. Then have your listener follow Ms. Marion's advice, and notice the difference in your own reaction to your listener.

Avoid staring fixedly at your customer, which can make him feel uncomfortable. Avoid assuming a rigid position. There is a happy medium. It is these "little things" that make a tremendous difference in the reactions customers have to us as individuals.

Comprehending what the customer is saying. If you don't follow your customer's reasoning or don't understand what she is saying, say so. There is no value in letting her talk on if you can't follow her. The very points you are missing may represent the premise upon which the balance of her views may be based. When it becomes necessary to stop her for clarification, assume this burden—don't place it upon her.

In interrupting, choose your expressions with care. Don't say, "I'm afraid that is not clear—can you go over that again?" Rather, say, "I'm sorry, but I think the point you are making is important and I want to be sure I understand it fully. Would you mind going over it again?"

Understanding the significance of what the customer is saying. It is not enough to understand what the customer is saying; it is necessary to understand *why* he is saying it. What is the stimulus that prompts his remarks? Does it provide a clue to the nature of his need—as *he* sees that need? Don't overwork this. Don't play amateur psychologist. Don't look for hidden meanings in everything he has to say. In the vast majority of instances, you will know your customer so well—or his need will be so obvious—that no analysis will be necessary.

There are other instances, however, when it is essential to know what is really on the customer's mind. This can sometimes be determined by asking yourself why he raises certain points, why he raises certain objections. This is especially true when you know there is no real validity to his objections or when you know the importance of the points he raises does not merit the fervor which accompanies his remarks.

Keep in mind that, though this may be somewhat of an oversimplification, your customer may feel that your proposition is going to cost him money (beyond its value to him), that it is going to make him look bad, that it is going to threaten his emotional security, or that it is going to give him additional work to do or require effort beyond that which he is willing to expend.

The nature of his remarks may reveal which of these is his "hot button" . . . his key motive for buying or resisting.

Seeking the reasons for the words the customer chooses. Your customer is a rational being, no matter how irrational he may appear to you. There are reasons, which are valid to him, for everything he says and does. By the same token, there are reasons for his choice of words and expressions, even though they may be subconscious reasons.

A columnist, for example, in attacking one of the giant companies of U.S. industry, chooses the word "sprawling." He does this to create an impression—an image—in the minds of his readers. The casual reader may take little notice of the word "sprawling," but its work has been done. On the other hand, the analytical reader may see the purpose of the columnist and, without finding supporting proof, may reject the attempts of the writer to brainwash him.

The one word "sprawling" can create the impression of a big, powerful, loosely controlled, octopus-like conglomerate. Your customer, too, will use words that reveal his true feelings and his prejudgments, as well as his desires. He will not do so with the same determined purpose of the columnist but will do so in spite of himself.

Be alert to your customer's choice of words... they can often tell you far more than he intends to convey. They can help you identify his needs for proof and assurance, and they can help you determine when his views have been colored by hearsay or by introjection (acting as though something done by you or your company to someone else had been done to him).

Listening with your eyes. You can listen to a customer who actually has little to say. The old adage applies, "What you are speaks so loudly I cannot hear a word you say."

Your customer may *profess* to have certain reasons for buying or for refusing to buy, or for buying from a certain firm, or for buying a certain product. But his real buying motives—or his motives for not buying—may be identified if you will but look for the significance of the actions he takes, the kinds of things (and people) with which he surrounds himself, the places he eats, what he wears, his hobbies, and so on. His behavior may supply you with an insight into his buying motives that he denies you through the spoken word. Be sure to look for a pattern and don't judge him by one or two isolated items or remarks.

Just as certain as the fact that you and I have identifiable patterns of behavior, so will the customer... and his behavior pattern offers him specific satisfactions. Identify the satisfactions he desires and you have your key to the kinds of things you must say and do to motivate him to buy.

Giving the customer free rein—don't interrupt her chain of thought. People talk about things of interest and significance to them. When we interrupt them—or worse yet, when we lead their thinking—we cause them to talk about things of interest to us. This deprives us of an opportunity to learn what is uppermost in their minds. Silence, on the other hand, is a vacuum; and two or more people, when gathered together, usually cannot tolerate this kind of vacuum. Communication becomes essential.

For example, a conference leader will ask a group a question. If it is difficult, it will plunge the group into silence while they grope for the answer. The untrained conference leader cannot stand the silence. He forgets that it is the group that "is on the spot"—not him. After a matter of 10 to 15 seconds, he will break the silence to "make the question easier," or he will provide the group with the answer. The result is that members of the group learn little or nothing because they have not been made to think. A worse result is that the conference leader has learned nothing about his group. Had he given them from 25 to 50 seconds, he would have found his group beginning to supply him with answers... even though they might be incorrect, and even from these he would have learned *something*. He forgets that members of his group would have felt compelled to speak, because they couldn't tolerate the silence either.

The professional salesman can learn a valuable lesson from this. First of all, asking a question is the most powerful tool you have. It can smoke out the customer. If he doesn't answer your question at once, wait. Don't take him off the hook by talking. Let him fill the void as he will if you remain silent. At the very least, wait nearly a minute, and rephrase your query if you feel that his failure to answer is due to an ambiguous or poorly worded question. Then, again, wait.

Showing respect when the customer talks. As has been pointed out previously, respect can be shown by giving the customer your full attention, which means looking at him. It can be demonstrated by mannerisms—for example, by the occasional nodding of your head (though not by an ingratiating and continuous bobbing of the head). It can be shown by such very brief comments as: "You have a point there"; "I see your point"; "That's an interesting point," etc. It can be shown even more effectively by the judicious use of additional questions which indicate to your customer that you have been listening, and they encourage him to keep talking and, in so doing, to reveal even more to you.

Avoiding mannerisms indicating disapproval or disagreement. There will be times when you really do not agree with what the customer seems to be saying. On such occasions, hold your reactions until he has finished! If you permit him to continue to talk, something he might say may indicate that you are in more agreement than you thought.

Keep in mind, always, that when you or I express or indicate disagreement, we are, in the customer's eyes, challenging his judgment and attacking his self-image. We should not be surprised when he reacts with anger, and when he freezes. When this happens, the flood of information we have been receiving stops... and we have a task of rebuilding the climate which we should have been nourishing. Facial expressions, signs of impatience, of irritation, or of challenge can create unnecessary roadblocks.

Looking at the situation pragmatically, you have nothing to gain by interrupting him. At that point, you are not going to change his mind. Furthermore, if something he has said irritates you your mental decks are not clear for action, and you are not going to be in a position to do your best thinking. While he continues to talk... listen, get your own emotions under control. Then, when he has finished—*really finished*—try to provide him with an escape hatch, so if he really is in error, he won't lose face!

We can easily *talk* ourselves out of a sale, but it is pretty difficult to envision a salesman *listening* himself out of a sale.

Compulsive talkers make poor salesmen and only serve to strengthen the stereotype that too many people have of salesmen. Why not try being a compulsive listener?

B19—Ask for his help, opinion or advice

This strategy is used primarily to establish rapport. It can also be used to get the customer talking. It can be used to cause the customer to overcome his own objections. For example, if he has raised certain objections to your proposal, you can sometimes react with, "I see what you mean"; "I'd appreciate your help on that"; "How do you suppose we can overcome that objection?"

Not for a moment do we suggest that all objections can be handled in this manner . . . but there are times when good judgment will indicate its applicability. The big question you might ask yourself is this, "How often have I tried this method, and how many times have I, instead, tried to argue him out of valid objections?" The chief merit of this strategy is found in the fact that it causes the customer to feel that *he* is dominating the selling situation (even though he is not), and, of course, it appeals to his needs for ego-gratification.

B20—Sympathize

There are some sales managers who will raise their hands in horror at the thought of anyone advising their salesmen to "sympathize" with the customer. They interpret this advice as meaning, "Yes, Mr. Kuniekoff, I can see your problem and I sympathize with it—this is no time for you to invest in our product," or, "I can see why you feel obligated to your present supplier, and this is no time for you to change."

This is *not* the type of "sympathy" we have in mind. One can express a sympathetic viewpoint without raising the flag of surrender. For example, "Mr. Kuniekoff, I can fully appreciate the situation you have described, and can, therefore, easily understand why you have felt as you have. Let's consider, for a moment, how we can overcome the problems you have raised!"

This approach is vastly different. By careful choice of words ("Why you have *felt* . . .") you give a transitory quality to his view—you indicate that there is possibility for change. You have also appealed to his self-image and have not attacked his judgment, his standards or (in the case of supposed obligations he feels he has) his ethics. More to the point, you have opened his mind. You have agreed he has a problem, but you have indicated that you have a possible solution. If he really *does* have a problem, he will want to know how to solve it.

CHAPTER 22

Compelling attention

The best sales presentation will be of no value if the salesman cannot hold the customer's attention. His interest may not be aroused, he may see no gain, he may be distracted or preoccupied, or his attention may be focused on problems which he regards as more important than the subject to which the salesman is addressing himself. It becomes obvious that the professional salesman must determine how to overcome other influences.

Our Sales Strategy Bank includes eight strategies that will assist the salesman in holding the attention of the customer, and it is with these that we will now concern ourselves.

C1—Provide him with a role

Jimmy Durante put it well when he said, "Everybody wants to get into the act!" Most of us are not content when we feel that the other fellow has complete control of a situation. Experiments conducted by a psychologist for a food company indicated that 11 times as many people *used* a recipe when they had been permitted to discuss it compared to a group that had to listen to a lecture on preparation of the food. The old Hawthorne experiment, conducted at Western Electric many years ago, showed that a group of employees would go along with a change when they felt that they had some voice in it . . . even though the actual change was made precisely as those who conducted the experiment had originally intended.

Customers will buy more readily when the salesman makes it possible for them to feel that they have participated in the decision. Some years ago, a door-to-door salesman visited us to sell a home fire-alarm system. Part of his presentation involved lighting a candle and placing it inside a miniature cardboard house set on the kitchen table. The demonstration was intended to dramatize the effects of superheated air. The salesman wisely asked us to light the candle. The experiment thus became partly *our* experiment, and we waited impatiently to see what would happen. Had he lit the candle himself, we would have waited to see "what gimmick he had up his sleeve."

This is not to say that the saleswoman must *always* provide her customer with a role. A given customer may have a pressing problem and be greatly concerned about it—so much so he will freely admit his inability to solve it. He will be delighted if the saleswoman simply shows him how to solve his problem. In such instances, the customer is content to play the role of listener.

There are, however, other situations where the customer considers himself well-informed or where he does not regard his need as pressing. Or, he may indeed not recognize that he even has a need. This customer finds it difficult to escape the feeling that when he is forced to be a listener or an observer, the salesman is ignoring his technical knowledge and capacities. Look for methods of providing him with a role. This may mean using other strategies that appear in the Sales Strategy Bank (see *B16*, for example, or *B19*).

Here are two examples of the customer being provided with a role:

"Mr. Aikens, how would you suggest that these items be displayed?" Here you have asked the customer to take the role of advisor. Perhaps you have a much better display idea than he has, but your goal is to sell the products—you can worry about how he displays them on future visits . . . first get the products in his establishment (if they are items he is to re-sell). Incidentally, he will work twice as hard to make his display idea succeed than he would work if the idea was yours.

"Mr. Canari, we don't want to exaggerate the amount of time this product will save you. I wonder if you would help us out by making some checks next week by keeping a record of the time used versus the time it takes with your present equipment? Then, we can talk over the results on my next visit, which will be a week from Monday, if that is convenient with you. We will both be in a position to estimate the time advantage this product will provide you."

In this last example, you accomplish five things: (1) You demonstrate your honesty; (2) you get the customer involved; (3) you get the "merchandise in his hands"; (4) you manage to set up a second appointment; (5) you show your regard for his time with the phrase, "if convenient with you."

The same technique can be used at a trade show (or elsewhere) as follows:

"Mr. Shapiro, all salesmen make claims for their product, and I wouldn't blame you for viewing my claims with some skepticism or doubt. But I'll tell you what... do you see that group of four men standing in front of that exhibit?" (Pause for agreement.) "Well, why don't you engage in a little product research yourself and ask their opinions of this product. Everyone of them is using it now. I'll introduce you if you prefer, although I would rather stay out of it so you can be sure you are getting unbiased opinions and that I am not influencing them."

Mr. Shapiro may decide not to visit the group because he feels you would not have made the recommendation unless you were certain of their reactions. If this *is* the case, your attempt to provide him with the role of product researcher has done its job anyway. Furthermore, he will be impressed with your suggestions and with your confidence in your product. Also, the very fact that he decides *not* to play the role gives him the feeling that he is not being dominated.

On the other hand, if he feels your claims are exaggerated, he may readily accept your suggestion if for no other reason than to call your "bluff" (as he may suspect it is just that). You have nothing to fear because unless you *were* certain of the favorable reactions of the four men you would not have made the offer.

Here is still another way to involve a customer in a role:

You have been attempting to sell Olson a product or service. Olson asks someone else in his firm to come in and hear your story. When Gregory, the second person, enters, you turn to Olson and say, "Mr. Olson, suppose you give Mr. Gregory the essence of what I have been saying to you... this will give me a chance to see if I have covered the subject thoroughly or if I have omitted something."

Don't use this strategy unless you *know* that Olson has been paying careful attention to you. If, in his narration to Gregory, Olson leaves out a point or two you had previously covered, don't tell him he skipped anything. Rather, say, "Here are two additional points that I may have passed over too lightly..." In this way, you take the blame and provide Olson with an escape hatch.

By using this technique, you have given Olson the job of being your attorney at the bar now presided over by Gregory. If Olson agrees, as he usually will, you have a rare opportunity to see which points he accents and which he neglects. This gives you clues as to what you must dwell upon if you are to sell Olson. It also lets you know if you have held his attention. It tells you which points you can play down. In this way, you will not overwhelm him with details that obviously have no interest for him (or he would have hit them harder when he relayed your presentation to Gregory).

You also are provided with an excellent opportunity to see if you missed any points in your presentation to Olson. Perhaps, most important of all, Olson, in relaying your remarks to Gregory, may sell himself... and Gregory will give greater credence to Olson's views than he would to yours. In addition, Gregory—in listening to Olson—may forget that it is your presentation Olson is expressing.

Don't provide this foregoing role to Olson if you know he has trouble expressing himself. This again emphasizes the professionalism of selling. A professional salesman doesn't use the same hackneyed technique on every customer. He sees every customer as a unique individual and fashions his strategies to meet the unique qualities of that customer.

"Here, Mr. Richeleau, *you* operate it!" With these words you assign the customer a role and get him in on the act. How many times have *you* been irritated when a salesman insisted on operating equipment in front of you and never gave you a chance to get the feel of the thing? Doesn't the astute car salesman sell more cars when he gets in the passenger's seat and suggests that *you* do the driving?

The foregoing examples are intended as thought-starters. Why not stop now and give some consideration to the means available to you, in your kind of selling, to provide a role to your customers?

C2—Use cause-and-effect examples

There is always the danger that your customer will feel she is going to lose face if she agrees with you. Let's be realistic... very often you are attempting to change her mind or to convince her that she should take actions that are at variance with her past actions. Even though you have no intention of making her past actions appear to have been ill-advised, she may feel that they have been something less than "smart." Unfortunately, though you wish to avoid putting her in that position, it is often essential that she see the error of her past views and/or actions.

It is under such conditions that the "cause-and-effect technique" will pay dividends. You may recall that the cause-and-effect method removes the customer from an uncomfortable situation by substituting the experience of some other person.

For example: Your customer, Atkinson, has been ill-advised and has been using an inferior material. You have a superior material that will meet his needs far more effectively but, unfortunately, it is necessary that he understand the material he has been using has been a poor choice.

You know Atkinson well enough to realize that if

you point this out to him, he is going to feel you are reflecting upon his judgment. You know that, as a result, he will take offense, and, feeling that his self-image is at stake, will resort to the use of every self-defense mechanism in the book—certainly including denial, aggression and rationalization. He will be so busy defending his past decision and buying practices that he won't hear a word you have to say. In other words, you won't have his attention... he will be totally involved with his efforts at self-delusion. This is a perfect setup for the cause-and-effect method.

Instead of pointing out to him that he has been using an inferior product (thereby threatening his emotional security), you slowly build up a story of another customer (with whom he can identify himself favorably) who has been using the same inferior material in a similar application. You stress the difficulties this other customer has been having or the losses he has sustained in terms of money or added and unnecessary effort. In doing this, you provide Atkinson with an escape hatch... you aid him in deluding himself that he has not been in error—or at the very least, his error is understandable and has been committed by other "equally intelligent people."

You might, as an example, say, "Richardson (your other customer) had been so up to his ears in work that he had not heard about the development of this new product so, naturally, he was using the material with which he was familiar. Consequently he ran into extra expense and had manufacturing difficulties (or whatever applies). Now that he has become acquainted with this new material he has adopted it, and here are the results he has been getting..." (You then recite the advantages that have accrued to the third party, Richardson, making sure you hit hard on those aspects which you know will be most meaningful to your present customer, Atkinson.)

Analyze the foregoing statements carefully. Note the use of the word "naturally." Almost alone, this word provides Atkinson with the escape hatch he needs. This one word implies that what happened to Richardson would have happened to *anyone* who was uninformed about the new material. Atkinson doesn't need your help to jump the gap between Richardson and himself. He can now feel that he has not been stupid to have used the inferior material; it was merely because he "has been so busy... after all, a fellow can't be up on *everything!*" This technique is called the cause-and-effect method because Richardson's action was the cause and his consequent operating problems represent the effect.

It should go without saying that you must know what your present customer, Atkinson, thinks about the customer whom you use as your "third party." If you must name the third party, be sure you select someone whom Atkinson regards as intelligent (not unlike himself). When possible, avoid naming the third party unless you are certain of the effect on your present customer.

C3—Use the testimonial

Use of the testimonial can be effective in attracting and holding the attention of the customer, but it requires more than a superficial knowledge of the customer. When we use a testimonial we are, in effect, saying to the customer, "I know you have respect for the ability (or knowledge, or judgment, or success) of Bill Blank (your testimonial), and you will be interested to know that he is using our product (or service." It becomes apparent, then, that you had better be certain that your customer really *does* hold in high regard the individual whom you are using as your testimonial.

We bolster our judgment, at times, by identifying ourselves with others who have made the same judgments. This is one of the 16 defense mechanisms (Identification). Be sure, then, that your customer does identify himself with the individual you are using as a testimonial.

In addition, there are these hazards to avoid:

1. Be sure that the person you have selected as a testimonial will support you if your customer contacts him and asks for an opinion.

2. Know your customer well enough to be certain that he will not become angry at your use of a testimonial. He can read into your action the conclusion that you are *not* giving him unique treatment but are regarding his problem as similar to other problems you have encountered in the past. All of us like to think we are unique (as we are), and we often like to think our problems are unique (nobody knows the trouble I've known). The customer who is insecure may place considerable faith in the testimonial you offer in the way of proof, but the customer who is secure may resent it.

Know the kind of people your customer respects before you decide whom you will use as a testimonial. Don't make the fatal error of thinking that because you regard the person you are offering as a testimonial as a person worthy of respect that your customer will view him in the same way. If you have not been able to determine this in advance, carefully feel out your customer by bringing in the name of the other person and watching his reactions.

C4—Emphasize his goals

If the customer's interest lags, focus his attention on what you know to be his goals—goals which, for the moment, he seems to have forgotten.

Example: "As I understand it, Mr. Ferranti, the main effort in your company this year is to try to reduce your maintenance costs; isn't that so?" (Pause.) "If that *is* the case, you will be interested in the fact that wherever this product has been used, maintenance costs have gone down by as much as X%... let me show you some figures."

While customer Ferranti may not have been interested in your product, his attention is awakened when he hears anything you have to say about "maintenance costs," because he *does* see this as a key problem. You have, also, in the foregoing example used the test question and awaited his nod of agreement—to be sure you have captured his attention. Questions, alone, can be effective tools for capturing attention. You have accomplished something equally important—you have demonstrated to the customer that you are aware of his problems, and this cannot help but impress him because he knows you have done your homework. Stating his problem for him also causes him to feel that you regard him as unique.

C5—Emphasize his objectives

This strategy becomes especially important when you are doing business with a customer who is short range in her thinking, and the product or service you are attempting to sell her provides long-range benefits. All of us, even if we are wise enough to have long-range objectives often forget them momentarily and concentrate on our immediate goals because they seem so much more important at the moment. Only reluctantly, when we are in a reflective frame of mind, do we transfer our attention to long-range objectives, which in the heat of action appear remote and of lesser importance. This becomes an even greater problem when you are up against a competitor who has demonstrated the prospects of immediate benefit to your customer. The customer figuratively will drool to obtain the immediate goal, even though this may be in some conflict with the attainment of his really more important long-range objectives. He will be so entranced with the prospects of immediate gain that he will not realize he is being inconsistent. Unfortunately, there will be sales lost because we have not been successful in causing the customer to keep his eye on his long-range objectives or because we have failed to show him the ultimate cost of satisfying what may be attractive, but transitory, goals.

We see this tendency in our own companies at times. To that point, we were in attendance at a departmental sales meeting some time ago, and the sales manager, addressing his men, said that they were all going to see a difference in him from now on: He was going to start building for the future and not be swayed by attractive short-range opportunities... and he wanted them to start thinking the same way.

He went on to add that from now on he was going to add only dealers who would improve the image and the prestige of his company in the marketplace. He devoted a full 30 minutes to this sound program and these lofty objectives. When he finished, one of his salesmen, with a smile of real appreciation, spoke up and said, "Gee, Mr. Habershaw, I'm sure glad to hear what you have just said... it sure makes me feel better. Now I know that I was 100% right in an action I took this week. You see, a couple of fellows whom I regard as sharp operators told me they would buy $100,000 of our stuff right now if we would give them a franchise. But I felt the way you do, and I turned them down because I thought they would be bad for our image in my area."

As the salesman finished, the sales manager was a sight to behold. He jumped up, in something akin to a frenzy, and shouted, "You did *what*? Who are they? What are their names? Quick!"

The entire sales meeting broke into roars of laughter at his unbelievable switch from objectives to attractive short-range goals. The sales manager broke into a grin as the reason for their mirth dawned upon him, but he continued, "Never mind, never mind... I know what town you had in mind, and I'll find out who they are." As the laughter rolled on unabated, he pulled a notebook from his pocket and jotted down a reminder.

It was obvious that this goal was so exciting that perhaps no one (other than *his* boss) could ever switch him back to his objectives. It is of interest to note that this individual resigned some weeks later and is now with a retail organization that thinks little of image importance but bases its planning on "how did we do at the cash register today?"

The point is obvious. If sales managers under pressure to "bring in a buck" switch with alacrity from long-range objectives, why should it surprise us to find customers, too, who think of the immediate future? As a professional salesman, your task may become one of alerting the customer to the long-range losses he will sustain if he falls prey to glamorous or even temporarily profitable short-range goals. Your job is to make the long-range objectives appear far more desirable.

Example: "I know that you want maximum profits this year, Mr. Vallejo, and I know that you must consider the effect that this investment will temporarily have on profits, but *just think* of what the impact on profits will be in two short years when your competition will have this same product operating and will be able to reduce their prices substantially. If you buy now, you are going to be in a position to dominate your market. You'll agree, I'm sure, that this is far more valuable to you than the temporary profits that may be neutralized in the next two years."

To make this strategy meaningful, it is suggested that you stop now and consider *your* product or service. Under what conditions could you encounter a customer whose concentration on short-range goals makes it difficult to see him when your product will bring him long-range benefits? What kinds of arguments can you have ready? Why not have some general answers ready that can be quickly adapted to each of your customers. (It might be added, at this point, that this is the way in which you can obtain maximum gain from the time you spend studying what is presented in this book on selling strategies. In each case, you should consider how you can use each strategy and try to think in terms of specific customers.)

C6—Omit something so he can inject his ideas

As has been said, one of the main reasons why it is sometimes difficult to capture and hold the customer's attention is because he feels "it's your act—not his." And not feeling involved, he feels no compulsion to listen.

Leaving something out of your presentation so the customer can inject one of his ideas is usually more effective (and safer) than *Strategy B16*, which calls for *being intentionally incorrect* in a statement so he can correct you. There is the danger in B16 that the customer's confidence in you can be shaken. That danger does not exist when you omit some point. He is usually so eager to fill the void you have left—so eager to "dominate the selling situation," even for a moment—that he will be thinking of what he is going to say rather than of what you left unsaid.

Except for 100% buyers' markets, customers always feel somewhat dominated in a selling situation. This they do not like—and when we offer them an opportunity to participate, they don't hesitate. Participating gives the customer a feeling of "co-authorship," and it is not unusual to find him selling himself in his efforts to sell you on his idea. This technique also enables you to use *Strategy B7—Compliment him*. ("That's a good point, and I'm glad you thought of it.")

Leaving something out of your presentation also fills the requirement for another strategy, *D9—Probe for reaction*. If what you have omitted is of some importance and the customer does not respond, you can be fairly sure he has not been giving you his attention or he has not understood the points you have been making. This is a signal to back up and clarify points you may have hit too lightly.

C7—Confuse him or use ambiguous questions

This strategy may prove effective under at least two conditions: (2) When the customer is a compulsive talker and won't give you a chance to make your points, or (2) when she has made up her mind to buy elsewhere and seems to have closed her mind. The task in both instances is to obtain her attention.

In the first instance, use of the ambiguous question or the difficult-to-answer question may be used to slow him down, to stop him or to arouse his curiosity—thus, gaining his attention. Then, when you have him listening, you may be able to present your selling points.

The ambiguous question should be designed so that it is either confusing and difficult to understand or designed to worry the person questioned. The phrasing should make him worry that you may be trying to put him on the spot, and he will slow down to consider the question's implications.

Example: "Excuse me for interrupting, Mr. Barmaier, but I want to be sure I am following you. Do I understand that you are saying it is the ethnic background of the customers in your area that makes them develop unusual buying patterns, or are you saying it is geographical considerations?" The customer stops at this point to consider the implications of the question (or just to unravel it). When he stops, the salesman grabs the ball and continues with, "The reason I am a bit confused on this is because we have customers in Worcester who sell to people with the same ethnic background as your customers. This product has gone over big with them because of a technique the dealers there have used."

The salesman has not caused the customer to lose face, but he has stopped him long enough to take over command of the situation. He has also made a statement calculated to arouse the curiosity of the customer sufficiently for him to inquire about what actions those Worcester dealers have taken. You have his attention. You had also better have some facts to support your claim.

Suppose your customer has decided against buying your product and is certain he is making the right move in buying from your competitor. You may be able to create uncertainty in his mind by bringing up issues that confuse him. This technique should not be used to sell a customer something he does not need. But it is ethical to use the technique when your customer is so sold on your competitor's product that he won't give you an opportunity to present your side of the case—that is, the virtues of your product.

Example: A salesman is attempting to sell a customer a car in the Buick category to a customer who is wedded to the idea of buying a foreign compact car. The salesman is making no headway because the customer keeps talking about the economy and maneuverability of the smaller car. Finally, the salesman says, "Have you thought about piston travel?" The question brings the customer to a screeching halt. He knows nothing about piston travel but figures this must be something he *should* know. His answer may be, "What about it?"

The salesman has now achieved his first objective... he has the customer's attention! He then goes on to explain the higher rpm at which the engine of the smaller car operates, its effect on rear-axle ratio —higher speed of movement of engine parts—consequently, faster wearing-out; hence cost; hence loss of the supposed economy. This device may not change the customer's mind, but it at least affords the salesman an opportunity to discuss the merits of his product, even though he had to come in through the back door.

To be effective, the device you use to temporarily confuse the customer must also be able to serve as a bridge to a discussion of the merits of your own product. It must also be beyond the threshold of the customer's knowledge so that at the very least it will arouse his curiosity.

C8—Interest him in more ambitious goals and objectives

In examining this strategy it might be well to define our terms. There is a terminology that is rapidly gaining use in industry. Goals are considered as the accomplishments we hope to achieve over the short range—the more immediate future. They may have target dates for achievement for the very immediate future up to two years or even more. Objectives are of longer-range nature and may be established for a period of from three to as long as 20 years. Both the target dates for goals and objectives will vary widely depending upon the nature of a customer's business. The point to keep in mind is that goals are more immediate in nature and objectives are more distant. Goals may be considered as benchmarks by which we measure our progress toward the attainment of the longer-range objectives and, obviously, they must be consistent with the nature of the objectives.

Many customers will hesitate before buying because their idea of a goal or objective is what they will do this afternoon or next week. The task often is to make the customer aware of his longer-range needs and the desirability of buying with those in mind. The salesman will often see the potential of a customer's business before the customer will be aware of it. Many customers have low levels of satisfaction, often because they are really unaware of either their own capabilities and potentials or unaware of the potentials their market offers them.

One of the greatest services you can provide as a salesman is to cause the customer to become aware of his real potential or to awaken that customer to business opportunities the customer has never considered.

How do you accomplish this? The following example may be one that *you* could not employ in your type of business but it illustrates the *type* of approach that may be employed. "I was thinking about you and your business the other day and I could not help but be tremendously impressed with what you have achieved. You started out with all kinds of competition. You've surpassed many of your competitors and some of them have even closed shop. Others have obtained even a greater market share. I can't help but wonder if you really realize how much ability it is obvious you have. Sure, I would agree, that some of your competitors may have gone even further but from my observations it isn't because they have greater ability than you have. As a matter of fact I'm certain that you are second to none in terms of ability and far out ahead of many of them. I think the reason some of them have gone further is because they have established some definite long-range objectives and planned to meet them. You have great potential to grow and I'd like to help you and, selfishly, to have my own business grow with you. I'd like to talk with you about a few things I think you might consider that could really help you to expand your business and profits even more. You have every right to be happy with the growth you have obtained. Wouldn't you be interested in looking into the future with me, to see how you can achieve even more, not only by not working harder, but probably by working even less?"

If you use this approach it is clear that you must have some concrete suggestions to make to excite her interest. Even if the customer's satisfaction with present growth or unawareness of her capabilities causes that customer to decline any offer to help attain greater growth, you have made real points with that individual because all of us "love" people who, in a sincere fashion, show us that they think we have the seeds of greatness in us.

The foregoing approach will be even more effective if you buttress your remarks with specific examples. For example when you stated that you had been impressed with the customer's achievements to date, it would have been even more effective if, at that point, you specifically stated some of the things the individual had accomplished. This distinguishes sincere admiration from flattery. It also causes the customer to realize you have really been thinking about him as an individual and have been studying the business.

One word of caution. Don't mention by name, if it can possibly be avoided, any of the competitors who have gone ahead of this customer or any of those whom you have said have "no more ability" than this customer has. You don't want this customer quoting you later. Keep your praise of this customer specific and your references to any of his competitors general—especially if they, too, are customers of yours.

CHAPTER 23

Overcoming objections

"Overcome his objections" is another major segment of the Sales Strategy Bank. The merits of a number of different strategies will be analyzed. The professional salesman will do well to remember them and then to decide which will be most effective, considering his own personality, that of the customer, and the nature of the situation in which he has met the customer. More to the point, the professional salesman must determine the basic reason for the customer's objection before deciding which of the strategies will be most effective.

Sales strategies are much like golf clubs; you play a better game if you have a wide selection of clubs available and if you decide which of the clubs will be most effective for each shot. A fly fisherman faces much the same problem. He has to find out the types of fish that are available and what their feeding habits are at the moment. A fly fisherman will maintain that fishing is, in many ways, scientific...so is the selection of sales strategies.

The questioning technique

In the Sales Strategy Bank, we have listed in D15 through D18 these four types of questions: The *direct* question, the *overhead* question, the *rhetorical* question, and the *provocative* question. To these we might add the *ambiguous* question. Inasmuch as this technique is not listed in the Sales Strategy Bank at this point but was covered in a previous section under C7, the reader is requested to refresh his memory by re-reading that section.

Questions are the most powerful tools of the professional salesman. Consider what they can do. They can:

1. Force the reluctant customer to talk.
2. Focus his attention on the facet of discussion which *you* wish to cover.
3. Demand proof (when she makes charges)... and the question, when used wisely, can demand proof in a manner that will not further alienate the customer.
4. Flatter the customer (by requesting his views or opinions).
5. Stop the over-talkative customer (by, for instance, posing difficult questions to answer).
6. Identify the spokesman of a group (by use of the "overhead" question).
7. Serve as a probe for reactions.
8. "Flush out" the real objections.
9. Force the customer to answer his own objections.
10. Cause the customer to feel that she is leading the discussion ("controlling" the selling situation).
11. Prevent us from talking too much.
12. Provide us with facts we need to know.
13. Force the other fellow to substitute reason for emotion.
14. Spotlight our own humility.
15. Appeal to the other fellow's ego.

The above list is not complete; others have compiled much longer lists of advantages of the questioning technique. It matters little whether there are 15 advantages or 50; the fact is that questions do have power and a multitude of uses. If questions have power, then we need to know as much as we can about their uses, their values, and how to select the right kind of question for each situation.

For those who are using this book as a vehicle for sales training meetings, it is suggested that men in the group practice using questions that can force the customer to talk...and to talk about the things you desire to learn. *To obtain maximum value* from the material, it is suggested that this practice be followed in testing out each of the strategies presented. It isn't enough, for example, to know that an effective strategy may be to *Tack with his stand (D8)*. The question is, "How do you do this?" It is suggested that the conference leader present a realistic problem situation from your kind of selling where the best strategy would be to "tack with his stand" and give your men an opportunity to demonstrate how they would do this. If they err, it costs you nothing. If they err in a real selling situation, it may cost you a sale.

267

D1—Give him a choice

"Don't fence me in" describes fairly well the fear that many customers have that they will be pushed into buying something they do not want or need. This is especially true of those selling situations where, until you entered the scene, the customer was perfectly satisfied with what he had. He may have lacked an awareness that he had a need or a problem; or, if he was aware of its existence, he did not place a high priority upon it. You have now alerted him to the need or have aroused a "want." Probably you have presented him with evidence in the form of a product or service that indicates his need can be filled or his problem can be solved. In other words, you have effectively changed his attitude. You now face the necessity to force a decision from him, and the decision must be favorable to your interest as well as to his.

One would expect that a rational being—once he had identified his problem and had a solution—would grasp that solution, regardless of its source. Your customer, today, may show no indication of being willing to do so. The key to his behavior is found in an understanding of what is meant by a "rational being."

Let us forsake dictionary definitions and examine the word "rational." A rational person, as we have been saying, has a reason for his every action and attitude. Consciously, or subconsciously, he measures the effect of each action against his goals and objectives, and against the needs of the self-image he is trying to project. If his judgment is good, he also measures the effect of each contemplated action against its impact on his long-range objectives. If he feels threatened in any way—economically or psychologically—he is going to resist.

Many of us are inclined to be goal-oriented (short-range). This is certainly true of the customer who hesitates to buy even after he *knows* your product will do the job he requires. By purchasing from you, he will attain his long-range objectives or will meet his short-range economic goals. These are business-oriented—to make a profit, to improve quality, to overcome an operating problem. He has, however, a short-range goal that may be uppermost in his mind and may warp his judgment; i.e., he does not like to be pushed around... he does not like having his mind made up for him. He sees, in you, a threat to his self-image.

In a vague sort of way, he senses that he is being "pushed around" when you attempt to force a decision from him or when he finds that in order to satisfy his business need he must accept "your" solution. Even though the decision is in his own business interests, he will resist. We say he is irrational because such an attitude is against his interests. This depends upon what interests you are talking about. His resistance may be against his business interests, but it is not against his psychological needs... the need in this case to consider himself as his own decision-maker. It is simply that these psychological needs are, for the moment, more important to him. It becomes obvious, then, that he does have a rational reason for his unfortunate attitude. We should not be surprised to find him hesitating to buy and even raising smoke screen objections.

The problem often boils down to this: The customer feels that he has had little to do with the solution you have provided... he feels trapped. Depending upon the nature of his unique psychic needs, he may feel that he is being deprived of his need for recognition, his need to feel needed, his need to dominate; or he may feel that his prestige is threatened. He has not been made part of the solution and senses he has been treated as part of the problem.

One method of making him feel that he has contributed to the solution is to provide him with a choice. Don't push the "school solution" down his throat. Don't indicate that there is only one way to solve his problem. Try to find a method of presenting your solution so that it offers him a choice—*even if it is only a choice of details.*

Most sales training texts mention "choice" only in reference to its effectiveness as one means of *closing* a sale. Actually, it is a tactic that can be used effectively throughout the selling situation. It can aid you, for instance, in determining whether you have held his interest and/or attention, and it serves as a vehicle for bringing the customer into the action throughout the sale.

As you approach the close of the sale, the choices should be made easier for the customer. They should, if possible, be choices you know he has probably made, mentally, already. If, as you approach the close of the sale, your customer resists your efforts, give him a choice and try to make the choice meaningful... it should be more than a choice, now, of relatively unimportant details. At the same time, as cautioned previously, the selection should not be so difficult to make that it will increase the difficulty of his problem and cause him to hesitate even more than he is doing at present.

In a retail selling situation, sales people are taught to provide a choice with such simple questions as: "Would you prefer blue or red?"; "Would you prefer delivery on Monday or Friday?"; "Would you prefer galvanized or stainless?" Such simple choices may work in a retail situation and undoubtedly do. In more difficult selling situations, however, the choice must represent more than a gesture. The more significant the choice, the more your customer will feel that he is controlling the situation and is not being pressured. The choice you enable him to make must cause him to feel "necessary" and must provide him with recognition.

Avoid giving her a choice that will represent a step backward in the sale. If it is clear that she is satisfied with one recommended action, don't confuse her by offering her an unnecessary choice. Keep your offer of a choice limited to something that will not create new problems.

Using choice to emphasize advantages

If the customer hesitates as you near the end of your presentation, use choice as a means of introducing comparisons. Offer him a choice between your proposal and something less effective...a selection that will immediately cause him to see the drawbacks of the other offering and will accent the advantages of your proposition.

Example: "Mr. Weber, you may want to prove to yourself that this proposition is really going to save you money. You might want to consider the possibility of using this product on one of your production lines while you continue to use the product you are now using on the other."

His reply may be, "No, that will never do...I'd prefer to use it on both lines if I am going to use it at all. Otherwise, I would multiply my service problems, require two sets of repair manuals and so forth ...No, that would cause more work."

You can respond with, "You have a good point there. As our figures indicate, you are losing money each day we delay. May I suggest that you give us your okay now so, as you suggest, we can put it in on both lines." This not only gives him the feeling he has had a choice, but it speeds you toward a rapid closing.

Note the salesman's choice of words:

"You have a point there" supplies recognition and agreement.

"Losing money each day" emphasizes the need for action.

"May I suggest" relieves the customer of the feeling he is being pressured.

"Each day 'we' delay"—not "each day *you* delay"—shares the onus for inaction.

"As you suggest" provides the customer with authorship of the idea, though he has really suggested no such thing...he has merely supplied a reason for not going ahead with the project on a 50–50 basis, but he *has* indicated that this is the only basis on which he would consider the purchase.

When you supply the customer with an alternative which he feels is unacceptable, he finds himself contrasting the alternative to your original offer. This is the technique of "illusion"...causing him to compare your offering to another in order to make yours look more acceptable—offering it in a different setting, as it were.

Offering a choice at various phases of the sale can gain a degree of commitment from the customer. The salesperson who sells your wife uses this technique in asking, "Do you prefer brown or black?" Thus, she obtains a partial commitment that at least one of her colors will be acceptable. In more sophisticated selling (if there really is a more difficult kind of selling), the choice should call for the exercise of a greater degree of judgment on the part of the customer. Here are some examples:

"Mr. Abercrombie, do you plan on servicing these yourself or having service supplied by an outside agency?" (Notice the assumption that he is going to buy..it's not "*Would* you plan..."; rather, it's "*Do* you plan...")

"Mr. Riley, your opinion on this point will be of help to us... which do you feel will be of greater advantage in your type of operation, having us design the primary for 220 volts or 440? We will want to work this out to your greater advantage."

"Joe, I know you people have some thoughts on the advantage of System A as compared to System B. In this instance, from our standpoint, we can supply it in either mode. Would you mind giving me your thinking on that?"

The foregoing may fall completely outside the sphere of your kind of selling. Put the principle to work. Develop a list of meaningful choices you can offer your customers and have them on tap, ready for immediate use. Plan on the specific words you are going to use, as has been done above. Keep the following factors in mind:

1. Don't make the choice so simple that it requires only a "Yes" or "No," a "Red" or "Black" reply. The choice should make the customer think and should enhance his ego.

2. Don't make the choice so difficult that it will dismay or confuse him, or cause him to think that the whole subject must be given much more thought before he can place an order.

3. Don't embarrass him by presenting a choice that is beyond his competence, thus causing him to lose face.

4. Don't present him with a choice that will cause him to feel he is going to have to take the blame if his selection is wrong. In other words, do not have it appear that you are trying to place the responsibility on him.

5. Have your choices indicate that you are aware of her problems and have a genuine desire to help.

6. When he gives you his answer—even if you knew what it was going to be—make meaningful comments that will indicate you have been waiting for his decision.

7. When you present her with a choice...*wait* and keep quiet...let her think without interruption. Don't assume because she does not reply at once that she has no intention of replying. Let her think it

through...she is on the spot—not you.

Choice releases the customer from the feeling that *you* are dominating the selling situation. It is a powerful selling tool. Use it with care and don't destroy its impact with over-usage. Don't employ choice so often that the customer will get the idea that you are not assisting him with his problem...after all, you are supposed to have some answers for him. Consider giving the advantages and disadvantages of the choices you present him and ask him to determine which are the more significant to him.

D2—Narrow down his choices

"I became so confused that I finally walked out without buying anything." So spoke a wife as she returned from a shopping trip. She had entered the store in a buying mood...she had *wanted* to buy. She had a need and wanted to fill it, yet she came home empty-handed. Why? Was it because she had not liked what she had seen? Quite the contrary, she liked *everything* that had been shown to her. The difficulty was that the salesman had not made it easy for her to choose.

This form of behavior is not limited to women. A man shopping for a suit or a tie may be heard to remark, "I like them all...I just can't decide among them. I'll be back with my wife." Maybe he will... but maybe he will buy from a salesman in another store who knows how to make choice easy.

These customers were confused by the *number* of choices available to them. This increases the difficulty of decision-making, and in the mind of the customer increases the chances of an unfortunate decision. The professional salesman, in any field, can learn from this. The customer is not really interested in the number of models you have to offer for his consideration. He is going to buy one. Show him all of the models that may meet his need, but watch his reactions and, as rapidly as possible, *narrow down his field of choice* by removing the other models from view, if possible. If the items you sell are too large for that, focus the customer's attention on the two that he seems to have preferred and don't mention the other models again.

Retail clothing salesmen who are effective follow this technique. They may show you a number of suits, but as they sense the focus of your key interest, they will either remove the other suits from view or toss them on top of one another on a nearby chair so they will not distract you. If you insist upon seeing one of the other suits he will quickly resurrect it from the pile, and nothing has been lost.

Decision-making is fatiguing, and when a customer buys, he is making a decision. Make it *easy* for him. Example: "Mr. Kelly, it seems to me that any one of these (products) will meet your needs, but as we have analyzed each of them, it appears that either of these two will be most effective. Which of these two appeals the most to you?"

If he disagrees and says, "I think this third model will also do the job," you nod in agreement and ask, "What features on that product appeal to you?" When he replies to this question, you have additional information that will enable you to identify his real buying motive. You can then concentrate on those features in which he is obviously most interested, and you can ease off on a discussion of other features which will only serve to make his choice more difficult.

D3—Provide solution

Essentially, a salewoman is in the business of providing solutions to problems, so it may seem strange that "Provide solution" should be listed as a separate strategy in the Sales Strategy Bank. The fact is that although your effort should *always* be to "provide a solution to the customer's problem," this is an end result to be sought rather than a specific strategic effort. We are now thinking of those situations where you must consider the wisdom of *overtly* providing the customer with a solution.

One would think that this would create no selling obstacle. Won't the customer be delighted to have his problem solved for him? Any experienced salesman knows that this is not always the case. Customers do not always grasp at the solutions you offer with the same zeal that a drowning man would grasp for your aid. Customers sometimes feel that we do not understand their problems and will not readily accept our solutions. Others who have failed to see a solution for their problem will not find it easy to believe that a solution is readily available that *they* did not recognize.

Other customers like to feel that *they* dominate the selling situation, and acceptance of your idea threatens "their control." As a result, they will resist your proposals. Often, customers have a high opinion of their own abilities and will feel that they will lose face if they accept ready-made solutions...this is especially true in those fields where you are dealing with customers who have problems of a highly technical nature.

There are, however, situations where the customer will welcome a ready-made solution. This is especially true when the customer is working against a deadline or is under heavy pressure to get a problem solved (in other words, where the fear of failure or fear of loss will be greater than his possible fear of losing face by accepting someone else's solution).

If you find it wise to overtly provide a solution to the customer's problem, keep these thoughts in mind:

1. Do not create the impression that your solution is akin to a "brilliant idea" that just occurred to you. Rather, indicate that it has been well planned . . . that you have given much time to the solution. (In effect: If the customer had only had as much time to give to this problem, *he* would have produced the solution.)

2. Another reason for creating the impression that your idea was not born on the spur of the moment is that your customer undoubtedly has been wrestling with the problem for a long time and will be inclined to be wary of solutions that did not occur to him. He will want plenty of assurance in the way of facts to convince him that your proposal is a sound one.

3. Even though your solution may completely solve his problem, it may be wise to present it as "a possible solution," as a "partial solution" or as "an approach to a solution." You can use such phrases as: "You may wish to modify this in some respects, but at least this is a base from which we can start . . ." In other words, it is often more important for him to feel that the idea is partly his than for you to attempt to get the credit. You often have a choice—get the order or get the credit.

D4—Provide alternative

When a customer requires an escape hatch—a means of saving face—it is sometimes wise to offer her an alternative. This is especially true when she has gone on record or has been adamant in her objections to one feature of our proposal. Provide her with an alternative to the objectionable feature.

Provide alternatives, also, to the customer who doesn't like to feel that you are doing his thinking for him and to the customer who resists what he regards as "high pressure." Alternate proposals may be given to the fellow who, you know, has to sell *his* boss, when you anticipate that the boss may give a rough reception to your basic proposal.

Alternate proposals are the answer to the lament, "Don't fence me in." They are the basic bread-and-butter tactics of the strategist who must be ready for each move of his customer or competitor. The military estimate of the situation used by every modern army and navy absolutely requires alternatives. Certainly selling is just as scientific as warfare. Both the military man and the professional salesman are dealing, primarily, with people. Weapons and products are merely means to an end—the solving of a problem or the breaking down of resistance. A nonprofessional can always be identified when he is heard to say, "There's just *one* way to handle that problem."

There are always choices of actions. It never ceases to amaze the author to watch groups of men discussing cases in sales training seminars. Often, a group will "home in" on a solution in their exploratory discussions. Then, when they return to the conference room to present their solution, facts are revealed that cause their solution to fall on its face, and the group sits there, dismayed, and with no alternative course of action. It is regrettable when this occurs in a sales training session, but it is *tragic* when it takes place in a real selling situation. It is the responsibility of every manager of salesmen, as he helps his men to develop themselves, to insist that his men enter every selling situation with alternatives ready. This can best be accomplished by considering in advance the alternatives that are available to customers and to competitors, and by not being caught unprepared to meet any of those alternatives.

D5—Compromise

In retail selling, there is an adage, "The customer is always ready to compromise." The question now facing us is: "Are *you?*"

We are mindful of the fact that there are areas in which compromise is "off limits." For example, a company may have a fixed policy that it will not sell products to customers and then permit those customers to place their trade name on the products. Included might be a whole gamut of items, ranging from tires to light bulbs. In such cases, companies have knowingly given up extremely large orders—and even large customers—simply because the policy makers had decided after much deliberation that more would be lost, over the long haul, than would be gained. In such situations, the salesman has neither the opportunity nor the authority to compromise. Neither does he have the right to complain when he loses an order because of the restrictions of a policy which his company has established after due regard for the temporary losses its enforcement may produce.

There are many other situations, however, where compromise is possible. All too often, salesmen give evidence of rigidity rather than flexibility. The customer may want something "just a bit different" in the way of color, delivery, terms, etc., but the salesman is unwilling to compromise.

Behind this unwillingness, usually, lurks an awareness on the part of the saleman that to accommodate the customer means that he, the salesman, is going to have to "battle" someone else in his own organization...the someone who is going to have to make the change the customer requests. In an effort to steer clear of this encounter—or to avoid a little extra work—the salesman will sometimes attempt to squeeze the customer into a mold and make him conform. The efforts of the salesman may be successful, but he has just lost a golden opportunity to prove to his customer that he is regarded as important and unique. In other words, the salesman has shown the customer that he, the salesman, is not truly customer-oriented.

We abhor rigidity in customers, yet we often expect them to tolerate rigidity in us. It is all so reminiscent of a civil service employee or the occasional military man who would resist new methods with the rigid posture epitomized in the retort, "It's against the regulations." This is the refuge of the witless, the unimaginative, the lazy, and the egocentric.

The professional salesman will *seek* opportunities to compromise when it is possible to do so, recognizing the great opportunity this provides to demonstrate his consideration for his customer as a unique individual or a unique organization. There comes to our mind the recollection of a land development organization in Florida that made a lasting impression upon us. This organization sold land and homes. As clients entered the development, they found a dozen model homes. The first sign the client encountered in a model informed him, "We will make any partition changes to suit your needs at no extra cost."

The customer was at once made to feel that he was not being forced to buy a home that had been produced for "just anyone." The homes were so well designed that most customers did not ask for changes, and the cost of alterations that were requested and made was negligible. It was not difficult to understand why this firm went on to greater success, with each of its clients becoming a walking advertisement.

Compare this to the attitude of a well-known foreign car manufacturer who refuses to equip his car with some accessory, or make a modest change in wiring of lights, for which the customer is willing to pay, because, "We do not put that accessory on cars being exported to the United States." The company, admittedly, is successful at this time, but you can be sure that customers have no affection for the manufacturer, nor will they exhibit any fidelity to that manufacturer when some competitor appears on the scene with a car that is equally as good.

When a company has a monopoly, it can—for the moment—reject compromise as a selling strategy, but it must do so with the awareness that it will eventually pay a price for this rigidity of philosophy. The same may be said for the salesman who refuses to compromise or who fails to see that opportunities for compromise are something to be sought rather than avoided.

We should remember, as has been stressed throughout this book, that we are selling not only the satisfaction of product or service needs—but we are selling the satisfaction of psychic needs as well. The customer's self-image is always present, and when we are willing to compromise, this attitude cannot help but make him feel important...and make him aware that *we feel he is important.*

D6—Talk benefits

"Talk benefits" is probably the most important of all the techniques listed in the Sales Strategy Bank. If our offering contains no advantage that is apparent to the customer, he will not buy...it's as simple as that.

If we can demonstrate that what he is required to spend on our offering is less than the gains he will obtain from ownership and/or use of our product or service, he will buy. "Gains" may be economic benefits, operational benefits, psychic benefits, or a combination of all three. These benefits may be there, but it is essential that he *knows* what they are and understands them.

Customers, we have stressed over and over, are unique, and rarely will *general* benefits satisfy them. That is why, although advertising is not only powerful and perhaps the greatest aid available to the salesman, it often will accomplish no more than to attract the attention of the customer and will require positive follow-up action on the part of a professional salesman to produce the sale...the buying action.

Before we talk benefits to a customer, we need to know what is important to him and that is a far cry from what would be important to *us* if we were the customer. We need to know his viewpoint, his concerns, his fears, his needs, his aspirations, and his problems, as he sees each of them.

For the sake of emphasis, let us repeat: The customer is asking these questions: (1) Are you going to make me rich? (2) Are you going to make me famous? (3) Are you going to make life easier for me? (4) Are you going to make me secure? In other words: Will I gain more than this costs me financially? Will I look good—or bad—if I buy (as seen through the eyes of those whose opinion concerns me most)? Is this going to give me problems or make it necessary for me to make more decisions? Will it threaten me in any way?

It should be obvious to us that no two customers will have the same "profile of needs." What will appeal to one customer will not only leave another customer cold, but it may bore him or even threaten him. One might ask, "Why not size up the *average* customer and prepare a presentation that will satisfy him?" We repeat, the answer is that you and I have never sold an "average" customer once in our lives. We don't sell average customers; we sell *unique* customers.

There are some door-to-door salesmen who have "canned presentations," and they know in advance that these presentations will not sell a certain percentage of their potential customers. This does not bother them. They have learned that this "pitch" will sell X% of their potential customers, and both they and their companies regard this as a satisfactory percentage.

A canned presentation is not professional selling—it's peddling. Because this method produces what is regarded by a company as a satisfactory profit, it is used. Such companies will probably never know how much *more* they could earn if their salesmen were really trained as professionals... how much *more* productive could be the time that is now consumed at those doors where the canned presentation fails to score... how much *more* they could reduce the turnover within their sales force, caused by men leaving simply because they ultimately realize that there is little challenge in this kind of selling.

In selling benefits, or advantages, keep the following thoughts in mind:

It is natural for us to think in terms of product or service features, but the customer does not always think in that language. We say, "It has this feature." He wonders, "What will that do for me?" We say, for example, "It is lightweight!" He wonders, "So what?"

A single feature may appeal to many customers—but for entirely different reasons *if* each sees the feature as an advantage from her own unique frame of reference. We could use any product as an example, from an automobile to a nutcracker. But, to demonstrate our point, let us use a product with which all of us, regardless of our field, have at least some familiarity—an electric motor, one that happens to be lightweight.

First of all, we might make the assumption that everyone would look favorably upon the fact that the motor is not heavy. Not so. One company produced a new motor made of a new lightweight steel. They immediately encountered vociferous objections from some established customers who equated *thickness* with strength. If the new steel was thin, ipso facto, the motor was "junk." The fact that the new steel was vastly superior to the older "heavy" metal was lost on them. In their case, what was really an advantage appeared as a disadvantage.

Obviously, then, if your company is introducing a new feature that represents *change*, let your customers know as far in advance as seems wise. Acquaint them with the nature of the change and show them why it is a change for the better from *their* point of view.

Take your own product and consider each of its features. Then take each feature and list all of the possible advantages it offers. Then, as you prepare to call on each customer, ask yourself which of these benefits will ring a bell with him. *Hit those benefits.*

Don't "play it safe" by listing all of the advantages. This may bore him—or worse yet, one of the supposed advantages may appear as a disadvantage to him. Itemize only those benefits that your study of the customer and his problem has indicated will be of value to him and will serve to impress him with your knowledge of his business—even if only subconsciously. You can always hold some other benefits in reserve in the event that you feel the need for more ammunition later during the selling situation.

Customers will look at a product from the standpoint of their economic, operational or psychic needs. What is an attractive feature to one customer may be of no importance to another.

Sell the advantage—not the feature. Interpret the feature in terms of its advantage to this customer.

1. To get and hold attention mention the advantage first, then prove it by citing the feature.

2. Concentrate on features that benefit this customer—and don't dilute their effectiveness by giving him a total run-down of features that offer him no advantage, though they may be most interesting to some other customer.

3. Hit hard on your most attention-compelling advantage first (except when you feel you will need this as a clincher).

4. Couple your advantages with losses she will sustain if she does not buy or if she buys elsewhere. But always conclude by restating your top advantage so that this is the last thought in her mind. By stating losses, you cause her to make a comparison and you accent the advantages.

As you have undoubtedly noted, we often take our examples from fields of selling that fall outside your scope of activity. We select a field of selling that is at least understandable to all of us. The real insight—the real learning—takes place when you, as the reader, ask yourself, "What is the *principle* that is being driven home, and how can I use this in my kind of selling?" This approach offers you a greater advantage, therefore, than if the examples were drawn from your field of selling, where you would be more inclined to question details.

To illustrate how to use product advantages (benefits), let's consider the following examples:

Throughout the nation there remain thousands of people who live in "fringe areas," lacking cable-TV facilities, where television signals are weak because of the distance from the transmitter or because of the interference of hills or other physical obstacles. Often, only high receiving antennas provide a solution to this problem, as you have undoubtedly noted in your own travels. Many of these people live in single-level homes where roof antennas may not do the job required and where towers must be erected.

Assume for the sake of this example that you are selling these towers. The tower you sell has a crank-down feature. The tower consists of two sections (or more), each 20 feet long, that telescope into the lower section as the antenna is lowered. A second crank enables the owner to tilt the tower over so that the top rests on the ground. These features will mean different things to different customers, as the following quotations indicate:

Customer A: "I can wind it down in a few minutes, and this is important because I live in an area where high winds and occasional tornadoes prevail."

Customer B: "We are exposed to salt water, so I have to paint the tower every year or so, and this makes the task infinitely easier."

Customer C: "I won't have to hire a TV repairman to change my antenna when it gets rusted or when improved models are available...I can do the job on the ground and save money."

Customer D: "I don't like my neighbors, and I won't have to be obligated to them or go to the expense of hiring help to erect the antenna...I can crank it up myself!"

Customer E: "Towers, at their best, are unsightly, and I can lower this one when necessary so it won't be so conspicuous." (He probably never will but likes to give himself this justification for buying something he wants to own.)

Customer F: "I'm afraid of heights and hate the expense of repairmen...I can crank this one down and work on it on the ground."

Customer G: "I don't see this crank-up feature as an advantage. Will it be as rugged as a fixed tower? Will the cables break when I'm cranking it up or down? Is there a device to stop it from accidentally telescoping?"

You could undoubtedly add to this list, but this should suffice to emphasize that a salesman is not really selling when he says to any one of these customers, "This is a two-sectional crank-up tower with tilt-over feature." Each of the customers may think, "So what!"...and Customer G may take a decidedly negative attitude..."What keeps it up?" Compare, then, the unimaginative "This is a two sectional crank-up tower with tilt-over feature" with any of the following presentations (upon which you can undoubtedly further improve):

To Customer A who lives in an area where high winds prevail: "Mr. Waldron, this tower can be cranked down and secured flat to the ground in three minutes. This will be important to you during the season of heavy winds."

Or: "No matter how hard the wind blows, Mr. Waldron, you'll feel safe with this tower because it cranks down in less than three minutes and in two minutes more you can have it secured on the ground."

Or: "You'll save money on this tower because you won't find it necessary to replace it as so many of your neighbors have had to do when theirs were damaged by heavy winds. With this tower, when heavy winds are predicted, you can crank the tower down and have it secured on the ground in under five minutes."

(a) *State the advantage,* (b) *identify the feature that produces the advantage, and* (c) *when necessary explain how this is accomplished.*

It will be of value to analyze the foregoing examples. If the salesman had started with "This is a crank-down tower," the value of the feature might have been entirely lost on the customer. In the first example, the salesman has mentioned the feature first ("Mr. Waldron, this tower can be cranked down...etc."). This is weak because the feature is mentioned first and the advantage last. Hit the advantage first in order to arouse attention. This has been accomplished in the last two of the three examples given. The first example is weak also in that the advantage is too general ("this will be important to you during the season of high winds"). Why will it be important? The last two examples, you will note, are specific and highlight the advantages to the customer.

How is the tower presented to Customer B who hates maintenance work? Here is one method: "If you dislike climbing as much as I do, you are going to be delighted with this crank-down feature. When this tower has to be painted, all you have to do is crank it down, and this is a job that even a child can do in three minutes. All of the painting can be accomplished with both feet on the ground."

Or: "Painting will keep this tower in excellent condition for years, and you are going to find painting *easy* because the tower can be cranked down in three minutes and painted in an hour, on the ground."

Or: "I've noticed how well you keep your place maintained, and you'll find it possible to keep this tower in first-class condition because it cranks down in three minutes, and you can paint it while it rests on the ground."

Analysis of these statements will be beneficial. You may suspect that your customer not only dislikes maintenance work but has the fear that many of us have of heights. In the first example, the salesman, without mentioning fear of climbing, establishes a bond between the customer and himself with, "If you dislike climbing as much as I do. . ." Fear is not mentioned, and the salesman merely indicates that he doesn't like the work involved in climbing.

In the same example, the salesman makes the job look easy with such phrases as, "even a child can do it in three minutes". . . "painted on the ground in less than an hour." Claims are made specific. As a matter of fact, the statement would be made stronger if the salesman had said ". . . in under four minutes." The phrase, "In three minutes," is almost a figure of speech and is often not taken literally by a customer, hence loses much of its impact. "In under four minutes" has greater validity because it sounds as though someone had actually timed-studied the operation.

Notice also that the salesman *assumes* he is going to get the order. He states, "You are going to find painting easy. . ." He doesn't say, "If you buy the tower, you will find painting easy." In the last example, the salesman appeals to the pride of his customer by commenting upon the fact that the customer keeps his place maintained well. By such personalized statements (if sincere), we show the customer that we see him as unique.

Selling the same tower to Customer C who is concerned with expense, we might use this approach: "You are going to save money on this tower, Mr. Gorman. As you know, a TV antenna usually needs replacement every few years or after severe storms. It is expensive sending a TV man up a stationary tower to replace the antenna. In under four minutes, this tower can be cranked down and a new antenna installed in a matter of minutes. As a matter of fact, it is a job you can do yourself, on the ground, and save the cost of the TV repairman. If you prefer using the TV man, you only pay him for a few minutes' work on the ground."

Again, the salesman assumes he is going to get the order with "You *are going* to save money. . ." Specific times are used, and the emphasis is entirely on saving money, which the salesman has deduced is the customer's prime concern. He does not dilute this by listing a number of other advantages or features which the customer might not regard as advantages, or in which he might have no interest.

Note, also, the use of the phrase, "*As you know,* a TV antenna usually needs replacement. . . etc." When you must "feed" information to a proud customer, use the powerful "as you know." If he does not "know," he will be flattered that you think he does. If he does "know," he will not be annoyed that you dwell on facts that would otherwise (in his opinion) insult his intelligence. *Don't abuse this phrase.* Vary it with such comments as, "As we both know. . ."; "as your experience has undoubtedly shown you. . ."

All of these examples recognize an unspoken objection; i.e., the probable expectation on the part of the customer that any large tower requiring cranking up and down probably represents much work. Each statement is designed to hit hard on the ease with which the cranking is accomplished. Actually, five minutes of cranking can be hard work, but "under four minutes" or even "only five minutes" is based on illusion. It is difficult to consider anything very fatiguing that can be accomplished in "only five minutes." (Try holding your arm outstretched "for only five minutes.") Yet, the salesman has been honest—it does not require great effort to crank the tower up or down.

In selling Customer D, who doesn't want to be obligated to neighbors, this approach might be used: "A great advantage of this tower is that you and your son will be able to assemble it in a matter of minutes, and once assembled, it only takes one person three to four minutes to crank it up to full height or to lower it when necessary."

Here, neighbors don't have to be mentioned because the advantage is so obvious to your customer that he will appreciate it at once. There are times when you will not want the customer to know that you are aware of his real objection or problem. In such cases, make the advantage so clear that he will see, by himself, that it solves his problem.

How would *you* phrase your remarks to the following customers who need your tower?

Customer E (Doesn't like unsightly towers.)
Customer F (Afraid of heights.)
Customer G (Fears mechanical failure of tower; i.e., afraid the cables that raise and lower the tower will snap when it is being cranked.)

The foregoing examples, developed around one product, will be more meaningful to you, if you now consider how you can present *your* product to some of *your* customers who have different reasons for buying or who have different objections. Select a feature of a product or service you sell and write down the actual words you would use in selling a specific customer. Then check what you have written against the ground rules which are now repeated for your convenience. Then be honest with yourself: Can you improve the words with a second writing? Words are powerful tools and we should choose them with care when we are trying to persuade others. Preparation is boring, we know, but it enables the professional saleswoman to operate at her peak performance. Here, then, are the ground

Product Feature-Advantage List

Customer	What is his key need?	What feature meets his need?	How can I express this in terms that will be most meaningful to *this* man?
Jones	1. A tower that will withstand high winds. 2. No interruption with his favorite programs.	Crank-down feature.	"You'll never miss a program because this tower can be quickly and easily lowered by a hand crank. Five minutes after a storm, you can have your set in operation while your neighbors are waiting for the repairman to fix or replace their antennas."
Brown	Tower that will stand up under conditions of salt air with minimum maintenance.	Galvanized steel.	"You'll never need to paint this tower in spite of your seaside location because, as you have undoubtedly noticed, it is galvanized and will resist corrosion."

rules... check your sentences against them:

1. Sell the advantage, don't sell the feature.
2. Interpret the feature in terms of the advantage to this customer.
3. OR: (a) State the advantage. (b) Identify the feature that produces the advantage. (c) When necessary, explain how it is accomplished.

We all depend upon memory too much and as a result we often forget key selling points, or we present them in something less than the most effective order, or we settle for a "general sales presentation" instead of the unique presentation the customer merits. To avoid these traps, the professional salesman should make a product feature-advantage list. He should write all the features and advantages in adjacent columns, and he should also list disadvantages which the customer may identify. And next to the disadvantages, he should itemize offsetting advantages or arguments he can use. Then he should, before each major call, draw up a list comparable to that which appears here. Such a schedule should be drawn up for each customer (not for two customers as shown in the exhibit). Be a word-smith and in the fourth column, list the very words you will use, after you have decided which will have maximum desirable psychological impact on your customer.

Keep in mind that your customer is less interested in what your product has (features) than in what it will do for him. So point out advantages, and back up each one with proof. ("You will get years of moisture-free operation because of this glazed finish.")

One important point: Don't overlook the importance of using the last column on the form to *explain* the advantage when you feel your customer may not see why a particular feature is an advantage. This is especially important when we are selling products that are highly "technical" to customers, who may not have the background to appreciate the advantages of your product's features. At the same time, feel your way carefully in doing so and avoid insulting a customer's intelligence or appearing to be patronizing. Your product may have features that produce tremendous benefits for your customer, and they may intrigue you... but if he doesn't understand how they will help him, they might as well not exist.

In conclusion, then:

1. Identify your customer's real reason for buying.
2. Identify those product features which fill that need.
3. Hit hard on the advantages and back up with proof.
4. Draw comparisons to emphasize advantages, when possible.
5. Don't bore or confuse him with a recital of features which might sell other customers but are of no interest to him.
6. Supply explanations when necessary.
7. Plan what you are going to say in advance and don't minimize the importance of selecting the right words and phrases for this customer.

D7—Draw comparisons

You may draw comparisons between your product and that of your competition. Comparisons may be made between the advantage of buying and the disadvantage of not buying. Comparisons may be drawn between the procedures your customer seems inclined to follow and those you believe he *should* follow.

Comparisons are helpful because they spotlight the advantages which your product will supply the customer. Without comparisons, the advantages you are prepared to supply may not become meaningful or even apparent to him. If you are shown a picture of the pyramids in Yucatan, you may remain unaware of their great height unless a human figure has been included in the photograph so that you can make a comparison between the known and the unknown.

You are selling a product to a customer, and it includes a new feature you sincerely believe will provide him with a benefit. The customer may not be technically informed, and the advantages of the new feature will be entirely lost upon him unless you make a comparison between the performance of your product versus that of a product which lacks this feature.

Comparisons must be made with caution. It will be wise to keep some of the following points in mind:

1. Under ordinary circumstances when you find it necessary to compare your product to that of a competitor's, don't give in to the temptation to "knock" the other product. Let the facts speak for themselves—and they will do so, eloquently, if you use this technique. Incidentally, we do not subscribe to the theory that it is always bad policy to point up the shortcomings of your competitor's product. Much depends upon the way in which it is accomplished.

For instance, a customer may ask your opinion of the competitor's offering. Or he may not know enough about the details—or technical aspects—of the competitor's product. If you do not bring certain shortcomings to the customer's attention, he may remain unaware of them. And you may, as a result, lose the order to a competitor who is offering a product inferior to that which you can provide. In such instances, you can use statements along the lines of the following:

"Under ordinary conditions, I find that it is unwise to criticize the offerings of a competitor. But in this instance, I feel that I have an obligation to point out one shortcoming of X's product. My job, as I see it, is to help you solve an operating problem, and I would be letting you down if I didn't make certain that all of the facts are available to you when you make your decision."

You then proceed to describe how the competitor's product or service will fail, in some specific respect, to meet the customer's problem. You may find it wise to indicate that the service or product being offered by the competitor is a good one except for this feature. Then, to conclude, be sure to use comparison, pointing out how your product is superior in this respect.

2. Your customer may have made prejudgments regarding the relative merits of your product and that of your competitor, and she may tend to favor your competitor's offering. Your success will depend upon your ability to change her mind—and making comparisons is one method of doing so. Remember, however, that when the customer holds an opinion that is adverse to yours, there are dangers implicit in such a situation. The implication is that you are right and she is wrong; in other words, her judgment has been poor. You must be certain that in changing her view, you do not attack the image she is always trying to project; i.e., being a person of good judgment and experience. Your approach must skirt criticism and must avoid creating a situation where she is forced to say, "I was wrong."

The task is to present, subtly, a comparison of the two offerings in such a manner that the customer is provided with an escape hatch. Naturally, the one best method is to make the comparison *before the customer has had a chance to go on record with his views*... then he is not in the position of being forced to change his mind openly and lose face. If he, unfortunately for you, has expressed an opinion in favor of a competitive offering, present your comparison, using phrases that "give him an out." For example, "I can understand your views, and I felt the same way until certain facts were brought to my attention..." Then continue to make the comparison.

You and I—and the customer—don't feel we have lost face if we have held erroneous beliefs, when other people "of quality" have also held those beliefs. In other words, what you are attempting to point out to him is that it is perfectly natural for a wise person to have been in error if he has been deprived of information.

Your comparisons must be meaningful to the customer, via the comparison route, that your product has features lacking in your competitor's. You must make certain that the customer understands the value of these features to him. Therefore, couple your comparisons with "reasons why"; i.e., explanations. In doing so, avoid any appearance of "talking down." Here is an example:

"This camera is equipped with a zoom lens, and you can take pictures that will range all the way from wide angle to telescopic views. This is accomplished in seconds by merely rotating the barrel. As you know, in cameras lacking this feature, you face the need to carry a number of lenses in a shoulder bag, and it takes so much time to change from one lens to another that good pictures are often lost. Before this zoom lens was available, I always had to carry a bag of lenses with me—and I'd like to have all of the good shots I lost while I was changing from one lens to another."

What has been accomplished in this one statement?

a. You have highlighted a feature of your product.

b. You have made its advantages meaningful to him.

c. You have avoided the possibility of causing him to lose face by using the phrase "as you know" and by pointing out that *you* had bought many lenses, yourself, before you learned about the new zoom lens. (This is especially important if you know he already owns a kit of lenses. You and I often make purchases before new product developments are available, and to obtain a new feature it means we must shelve previously purchased items. This always causes us some mental pain and a feeling of guilt if we invest in the product while the old product is still usable. However, the pain is greatly eased if we know that many others have had to take the same action. This has been accomplished by the salesman in the example above.)

d. You have drawn a comparison—favorable to you—between your product and your competitor's, without once having mentioned the competitor.

e. You have caused the feature of your product to appear highly desirable to him. In doing so, you explained its operation—and in such a manner that you did not appear to be patronizing.

f. *You have accomplished all of this in about 28 seconds.* (Time it!)

Note: If the foregoing material is to be of maximum help to you, it is important that you select one of the features of a product or service that you sell and then apply this formula. As you design your sentence, the tremendous power of words and the importance of their selection will become apparent to you. Words are your finest selling tool; use the right ones for each selling situation. To sales managers who are using this book as the vehicle for training sessions, it is urged that each man be assigned, for this session, the responsibility for bringing to the conference two or three examples of actual sentences he can use for comparing his product to others.

D8—Tack with his stand

In our comments on the use of comparisons in the previous strategy, we noted how much more difficult the selling situation becomes if the customer has been permitted to go on record with a flat statement of his views when those views are adverse to our position. It is also apparent that we cannot always stop a customer from taking a determined stand. He has sometimes made prejudgments before we even enter the selling situation.

It is equally important that we carefully consider the advisability of making flat statements or claims before we know how the customer feels or what the main focus of his interest is—his reason for buying or for being opposed. Whenever either party in a selling situation takes an inflexible stand, the chances of establishing the essential rapport is threatened, as is the successful closing of a sale.

We have already discussed the need to provide the customer with an escape hatch. It is just as important that you do not maneuver yourself into a position far out on the limb from which there is no escape. Avoid dogmatic statements, and avoid an expression of views that is so flatly stated that you have no opportunity to maneuver or to agree, even in part, with contrary views that may be expressed by the customer.

In *Strategy B9 (Emphasize mutual interests)*, stress was placed on finding areas of agreement with the customer. The more such areas you can find, the better are your chances of closing the sale successfully. This is because we tend to have greater confidence in people who agree with us on many things.

Earlier we discussed Gestalt—our tendency to see the part as the whole. This tendency applies here also. If a person is inclined to agree with us on many things, we see him as a whole; and because his agreement flatters us, we tend to give credence to views he expresses on *other* subjects. No better reason exists for avoiding expressions of opinion on minor points where we do not agree with the customer. Gestalt works both ways. The customer will tend to view us unfavorably in proportion to the number of contrary views we hold...so keep them to a minimum.

There will be situations where principle is involved, and you cannot find it possible to agree with the customer on some premise he has stated. The trouble with many of us is that we use "principle" as an excuse for a lack of self-control that leads us into making a major issue of something that was really unimportant. In most instances, it is not too difficult to make unimportant compromises, via silence, with your customer in order to accomplish your major goals and long-range objectives.

In order to avoid unnecessary clashes of opinion, we tack with his stand. "Tack" is used in the nautical sense. The dictionary defines "tacking" as sailing before the wind instead of in a direct, straight, course. The dictionary states, "the speed attained is often greater than in a straight running."

In other words, it is not important that you gain your customer's agreement on every point you make. It *is* important that you gain her agreement on major points. Therefore, don't challenge a premise stated by the customer unless acceptance of that premise would undermine your main thesis. Let's clarify that point by applying this technique to selling an object with which we all have at least some familiarity... the selling of a camera.

Let's assume the salesman is attempting to sell a camera that has an interchangeable lens. Each lens is equipped with a bayonet-type base. It is a simple matter to change from one lens to another by simply twisting the lens that is in the camera a quarter of an inch and pulling it out of the bayonet socket, then inserting the other lens. The customer is interested in the camera and has commented favorably upon many of its features... but now he offers his view:

"Of course, there *is* one thing wrong with this camera. It has a bayonet-type mounting, and lenses that screw in are much better because there is no chance of wear and your pictures will always be in sharper focus."

When the customer offers this view, the salesman has a choice of actions. For example, he could point out that the "wear and tear" on a lens mounting is infinitesimal, and in decades of use you would never see the *slightest* change in the accuracy of focus as a result of this type of mounting. To point this out to the customer would serve only to irritate him and might lose the sale. To agree—without qualification—might also lose the sale. The professional salesman decides that this is not a key point, and he is not going to permit it to become one. After all, the camera *does* have a bayonet mounting lens and this can't be changed. The salesman would prefer to ignore the comment but, instead, recognizes it for the moment and then diverts the customer's attention to other features (in other words, he "tacks"). For example, he might say:

"I certainly agree with you in theory because there probably is more wear in this type of mounting. I would, however, like to point out one advantage of the bayonet mounting that will be important to you. With a screw-in type lens, it takes much longer to change lenses, and even good photographers lose pictures as a result. You've probably already had that experience—I know I have. With the bayonet mounting, you are going to find that you can change lenses quickly. And you'll rarely lose a picture because you have the wrong lense in the camera when a picture unexpectedly pops up. You are going to find the possible disadvantages will be outweighed by this advantage which this camera is going to provide you."

What has been accomplished in this comment?

a. A degree of credibility has been accorded the customer's expressed point of view. His comment has not been ignored, disclaimed, debated or scoffed at.

b. The salesman has tempered the effect of the complaint. (He could have done so even more effectively has he added, "While there may be more wear in this type of mounting, I know you agree that there won't be appreciable wear for years.")

c. He then placed the customer's attention on an advantage that outweighs the claimed disadvantage... he turns an objection into an advantage.

d. He has used, effectively, the technique of "comparison" discussed earlier.

e. He flatters the customer when he states that the customer has probably had the same troubles experienced by (other) "good photographers."

f. When he states that the customer has probably had the experience of losing pictures in the past, he takes the customer off the spot by adding, "I know *I* have."

g. He emphasizes the losses the customer will suffer if he chooses the wrong camera, and then makes certain that he closes on a positive note by leaving advantages uppermost in the customer's mind.

h. Notice how the salesman assumes throughout his remarks that the customer is going to buy the camera. He accomplished this with such phrases as, "you *are going to find*"; "one advantage... *will be* important to you"; "this advantage... *is going* to provide you."

The customer's arguments should not be ignored, brushed aside or ridiculed. They should be recognized, and agreed with in part or in principle (if possible), but then, by use of "comparison," they should be made to appear unimportant as compared with the advantages of choosing our product or service.

D9—Use probe for reaction

As a selling situation unfolds, conditions frequently arise making it desirable, even necessary, to identify the nature of the customer's probable reaction to a contemplated action on your part, or to determine what his reaction has been, up to this point, to your presentation. The following situations are typical of what we have in mind:

a. The customer has witnessed a demonstration of your product, and you wish to learn what it is about your product that has impressed him the most or you wish to identify possible objections which the demonstration may have caused to arise in his mind.

b. You know he has had real difficulties with your company or product since your last visit, and you need to determine how she feels now.

c. You are considering using the name or experience of a third party as a testimonial, and you, first, want to know how he regards this third person.

d. You would like to ask for the order at this point, but you want to determine whether or not he is satisfied with what he has seen and heard up to this point.

e. Your product has many advantages, but you want to determine which will be of greatest interest to him so that you will not bore him—or even lose him—by reciting all of the advantages or by claiming that a feature is an advantage when, from his frame of reference, he will see it as a disadvantage. In other words, you would like to find out if he wants you to "make him rich"...."make him famous"...."protect his status quo (security)"... or "make life easier for him."

These are a cross-section of the many situations in which a "probe for reaction" will be helpful strategy. Because the specific goal of each of the five situations described above will be different, the actual wording of the probe may vary. There are, however, certain basic considerations to keep in mind when you find there is a need to probe for reaction:

1. You seek his reactions to a demonstration you have just made:
a. The probe should be worded so as to focus his attention on the points that you hope made the most favorable impression upon him. The purpose of the probe in this situation is to aim a spotlight on those points before he can open up a discussion on the points he found less favorable or even objectionable.

If you can cause your customer to talk positively (versus negatively) about the attractive features of the demonstration, he will, to a degree, stimulate his own desire to buy. If he sells himself on the good points of your product, he will *want* to find ways of overcoming those points he found objectionable because he will not want to lose the advantages. Contrary-wise, if he is given an opportunity to talk about objectionable features first, he may not develop the desire to buy. He will have been permitted to think about the disadvantages, and as a result he may not see any advantages.

Example of how to establish a favorable focus: "Mr. Henry, I would be interested in knowing what features impressed you the most during this demonstration?"

This simple, straightforward question almost forces the customer to seek good features and advantages. If he is one of those real obstreperous people, he may answer, "I didn't see *any*!" If this is his answer, nothing has been lost because at this point he is unsold in any event. Now you can probe for the reason for his objections with, "I'm sorry to hear that—what did you find most objectionable?" As you listen to his reply, you may find that his disagreement is the result of poor communication or misunderstanding. At least you have spotlighted the nature of your problem.

When the probe brings forth the favorable reply you hope for, you have identified what is of primary interest to him, and you can now expand upon those features.

b. The probe should be worded as to avoid producing complaints which you prefer to avoid or which you are unprepared to answer. However, it is not wise to avoid objections which you know he has and which are important to him, hence vital to the success of the sale. They are not just "going to go away" if ignored. You must smoke them out via the probe for reaction. Presale preparation and product knowledge will now stand you in good stead. You should know your product (and your competitor's) so well that you will have considered the possibility of every objection which he could raise, and you will have developed answers to those objections. Now that you are in the middle of the selling situation, you must determine whether he is harboring unspoken objections. Only the ill-prepared salesman avoids serious objections. They must be dispelled with convincing facts.

2. Your customer may have had a disagreeable experience with your company since your last call, and you will want to know how it has affected her attitude of the moment. This is an occasion where the direct probe is not recommended. It is far better to act as though nothing had happened and to try to focus her attention on something you have to offer that is of interest to her now. This is still a kind of probe in that it enables you to watch her reactions carefully. If you have something to offer that is definitely of interest to her, concentrate on that point. If the advantages to her are great, she may change her attitude. If she continues to be hostile, you must face up to it. References to her past difficulties with your company may simply re-awaken hostilities that have abated since the unfortunate event occurred.

3. You would like to use a third person as a testimonial, but you wonder how he regards that individual. Have you ever had either of these experiences?

1. You had two friends, each of whom you liked very much. You brought them together, thinking that because you liked each of them, they would like one another... only to find that they did not.

2. You arranged a party, carefully selecting people you were certain would be compatible because you like all of them, only to find some of them intensely disliking some members of the group.

You may respect one customer and use him as a testimonial, only to find the customer you are trying to influence not only fails to share your high opinion but actually disdains the other?

How do you avoid the problem?

Here is one method you might employ:

Don't wait until you are about to use the name of the third person as a testimonial, or you may find yourself in an uncomfortable position. Example: "Do you know Al Simpson of Moravia Brass?" The customer replies, "Yeah, he's a real screwball... why do you ask?" In this instance, you are left with the lame reply, "Oh, nothing... his name came up in a conversation the other day, and I just wondered if you knew him." All that can be said for this approach is that it leaves you in no doubt about the advisability of forgetting Al Simpson as a testimonial with *this* customer.

Naturally, the best bet is to know in advance how your customer feels about various people whom you might wish to use as testimonials. The opportunity is often presented to ask such questions as, "Whom do you regard as the leaders in your field?" If he mentions Al Simpson, you can be nearly certain that the use of Al Simpson as a testimonial will be in order. If he does not mention Al Simpson, you can always say, "There are two or three others in the field, as I recall. There's Bevins, Simpson and Sandres. Do you know them?" Then listen to his reaction.

Don't use Simpson's name immediately after you have learned that your customer regards him well; your reason for asking will be transparent to the customer. Wait until the appropriate time in the sale and then say, "Oh, Al Simpson's name came up earlier in our conversation. Did you know that he is using our service?"

4. The most common need for the probe for reaction is when the professional saleswoman wishes to check the progress she has made in the selling situation. The rhetorical question often proves to be the most effective in such instances. This is because it does not necessarily require an answer, yet it provides you with an opportunity to look for quick reactions. For example:

"I'm sure, Mr. Eldridge, that the time-saving features of this installation are attractive to you." Then you pause only momentarily to watch for a reaction. You may get a negative response—and if such is the case, the probe has done its job. You know that you should not proceed further until you have overcome any disagreement or misunderstanding that exists at the moment. If you proceed, he may not hear a word you say... he will still be thinking about those portions of your presentation that left him dissatisfied. If, however, he merely shakes his head, or seems to agree with you by his attitude, you can proceed with your presentation.

The probe for reaction can be made to do double duty. It can provide you with a clue to his reaction, as we have seen, but it can also help you drive home a selling point. This is accomplished by a thought-stimulating type of query, such as the following:

"You can see the quality that has been built into this product, Mr. Emery. Can you really afford to push your present equipment to get the last mile out of it? I'm sure you see that there is greater economy in replacing it now before breakdowns cost lost time..."

Or: "We all share the understandable tendency to hold off on making capital investments until the last minute... and often it is wise to do so... but I am sure you agree that installing this equipment now is going to save you costly breakdowns... hence less of an expense than if we wait until later in the year. Don't you agree, Mr. Bond?"

In the foregoing example, we have welded together a number of strategies from the Sales Strategy Bank:

1. "We all share the understandable tendency" —this phrase provides him with an escape hatch (you have suggested that he has a tendency to hold off making wise investments, but you "share the guilt" by pointing out that you do also). This must be used with care with the customer who regards himself as above other people and above you. He won't relish your putting yourself on the same plane he feels he occupies. In dealing with such individuals, you can say, "Many key executives hesitate to make capital investments... they share an understandable tendency... to put off making capital investments," etc. Again, if he is *not* a key executive, substitute some other term he will look upon as favorably such as, "many men in your position."

2. The statements provided in the example above also use other techniques which apear in the Sales Strategy Bank; i.e., *(F4) Emphasize cost of not buying; (E4) Explore cost of wrong decision;* and *(D6) Emphasize benefits.*

The foregoing material represents only a few sample statements that can be made in probing for a reaction. Develop your own list, and don't use the same statement over and over so that your customer becomes annoyed or that your purpose becomes transparent.

D10—Divert him

There are occasions when a customer will concentrate on something that will, at the very least, impede the progress of the sale. He may have some pet peeve that he will belabor. He may have raised some point, which, for the moment, you would prefer to avoid. It is at times like these that you can effectively employ the technique D10—Divert Him!

You accomplish this by the use of questions. Just so long as he must answer *your* questions, you will be directing the course of the selling situation, and he will be in no position to talk about the things he wants to talk about! This would seem to be in direct conflict with the advice that has been repeated over and over in this book; i.e., "Let *him* talk, and you listen." There are times, however, when it will be necessary to have the customer do the listening. And when he seems determined to embark upon a discussion which impedes or threatens the progress of the sale, you must employ techniques that will compel his attention.

Once again we will see the wisdom of pre-sale preparation. If you do an effective job of preparing for the sale, you may identify some of the pet peeves which the customer has, or you may be able to predict that he will want to discuss poor experiences he has had with your company or your product. Before the sale, with emotional conditions absent, you will be in a much better position to plan the type of questions you will employ to divert his attention away from these critical areas. You should not wait until you are actually *in* the selling situation before you plan your strategy.

Being realistic, we must agree that there *will* be times when you cannot predict what avenues the customer will seek to explore. It is at times like these, when he "strays from the course you have plotted," that you should employ the use of questions in order to divert his attention away from those nonproductive areas and back to the points you are seeking to establish. *Questions* are your most effective selling techniques. A question always puts the listener "on the spot" and enables you to control the situation for the moment. At least, while the person being questioned ponders his reply, you have been successful in directing his attention to an area other than the one he would prefer to examine.

If a customer is truly angry about some condition, don't think that by simply asking a question you are going to be able to divert his attention. You may find it wise to couple your query with another strategy . . . such as, find a point of agreement, or show respect for his arguments . . . and *then* pose the question. For example (in speaking to the customer who has raised a point which you don't wish to discuss at this juncture of the selling situation): "Bill, you certainly have a point there, and I don't think it should be ignored. If you don't mind, though, I'd like to hold it for discussion in a few minutes. Right now, I'm wondering about another problem that you have—and one upon which I think I can give you some help. Bill, am I right in assuming that your company is having some difficulty with . . ." (Here, you raise some problem that your pre-sale preparation has shown you he has, and for which you can offer him some relief.) In other words, divert his attention from the subject you wish to avoid to a subject upon which you feel you can take positive action.

There are times when diverting the customer calls for frightening him. We recall such a situation. A customer had failed to follow certain installation instructions that had been given him by the salesman. As a result, he had difficulties and called the salesman in a fury. He immediately denied that he had even been instructed regarding the installation procedures. The salesman knew that a letter containing such instructions was in the customer's file, but he also knew that proving the customer to be wrong would avail the salesman nothing. He realized that the customer, too, knew that he was wrong and that he was seeking an escape hatch by blaming the salesman.

Instead of discussing who was right and who was wrong, the salesman diverted the customer's attention to a fear that was greater than the fear of having been wrong. He said, "Look, Ed, the point is that right now you are in trouble with your boss, and I know how to get you out of trouble on this installation." The customer made a speedy about-face and asked anxiously, *"How?"*

The principle is that, when possible, if your customer is concerned about one fear (and a fear about which you can do nothing), focus his attention on a greater cause for anxiety about which you *can* do something.

Keep in mind, too, the point that has been repeated here so often. Your customer falls into one or more of these categories: He wants to be rich, or he wants to be famous; he wants to protect his security, or he wants life to be easier. You should know him well enough to gauge which of these is paramount in his mind at this time. If he is angry because he sees one of these needs threatened, you must employ questions or statements that will direct his attention to one of the other areas and make it, for the moment, seem more desirable than the need that has until now been compelling his attention.

For example, your customer may be objecting to your "high price" . . . in other words you are threatening, to a degree, his "desire to be rich." The facts of the situation may enable you to divert his attention with remarks such as these: "Jerry, I can appreciate your feelings about the price, but I also am aware, Jerry, that you are one of the busiest individuals I know, with great demands upon your time. Have you considered how valuable your time is, Jerry, and how much of it will be made available to you, if you use our service? Jerry, you *deserve* a little relaxation . . . *you've earned it* . . . and life will be a lot easier for you if you employ this service. Jerry, how much of your time is now spent on this activity?"

In this simple statement, you have accomplished many things:

1. You have dignified his complaint ("I can appreciate your feelings about the price").

2. You have appealed to his ego ("You are one of

the busiest individuals I know").

3. You have transferred his attention to something about which you know he feels strongly...the demands upon his time. In doing so you have employed questions: "Have you considered how valuable your time is?"...."How much of it will be made available to you?"

4. You have employed one of the most powerful tools in the Sales Strategy Bank ("You deserve it... you've earned it"). Busy men usually are afflicted with consciences that drive them relentlessly. They often ache for a few hours' respite but feel guilty when they are not working. They grasp avidly for the opportunity to relax without feelings of guilt that is made available to them through the comforting phrase, "You deserve it...you've earned it." This phrase is a powerful technique for diverting a customer's attention from the things his conscience is demanding.

5. Finally, you have forced the customer to consider the avenue you wish to explore when you have ended with a direct question ("Jerry, how much of your time is now spent on this activity?").

In summary, the formula that will often divert the customer is: (a) Find a point of agreement; (b) focus his attention on something that is basically of greater interest and importance to him than the subject he has raised; and (c) force him to consider this new subject by raising a direct question that is sufficiently complex to cause him to ponder his reply. The more the question causes him to stop and consider, the more likely it is that he will forget his pet peeve. The question employed in the example above ("...how much of your time is spent on this activity?") can't be answered without thought. And consideration of this question leaves no room in the customer's mind for other topics which you prefer to avoid.

D11—Rephrase and overstate his objections

Once your customer has raised an objection, your task is either to eliminate the objection if it is not valid; or, if it *is* valid, to minimize its importance. One technique for accomplishing the latter is to restate her objection and, in so doing, to *overstate* it. The customer's normal reaction will be to deny authorship of the objection as you have stated it. Then, in restating her objection, she will usually tone it down a bit to emphasize that you have exaggerated what she said. You now have something with which you can cope much more easily.

Often, you and I—and the customer—in voicing complaints, *do* tend to exaggerate or to make broad, generalized charges. These are difficult for a salesman to handle. He needs something specific. In restating our objection and overstating it, he forces us to deny that we made such an all-encompassing charge. In restating the complaint, we tend to become more specific...and this is just what the salesman needs.

Here is an example illustrating how to use this strategy:

Customer: "Your product would suit me just fine except for one thing. We can't always control the size of raw materials that our vendors supply us. So, we have to have equipment that provides safety against overload, and I don't think that your equipment makes sufficient provisions for overload."

Salesman: "I can see why provision for overload is important to you in this type of application, and I agree, too, that this should be a key consideration in your buying decision. Now, as I understand it, you feel that we have not factored in *any* provision for this kind of protection in our product...Is that correct, sir?"

Customer: "Well...no, I don't say *that*. I'm just saying that before I buy, I'm going to be sure that the equipment I select makes sufficient provision for overload...in the case of our application, say, about 10%. And I felt your product might not provide that much protection."

Salesman: "Oh, I see. Well, our problem is easily resolved, Mr. Roberts...it's just a matter of seeing what 'enough protection' is in the case of your application. Let me show you how we have provided for overload in this design so that 10% or even 15% overload will never present you with a problem."

By this device, the salesman has reduced the severity of the customer's objection. Originally, the customer intimated that he did not feel the salesman's product provided *any* protection against overload. The salesman overstated the objection, and the customer hastened to rephrase his objection by reducing it to a "feeling" that there might not be *enough* protection against overload. This is considerably easier to handle, especially when the salesman has been successful in having a specific overload percentage stated by the customer.

It will be noted that this device also provides the salesman with an opportunity to "fence"—to obtain a bit of time to consider the best way to handle the objection.

There are a number of reasons why overstating the objection works. It sometimes causes the customer to feel that you are, perhaps, trying to "trap" him by attributing statements to him that he cannot substantiate. He, therefore, hastens to reduce the scope of his objection so that he will be in a position to defend his charge. In doing so, he moderates the severity of his objection and makes it easier for you to handle it.

Another reason for the customer's fast reaction to an exaggeration of his objection is found in the basic desire that most people have to be "fair." He does not want to be placed in the position of making unfair criticisms of your product—and wide, sweeping generalities have a tendency to be "unfair." Example:

Customer: "I have no objection to your product; it's just that I'm a small businessman, and you people are big. I think I'd be considered insignificant compared to your big customers, and any problems I would have wouldn't receive attention."

Salesman: "As I understand it, Mr. Smedley, you feel we wouldn't be interested in your business?"

Customer: "No, no, I didn't say that! It's just that service would be pretty important to me. I can't afford delays due to shutdowns when equipment problems arise. Big outfits can probably take that sort of thing in stride. I'd have to be sure I would get fast service."

Salesman: "Mr. Smedley, I'm happy that this is the concern you have. The fact that I am here shows you we want your business. We know that to *keep* your business we must give you prompt, efficient service. I can readily understand your desire to avoid lost time and I'd like to explain our service arrangements. As you know, we have thousands of customers, and there just aren't that many big customers. We have been successful because the solid foundation upon which our company has been built and continues to grow is the business that comes to us from businessmen who are not what you or I would call 'big'. Let me show you how we service those businesses." (Then, explain your service procedures.)

Here the customer has been made to back-track in his original, sweeping objection. The salesman asked the customer if he thought they were not interested in his business. The customer knows you *are* interested or you wouldn't be calling upon him. Therefore, he suspects some kind of trap is being set, and he hastens to modify his original objection. Now, instead of concentrating on *size*, he focuses on his need for fast service, his need for protection against lost time. This is something to which the salesman can readily respond. It also spotlights the customer's real problem and shows the salesman the direction that his presentation should take. He will concentrate on service and will not waste valuable time on aspects of his proposal that are not of key concern to his customer.

D12—Get him to talk

In the earlier phase of this book, we discussed the Anatomy of a Sale, and at that time the strategy "*Get him to talk*" was described in detail. A lengthy expansion here would be redundant, other than to examine the reasons *why* this strategy is effective. These include:

1. Given free rein, people will talk about things of interest to them. If you do not steer the conversation, your customer will reveal many of her aspirations, problems and opinions—and these provide you with valuable information.

2. A customer may give you a "smoke screen objection" to hide his real objections (which may be socially unacceptable; i.e., may not make him look good). If you can get him talking and keep him talking, the chances are that he will reveal his real objections.

3. People with problems like listeners. Give your customer an opportunity, and he will unravel his problems before your eyes. Keep in mind that what he really wants is a good listener. Frequently customers do not want advice even when they seem to be asking for it. They usually want someone else to confirm the wisdom of the course of action they already have contemplated. Therefore, if your customer has dwelled on some problem that is *not* connected with your product or service, and he finally asks you, "What would you do if you were in my place?" avoid the tendency to play Solomon. Here are the actions you can take, *safely*:

1. "Well, Bill, what are some of the actions you have considered taking?" Don't be surprised to have him respond, with some spirit, "I'll tell you what I'm going to do...here's what I'm going to do...etc." He may finish with, "What do you think about that?" In this event, your best reply may be "Bill, that's certainly *one* action that's available to you."

2. "We might explore some possible actions, Bill. What is the first course of action that occurs to you?"

3. "Perhaps, Bill, to answer your question, we could weigh the advantages and disadvantages of each action you have contemplated. What's the first action we should consider?" You are probably giving him maximum help when you aid him in approaching his problem in this scientific manner. If you feel that the action he seems determined to take is going to add to his troubles (and affect your business opportunities with him), the most you should do is to try to steer him toward method 3. This technique will usually help him discover the inadvisability of the proposed action.

Above all, getting him to talk presents you with the most effective way of establishing rapport. Being a good listener is still one of the highest forms of flattery. It goes without saying (or should) that when you get him talking, you listen—*really* listen.

D13—Appeal to fairness

Most of us hate to be stigmatized by the charge that we are "unfair." This is basic to human nature and starts in our childhood when we are sure to hear the colloquial children's cry, "No fair," shouted by some youngster who feels he has been subjected to unfair treatment by his peers or elders.

A customer may be unfair for many reasons. She may hold some old grievance, real or fancied, against you or your company. She may blame you for some action taken by some other individual or department in your company. She may blame you for something that she has picked up along the "hearsay route." The list of reasons for her attitude may be endless. The question is, what can be done about her attitude?

First of all, we must keep in mind, that we use defense mechanisms when we sense a threat (against which we must *defend* ourselves), or we see no gain in following the suggestions of a salesman. We may take actions against him to protect ourselves that are essentially unfair to him. Often, we are not really aware that we are being unfair. And when our customer treats us unfairly, he often is not really aware of that fact either. He is so intent upon protecting his self-image that he gives little thought to the techniques he employs. He may be so angry that he seems to be something less than rational.

The direct approach, with an appeal to his pride, is often the most effective counter-action. Here are three examples:

1. Mr. Harrington, I'd like to appeal to your sense of fairness. I realize you did have an unfortunate experience with our company some time ago, but don't you honestly think I deserve a fair chance to prove myself to you?"

This statement has the weakness that it leaves the door open for a "No!" Therefore, it may be more effective if it were phrased: "You did have an unfortunate experience with our company some time ago, and I know that you'll agree that we deserve a fair chance to prove ourselves to you."

Better, perhaps, is the more direct appeal:

2. "Mr. Babcock you have a reputation for being demanding but fair. You are concerned—and quite likely, justifiably so—with something that happened in the past. I'm sure, though, that your sense of fair play will cause you to agree that we shouldn't be given a 'life sentence', but should have a chance to prove ourselves." (The "life sentence" remark must be accompanied by a smile so that it does not sound like an indictment.)

3. "Mr. Osborne, we are a large company and we have many departments and individuals. I'm sure that occasionally one department or some one individual may take an action that understandably irritates a customer. However, you are a fair person, and I'm sure that you won't want to penalize me for something that involved people I don't even know. I'd appreciate a fair chance to show you that I am as interested in your welfare as in my own."

D14—Fence

Customers throw totally unexpected objections our way, and no matter how well prepared we may be, we are momentarily at a loss for an answer. Succumbing to the temptation to give off-the-cuff answers may be highly dangerous. What we need is time to think. This is where the strategy "Fence" can be most helpful. Example:

Customer: "That horizontal chamber is going to be very hard to reach when maintenance is necessary."

Salesman: "What type of maintenance did you have in mind, Mr. Chambers?"

The salesman knows very well what kind of maintenance is going to be necessary, but he needs time to marshal his thoughts and to respond to what is, essentially, an objection to his product. The customer, however, cannot question the sincerity of the question because there are many types of maintenance necessary on the product being offered for sale—electrical, mechanical, plumbing, painting, etc. The innocence of the question throws the customer for a momentary loss, and he has to stop and think of just what kind of maintenance he *did* have in mind.

The "Fence" has added value in that it often enables the salesman to "explode" objections. In the example just given, the customer is stopped by the question. Now, as he thinks it through, he may realize that there will be no real maintenance problems. He is then forced to respond to the question with, "Well, I suppose there are not real maintenance problems." By this time, the salesman, who has had time to think, has arrived at the same opinion but is not put in the position of causing the customer embarrassment by pointing it out to him. Here are two examples:

Example 1

Customer: "Your product is all right but your terms are out of the question."

Salesman: "What terms would you find more realistic, Mr. Van Zandt?"

Here, the customer is again on the spot. Remember the old Chinese proverb: A fool can ask a question it would take 10,000 wise men to answer. While Mr. Van Zandt is groping for an answer to the unexpected query, the salesman has time to formulate a reply to the customer's apparent objection.

Example 2
Customer: "We've decided to advertise only in the Sunday edition... we think this will produce better results."
Salesman: "That's interesting, Mr. Eggers. I'd appreciate knowing why you feel that way. It's a viewpoint I'd like to explore for a moment with you."

Notice in this example that the salesman did not say, "It's a *decision* I'd like to explore with you." He is not going to admit that a decision has been made ... he wants to keep it in the form of a viewpoint because viewpoints are easier to change than announced decisions.

Notice, also, that the salesman does not issue his question as a challenge. He phrases it as a favor he would like granted. Now, while the customer tries to answer his question, the salesman has an opportunity to see the real nature of his problem and prepare a rebuttal. Note also that the salesman would like to explore the viewpoint *for a moment*, leaving the customer feeling that there will be no long, drawn-out debate which would use up his time.

The "Fence" is a powerful strategy because:

1. It places the customer on the defensive.
2. It gives you a chance to think and prepare something better than an off-the-cuff reply.
3. It forces the customer to narrow down his objection.
4. It aids you in identifying your problem.
5. It forces the customer to talk... you listen and gain facts.
6. It often exposes the fact that the objection is without validity, which the customer discovers for himself when he is forced to amplify or explain his objection.
7. Even when you subsequently lose the order, you identify the real nature of your problem and are ready to meet it head-on in the next sale by using Strategies *E10—Anticipate, raise and answer objections* and *A12—Identify his possible sources of resistance.*
8. When you use the "Fence," it gives the customer the feeling that he is dominating the selling situation because *he* is doing the talking. Actually, you are controlling the situation because you are putting him in a position where he must be responsive to *your* question.

D15—Use the direct question

This type of question usually demands an answer. It can be used on the individual who you believe is giving you a smoke screen objection to hide his real objection. It can be used as a "last resort" method on the chap who permits his personal animosity for you or your company to cloud his judgment. It can be used to demand proof. It can be used to cause the customer to substitute reason for emotion. The direct question is often referred to as the "factual" question. In this form it starts with "who—what—where—when—why—which—and—how." For example:

"Who told you that?" ("Who told you this motor cannot be used in a vertical position?")

"Why do you feel that way, Mr. Jensen?" ("Why do you feel we have been unfair to you?") ("Why don't you like us?") You'll recall that this latter question brought back a customer who had purchased elsewhere for 20 years, simply because no one had ever considered asking him this very simple question. Once he had an opportunity to answer it and get the bile out of his system, he returned to the original supplier who he thought had treated him badly 20 years previously.

"You feel this will be difficult to maintain? *What* maintenance do you have in mind?"

"The product you are considering buying is a good one, but the manufacturer doesn't have service facilities within 1,000 miles of here. *Where* will you obtain your service, sir?"

"If you don't make this investment now, how do you intend to meet the power demands of the three new manufacturers who plan on building here within the year?"

"If you feel this is not the time to buy, *when* do you feel will be the right time to make this investment?" Or: "*Why* do you feel this is an inappropriate time to buy?"

"If you are not satisfied with our product, which of those available do you feel *will* give you what you need?" (Followed by "*Why* do you feel that way?")

It is obvious from the tone of this type of question that it must be used with extreme care. It can cause the customer to feel fenced in. It can irritate him. It can cause him to think you are brash (especially if you are younger than he is). It can cause the customer to lose face when he cannot come up with a good answer. However, it *can* resolve points at issue. It can force showdowns when showdowns are necessary. It can shock the customer into reality. It can force him to think through his own problem. It has so much value that it should not be ruled out because of its hazards.

The direct or factual question requires great emphasis on your own mood and manner. Because it *is* direct, it must not be used as a threat or a challenge. It must be accompanied by a facial expression and a tone of voice that implies that you are honestly perplexed and seeking information. It must not be made to appear defiant. Often, it must be considered as a "last resort" method.

D16—Use the overhead question

Committee-type buying is on the increase in many types of selling situations. Even when a formal committee has not been constituted, the professional saleswoman often encounters situations involving two or more people in the customer's organization. The saleswoman faces the real problem of determining the identity of the key influence on the decision to buy—or *not* to buy. The overhead question can be very helpful in this type of situation. When we use this type of query, we throw our question to the entire group, be it two men or ten.

In asking the overhead question, be sure that you do not fasten your gaze on any one individual or it will possibly lose its entire value...you don't want to force the wrong man to reply simply because you are looking at him and he thinks he must reply. After throwing your question to the group, you wait for a reply, and you try to see who seems to be the leader.

Avoid asking questions that are tied to one function. In other words, if your audience consists of men from different functions of the customer's business, you should avoid asking a question that will fall almost exclusively within the scope of one individual's work. Obviously, he would answer the question, and you would still not know the identity of the true decision-maker.

One such overhead question will hardly indicate the identity of the decision-maker. But if you toss out a number of queries and one woman seems to be first to reply, you are certainly entitled to draw assumptions from this fact. When asking an overhead question, notice whether or not the eyes of your audience seem to turn to one individual, as though waiting to see what *his* answer will be.

Don't jump to the conclusion that the real spokesman must be the fellow with the organizational rank or the person who is most vocal. If you use questions that pertain to customer's policy, even a blabbermouth will hesitate before he replies. The noisy individual will not hesitate to speak for himself—but if you have tossed out a "policy question," he will not be so fast to answer for the entire group.

Often the fellow with the most rank is wise enough to know that his subordinates are better qualified to make certain types of decisions. This is often especially true of *buying decisions*. He will be slow to overrule a subordinate who has nominal responsibility for buying. If your customer's organization has been afflicted by the "Peter principle" (where men rise above the threshold of their competence), the ranking man may be the last man who will make the decision. (Parenthetically, we don't believe that the Peter principle is found too often in industrial or commercial organizations, so it is a mistake to assume that the ranking man is not well-informed.)

Another hazard of directing remarks or questions specifically to the ranking man present is that if he does delegate authority to buy to others in the group, they may well become antagonistic to you.

Here is an example of the overhead question:

"Gentlemen, what is it that you are *primarily* looking for in this purchase?"

This is a policy type of question. A purchasing agent *might* reply, "Value for the dollar." An engineer *might* say, "Quality." With more than one function present, the noisy chap will think twice before he replies for the group. This type of question, it will be noted, has another value...it may smoke out the key buying motive, thus enabling you to decide what features or values to emphasize.

D17—Use the rhetorical question

The rhetorical question requires no reply. It is spoken and is followed by an extremely short pause—just long enough to identify reactions. It is used to leave the impression that the point under discussion has been resolved. It can be utilized, as has been seen, as a probe for reaction.

The rhetorical question is of real value when the sale has bogged down on some minor point in which minor conflict has been created, and the salesman wants to get by this point.

Example: "I'm sure that we are in agreement that there will be no difficulty in handling this minor problem...(during installation, at the proper time, etc.)...(pause)...but I would like to discuss shipment requirements with you."

Example: "I'm sure that we are all in agreement that this (product) has so many advantages that this minor problem will be solved...(pause to identify reactions, or to wait for objections)...now, if I may, I'd like to show you an additional feature that has real significance."

D18—Use the provocative question

From the very nature of its name, this question is a "last resort" technique. The late Elmer Wheeler used to refer to "Dangerous Selling"...but *effective* selling when there is nothing to lose.

Recently, during a European Institute we conducted on professional salesmanship, we encountered an excellent example of dangerous selling where the provocative question was used. Space requirements do not permit a discussion of the entire selling situation involved, but the key points were these: The salesman was selling oil for heating a large dairy. The dairy owner was being accused by the government of having a high water content in his milk. At the same time, the dairyman experienced real difficulty with his heating plant. He called the oil company, and they passed the buck to the burner people. The burner people made their inspection and blamed the oil company.

By this time, the customer was enraged. The oil salesman made a check of the oil with the aid of his lab people and found water in the oil. He reported this to the customer who roared in his fury and brandished a letter in the air which he announced he was going to mail in five minutes, cancelling the oil contract and giving it to a competitor of the salesman. The salesman asked for a chance to pump out the tanks and put in fresh oil. The customer roared again that they could pump out his tanks all right, but it wasn't *their* oil that would go in. He ended with a shouted question, "How do you guys get water in your oil anyway?"

The salesman looked at him and calmly asked, "How do you get water in your milk?"

The customer stood transfixed, with the letter still in his outstretched hand, mouth wide open. Then with a roar of laughter, he tore up the letter and said, "Hurry up and change that oil."

No, this is not being recommended as a standard operating procedure when you are up against an irate customer. *This* salesman knew his customer and knew the situation. He felt he had nothing left to lose, and only this provocative type of question could make the customer see how unfair he was being. It happened to work.

Less dramatic but equally effective examples of the use of the provocative question are these:

"Do you mean to say that you can't see what you stand to lose by not buying now?"

"Are you saying you're not interested in more business?"

"You have a reputation for making top-quality products. Do you mean to say that you would jeopardize that reputation by using low-quality components just because they are cheap?"

"You have agreed that our service has been without fault. Do I understand that you would even think of taking a chance of losing that kind of service, in an operation of your size, because someone else has offered you a discount as small as the one you have mentioned?"

If you examine these questions, you will ascertain that in addition to their irritation factor they have a redeeming quality...they either hint strongly of customer gain or loss, or they reflect favorably in some way upon the customer or his reputation. With care, you can fashion questions that will be infinitely better than those contained in the examples. But this should be done before you confront the customer, before you enter the selling situation. Consider, in advance, the problems you may encounter and the courses of action you intend to pursue. Then, if they all fail, have your provocative question already framed, ready for use.

Before you consider using the provocative question, it is essential that you know your customer well (or that you know you have nothing to lose because of the strong stand he has taken). It is also imperative that you weigh the effect of this type of question on your long-range relationship with the customer. Use of this question may get today's order, but as the customer thinks it over, after the sale, what may be the effect on the lasting relationship you seek to build? It has to be recognized that use of the provocative question throws caution to the winds and risks the danger of causing the customer to lose face. The gains you have to offer must be extremely attractive, or he won't "take" this kind of question without becoming angry.

Make certain, too, that you do not use the provocative question because you can't think of anything else to do. It really *must* be the *last* resort.

Some additional types of questions

There are four others with which it will pay you to become acquainted:

The leading question
The alternative question
The ambiguous question
The re-directed question

An example of the *leading question* is any of the following:

"You are primarily interested in economy, isn't that so?"..."Isn't it fair to say that this feature will meet your production requirements?"..."Wouldn't this be a good system to use?"

Your question indicates the answer you hope to get. If it is successful, you can move on to your next point. If you receive a negative reply, it may smoke out the true nature of the customer's objection, especially if you follow up the negative reply with a factual question.

An example of the *alternative question* is: "What is your primary concern: production, quality or cost?" (Hopefully, the customer will not respond, "All three.") It does, however, force an answer, and it gives the customer alternatives. Generally, the alternative question is designed to give the customer a choice of a "Yes" or "No" answer. In any event, it supplies you with information you need.

The *ambiguous question* was also mentioned briefly in a previous chapter. Remember it is used primarily to slow down an overtalkative customer. While he takes time to unravel the question, to determine what it is you are asking him, you have time to plan your next tactic.

The *re-directed question* is especially helpful when your customer is represented by more than one individual present, although it can be used effectively with a single customer representative.

If more than one are present and the customer asks you a tough question, you can say, "Before I answer that question, let's get Bill's thoughts on that...what do you think, Bill?"

While Bill is groping for an answer, you have time to think. Furthermore, Bill may come up with a better answer than you can provide.

Even if you are dealing with a single customer who asks a tough question, you can use the re-directed question: "That's a good question, but before I answer it, I'd like to hear your thinking on it...what would you suggest?"

A classic example of this occurred when a manufacturing man present asked, "Yeah, what are we going to do about stock storage if we go along with this idea?" (The question was asked with a tone of belligerence because the salesman's proposal meant a problem for the chap from manufacturing.)

The salesman replied, "That is something that must be considered. Do you have any suggestions on how storage could be handled, Ed?" Whereupon the manufacturing man told exactly how storage facilities could be arranged.

Often people will raise questions of this type because they want to have a voice in the solution, or they feel their views have not been sought or simply because they want everyone present to know that the proposal is going to require their cooperation and require them to do more work. Using the re-directed question satisfies any of these needs.

Questions can be your most powerful selling tool. They get you off the spot, they produce information you need and they let you continue to control the selling situation. Study them and produce questions that will work best in your kind of selling.

D19—Scare him!

Scare him! This strategy has the overtones of manipulation. The professional salesman motivates —he does not manipulate. There is a difference, a marked difference. We think of manipulating as getting the other fellow to do something in *our* selfish interests. We think of motivating as getting the other chap to do something in our *mutual* interests. It is only under such conditions that this strategy, "Scare him!" should be used.

Let's look at it this way. Often, his objections (spoken or unspoken) are rooted in his fear that if he follows your advice, it will (1) cost him money, (2) make him look badly, (3) threaten his security—psychological or economic—or (4) put him to inconvenience; i.e., make life more difficult for him. In other words, "fear" is the basic cause of his resistance.

When a man resists our suggestions because of fear, we have two strategies available to us. We can show him tremendous benefits that will be so important to him that they will overcome his fear. Or, we can emphasize losses he will sustain if he makes the wrong decision. In other words, if his actions have been dictated by fear, we can pose a greater fear.

For example: Your customer is about to make a hasty decision. You know that he intends to buy from your competitor, and he wants to place the order at once. You also know that your competitor's salesman is on the way to see him. You want a "last look." You can use such statements as : "Joe, I know you'd like to get this decision off your mind before you leave for the weekend. But...Joe, if you make a decision before I can come in here Monday with what I am sure is a terrific price, this may be the most costly decision you have ever made."

In this case, your customer is going to think twice before he makes a speedy decision. The fear of losing out on a good price may cause him to delay his decision until Monday. If, on Monday, you have no better price than your competitor offered him on Friday, you may still lose the order, but you did gain time and a second opportunity to bid.

Many marketing strategies have been based on this principle. One company has tied up its market in a knot for months. They have been advertising a new product and making great claims for it. It is a product that is quite expensive, and a consumer doesn't replace it for years. As a result, many consumers have resisted the overtures of other salesmen week after week until they see this "super model" that Company X will be out with any day now. *Fear* of losing out on something good has caused them to wait.

Obviously, fear can also be used to cause a customer to come to a decision. If he has been procrastinating because he is undecided, the real cause of his delay is fear that he will not make the best decision. You have a choice: You can overwhelm him with proof, assurances, demonstrations, testimonials and guarantees. These may relieve him of his fear and cause him to take action. (This course of action should usually be taken before the decision is made to use the "Scare him!" tactic.)

When such strategies do not produce results, you then might turn to arousing fear. For example: "Joe, I know that you've heard this story many times, but I really am worried about a price increase, and I'd hate to see you delay and then find the price has gone up." Or: "Joe, you know that installation costs are a big factor, and installation costs soar once cold weather sets in, simply because outside work takes more time. If you delay in making the decision now, it will be far more costly later on. Why not save these extra costs by placing your order now?"

Consider your own product or service. What problems *could* be created by delay on the customer's part? How could these problems cost him money, deprive him of a chance to "look good," make life more difficult for him, or threaten his security in any way? How could you weave these threats into a sequence that would cause him to take action and abandon his tactic of delay? Be sure that your "scare" tactics are based on legitimate threats to him that he may have overlooked or discounted. Use this tactic sparingly so that it will be meaningful when you find it necessary to utilize it.

D20—Let him do you a favor

Reference was made to this strategy in Part Two. None of us like to feel obligated to another person. There are times when a salesman will make every effort to cause the customer to feel some obligation for an unusual service the salesman has provided. There are occasions, however, when the salesman may detect annoyance on the part of the customer when the latter feels obligated. When this condition arises, the salesman can use strategy D20, Let him do you a favor. The purpose of this strategy is to cause the customer to feel that the books are balanced in terms of who is obligated to whom. Once this is accomplished the salesman must move rapidly to emphasize the benefits of his product or service in order to give the customer a valid reason for buying the product being offered. The point is that all of us like to feel that we have freedom of choice and there are customers who will feel that they do not have that freedom when they feel obligated to buy solely because of some service previously rendered by the salesman.

The nature of the favor requested of the customer will vary widely from one type of selling to another. The salesman may simply ask for an introduction to another customer whom the customer of the moment knows well. If the customer is especially qualified technically, the salesman may ask for his advice on some problem the salesman has in another selling situation unrelated to the customer's business but within the scope of his expertise. Or the favor may be something as simple as asking the customer to recommend the saleswoman for membership in a professional organization. When you ask a favor of another individual you are also inhancing that person's self-image as a person of influence or you are recognizing his expertise.

CHAPTER 24

Overcoming the customer's fears

Some time ago we were asked why we did not list "fear" as an obstacle. The answer is that there are many kinds of fear, and they certainly cannot be given the same treatment. For example, there is the fear of "looking foolish," the fear of losing money, the fear of making an error, the fear of being manipulated, the fear of being used as a scapegoat or guinea pig. The list is a long one, and the first task of the salesman is to identify the kind of fear and to determine its cause. Only then can he decide which course of action will be most effective.

E1—Provide support

There are times when the individual to whom you must make your presentation must, in turn, sell to some other individual in her organization. In other words, the woman to whom you have access may not be the ultimate decision-maker in her company, but she is an influence on that decision-maker. Your contact may be quite willing to buy from you but she may entertain some qualms (fears) relative to her ability to justify the purchase to "higher-ups" in her organization. Your task is twofold. You must provide her with evidence that supports her decision to buy (or to recommend buying), and you must provide your contact with this evidence in a form that will enable her to transmit the information to those whom she must convince.

The situation may be further complicated if your contact does not want you to know that he lacks the last word in decision-making in his company. Therefore, your problem has at least three facets:

1. You cannot let him know that you know he is something less than a free agent.

2. You can't feed him information which he already possesses without having him feel you are "talking down to him." (if, for example, your customer is already familiar with certain technical advantages of your product, and you start to enumerate these, he will retort with an impatient or annoyed, "I *know* that!" Yet, somehow, you *must* cause him to keep these points in mind because he is going to need to bring them to the attention of the people he must sell in his own organization.)

3. You must decide what kinds of information are going to impress these ultimate buyers (they may not be the same points that will impress your contact).

Let's look at this example:

Your customer, Eric Jordan, is a plant maintenance engineer. The product you are trying to sell him will make his work easier, it will free him from some of the maintenance duties that have taken much of his time. It will almost eliminate the need for training his men...a need that has demanded much of his time in the past and has resulted in errors on the part of his men. This new product, in a word, is going to make life easier for Jordan and make him look good in the eyes of his superiors.

All of these advantages must be pointed out to your customer, Jordan. However, these same advantages which are so important to Jordan might be of no interest to the men who are going to make the buying decision—the men whom you cannot contact but who must be sold by your contact, Jordan. These people may be far more concerned with the price of your product, with whether or not it is self-liquidating, whether it will help them meet production schedules by eliminating breakdowns, etc.

Your contact, Jordan, wants to think that he makes the buying decision...yet you know that he cannot decide until he gets his O.K. from above. You can follow this course of action: Tell Jordan what advantages the product has for him. Leave him, in writing, the other advantages that will impress the people to whom Jordan has to sell your product. Undoubtedly, he will paraphrase these advantages in passing them along. You couldn't care less...just so long as he *does* pass these advantages on to the people who will find them of interest.

The benefits of this procedure are that you don't annoy Jordan with a long presentation, yet you have made available all of the facts to him. You have done so in a manner that will prevent him from forgetting any of them and without betraying to Jordan the fact that you know he is not the key decision-maker. You have also made it possible for Jordan to look good in the eyes of his superiors.

You may "provide support" in many ways. Your

contact may not care whether you know that he has to go "higher" for the purchasing decision. He may be very busy, or he may not have the talent for preparing a presentation which he must ultimately make to the fellows "upstairs" if he is going to get the O.K. to buy your product. In these cases, you can often provide support by preparing charts or demonstrations for him. You can obtain foolproof statistics and testimonials (which will impress the ultimate decision-makers), you can undertake research for him, and you can even write his "speech" for him.

Anticipate the arguments that the ultimate decision-makers will throw at your contact. Write them out and then supply the answers to each of these arguments. Make this list available to your contact and do so far enough in advance so that he will have time to become familiar with the arguments for and against your product.

If this material is to have maximum value for you, stop now and think of your product and your customers. What kind of support can you offer your customer? How can you help him justify his decision to buy? There is much loose talk in marketing circles about being "customer-oriented." We are customer-oriented only when we *want* to know the customer's problems and when we *want* to give him such support as we can in aiding him in the solution of those problems.

To be able to feel the customer's "hurts" and
to *want* to feel those hurts,
To be able to sense his needs, and to *want*
to fill those needs...
this is the root of true professional selling.

And that's a long cry from "peddling" and order-taking.

E2—Provide assurance or proof

Customers often *want* to buy, but they lack confidence that the purchase will really satisfy their needs. Or they worry that by buying your product or service, they will pass up a chance to buy superior products or services from some other source. Even if they are "sold" on what you sell, they may lack confidence in their own ability to select the best offering from the choice you may have made available to them.

This feeling may stem from a lack of technical knowledge, from fear that they lack all the facts, from their fear that they may be expecting too much from your product, from fear that you may not really understand their problem, or from fear that they are placing too much belief in your claims. The task facing you, as a salesman, is to provide assurance and to build the customer's confidence in his ability to make the right decision. Consider the wide choice of actions open to you (all of which will be covered in this or subsequent articles):

1. *Use the cause and effect method (C2)*
2. *Demonstrate (E3)*
3. *Use testimonials (C3)*
4. *Anticipate objections and answer them (E10)*

Don't be "checklist deep." In other words, analyze these four courses of action. "Demonstrate," for example, may extend from an actual demonstration of a product to showing the customer other installations of your product, or even to bringing him to the factory.

Keep in mind that "fear" is not solely psychological. It may be a very pragmatic economic fear of losing money. When such is the case, appeals to prestige and pride may fall flat. You are going to be forced to provide the customer with irrefutable proof that you are asking him to make a judicious investment. That is not to say that you cannot couple this approach with appeals to pride, prestige, and other emotional movers.

A customer, for instance, may balk when you attempt to convince him that he should anticipate his long-range needs and order in large quantities. He may *want* to do this but also may have a variety of anxieties. He may fear obsolescence (in some kinds of products); he may fear that the price will drop after he buys; or he may fear that he will experience a drop in his own business volume and end up in an over-stocked condition. Your task may be to confront him with evidence that other customers with conditions comparable to his (location, volume, clientele, space and so forth) have benefited by buying in the qualities you recommend. Here is an example:

"Mr. Barhydt, I know that you want to save the money that this quantity buying will make possible, and I can readily understand it if you wondered whether you were being asked to stock up with too much merchandise. You are familiar with Mackintosh's over in Dellwood, and you know the kind of business they do. Certainly, you have been every bit as successful as Mackintosh, and I am sure that you will agree that, in many respects, business conditions here and in Dellwood have much in common.

"Mackintosh put in this plan this year and it has saved them real money. They have been able to move the merchandise without difficulty. This product appeals to people in these communities, and the mass display made possible by ordering in these quantities really attracts attention and creates sales. You've been able to accomplish everything that Mackintosh has accomplished—and I know you will now. Incidentally, if you would prefer to get this evidence directly from Mackintosh, please check with them.

"Mr. Barhydt, my own success depends entirely upon your success, and I'm not about to recommend anything that isn't going to be good for both of us. Leaving you in an overstocked condition would understandably destroy your confidence in me—and I'm not about to lose that confidence. This is going to save you money and move merchandise for both of us... there's just no doubt about it!"

Now, let's analyze the foregoing: You started out by attacking the problem directly and by emphasizing the advantage that you have found is uppermost in his mind—saving (therefore, making) money.

You did not mention that he had "fears" (no one likes to be accused of "being afraid"). Rather, you said, "I can readily understand it if you *wondered...*" By this device you indicate that if you were in his position, you, too, might entertain some misgivings. This provides him with an escape hatch. In other words, he doesn't lose face with himself; i.e., you have just intimated that any "normal" person would have misgivings (fears that are nicely disguised). You have also caused the customer to see that you have empathy; i.e., that you are looking at the sale from his standpoint.

You get the objection out into the open with the statement, "...you wonder whether you are being asked to stock up with too much merchandise." If this is not an objection, he will probably say so.

You appeal to pride with, "You have been every bit as successful as Mackintosh." You also do *not* overdo it with, "You've been more successful than Mackintosh." You don't want it said later (to Mackintosh) that you made such a comparison. You indicate your own confidence with phrases such as, "I'm sure you'll agree"...."There's just no doubt about it." Confidence begets confidence.

"Doing business here and in Dellwood have much in common." Here, you avoid the possible argument or retort that might come your way if you had said, "Doing business here and in Dellwood are the same." He must agree that there are some things that the two business communities have in common, and you made sure that you did not select for comparison a community that was vastly different.

"Mackintosh put the plan in *this year.*" You selected another client who had recent or current experience. This made the comparison timely yet not so recent that the customer would feel the value or weakness of the plan would not yet have been demonstrated. This means that in your pre-sale preparation you must select your "comparison client" carefully and honestly. Then you again hit his area of key interest with, "...it has saved them real money!" You again identify the key problem in his mind and point out that experience in a comparable organization indicates that it is really *not* a problem with, "They have been able to move the merchandise without difficulty."

"Mass display"...you show *how* the problem was avoided or overcome.

"You've been able to accomplish everything that Mackintosh has accomplished—and I know you will now." Again, the appeal to pride is coupled with the definite "I *know*" that serves as evidence of your own confidence. Your suggestion that he contact Mackintosh directly provides your customer with the assurance that you believe Mackintosh's experience will bear you out. This increases his own confidence.

Perhaps the next point is your strongest sales prescription in the foregoing example. You stress that you are no more anxious to lose money than he is and that your own future success is completely dependent upon his ability to make money on your proposal. Isn't this the same argument you and I give ourselves when we find ourselves in a plane in rough weather? We keep reminding ourselves that the pilot wants to get down and see his family just as much as we do...and so does the stewardess. Just look at her—serene, confident and untroubled. The result is that we settle back and enjoy at least a greater degree of freedom from concern. Their attitude gives us assurance because we see there is a communion of interest.

See ways of getting across to the customer that it is in your selfish interests to make sure he is given the best advice. Even if you are selling "a one-shot deal," keep in mind—and in his mind—that his reaction can affect your reputation!

E3—Demonstrate

The demonstration can be one of the most effective selling tools if it is used with an understanding of its values and disadvantages. Above all, check out the demonstration before you use it. Recently, we had an experience, personally—and heard of another from the ill-fated salesman—that bears out the importance of this warning. We were considering the purchase of a rather expensive closed-circuit television system for our Professional Sales Situation Management Institute. Two salesmen called upon us. They were enthusiastic over their equipment, but when they demonstrated it a series of black bars followed one another in a procession across the face of the TV tube. After numerous attempts to eliminate this disconcerting feature, we were informed that this was due to the need for a "simple" adjustment, and, "of course, it would not be present" in the equipment we would be supplied. Our unspoken reaction was "maybe so" (and "maybe not"), and "if it's so simple, why don't they fix it?" Result: no sale.

In another instance, a salesman attempting to sell a client on a time-sharing computer plan brought his console into the customer's office. The customer had approved the appointment and had his entire staff on hand. For certain technical reasons, the console would not function on the customer's equipment. The customer smiled and said he "knew how it was with demonstration models" and agreed to have a second showing. The second showing was made, but the customer did *not* bring his staff, and again the equipment failed to work. The salesman tells us that the defect is now remedied and although the customer agreed to have a third demonstration, for some reason or other he had been constantly too busy to find the time.

The demonstration will be found to be effective under the following conditions:

1. When the customer cannot be made fully aware of the advantages of your product solely by the spoken or written word.

2. When you know that the impact will be pronounced if the customer sees the product in operation.

3. When you feel that she does not completely accept your claims. (When proof is needed.)

4. When you want to have the customer become involved—to participate.

5. When the operation of your product is complex, and his technical knowledge is limited. The demonstration provides you with an opportunity to build up your presentation, point-by-point at a speed he can absorb, and provides him with an opportunity to ask questions.

6. When a comparison of your product and that of your competitor is possible. This enables you to avoid "knocking" competition...the customer can see for herself.

7. When you desire an opportunity to bring some of his people "in on the act." This presupposes that you are certain they will be "sold" (because it has real advantages to them) and they, in turn, will help sell their boss—your customer.

8. When your customer has been attracted to your product by an advertisement and wants to *see* if it lives up to the claims made for it.

9. When you want the opportunity to bring in some of your own people who may be more technically knowledgeable than you are.

10. When your product is so revolutionary that one must see it to believe it.

11. When you suspect that you do not have the full confidence of your customer because of your age, relative inexperience (compared to his), and you need a device that will, on its own, earn his confidence.

12. When your product worries the customer because, without seeing it, he believes it will be too complicated for his people to operate or maintain.

All of this assumes that your products or services are such that they lend themselves to demonstration. You may sell products that cannot be transported because they are too bulky, heavy or cannot be demonstrated because they are built on order. Demonstration, however, is still possible:

You may be able to bring him to the factory or location where similar products are in operation.

You may be able to bring him to the factory or display room.

You may be able to utilize models, sketches, films, visuals or mock-ups with moving parts made of cardboard or similar materials.

Considerations to keep in mind

1. Be sure the customer is "positioned" where he can see what you are demonstrating. This sounds rather basic, yet salesmen often fail to take the time to make certain the customer's view is unobstructed. They will refer to a product feature without being sure the customer is in a position to appreciate the feature. For instance, a salesman may put his finger in an aperture and state that the highly polished finish will make it difficult for rust or corrosion to start. The salesman may never think of asking the customer to place his finger in the same aperture and "feel" the finish. (Appeal to as many of the customer's senses as possible.) When the salesman fails in this respect, the customer can only accept the word of the salesman that there really is a highly polished finish, and the full value of the demonstration is lost.

2. If you are demonstrating a device that has moving parts, make sure you have checked its operation under the conditions that will prevail during the demonstration. If you are using a projector, have a spare projection lamp with you. If your product requires an external electric supply, make certain in advance that the appropriate voltage and current supplies are available. Don't have your demonstration fall on its face because of performance failures that could have been caught in advance.

3. If your demonstration requires that you use other people to operate the equipment, be sure they have been instructed in advance to let you do the talking. Also, if questioned, make certain that they pause long enough before answering the customer to permit you to supply the answer if you prefer to do so. These people are well-intentioned and seek to be helpful, but the comments they can make, without an understanding of the customer as a unique being, can retard or lose a sale.

4. Build up to the demonstration. Rely on the old adage, "Tell 'em what you're going to tell 'em, tell 'em, and tell 'em what you told 'em!" In other words, precondition the customer to look for the things you wish her to see. After the demonstration, summarize the key points you wish her to keep in mind. Never, but never, show a film, without following this prescription.

5. Gear your presentation to his threshold of knowledge. Don't bore him or insult his intelligence with details of which he is already aware. Avoid creating the impression that he is being subjected to a canned performance and that come hell or high water you are going to cover every point. No matter how often you have made the demonstration, constantly make meaningful remarks that tie in the demonstration to what he is certain to consider his "unique" problem. ("Nobody knows the troubles I've seen" are more than words of an old spiritual . . . they are a good description of most of us when we are customers.)

6. Try to make the demonstration appear as spontaneous as possible. Watch for opportunities to respond to a question with such statements as, "Here—let me show you!" Then go into your demonstration. Customers, even as you and I, wish to be regarded as unique individuals (which they are), and they want to be treated in a unique manner. They will respond more favorably to a demonstration if they feel it was designed especially, if not solely, for them.

7. Get reactions. Don't lose the customer because you are so engrossed in the demonstration that you forget you are in the middle of a selling situation.

8. Get the customer in on the act when possible. Let him operate the device. Find ways of involving him personally in the demonstration. Avoid giving him roles to play, however, that may embarrass him or cause him to feel that your equipment is too complicated. Don't overlook the fact that many key executives are all thumbs when it comes to using their hands; and if they find a product difficult to operate, they may forget that it will be easy for the men and/or women who will actually operate it.

9. Capitalize on the demonstration. Emphasize the advantages that the demonstration has revealed. Get the customer to agree that the advantages are there. Then move in for a close and don't forget to ask for the order.

10. Finally, don't just "hope" that the demonstration was significant to him. Make sure it is. Help him bridge the gap between the demonstration and his own problems. Hit hard on the advantages, gains and benefits to him. Talk about the solution of his problems and the fulfillment of his needs.

Don't concentrate on the product, concentrate on what it will do for him. For example: *don't* talk about "how quietly she runs" without converting this into a benefit. *Do* point out how the quiet motor means lower noise level, less employee fatigue, more production, etc. Obviously, in this example, we may have listed benefits that apply to all motor sales. If you were selling motors, and the fact was that a quiet motor would not be an advantage to your customer, you obviously would not single out this feature. You would not dilute the balance of your demonstration by accenting something of no interest to this specific customer. (It might well be a point you would hit hard with another customer.) Concentrate on extracting from the demonstration the points that do mean a gain for your customer.

To many people regard salesmen as fast talkers —something that simply is not true. But why not let a demonstration do your talking for you? It cannot help but be more effective because it can certainly appeal to more of the customer's senses, and therefore has a greater impact than the spoken word alone.

E4—Explore cost of wrong decision

We have treated this strategy, to some degree, in previous elements of this book (*D4—Provide alternatives*), but the following thoughts will prove to be of additional value.

When a customer is going to act precipitously, there are two methods that can be employed to cause him to "slow down." One is to inject what amounts to a threat, and the other is to show him that greater benefits can be obtained by waiting. The word "threat" that we have employed can leave the wrong impression unless we explain what is meant. By threat, we mean that the customer must be made to see that he will, in some way, suffer losses if he acts too quickly. This can vary all the way from economic losses to "losses" represented by buying from another supplier before you come out with a new product (soon to be available) that will give him greater satisfaction. The second course of action, obviously, is to draw comparisons between the benefits he will gain by buying from you versus the benefits he will gain by buying from a competitor.

A third course of action, which really embraces both of those we have discussed, is to make the customer consider the various alternatives open to him, so that he will see for himself the comparative losses and benefits of the various courses of action. This technique gives him the impression that you are reasoning your way through these other avenues with him, rather than pointing with alarm at losses he will sustain or exaggerating the benefits he will gain by buying from you. For example:

"Mr. Taylor, let's take just a minute or two to think our way through some of the alternatives that are available to you and see what is involved." Then, without pausing, select the most vulnerable of the alternatives and lead him into a discussion of its disadvantages (as compared to the advantages available through the purchase of your product).

If he has given evidence that he favors a course of action that is against your interests, it will be wise to agree that his preference does have advantages. For example, let's assume that you are selling transistors and he wants to buy a specific transistor from your competitor...a transistor that he favors because it will provide considerable power. Your transistor provides less power but furnishes a much better quality output. Here is what you might say:

"Mr. Anderson, I agree that use of the transistor you have in mind will provide you with plenty of power. Your company, however, has an enviable reputation for top quality, and I think that before you make a decision, we should explore some of the disadvantages involved in the use of the type of transistor you have in mind...."

In this approach, you have not antagonized him. You have found a point upon which you can agree with him. You have appealed to his ego. By doing these things, you have opened his mind so that he will really consider some of the drawbacks of the product that he may not have considered before.

Another example: Your customer is stalling on reordering either because he thinks he can squeeze by on what he has, or he dislikes giving up more space for storage, or he doesn't want money tied up in a larger inventory. Whatever his reasons are, they have validity to him. If you are going to get him to consider alternatives, you must first open his mind. This is accomplished by finding a point of agreement. For instance:

"Mr. Allan, I certainly agree with your views on inventory control...the success of your business is evidence of the approach you have used for this problem. I think much of your success has been due to the wisdom of your policy of having both depth and breadth in your line and maintaining minimum stocks. It's solely for that reason that I would like to explore with you some of the hazards that may be involved if you were to postpone replenishing your stocks at this time...."

Notice that in each example the use of the "Yes, but..." approach has been avoided. In essence, the "Yes, *and*..." approach has been used.

E5—Remove him from the situation with third party examples

You will encounter customers who fear to make decisions or fear to reverse their previous decisions because they do not want to lose face. They feel that the image they are trying to project will, in some way, be adversely affected. To persuade this customer, you must "remove her from the situation." In other words, you must cause her to see that personal considerations are not involved. You must cause her to see that the situation she faces is merely one of making a buying decision to solve an operating problem (or fill a product need).

There are a number of ways in which this can be accomplished. One is to present your solution as the "standard solution" for this problem and one that is universally used by people of good judgment. Another technique is to focus his attention on the operating problem (or his product need) and emphasize this to such a degree that he "forgets himself."

When it is not possible to use either of these solutions, use the third party method. You select an absent third party, whom you know the customer respects, and you use this type of argument: "As you probably know, Jim Barnett had this same operating problem that you have. He had the same understandable reservations that you have, too. Jim changed his mind, however, when he came into possession of certain facts of which he had been unaware. Here, let me show you what changed Jim's mind...."

By this device you have, in effect, taken the customer out of the situation psychologically. You have concentrated his attention upon someone else who has had the same problem he has had and has solved it with your product. It should go without saying that you make certain that your customer holds the individual whom you have used as an example in high regard. You have also emphasized that this third party previously held the same reservations (fears) that are held by your customer.

Your customer considers that he has good judgment and he does not like to have it challenged. By using the third party example, and by using as that example a person known to have good judgment, you have protected your customer's self-image. You then point out that the third party changed his mind only after he came into possession of new facts. Your customer now is curious to know what those new facts are. You have now completely removed him from the situation (in a psychological sense) and have concentrated his attention upon his product problem.

If your customer is determined to take an action that you know is "wrong" for him, you can also use the third person method by pointing out the losses that this individual suffered by taking the action the customer has in mind.

E6—Appeal to his pride

Appeals to pride should not be used lightly. They should be employed only when the customer's pride can be aroused or when you know that his key motive for buying is to satisfy his pride. It is a strategy that should be used with care because it smacks of flattery. If it is seen by the customer as flattery, it can antagonize him. Flattery indicates to the person we flatter that we find him so transparent that we know what will motivate him.

It is always wise to "praise the work and not the worker." In other words, compliment something he has done, something he stands for, or commend his organization. "Beware of Greeks bearing gifts" epitomizes the reaction of the customer who senses that he is being "buttered up." If you have made a sales presentation that has been based solidly on product features and customer gains, and you sense that for some reason the customer has a fear of buying, you can often take his mind off that fear by arousing his pride, if you do so subtly. For example:

"Your products have a reputation for quality, and using these components is going to enhance that reputation." Or "With the reputation your company has for quality, I'm certain you won't want to use any components that will not measure up to the standard of your finished product."

To make this strategy meaningful, it is suggested that the reader consider the product she sells, select a specific customer, and then consider what words she could use to appeal to that customer's pride.

E7—Appeal to his prestige

There is a fine line between pride and prestige. We might think of pride as the desire the customer has to maintain self-respect. Let us think of prestige (as used here) to indicate the desire the customer has to appear favorably in the eyes of others—to impress them, to be admired, to maintain status. For example:

"I noticed, Mr. Simone, that you were on the program of the Regional Conference again. Over the years, you have established a reputation for progressive methods and for being sensitive to new trends. That is why I am certain you will want to be one of the first to cash in on the market that is represented by the migration that is taking place in the area you serve . . . these new people are going to be looking for this product on your shelves."

As in the case of appeals to pride, there are factors that must be considered in using appeals to prestige:

1. When the sale can be made without this appeal, do not use it. There is always the danger that the customer will see it as flattery.

2. There must be the basis for this appeal—and if it is used, it must be used sincerely. However, this appeal to prestige can be employed even if the customer does not enjoy prestige, *if* he *thinks* he has, or if you know that he wants prestige. You are then addressing the "me he wishes he were" or the image he is trying to project.

3. As in the case appealing to pride, you must be convinced in your own mind that your customer really does *need* that which you are asking him to buy. In other words, taking advantage of a customer's vanity in order to make a sale is not sound, long-range selling, and it is not customer-oriented. In the long run, it will cost the salesman a customer if he capitalizes upon this human weakness to make quick sales. It is this sort of thing that has brought selling a disrepute that professional salesmen find is one of the greatest obstacles they must overcome.

E8—Appeal to the me he wants to project

Probably every word and act of ours that we plan in advance has its roots in our self-image. As a matter of fact, even the actions that we take impulsively are usually dictated by the needs of our self-image. As we have seen, our self-image gives our behavior a pattern—a "continuity of personality"—or as the police would put it, "an MO" (modus operandi).

"Our kind of person" will take predictable actions in a given set of circumstances. The individual who has studied us will be able to predict that behavior with an amazing degree of accuracy. He will also be able to determine what approaches to use to cause us to act as he wants us to act. He knows that we will not do anything that is going to threaten the attainment of "the me we are trying to project," and he knows we will embrace any course of action that will enhance that image.

So it is with your customer. If you have really studied your customer in advance and have allowed him to do most of the talking, you have probably been able to see a pattern in his behavior that reveals how he wants others to see him and how he regards himself. His actions will be based on his desire to project that image. Therefore, you can determine what product advantages to emphasize and what selling arguments will be most effective with this unique customer.

Buying motives have their genesis in the customer's operating or product needs, but the customer usually has some flexibility in deciding from whom to buy, or which of the many products available will best satisfy his product needs. He will also have "psychic needs" (need to look smart, need to dominate, need to be secure, etc.). Once he knows that a number of products will meet his product needs, all other things being equal, he will buy the one that gives him maximum psychic satisfaction. Keep in mind that it may not be the *product* that supplies these psychic needs... it may be the salesman who furnishes them, by making the customer feel important or secure, or however the salesman has decided this unique customer wants to feel.

One proof of this is found in the fact that you, in your capacity of a customer, have encountered certain people from whom you will buy and other saleswomen whom you prefer to avoid. Ask yourself why. You can be sure that if their products are comparable, the difference in your attitude is found in the attitude of the saleswoman toward you...how she recognizes you.

While it is true that we generally prefer a salesman who knows his product—and we certainly avoid the chap who is not informed—it is also true that the well-informed salesman can lose many customers simply because he makes his customers feel ignorant by comparison, or he makes them feel inadequate or uncomfortable. The well-informed salesman has to know how to convey his product knowledge without threatening the self-image of the customer. The principle is to keep aware of the fact that selling is not only satisfying product needs, it is also recognizing and satisfying the unique psychological needs of this customer. (Review Part One of this book, which covers, in detail, the Psychological Aspects of Selling.)

E9—Give him recognition

Too many saleswomen are not satisfied with getting an order, they also want credit for having solved the customer's problem. The result is that they lose out on both counts.

Whenever possible, make information available to your customer in such a subtle manner that he uses it to solve his own problem. You may be sure that subconsciously he knows that without the information you made available to him he would not have solved the problem. Not necessarily out of gratitude (which is a rare commodity), but out of his own enlightened self-interest, he will continue to buy from you because again, subconsciously, he knows he needs you.

Salesmen have been known to make another major mistake that costs sales. Example: The salesman has just solved the customer's problem. Then the customer's boss enters the picture, and the salesman—wanting to impress someone "higher-up"—lets the boss know that he, the salesman, came up with the solution. Is there really any need to discuss subsequent developments? Recently a vice president of purchasing very bluntly and honestly told us that what he looks for in a salesman is *"the fellow that will make me look important upstairs."* The obvious antidote is to grasp every opportunity to make your customer look good by passing the credit on to him.

Find ways of including your customer in the solution. If your proposal is to be in writing, use such phrases as: "As you pointed out"..."As you suggested"..."In keeping with the observation you made"..."As you emphasized." Accent his "you" and play down your "I." Identify him as much as possible with the solution you have developed. To the extent that it is possible, incorporate his ideas into your solution. This act on your part will speak more convincingly than all of your words.

E10—Anticipate, raise and answer objections

If you know that your customer is going to raise an objection, don't assume that by ignoring it, it will go away by itself. If you wait until he raises the objection and then supply him with an answer, he may feel that you have given him an off-the-cuff reply and that you don't really consider his objection very important. On the other hand, if *you* raise the objection before he does, and if you then answer it, the following advantages may result:

1. The customer feels you have thought her

Another reason for the customer's fast reaction to an exaggeration of his objection is found in the basic desire that most people have to be "fair." He does not want to be placed in the position of making unfair criticisms of your product—and wide, sweeping generalities have a tendency to be "unfair." Example:

Customer: "I have no objection to your product; it's just that I'm a small businessman, and you people are big. I think I'd be considered insignificant compared to your big customers, and any problems I would have wouldn't receive attention."

Salesman: "As I understand it, Mr. Smedley, you feel we wouldn't be interested in your business?"

Customer: "No, no, I didn't say that! It's just that service would be pretty important to me. I can't afford delays due to shutdowns when equipment problems arise. Big outfits can probably take that sort of thing in stride. I'd have to be sure I would get fast service."

Salesman: "Mr. Smedley, I'm happy that this is the concern you have. The fact that I am here shows you we want your business. We know that to *keep* your business we must give you prompt, efficient service. I can readily understand your desire to avoid lost time and I'd like to explain our service arrangements. As you know, we have thousands of customers, and there just aren't that many big customers. We have been successful because the solid foundation upon which our company has been built and continues to grow is the business that comes to us from businessmen who are not what you or I would call 'big'. Let me show you how we service those businesses." (Then, explain your service procedures.)

Here the customer has been made to back-track in his original, sweeping objection. The salesman asked the customer if he thought they were not interested in his business. The customer knows you *are* interested or you wouldn't be calling upon him. Therefore, he suspects some kind of trap is being set, and he hastens to modify his original objection. Now, instead of concentrating on *size*, he focuses on his need for fast service, his need for protection against lost time. This is something to which the salesman can readily respond. It also spotlights the customer's real problem and shows the salesman the direction that his presentation should take. He will concentrate on service and will not waste valuable time on aspects of his proposal that are not of key concern to his customer.

D12—Get him to talk

In the earlier phase of this book, we discussed the Anatomy of a Sale, and at that time the strategy *"Get him to talk"* was described in detail. A lengthy expansion here would be redundant, other than to examine the reasons *why* this strategy is effective. These include:

1. Given free rein, people will talk about things of interest to them. If you do not steer the conversation, your customer will reveal many of her aspirations, problems and opinions—and these provide you with valuable information.

2. A customer may give you a "smoke screen objection" to hide his real objections (which may be socially unacceptable; i.e., may not make him look good). If you can get him talking and keep him talking, the chances are that he will reveal his real objections.

3. People with problems like listeners. Give your customer an opportunity, and he will unravel his problems before your eyes. Keep in mind that what he really wants is a good listener. Frequently customers do not want advice even when they seem to be asking for it. They usually want someone else to confirm the wisdom of the course of action they already have contemplated. Therefore, if your customer has dwelled on some problem that is *not* connected with your product or service, and he finally asks you, "What would you do if you were in my place?" avoid the tendency to play Solomon. Here are the actions you can take, *safely:*

1. "Well, Bill, what are some of the actions you have considered taking?" Don't be surprised to have him respond, with some spirit, "I'll tell you what I'm going to do... here's what I'm going to do... etc." He may finish with, "What do you think about that?" In this event, your best reply may be "Bill, that's certainly *one* action that's available to you."

2. "We might explore some possible actions, Bill. What is the first course of action that occurs to you?"

3. "Perhaps, Bill, to answer your question, we could weigh the advantages and disadvantages of each action you have contemplated. What's the first action we should consider?" You are probably giving him maximum help when you aid him in approaching his problem in this scientific manner. If you feel that the action he seems determined to take is going to add to his troubles (and affect your business opportunities with him), the most you should do is to try to steer him toward method 3. This technique will usually help him discover the inadvisability of the proposed action.

Above all, getting him to talk presents you with the most effective way of establishing rapport. Being a good listener is still one of the highest forms of flattery. It goes without saying (or should) that when you get him talking, you listen—*really* listen.

the busiest individuals I know").

3. You have transferred his attention to something about which you know he feels strongly...the demands upon his time. In doing so you have employed questions: "Have you considered how valuable your time is?"...."How much of it will be made available to you?"

4. You have employed one of the most powerful tools in the Sales Strategy Bank ("You deserve it... you've earned it"). Busy men usually are afflicted with consciences that drive them relentlessly. They often ache for a few hours' respite but feel guilty when they are not working. They grasp avidly for the opportunity to relax without feelings of guilt that is made available to them through the comforting phrase, "You deserve it...you've earned it." This phrase is a powerful technique for diverting a customer's attention from the things his conscience is demanding.

5. Finally, you have forced the customer to consider the avenue you wish to explore when you have ended with a direct question ("Jerry, how much of your time is now spent on this activity?").

In summary, the formula that will often divert the customer is: (a) Find a point of agreement; (b) focus his attention on something that is basically of greater interest and importance to him than the subject he has raised; and (c) force him to consider this new subject by raising a direct question that is sufficiently complex to cause him to ponder his reply. The more the question causes him to stop and consider, the more likely it is that he will forget his pet peeve. The question employed in the example above ("...how much of your time is spent on this activity?") can't be answered without thought. And consideration of this question leaves no room in the customer's mind for other topics which you prefer to avoid.

D11—Rephrase and overstate his objections

Once your customer has raised an objection, your task is either to eliminate the objection if it is not valid; or, if it *is* valid, to minimize its importance. One technique for accomplishing the latter is to restate her objection and, in so doing, to *overstate* it. The customer's normal reaction will be to deny authorship of the objection as you have stated it. Then, in restating her objection, she will usually tone it down a bit to emphasize that you have exaggerated what she said. You now have something with which you can cope much more easily.

Often, you and I—and the customer—in voicing complaints, *do* tend to exaggerate or to make broad, generalized charges. These are difficult for a salesman to handle. He needs something specific. In restating our objection and overstating it, he forces us to deny that we made such an all-encompassing charge. In restating the complaint, we tend to become more specific...and this is just what the salesman needs.

Here is an example illustrating how to use this strategy:

Customer: "Your product would suit me just fine except for one thing. We can't always control the size of raw materials that our vendors supply us. So, we have to have equipment that provides safety against overload, and I don't think that your equipment makes sufficient provisions for overload."

Salesman: "I can see why provision for overload is important to you in this type of application, and I agree, too, that this should be a key consideration in your buying decision. Now, as I understand it, you feel that we have not factored in *any* provision for this kind of protection in our product...Is that correct, sir?"

Customer: "Well...no, I don't say *that*. I'm just saying that before I buy, I'm going to be sure that the equipment I select makes sufficient provision for overload...in the case of our application, say, about 10%. And I felt your product might not provide that much protection."

Salesman: "Oh, I see. Well, our problem is easily resolved, Mr. Roberts...it's just a matter of seeing what 'enough protection' is in the case of your application. Let me show you how we have provided for overload in this design so that 10% or even 15% overload will never present you with a problem."

By this device, the salesman has reduced the severity of the customer's objection. Originally, the customer intimated that he did not feel the salesman's product provided *any* protection against overload. The salesman overstated the objection, and the customer hastened to rephrase his objection by reducing it to a "feeling" that there might not be *enough* protection against overload. This is considerably easier to handle, especially when the salesman has been successful in having a specific overload percentage stated by the customer.

It will be noted that this device also provides the salesman with an opportunity to "fence"—to obtain a bit of time to consider the best way to handle the objection.

There are a number of reasons why overstating the objection works. It sometimes causes the customer to feel that you are, perhaps, trying to "trap" him by attributing statements to him that he cannot substantiate. He, therefore, hastens to reduce the scope of his objection so that he will be in a position to defend his charge. In doing so, he moderates the severity of his objection and makes it easier for you to handle it.

directing the course of the selling situation, and he will be in no position to talk about the things he wants to talk about! This would seem to be in direct conflict with the advice that has been repeated over and over in this book; i.e., "Let *him* talk, and you listen." There are times, however, when it will be necessary to have the customer do the listening. And when he seems determined to embark upon a discussion which impedes or threatens the progress of the sale, you must employ techniques that will compel his attention.

Once again we will see the wisdom of pre-sale preparation. If you do an effective job of preparing for the sale, you may identify some of the pet peeves which the customer has, or you may be able to predict that he will want to discuss poor experiences he has had with your company or your product. Before the sale, with emotional conditions absent, you will be in a much better position to plan the type of questions you will employ to divert his attention away from these critical areas. You should not wait until you are actually *in* the selling situation before you plan your strategy.

Being realistic, we must agree that there *will* be times when you cannot predict what avenues the customer will seek to explore. It is at times like these, when he "strays from the course you have plotted," that you should employ the use of questions in order to divert his attention away from those nonproductive areas and back to the points you are seeking to establish. *Questions* are your most effective selling techniques. A question always puts the listener "on the spot" and enables you to control the situation for the moment. At least, while the person being questioned ponders his reply, you have been successful in directing his attention to an area other than the one he would prefer to examine.

If a customer is truly angry about some condition, don't think that by simply asking a question you are going to be able to divert his attention. You may find it wise to couple your query with another strategy...such as, find a point of agreement, or show respect for his arguments...and *then* pose the question. For example (in speaking to the customer who has raised a point which you don't wish to discuss at this juncture of the selling situation): "Bill, you certainly have a point there, and I don't think it should be ignored. If you don't mind, though, I'd like to hold it for discussion in a few minutes. Right now, I'm wondering about another problem that you have—and one upon which I think I can give you some help. Bill, am I right in assuming that your company is having some difficulty with..." (Here, you raise some problem that your pre-sale preparation has shown you he has, and for which you can offer him some relief.) In other words, divert his attention from the subject you wish to avoid to a subject upon which you feel you can take positive action.

There are times when diverting the customer calls for frightening him. We recall such a situation. A customer had failed to follow certain installation instructions that had been given him by the salesman. As a result, he had difficulties and called the salesman in a fury. He immediately denied that he had even been instructed regarding the installation procedures. The salesman knew that a letter containing such instructions was in the customer's file, but he also knew that proving the customer to be wrong would avail the salesman nothing. He realized that the customer, too, knew that he was wrong and that he was seeking an escape hatch by blaming the salesman.

Instead of discussing who was right and who was wrong, the salesman diverted the customer's attention to a fear that was greater than the fear of having been wrong. He said, "Look, Ed, the point is that right now you are in trouble with your boss, and I know how to get you out of trouble on this installation." The customer made a speedy about-face and asked anxiously, *"How?"*

The principle is that, when possible, if your customer is concerned about one fear (and a fear about which you can do nothing), focus his attention on a greater cause for anxiety about which you *can* do something.

Keep in mind, too, the point that has been repeated here so often. Your customer falls into one or more of these categories: He wants to be rich, or he wants to be famous; he wants to protect his security, or he wants life to be easier. You should know him well enough to gauge which of these is paramount in his mind at this time. If he is angry because he sees one of these needs threatened, you must employ questions or statements that will direct his attention to one of the other areas and make it, for the moment, seem more desirable than the need that has until now been compelling his attention.

For example, your customer may be objecting to your "high price"...in other words you are threatening, to a degree, his "desire to be rich." The facts of the situation may enable you to divert his attention with remarks such as these: "Jerry, I can appreciate your feelings about the price, but I also am aware, Jerry, that you are one of the busiest individuals I know, with great demands upon your time. Have you considered how valuable your time is, Jerry, and how much of it will be made available to you, if you use our service? Jerry, you *deserve* a little relaxation ...you've earned it...and life will be a lot easier for you if you employ this service. Jerry, how much of your time is now spent on this activity?"

In this simple statement, you have accomplished many things:

1. You have dignified his complaint ("I can appreciate your feelings about the price").

2. You have appealed to his ego ("You are one of

but actually disdains the other?

How do you avoid the problem?

Here is one method you might employ:

Don't wait until you are about to use the name of the third person as a testimonial, or you may find yourself in an uncomfortable position. Example: "Do you know Al Simpson of Moravia Brass?" The customer replies, "Yeah, he's a real screwball... why do you ask?" In this instance, you are left with the lame reply, "Oh, nothing... his name came up in a conversation the other day, and I just wondered if you knew him." All that can be said for this approach is that it leaves you in no doubt about the advisability of forgetting Al Simpson as a testimonial with *this* customer.

Naturally, the best bet is to know in advance how your customer feels about various people whom you might wish to use as testimonials. The opportunity is often presented to ask such questions as, "Whom do you regard as the leaders in your field?" If he mentions Al Simpson, you can be nearly certain that the use of Al Simpson as a testimonial will be in order. If he does not mention Al Simpson, you can always say, "There are two or three others in the field, as I recall. There's Bevins, Simpson and Sandres. Do you know them?" Then listen to his reaction.

Don't use Simpson's name immediately after you have learned that your customer regards him well; your reason for asking will be transparent to the customer. Wait until the appropriate time in the sale and then say, "Oh, Al Simpson's name came up earlier in our conversation. Did you know that he is using our service?"

4. The most common need for the probe for reaction is when the professional saleswoman wishes to check the progress she has made in the selling situation. The rhetorical question often proves to be the most effective in such instances. This is because it does not necessarily require an answer, yet it provides you with an opportunity to look for quick reactions. For example:

"I'm sure, Mr. Eldridge, that the time-saving features of this installation are attractive to you." Then you pause only momentarily to watch for a reaction. You may get a negative response—and if such is the case, the probe has done its job. You know that you should not proceed further until you have overcome any disagreement or misunderstanding that exists at the moment. If you proceed, he may not hear a word you say... he will still be thinking about those portions of your presentation that left him dissatisfied. If, however, he merely shakes his head, or seems to agree with you by his attitude, you can proceed with your presentation.

The probe for reaction can be made to do double duty. It can provide you with a clue to his reaction, as we have seen, but it can also help you drive home a selling point. This is accomplished by a thought-stimulating type of query, such as the following:

"You can see the quality that has been built into this product, Mr. Emery. Can you really afford to push your present equipment to get the last mile out of it? I'm sure you see that there is greater economy in replacing it now before breakdowns cost lost time..."

Or: "We all share the understandable tendency to hold off on making capital investments until the last minute... and often it is wise to do so... but I am sure you agree that installing this equipment now is going to save you costly breakdowns... hence less of an expense than if we wait until later in the year. Don't you agree, Mr. Bond?"

In the foregoing example, we have welded together a number of strategies from the Sales Strategy Bank:

1. "We all share the understandable tendency" —this phrase provides him with an escape hatch (you have suggested that he has a tendency to hold off making wise investments, but you "share the guilt" by pointing out that you do also). This must be used with care with the customer who regards himself as above other people and above you. He won't relish your putting yourself on the same plane he feels he occupies. In dealing with such individuals, you can say, "Many key executives hesitate to make capital investments... they share an understandable tendency... to put off making capital investments," etc. Again, if he is *not* a key executive, substitute some other term he will look upon as favorably such as, "many men in your position."

2. The statements provided in the example above also use other techniques which apear in the Sales Strategy Bank; i.e., *(F4) Emphasize cost of not buying; (E4) Explore cost of wrong decision;* and *(D6) Emphasize benefits.*

The foregoing material represents only a few sample statements that can be made in probing for a reaction. Develop your own list, and don't use the same statement over and over so that your customer becomes annoyed or that your purpose becomes transparent.

D10—Divert him

There are occasions when a customer will concentrate on something that will, at the very least, impede the progress of the sale. He may have some pet peeve that he will belabor. He may have raised some point, which, for the moment, you would prefer to avoid. It is at times like these that you can effectively employ the technique D10—Divert Him!

You accomplish this by the use of questions. Just so long as he must answer *your* questions, you will be

c. You are considering using the name or experience of a third party as a testimonial, and you, first, want to know how he regards this third person.

d. You would like to ask for the order at this point, but you want to determine whether or not he is satisfied with what he has seen and heard up to this point.

e. Your product has many advantages, but you want to determine which will be of greatest interest to him so that you will not bore him—or even lose him—by reciting all of the advantages or by claiming that a feature is an advantage when, from his frame of reference, he will see it as a disadvantage. In other words, you would like to find out if he wants you to "make him rich"..."make him famous"..."protect his status quo (security)"... or "make life easier for him."

These are a cross-section of the many situations in which a "probe for reaction" will be helpful strategy. Because the specific goal of each of the five situations described above will be different, the actual wording of the probe may vary. There are, however, certain basic considerations to keep in mind when you find there is a need to probe for reaction:

1. You seek his reactions to a demonstration you have just made:
a. The probe should be worded so as to focus his attention on the points that you hope made the most favorable impression upon him. The purpose of the probe in this situation is to aim a spotlight on those points before he can open up a discussion on the points he found less favorable or even objectionable.

If you can cause your customer to talk positively (versus negatively) about the attractive features of the demonstration, he will, to a degree, stimulate his own desire to buy. If he sells himself on the good points of your product, he will *want* to find ways of overcoming those points he found objectionable because he will not want to lose the advantages. Contrary-wise, if he is given an opportunity to talk about objectionable features first, he may not develop the desire to buy. He will have been permitted to think about the disadvantages, and as a result he may not see any advantages.

Example of how to establish a favorable focus: "Mr. Henry, I would be interested in knowing what features impressed you the most during this demonstration?"

This simple, straightforward question almost forces the customer to seek good features and advantages. If he is one of those real obstreperous people, he may answer, "I didn't see *any*!" If this is his answer, nothing has been lost because at this point he is unsold in any event. Now you can probe for the reason for his objections with, "I'm sorry to hear that—what did you find most objectionable?" As you listen to his reply, you may find that his disagreement is the result of poor communication or misunderstanding. At least you have spotlighted the nature of your problem.

When the probe brings forth the favorable reply you hope for, you have identified what is of primary interest to him, and you can now expand upon those features.

b. The probe should be worded as to avoid producing complaints which you prefer to avoid or which you are unprepared to answer. However, it is not wise to avoid objections which you know he has and which are important to him, hence vital to the success of the sale. They are not just "going to go away" if ignored. You must smoke them out via the probe for reaction. Presale preparation and product knowledge will now stand you in good stead. You should know your product (and your competitor's) so well that you will have considered the possibility of every objection which he could raise, and you will have developed answers to those objections. Now that you are in the middle of the selling situation, you must determine whether he is harboring unspoken objections. Only the ill-prepared salesman avoids serious objections. They must be dispelled with convincing facts.

2. Your customer may have had a disagreeable experience with your company since your last call, and you will want to know how it has affected her attitude of the moment. This is an occasion where the direct probe is not recommended. It is far better to act as though nothing had happened and to try to focus her attention on something you have to offer that is of interest to her now. This is still a kind of probe in that it enables you to watch her reactions carefully. If you have something to offer that is definitely of interest to her, concentrate on that point. If the advantages to her are great, she may change her attitude. If she continues to be hostile, you must face up to it. References to her past difficulties with your company may simply re-awaken hostilities that have abated since the unfortunate event occurred.

3. You would like to use a third person as a testimonial, but you wonder how he regards that individual. Have you ever had either of these experiences?

1. You had two friends, each of whom you liked very much. You brought them together, thinking that because you liked each of them, they would like one another... only to find that they did not.

2. You arranged a party, carefully selecting people you were certain would be compatible because you like all of them, only to find some of them intensely disliking some members of the group.

You may respect one customer and use him as a testimonial, only to find the customer you are trying to influence not only fails to share your high opinion

In other words, it is not important that you gain your customer's agreement on every point you make. It *is* important that you gain her agreement on major points. Therefore, don't challenge a premise stated by the customer unless acceptance of that premise would undermine your main thesis. Let's clarify that point by applying this technique to selling an object with which we all have at least some familiarity... the selling of a camera.

Let's assume the salesman is attempting to sell a camera that has an interchangeable lens. Each lens is equipped with a bayonet-type base. It is a simple matter to change from one lens to another by simply twisting the lens that is in the camera a quarter of an inch and pulling it out of the bayonet socket, then inserting the other lens. The customer is interested in the camera and has commented favorably upon many of its features... but now he offers his view:

"Of course, there *is* one thing wrong with this camera. It has a bayonet-type mounting, and lenses that screw in are much better because there is no chance of wear and your pictures will always be in sharper focus."

When the customer offers this view, the salesman has a choice of actions. For example, he could point out that the "wear and tear" on a lens mounting is infinitesimal, and in decades of use you would never see the *slightest* change in the accuracy of focus as a result of this type of mounting. To point this out to the customer would serve only to irritate him and might lose the sale. To agree—without qualification—might also lose the sale. The professional salesman decides that this is not a key point, and he is not going to permit it to become one. After all, the camera *does* have a bayonet mounting lens and this can't be changed. The salesman would prefer to ignore the comment but, instead, recognizes it for the moment and then diverts the customer's attention to other features (in other words, he "tacks"). For example, he might say:

"I certainly agree with you in theory because there probably is more wear in this type of mounting. I would, however, like to point out one advantage of the bayonet mounting that will be important to you. With a screw-in type lens, it takes much longer to change lenses, and even good photographers lose pictures as a result. You've probably already had that experience—I know I have. With the bayonet mounting, you are going to find that you can change lenses quickly. And you'll rarely lose a picture because you have the wrong lense in the camera when a picture unexpectedly pops up. You are going to find the possible disadvantages will be outweighed by this advantage which this camera is going to provide you."

What has been accomplished in this comment?

a. A degree of credibility has been accorded the customer's expressed point of view. His comment has not been ignored, disclaimed, debated or scoffed at.

b. The salesman has tempered the effect of the complaint. (He could have done so even more effectively has he added, "While there may be more wear in this type of mounting, I know you agree that there won't be appreciable wear for years.")

c. He then placed the customer's attention on an advantage that outweighs the claimed disadvantage... he turns an objection into an advantage.

d. He has used, effectively, the technique of "comparison" discussed earlier.

e. He flatters the customer when he states that the customer has probably had the same troubles experienced by (other) "good photographers."

f. When he states that the customer has probably had the experience of losing pictures in the past, he takes the customer off the spot by adding, "I know *I* have."

g. He emphasizes the losses the customer will suffer if he chooses the wrong camera, and then makes certain that he closes on a positive note by leaving advantages uppermost in the customer's mind.

h. Notice how the salesman assumes throughout his remarks that the customer is going to buy the camera. He accomplished this with such phrases as, "you *are going to find*"; "one advantage... *will be* important to you"; "this advantage... *is going to* provide you."

The customer's arguments should not be ignored, brushed aside or ridiculed. They should be recognized, and agreed with in part or in principle (if possible), but then, by use of "comparison," they should be made to appear unimportant as compared with the advantages of choosing our product or service.

D9—Use probe for reaction

As a selling situation unfolds, conditions frequently arise making it desirable, even necessary, to identify the nature of the customer's probable reaction to a contemplated action on your part, or to determine what his reaction has been, up to this point, to your presentation. The following situations are typical of what we have in mind:

a. The customer has witnessed a demonstration of your product, and you wish to learn what it is about your product that has impressed him the most or you wish to identify possible objections which the demonstration may have caused to arise in his mind.

b. You know he has had real difficulties with your company or product since your last visit, and you need to determine how she feels now.

problem through and that your answer is well considered.

2. He feels that you are really concerned with his welfare because you have considered some of the problems he will face, and you have produced a solution.

3. He gets the feeling that you are not trying to "snow him" and that you are willing to admit that your product may have some shortcomings. This will also tend to lend greater credence to the claims you do make for your product.

4. It builds her confidence in you.

Example: "Mr. Graham, our use of a vertical chamber instead of the orthodox horizontal chamber does, admittedly, introduce one problem. It may be more difficult to maintain. However, Mr. Graham, compare the figures on maintenance cost with the savings you make on valuable floor space, and you will find a significant balance in your favor by using the vertical chamber. Furthermore, because our engineers knew that maintenance would not be as easy, they insisted on extremely critical tests and intense inspection procedures at the factory. As a result, we actually can show reduced maintenance cost. So, what at first appears to be a disadvantage is really an advantage to you.

"I would be delighted to arrange a visit to the plant for you so that you can witness these additional quality control measures for yourself. Or you may prefer to have me arrange visits for you to other installations where the vertical chamber has been in use for some time."

By admitting the existence of the objection, the salesman has opened the customer's mind so that he will listen to the advantages. He has also supplied assurance by his twin offers of a factory visit or visits to installations using the product.

E11—Stimulate recall

Incidentally, the Sales Strategy Bank should not be considered as a static device. You will develop other strategies and they should be added to the Bank.

The influences on recall were discussed at length in Part One. There we discussed why the customer often remembers things you wish he would forget. He forgets things you wish he would remember, and he remembers experiences which you both shared, in a manner that is quite different from your own recall.

When a customer is making the going difficult because he entertains fears of one kind or another, it may be to your advantage to cause him to remember past experiences. He may, on past occasions, have had similar fears when you attempted to sell him other products. Recall these experiences to his mind. Cause him to see that just as his past fears were groundless, so are his present fears. In doing this, be sure to protect his self-image. For example:

"Joe, I can easily understand your concern. If I were in your position I would feel the same way. You will recall, Joe, that some of this same concern existed last year when it was decided to install equipment X. It seemed reasonable then to be concerned, but you'll remember how well the application worked out. You can see how much both of these applications have in common, and that there is really no need for any concern."

In this manner, you have not made him look badly (for entertaining fears or concern). You have made him recall that he had similar fears on a past occasion and that they proved to be groundless. You have pointed out the similarities in both situations, and you have provided him with assurance. Naturally, the example above is a skeletonized version (as all of our examples must be). In an actual selling situation, you would be much more specific and would actually spell out the nature of the concerns he had on the previous occasion and you would specify the exact similarities between today's selling situation and the one you have recalled.

Another approach in which his recall is refreshed is found in this example:

"You were absolutely right, Joe, when you decided to expand two years ago. You were one of the first to see the possibilities of additional business in the suburbs. Now, we see a similar situation approaching."

Note that you do not have to spell this out for the customer. Let "Joe" revel for a moment in the fact that his previous wisdom has been recognized, and he will bridge the gap between the situation that existed two years ago and the situation he now faces.

E12—Provide guarantee

There are customers who will hesitate to buy or even resist your efforts because of fear or because of past experiences. Fear is often an obstacle that underlies the resistance of a customer who believes that a saleswoman has only one goal; i.e., to get a sale, and really is not concerned about the customer's welfare. This is often a fear of obsolescence, a fear that prices will drop after the purchase is made, fear that this is not really the best product or service for the customer's needs, or a fear that once the purchase has been made, satisfactory service will not be provided. You can add to the list of fears by asking yourself why *you*, in your role as a customer, have been slow to buy items *you* needed. Some of your customers will entertain the same fears that caused you to delay in making a buying decision.

Past experiences also cause the customer to delay. One customer resisted the efforts of a salesman because ten years previously the salesman's company had stopped making a product the customer required and left him high and dry. This customer had also been unable to receive adequate post-sales service because the supplier had, at the same time, turned over servicing of the product line to a distributor who had done an especially poor job. To put it mildly, the customer had a dim view of the supplier who had now re-entered the market. The fact is that this supplier has made a management decision to stay in the market permanently and to provide excellent service. The same reluctance to buy may have stemmed from a previous experience, where the customer, after buying, found that a large reduction in price had taken place in what the customer considered was a short time after the purchase at the higher price had taken place.

In another instance a supplier, having put into effect cost reduction methods and automatic equipment, was able to offer the customer a price 25% under that which he had been paying another supplier for ten years. Instead of being delighted, the customer became aggressive. The reason was that he had been made to look like a fool in the eyes of his top management. The question was being raised, "Why have you been letting our present supplier bleed us all these years when he, too, could have been giving us this low price?" The fact was that the present supplier felt he had a captive market and never lowered prices. Now, this customer is not only angry with the new price, but states that he is certain the salesman's company is just trying to get a foot in the door with an impossibly low price and would raise it as soon as the customer got locked in with this product.

What to do in any of these instances? Provide guarantees whenever possible. Granted that it is not always possible to do so in the case of sustaining a current price, the salesman should, after identifying the underlying reason for the customer's hesitation to buy, contact his company and determine what kinds of guarantees can be offered and which can be presented to this potential customer in writing. If the guarantee is verbal, the customer is inclined to wonder what will happen if this present salesman is replaced by the supplier with another salesman who was not present when the present guarantee was offered. Guarantees represent one of the best means of providing the assurance that these types of customers, represented in the foregoing examples, are going to require before they will buy.

E13—Offer additional services

When a customer has any of the fears previously covered—or others—one way to dilute these fears is to offer additional services that cost the customer nothing, and often cost the salesman's company, or the salesman, nothing except a very small amount of time. This offer and provision of extra free services accomplishes three valuable things: (1) it serves to dilute the fears; (2) it causes the customer to feel that greater value is being received; and (3) it establishes in the mind of the customer the feeling that this salesman, and the supplier, are sincerely interested in helping the customer.

These additional services may range from something the saleswoman may be able to do as an individual (such aids as helping the customer train his own sales staff, aiding with displays, and similar forms of help) to help that the saleswoman's company can provide. This last is especially easy to provide if you, as the salesman, are working for a large corporation. There is just no end to the type of help that can be provided. This help can include such forms as advice and assistance from your company in record-keeping, record and report forms, accounting, sales training, advertising, ad infinitum.

In reference to the last paragraph, this applies primarily when your customer's organization is considerably smaller than your own. Customers who are as large or larger than your own company may not take kindly to offers of these last types of aid because the inference may be that you feel your methods are better than theirs. The smaller organization, however, lacking the facilities of your own company, will often be delighted to accept the help offered. Even if your offer is not accepted, you have established yourself in the mind of the customer as a person who cares about that individual. It also sets you apart from the competitor who concentrates solely upon getting this customer's business. If the customer had previously entertained some fears about your company these will often be allayed or eliminated. The basis for this is that the customer, seeing you as an individual who is sincerely interested in her problems, will be equally helpful if, subsequently, problems arising from your product or service occur.

Concluding our treatment of this strategy, we are impelled to point out that we are constantly amazed that so many large corporations that have great expertise in the fields of finance, personnel management, market research and the like, seem to be oblivious to the great advantage this can give them over smaller competitors, through the opportunity it provides these suppliers to offer free aid and assistance above and beyond the products and services they sell.

Summary

This completes our treatment of the strategies that may be used to overcome the customer's fears. It will be well to repeat these most important points:

1. Customers, as a rule, will not reveal their fears. You must identify them.

2. Fears will vary widely and will include fear of looking badly, fear of losing money, fear of making a wrong decision, fear of the work and/or expense that may be involved to implement his decision...to name just a few.

3. It is essential that the professional salesman identify the specific kind of fear and then use a strategy that will neutralize that fear. This means that the salesman must develop an ability to empathize with the customer (and this must not be allowed to degenerate into sympathy). You must ask yourself, "If I were the customer, what fear would I have, and what could someone do to me to overcome my fears... that's what I must do with this customer."

CHAPTER 25

Forcing a decision

As you will have noticed, the Sales Strategy Bank consists of seven sections. Individual strategies have been assigned to the grouping to which they seem most appropriate. Thus far we have covered the following groupings:

    Strategies to be followed in preparing for the sale.
    Strategies that are designed to create a good climate.
    Strategies that compel attention.
    Strategies that are designed to overcome objections.
    Strategies that will allay fears.

We now turn our attention to those strategies that are designed to force the customer to a decision.

As we have seen, customers will delay making a decision for many reasons. They may not have complete confidence in our solution to their problem, they may have other demands on their time and/or funds, they may not be truly aware of the nature of their problem... or countless other factors may cause them to delay in making the buying decision. In previous chapters, we have discussed the many strategies that can be employed to overcome each of these obstacles.

These strategies do not always produce the desired result. What then? It may be time to live dangerously, to resort to what the late Elmer Wheeler, one of the great sales trainers of all time, would undoubtedly have dubbed "dangerous selling." In a word, it may be that the only course of action left is to bypass the buyer.

F1—Bypass him

Before embarking on this admittedly dangerous course of action, it will be well to discuss its hazards, its values, and methods by which it may be accomplished.

We recall an instance in Arizona where a company was giving one third of its business to a local concern. The vendor objected, stating that because he was local he should be given *all* the business. The chap who had the authority to place the business explained that in the market in which the vendor served, this customer would never place all of its business with one supplier. A few days later, the salesman invited the buyer to lunch. When he arrived at the restaurant, the buyer was amazed to find that the vendor had also invited the buyer's immediate superior and another key executive who had the power of veto over the buyer.

During the course of the luncheon, the supplier made a determined drive for all of the business, pointing out that he was not getting 100% of the business simply because the buyer had decided it would be his policy to spread the business. It became quickly apparent to the buyer's superior and the other executive that the vendor was trying to "box in" the buyer. The result was that they became incensed. Before the lunch was over, the vendor had lost even the third of the business he had enjoyed to date.

This, as is the case with all of our examples, is a real instance of the hazards of bypassing the buying influence. There are other times when it is the wisest strategy to use, but let's consider the following points.

Let's assume that you *know* the buyer is corrupt and is buying from another source because he is being paid off. Let's hope that your own company's policy, and your own personal policy, is to refuse to obtain business via this route. If so, you may be thoroughly aware that it is going to be difficult, if not impossible, to obtain his business. The temptation is to go over his head, with your proof, to his top management.

We have seen that done and have seen buyers fired as a result. However, often a strange thing happens. The company stops purchasing from the vendor that has been "buying" the business through bribes, but it gives the business to a *third party* rather than to the individual who called the situation to its attention. The full answer to why this is so will be found in Part One.

Briefly, the complainant has caused the company to lose face, to be embarrassed... the corporate image they are striving to project has been damaged. To do business now with the individual who brought the matter to their attention would be to constantly remind themselves of a situation they wish to forget as rapidly as possible.

Look at it this way: You know Mr. A and Mr. B. Mr. A is a good friend of yours. He often tells you that Mr. B is a fine fellow. You put up with this for a long time until you finally let your friend A know that B has been talking about him in uncomplimentary terms and has been working against A's interests. What happens? Does A thank you profusely for causing his judgment to appear to have been bad? He does not! First he attacks the sources of your information and shows great irritation. Later, he will check on B, now that his suspicions have been aroused, but you can be certain that—except in rare cases—he won't thank you for making him look foolish.

What's the point? The point is that corporations, companies, or firms—are simply groups of individuals, and they all have images, both personal and corporate, that they are trying to project. They will, therefore, act almost exactly as you and I will act when our self-image is being attacked, or when we are being made to appear foolish or inept.

Certainly if you have indisputable proof that a purchasing decision-maker is being "bought," you know that your selling efforts will probably be fruitless, and you have a moral right to have this brought to the attention of her superiors. You can be sure that they will take no action unless the proof is really there. At the same time, note that we said... "have it brought to their attention." That is a far cry from *bringing* it to their attention.

Let's not jump to the conclusion when we have failed to sell the buyer that he is corrupt. He may have excellent reasons—in his mind—for not switching suppliers. If you *know* that your product or service is superior to that which he is buying and you feel, strongly, that his intransigence is not only costing you an order but is depriving his company of real benefits, you may bypass him as a last resort. This can sometimes be accomplished by making certain that people at higher levels come into possession of facts which you know their buyer has withheld from them. Your hope is that they may become so enthusiastic that they will put pressure on him to buy from you. Even then you stand the chance of getting the order but earning his enmity, especially if he connects you with the pressure. Or... you may get *this* order and lose future business if the buyer has his way.

Often a buyer will refuse to buy from a saleswoman simply because he feels he is being measured on how well he keeps costs down. Your product or service may cost more, so you are automatically excluded from consideration. In such cases, it becomes essential to bypass him...not by "going upstairs" over his head, but by going around him to the people who will actually use your product. The methods of doing this will vary as widely as there are course participants. In some companies, the buyers do not object to saleswomen seeing engineers, factory personnel or the workers who use the products. In others, the buyer stands like a colossus barring the way.

Study the organization to see how you can get around this roadblock. Determine whether there are methods of getting to know some of the people you must reach, off the job. Can you contact them with circulars, brochures or other forms of technical information through the mails, so that they will become interested and specify or at least suggest your product? This is the "pull-through" method.

Another method of bypassing that often proves effective and less hazardous is to obtain help from "upstairs" in your own company. Often, people at higher levels in your organization know some of the key people in the buyer's company and can do much of the selling job for you. But don't make the error of thinking this cancels all hazards.

We recall the true story of two executives travelling by train. The train stopped at a remote station on a hot day, and the two executives found no water fountain available in the station. Knowing the president of the railroad, they called it to his attention and to their delight were told that they would get an order for 400 coolers. After two months passed with no order appearing, they again visited the railroad president. He stated emphatically that they *must* have received the order because he had told the purchasing agent to place the business with them. All three dropped in on the purchasing agent. When the president asked him what had happened, the P.A. blandly said he had placed the business elsewhere at a lower price because he "was following the president's orders to seek the lowest price consistent with quality." Period! It was obvious that this P.A. was irritated at the pressure, or at least he would have asked them if their first price was their lowest price.

If you *are* successful in getting business via the bypass route, it goes without saying that it will be wise never to bring up the matter in the buyer's presence.

It would seem from the foregoing that we are advising you never to take this bypass route. Not so. When everything else has failed, when you know you have the best product, when some one individual is standing between you and the order, and when no other department of your company will be adversely affected, it may well be that the bypass route is the only open road to an order... if it will not adversely affect future orders.

F2—Shock or antagonize him

Once again, before employing this strategy, consider its effect upon your long-range relationship with the buyer. Is it worth gaining this one sale and losing her business? Will you get this order and put other components of your company at a great disadvantage in their dealings with this customer?

This strategy has much in common with the controversial question discussed in a previous chapter; i.e., "Do you mean to say that you can't see what you stand to lose by not buying now?" Note that this treatment has an irritating quality that will compel attention, but it also includes the strong hint of benefits or losses to the customer.

Take the case of a customer who simply refuses to listen to reason. He may insist that your product will not do what you claim for it, or he may simply act totally disinterested. You have exhausted every avenue, and you realize that unless you can get him to listen, you have no chance of getting an order. You decide to take the risk that is inherent in shocking or antagonizing him.

If feasible, you may *prove* his statements to be wrong by demonstrating your product. You may show him records that will *prove* him wrong. No matter what means you use, your intention is to devastate his arguments with irrefutable proof. This is certainly the last resort method because you and I know that we don't take kindly to being *proven* wrong. This means you had better know your customer and be fairly sure you are right in your assessment of his probable reaction.

Don't think for a minute that all this strategy requires is the ability to use a verbal blunderbuss... to be able to be insulting. Once you have shocked or antagonized this customer, the basic technique has done its job—it has compelled attention—and you must be quick to take him off the spot by rapidly changing your mood and manner. Your "attack" should consist of something far more than an outright insult. Here is an example:

(The salesman has been working on the customer, Mr. Moye, with no effect. He decides to try the last resort method.)

Salesman: "Mr. Moye, I'm amazed that you seem to be against progress."

Moye: "What do you mean *progress*—what has progress got to do with it... I'm just not interested in what you have to sell."

Salesman: "Well, if cutting your costs by 15% and increasing production by an equal amount isn't progress, then I just don't know progress when I see it."

Moye: "Just how do you think this thing will cut my costs by anything like 15%?"

For the best part of 30 minutes, the salesman had been trying to focus Mr. Moye's attention on the cost-saving possibilities of the device he is selling. Moye hadn't heard a word simply because he wasn't interested. Now that Moye has been charged with being "against progress," he becomes very much interested. He is going to make the salesman explain or apologize. (You and I hate to be told we are unfair, against progress, getting old, getting fat... or a number of other things—all of which are, of course, totally false.) It doesn't take much to irritate us at times or to get our attention. Simply have some friend (?) look at you quizzically and ask, "Have you put on a pound or two?" and see whether or not he gets your attention. The point being made is that to use this strategy you don't have to be abusive.

The salesman in our example must now move rapidly into a presentation that converts the irritating charge into a parade of important advantages to Mr. Moye... and he had better be prepared to back up the claims he implied in his attack. His entire mood and manner must change quickly. He must now concentrate on every argument at his disposal so Moye will forget the "insult" and focus his attention on what this product or service will do for him.

You can be insulting by degrees and, as has been said so often, you'd better know your customer. If he is an old acquaintance, you might have gotten away with: "Joe, I'm beginning to think your business methods are old fogeyish." The hazard you run in being this blunt is that you may anger him so much that he will close his mind, completely, to anything you have to say... but it *can* be done:

Salesman: "I'm beginning to think your methods are old fogeyish!"

Moye: "What d'ya mean... old fogeyish, we're as progressive as anyone you call on!"

Salesman: "Well, I don't think passing up opportunities to cut costs by 15% is progressive."

Moye: "Just how do you propose to cut my costs by 15%?"

This is just the challenge the salesman wants. Moye issues the challenge to put the salesman on the spot... and this is just the spot the salesman has been seeking.

Consider this example:

(Jerry Calhoun has been trying to sell Jim Stewart, a store owner, on the idea of installing a display kitchen. He has had no luck in getting, let alone holding, Stewart's attention. Stewart's business has real potential but his volume, up to now, has been unsatisfactory. The salesman, Calhoun, is convinced that it is due to Stewart's poor display techniques.)

Calhoun: "Jim, you're just lazy!"

Stewart: "What are you talking about... lazy? I put in more hours here in a day than you put in, in a week. Just what do you mean by that crack?"

Calhoun: "Jim, any dealer who is as progressive as you are, and is as up to date on new trends as you are, simply couldn't have any logical reason for refusing to make the necessary layout changes that this model kitchen would require."

Stewart: "It has nothing to do with being lazy.

We've got other things to do with our time and our selling space."

Calhoun: "Maybe, Jim, but can you show me anything else you have to do with your time or your space that will increase the sale of your appliances by 30%?"

Stewart: (with belligerence) "Just how do you figure it will do that?"

Again—just the invitation the salesman has been seeking! Now, he had better be ready with figures, facts and other proof to show that his claims are realistic and attainable. In other words, the intent to antagonize must have something behind it. An insult, born of desperation, will only lose you a customer—as it should if you employ it for no sound reason. *The "attack" is intended to excite attention*, and once the salesman has that attention he had better be ready to direct it into positive channels.

Examine, again, our last example. Notice how quick the salesman was to change from raw attack to compliments. Once he had Stewart's attention, he started to talk about how "progressive" Stewart was...to take some of the fire out of the remark about Stewart's being lazy. Then he quickly hit hard on something he knew would be important to this customer...an increase in business.

F3—Show him cost of procrastination is higher than cost of decision

This is an interesting strategy because in certain circumstances exactly the opposite strategy may be used. If, for example, your customer is about to make a *fast* decision that is adverse to your interests, you must show him that it will be more costly to act than to wait.

But, first, let us take up F3, the strategy we use when the customer delays making a decision.

To understand others, we must first understand ourselves. Why do you and I often procrastinate... delay in making a decision? Here are a few reasons to which, in honesty, we must plead guilty:

1. "It will be easier to decide tomorrow." (It never is—it is almost always harder.)

2. "If we wait, maybe the situation will clear up and we won't have to decide." (It hardly ever does ...it usually gets worse.)

3. We see the benefits and losses of opposing courses of action as being so balanced that we just can't decide which action to take.

4. We fear the results of a bad decision. You can add some other reasons of your own to this list.

When the customer procrastinates, you have two courses of action open to you. Let's say that you are attempting to cause him to choose course A, and he can't decide between A and B (the latter which you oppose). You must either show him tremendous benefits of A, or you must cause him to see that he will lose more by his delay than he will gain by continuing to delay.

The customer is going to remain on dead center just so long as he cannot see one course of action as being preferable to the other. To choose the proper argument for budging your customer means that you must know him. You must know him so well that you will be able to predict how he is going to react.

For example, we feel that a friend of ours does not know us very well. He sells investment funds, and his approach, every time, is to convince us that we must move fast because this is a very limited issue. The result is that we dig in our heels and refuse to buy at all. This is an attack on our self-image. "Does he think he can pull that 'high-pressure' stuff on us?"

So know your customer and, above all, don't use persuasive methods that are not true. If you intend to scare him into action with a threat of imminent price increases, be ready to show him supporting proof or he will only resent what he will regard as crude attempts to pressure him. *Know him.* Does he want to be first in the market? Then cause him to see that he won't be first if he doesn't buy now. Are there really chances that there will be a shortage of that which you sell if he doesn't buy now? If so, tell him, but support your comments with proof. If possible, show him what he has already lost through delay, and cause him to see that the decision will be more difficult to make with each passing day. Don't let him think it is going to be easier.

Now, let's look at the opposite situation. Here we have a customer who is going to take a fast action, and it is in your interest to have him wait. Let us take a hypothetical situation. Your customer, you sense, is going to buy from one of your competitors with whom you know he is about to leave on a cruise. He is known to act impulsively, and you know he wants to make the decision while on the cruise. He feels he will get the best "deal" in the setting of camaraderie that exists on the cruise. You had been invited but you couldn't get away (and now it's too late). Now, he is going with one of your competitors. What do you do?

"Joe, I hope you have a wonderful cruise. But, Joe, I have reason to know that we are going to be able to make you an extremely attractive offer, and if you make a decision before you come back from this cruise, it may be the most expensive cruise anyone has taken in years. I'd strongly recommend that you consider the offer we will be able to make when you return."

In examining this action, you may say, "But what has that gained you... you still don't have the order, and you may not get it when he returns." That's quite right... but we have gained time, and we have removed him from the atmosphere of the cruise where our competitor may otherwise have been able to close the order.

You may ask, "Will the competitor stand still for this?" He won't have any choice. If you know your customer Joe, and you know he is concerned about costs, he is going to convince your competitor that they should both relax and enjoy the cruise and discuss business the "day after" they return. In the meantime, he will be planning on seeing you the *day* he gets back to see what your offer consists of.

Who knows, too, he may have a perfectly rotten time on the cruise with your competitor. In any event, you can't be worse off than you are now, with your customer planning to take fast action. You speed up the procrastinating customer with the great benefits that will be his if he acts now (or the great losses he will suffer if he waits). You slow down the customer who is impulsive... or the customer who is going to make a fast decision to buy before you can get your presentation ready, by showing him the losses he will suffer if he moves too rapidly... or the gains he will make if he waits.

Remember that the gains and losses do not have to be financial. You can often cause the customer to act quickly... or slowly... as you desire, by showing him the impact that will be made on the image he is trying to project—how he will look good or bad, as the case may be.

You and I—and your customer—react in a rational manner to specific stimuli. We interpret these stimuli in terms of how they will make us look, feel or benefit. If someone else wants to change our attitude, they must inject stimuli into our frame of reference that neutralize or negate the effect of the original stimuli.

To change your customer, you must make an attempt to determine how and why he feels as he does. He either wants to be rich or famous, or he wants to be secure, or he wants life to be easier. *Which of these needs is most important to your customer at this point in time?* If he is delaying, it is because he sees no chance of satisfying that need which is most important to him or he sees that need threatened.

If he is about to act precipitously, it is because he feels that his most urgent need will be satisfied by fast action.

You have the choice in either case of showing him that procrastination or fast action is against his best interests. As we have seen, you can do this by injecting fear... or concern... or by making him think of other needs that he has that are even more compelling... which the course of action you are urging (be it delay or fast action) will make possible for him to attain.

Selling is persuading. Persuading consists, primarily, of showing the other fellow that a given course of action will satisfy those needs upon which he sets the highest priority... or it means convincing him that he has other needs that deserve even higher priority—needs, incidentally, which our product and/or service can supply.

F4—Emphasize cost of not buying

Every experienced salesman knows the need to emphasize selected benefits that will appeal to the customer and produce a favorable buying decision. There are times when even these benefits leave the customer undecided. What to do? At that point it may be necessary to emphasize what it will cost the customer if the decision to buy is delayed. These loses must be presented with great care so that the customer does not get the feeling that "high pressure" is being exerted. It may well be that the losses you have in mind are legitimate losses but they must be made believable.

For example, saleswomen often try to force a favorable buying decision by pointing out that price increases are in the offing and delay in buying will result in added expense for the customer. Customers have had this pressure used on them for years. If, indeed, price increases are imminent, the customer should be presented with facts that these increases are really due to take place.

Another time-worn device used by some salesmen is to warn the customer that supply is short and if a decision to buy now is not forthcoming, it may not be possible to supply the customer if the order is placed later. All too often the customer sees this as pressure and suspects that you are really up to your ears in unsold products. Once again, if you are making a legitimate claim of probable short supply, present facts that prove it. It is essential to avoid creating the impression that scare tactics are being employed.

Fear of loss can be as powerful an incentive to buy in many selling situations as a parade of benefits no matter how attractive they may be. If the customer appears to be reacting favorably as buying benefits are presented, do not discuss the losses that will be sustained if the order is not placed now. Presentation of losses should be held in reserve and used only when the customer seems to be undecided.

In presenting potential losses use what we might call the "sandwich" technique. Mention the key benefits first, then sandwich in the losses and finish with benefits. For instance: "Mr. Brown, I know you agree that placing an order now is going to produce some real benefits to your organization (mention a chosen few briefly). I'm sure you've noticed in the financial news that our industry is facing new wage negotiations and with inflation I'm sure you will agree that some increases in labor costs are certainly going to take place and, obviously, they are going to be factored into price. I want you to get the best price possible and that's why I would sincerely advise that you place your order now, so you can avoid paying more in the near future. And, of course, buying now means that you will be able to start obtaining (mention the key benefit again) at the earliest possible time."

Using that method, the customer sees the possible loss. He has no feeling that high pressure or scare tactics are being exerted, and the last thought left in the customer's mind is the attractive benefit of immediate gain. Possible and likely losses have not been used as a blunderbuss and the customer has not been irritated. Your approach has been one of sound reasoning.

CHAPTER 26

Post-sale actions

Post sales service is step one to the next sale. It causes the customer to see that the salesman is doing more than merely trying to sell the products or services the supplier seeks to provide. It convinces the customer that the salesman is sincerely interested in aiding the customer in solving a problem or in filling a need.

In many instances it might be said that a sale has not really been completed until the second sale is made. In most businesses the real profit comes not from the first sale but in the subsequent business the customer can represent.

In all too many cases salesmen feel that their obligation ceases when the product is delivered or the service is provided. In one instance a salesman was telling us that because his company consistently used the wrong name of the customer's company and enraged the customer, unless "his company" did something to correct this, the salesman was about to lose a most important account. As we listened to this complaint we could not help but wonder, why, in the time it took the salesman to tell us about this situation, he could not have called someone in his office at headquarters and asked them to get out to the individual who operated the equipment that addressed the mail and have them change the address.

Probably by this time every reader who has had difficulty with a company in his own capacity as a customer has been told, "It was a computer error." Computers do not make mistakes but people who feed information into the computer do make mistakes. The customer is not interested in *how* the errors occur, but she is very much interested in having the errors stopped. This is just one of the areas in which professional salesmen have a continuing obligation.

Fortunately, every saleswoman is at times a customer. In our lives as customers we know what irritates us. These same irritations are experienced by our customers. One of the key sources of irritation is often the failure of a saleswoman who has sold us a product to show any continuing interest in the solution of our problem or the fulfillment of our need by simply contacting us to learn how the product is performing.

One of the major objectives of any salesman should be to convince the customer that the salesman and the company selling the product are both really customer-oriented.

Because we claim that this is one of the most important elements of a sale it may be wondered why more space has not been given to the subject. The answer is that the forms of post-sales service vary as widely as do the selling fields represented by the readers of this book. The fact is that all that is needed is for every salesman to develop a mental outlook and concept of the job of selling that is based on an awareness of the true nature of the selling profession; i.e., a continuing interest in the welfare of the people who buy. This should be based on an awareness that while satisfaction can be obtained from receiving the customer's cash, an equal, if not greater satisfaction, can be the reward when a lasting relationship is created by causing the customers to see that professional selling is a field of continuing service for, and empathy with, the customer.

G1—Show continued interest

Step one of a second sale may well be the actions you take in following up the initial success of the first sale. The professional salesman can often accomplish more after the sale than at any other time. Because he has the first sale "in his pocket," so to speak, his motives are not suspect and it is easier to make the customer understand that he, the salesman, is *really* sincerely interested in helping the customer, and, furthermore, is interested in him as an individual.

Look at it this way. After you purchased your last car:

1. Did the salesman call you a week or two later to inquire about your reactions to the automobile? If, for instance, you had expressed a desire to obtain a car that would provide you with good gas mileage, did he inquire as to whether the one he sold to you was living up to expectations?

2. Did he call you sometime later to remind you that your guarantee period was about to elapse, and if you had any problems this would be the time to bring in the car and have the difficulties remedied?

3. Did he call you some months later to inquire

about how many miles you had driven the car, and then to suggest that you had better rotate your tires in order to insure maximum mileage?

When did you hear from the automobile salesman the next time? Was it at Christmas when you perhaps received a stereotyped greeting card that he sent to all of his customers? Or, was it the following year when he sent you a card suggesting that you might care to "stop in and see our new models"? Do you really feel today that he was as interested in solving your unique transportation problems as he was in making another sale and obtaining his commission? Half of us—perhaps more—go back again to the same salesman, not because we are sold on him, but because we feel that the rest of the automobile salesmen are probably just as bad. And, at least, he knows us... or we like the kind of car he sells.

Sale No. 2 starts with after-the-sale interest. Here is one area where you can out-perform your competitor at little or no cost because the chances are that your competitors are like the automobile salesman. Any interest which you demonstrate in the continued welfare of the customer is going to provide you with what may be a vital edge the next time you compete for his business.

It was suggested to the owner of a men's furnishings store that he try this with his regular customers (who get the feeling, at times, that they are being taken for granted). It was recommended that he call the customer on the phone, some days after a sale he had made, and ask whether "dad liked the new shirt."

The store owner grasped at the idea and excitedly said, "Great—I'll do it. And, at the same time, I can tell them what I have that is new and will be of interest to them." This owner thus demonstrated that he had missed the point completely. His after-the-sale call should not mention new business at all! The entire effort should be to show the customer that you have no "irons in the fire"... no ulterior, selfish motives. You should indicate that you really want to determine whether the item you sold him did solve his problem and bring him satisfaction.

As it is now, all too many customers feel that salesmen never do anything without an ulterior motive. If we can prove to them that we are sincerely and deeply interested in them and their welfare, the next order will be more easily obtained.

G2 through G7

G2 through G7 on the Sales Strategy Bank are really test questions that you should use as a checklist after a sale. Let's review them:

G2—Make a return call.
G3—Check operation, use or installation of that which you sold him.
G4—Has his problem really been solved?
G5—Have new problems been created? What can you do to assist him on these?
G6—Is he satisfied in all respects?
G7—Have all of your promises been kept by yourself or by your company?

By coincidence, our typewriter gives us all kinds of problems, the details of which we will not impose upon you. The fact is that we have just called the dealer for the third time. He is ready to rectify the difficulty, but we have no feeling of confidence, nor do we have any feeling of gratification. After all, although he knew we were experiencing difficulties, he did not show enough interest in our problem to call to inquire whether or not the problems had been corrected.

Imagine the resistance he is going to meet the next time he tries to sell us office equipment. What could have been a great advantage to him; i.e., the demonstration of sincere interest, is not going to be working for him. He will, in fact, have even greater difficulty in getting us to believe that he is interested —*truly* interested—in future problems we will have that the purchase of additional office equipment could rectify.

CHAPTER 27

Self-fulfillment

Self-fulfillment should be the goal of every professional salesman. True, it will never be fully attained, but the closer one moves toward that objective, the greater are the rewards, the greater is the satisfaction, and the greater is the self-respect.

While few men, if any, ever reach true self-fulfillment, most of us strive at the very least for self-improvement. Self-improvement is possible only if we are really willing to engage in introspection—self-analysis—with the intention of identifying our strong points and capitalizing upon them, and with an equally firm intention of identifying our weak points and taking positive, planned action to strengthen our performance where the need is indicated.

The best time for introspection is right after we have been engaged in a selling situation. Win or lose, the professional salesman should want to know the reason for his success or his failure. If he recognizes why he was successful, he can capitalize upon this knowledge in similar selling situations in the future. If he can determine the causes of his failure, he will be able to take appropriate action to make sure that his mistakes are not repeated in the future.

The difficulty with self-analysis is that we are all prone to rationalize... to give ourselves excuses for having taken actions that resulted in a lost sale. These excuses may bring us a sense of relief, but they don't correct or change bad habits, and they certainly don't bring us more business. All of us have the tendency to view those who have been more successful than we have with a touch of envy, and we attribute their success to "luck." Probably, more often than not, the reason for their success was that they faced up to the need for self-analysis, engaged in introspection, and took corrective actions.

Obviously, the identification of one's faults is facilitated if the opinion of others is sought. For example, if a friend has been with you on a sales call, you will find it of real value, whether you made the sale or lost it, to solicit his honest opinion of your actions and strategy during the call. Then, if you find his opinion is less than laudatory, don't permit yourself to fall into the trap of starting to defend yourself. Remember, it isn't what you thought you said and/or did, it is what others who observed you thought you said and did. Your actions and words were intended to influence those who observed you or listened to you. If you failed to influence them (as you wanted them influenced), it will avail you nothing to fight the problem and figuratively to curse their stupidity. What you must do is learn what actions you take that fail to influence people favorably.

For example, after a lost sale, it is not unusual for a professional saleswoman to lose a bit of self-assurance. This is a little bit like dying. Self-confidence is a saleswoman's number one selling tool. To lose even a portion of our self-assurance can cost us future sales. So that the situation "won't be a total loss," we often call upon "Good Old Joe" after losing a sale. He is the customer who always gives us an order, albeit a small one. Getting an order from "Good Old Joe" reinforces our faith in ourselves and convinces us that the reason for losing the other order, earlier in the day, is that the prospective customer was simply a "jerk that nobody could sell." This may be good for restoring self-confidence and self-esteem, but it doesn't produce new sales.

The action we should have taken was to stand on the curb, after losing the sale, and ask ourselves such questions as these:

1. What principle of human behavior did I ignore or trample in my attitude or actions toward the customer?

2. What intelligence (information made meaningful) did I fail to gather on:

My product?
My competitor's product?
The market situation?
My competitor's behavior pattern?
My customer's real need or want?
My customer's economic situation?
My customers psychological needs?
My customer's objectives (long range) and his goals (short range)?
My customer's problems?

These are, by no means, the full range of essential elements of intelligence, but they will do, as a starter, for some meaningful—and profitable—self-analysis.

And let us not think that we can benefit only

from identifying our weaknesses. Learning the nature of our strengths and then reinforcing them can be equally helpful in attaining self-improvement. After the successful sale—instead of asking the boys to "belly up to the bar, the drinks are on me"—stop for a minute and conduct this same curbstone self-analysis. Ask yourself what principle of human behavior you identified in the selling situation. Determine what information it was that you gained prior to the sale that led you to the employment of a successful strategy. Use the same list of questions we have recommended for self-analysis after a lost sale.

All of this is not to say that you should pass up an order from "Good Old Joe" after the lost sale. Getting Joe's order will be good for your morale—but guard against the tendency, then, to invite "Good Old Joe" out for lunch to celebrate the small order he has given you and to erase from your mind the failure of earlier in the day.

The chap you should probably take out for lunch is the fellow who refused to give you an order. Get closer to him and learn more about him as an individual, learn more about his problems and his needs—psychic as well as operating needs and problems. Then ask yourself what you can do to help him in solving any of these problems. While your competitor may be concentrating on the customer's product needs and problems, you can consider these, too. But you can go your competitor one step better by identifying and satisfying the needs of the customer's self-image...what are commonly called his psychic needs.

Then examine both strategy and tactical plan to identify any instance where your plan might possibly include one or more of the errors found in (3), (4) or (5) of the left column. If you find any such possibilities, change your plan of approach so that they will be eliminated.

To prove to yourself the value of this chart, take a moment or two now, to review certain selling successes and failures you have experienced in the past. Ask yourself which of the motivators in (1) and (2) you used and were responsible for your success. Ask yourself which of the de-motivators in (3), (4) and (5) were used and to which you can now attribute your failure.

Finally, after every selling experience, ask yourself these same questions, using the chart so that you can continue to use the motivators and to become so familiar with the de-motivators that you will instinctly steer clear of them in your daily selling efforts and improve your percentage of successful selling experiences.

Highly predictable behavior patterns

The following material may be some of the most valuable in this book and it merits your close examination. You note that the customer's possible reactions are in the right column, ranging from what you desire (Full acceptance) to what you certainly do *not* want... Rejection and attack. In the left column you will find the actions you must take in order to obtain acceptance and those which will bring something ranging from no action on the part of the customer, through total rejection, to aggression.

Whenever you face a problem selling situation and you have planned your overall strategy and tactics, refer to this chart. Ask yourself if you have made certain you have included all of the motivators found in (1) and (2) of the left column. If you have missed any that you believe would be appropriate and would produce the customer reactions you desire, make sure you include them now.

Highly Predictable Behavior Patterns

These apply in any person-to-person situation, whether it be in selling products or in managing or persuading people

If you take these actions: | **Your customer will probably act this way:**

1. Offer and proof of strong advantage:
   (1) Wealth—offer financial gain; (2) Famous—enhance his self-image; (3) Insure his security; (4) Make his life easier; (5) Give him greater power of authority.
   NOTE: It's absolutely essential to determine which one of these is the customer's key desire and to concentrate on it!

   Full acceptance (1A)

2. Show understanding of his problem.
   Show respect for him and for his ideas.
   Find point of mutual agreement.
   Share credit with him when you solve his problem.
   Listen—really listen—when he talks.

   Partial acceptance—(2A)
   Exhibits willingness to listen

3. Offer dubious advantages.
   Exhibit uncertainty.
   Offer no proof of gain.
   Don't answer his objections satisfactorily.
   Cause him to sense possible threat to any of No. 1 through 5 (above), with no offsetting gain.
   Fail to identify which one of No. 1 to 5 (above) is uppermost in customer's mind!

   No action—Dead center (3A)
   Delay decision—procrastinate.
   Diversion—change subject.
   Evasion—will try to avoid you.
   Will try to discourage you.
   Exhibit apathy—show no interest.
   Resorts to rationalization or repression.
   Exhibit suspicion of your motives.
   Submits—cancels later.
   Attention will wander.
   Offers smoke screen objections.

4. Fail to anticipate objections.
   Concentrate on your proposal and ignore what he sees as his real problems.
   Fail to listen.
   Fake—bluff your way through.
   Fail to identify his problem.
   Be unprepared.
   Ignore hearsay to which he has been exposed.
   Ignore his self-image and its needs
   Ignore his habits and customs.
   Be unethical, immoral, or illegal.
   Ignore his objections.
   Argue with him or interrupt him.

   Total rejection (4A)

5. Present definite threat in any of these forms:
   (1) Economic loss; (2) loss of face; (3) emotional insecurity—discredit him; (4) cause him inconvenience or more work; (5) reduce his power of authority.

   Rejection and attack! (5A)

When you have your plan of action ready, test it against these general guidelines!

APPENDIX

Essentials for Successful Professional Selling Performance

Appendix

Essentials for successful professional selling performance

Now that you have finished reading the book, the following appendix provides you with a summary of 32 key points and a self-examination that will serve to point out the areas in which your strength lies and also to indicate those areas in which you realize you should endeavor to improve. To gain maximum value from this self-examination resist the normal human tendency to rate yourself as you would like to be. Rate yourself as you know you really are.

1. **Self-confidence**

   This is the sine qua non. Without it, no real success is possible. It can be developed when we observe the points listed below.

2. **Determine to be the best in the world among those working in your activity**

   As you read this, *someone, somewhere*, is. Why can't it be *you*?

   In the 1950's a major magazine said that General Electric's marketing organization was the best in the world at the time. There was a reason for this. Each person assigned to Marketing Services was told, "You will be the best in the world in your activity."

3. **Know your product**

   Whether you sell a product or a service, it is essential that you know it thoroughly. This includes knowing what it will do, what it won't do, how to use it, and being able to answer any reasonable question a customer may ask about any product you sell. It means *never* bluffing. It means when you don't have the information requested, promising to obtain it... and then keeping that promise.

4. **Know competitive products equally well**

   You should know the benefits and disadvantages of the product or service you sell, but you should also have equally complete information relative to that which competition has to offer. It means you must be able to draw honest comparisons between your offerings and those of your competitors so that you may emphasize the features and customer benefits of your product.

5. **Believe in your product and exhibit that belief**

   Hell must be a place where salesmen are required to sell products they don't believe in. Enthusiasm and honest conviction are contagious and will "rub off" on your customer.

6. **Identify your customers' problems and have a burning desire to help solve those problems**

   Our effort is all too often viewed by the customer as a "burning desire" to get the order and the customer's money, rather than as a determination to help the customer solve a problem, fill a need or to prevent future problems from occurring.

   Review your own past experience when *you* were a customer. How many times have you felt that the saleswoman was *really* concerned about your problem? Is it not the exception, rather than the rule, to encounter truly helpful service.

   Can't you think of a source of your own supplies to which you return over and over because the salespeople there keep their eyes on their "FHI"—give Friendly, Helpful, and Interested service?

   It is undeniably true that some customers will take advantage of good service. You may give them the help and advice they require and then see them buy elsewhere on the basis of price. Our own experience indicates that this behavior is engaged in by about 2% of customers. 98% is a pretty desirable batting average.

You are primarily in the *business of selling solutions to problems*. Showing the customer that your key interest is in helping him or her will bring the orders.

7. Love your work

This may sound "corny," but studies show that 70% of employed men are *not* in the occupational field of their first choice. With the barriers that have been erected against women, the figure, in their case, will undoubtedly be even higher.

Many of us, having been denied the opportunity to engage in the field—or the work—that was our first choice, turn to a field of endeavor that promises to put food on the table and is one that we can tolerate or one that will provide some satisfactions.

This frame of mind prevents us from giving the maximum effort to the job and affects our progress. If salesmen would understand the importance of their work...its contribution to the economy...the chance that it provides to *help others* and to understand the true professional nature of selling versus order-taking, they not only would get "more fun out of the job," but they would achieve maximum success. Take a new, hard look at your job and fall in love with it.

It can't be denied that we do best when we are engaged in activities we like. It is, therefore, self-defeating to look upon the job of selling as something "we *have to do*." Loving your work is not "corny," it's pragmatic and rewarding.

8. Anticipate objections

Pan American Airways says, "We want pilots that fly out in front of our planes." In other words, pilots that anticipate problems before they occur and then take action to avoid them.

Professional salespeople ask themselves, "If I were the customer, what objections would I have to this product or service?" In other words, they try to determine what objections they *may* encounter, and then decide, in advance, what they will say and/or do *if* any of these objections arise or if problems occur.

Obviously, the quality of their actions is going to be incomparably more effective simply because they have given these anticipated objections sound and unhurried consideration outside of the emotional situation that a sale often represents.

This type of professionalism has other advantages: (1) It causes us to be customer-oriented ...to think in terms of the customer's problems or needs. (2) It means that we are not *surprised* when an objection is raised that we had failed to anticipate...we aren't forced to produce an off-the-cuff reply.

Note: As a companion thought, when you do have an answer ready because you anticipated the objection, don't give the customer a fast reply, even though it is correct, or you will cause the customer to believe that you haven't even considered the objection being raised and that you truly are giving an off-the-cuff reply. Present your answer to the customer in a measured, thoughtful manner.

9. Learn to question

Questions are the most powerful tools you have. Consider what they accomplish:

- They obtain information.
- They keep you off the defensive.
- They start discussion, or can stop it.
- They demand proof. ("Can you give me an example of the problems you have had with similar products?")
- They help identify the customer's *real need*, be it product, economic, or psychic-oriented.
- They force the silent type of customer to talk.
- They can cause the customer to realize that you really are interested in his or her problems.
- They give you information that aids you in knowing which of your products or services will be most acceptable to the customer.

There are ten types of questions—rhetorical, direct, overhead, and so on. Learn not only what they are but what each of them accomplishes and when to use them. Perhaps the greatest advantage of using the questioning technique is that it forces us to listen.

Don't make your questions ambiguous... make them clear and open to one interpretation. Ask questions that require factual answers... facts that will help you identify the customer's primary buying motive and will enable you to select the product or service features that satisfy the customer's need or want.

10. Listen—really listen

No one ever hated a listener. We have yet to hear someone criticized with the words, "He listens too much!"

Even experienced salespeople often *talk* themselves out of a sale...more on that later. When *you* listen and the *customer* does the talking, the customer feels that he or she is dominating the selling situation...a feeling the customer likes.

Actually, it is *you* who are dominating the selling situation because listening, coupled with adroit questioning, will channel the customer's remarks into the areas you desire to explore.

A formula for successful selling is, *Be dominantly persuasive while enhancing the status needs of the customer.* The customer's basic need to dominate the selling situation is enhanced when the customer is talking.

Getting the customer to talk is important, but of even greater importance is developing an ability to listen...to listen not just to the customer's words, but to understand what is being said.

Good listening requires:

- An ability to control the tendency we all have to be planning what *we* are going to say when the customer stops talking.
- Being aware that when the customer is talking, we are getting information we need.
- Showing the customer you are listening. (More on this next under "Body Language.") One of the best methods for convincing the customer that you are listening is, when the customer stops talking, to repeat, in brief, something the customer has said...to agree with it when possible.

If you cannot agree, you can use such statements as, "That's an interesting point you made when you said..." or "I can understand why you feel as you do when you said..."

- Looking at the customer when he or she is talking. Don't allow yourself to be distracted by anything else.

Listening provides you with an opportunity to satisfy the customer's number-one need; i.e., to feel *comfortable in your presence and to be regarded as important and to have his or her needs to be seen as important and as unique.*

The customer's problem may be "old hat" to you...you may have encountered it many times, but keep in mind that to the customer it is *new.*

When it is an old problem to you, you may be tempted to "half-listen" and thereby miss something that will be important to you when you decide how to handle this customer's problem.

11. Watch your body language

We hear much recently about watching the customer's body language...observing his or her reactions to what you are saying.

It *is* important but much of what we are being told about the significance of the customer's physical actions...the position of his arms, legs, etc. is unadulterated rubbish. It makes an interesting, humorous, and entertaining speech, but it is dangerous because body movements can be misleading.

You don't need a course to be able to understand the customer's body language. All you need to do is to *listen with your eyes,* and use common sense.

You and I are very familiar with facial and hand actions that represent skepticism, disagreement, or agreement on the customer's part, if we will but observe them and act accordingly.

What is of great importance is to watch your own body language because the customer is observing you, as you observe the customer.

We have been discussing good listening. As you listen, if you are seated, face the customer, sit well forward on your chair, bent slightly toward the customer, keep your hands loosely folded so they won't distract the customer, and while not *staring* at the customer, *do* look at him or her.

By doing so, you are paying the customer the highest compliment possible...you are giving the customer your full attention.

Don't interrupt the customer...ever. If it is absolutely necessary that you do so, be sure to say, "Pardon me for interrupting, but..." This is so fundamental that we almost hesitate to mention it, but just for a moment, consider the number of times that *you* have been interrupted and been made furious as a result.

If you are with a customer and the customer is in the middle of a sentence and a third party starts to talk, stop him or her at once. This can be accomplished while saving the face of all concerned by such statements as:

"Excuse me (name of person interrupting), but I want to make sure I understand what (name of customer) is saying...just a minute, and I'll be with you."

This is of major importance. If you should answer the person doing the interrupting while the customer is talking or waiting for a reply from you, you have just told the customer that he or she is less important than the interrupting individual—a sure recipe for a lost sale or strained relationship. *No one—even your boss—is as important as the customer.* That customer has the money that both you and the boss need to succeed in business.

Keep in mind that you and I are walking mirrors. By our body language, we demonstrate to the customer how important or unimportant we consider him or her.

12. Don't stereotype

Don't stereotype by physical appearance, clothing, race, color, creed, nationality, educational background, age, youth, and the like.

Let's see how stupid it is to use such superficial indices.

Probably we and the British are, in one respect, the worst of snobs. We tend to consider a person speaking English with an accent as something less than intelligent.

Ask yourself, "How many languages do I speak *fluently* in addition to English?"

Remember this customer with an accent speaks at least one other language fluently and speaks enough English to make himself or herself understood.

Ask the question above in almost any European country, and the answer will almost certainly be, "I speak two, three, four, or as many as seven other languages, in addition to my own."

We rationalize by saying, "Oh, they live in such close proximity that they must learn other languages." Rubbish again. We have met Europeans who have never been more than 18 kilometers away from their hometown that spoke English almost flawlessly...and often many more languages.

The reason? They start in the first grades of school studying other languages.

Or take education. We assume that if our customer is college-educated, he or she will be intelligent, and if they did not attend college, their comprehension level will be low.

Under that system, Harry Truman would never have been President because he did not go to college, and Charles Wilson would never have been Chairman of the Board of General Electric for the same reason...and the list is endless.

Judging by the way the customer is dressed? We may tend to equate this with buying power. This can be a costly error. Our files are loaded with examples of people who dressed poorly who had more cash resources than we will ever see, let alone have.

Keep in mind that those with money can afford to dress any way they care to. It is often those who have little money that dress well to project the image of the way they want to be seen... "the me they wish they were."

We often see age as synonymous with wisdom and experience. An answer to that was given to us by the late Norris Hyett, professor at the University of Texas, who said, "We have some people whose epitaph on their tombstones should be, "Buried at 70, died at 30."

Some older people had their value systems locked in at 30, and lacking what Dr. Massey, of the University of Colorado, calls a "Significant Emotional Event," will never change.

We often equate youth with a desire to explore new ideas. There are *some* young people who are so insecure that they will run for the hills when confronted with any new idea, product, service, or method.

We will not take the time to explore the rest of the stereotypes listed at the start of this unit. This can be left to you. Examine them again, and see how many exceptions you can remember to the "rule"(?).

Probably the best guard against stereotyping is to ask yourself, "How would I like to be typed because I fall into any one of the groups mentioned in the list of stereotypes?"

And the list of stereotypes presented is far from complete. Stereotyping is the refuge of the witless...an excuse for thinking.

Above all, it robs your customer of his or her birthright—the right to be seen as a unique individual, a right that he or she will not surrender, and it will cost any of us that attempt to steal that right...sales.

13. Identify the customer's key area of interest

There are over 150 basic buying motives, but they all fall into one of the following categories:

a. Are you going to make me rich?

b. Are you going to make me famous?

c. Are you going to make life easier for me?

d. Are you going to make me more secure?

e. Are you going to give me more power or authority?

It is of primary importance that you identify the key area of interest of the customer because only then will you be able to decide which benefits or features of your product or service will be of compelling interest to that customer.

You will also know which of these benefits will be of no interest or will bore, even irritate, or "turn off" the customer.

Don't look at these five motivations narrowly.

"Are you going to make me rich" really means, "Am I going to make money or lose money if I follow your advice?"

"Are you going to make me famous" doesn't mean, "Are you going to put me in your national advertising?" It means, "If I follow your advice, how will it make me look in the eyes of the people whose opinion is important to me?"

These "people" can mean the customer's wife or husband, the boss, the peers, other people engaged in a similar business to one in which the customer is engaged. It means, "Will I look wise, or will I appear to have been 'taken in'?"

"Will it make life easier" means, "If I do what you want me to do, will it add to my workload, or decrease it?" "Will it mean I have to make decisions, abandon a comfortable habit I have always followed?" "Will it mean I have the tough and/or distasteful task of selling other people in my organization on the idea, if I buy from you and follow your advice?"

Salespeople have often discussed a new product and found the customer to be enthused. A month later, the customer had totally lost interest. Why? Because after the salesperson left, the customer broached the idea to people in his organization who would have to use the product or service and found them rebellious. The customer's attitude then became, "I need this trouble like a hole in the head" and the sale was lost.

"Are you going to make me more secure?"

We could forget this one without damage because when we talk about "making me rich" we are talking about financial security. When we talk about "making me famous," we are talking about psychic security.

"Are you going to give me more power or authority?" You will find that, too often, when you are trying to sell a service or a product that will result in labor savings, the customer (or the influences on the customer's buying decision) will oppose the idea.

They will see it as a threat...a threat to their need for feeling needed. ("If we buy this, then *anyone* can do my job!")

In some organizations, the manager's importance is based upon the number of people they supervise. Anything that will reduce that number may be seen as a threat because they are thinking of their own importance rather than of the welfare of their over-all organization.

In each of these cases, where the customer sees a threat, you must identify and present *features and benefits that satisfy the customer's needs and present no threats.*

When the customer senses any threat, you must present other gains that more than offset the threat.

It is of major importance that the salesman keeps in mind that these five basic needs of the customer do not remain static. They change as the customer's financial position or authority change. Because price, for example, may have been the customer's key criterion in the past does not mean it will be today. Keep current with the customer's present compelling requirement by getting the customer to talk... listen and then address yourself to the customer's current need.

14. Avoid projection

One of the most grievous sins in selling is to resort to projection, that is to say, "If *I* were the customer, this is what would interest or appeal to *me*, therefore, this is what will interest the customer."

You are not the customer.

Sales are lost so often when the salesman hits hard on features and benefits that interest the salesman, but are of little or no interest to the customer and may even destroy his or her buying mood. These features may even be seen negatively.

An example: A realtor constantly mentions the closet space available in every room and loses the interest of the customer. The realtor has a wife and five children, and closet space is at a premium in his or her home. The customer lives alone with his or her spouse, and is primarily interested in kitchen facilities. Closet space is not a problem.

Or a saleswoman constantly emphasizes appearance of a stereo system, whereas the customer is looking for power output.

Ask yourself, how, in the kind of selling in which you are engaged, you could find yourself emphasizing a feature that appeals to you but that might be of little interest to the customer.

15. Treat the customer as unique

Customers dislike canned sales presentations. Right or wrong, they often believe that no one has ever had the problems they have.

Before you use any other customer as an example or as a testimonial, be sure you have determined what *this* customer thinks about the customer you intend to use as a testimonial. If you fail to do so, you may cause the customer to think, "Does this salesperson see me as being like *that* customer?"

16. Don't hesitate to say, "I'm wrong!"

We often hesitate to say to the customer,

"You're right—I'm wrong!" We think it is a sign of weakness or that this admission will destroy our credibility in the eyes of the customer.

Nothing could be further from the truth.

Think of how you regard the exceptional person who says to you, in the middle of an argument, "Wait—I never thought of that, you're right, I'm wrong!"

We think this kind of person is great. Our own ego has been inflated, and he or she has shown himself or herself to be open-minded.

This is just the way we will be seen when we do not hesitate to admit we are wrong. Only the weak are afraid to make this admission. The strong know they are right most of the time, and they don't hesitate to admit when they are wrong.

Instead of being the most difficult words that come to the tongue of a human, they can be the easiest and the most productive in building the esential rapport with the customer.

17. Ask for the order

It is amazing how often orders are lost solely because the salesman fails to actually ask for the order. Our book examines the reasons for this but suffice it to say here, it is a problem that can be overcome.

Too often, we fear a negative answer. A "No" is not necessarily the end of the sale. Trial closes are just that! If we find the customer is not sold, we then determine what it is that we have not done by asking the customer.

Often the customer is on the verge of buying, but is hesitating, and a direct request for the order will force the customer to decide now.

18. Treat everyone as though that person is the chief executive of the customer's organization

At a meeting with a customer's organization, a salesman was talking in an informal party before the meeting started. Later someone said, "Do you know you were just talking to the chairman of the board?"

The salesman was aghast and wondered, "What did I say?" This problem will never arise if, in talking to *anyone* in the customer's organization, we treat him or her in exactly the same fashion we would treat the CEO of the organization.

19. Don't treat the customer as the person he or she "used to be"

We often become so familiar with a customer that we do not see the day-to-day changes that have taken place in that customer's intellectual growth.

This is like the manager who is told one of his or her people is doing a great job. The manager replies, "Yes, isn't Danny doing fine for a mechanic?"

Danny started out with the organization as a mechanic ten years ago. Last year he went into sales, and is doing the best job in the sales department, but the manager will always see Danny as the mechanic.

We may get away with that with our employees, but never with customers. We must see each of them as they are today. Keep current with the customer's growth, self-image, and aspirations.

20. Don't think the customer only likes you

Another serious error we can make. The customer likes us, therefore he won't like other salesmen in other companies just as well. He may like us and others, too. This points up the need for constantly earning the customer's faith by going all-out for him or her.

It also should emphasize the hazards of depending too much upon past relationships. If the customer likes others as much as he likes you, the antidote is to find ways of giving that customer something more than others give him.

One way to do this is to identify problems your customer has that have nothing to do with your product or service, and trying to find ways you can help him solve those problems.

If you do this, you will find that, in his mind, there are few salesmen like you.

21. Post-sales service

When you bought your last car—or anything for that matter, when did the salesperson who made the sale get in touch with you again?

The answer is: When that salesperson had something new to sell you.

How seldom many salespeople call the customer or visit the customer again to see if that which they sold had cured the customer's problem or had filled a need.

Too many saleswomen fear to do this for fear the customer may have complaints. Well, if so, the complaints are not going to disappear.

Showing a continued interest in the customer and in his or her problems will set you apart from the order-takers. This applies to those selling at retail as well as those in heavy industry.

It established you in the customer's mind as someone who was not merely seeking an order but was trying to solve or prevent problems. It is step one in sale two. It is in the continued business of the customer that the most meaningful profit is obtained.

There is probably nothing more important than this step.

22. **Don't use the 80%–20% concept**

    Many sales managers continuously drive home this point: 80% of the business comes from 20% of our customers, and that's the place to spend your time.

    Sounds logical, but if interpreted literally, may spell lost business. It is short-range thinking. The fact is, growth companies are small companies, for the most part.

    When they grow, they will remember the suppliers who gave them service and treated them as though they were important when they were small...before they started to grow.

    They will also remember those who paid little attention to them when they were small.

    Certainly, we should give a larger share of our time to the 20% that gives us that 80% of our business, but an effective salesman will analyze his small customers and identify those with growth potential, and give them sufficient attention and assistance to insure that they will be remembered when the customer achieves growth.

23. **Overcome the price obstacle**

    Price is, undeniably, sometimes the obstacle you have to overcome, but there are many times when we make price the obstacle when it is no obstacle.

    The case of Dr. Bakeland is a good example. He was an impoverished Belgian inventor. He invented Bakelite and Velox film. He came to the United States and obtained an appointment with Mr. Eastman, of Eastman Kodak, to sell his Velox film invention.

    On the way to the appointment, he worried constantly over whether he should ask as much as $50,000 for the invention. During the meeting with Eastman, all that was discussed were the attributes of the invention. Suddenly Eastman said, "Young man, will you accept one million dollars for this invention?"

    Bakeland smothered a gasp and agreed. Later, he said, "Mr. Eastman, do you know that I came here hoping to receive $50,000?" Whereupon Eastman retorted, "That's all right young man, *we* came here prepared to offer you three million!"

    Why *do* we often *create* a price problem? Here are only some of the reasons:

    - We forget that almost always, the customer knows *about* what he or she is going to pay. The price will not be a complete surprise.
    - We sometimes are asking for more money for some products than we will ever hold in our hand at one time in our lives, and we are over-impressed to the point that we fear we are over-priced.
    - We forget that this same customer whom we are trying to sell, buys other products or services to fill other needs he or she has that cost *many* times as much as the price we are asking for our product.
    - What seems to us like much money, from our own economic level, may represent a very small amount from the customer's point of view because he or she lives at a higher economic level than we do.
    - We forget also that price is uppermost in *our* mind, but in the customer's mind, very often, the absolute need to solve a problem, satisfy a desire, or fill a need may outweigh price.
    - We often worry because we know that competition is offering a product at a lower price than we are going to ask the customer to pay. What then?

    Then we must know that competitive offering thoroughly. Is the customer about to compare apples with oranges? Is the competitive offering *really* the same as that which we supply?

    Does our product have features which are lacking in the competitive product? Does the competitor really have the product on hand to supply? (It's *easy* to offer a low price on something you don't have in stock...and you and I know it is done every day...then, when delivery is possible, "rising costs have raised the price.")

    The antidote for this problem that *we* create without the help of the customer is to see price from the viewpoint of the customer, considering the points raised above (•).

    But you ask, what do you do when your price for the same product or service is higher than that being asked by competition?

    There are a number of things to do:

    Sell yourself, sell your company, and sell your product or service and in that order.

    This means showing greater interest in the customer as an individual, and in his or her problems, than the competitor who may be concentrating on product and price. It means going out of your way to identify other problems the customer has on which you can provide assistance.

Analyze *why* the competitor can offer the same product at a lower price.

You *do* have a problem if you find that the reason for your high price is that (1) your organization is trying to make an exorbitant profit, (2) your competitor is merchandising or manufacturing more efficiently, or (3) your competitor is using this product or service as a "loss-leader," or offering an "introductory offer."

In such cases, you may not be able to overcome the price problem, especially in the case of (1) above. In the other two instances, you must place greater emphasis on the service, especially post-sale service, you offer.

But, if your analysis of your higher price versus the competitive price indicates that you are offering more, then the task is to drive this home to the customer.

Point out that your profit is about the same as the competitor's, and if you are asking a higher price, then it is because you are offering more to the customer. Then identify that which you are providing (lacking in the competitive offer) and cause the customer to feel that this "extra" is so important that he or she must have it.

In other words, justify price by *facts*. This is no Pollyanna approach, but it calls for something more than order-taking, it calls for professionalism and it is rewarding.

Two nationally known sales trainers offer an excellent way to overcome price when it *is* an obstacle. Mack Hanan calls it the Presumption of Exclusivity, referred to by Charles Herrmann as An Image of Difference. How does it work? Like this:

You determine what this specific customer wants from your product or service more than anything else, and then you identify the feature of your product or service that supplies satisfaction for that want or need.

Then you concentrate on that feature, hitting it over and over again, until the customer feels that it is only from you that the satisfaction required can be obtained.

Some might say this is an "immoral" technique because competition has exactly the same feature to offer. Nothing could be further from the truth. It is the most moral type of selling possible. It means that you have treated this customer as a *unique individual with a unique problem or need*. It means that you have been sufficiently interested in this customer and his or her problem to identify that which is most important to him or her. You have not given a "canned presentation" that could be given to *any* customer.

24. Romance your offer

In Essential No. 5, it was stated, "Believe in your product and exhibit that belief." It is equally important that you cause your customer to share that belief. How to do it? Among other ways, here are a few to consider:

- Get the merchandise in the customer's hands. "A piece of merchandise in the customer's hands is worth two on the shelf."

In one large retail organization, we took men's socks out from a glassed-in counter and placed them on aisle tables...sales soared 300% and *remained at that level.*

If you were selling automobiles, you would certainly want to get the potential customer *in that car and to drive it.* Which raises another point:

- Get the customer involved! Of course, this is what you are doing when you get the merchandise in his or her hands. If yours is a product that can be demonstrated, if you must make the first demonstration, be sure to have the customer participate or take part in the second demonstration...show how easy and pleasant it is to use. (Isn't this one reason why perfume counters have samples for the customers to use? You could talk all day about how pleasant a perfume may be, but it isn't until that customer puts a drop on her hand and tries it that a sale is possible.)

But you may say, there is just *no way* I can get a steam turbine, a power transformer, or even a large copying machine in the customer's hands, or "let him or her operate it!" In fact, you're right. In application of the technique, that's wrong.

Here are some ways you can "romance a product" even if it is of such a nature that it can't be operated (and sometimes even seen) by the customer.

- In retailing, we would urge salesmen selling anything from furniture to dresses to use such techniques as: As the salesman fondly rubs the finish of a tabletop, he would say to the customer, "Just feel that beautiful finish"...or "Just sit in that chair and feel how relaxing it is"...or "Just feel this fabric; it's the latest Italian creation."

We have had a salesman selling us an offset machine, say, "Just feel the thickness and the quality of the steel they have put into this machine." Before we realized it, we were stroking the metal side of the machine. Then the salesman added, "They have put that quality into every part of the machine." We believed it, and bought the machine. Fortunately, the salesman had been truthful.

One sales engineer involved in selling heavy electrical apparatus to utility companies, fell so much in love with his products that as he drove across the countryside, he could never pass an electrical sub-station without getting out of his car and taking a picture. He thought (and we agree) that they are beautiful to behold.

Then he would show some of these photographs to his customers, and his enthusiasm was contagious.

It isn't enough to believe in your product, you must cause your customer to share that feeling.

Even if you are only trying to upgrade a customer from 60 lb paper to 80 lb, it's possible to use these techniques. "Just *feel* the thickness of this paper!" Or, "Just *feel* this finish!"

The foregoing is just a once-over-lightly of methods of conveying enthusiasm to a customer. The challenge to you is to consider the items or services *you* sell and ask yourself how you can romance *them*.

If you must depend upon brochures, then use them effectively. Above all, after you place something in the hands of the customer to read, *stop talking while he or she reads it!* Customers don't have two heads...one to use to read your material, and another to listen to you.

25. Pre-sale preparation

Admittedly, retail selling, or across-the-counter selling in such activities as an electrical wholesaling operation are among the most difficult kinds of selling. Reason for this? These salesmen—except in the case of repeat customers—have the least opportunity to get to know the customer as an individual. They *must* learn how to use questions and get the customer talking, as said before. They have less opportunity for pre-sale preparation.

We have found that more sales are lost for two reasons than all other reasons combined. (1) Lack of professional pre-sale preparation, and (2) a failure to understand the importance of satisfying the psychic needs of the customer (how he or she wants to be seen and whom they are most anxious to impress or whose reactions concern them the most).

Even retail sales personnel can engage in *some* pre-sale preparation by, as said earlier, *anticipating* some of the objections customers may raise and having a battery of answers ready.

A general outline of what constitutes pre-sale planning incorporates these factors:

a. Identify the customer. Who are the other key influences behind the decision to buy?

b. (If a company is involved.) What is the customer's long-range objective for his or her company that can be affected by this purchase?

c. (Again, if a company is involved.) What is the company's immediate goal that this purchase can affect.

d. What is this customer's long-range objective, how does he or she want to be seen in the future, and who does he or she want to impress over the "long haul"?

e. What is the customer's immediate goal? What psychic need (if any) in addition to product or economic satisfaction does he or she seek *now*?

f. Which of these (b through e, above) is of greatest interest to the customer *now*—that's the need I must satisfy if I am to make a sale.

g. What is my long-range objective with this customer (if a possible repeat customer)?

h. What is my immediate goal with this customer?

i. What is the problem? (What stands between me and a successful sale?) If I identify the wrong problem, I can produce a brilliant solution but no sale.

j. How many alternatives do I have?

In (j) we find the source of many lost sales. We are, often, so certain that the strategy and tactics we intend to employ will sell the customer that we are caught totally unprepared when that customer does not react "as he or she was supposed to react."

IT IS ABSOLUTELY ESSENTIAL THAT WHEN WE PLAN OUR SALES STRATEGY WE CONSIDER WHAT THE CUSTOMER AND THE COMPETITION CAN DO IN ADDITION TO WHAT WE THINK THEY WILL DO, AND TO HAVE ALTERNATE STRATEGIES AND TACTICS READY TO BE USED... AND NEVER, BUT NEVER, BE SURPRISED AND CAUGHT UNPREPARED!

k. Select a strategy, as an opener, that promises the greatest possibility of success...but have the others ready.

26. Assume you have the order

Give the impression you are going to get the order! Avoid such phrases as, "*If* you decide to buy, *would* you want...etc."

Use such phrases as, "*When* you use this, you *will* want to...etc."

Avoid words and phrases that create the impression that the customer may decide not to buy. Be a word-smith. Develop positive-sounding phrases and words in advance that you can employ in all sales situations. Test them in advance.

27. Consider your appearance

Your decision on the kinds of clothing you should wear is going to vary widely according to the area in which you sell and according to the kinds of people you sell. Consider the following:

- In the oil fields, in many instances, a salesman had best not come on the scene dressed like an executive from Dallas or Tulsa. He had best be in clothes (neat, always) that will cause customers to see him as "one of us."

This, however, is no set rule.

There will be men in some oil fields that will think, "Who does this guy think we are, coming here dressed in khaki...what's he trying to prove? I'll bet he wouldn't call on a Dallas executive dressed that way—who does he think we are?"

So it is obvious that the salesperson had better know his customer.

- In some high-fashion stores, the perennial problem exists of trying to convince salespeople that they should not dress better than their customers.
- A growing problem in retail stores seems to be the failure to have salespeople wear some kinds of readily recognizable indication of their identity. Customers are constantly embarrassed when they must approach someone and ask, "Do you work here?" Embarrassed customers are not happy customers.
- Beards, long hair, too-casual clothing? What to do?

One executive told us that it was imperative that his male sales personnel have "long hair" (styled). He pointed out that the advertising people they call upon, as a rule, also have that type of hair-styling. He went on to point out that the only salesman he had with a "crew-cut" could be used only in selling to the automobile headquarters in Detroit, where he would be accepted.

We don't vouch for the reliability of that viewpoint, but it does indicate that there is a problem.

One group asked us what we thought about "long hair." Their manager interrupted to say, "How you wear your hair is of no concern to me, unless it gives you a problem with our customers. *Then* you have a problem with *me*."

There followed some grousing about being forced to "compromise."

Let's face it, if you and I are going to sell, we must compromise. What is really new about that? Many of us hate neckties, but we know that in certain situations we have no choice.

If we are asked to go to a restaurant in the summer, our first question may be, "Are jackets and ties required?"

If the reply is "yes" we compromise, at least if we want to eat in that restaurant.

This is one that you will have to answer for yourself. It just doesn't make much sense, to us, for a salesperson to stand on his or her "rights" and dress as he or she pleases, if it is going to cost sales and the economic security those sales produce.

Attractively trimmed beards and styled hair in many instances are accepted by almost everyone. There are many customers who equate far-out beards, unattractive hair-styles, unusual mustaches with poor judgment and bad taste, and it has an effect upon their appraisal of the judgment of the salesman on the professional aspects of the job.

The same may be said of female sales personnel in the matter of jeans, and other articles of clothing. When to wear them, when to wear "conservative" clothing...these are questions each saleswoman must ask herself.

The point is, sales will be lost, no matter how good your product or service may be, if the salesman takes the attitude, "I'll wear my hair the way I want to wear it, and I'll wear whatever I feel like wearing."

That attitude will exact a price.

28. Avoid anger

Anger blows out the lamp of the mind. One salesman said, "I save my anger for those things worth getting angry about, and those are the very things I have found anger won't help."

There *definitely are* customers who have the ability to irritate us, annoy us, and make us angry. But we must agree that to the degree to which we allow ourselves to become angry—to that degree—our thinking ability is impaired.

When you and I allow ourselves to become angry, we *are letting the other person pull our strings.*

Anyone can permit themselves to become angry. Only the professional knows how to overcome anger.

Becoming angry only injects further acrimony into the selling situation. There *are* people who are habitually "mean," but consider the satisfaction that can be obtained when you manage to make a successful sale to one of these people.

29. **Listen with your eyes**

    The psychologist tells us that "we don't like our behavior to be transparent."

    In spite of this, without realizing it, we make ourselves transparent to those who listen with their eyes and ears.

    It's vital, in many selling situations, that we determine what image the customer is trying to project—how that person wants to be seen.

    They won't admit it, but they can show it... they show it by such things as:
    - The kinds of artifacts displayed in their office and home.
    - The clothes they wear.
    - Their hairstyles (the chap, for example, who has passed his fiftieth birthday and wears long hair or exaggerated "mutton-chops" and wants to be seen as young—or wants to see himself that way).
    - The cars they drive.
    - The neighborhood they live in.
    - The names and places they "drop."
    - The people with whom they associate.
    - The lapel buttons they wear.
    - The subjects they like to discuss.
    - The rings and/or jewelry they wear.

    These are all windows to their self-image.

30. **Seek a pattern**

    Don't make the mistake of judging any customer by any one or two of these revealing indicators. *Look for a pattern in their possessions, conversation, and behavior.*

    Any single one or two of these points above may be totally meaningless. Someone else in the family may have selected the car or the house, or the expensive watch may have been a gift.

    But look for a pattern. Especially be alert for the subjects the customer selects to talk about when he or she is given free rein to talk.

    Be aware of the people the customer discusses in a favorable light...this may be the way the customer wants to be seen.

    You may know the customer's product needs and economic needs and limitations, but the task often remains to identify his or her psychic needs.

NOW ABOUT YOU!

31. **Put in writing your long-range personal objectives and your short-range personal goals**

    Set high thresholds of satisfaction. Don't underestimate your potential.

    If we were to say that you may, some day, be the chief executive officer of a major company, you would probably reject that as being impossible.

    Yet, one day, the present chief executive officer of every company was 25 years old. They weren't "hatched out of an egg" at 50.

    Years ago, if you had told most of these men and women that they would rise to the heights, they would have rejected this as impossible.

    But they did it. How?

    By having faith in themselves and setting goals and objectives! By combining them with an action plan.

    What *is* the secret of success? Just this.

    a. Engage in regular, periodic and scheduled *self-analysis*. It should consist of:
       - What I am versus what I must become (my habits and attitudes).
       - What I know versus what I must learn.
       - What I can do versus what skills I must acquire.

    On the first of these (habits-attitudes), check your opinion of yourself in these areas with the opinions of others who know you.
    SELF-ANALYSIS MAKES POSSIBLE SELF-AWARENESS (*what am I really like?*), WHICH LEADS TO SELF-DEVELOPMENT, WHICH LEADS US TO SOMETHING CLOSER TO SELF-FULFILLMENT.

    b. Don't underestimate yourself. You might be surprised to find you have a higher IQ than the president of a major organization, but keep in mind that although "IQ" may be considered the capacity to learn, we prefer to think of IQ as not what you know, but what you *do* with what you know...and he or she who doesn't do with what is known is no better off than the individuals who "didn't know" to begin with.

    c. Don't set unrealistic objectives at the start. They can always be upgraded later. Your objectives should be from five to 15 years ahead. You can set intermediate objectives for the period two to five years ahead.

    d. Set goals for the next two years...they can be more flexible. Set goals tough enough to make them attainable—goals you can get

your arms around, but only by stretching.
e. Check yourself against the goals every six months.
f. Establish action plans, and make them measurable.
g. *Let your immediate manager know what both your goals and objectives are—especially those that embrace the next five years!*

If he or she doesn't know what they are, how can he or she help you attain them?

We agree that your boss should have asked you and should know what they are, but our studies indicate that at least 95% of the managers do *not* know and have *not* asked... simply because they were so preoccupied with their business objectives, and in some instances, so preoccupied with the attainment of their own goals.

h. Don't be satisfied by assuring yourself that you have done pretty well by comparing yourself to others that came from your neighborhood, or were in your class at high school or the university.

Compare your progress to what it can be! Keep in mind that many of them may be comparing themselves to *you*, and if they have kept pace with you, they, too, have set low thresholds of satisfaction. They have not engaged in self-analysis and set goals and objectives.

32. Take another look at selling as a profession

A salesman today must be a psychologist, an economic advisor, an expert in his or her field, knowledgeable about products and services and how to use them.

The salesmen today must be seen as the intelligence gathering network of the organization. They are closer to the customer than any single individual.

They often know competitive plans before anyone in their own organization. They are privy to the reactions, attitudes, needs and wants of their customers.

They know whether the product line now sold is in tune with those needs and wants, and whether it anticipates *future* customer needs.

In other words, the saleswomen today must be seen—and view themselves—as marketeers.

We measure our country's stability and potential (rightly or wrongly) in terms of GNP—Gross National Product.

That GNP depends more upon what we *sell* than any other factor. We can have the best engineering, manufacturing, finance, and personnel in the world, but, in the final analysis, all of those people, and the nation, depend upon what we sell.

Arthur ("Red") Motley said it effectively and succinctly: *"Nothing happens until we sell something!"*

If every salesman would but realize the chain of events that he affects when a sale is made, pride in the profession would soar.

It is from sales that profit accrues, and from profits that taxes are obtained, and from taxes that our schools, hospitals, national defense, and all the other essentials for the strength of this government are provided.

And it all starts with you.

33. Now measure yourself

So, why not, as we said in Essential No. 2, be the best in the world in the activity in which you are engaged? As a start, how about checking yourself on each of these 33 points so you can get moving.

The first step is to use the next page.

How Am I Doing?

1—Good, no improvement necessary. **2**—Good, but some improvement necessary.
3—OK, but much improvement necessary. **4**—Weak, immediate action needed.

		1	2	3	4
1.	Do I have sufficient self-confidence?				
2.	Am I determined to be the best in the world on this job?				
3.	Do I have sufficient knowledge of the things I sell?				
4.	Do I know competitive products sufficiently well?				
5.	Do I believe in and exhibit my faith in products I sell?				
6.	Do I really have a burning desire to help my customer?				
7.	Have I analyzed my job and realized how I can really love it?				
8.	Do I anticipate customer objections and ready myself to handle them?				
9.	Do I know the types of questions and how to employ them?				
10.	Do I really listen when the customer (or anyone) is talking?				
11.	Do I observe my customer's body language and watch my own?				
12.	Do I tend to stereotype customers . . . to group them?				
13.	Do I try to identify the customer's key areas of interest?				
14.	Do I avoid projection ("I like it, so the customer will")?				
15.	Do I treat the customer as unique (or just like others)?				
16.	Do I avoid ever saying "You're right—I'm wrong"?				
17.	Do I fail to ask for the order, afraid the answer will be "No"?				
18.	Do I treat everyone as though he/she were the CEO of the company?				
19.	Do I tend to see people as they "used to be"?				
20.	Do I think the customer likes me, so he won't like competitors?				
21.	Do I show a continuing interest in the customer *after the sale*?				
22.	Do I ignore or give insufficient attention to small customers?				
23.	Do I create price as an obstacle when it may be no obstacle?				
24.	Do I "romance" my products or services?				
25.	Do I engage in really professional pre-sale preparation?				
26.	Do I assume I have the order, or do I act uncertain?				
27.	Do I consider my appearance and its effect on the customer?				
28.	Do I avoid anger, or do I become irritated and annoyed?				
29.	Do I listen with my eyes and to the implications of what the customer likes to discuss?				
30.	Do I look for a pattern in the customer's behavior, or do I tend to base my opinions on one thing about him or her?				
31.	Have I set objectives and goals for myself, with dates and action plans?				
32.	Have I failed to realize the importance of this profession?				
33.	Have I just glanced at this self-analysis form, or have I really been willing to analyze myself and get started?				

*The power to sense another's hurt,
and to want to feel that hurt...*

*The power to sense another's need,
and to want to fill that need...*

This is the root of true professional selling!

Elements in the Anatomy of a Sale—Plus Possible Obstacles

Pre-sale preparation

Too sure
Lack of time
Cold call
Customer has "new rep"
New type of account
Depends on predecessor
Deprecates competition
Overlooks new competition
No sources of information
Depends on reputation
Take customer for granted
One-sale oriented
Product oriented
Misinterprets info.—doesn't identify a behavior pattern

Contact

Time
Priority
Recall
Competition
Mannerisms (our)
Atmosphere
Habits (customer's)
Timing (our)
Predecessor (our)
Organization structure
Personalities
Appearance (our)
Prejudice
Voice (our)
Age (our)
Interruptions
Customer has other problems
Personal worries (customer's)
Health (customer's)
Policy (both)
Trade relations
"Golf today"
Our own prejudice
Defeatism (our)
Secretaries (customer's)

Gain customer's confidence

Appearance (our)
Our own lack of confidence
Habits (our)
Mannerisms (our)
Canned presentation
Images customer has of us
of our company
of our products
Customer's self-image
Hearsay
Customer's image of our competitor
Age (our)
Voice (our)
Customer's off-beat outlook
Our reaction to customer's behavior

Identify or establish need

Unaware of need
Doesn't comprehend
Poor judgment
Ignorance
Priority
Disinterest
Doesn't want to see need
Customer is confused

Establish satisfactions

Price high
Unattractive terms
Limited use
Delivery slow
Union opposed
Installation costly
Space problem
Obsolescence
Price changes
Past performance
Service problems
Intangible benefits
Remote benefits
Customer is confused
Doesn't understand our solution

Gain agreement

Smoke screen objections
High cost in psychic needs
Won't buy your premise
Wants to avoid obligation
Has other obligations now
Influence of conditioners
Influence of experiences
Hearsay
Demands a concession
Fears
Hearsay

Gain preference

Customer has obligations
Customer's psychic needs satisfied by competition
Competitor is local
Competitor has only one salesman for customer to deal with
Customer has prejudice
Experience with other depts.
We threaten customer's psychic needs
Customer's habits
Compares our product to others
Spreads business
Influence of outsiders

Establish priority

Fear of business climate
Fear of higher authority
Fear of failure
Won't plan ahead
Other needs seem greater
"Can get by with what I have"

Ask for the order

Front office jitters
Not expecting order
What to say?
Buyer's attitude
Is customer ready yet?
Size of order jitters
Impressed by customer's rank
Fear of turndown
Wants discount
Wants preference

Take the order

Hates detail (either party)
Afraid to push "luck"
Doesn't know tech. details
Afraid to lose control

Work for continuance

Is sale oriented
Forgets promises
Has no follow up plan
Doesn't see full customer potential
Geographical problems
Poor communication with factory or office